MULTIVARIABLE
Calculus

MULTIVARIABLE
Calculus

Donald Trim
University of Manitoba

Prentice Hall Canada Inc.
Scarborough, Ontario

Canadian Cataloguing in Publication Data
Trim, Donald W.
Multivariable calculus

ISBN 0–13-015959–X

1. Calculus.   I. Title.
QA303.T76 1993   515   C93–093111-4

© 1993 Prentice-Hall Canada Inc., Scarborough, Ontario
ALL RIGHTS RESERVED

No part of this book may be reproduced in any form without permission in writing from the publisher.

Prentice-Hall, Inc., Englewood Cliffs, New Jersey
Prentice-Hall International, Inc., London
Prentice-Hall of Australia, Pty., Ltd., Sydney
Prentice-Hall of India Pvt., Ltd., New Delhi
Prentice-Hall of Japan, Inc., Tokyo
Prentice-Hall of Southeast Asia (Pte.) Ltd., Singapore
Editora Prentice-Hall do Brasil Ltda., Rio de Janeiro
Prentice-Hall Hispanoamericana, S.A., Mexico

ISBN 0–13-015959–X

Acquisitions Editor: Jacqueline Wood
Developmental Editor: David Jolliffe
Copy Editor: Santo D'Agostino
Production Editor: Kelly Dickson
Production Coordinator: Florence Rousseau
Interior and Cover Design: Full Spectrum Art
Cover Image Credit: Mark Tomalty/Masterfile
Page Layout: Pronk&Associates

1 2 3 4 5 AGC 97 96 95 94 93

Printed and bound in the U.S.A. by Arcata Graphics

Contents

Preface xi

Chapter 11 ■ Infinite Sequences and Series 507
Section 11.1 Infinite Sequences of Constants 508
Section 11.2 Applications of Sequences 518
Section 11.3 Infinite Series of Constants 524
Section 11.4 Nonnegative Series 531
Section 11.5 Further Convergence Tests for Nonnegative Series 537
Section 11.6 Series with Positive and Negative Terms 542
Section 11.7 Approximating the Sum of a Series 547
Section 11.8 Power Series and Intervals of Convergence 551
Section 11.9 Taylor's Remainder Formula and Taylor Series 560
Section 11.10 Taylor Series Expansions of Functions 569
Section 11.11 Sums of Power Series and Series of Constants 580
Section 11.12 Applications of Taylor Series and Taylor's Remainder Formula 583
Summary 591
Review Exercises 593

Chapter 12 ■ Vectors and Three-dimensional Analytic Geometry 595
Section 12.1 Rectangular Coordinates In Space 595
Section 12.2 Curves and Surfaces 599
Section 12.3 Vectors 608
Section 12.4 The Scalar Product 624
Section 12.5 The Vector Product 636
Section 12.6 Differentiation and Integration of Vectors 648
Section 12.7 Parametric and Vector Representations of Curves 654
Section 12.8 Tangent Vectors and Lengths of Curves 659
Section 12.9 Normal Vectors, Curvature, and Radius of Curvature 666
Section 12.10 Displacement, Velocity, and Acceleration 675
Summary 686
Review Questions 688

Chapter 13 ■ Differential Calculus of Multivariable Functions 691
Section 13.1 Multivariable Functions 691
Section 13.2 Limits and Continuity 697
Section 13.3 Partial Derivatives 701
Section 13.4 Gradients 704
Section 13.5 Higher-Order Partial Derivatives 708
Section 13.6 Chain Rules for Partial Derivatives 712
Section 13.7 Implicit Differentiation 724
Section 13.8 Directional Derivatives 731
Section 13.9 Tangent Lines and Tangent Planes 737
Section 13.10 Relative Maxima and Minima 744

Section 13.11 Absolute Maxima and Minima   755
Section 13.12 Lagrange Multipliers   764
Section 13.13 Differentials   773
Section 13.14 Taylor Series for Multivariable Functions   776
Summary   778
Review Exercises   779

Chapter 14 ■ Multiple Integrals   781

Section 14.1 Double Integrals and Double Iterated Integrals   781
Section 14.2 Evaluation of Double Integrals by Double Iterated Integrals   786
Section 14.3 Areas and Volumes of Solids of Revolution   793
Section 14.4 Fluid Pressure   799
Section 14.5 Centres of Mass and Moments of Inertia   803
Section 14.6 Surface Area   809
Section 14.7 Double Iterated Integrals in Polar Coordinates   815
Section 14.8 Triple Integrals and Triple Iterated Integrals   823
Section 14.9 Volumes   830
Section 14.10 Centres of Mass and Moments of Inertia   834
Section 14.11 Triple Iterated Integrals in Cylindrical Coordinates   839
Section 14.12 Triple Iterated Integrals in Spherical Coordinates   844
Section 14.13 Derivatives of Definite Integrals   852
Summary   859
Review Exercises   860

Chapter 15 ■ Vector Calculus   863

Section 15.1 Vector Fields   863
Section 15.2 Line Integrals   874
Section 15.3 Line Integrals Involving Vector Functions   881
Section 15.4 Independence of Path   888
Section 15.5 Energy and Conservative Force Fields   897
Section 15.6 Green's Theorem   900
Section 15.7 Surface Integrals   907
Section 15.8 Surface Integrals Involving Vector Fields   914
Section 15.9 The Divergence Theorem   919
Section 15.10 Stokes's Theorem   928
Section 15.11 Flux and Circulation   935
Summary   938
Review Exercises   940

Chapter 16 ■ Differential Equations   943

Section 16.1 Introduction   943
Section 16.2 Separable Differential Equations   948
Section 16.3 Linear First-Order Differential Equations   953
Section 16.4 Second-Order Equations Reducible to Two First-Order Equations   958
Section 16.5 Newtonian Mechanics   963
Section 16.6 Population Dynamics   968
Section 16.7 Linear Differential Equations   972
Section 16.8 Homogeneous Linear Differential Equations   975
Section 16.9 Homogeneous Linear Differential Equations with Constant Coefficients   979
Section 16.10 Nonhomogeneous Linear Differential Equations with Constant Coefficients   984

Section 16.11 Applications of Linear Differential Equations 998
Summary 1007
Review Exercises 1008

Appendix D ■ Answers to Even-numbered Exercises 1011

Appendix E ■ Additional Exercises 1029

Appendix F ■ Determinants 1041

Index 1051

Donald W. Trim received his honours degree in mathematics and physics in 1965. Masters and doctorate degrees followed with a dissertation in general relativity. As a graduate student at the University of Waterloo, he discovered that teaching was his life's ambition. He quickly became known for his teaching, and in 1971 he was invited to become the first member of the Department of Applied Mathematics in the Faculty of Science at the University of Manitoba. In 1976, after only five years at the university, he received one of the university's highest awards for teaching excellence, based on submissions by faculty and graduating students. At the University of Manitoba no individual can receive the other award within ten years of receiving the first. In 1988, Professor Trim received the second award. In addition, the Faculty of Engineering presented him with a gold replica of the engineer's iron ring in appreciation for his service.

Professor Trim's skills in teaching are reflected in the notes and books that he has written. These include an Introduction to Applied Mathematics, an Introduction to the Theory and Applications of Complex Functions, and Applied Partial Differential Equations, and this calculus text. All have received most favourable reviews from students.

Professor Trim is a member of the Canadian Applied Mathematics Society and the Mathematical Association of America. His present interests are in partial differential equations and mathematical education.

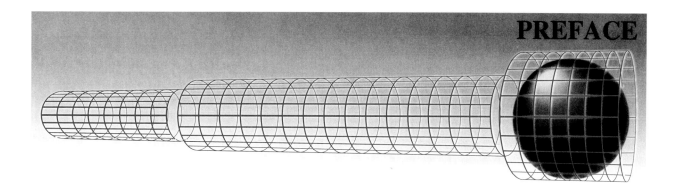

PREFACE

To the Instructor

Approach

This calculus book evolved from a set of notes developed by the author over a number of years of teaching first- and second-year calculus, and an earlier version of the text. For the most part our approach is intuitive, making free use of goemetry and familiar physical settings to motivate and illustrate concepts.

Order of Topics and Pedagogy

The order of topics in the book is fairly standard with a few exceptions. We indicate some of these differences below as well as some of our thoughts on the best way to approach multivariable calculus.

1. Although "infinite series" is a topic in single-variable calculus, it is often taught in more advanced courses, and has therefore been included in the text. In discussions of series of constants we apply the limit ratio test only to series with positive terms. It is not used directly as a test for convergence of series with positive and negative terms. Many students have difficulty with the convergence and divergence of series. Let us take it one step at a time. First discuss convergence of series of positive terms using the comparison, limit comparison, integral, limit ratio, and limit root tests. Then introduce series with positive and negative terms, and use the above tests as tests for absolute convergence.

2. The basis for power series expansions is Taylor's remainder formula; when remainders approach zero, the Taylor series converges to the function. This is discussed first. It then becomes a matter of illustrating that various techniques exist to find power series (such as integrating and differentiating known series), and these techniques conveniently avoid remainders.

3. Many students take entire courses in differential equations, once again a topic in single-variable calculus. For those who need only an introduction to the topic and knowledge of some of the more important techniques for solving differential equations we have included Chapter 16. There are sections on separable, linear first order, and second order equations easily reduced to first order equations; the exercises introduce (first order) homogeneous and Bernoulli equations. Considerable emphasis is placed on the important topic of linear differential equations (four sections).

Applications to Newtonian mechanics, population dynamics, vibrating mass-spring systems, and electric circuits are discussed in detail. Many other applications are are introduced through examples and exercises.

4. Three-dimensional analytic geometry and vectors in Chapter 12 provide the tools for multivariable calculus. We stress the value of drawing curves and surfaces in space, using the curve sketching tools learned in single-variable calculus. Such diagrams are essential to the evaluation of double, triple, line, and surface integrals, and to an appreciation of many of the ideas of differential calculus. Vectors are handled both algebraically and geometrically, since neither approach is satisfactory by itself. Every algebraic definition is interpreted geometrically, and every geometric definition is followed by an algebraic equivalent. Differentiation and integration of vectors dependent on a single parameter lead to discussions of tangent and normal vectors to curves, curvature and arc length, and three-dimensional kinematics.

5. Gradients are useful in many areas of applied mathematics. Introducing them in the context of an application leads students to associate them with that particular application. For this reason we introduce gradients in Section 13.4, and apply them to directional derivatives and normal vectors to curves and surfaces in Sections 13.8 and 13.9.

6. Chain rules for composite, multivariable functions are endless in variations and applications. We show students how to appreciate each term in a chain rule as a contribution of particular variables to the overall rate of change of a function, and then provide them with a schematic diagram to handle the most complicated functional situations.

7. Chain rules are used to find systems of equations for partial derivatives of implicitly defined functions. Cramer's rule and Jacobians facilitate solving the equations.

8. Relative extrema and absolute extrema for multivariable functions are discussed in separate sections. Most applied extrema problems involve absolute extrema; surface sketching uses relative extrema. They are different, and they are used differently; let the student see them "one-at-a-time". Lagrange multipliers provide an important alternative for constrained extrema problems.

9. Our definitions of double and triple integrals as " limit-summations" is completely analogous to the definition of the definite integral; evaluation by double and triple iterated integrals is geometric. Through representative boxes, rectangles, strips, and columns, the student visualizes the summmation process and affixes appropriate limits to integrals. There is no algebraic manipulation of inequalities; a thoughtfully prepared diagram does it all. The geometric approach also helps students to visualize integrations in polar, cylindrical, and spherical coordinates.

10. One of the biggest difficulties for students is to decide whether a definite, double, triple, line, or surface integral should be used on a given problem. We suggest that they ask what it is that they are integrating over, not what they are finding. To integrate over an area in a coordinate plane, use a double integral; over a volume, use a triple integral; along a curve, use a line integral; and over a surface, use a surface integral. It seems natural then to find volumes in space with triple integrals rather than double integrals.

11. Students believe that topics with very few definitions are inherently simpler than those with many definitions. With this in mind, we tell students that there is one kind of line integral, $\int_C f(x,y,z)\,ds$ and that integrals of the form $\int_C P\,dx + Q\,dy + R\,dz$ are a special case when $f(x,y,z)$ is the tangential component of some vector function defined along C. They can all be evaluated by substituting from parametric equations of the curve. Alternatively, it may be expedient to use Green's or Stokes's thereom

or to determine whether the integral is independent of path.
12. Likewise, there is only one surface integral $\iint_S f(x,y,z)\,dS$, but in many applications, $f(x,y,z)$ is the normal component of some vector function defined on S.
13. Appendix E contains 262 additional problems for the entire book. Students should use them in preparation for their final exam. We have provided answers to only the first 123 even-numbered problems. The last few are more challenging. We usually throw some of them out to the class, one at a time, for a bonus of 3, 4, or 5 marks. Perhaps the first correct solution, or the best solution, earns the bonus. This often sparks the interest of the better student.

Supplements

Two supplements have been prepared to aid instructors and students. Available to students is a manual containing detailed solutions to all even-numbered exercises. For instructors there is a manual with answers to odd-numbered exercises. Any errors in the text or supplements are the responsibility of the author, and we would appreciate having them brought to our attention.

Acknowledgements

I wish to express my appreciation to the staff at Prentice Hall Canada Inc. for a superb effort in editing, design and illustration. Special thanks to J. Wood, D. Jolliffe, S. D'Agostino, and K. Dickson for their suggestions and encouragement.
Thanks also to Panusunan Siregar and Karen J. Hand, University of Guelph, who proofed the calculations.

To the Student

In multivariable calculus we apply what you have learned in single-variable calculus to functions of more than one variable. We have limits, derivatives, and integrals of functions of two, three, or more variables and these are based on analogous definitions for single-variable functions. It follows therefore that a prerequisite for your studies here is successful completion of a course in single-variable calculus. Even so, it may be necessary for you to review some topics in single-variable work that perhaps were not mastered as well as they might have beeen, or perhaps need refreshing. Since our approach to multivariable calculus is based on our approach to single-variable calculus, it would be advantageous if your instructor and/or library had our single-variable text available for your reference.

Our approach is very intuitive, making use of three-dimensional diagrams to illustrate ideas and solve problems. Make a habit of making diagrams. A picture, no matter how rough, is invaluable in giving you a "feeling" for what is going on. It displays the known facts surrounding the problem; it permits you to see what the problem really is, and how it relates to the known facts; and it often suggests that all-important first step. The curve sketching skills that you develop in single-variable calculus will pay dividends here. They will be invaluable in drawing surfaces in space.

We place considerable emphasis on problem solving, and we do this because you will never learn calculus by only going to class and/or reading the textbook. You must involve yourself in the subject, and the best way to do this is with problems. Only when you solve problems do you begin to think about what your instructor has said and what you have read, and then you begin to understand. The exercises in each section begin with routine, drill-type problems designed to reinforce fundamentals. If you have difficulties with early exercises then you have not understood the fundamentals, and you should therefore review material in that section. In later exercises we begin to make you think. Some of the exercises are of the same type as were found earlier, but with more involved calculations; some require you to think about a concept from a different point of view; others may draw on ideas from previous sections; still others may do all of these. It is in these exercises that you find most of the applications, for only when you have mastered fundamentals can you begin to apply calculus to problems from other fields. Certain exercises are marked with a calculator icon to indicate the need for an electronic calculator. There are many other exercises for which a calculator could eliminate laborious hand calculations and use of trigonometry and logarithm tables, but exercises marked with the icon definitely require a calculator. In addition, a calculator is essential for those sections that deal with approximations and iterations (Sections 11.2, 11.7, and 11.12).

At the end of each chapter there is a brief summary of the results in that chapter and a set of review exercises. Don't underestimate the value of these exercises; they test your understanding of the material in the entire chapter. Section exercises develop expertise in small areas; chapter exercises test knowledge of major blocks of subject matter. In addition, Appendix E contains a set of exercises on the entire book. Use these exercises as a review when you have completed your calculus studies, or in preparation for you final examination.

Answers to even-numbered problems can be found in Appendix D. In addition, a student supplement containing detailed solutions to all even-numbered problems is available.

Throughout the book you will encounter an icon to warn you about pitfalls that frequently entrap students. These pitfalls may be inappropriate calculations or misinterpretations of ideas. Pay close attention to each one.

The key word in our approach to calculus is "think". Above all else, we hope that you will train your mind to think and to think logically. We want you to learn how to organize facts and interpret them mathematically. If there is a problem, we want you to be able to decide exactly what that problem is, and finally, to produce a step-by-step procedure by which to solve it. Do not, therefore, read this book expecting formulas that can be used to solve all problems; we won't give them to you. We will give you a few formulas and, we hope, a great deal of insight into the ideas surrounding these formulas. Then you begin to think for yourself. You use what you know to solve the first problem. Having gained from that experience, you then move on to the next problem. In this way, your calculus studies will be a true learning experience—not an exercise in memorization and regurgitation of facts and formulas, but a developmental period during which you both learn calculus and increase your mental capacity. When you reach the end of the road and successfully complete your calculus studies, compare how much you knew at the beginning to how much you know at the end. Congratulate yourself on accomplishing a great deal, and mostly through your own efforts.

Winnipeg, Canada
October, 1992

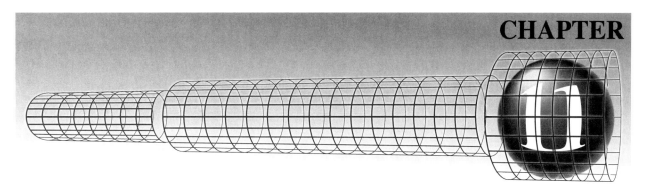

CHAPTER

10

Infinite Sequences and Series

Sequences and series play an important role in many areas of applied mathematics. Sequences were first encountered in Chapter 4, although we did not use the term "sequences" at the time. In Section 4.1, Newton's iterative procedure was used to develop a set of numbers $x_1, x_2, x_3,\ldots$ to approximate a root of an equation $f(x) = 0$. The first number is chosen as some initial approximation to the solution of the equation, and subsequent numbers, defined by the formula

$$x_{n+1} = x_n - \frac{f(x_n)}{f'(x_n)},$$

are better and better approximations. This ordered set of numbers is called a sequence. Each number in the set corresponds to a positive integer, and each is calculated according to a stated formula.

A series is the sum of the numbers in a sequence. If the numbers are $x_1, x_2, x_3,\ldots$, the corresponding series is denoted symbolically by

$$\sum_{n=1}^{\infty} x_n = x_1 + x_2 + x_3 + \cdots,$$

where the three dots indicate that the addition is neverending. It is very well to write an expression like this, but it does not seem to have any meaning. No matter how fast we add, or how fast a calculator adds, or even how fast a supercomputer adds, an infinity of numbers can never be added together in a finite amount of time. We shall give meaning to such expressions, and show that they are really the only sensible way to define many of the more common transcendental functions such as trigonometric, exponential, and hyperbolic.

The first two sections of this chapter are devoted to sequences and the remaining ten to series. This is not to say that series are more important than sequences; they are not. Discussions on series invariably become discussions on sequences associated with series. We have found that difficulties with this chapter can almost always be traced back to a failure to distinguish between the two concepts. Special attention to the material in Sections 11.1 and 11.2 will be rewarded, while a cursory treatment of these sections leads to confusion in later sections.

SECTION 11.1

Infinite Sequences of Constants

Sequences are defined as follows.

Definition 11.1

An **infinite sequence of constants** is a function $f(n)$ whose domain is the set of positive integers.

For example, when $f(n) = 1/n$, the following numbers are associated with the positive integers,

$$1, \frac{1}{2}, \frac{1}{3}, \frac{1}{4}, \ldots.$$

The word "infinite" simply indicates that an infinity of numbers is defined by the sequence as there is an infinity of positive integers, but it indicates nothing about the nature of the numbers. Often we write the numbers $f(n)$ in a line separated by commas,

$$f(1), f(2), \ldots, f(n), \ldots \qquad (11.1a)$$

and refer to this array as the sequence rather than the rule by which it is formed. Since this notation is somewhat cumbersome, we adopt a notation similar to that used for the sequence defined by Newton's iterative procedure. We set $c_1 = f(1)$, $c_2 = f(2)$, ..., $c_n = f(n)$, ..., and write for 11.1a

$$c_1, c_2, c_3, \ldots, c_n, \ldots. \qquad (11.1b)$$

The first number c_1 is called the first **term** of the sequence, c_2 the second term, and for general n, c_n is called the nth term (or general term) of the sequence. For the above example, we have

$$c_1 = 1, \quad c_2 = \frac{1}{2}, \quad c_3 = \frac{1}{3}, \quad \text{etc.}$$

In some applications, it is more convenient to define a sequence as a function whose domain is the set of all integers larger than or equal to some fixed integer N, and N can be positive, negative, or zero. Indeed, in Section 11.8, we find it convenient to initiate the assignment with $N = 0$. For now we prefer to use Definition 11.1 where $N = 1$, in which case we have the natural situation where the first term of the sequence corresponds to $n = 1$, the second term to $n = 2$, etc.

EXAMPLE 11.1

The general terms of four sequences are:

(a) $\dfrac{1}{2^{n-1}}$ (b) $\dfrac{n}{n+1}$ (c) $(-1)^n |n - 3|$ (d) $(-1)^{n+1}$

Write out the first six terms of each sequence.

SOLUTION The first six terms of these sequences are:

(a) $1, \dfrac{1}{2}, \dfrac{1}{4}, \dfrac{1}{8}, \dfrac{1}{16}, \dfrac{1}{32}$;

(b) $\dfrac{1}{2}, \dfrac{2}{3}, \dfrac{3}{4}, \dfrac{4}{5}, \dfrac{5}{6}, \dfrac{6}{7}$;

(c) $-2, 1, 0, 1, -2, 3$;

(d) $1, -1, 1, -1, 1, -1$.

The sequences in Example 11.1 are said to be defined **explicitly**; we have an explicit formula for the nth term of the sequence in terms of n. This allows easy determination of any term in the sequence. For instance, to find the one-hundredth term, we simply replace n by 100 and perform the resulting arithmetic. Contrast this with the sequence in the following example.

EXAMPLE 11.2

The first term of a sequence is $c_1 = 1$ and every other term is to be obtained from the formula
$$c_{n+1} = 5 + \sqrt{2 + c_n}, \quad n \geq 1.$$
Calculate c_2, c_3, c_4, and c_5.

SOLUTION To obtain c_2 we set $n = 1$ in the formula,
$$c_{1+1} = c_2 = 5 + \sqrt{2 + c_1} = 5 + \sqrt{2 + 1} = 5 + \sqrt{3} = 6.732.$$

To find c_3, we set $n = 2$,
$$c_3 = 5 + \sqrt{2 + c_2} = 5 + \sqrt{2 + \left(5 + \sqrt{3}\right)} = 5 + \sqrt{7 + \sqrt{3}} = 7.955.$$

Similarly,
$$c_4 = 5 + \sqrt{2 + c_3} = 5 + \sqrt{7 + \sqrt{7 + \sqrt{3}}} = 8.155, \quad \text{and}$$

$$c_5 = 5 + \sqrt{2 + c_4} = 5 + \sqrt{7 + \sqrt{7 + \sqrt{7 + \sqrt{3}}}} = 8.187.$$

■

When the terms of a sequence are defined by a formula such as the one in Example 11.2, the sequence is said to be defined **recursively**. Note that the terms for a sequence obtained from Newton's iterative procedure are so defined. To find the 100th term of a recursively defined sequence, we must know the 99th; to find the 99th, we must know the 98th; to find the 98th, we need the 97th; and so on down the line. In other words, to find a term in the sequence, we must first find every term that precedes it. Obviously it is much more convenient to have an explicit definition for c_n in terms of n, but this is not always possible. It can be very difficult to find an explicit formula for the nth term of a sequence which is defined recursively.

Sometimes it is impossible to give any algebraic formula for the terms of a sequence. This is illustrated in the following example.

EXAMPLE 11.3

The nth term of a sequence is the nth prime integer (greater than 1) when all such primes are listed in ascending order. List its first ten terms.

SOLUTION The first ten terms of the sequence are
$$2, 3, 5, 7, 11, 13, 17, 19, 23, 29.$$

No one has yet discovered a formula for the prime integers, and it is therefore impossible to express terms of the sequence explicitly or recursively. In spite of this, the sequence is well-defined; and with enough perseverence, and perhaps a high-powered computer, we could find any number of terms in this sequence. ∎

When the general term of a sequence is known explicitly, any term in the sequence is obtained by substituting the appropriate value of n. In other words, the general term specifies every term in the sequence. We therefore use the general term to abbreviate the notation for a sequence by writing the general term in braces and using this to represent the sequence. Specifically, for the sequence in Example 11.1(a), we write

$$\left\{\frac{1}{2^{n-1}}\right\}_1^\infty = 1, \frac{1}{2}, \frac{1}{4}, \frac{1}{8}, \ldots, \frac{1}{2^{n-1}}, \ldots,$$

where 1 and ∞ indicate that the first term corresponds to the integer $n = 1$, and that there is an infinite number of terms in the sequence. In general, we write

$$\{c_n\}_1^\infty = c_1, c_2, c_3, \ldots, c_n, \ldots. \qquad (11.2)$$

If, as is the case in this section, the first term of a sequence corresponds to the integer $n = 1$, we abbreviate the notation further and simply write $\{c_n\}$ in place of $\{c_n\}_1^\infty$.

Since a sequence $\{c_n\}$ is a function whose domain is the set of positive integers, we can represent $\{c_n\}$ graphically. The sequences of Example 11.1 are shown in Figure 11.1.

FIGURE 11.1

(a)

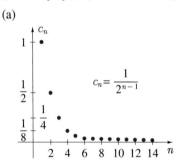

(b)

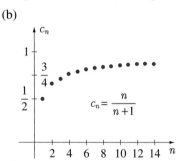

(c)

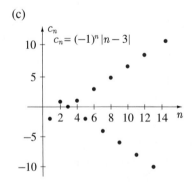

(d)

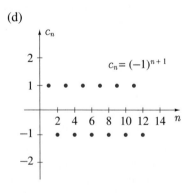

In most applications of sequences we are interested in a number called the **limit** of the sequence. Intuitively, a number L is called the limit of a sequence $\{c_n\}$ if, as we go farther and farther out in the sequence, the terms get arbitrarily close to L, and stay close to L. If such a number L exists, we write

$$L = \lim_{n \to \infty} c_n, \qquad (11.3)$$

and say that the sequence $\{c_n\}$ **converges** to L. If no such number exists, we say that the sequence does not have a limit, or that the sequence **diverges**.

For the sequences of Example 11.1, it is evident that:

(a) $\lim\limits_{n \to \infty} \dfrac{1}{2^{n-1}} = 0$;

(b) $\lim\limits_{n \to \infty} \dfrac{n}{n+1} = 1$;

(c) $\lim\limits_{n \to \infty} (-1)^n |n-3|$ does not exist;

(d) $\lim\limits_{n \to \infty} (-1)^{n+1}$ does not exist;

and in Example 11.3,

$$\lim_{n \to \infty} c_n \text{ does not exist.}$$

Note how the points on the graphs in Figures 11.1(a) and (b) cluster around the limits 0 and 1 as n gets larger and larger. No such clustering occurs in the remaining two figures.

It is usually, but not always, easy to determine whether an explicitly defined sequence has a limit, and what that limit is. It is like finding the limit of a function $f(x)$ as $x \to \infty$. Seldom is this the case, however, for recursive sequences. For instance, it is not at all clear whether the recursive sequence of Example 11.2 has a limit. In spite of the fact that differences between successive terms are approaching zero, and terms of the sequence are therefore getting closer together, the sequence might not have a limit. This is explored further in Exercise 70 at the end of this section. Likewise for recursive sequences defined by Newton's iterative procedure, we pointed out in Section 4.1 that sometimes the sequence does not converge, but gave no conditions that would ensure that it does. What we need, then, are criteria by which to determine whether a given sequence has a limit. The following two definitions lead to Theorem 11.1, which can be very useful in determining whether a recursive sequence has a limit.

Definition 11.2

A sequence $\{c_n\}$ is said to be

(i) **increasing** if $\quad c_{n+1} > c_n \quad$ for all $n \geq 1$; $\quad$ (11.4a)
(ii) **nondecreasing** if $\quad c_{n+1} \geq c_n \quad$ for all $n \geq 1$; $\quad$ (11.4b)
(iii) **decreasing** if $\quad c_{n+1} < c_n \quad$ for all $n \geq 1$; $\quad$ (11.4c)
(iv) **nonincreasing** if $\quad c_{n+1} \leq c_n \quad$ for all $n \geq 1$. $\quad$ (11.4d)

If a sequence satisfies any one of these four properties, it is said to be **monotonic**.

Definition 11.3

A sequence $\{c_n\}$ is said to have an **upper bound** U (be bounded above by U) if

$$c_n \leq U \qquad (11.5)$$

for all $n \geq 1$. It has a **lower bound** V (is bounded below by V) if

$$c_n \geq V \qquad (11.6)$$

for all $n \geq 1$.

Note that if U is an upper bound for a sequence, then any number greater than U is also an upper bound. If V is a lower bound, so too is any number smaller than V.

For the sequences of Example 11.1, we find that $\{1/2^{n-1}\}$ is decreasing and has an upper bound $U = 1$ and a lower bound $V = 0$; $\{n/(n+1)\}$ is increasing and has an upper bound $U = 5$ and a lower bound $V = -2$; $\{(-1)^n|n-3|\}$ is not monotonic and has neither an upper nor a lower bound; and $\{(-1)^{n+1}\}$ is not monotonic and has an upper bound $U = 1$ and a lower bound $V = -3$.

The sequence of Example 11.3 is increasing and has a lower bound $V = 2$, but has no upper bound (see Exercise 81).

The recursive sequence of Example 11.2 will be discussed in detail in Example 11.4. For a complete discussion of this sequence, however, we require the following theorem.

Theorem 11.1 *A bounded, monotonic sequence has a limit.*

To expand on this statement somewhat, consider a sequence $\{c_n\}$ whose terms are illustrated graphically in Figure 11.2(a). Suppose that the sequence is increasing and therefore monotonic, and that U is an upper bound for the sequence. We have shown the upper bound as a horizontal line in the figure; c_1 is a lower bound. Our intuition suggests that because the terms in the sequence always increase, and they never exceed U, the sequence must have a limit. Theorem 11.1 confirms this. The theorem does not suggest the value of the limit, but obviously it must be less than or equal to U.

Similarly, when a sequence is decreasing or nonincreasing and has a lower bound V (Figure 11.2(b)), it must approach a limit that is greater than or equal to V.

FIGURE 11.2

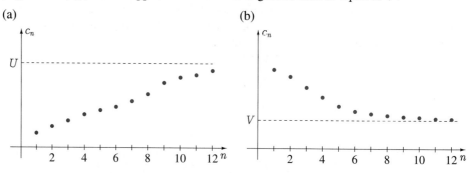

Another way of stating Theorem 11.1 is as follows.

Corollary A monotonic sequence has a limit if and only if it is bounded.

We are now prepared to give a complete and typical discussion for the recursive sequence of Example 11.2.

EXAMPLE 11.4 Find the limit of the sequence of Example 11.2, if it has one.

SOLUTION The first five terms of the sequence are

$$c_1 = 1, \quad c_2 = 6.732, \quad c_3 = 7.955, \quad c_4 = 8.155, \quad c_5 = 8.187.$$

They suggest that the sequence is increasing; that is, $c_{n+1} > c_n$. To prove this we use mathematical induction (see Appendix C). Certainly the inequaltiy is valid for $n = 1$ since $c_2 > c_1$. Suppose k is an integer for which $c_{k+1} > c_k$. Then

$$2 + c_{k+1} > 2 + c_k,$$

from which
$$\sqrt{2 + c_{k+1}} > \sqrt{2 + c_k}.$$

It follows that
$$5 + \sqrt{2 + c_{k+1}} > 5 + \sqrt{2 + c_k}.$$

The left side is c_{k+2} and the right side is c_{k+1}. Therefore we have proved that $c_{k+2} > c_{k+1}$. Hence, by mathematical induction $c_{n+1} > c_n$ for all $n \geq 1$. Since the sequence is increasing, its first term $c_1 = 1$ must be a lower bound. Certainly any upper bound, if one exists, must be at least 8.187 (c_5). We can take any number greater than 8.187 and use mathematical induction to test whether it is indeed an upper bound. It appears that $U = 10$ might be a reasonable guess for an upper bound for this sequence, and we verify this by induction as follows. Clearly $c_1 < 10$. We suppose that k is some integer for which $c_k < 10$. Then

$$2 + c_k < 12,$$

from which
$$\sqrt{2 + c_k} < \sqrt{12}.$$

Thus,
$$c_{k+1} = 5 + \sqrt{2 + c_k} < 5 + \sqrt{12} < 10.$$

By mathematical induction, then $c_n < 10$ for $n \geq 1$.

Since the sequence is monotonic and bounded, Theorem 11.1 guarantees that it has a limit, call it L. To evaluate L, we take limits on each side of the equation defining the sequence recursively:

$$\lim_{n \to \infty} c_{n+1} = \lim_{n \to \infty} \left(5 + \sqrt{2 + c_n}\right).$$

It is important to note that this cannot be done until the conditions of Theorem 11.1 have been checked. Since the terms c_n of the sequence approach L as $n \to \infty$, it follows that $5 + \sqrt{2 + c_n}$ approaches $5 + \sqrt{2 + L}$. Furthermore, as $n \to \infty$, c_{n+1} must also approach L. Do not make the mistake of saying that c_{n+1} approaches $L + 1$ as $n \to \infty$. Think about what $\lim_{n \to \infty} c_{n+1}$ means. We conclude therefore that

$$L = 5 + \sqrt{2 + L}.$$

If we transpose the 5 and square both sides of the equation, we obtain the quadratic equation
$$L^2 - 11L + 23 = 0,$$

with solutions
$$L = \frac{11 \pm \sqrt{29}}{2}.$$

Only the positive square root satisfies the original equation $L = 5 + \sqrt{2 + L}$ defining L, so that $L = (11 + \sqrt{29})/2$. The other root, $(11 - \sqrt{29})/2 \approx 2.8$ can also be eliminated on the grounds that all terms beyond the first are greater than 6. ∎

Now that we know what it means for a sequence to have a limit, and how to find limits, we can be more precise. To give a mathematical definition for the limit of a sequence, we start with our intuitive description and make a succession of paraphrases, each of which is one step closer to a precise definition:

A sequence $\{c_n\}$ has limit L if its terms get arbitrarily close to L, and stay close to L, as n gets larger and larger.

A sequence $\{c_n\}$ has limit L if its terms can be made arbitrarily close to L by choosing n sufficiently large.

A sequence $\{c_n\}$ has limit L if the difference $|c_n - L|$ can be made arbitrarily close to 0 by choosing n sufficiently large.

A sequence $\{c_n\}$ has limit L if given any real number $\epsilon > 0$, no matter how small, we can make the difference $|c_n - L|$ less than ϵ by choosing n sufficiently large.

Finally, we arrive at the following definition.

Definition 11.4

A sequence $\{c_n\}$ has **limit** L if for any given $\epsilon > 0$, there exists an integer N such that for all $n > N$
$$|c_n - L| < \epsilon.$$

This definition puts in precise terms, and in the simplest possible way, our intuitive idea of a limit. For those who have studied Section 2.5, note the similarity between Definition 11.4 and Definition 2.2. For a better understanding of Definition 11.4, it is helpful to consider its geometric interpretation. The inequality $|c_n - L| < \epsilon$, when written in the form $L - \epsilon < c_n < L + \epsilon$ is interpreted as a horizontal band of width 2ϵ around L (Figure 11.3). Definition 11.4 requires that no matter how small ϵ, we can find a stage, denoted by N, beyond which all terms in the sequence are contained in the horizontal band. For the sequence and ϵ in Figure 11.3, N must be chosen as shown. For the same sequence, but a smaller ϵ, N must be chosen correspondingly larger (Figure 11.4). Proofs of many theorems in this chapter require a working knowledge of this definition. As an example, we use it to verify that a sequence cannot have two limits, a fact that we have implicitly assumed throughout our discussions.

FIGURE 11.3

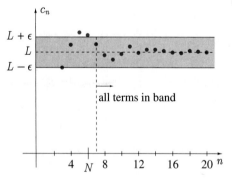

all terms in band

FIGURE 11.4

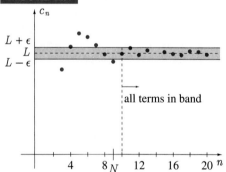

all terms in band

Theorem 11.2

A sequence can have at most one limit.

Proof We prove this theorem by showing that a sequence cannot have two distinct limits. Suppose to the contrary that a sequence $\{c_n\}$ has two distinct limits L_1 and L_2, where $L_2 > L_1$, and let $L_2 - L_1 = \delta$. If we set $\epsilon = \delta/3$, then according to Definition 11.4, there exists an integer N_1 such that for all $n > N_1$

$$|c_n - L_1| < \epsilon = \delta/3;$$

that is, for $n > N_1$, all terms in the sequence are within a distance $\delta/3$ of L_1.

But since L_2 is also supposed to be a limit, there exists an N_2 such that for $n > N_2$, all terms in the sequence are within a distance $\epsilon = \delta/3$ of L_2:

$$|c_n - L_2| < \epsilon = \delta/3.$$

But this is impossible (Figure 11.5), if L_1 and L_2 are a distance δ apart. This contradiction therefore implies that L_1 and L_2 are the same; that is, $\{c_n\}$ cannot have two limits.

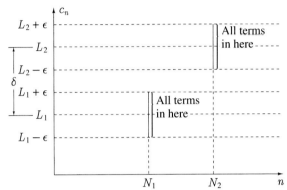

FIGURE 11.5

The following theorem, which states some of the properties of convergent sequences, can also be proved using Definition 11.4. Only the first two parts, however, are straightforward (see Exercises 71, 78, and 79).

Theorem 11.3

If sequences $\{c_n\}$ and $\{d_n\}$ have limits C and D, then:

(i) $\{kc_n\}$ has limit kC if k is a constant;
(ii) $\{c_n \pm d_n\}$ has limit $C \pm D$;
(iii) $\{c_n d_n\}$ has limit CD;
(iv) $\{c_n/d_n\}$ has limit C/D provided that $D \neq 0$, and none of the $d_n = 0$.

EXERCISES 11.1

In Exercises 1–14 state (without proof) whether the sequence is monotonic, and whether it has an upper bound, a lower bound, and a limit.

1. $\left\{\dfrac{1}{n}\right\}$
2. $\{3^n + 1\}$
3. $\{3\}$
4. $\left\{\left(\dfrac{3}{4}\right)^{n+1}\right\}$
5. $\left\{\left(\dfrac{4}{3}\right)^{n+1}\right\}$
6. $\left\{\left(-\dfrac{15}{16}\right)^{n+5}\right\}$
7. $\left\{\sin\left(\dfrac{n\pi}{2}\right)\right\}$
8. $\left\{\dfrac{n}{n^2 + n + 2}\right\}$
9. $\left\{\dfrac{(-1)^n}{n}\right\}$
10. $\{\operatorname{Tan}^{-1} n\}$
11. $\left\{\dfrac{\ln n}{n^2 + 1}\right\}$
12. $\left\{(-1)^n \sqrt{n^2 + 1}\right\}$
13. $c_1 = 2$, $c_{n+1} = \dfrac{2}{c_n - 1}$, $n \geq 1$
14. $c_1 = 4$, $c_{n+1} = -\dfrac{c_n}{n^2}$, $n \geq 1$

In Exercises 15–34 determine whether the statement is true or false. Verify any statement that is true, and give a counter example to any statement that is false.

15. A sequence can be increasing and nondecreasing.
16. A monotonic sequence must be bounded.
17. An increasing sequence must have a lower bound.

18. A decreasing sequence must have a lower bound.

19. An increasing sequence with a lower bound must have a limit.

20. An increasing sequence with an upper bound must have a limit.

21. If a sequence diverges, then either $\lim_{n\to\infty} c_n = \infty$, or, $\lim_{n\to\infty} c_n = -\infty$.

22. A sequence cannot be both increasing and decreasing.

23. A sequence can be both nonincreasing and nondecreasing.

24. If a sequence is monotonic and has a limit, it must be bounded.

25. If a sequence is bounded and has a limit, it must be monotonic.

26. If a sequence is bounded, but not monotonic, it cannot have a limit.

27. If a sequence has a limit, it must be bounded.

28. If a sequence is not monotonic, it cannot have a limit.

29. If all terms of a sequence $\{c_n\}$ are less than U, and $L = \lim_{n\to\infty} c_n$ exists, the L must be less than U.

30. A sequence $\{c_n\}$ has a limit if and only if $\{c_n^2\}$ has a limit.

31. A sequence $\{c_n\}$ of positive numbers converges if $c_{n+1} < c_n/2$.

32. A sequence $\{c_n\}$ of numbers converges if $c_{n+1} < c_n/2$.

33. If an increasing sequence has a limit L, then L must be equal to the smallest upper bound for the sequence.

34. If an infinite number of terms of a sequence all have the same value a, then either a is the limit of the sequence or the sequence has no limit.

In Exercises 35–48 discuss, with proofs, whether the sequence is monotonic, and whether it has an upper bound, a lower bound, and a limit.

35. $\left\{\dfrac{n+1}{2n+3}\right\}$

36. $\left\{\dfrac{2n+3}{n^2-5}\right\}$

37. $\left\{\dfrac{n^2+5n-4}{n^2+2n-2}\right\}$

38. $\{ne^{-n}\}$

39. $\left\{\dfrac{\ln n}{n}\right\}$

40. $\left\{\dfrac{n}{n+1}\operatorname{Tan}^{-1} n\right\}$

41. $c_1 = 1$, $c_{n+1} = \dfrac{1}{10}(c_n^3 + 12)$, $n \geq 1$

42. $c_1 = 0$, $c_{n+1} = \dfrac{1}{12}(c_n^4 + 5)$, $n \geq 1$

43. $c_1 = 3$, $c_{n+1} = \sqrt{5 + c_n}$, $n \geq 1$

44. $c_1 = 1$, $c_{n+1} = \sqrt{5 + c_n}$, $n \geq 1$

45. $c_1 = 5$, $c_{n+1} = 1 + \sqrt{6 + c_n}$, $n \geq 1$

46. $c_1 = 3$, $c_{n+1} = 1 + \sqrt{6 + c_n}$, $n \geq 1$

47. $c_1 = 1$, $c_{n+1} = 4 - \sqrt{5 - c_n}$, $n \geq 1$

48. $c_1 = 4$, $c_{n+1} = 4 - \sqrt{5 - c_n}$, $n \geq 1$

In Exercises 49–53 find an explicit formula for the general term of the sequence. In each case assume that the remaining terms follow the pattern suggested by the given terms.

49. $\dfrac{1}{2}, \dfrac{3}{4}, \dfrac{7}{8}, \dfrac{15}{16}, \dfrac{31}{32}, \ldots$

50. $4, \dfrac{7}{4}, \dfrac{10}{9}, \dfrac{13}{16}, \dfrac{16}{25}, \dfrac{19}{36}, \ldots$

51. $\dfrac{\ln 2}{\sqrt{2}}, -\dfrac{\ln 3}{\sqrt{3}}, \dfrac{\ln 4}{\sqrt{4}}, \dfrac{-\ln 5}{\sqrt{5}}, \ldots$

52. $1, 0, 1, 0, 1, 0, \ldots$

53. $1, 1, -1, -1, 1, 1, -1, -1, \ldots$

54. Verify that the sequence $\{n^{1/n}\}$ is decreasing for $n \geq 3$ by showing that the function $f(x) = x^{1/x}$ has a negative derivative for $x \geq 3$. Illustrate this result graphically. What is the limit of this sequence?

55. A sequence $\{c_n\}$ is defined by $c_n = 3n/(n^2 - 5n + 1)$.

(a) Consider the following proof that the sequence is decreasing: $c_{n+1} < c_n$ if and only if

$$\dfrac{3(n+1)}{(n+1)^2 - 5(n+1) + 1} < \dfrac{3n}{n^2 - 5n + 1}.$$

But this is true if and only if

$$3(n+1)(n^2 - 5n + 1) < 3n(n^2 - 3n - 3) \quad \text{or}$$
$$n^3 - 4n^2 - 4n + 1 < n^3 - 3n^2 - 3n;$$

that is, if and only if

$$1 < n^2 + n.$$

Since this inequality is true for all $n \geq 1$, the sequence is decreasing. Do you see any error in this proof?

(b) Calculate the first five terms in the sequence. What do you now conclude about the argument in (a)?

In Exercises 56–59 show how L'Hôpital's rule (Theorem 4.6, Section 4.9) can be used to evaluate the limit for the sequence.

56. $\left\{\dfrac{\ln n}{\sqrt{n}}\right\}$ 57. $\left\{\dfrac{n^3+1}{e^n}\right\}$

58. $\left\{n\sin\left(\dfrac{4}{n}\right)\right\}$ 59. $\left\{\left(\dfrac{n+5}{n+3}\right)^n\right\}$

In Exercises 60–68 discuss, with all necessary proofs, whether the sequence is monotonic, is bounded, and has a limit.

60. $c_1 = 2$, $c_{n+1} = \dfrac{1}{3-c_n}$, $n \geq 1$

61. $c_1 = 1$, $c_{n+1} = \dfrac{1}{4-2c_n}$, $n \geq 1$

62. $c_1 = 1$, $c_{n+1} = \dfrac{7}{16-8c_n^2}$, $n \geq 1$

63. $c_1 = 0$, $c_{n+1} = \dfrac{7}{16-8c_n^2}$, $n \geq 1$

64. $c_1 = 1$, $c_{n+1} = \dfrac{4c_n}{4+c_n}$, $n \geq 1$

65. $c_1 = 4$, $c_{n+1} = \dfrac{3c_n}{2+c_n}$, $n \geq 1$

66. $c_1 = 2$, $c_{n+1} = \dfrac{2c_n^2}{3+c_n}$, $n \geq 1$

67. $c_1 = \dfrac{3}{2}$, $c_{n+1} = \dfrac{c_n+2}{4-c_n}$, $n \geq 1$

68. $c_1 = 0$, $c_{n+1} = \dfrac{3-c_n}{5-2c_n}$, $n \geq 1$

69. Show that the sequence

$$c_1 = 1, \quad c_{n+1} = \dfrac{1}{4-c_n-c_n^2}, \quad n \geq 1$$

is monotonic and bounded. Find an approximation to its limit accurate to 5 decimals.

70. (a) Prove that if a sequence converges, then differences between successive terms in the sequence must approach zero.

    (b) The following example illustrates that the converse is not true. Show that differences between successive terms of the sequence $\{\ln n\}$ approach zero, but the sequence itself diverges.

71. Prove parts (i) and (ii) of Theorem 11.3.

72. Find bounds for the sequence

$$c_1 = 1, \quad c_{n+1} = \dfrac{1+c_n}{1+2c_n}, \quad n \geq 1.$$

73. Determine whether the following sequence is monotonic, has bounds, and has a limit,

$$c_1 = -30, \quad c_2 = -20, \quad c_{n+1} = 5 + \dfrac{c_n}{2} + \dfrac{c_{n-1}}{3}, \quad n \geq 2.$$

74. A sequence $\{c_n\}$ has only positive terms. Prove that:
    (a) if $\lim_{n\to\infty} c_n = L < 1$, there exists an integer N such that for all $n \geq N$

    $$c_n < \dfrac{L+1}{2}.$$

    (b) if $\lim_{n\to\infty} c_n = L > 1$, there exists an integer N such that for all $n \geq N$,

    $$c_n > \dfrac{L+1}{2}.$$

    These results are used in Theorem 11.9, Section 11.5.

75. Prove that if $\lim_{n\to\infty} c_n = L$, there exists an integer N such that for all $n \geq N$,

    $$c_n < L+1.$$

    This result is used in Theorem 11.8, Section 11.4.

76. The *Fibonacci sequence* found in many areas of applied mathematics is defined by

    $$c_1 = 1, \quad c_2 = 1, \quad c_{n+1} = c_n + c_{n-1}, \quad n \geq 2.$$

    (a) Evaluate the first ten terms of this sequence.
    (b) Is the sequence monotonic, is it bounded, and does it have a limit?
    (c) Prove that

    $$c_n^2 - c_{n-1}c_{n+1} = (-1)^{n+1}, \quad n \geq 2.$$

    (d) Verify that an explicit formula for c_n is

    $$c_n = \dfrac{1}{\sqrt{5}}\left[\left(\dfrac{1+\sqrt{5}}{2}\right)^n - \left(\dfrac{1-\sqrt{5}}{2}\right)^n\right].$$

    (e) Define a sequence $\{b_n\}$ as the ratio of terms in the Fibonacci sequence

    $$b_n = \dfrac{c_{n+1}}{c_n}.$$

    Is this sequence monotonic? Does it have a limit?

77. Prove that the sequence

    $$c_1 = d, \quad c_{n+1} = \sqrt{a+2bc_n}, \quad n \geq 1$$

(a, b, and d all positive constants) is increasing if and only if $d < b + \sqrt{a + b^2}$. What happens when $d = b + \sqrt{a + b^2}$?

78. In this exercise we outline a proof of part (iii) of Theorem 11.3.

(a) Verify that
$$|c_n d_n - CD| \leq |c_n||d_n - D| + |D||c_n - C|.$$

(b) Show that given any $\epsilon > 0$, there exist positive integers N_1, N_2, and N_3 such that

$$|c_n| < |C| + 1 \quad \text{whenever } n > N_1,$$
$$|d_n - D| < \frac{\epsilon}{2(|C| + 1)} \quad \text{whenever } n > N_2,$$
$$|c_n - C| < \frac{\epsilon}{2|D| + 1} \quad \text{whenever } n > N_3.$$

(c) Use these results to prove part (iii) of Theorem 11.3.

79. In this exercise we outline a proof for part (iv) of Theorem 11.3.

(a) Verify that when $d_n \neq 0$ and $D \neq 0$,

$$\left|\frac{c_n}{d_n} - \frac{C}{D}\right| \leq \frac{|c_n - C|}{|d_n|} + \frac{|C||d_n - D|}{|D||d_n|}.$$

(b) Show that given any $\epsilon > 0$, there exist positive integers N_1, N_2, and N_3 such that

$$|d_n| > \frac{|D|}{2} \quad \text{whenever } n > N_1,$$
$$|c_n - C| < \frac{\epsilon|D|}{4} \quad \text{whenever } n > N_2,$$
$$|d_n - D| < \frac{\epsilon|D|^2}{4|C| + 1} \quad \text{whenever } n > N_3.$$

(c) Now prove part (iv) of Theorem 11.3.

80. Find an explicit formula for the recursive sequence
$$c_1 = 1, \quad c_2 = 2, \quad c_{n+1} = \frac{c_n + c_{n-1}}{2}, \quad n \geq 2.$$

81. Justify the statement that the sequence of Example 11.3 has no upper bound.

SECTION 11.2

Applications of Sequences

When a sequence is monotonic and bounded, it has a limit. Verification that a recursive sequence is monotonic and bounded usually requires mathematical induction. However, when the sequence arises in an application, the derivation of the sequence may make it clear that the sequence has a limit, and it may therefore be possible to proceed directly to its evaluation without verifying monotony and bounds. Such is often the case in the field of numerical analysis, wherein recursive sequences commonly arise in the form of iterative procedures. We have already encountered Newton's iterative procedure as one example. When this method is applied to the equation

$$f(x) = x^3 - 3x + 1 = 0$$

with an initial approximation $x_1 = 0$ to find the root between 0 and 1, the sequence obtained is

$$x_1 = 0, \quad x_{n+1} = x_n - \frac{f(x_n)}{f'(x_n)} = x_n - \frac{x_n^3 - 3x_n + 1}{3x_n^2 - 3}, \quad n \geq 1.$$

The first three terms of this sequence are illustrated in Figure 11.6, and the tangent line construction by which they are obtained makes it clear that the sequence must converge to the solution of the equation between 0 and 1. In other words, it is not necessary for us to analyze monotony and bounds for the recursive sequence algebraically; they are obvious geometrically. The sequence is increasing and has bounds $U = 1$ and $V = 0$.

To find the limit of a recursive sequence in Section 11.1, we adopted the procedure of letting n become very large on both sides of the recursive definition for the sequence (see Example 11.4). This does not work here as we are simply led back to the equation

$x^3 - 3x + 1 = 0$. Try it. Instead we evaluate terms of the sequence algebraically until they repeat

$x_1 = 0$, $x_2 = 1/3$, $x_3 = 0.347222$, $x_4 = 0.34729635$, $x_5 = 0.347296355$, $x_6 = 0.347296355$.

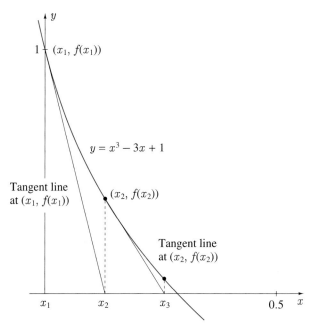

FIGURE 11.6

At this point we might be led to conclude that the required limit and solution of $x^3 - 3x + 1 = 0$ is $x = 0.347296355$. But realizing that our calculations are calculator dependent, we should at least be skeptical of the last digit on the display of our calculator. Suppose the real question were to find the solution accurate to six decimals. This could be accomplished by calculating

$$f(0.3472955) = 2.3 \times 10^{-6} \quad \text{and} \quad f(0.3472965) = -3.8 \times 10^{-7}.$$

Since one function value is positive and the other is negative, the intermediate value theorem (see Theorem 4.1) guarantees that the root is between 0.3472955 and 0.3472965. Hence, to six decimals, the solution is 0.347296.

Figure 11.6 illustrated that the sequence defined by Newton's iterative procedure must converge to a root of $f(x) = x^3 - 3x + 1 = 0$, and therefore it was unnecessary to verify existence of a limit of the sequence by proving that the sequence was monotonic and bounded. Are we proposing that a graph of the function $f(x)$ should always be drawn when using Newton's method to solve equations? Not really, although we are of the philosophy that pictures should be drawn whenever possible. Numerical analysts have proved that under very mild restrictions, the sequence defined by Newton's method always converges to a root of the equation provided the initial approximation is sufficiently close to that root. See Exercise 33 for further discussion of this point.

Another way to solve for the same root of $x^3 - 3x + 1 = 0$ is to rewrite the equation in the form

$$x = \frac{1}{3}(x^3 + 1),$$

and use this to define the following recursive sequence

$$x_1 = 0, \quad x_{n+1} = \frac{1}{3}(x_n^3 + 1), \quad n \geq 1.$$

It is straighforward to show that this sequence is bounded above by 1 and below by 0 and that it is increasing. According to Theorem 11.1, it therefore has a limit L, which must lie in the interval $0 \le L \le 1$. To find L, we set

$$\lim_{n \to \infty} x_{n+1} = \frac{1}{3} \lim_{n \to \infty} (x_n^3 + 1) \quad \text{or}$$

$$L = \frac{1}{3}(L^3 + 1).$$

But this implies that L is the required root of the original equation. In other words, we have defined a recursive sequence $\{x_n\}$ in such a way that its limit is the required solution of $x^3 - 3x + 1 = 0$. The first ten terms of the sequence are:

$x_1 = 0$, $\qquad\qquad x_6 = 0.34729353$,
$x_2 = 1/3$, $\qquad\quad x_7 = 0.34729601$,
$x_3 = 0.345679$, $\quad x_8 = 0.34729631$,
$x_4 = 0.34710219$, $x_9 = 0.34729635$,
$x_5 = 0.34727295$,

which leads us again to the conclusion that an approximation to the root of $x^3 - 3x + 1 = 0$ between 0 and 1 is $x = 0.347296$.

This method of finding the root of an equation is often called the **method of successive substitutions**. Exercises 21–32 provide more opportunities to practise this method.

In Section 3.9 and Appendix B, we essentially defined the number e as the limit of the sequence

$$e_n = \left(1 + \frac{1}{n}\right)^n, \qquad (11.7)$$

but gave no proof that the sequence does indeed converge. In spite of the fact that the sequence is defined explicitly, it is far from trivial to prove that it has a limit. To do so, we first find an expanded expression for e_n by means of the binomial theorem,

$$e_n = \left(1 + \frac{1}{n}\right)^n = \sum_{r=0}^{n} \binom{n}{r}\left(\frac{1}{n}\right)^r$$

$$= 1 + \sum_{r=1}^{n} \frac{n(n-1)\cdots(n-r+1)}{r! \, n^r},$$

where we have used the expression $\binom{n}{r} = n(n-1)\cdots(n-r+1)/r!$ for binomial coefficients. Now in the product $n(n-1)\cdots(n-r+1)$, there are r factors. If each is divided by one of the n's in the denominator, e_n can be expressed in the form

$$e_n = 1 + \sum_{r=1}^{n} \frac{1}{r!}\left(\frac{n}{n}\right)\left(\frac{n-1}{n}\right)\left(\frac{n-2}{n}\right)\cdots\left(\frac{n-r+1}{n}\right)$$

$$= 1 + \sum_{r=1}^{n} \frac{1}{r!}(1)\left(1 - \frac{1}{n}\right)\left(1 - \frac{2}{n}\right)\cdots\left(1 - \frac{r-1}{n}\right). \qquad (11.8)$$

To show that sequence 11.7 has a limit, we verify that it is increasing ($e_{n+1} > e_n$) and bounded ($2 < e_n < 3$). The expanded form for e_{n+1} is obtained by replacing n by $n+1$ in 11.8,

$$e_{n+1} = 1 + \sum_{r=1}^{n+1} \frac{1}{r!}(1)\left(1 - \frac{1}{n+1}\right)\left(1 - \frac{2}{n+1}\right)\cdots\left(1 - \frac{r-1}{n+1}\right).$$

If we drop the last term from this summation (when $r = n + 1$), the right side becomes smaller. In addition, if the $(n + 1)$'s in the denominator are all replaced by n's, each factor decreases. Thus we may write

$$e_{n+1} > 1 + \sum_{r=1}^{n} \frac{1}{r!}(1)\left(1 - \frac{1}{n}\right)\left(1 - \frac{2}{n}\right)\cdots\left(1 - \frac{r-1}{n}\right)$$

But this is e_n (see 11.8). In other words, we have verified that $e_{n+1} > e_n$, and the sequence is increasing.

Since the sequence is increasing, it is bounded below by its first term $e_1 = 2$. To verify that 3 is an upper bounded, we use expression 11.8 to write

$$e_n = 1 + \sum_{r=1}^{n} \frac{1}{r!}(1)\left(1 - \frac{1}{n}\right)\left(1 - \frac{2}{n}\right)\cdots\left(1 - \frac{r-1}{n}\right)$$

$$< 1 + \sum_{r=1}^{n} \frac{1}{r!} \qquad \text{(since all omitted factors are less than one)}$$

$$= 1 + \left(1 + \frac{1}{2!} + \frac{1}{3!} + \cdots + \frac{1}{n!}\right)$$

$$< 1 + \left(1 + \frac{1}{2} + \frac{1}{2^2} + \frac{1}{2^3} + \cdots + \frac{1}{2^{n-1}}\right).$$

A little experimentation shows that the sum of the terms in the parentheses is $2 - 1/2^{n-1}$, and therefore

$$e_n < 1 + 2 - \frac{1}{2^{n-1}} = 3 - \frac{1}{2^{n-1}} < 3.$$

With an upper bound established, Theorem 11.1 now guarantees that the sequence $\{e_n\}$ has a limit. The limit can be approximated by setting n in $(1 + 1/n)^n$ equal to a very large number. For example, with $n = 10^6$, we obtain $e_{1\,000\,000} = 2.71828047$. Unfortunately, we have no idea how accurate this approximation is for e. A superior method for approximating e using infinite series is discussed in Section 11.7, and with this method we can make definitive statements about the accuracy of the approximation.

EXERCISES 11.2

In Exercises 1–12 use Newton's iterative procedure with the given initial approximation x_1 to define a sequence of approximations to a solution of the equation. Determine graphically whether the sequence is monotonic, is bounded, and has a limit. Approximate any limit that exists to seven decimals.

1. $x_1 = 1$, $\quad x^2 + 3x + 1 = 0$
2. $x_1 = -1$, $\quad x^2 + 3x + 1 = 0$
3. $x_1 = -1.5$, $\quad x^2 + 3x + 1 = 0$
4. $x_1 = -3$, $\quad x^2 + 3x + 1 = 0$
5. $x_1 = 4$, $\quad x^3 - x^2 + x - 22 = 0$
6. $x_1 = 2$, $\quad x^3 - x^2 + x - 22 = 0$
7. $x_1 = 2$, $\quad x^5 - 3x + 1 = 0$
8. $x_1 = 1$, $\quad x^5 - 3x + 1 = 0$
9. $x_1 = 0$, $\quad x^5 - 3x + 1 = 0$
10. $x_1 = 4/5$, $\quad x^5 - 3x + 1 = 0$
11. $x_1 = 0.85$, $\quad x^5 - 3x + 1 = 0$
12. $x_1 = -2$, $\quad x^5 - 3x + 1 = 0$
13. (a) Use Newton's iterative procedure with $x_1 = 1$ to approximate the root of $x^4 - 15x + 2 = 0$ between 0 and 1.

    (b) Use the method of successive substitutions with $x_1 = 1$ and $x_{n+1} = (x_n^4 + 2)/15$ to approximate the root in (a).

(c) Use Newton's method to approximate the root between 2 and 3.

(d) What happens if the sequence in (b) is used to approximate the root between 2 and 3 with $x_1 = 2$ and $x_1 = 3$?

14. A superball is dropped from the top of a building 20 m high. Each time it strikes the ground, it rebounds to 99% of the height from which it fell.
    (a) If d_n denotes the distance traveled by the ball between the nth and $(n+1)^{\text{th}}$ bounces, find a formula for d_n.
    (b) If t_n denotes the time between the nth and $(n+1)^{\text{th}}$ bounces, find a formula for t_n.

15. A dog sits at a farmhouse patiently watching for his master to return from the fields. When the farmer is 1 km from the farmhouse, the dog immediately takes off for the farmer. When he reaches the farmer, he turns and runs back to the farmhouse, whereupon he again turns and runs to the farmer. The dog continues this frantic action until the farmer reaches the farmhouse. If the dog runs twice as fast as the farmer, find the distance d_n run by the dog from the point when he reaches the farmer for the nth time to the point when he reaches the farmer for the $(n+1)^{\text{th}}$ time. Ignore any accelerations of the dog in the turns.

16. The equilateral triangle in Figure 11.7(a) has perimeter P. If each side of the triangle is divided into three equal parts, an equilateral triangle is drawn on the middle segment of each side, and the figure transformed into Figure 11.7(b), what is the perimeter P_1 of this figure? If each side of Figure 11.7(b) is now subdivided into three equal portions and equilateral triangles are similarly constructed to result in Figure 11.7(c), what is the perimeter P_2 of this figure? If this subdivision process is continued indefinitely, what is the perimeter P_n after the nth subdivision? What is $\lim_{n \to \infty} P_n$?

FIGURE 11.7

(a)   (b)   (c)

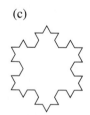

17. A stone of mass 100 gm is thrown vertically upward with speed 20 m/s. Air exerts a resistive force on the stone proportional to its speed, and has magnitude 1/10 N when the speed of the stone is 10 m/s. It can be shown that the height y (in metres) above the projection point attained by the stone is given by

$$y = -98.1t + 1181\left(1 - e^{-t/10}\right),$$

where t is time (measured in seconds with $t = 0$ at the instant of projection).

(a) The time taken for the stone to return to its projection point can be obtained by setting $y = 0$ and solving the equation for t. Do so (correct to 2 decimals).

(b) Find the time for the stone to return if air resistance is ignored.

18. When the beam in Figure 11.8 vibrates vertically, there are certain frequencies of vibration, called *natural frequencies of the system*. They are solutions of the equation

$$\tan x = \frac{e^x - e^{-x}}{e^x + e^{-x}}$$

divided by 20π. Find the two smallest natural frequencies.

FIGURE 11.8

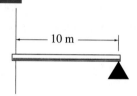

19. Consider the sequence

$$c_1 = 1, \quad c_{n+1} = \frac{1}{2}\left(c_n + \frac{a}{c_n}\right), \quad n \geq 1,$$

where a is any number greater than 1.
(a) Show that $\sqrt{a} < c_n < a$ for $n \geq 2$.
(b) Prove that the sequence is decreasing for $n \geq 2$.
(c) Verify that the sequence converges to $\sqrt{a}$. This result establishes an iterative procedure for finding the square root of real numbers. The sequence converges to $\sqrt{a}$ for $0 < a < 1$ also.
(d) Show that this sequence can be obtained from Newton's iterative procedure for the solution of the equation $x^2 = a$.

20. If A_n is the area of the figure with perimeter P_n in Exercise 16, find a formula for A_n.

To solve an equation $f(x) = 0$ by the method of successive substitutions, we first rearrange the equation into the form $x = g(x)$. We then define a recursive sequence by choosing some initial approximation x_1 and setting

$$x_{n+1} = g(x_n), \quad n \geq 1.$$

Depending on the choice of $g(x)$, this sequence may or may not converge to a root of the equation. In Exercises 21–25 illustrate that the suggested rearrangement of the equation $f(x) = 0$, along with the initial approximation x_1, leads to a sequence which converges to a root of the equation. Find the root accurate to 4 decimals.

21. $f(x) = x^2 - 2x - 1$; $\quad x_1 = 2$; $\quad$ use $x = 2 + 1/x$

22. $f(x) = x^3 + 6x + 3$; $\quad x_1 = -1$; $\quad$ use $x = -(1/6)(x^3 + 3)$

23. $f(x) = x^4 - 120x + 20$; $\quad x_1 = 0$; $\quad$ use $x = (x^4 + 20)/120$

24. $f(x) = x^3 - 2x^2 - 3x + 1$; $\quad x_1 = 3$; $\quad$ use $x = (2x^2 + 3x - 1)/x^2$

25. $f(x) = 8x^3 - x^2 - 1$; $\quad x_1 = 0$; $\quad$ use $x = (1/2)(1 + x^2)^{1/3}$

In Exercises 26–31 find a rearrangement of the equation which leads, through the method of successive substitutions, to a 4 decimal approximation to the root of the equation.

26. $x^3 - 6x^2 + 11x - 7 = 0$ $\quad$ between $x = 3$ and $x = 4$

27. $x^4 - 3x^2 - 3x + 1 = 0$ $\quad$ between $x = 0$ and $x = 1$

28. $x^4 + 4x^3 - 50x^2 + 100x - 50 = 0$ $\quad$ between $x = 0$ and $x = 1$

29. $\sin^2 x = 1 - x^2$ $\quad$ between $x = 0$ and $x = 1$

30. $\sec x = 2/(1 + x^4)$ $\quad$ between $x = 0$ and $x = 1$

31. $e^x + e^{-x} - 10x = 0$ $\quad$ between $x = 0$ and $x = 1$

32. (a) Show that if α is the only root of the equation $x = g(x)$ and Figure 11.9(a) is a graph of $g(x)$, then Figure 11.9(b) exhibits geometrically the sequence of approximations of α determined by the method of successive substitutions.

FIGURE 11.9

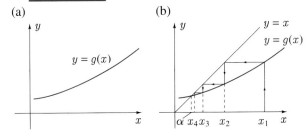

(b) Illustrate graphically how the sequence defined by successive substitutions converges to the root $x = \alpha$

for the equation $x = g(x)$ if $g(x)$ is as shown in Figure 11.10.

FIGURE 11.10

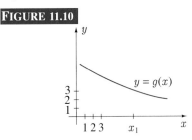

(c) Illustrate graphically that the method fails for the function $g(x)$ in Figure 11.11.

FIGURE 11.11

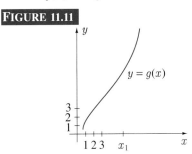

(d) Based on the results of parts (a)–(c), what determines success or failure of the method of successive substitutions? Part (e) provides a proof of the correct answer.

(e) Prove that when $x = g(x)$ has a root $x = \alpha$, the method of successive substitutions with an initial approximation of x_1 always converges to α if $|g'(x)| \leq a < 1$ on the interval $|x - \alpha| \leq |x_1 - \alpha|$. In other words, when an equation $f(x) = 0$ is rearranged into the form $x = g(x)$ for successive subsitutions, success or failure depends on whether the derivative of $g(x)$ is between -1 and 1 near the required root.

33. Suppose that $f''(x)$ exists on an open interval containing a root $x = \alpha$ of the equation $f(x) = 0$. Use the result of Exercises 32 to prove that if $f'(\alpha) \neq 0$, Newton's iterative sequence always converges to α provided the initial approximation x_1 is chosen sufficiently close to α. (Hint: First use Exercise 32(e) to show that Newton's sequence converges to α if on the interval $|x - \alpha| \leq |x_1 - \alpha|$, $|ff''/(f')^2| \leq a < 1$.) In actual fact, Newton's method usually works even when $f'(\alpha) = 0$.

SECTION 11.3

Infinite Series of Constants

If $\{c_n\}$ is an infinite sequence of constants, an expression of the form

$$c_1 + c_2 + c_3 + \cdots + c_n + \cdots \quad (11.9)$$

is called an **infinite series**, or simply a series. We often use sigma notation to represent a series:

$$\sum_{n=1}^{\infty} c_n = c_1 + c_2 + c_3 + \cdots + c_n + \cdots . \quad (11.10)$$

By initiating the sigma notation with $n = 1$, we once again have the convenient situation in which $n = 1$ yields the first term of the series, $n = 2$ the second, etc. We usually abbreviate the notation further by writing $\sum c_n$ in place of $\sum_{n=1}^{\infty} c_n$ whenever the lower limit of summation is chosen as unity.

From the sequences of Example 11.1 we may form the following series.

EXAMPLE 11.5

Write out the first six terms of the following series:

(a) $\sum_{n=1}^{\infty} \dfrac{1}{2^{n-1}}$ (b) $\sum_{n=1}^{\infty} \dfrac{n}{n+1}$ (c) $\sum_{n=1}^{\infty} (-1)^n |n-3|$ (d) $\sum_{n=1}^{\infty} (-1)^{n+1}$

SOLUTION The first six terms of these series are shown below:

(a) $\sum_{n=1}^{\infty} \dfrac{1}{2^{n-1}} = 1 + \dfrac{1}{2} + \dfrac{1}{4} + \dfrac{1}{8} + \dfrac{1}{16} + \dfrac{1}{32} + \cdots$

(b) $\sum_{n=1}^{\infty} \dfrac{n}{n+1} = \dfrac{1}{2} + \dfrac{2}{3} + \dfrac{3}{4} + \dfrac{4}{5} + \dfrac{5}{6} + \dfrac{6}{7} + \cdots$

(c) $\sum_{n=1}^{\infty} (-1)^n |n-3| = -2 + 1 - 0 + 1 - 2 + 3 - \cdots$

(d) $\sum_{n=1}^{\infty} (-1)^{n+1} = 1 - 1 + 1 - 1 + 1 - 1 + \cdots$

Perhaps the most frequently encountered series in applications is the **geometric series**; a series of the form

$$a + ar + ar^2 + ar^3 + \cdots .$$

Each term in a geometric series is obtained by multiplying the term that came before it by the same constant r. The number $a \neq 0$ is called the first term of the series and r is called the **common ratio**. In sigma notation,

$$\sum_{n=1}^{\infty} ar^{n-1} = a + ar + ar^2 + ar^3 + \cdots . \quad (11.11)$$

The series in parts (a) and (d) of Example 11.5 are geometric. The first has common ratio $1/2$ the second has common ratio -1.

Another important series is

$$\sum_{n=1}^{\infty} \frac{1}{n} = 1 + \frac{1}{2} + \frac{1}{3} + \frac{1}{4} + \cdots. \qquad (11.12)$$

It is called the **harmonic series**. We shall see it many times in our discussions.

It may be quite convenient to write expressions like 11.9 and say that they are infinite series, but since we can add only finitely many numbers, 11.9 is as yet meaningless. To illustrate how we may define a meaning for such a sum, consider the geometric series of Example 11.5(a),

$$\sum_{n=1}^{\infty} \frac{1}{2^{n-1}} = 1 + \frac{1}{2} + \frac{1}{4} + \frac{1}{8} + \cdots.$$

If we nonchalantly start adding terms of this series together, a pattern soon emerges. Indeed, if we denote by S_n the sum of the first n terms of this series, we find that

$$S_1 = 1,$$
$$S_2 = 1 + \frac{1}{2} = \frac{3}{2}, \qquad \text{(sum of first two terms)}$$
$$S_3 = 1 + \frac{1}{2} + \frac{1}{4} = \frac{7}{4}, \qquad \text{(sum of first three terms)}$$
$$S_4 = 1 + \frac{1}{2} + \frac{1}{4} + \frac{1}{8} = \frac{15}{8}, \qquad \text{(sum of first four terms)}$$
$$S_5 = 1 + \frac{1}{2} + \frac{1}{4} + \frac{1}{8} + \frac{1}{16} = \frac{31}{16}, \qquad \text{(sum of first five terms)}$$

and so on. By careful examination of the pattern, we can see that the sum of the first n terms of the series is given by the formula

$$S_n = 2 - \frac{1}{2^{n-1}}.$$

As we add more and more terms of this series together, the sum S_n gets closer and closer to 2. It is always less than 2, but S_n can be made arbitrarily close to 2 by choosing n sufficiently large. If this series is to have a sum, the only reasonable sum is 2. In practice, this is precisely what we do; we define the sum of the series $\sum 1/2^{n-1}$ to be 2.

Let us now take this idea and define sums for all infinite series. We begin by defining a sequence $\{S_n\}$ as follows:

$$S_1 = c_1,$$
$$S_2 = c_1 + c_2,$$
$$S_3 = c_1 + c_2 + c_3,$$
$$\vdots$$
$$S_n = c_1 + c_2 + c_3 + \cdots + c_n,$$
$$\vdots$$

It is called the **sequence of partial sums** of the series $\sum c_n$. The nth term of the sequence $\{S_n\}$ represents the sum of the first n terms of the series. If this sequence has a limit, say, S, then the more terms of the series that we add together, the closer the sum gets to S. It seems reasonable, then, to call S the sum of the series. We therefore make the following definition.

Definition 11.5 Let $S_n = \sum_{k=1}^{n} c_k$ be the nth partial sum of a series $\sum_{n=1}^{\infty} c_n$. If the sequence of partial sums $\{S_n\}$ has limit S,

$$S = \lim_{n \to \infty} S_n,$$

we call S the sum of the series and write

$$\sum_{n=1}^{\infty} c_n = S;$$

if $\{S_n\}$ does not have a limit, we say that the series does not have a sum.

If a series has sum S, we say that the series **converges** to S, which means that its sequence of partial sums converges to S. If a series does not have a sum, we say that the series **diverges**, which means that its sequence of partial sums diverges.

According to this definition, the series $\sum 1/2^{n-1}$ of Example 11.5(a) has sum 2. Since every term of the series $\sum n/(n+1)$ in Example 11.5(b) is greater than or equal to $1/2$, it follows that the sum of the first n terms is $S_n \geq n(1/2) = n/2$. As the sequence of partial sums is therefore (increasing and) unbounded, it cannot possibly have a limit (see the corollary to Theorem 11.1). The series does not therefore have a sum; it diverges. Examination of the first few partial sums of the series $\sum (-1)^n |n-3|$ in Example 11.5(c) leads to the result that for $n > 1$,

$$S_n = \begin{cases} \dfrac{n-4}{2} & \text{if } n \text{ is even} \\ \dfrac{1-n}{2} & \text{if } n \text{ is odd.} \end{cases}$$

Since the sequence $\{S_n\}$ does not have a limit, the series diverges. The partial sums of the series $\sum (-1)^{n+1}$ are

$$S_1 = 1, \quad S_2 = 0, \quad S_3 = 1, \quad S_4 = 0, \quad \cdots.$$

Since this sequence does not have a limit, the series does not have a sum.

Consider now geometric series 11.11. If $\{S_n\}$ is the sequence of partial sums for this series, then

$$S_n = a + ar + ar^2 + \cdots + ar^{n-1}.$$

If we multiply this equation by r,

$$rS_n = ar + ar^2 + ar^3 + \cdots + ar^n.$$

When we subtract these equations, the result is

$$S_n - rS_n = a - ar^n.$$

Hence, for $r \neq 1$, we obtain

$$S_n = \frac{a(1-r^n)}{1-r}.$$

Furthermore, when $r = 1$,

$$S_n = na,$$

and therefore the sum of the first n terms of a geometric series is

$$S_n = \begin{cases} \dfrac{a(1-r^n)}{1-r} & r \neq 1 \\ na & r = 1. \end{cases} \qquad (11.13\text{a})$$

To determine whether a geometric series has a sum, we consider the limit of this sequence. Certainly, $\lim_{n \to \infty} na$ does not exist, unless trivially $a = 0$. In addition, $\lim_{n \to \infty} r^n = 0$ when $|r| < 1$, and does not exist when $|r| > 1$. Thus, we may state that

$$\lim_{n \to \infty} S_n = \begin{cases} \dfrac{a}{1-r} & |r| < 1 \\ \text{does not exist} & |r| \geq 1. \end{cases}$$

The geometric series therefore has sum

$$\sum_{n=1}^{\infty} ar^{n-1} = \frac{a}{1-r} \qquad (11.13\text{b})$$

if $|r| < 1$, but otherwise diverges.

Next we consider harmonic series 11.12. This series does not have a sum, and we can show this by considering the following partial sums of the series:

$$S_1 = 1,$$
$$S_2 = 1 + \frac{1}{2} = \frac{3}{2},$$
$$S_4 = S_2 + \frac{1}{3} + \frac{1}{4} > \frac{3}{2} + \frac{1}{4} + \frac{1}{4} = \frac{4}{2},$$
$$S_8 = S_4 + \frac{1}{5} + \frac{1}{6} + \frac{1}{7} + \frac{1}{8} > \frac{4}{2} + \frac{1}{8} + \frac{1}{8} + \frac{1}{8} + \frac{1}{8} = \frac{5}{2},$$
$$S_{16} = S_8 + \frac{1}{9} + \frac{1}{10} + \cdots + \frac{1}{16} > \frac{5}{2} + \frac{1}{16} + \frac{1}{16} + \cdots + \frac{1}{16} = \frac{6}{2}.$$

This procedure can be continued indefinitely, and shows that the sequence of partial sums is unbounded. The harmonic series therefore diverges (see once again the corollary to Theorem 11.1).

In each of the above examples, we used Definition 11.5 for the sum of a series to determine whether the series converges or diverges; that is, we formed the sequence of partial sums $\{S_n\}$ in order to consider its limit. For most examples, it is either too difficult or impossible to evaluate S_n in a simple form, and in such cases consideration of the limit of the sequence $\{S_n\}$ is impractical. Consequently, we must develop alternative ways to decide on the convergence of a series. We do this in Sections 11.4–11.6.

We now discuss some fairly simple but important results on convergence of series. If a series $\sum c_n$ has a finite number of its terms altered in any fashion whatsoever, the new series converges if and only if the original series converges. The new series may converge to a different sum, but it converges if the original series converges. For example, if we double the first three terms of the geometric series in Example 11.5, but do not change the remaining terms, the new series is

$$2 + 1 + \frac{1}{2} + \frac{1}{8} + \frac{1}{16} + \cdots .$$

It is not geometric. But its nth partial sum, call it S_n, is very closely related to that of the geometric series, call it T_n. In fact for $n \geq 4$, we can say that $S_n = T_n + \frac{7}{4}$ ($\frac{7}{4}$ is the total change in the first three terms). Since $\lim_{n \to \infty} T_n = 2$, it follows that

$\lim_{n\to\infty} S_n = 2 + \frac{7}{4} = \frac{15}{4}$; that is, the new series converges, but it has a sum which is $\frac{7}{4}$ greater than that of the geometric series.

On the other hand, suppose we change the first 100 terms of the harmonic series, which diverges, to 0, but leave the remaining terms unaltered. The new series is

$$\underbrace{0 + 0 + \cdots + 0}_{100 \text{ terms}} + \frac{1}{101} + \frac{1}{102} + \cdots.$$

We know that the sequence of partial sums $\{T_n\}$ of the harmonic series is increasing and unbounded. The nth partial sum S_n of the new series, for $n \geq 100$, is equal to $T_n - k$, where k is the sum of the first 100 terms of the harmonic series. But because $\lim_{n\to\infty} T_n = \infty$, so also must $\lim_{n\to\infty} S_n = \infty$; that is, the new series diverges.

The following theorem indicates that convergent series can be added and subtracted, and multiplied by constants. These properties can be verified using Definition 11.5.

Theorem 11.4 If series $\sum_{n=1}^{\infty} c_n$ and $\sum_{n=1}^{\infty} d_n$ have sums C and D, then:

$$\text{(i)} \quad \sum_{n=1}^{\infty} kc_n = kC \qquad \text{(when } k \text{ is a constant);} \qquad (11.14\,\text{a})$$

$$\text{(ii)} \quad \sum_{n=1}^{\infty} (c_n \pm d_n) = C \pm D. \qquad (11.14\,\text{b})$$

Our first convergence test is a corollary to the following theorem.

Theorem 11.5 If a series $\sum_{n=1}^{\infty} c_n$ converges, then $\lim_{n\to\infty} c_n = 0$.

Proof If series $\sum c_n$ has a sum S, its sequence of partial sums

$$S_1, S_2, \cdots, S_n, \cdots$$

has limit S. The sequence

$$0, S_1, S_2, \cdots, S_{n-1}, \cdots$$

must also have limit S; it is the sequence of partial sums with an additional term equal to 0 at the beginning. According to Theorem 11.3, if we subtract these two sequences, the resulting sequence must have limit $S - S = 0$; that is,

$$S_1 - 0, S_2 - S_1, \cdots, S_n - S_{n-1}, \cdots \to 0.$$

But $S_1 - 0 = S_1 = c_1$, $S_2 - S_1 = c_2$, etc; that is, we have shown that $\lim_{n\to\infty} c_n = 0$.

Theorem 11.5 states that a necessary condition for a series $\sum c_n$ to converge is that the sequence $\{c_n\}$ of its terms must approach zero. What we really want are sufficient conditions to guarantee convergence or divergence of a series. We can take the contrapositive of the theorem and obtain the following.

Corollary (n^{th} term test) If $\lim_{n\to\infty} c_n \neq 0$, or does not exist, then the series $\sum_{n=1}^{\infty} c_n$ diverges.

This is our first convergence test, the nth term test. It is in fact a test for divergence rather than convergence, stating that if $\lim_{n\to\infty} c_n$ exists and is equal to anything but zero, or the limit does not exist, then the series $\sum c_n$ diverges. Note well that the nth term test never indicates that a series converges. Even if $\lim_{n\to\infty} c_n = 0$, we can conclude nothing about the convergence or divergence of $\sum c_n$; it may converge or it may diverge. For example, the harmonic series $\sum 1/n$ and the geometric series $\sum 1/2^{n-1}$ both satisfy the condition $\lim_{n\to\infty} c_n = 0$, yet one series diverges and the other converges. The nth term test therefore may indicate that a series diverges, but it never indicates that a series converges.

In particular, both series $\sum (-1)^n |n-3|$ and $\sum (-1)^{n+1}$ of Example 11.5 diverge by the nth term test.

To understand series, it is crucial to distinguish clearly among three entities: the series itself, its sequence of partial sums, and its sequence of terms. For any series $\sum c_n$, we can form its sequence of terms $\{c_n\}$ and its sequence of partial sums $\{S_n\}$. Each of these sequences may give information about the sum of the series $\sum c_n$, but in very different ways.

The sum of the series is defined by its sequence of partial sums in that $\sum c_n$ has a sum only if $\{S_n\}$ has a limit. The sequence of partial sums therefore tells us definitely whether the series converges or diverges, provided we can evaluate S_n in a simple form.

The sequence of terms $\{c_n\}$, on the other hand, may or may not tell us whether the series diverges. If $\{c_n\}$ has no limit or has a limit other than zero, we know that the series does not have a sum. If $\{c_n\}$ has limit zero, we obtain no information about convergence of the series, and must continue our investigation.

EXERCISES 11.3

In Exercises 1–10 determine whether the series converges or diverges. Find the sum of each convergent series. To get a feeling for a series, it is helpful to write out its first few terms. Try it.

1. $\sum_{n=1}^{\infty} \dfrac{n+1}{2n}$

2. $\sum_{n=1}^{\infty} \dfrac{2^n}{5^{n+1}}$

3. $\sum_{n=1}^{\infty} \cos\left(\dfrac{n\pi}{2}\right)$

4. $\sum_{n=1}^{\infty} \left(\dfrac{n}{n+1}\right)^n$

5. $\sum_{n=1}^{\infty} \dfrac{7^{2n+3}}{3^{2n-2}}$

6. $\sum_{n=1}^{\infty} \dfrac{7^{n+3}}{3^{2n-2}}$

7. $\sum_{n=1}^{\infty} \sqrt{\dfrac{n^2-1}{n^2+1}}$

8. $\sum_{n=1}^{\infty} \dfrac{\cos(n\pi)}{2^n}$

9. $\sum_{n=1}^{\infty} \dfrac{4^n + 3^n}{3^n}$

10. $\sum_{n=1}^{\infty} \mathrm{Tan}^{-1} n$

In Exercises 11–14 we have given a repeating decimal. Express the decimal as a geometric series and use formula 11.13b to express it as a rational number.

11. $0.666666\ldots$
12. $0.1313131313\ldots$
13. $1.347346346346\ldots$
14. $43.020502050205\ldots$

In Exercises 15–17 complete the statement and give a short proof to substantiate your claim.

15. If $\sum c_n$ and $\sum d_n$ converge, then $\sum (c_n + d_n) \ldots$

16. If $\sum c_n$ converges and $\sum d_n$ diverges, then $\sum (c_n + d_n) \ldots$

17. If $\sum c_n$ and $\sum d_n$ diverge, then $\sum (c_n + d_n) \ldots$

In Exercises 18–21 determine whether the series converges or diverges. Find the sum of each convergent series.

18. $\sum_{n=1}^{\infty} \dfrac{2^n + 3^n}{4^n}$

19. $\sum_{n=1}^{\infty} \dfrac{3^n - 1}{2^n}$

20. $\sum_{n=1}^{\infty} \dfrac{n^2 + 2^{2n}}{4^n}$

21. $\sum_{n=1}^{\infty} \dfrac{2^n + 4^n - 8^n}{2^{3n}}$

22. Find the sum of the series

$$\sum_{n=1}^{\infty} \dfrac{1}{n(n+1)}.$$

Hint: Use partial fractions on the nth term and find the sequence of partial sums.

23. Find the total distance travelled by the superball in Exercise 14 of Section 11.2 before it comes to rest.

24. Find the time taken for the superball in Exercise 14 of Section 11.2 to come to rest.

25. What distance does the dog run from the time when it sees the farmer until the farmer reaches the farmhouse in Exercise 15 of Section 11.2?

26. Find a simplified formula for the area A_n in Exercise 20 of Section 11.2. What is $\lim_{n\to\infty} A_n$?

27. According to equation 11.13a, the nth partial sum of the geometric series $1 + r + r^2 + r^3 + \cdots$ is $S_n = (1 - r^n)/(1 - r)$. If T_n denotes the nth partial sum of the series

$$1 + 2r + 3r^2 + 4r^3 + \cdots,$$

show that
$$T_n - S_n = r(T_n - nr^{n-1}).$$

Solve this equation for T_n and take the limit as $n \to \infty$ to show that

$$\sum_{n=1}^{\infty} nr^{n-1} = \frac{1}{(1-r)^2}, \quad |r| < 1.$$

In Exercises 28–31 use the result of Exercise 27 to find the sum of the series.

28. $\dfrac{1}{2} + \dfrac{2}{2^2} + \dfrac{3}{2^3} + \dfrac{4}{2^4} + \cdots$

29. $\dfrac{2}{5} + \dfrac{4}{25} + \dfrac{6}{125} + \dfrac{8}{625} + \cdots$

30. $\dfrac{2}{3} + \dfrac{3}{27} + \dfrac{4}{243} + \dfrac{5}{2187} + \cdots$

31. $\dfrac{12}{5} + \dfrac{48}{25} + \dfrac{192}{125} + \dfrac{768}{625} + \cdots$

32. Two people flip a single coin to see who can first flip a head. Show that the probability that the first person to flip wins the game is represented by the series

$$\frac{1}{2} + \frac{1}{8} + \frac{1}{32} + \cdots + \frac{1}{2^{2n-1}} + \cdots.$$

What is the sum of this series?

33. Two people throw a die to see who can first throw a six. Find the probability that the person who first throws wins the game.

34. One of *Zeno's paradoxes* describes a race between Achilles and a tortoise. Zeno claims that if the tortoise is given a head start, then no matter how fast Achilles runs, he can never catch the tortoise. He reasons as follows: In order to catch the tortoise, Achilles must first make up the length of the head start. But while he is running this distance, the tortoise "runs" a further distance. While Achilles makes up this distance, the tortoise covers a further distance, and so on. It follows that Achilles is always making up distance covered by the tortoise, and therefore can never catch the tortoise. If the tortoise is given a head start of length L and Achilles runs $c > 1$ times as fast as the tortoise, use infinite series to show that Achilles does in fact catch the tortoise and that the distance he covers in doing so is $cL(c-1)^{-1}$.

35. Find the time between 1:05 and 1:10 when the minute and hour hands of a clock point in the same direction
    (a) by reasoning with infinite series as in Exercise 34; and
    (b) by finding expressions for the angular displacements of the hands as functions of time.

36. Repeat Exercise 35 for the instant between 10:50 and 10:55 when the hands coincide.

37. A child has a large number of identical cubical blocks, which she stacks as shown in Figure 11.12. The top block protrudes $\frac{1}{2}$ its length over the second block, which protrudes $\frac{1}{4}$ its length over the third block, which protrudes $\frac{1}{6}$ its length over the fourth block, etc. Assuming the centre of mass of each block is at its geometric centre, show that the centre of mass of the top n blocks lies directly over the edge of the $(n+1)^{\text{th}}$ block. Now deduce that if a sufficient number of blocks is piled, the top block can be made to protrude as far over the bottom block as desired without the stack falling.

FIGURE 11.12

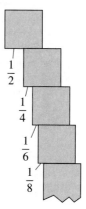

38. It is customary to assume that when a drug is administered to the human body it will be eliminated exponentially; that is, if A represents the amount of drug in the body, then

$$A = A_0 e^{-kt},$$

where $k > 0$ is a constant and A_0 is the amount injected at time $t = 0$. Suppose n successive injections of amount A_0 are administered at equally spaced time intervals T, the first injection at time $t = 0$.

(a) Show that the amount of drug in the body at time t between the nth and $(n+1)^{\text{th}}$ injection is given by

$$A_n(t) = A_0 e^{-kt}\left[\frac{1 - e^{knT}}{1 - e^{kT}}\right], \quad (n-1)T < t < nT.$$

(b) Sketch graphs of these functions on one set of axes.

(c) What is the amount of drug in the body immediately after the nth injection for very large n; that is, what is
$$\lim_{n \to \infty} A_n((n-1)T) ?$$

39. Prove the following result: If $\sum_{n=1}^{\infty} c_n$ converges, then its terms can be grouped in any manner, and the resulting series is convergent with the same sum as the original series.

SECTION 11.4

Nonnegative Series

In Section 11.3 we derived the nth term test, a test for divergence of a series. In order to develop further convergence and divergence tests, we consider two classes of series:

1. Series with terms that are all nonnegative;
2. Series with both positive and negative terms.

A series with terms that are all nonpositive is the negative of a series of type 1, and therefore any test applicable to series of type 1 is easily adapted to a series with nonpositive terms.

Definition 11.6

A series $\sum_{n=1}^{\infty} c_n$ is said to be **nonnegative** if each term is nonnegative: $c_n \geq 0$.

For example, the harmonic series $\sum 1/n$ and the geometric series $\sum 1/2^{n-1}$ are both nonnegative series. We have already seen that both series have the property that $\lim_{n \to \infty} c_n = 0$, yet the harmonic series diverges and the geometric series converges. Examination of the terms of these series reveals that $1/2^{n-1}$ approaches 0 much more quickly than does $1/n$. In general, whether a nonnegative series does or does not have a sum depends on how quickly its sequence of terms $\{c_n\}$ approaches zero. The study of convergence of nonnegative series is an investigation into the question "How fast must the sequence of terms of a nonnegative series approach zero in order that the series have a sum?" This is not a simple problem; only a partial answer is provided through the tests in this section and the next. We begin with the integral test.

Theorem 11.6

(Integral test) Suppose the terms in a series $\sum_{n=1}^{\infty} c_n$ are denoted by $c_n = f(n)$ and $f(x)$ is a continuous, positive, decreasing function for $x \geq 1$. Then the series converges if and only if the improper integral $\int_1^{\infty} f(x)\, dx$ converges.

Proof Suppose first that the improper integral converges to value K. We know that this value can be interpreted as the area under the curve $y = f(x)$, above the x-axis, and to the right of the line $x = 1$ in Figure 11.13. In Figure 11.14 the area of the nth rectangle is $f(n) = c_n$, and clearly it is less than the area under the curve from $x = n-1$ to $x = n$.

FIGURE 11.13

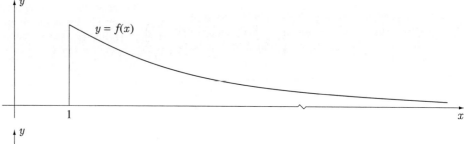

FIGURE 11.14

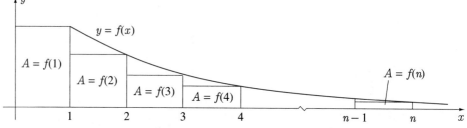

Now, the nth partial sum of the series $\sum c_n$ is

$$\begin{aligned}
S_n &= c_1 + c_2 + \cdots + c_n \\
&= f(1) + f(2) + \cdots + f(n) \\
&< f(1) + \int_1^2 f(x)\,dx + \int_2^3 f(x)\,dx + \cdots + \int_{n-1}^n f(x)\,dx \\
&= f(1) + \int_1^n f(x)\,dx \\
&< f(1) + \int_1^\infty f(x)\,dx \\
&= f(1) + K.
\end{aligned}$$

What this shows is that the sequence $\{S_n\}$ is bounded (since $f(1) + K$ is independent of n). Because the sequence is also increasing, it follows by Theorem 11.1 that it must have a limit; that is, the series $\sum c_n$ converges.

Conversely, suppose now that the improper integral diverges. This time we draw rectangles to the right of the vertical lines at $n = 1, 2, \ldots$ (Figure 11.15). Then

$$\begin{aligned}
S_n &= c_1 + c_2 + \cdots + c_n \\
&= f(1) + f(2) + \cdots + f(n) \\
&> \int_1^2 f(x)\,dx + \int_2^3 f(x)\,dx + \cdots + \int_{n-1}^n f(x)\,dx \\
&= \int_1^n f(x)\,dx.
\end{aligned}$$

Since $\lim_{n\to\infty} \int_1^n f(x)\,dx = \infty$, it follows that $\lim_{n\to\infty} S_n = \infty$, and the series therefore diverges.

FIGURE 11.15

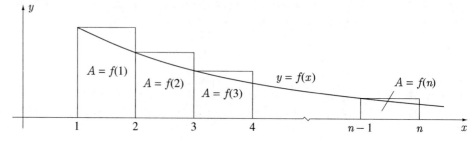

To use the integral test it is necessary to antidifferentiate the function $f(x)$, obtained by replacing n's in c_n with x's. When the antiderivative appears obvious, and other conditions of Theorem 11.6 are met, the integral test may be the easiest way to decide on convergence of the series; when the antiderivative is not obvious, it may be better to try another test.

EXAMPLE 11.6

Determine whether the following series converge or diverge:

$$\text{(a)} \sum_{n=1}^{\infty} \frac{1}{n^2+1} \qquad \text{(b)} \sum_{n=2}^{\infty} \frac{1}{n \ln n} \qquad \text{(c)} \sum_{n=1}^{\infty} n e^{-n}$$

SOLUTION (a) Since $f(x) = 1/(x^2+1)$ is continuous, positive, and decreasing for $x \geq 1$, and

$$\int_1^{\infty} \frac{1}{x^2+1}\,dx = \left\{\text{Tan}^{-1} x\right\}_1^{\infty} = \frac{\pi}{2} - \frac{\pi}{4} = \frac{\pi}{4},$$

it follows that the series $\sum 1/(n^2+1)$ converges.

(b) Since this series begins with $n = 2$, we modify the integral test by considering the improper integral of $f(x) = 1/(x \ln x)$ over the interval $x \geq 2$. Since $f(x)$ is positive, continuous, and decreasing for $x \geq 2$, and

$$\int_2^{\infty} \frac{1}{x \ln x}\,dx = \left\{\ln(\ln x)\right\}_2^{\infty} = \infty,$$

it follows that $\sum_{n=2}^{\infty} 1/(n \ln n)$ diverges.

(c) Since xe^{-x} is continuous, positive, and decreasing for $x \geq 1$, and

$$\int_1^{\infty} x e^{-x}\,dx = \left\{-xe^{-x} - e^{-x}\right\}_1^{\infty} = \frac{2}{e},$$

the series converges. ∎

We mentioned before that two very important series are geometric series and the harmonic series. Convergence of geometric series was discussed in Section 11.3. The harmonic series belongs to a type of series called *p*-**series** defined as follows

$$\sum_{n=1}^{\infty} \frac{1}{n^p} = 1 + \frac{1}{2^p} + \frac{1}{3^p} + \cdots . \qquad (11.15)$$

When $p = 1$, the *p*-series becomes the harmonic series which we know diverges. The series clearly diverges for $p < 0$ also. Consider the case when $p > 0$, but $p \neq 1$. The function $1/x^p$ is continuous, positive, and decreasing for $x \geq 1$, and

$$\int_1^{\infty} \frac{1}{x^p}\,dx = \left\{\frac{1}{-(p-1)x^{p-1}}\right\}_1^{\infty} = \begin{cases} \infty & p < 1 \\ \dfrac{1}{p-1} & p > 1. \end{cases}$$

According to the integral test, the *p*-series converges when $p > 1$ and diverges when $p < 1$. Let us summarize results for geometric and *p*-series:

$$\begin{matrix} \text{Geometric Series} & & \text{p-series} \end{matrix}$$

$$\sum_{n=1}^{\infty} ar^{n-1} = \begin{cases} \dfrac{a}{1-r} & |r| < 1 \\ \text{diverges} & |r| \geq 1 \end{cases} \qquad \sum_{n=1}^{\infty} \dfrac{1}{n^p} = \begin{cases} \text{converges} & p > 1 \\ \text{diverges} & p \leq 1 \end{cases} \quad (11.16)$$

It is unfortunate that no general formula can be given for the sum of the p-series when $p > 1$. Some interesting cases which arise frequently are cited below:

$$\sum_{n=1}^{\infty} \frac{1}{n^2} = \frac{\pi^2}{6}, \qquad \sum_{n=1}^{\infty} \frac{1}{n^3} = 1.202\,056\,903\,1, \qquad \sum_{n=1}^{\infty} \frac{1}{n^4} = \frac{\pi^4}{90}. \quad (11.17)$$

Many series can be shown to converge or diverge by comparing them to known convergent and divergent series. This is the essence of the following two tests.

Theorem 11.7

(Comparison Test)   If $0 \leq c_n \leq a_n$ for all n and $\sum_{n=1}^{\infty} a_n$ converges, then $\sum_{n=1}^{\infty} c_n$ converges. If $c_n \geq d_n \geq 0$ for all n and $\sum_{n=1}^{\infty} d_n$ diverges, then $\sum_{n=1}^{\infty} c_n$ diverges.

Proof   Suppose first that $\sum a_n$ converges and $0 \leq c_n \leq a_n$ for all n. Since $a_n \geq 0$ the sequence of partial sums for the series $\sum a_n$ must be nondecreasing. Since the series converges, the sequence of partial sums must also be bounded (corollary to Theorem 11.1). Because $0 \leq c_n \leq a_n$ for all n, it follows that the sequence of partial sums of $\sum c_n$ is nondecreasing and bounded also. This sequence therefore has a limit and series $\sum c_n$ converges. A similar argument can be made for the divergent case.

Theorem 11.7 states that if the terms of a nonnegative series $\sum c_n$ are smaller than those of a known convergent series, then $\sum c_n$ must converge; if they are larger than a known nonnegative divergent series, then $\sum c_n$ must diverge.

In order for the comparison test to be useful, we require a catalogue of known convergent and divergent series with which we may compare other series. The geometric and p-series in equation 11.16 are extremely useful in this respect.

EXAMPLE 11.7

Determine whether the following series converge or diverge:

(a) $\displaystyle\sum_{n=2}^{\infty} \frac{\ln n}{n}$ \qquad (b) $\displaystyle\sum_{n=1}^{\infty} \frac{2n^2 - 1}{15n^4 + 14}$ \qquad (c) $\displaystyle\sum_{n=1}^{\infty} \frac{2n^2 + 1}{15n^4 - 14}$

SOLUTION   (a) For this series, we note that when $n \geq 3$,

$$\frac{\ln n}{n} > \frac{1}{n}.$$

Since $\sum_{n=3}^{\infty} 1/n$ diverges (harmonic series with first two terms changed to zero), so does $\sum_{n=3}^{\infty} (\ln n)/n$ by the comparison test. Thus $\sum_{n=2}^{\infty} (\ln n)/n$ diverges. This can also be verified with the integral test.

(b) For the series $\sum (2n^2 - 1)/(15n^4 + 14)$, we note that

$$\frac{2n^2-1}{15n^4+14} < \frac{2n^2}{15n^4} = \frac{2}{15n^2}.$$

Since $\sum 2/(15n^2) = (2/15)\sum 1/n^2$ converges (p-series with $p=2$), so does the given series by the comparison test.

(c) For this series we note that the inequality

$$\frac{2n^2+1}{15n^4-14} \leq \frac{3}{n^2}$$

is valid if and only if

$$45n^4 - 42 \geq 2n^4 + n^2.$$

But this inequality is valid if and only if

$$n^2(43n^2 - 1) \geq 42,$$

which is obviously true for $n \geq 1$. Consequently,

$$\frac{2n^2+1}{15n^4-14} \leq \frac{3}{n^2},$$

and since $\sum 3/n^2 = 3\sum 1/n^2$ converges, so too does the given series. ∎

Two observations about Example 11.7 are worthwhile:

1. To use the comparison test we must first have a suspicion as to whether the given series converges or diverges in order to discuss the correct inequality; that is, if we suspect that the given series converges, we search for a convergent series for the right-hand side of the inequality $\leq$, and if we suspect that the given series diverges, we search for a divergent series for the right-hand side of the opposite inequality $\geq$.

2. Recall that when the terms of a nonnegative series approach zero, whether the series has a sum depends on how fast these terms approach zero. The only difference in the nth terms of Examples 11.7(b) and (c) is the position of the negative sign. For very large n this difference becomes negligible since each nth term can, for large n, be closely approximated by $2/(15n^2)$. We might then expect similar analyses for these examples, and yet they are quite different. In addition, it is natural to ask where we obtained the factor 3 in part (c). The answer is, "by trial and error".

Each of these observations points out weaknesses in the comparison test, but these problems can be eliminated in many examples with the following test.

Theorem 11.8

(Limit Comparison Test) If $0 \leq c_n$ and $0 < b_n$, and

$$\lim_{n \to \infty} \frac{c_n}{b_n} = l, \quad 0 < l < \infty, \tag{11.18}$$

then series $\sum_{n=1}^{\infty} c_n$ converges if $\sum_{n=1}^{\infty} b_n$ converges, and diverges if $\sum_{n=1}^{\infty} b_n$ diverges.

Proof Suppose that series $\sum b_n$ converges. Since sequence $\{c_n/b_n\}$ converges to l, we can use the result of Exercise 75 in Section 11.1 and say that for all n greater than or equal to some integer N,

$$\frac{c_n}{b_n} < l + 1, \quad \text{or} \quad c_n < (l+1)b_n.$$

Since the series $\sum_{n=N}^{\infty}(l+1)b_n = (l+1)\sum_{n=N}^{\infty} b_n$ converges, so also must $\sum_{n=N}^{\infty} c_n$ (by the comparison test). Hence, $\sum_{n=1}^{\infty} c_n$ converges. A similar argument can be made for the divergent case.

If $\lim_{n \to \infty} c_n/b_n = l$, then for very large n we can say that $c_n \approx lb_n$. It follows that if $\{b_n\}$ approaches zero, $\{c_n\}$ approaches zero $1/l$ times as fast as $\{b_n\}$. Theorem 11.8 implies then that if the sequence of nth terms of two nonnegative series approach zero at proportional rates, the series converge or diverge together.

To use the limit comparison test, we must find a series $\sum b_n$ so that the limit of the ratio c_n/b_n is finite and greater than zero. To obtain this series, it is sufficient in many examples to simply answer the question, "What does the given series really look like for very large n?". In both Examples 11.7(b) and (c) we see that for large n

$$c_n \approx \frac{2n^2}{15n^4} = \frac{2}{15n^2}.$$

Consequently, we calculate in Example 11.7(c) that

$$l = \lim_{n \to \infty} \frac{\frac{2n^2+1}{15n^4-14}}{\frac{2}{15n^2}} = \lim_{n \to \infty} \frac{n^2\left(2 + \frac{1}{n^2}\right)}{n^4\left(15 - \frac{14}{n^4}\right)} \cdot \frac{15n^2}{2} = 1.$$

Since $\sum 2/(15n^2)$ converges, so too does $\sum (2n^2+1)/(15n^4-14)$.

EXAMPLE 11.8

Determine whether the following series converge or diverge:

(a) $\displaystyle\sum_{n=1}^{\infty} \frac{\sqrt{n^2+2n-1}}{n^{5/2}+15n-3}$ (b) $\displaystyle\sum_{n=1}^{\infty} \frac{2^n+1}{3^n+5}$

SOLUTION (a) For very large n, the nth term of the series can be approximated by

$$\frac{\sqrt{n^2+2n-1}}{n^{5/2}+15n-3} \approx \frac{n}{n^{5/2}} = \frac{1}{n^{3/2}}.$$

We calculate therefore that

$$l = \lim_{n \to \infty} \frac{\frac{\sqrt{n^2+2n-1}}{n^{5/2}+15n-3}}{\frac{1}{n^{3/2}}} = \lim_{n \to \infty} \frac{n\sqrt{1 + \frac{2}{n} - \frac{1}{n^2}}}{n^{5/2}\left(1 + \frac{15}{n^{3/2}} - \frac{3}{n^{5/2}}\right)} \cdot n^{3/2} = 1.$$

Since $\sum 1/n^{3/2}$ converges (p-series with $p = 3/2$), so too does the given series by the limit comparison test.

(b) For large n,
$$\frac{2^n+1}{3^n+5} \approx \left(\frac{2}{3}\right)^n.$$

We calculate therefore that
$$l = \lim_{n\to\infty} \frac{\frac{2^n+1}{3^n+5}}{\left(\frac{2}{3}\right)^n} = \lim_{n\to\infty} \frac{2^n\left(1+\frac{1}{2^n}\right)}{3^n\left(1+\frac{5}{3^n}\right)} \cdot \frac{3^n}{2^n} = 1.$$

Since $\sum (2/3)^n$ converges (a geometric series with $r = 2/3$), the given series converges by the limit comparison test. ■

EXERCISES 11.4

In Exercises 1–22 determine whether the series converges or diverges.

1. $\sum_{n=1}^{\infty} \dfrac{1}{2n+1}$
2. $\sum_{n=1}^{\infty} \dfrac{1}{4n-3}$
3. $\sum_{n=1}^{\infty} \dfrac{1}{2n^2+4}$
4. $\sum_{n=1}^{\infty} \dfrac{1}{5n^2-3n-1}$
5. $\sum_{n=2}^{\infty} \dfrac{1}{n^3-1}$
6. $\sum_{n=4}^{\infty} \dfrac{n^2}{n^4-6n^2+5}$
7. $\sum_{n=1}^{\infty} \dfrac{1}{(2n-1)(2n+1)}$
8. $\sum_{n=1}^{\infty} \dfrac{n-5}{n^2+3n-2}$
9. $\sum_{n=2}^{\infty} \dfrac{1}{\ln n}$
10. $\sum_{n=1}^{\infty} n^2 e^{-2n}$
11. $\sum_{n=2}^{\infty} \dfrac{\sqrt{n^2+2n-3}}{n^2+5}$
12. $\sum_{n=1}^{\infty} \dfrac{\sqrt{n+5}}{n^3+3}$
13. $\sum_{n=1}^{\infty} \sqrt{\dfrac{n^2+2n+3}{2n^4-n}}$
14. $\sum_{n=2}^{\infty} \dfrac{1}{n^2 \ln n}$
15. $\sum_{n=1}^{\infty} \dfrac{1}{2^n} \sin\left(\dfrac{\pi}{n}\right)$
16. $\sum_{n=1}^{\infty} \dfrac{\sqrt{n^2+1}}{n^3} \operatorname{Tan}^{-1} n$
17. $\sum_{n=1}^{\infty} \dfrac{2^n+n}{3^n+1}$
18. $\sum_{n=2}^{\infty} \dfrac{1+\ln^2 n}{n\ln^2 n}$
19. $\sum_{n=1}^{\infty} \dfrac{1+1/n}{e^n}$
20. $\sum_{n=1}^{\infty} \dfrac{\ln(n+1)}{n+1}$
21. $\sum_{n=1}^{\infty} n e^{-n^2}$
22. $\sum_{n=2}^{\infty} \dfrac{1}{n\sqrt[3]{\ln n}}$

In Exercises 23–25 find values of p for which the series converges.

23. $\sum_{n=2}^{\infty} \dfrac{1}{n^p \ln n}$
24. $\sum_{n=2}^{\infty} \dfrac{1}{n(\ln n)^p}$
25. $\sum_{n=2}^{\infty} \dfrac{1}{(\ln n)^p}$

SECTION 11.5

Further Convergence Tests for Nonnegative Series

In this section we consider two additional tests to determine whether nonnegative series converge or diverge. The first test, called the limit ratio test, indicates whether a series $\sum c_n$ resembles a geometric series for large n.

Theorem 11.9

(Limit Ratio Test) *Suppose $c_n > 0$ and*

$$\lim_{n \to \infty} \frac{c_{n+1}}{c_n} = L. \qquad (11.19)$$

Then,

(i) $\sum_{n=1}^{\infty} c_n$ *converges if* $L < 1$;

(ii) $\sum_{n=1}^{\infty} c_n$ *diverges if* $L > 1$ *(or if* $\lim_{n \to \infty} c_{n+1}/c_n = \infty$);

(iii) $\sum_{n=1}^{\infty} c_n$ *may converge or diverge if* $L = 1$.

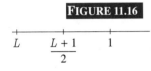

FIGURE 11.16

Proof (i) By the result of part (a) in Exercise 74 of Section 11.1 (see also Figure 11.16), we can say that if $\lim_{n \to \infty} c_{n+1}/c_n = L < 1$, there exists an integer N such that for all $n \geq N$,

$$\frac{c_{n+1}}{c_n} < \frac{L+1}{2}.$$

Consequently,

$$c_{N+1} < \left(\frac{L+1}{2}\right) c_N;$$

$$c_{N+2} < \left(\frac{L+1}{2}\right) c_{N+1} < \left(\frac{L+1}{2}\right)^2 c_N;$$

$$c_{N+3} < \left(\frac{L+1}{2}\right) c_{N+2} < \left(\frac{L+1}{2}\right)^3 c_N;$$

and so on. Hence,

$$c_N + c_{N+1} + c_{N+2} + \cdots < c_N + \left(\frac{L+1}{2}\right) c_N + \left(\frac{L+1}{2}\right)^2 c_N + \cdots.$$

Since the right-hand side of this inequality is a geometric series with common ratio $(L+1)/2 < 1$, it follows by the comparison test that $\sum_{n=N}^{\infty} c_n$ converges. Therefore, $\sum_{n=1}^{\infty} c_n$ converges also.

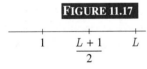

FIGURE 11.17

(ii) By part (b) in Exercise 74 of Section 11.1 (see also Figure 11.17), if $\lim_{n \to \infty} c_{n+1}/c_n = L > 1$, there exists an integer N such that for all $n \geq N$,

$$\frac{c_{n+1}}{c_n} > \frac{L+1}{2} > 1.$$

When $\lim_{n \to \infty} c_{n+1}/c_n = \infty$, it is also true that for n greater than or equal to some N, c_{n+1}/c_n must be greater than 1. This implies that for all $n > N$, $c_n > c_N$, and therefore

$$\lim_{n \to \infty} c_n \neq 0.$$

Hence $\sum c_n$ diverges by the nth term test.

(iii) To show that the limit ratio test fails when $L = 1$, consider the two p-series $\sum 1/n$ and $\sum 1/n^2$. For each series, $L = 1$, yet the first series diverges and the second converges.

In a nonrigorous way we can justify the limit ratio test from the following standpoint. If $\lim_{n \to \infty} c_{n+1}/c_n = L$, then for large n, each term of series $\sum c_n$ is essentially L times the term before it; that is, the series resembles a geometric series with common ratio L. We would expect convergence of the series if $L < 1$ and divergence if $L > 1$. We might also anticipate some indecision about the $L = 1$ case, depending on how this limit is reached, since this case corresponds to the common ratio which separates convergent and divergent geometric series.

EXAMPLE 11.9

Determine whether the following series converge or diverge:

(a) $\sum_{n=1}^{\infty} \frac{2^n}{n^4}$ (b) $\sum_{n=1}^{\infty} \frac{n^{100}}{1 \cdot 3 \cdot 5 \cdot \cdots \cdot (2n-1)}$ (c) $\sum_{n=1}^{\infty} \frac{n^n}{n!}$

SOLUTION (a) For the series $\sum 2^n/n^4$,

$$L = \lim_{n \to \infty} \frac{\frac{2^{n+1}}{(n+1)^4}}{\frac{2^n}{n^4}} = \lim_{n \to \infty} 2 \left(\frac{n}{n+1}\right)^4 = 2,$$

and the series therefore diverges by the limit ratio test.

(b) For this series,

$$L = \lim_{n \to \infty} \frac{\frac{(n+1)^{100}}{1 \cdot 3 \cdot 5 \cdot \cdots \cdot (2n+1)}}{\frac{n^{100}}{1 \cdot 3 \cdot 5 \cdot \cdots \cdot (2n-1)}} = \lim_{n \to \infty} \left(\frac{n+1}{n}\right)^{100} \frac{1}{2n+1} = 0,$$

and the series therefore converges by the limit ratio test.

(c) Since

$$L = \lim_{n \to \infty} \frac{\frac{(n+1)^{n+1}}{(n+1)!}}{\frac{n^n}{n!}} = \lim_{n \to \infty} \left(\frac{n+1}{n}\right)^n = e > 1$$

(see equation 3.29b), the series $\sum n^n/n!$ diverges. ∎

We have one last test for nonnegative series.

Theorem 11.10

(Limit Root Test) Suppose $c_n \geq 0$ and

$$\lim_{n \to \infty} \sqrt[n]{c_n} = R. \qquad (11.20)$$

Then,

(i) $\sum_{n=1}^{\infty} c_n$ converges if $R < 1$;
(ii) $\sum_{n=1}^{\infty} c_n$ diverges if $R > 1$ (or if $\lim_{n \to \infty} \sqrt[n]{c_n} = \infty$);
(iii) $\sum_{n=1}^{\infty} c_n$ may converge or diverge if $R = 1$.

Proof (i) By the result of part (a) in Exercise 74 of Section 11.1, if $\lim_{n \to \infty} \sqrt[n]{c_n} = R < 1$, there exists an integer N such that for all $n \geq N$,

$$\sqrt[n]{c_n} < \frac{R+1}{2}.$$

Consequently,

$$\sqrt[N]{c_N} < \frac{R+1}{2}, \quad \text{or} \quad c_N < \left(\frac{R+1}{2}\right)^N;$$

$$\sqrt[N+1]{c_{N+1}} < \frac{R+1}{2}, \quad \text{or} \quad c_{N+1} < \left(\frac{R+1}{2}\right)^{N+1};$$

$$\sqrt[N+2]{c_{N+2}} < \frac{R+1}{2}, \quad \text{or} \quad c_{N+2} < \left(\frac{R+1}{2}\right)^{N+2};$$

and so on. Hence,

$$c_N + c_{N+1} + c_{N+2} + \cdots < \left(\frac{R+1}{2}\right)^N + \left(\frac{R+1}{2}\right)^{N+1} + \left(\frac{R+1}{2}\right)^{N+2} + \cdots.$$

Since the right-hand side of this inequality is a convergent geometric series, the left-hand side must be a convergent series also by the comparison test. Thus, $\sum_{n=N}^{\infty} c_n$ converges, and so also must $\sum_{n=1}^{\infty} c_n$.

(ii) When $\lim_{n \to \infty} \sqrt[n]{c_n} = R > 1$, there exists an integer N such that for all $n \geq N$,

$$\sqrt[n]{c_n} > \frac{R+1}{2} > 1 \quad \text{or} \quad c_n > 1,$$

as in part (b) of Exercise 74 in Section 11.1. When $\lim_{n \to \infty} \sqrt[n]{c_n} = \infty$, it is also true that for n greater than or equal to some N, $\sqrt[n]{c_n}$ must be greater than 1. But it now follows that $\lim_{n \to \infty} c_n \neq 0$, and the series diverges by the nth term test.

(iii) To show that the test fails when $R = 1$, we show that $R = 1$ for the two p-series $\sum 1/n$ and $\sum 1/n^2$, which diverges and converges respectively. For the harmonic series, let $R = \lim_{n \to \infty} (1/n)^{1/n}$. If we take logarithms, then

$$\ln R = \ln \left[\lim_{n \to \infty} \left(\frac{1}{n}\right)^{1/n}\right] = \lim_{n \to \infty} \ln \left(\frac{1}{n}\right)^{1/n} = \lim_{n \to \infty} \frac{1}{n} \ln \left(\frac{1}{n}\right) = -\lim_{n \to \infty} \frac{1}{n} \ln n$$

$$= -\lim_{n \to \infty} \frac{\frac{1}{n}}{1} \quad \text{(by L'Hôpital's rule)}$$

$$= 0.$$

Hence, $R = 1$ for this divergent series. A similar analysis gives $R = 1$ for the convergent series $\sum 1/n^2$.

EXAMPLE 11.10

Determine whether the following series converge or diverge:

$$\text{(a)} \quad \sum_{n=1}^{\infty} \left(\frac{n+1}{n}\right)^{n^2} \qquad \text{(b)} \quad \sum_{n=1}^{\infty} \frac{n}{(\ln n)^n}$$

SOLUTION (a) Since

$$R = \lim_{n \to \infty} \left[\left(\frac{n+1}{n}\right)^{n^2}\right]^{1/n} = \lim_{n \to \infty} \left(\frac{n+1}{n}\right)^n = e > 1,$$

the series diverges by the limit root test.

(b) For this series,

$$R = \lim_{n\to\infty} \left[\frac{n}{(\ln n)^n}\right]^{1/n} = \lim_{n\to\infty} \frac{n^{1/n}}{\ln n}.$$

If we set $L = \lim_{n\to\infty} n^{1/n}$, then

$$\ln L = \ln\left(\lim_{n\to\infty} n^{1/n}\right) = \lim_{n\to\infty} \frac{1}{n} \ln n$$

$$= \lim_{n\to\infty} \frac{\frac{1}{n}}{1} \qquad \text{(by L'Hôpital's rule)}$$

$$= 0.$$

Thus, $L = 1$, and it follows that

$$R = \lim_{n\to\infty} \frac{n^{1/n}}{\ln n} = 0.$$

The series therefore converges. ∎

We have developed six tests to determine whether series converge or diverge:

n^{th} term test

integral test

comparison test

limit comparison test

limit ratio test

limit root test

The form of the nth term of a series often suggests which test should be used. Keep the following ideas in mind when choosing a test:

1. The limit ratio test can be effective on factorials, products of the form $1 \cdot 3 \cdot 5 \cdot \ldots \cdot (2n-1)$, and constants raised to powers involving n (2^n, 3^{-n}, etc.)
2. The limit root test thrives on functions of n raised to powers involving n (see Example 11.10).
3. The limit comparison test is successful on rational functions of n, and fractional powers as well ($\sqrt{n}$, $\sqrt[3]{n/(n+1)}$, etc.)
4. The integral test can be effective when the nth term is easily integrated. Logarithms often require the integral test.

By definition, a series $\sum c_n$ converges if and only if its sequence of partial sums $\{S_n\}$ converges. The difficulty with using this definition to discuss convergence of a series is that S_n can seldom be evaluated in a simple form, and therefore consideration of $\lim_{n\to\infty} S_n$ is impossible. The above tests have the advantage of avoiding partial sums. On the other hand, they have one disadvantage. Although they may indicate that a series does indeed have a sum, the tests in no way suggest the value of the sum. The problem of calculating the sum often proves more difficult than showing that it exists in the first place. In Sections 11.7 and 11.11 we discuss various ways to calculate sums for known convergent series.

EXERCISES 11.5

In Exercises 1–20 determine whether the series converges or diverges.

1. $\sum_{n=1}^{\infty} \frac{e^n}{n^4}$

2. $\sum_{n=1}^{\infty} \frac{1}{n!}$

3. $\sum_{n=1}^{\infty} \frac{n^3}{2^n}$

4. $\sum_{n=1}^{\infty} \frac{1}{n^n}$

5. $\sum_{n=1}^{\infty} \frac{(n-1)(n-2)}{n^2 2^n}$

6. $\sum_{n=1}^{\infty} \frac{(2n)!}{(n!)^2}$

7. $\sum_{n=1}^{\infty} \frac{\sqrt{n+1}}{n^{n+1/2}}$

8. $\sum_{n=1}^{\infty} \frac{3^{-n} + 2^{-n}}{4^{-n} + 5^{-n}}$

9. $\sum_{n=1}^{\infty} \frac{e^{-n}}{\sqrt{n+\pi}}$

10. $\sum_{n=1}^{\infty} \frac{2 \cdot 4 \cdot \dots \cdot (2n)}{4 \cdot 7 \cdot \dots \cdot (3n+1)}$

11. $\sum_{n=1}^{\infty} \frac{n^{n-1}}{3^{n-1}(n-1)!}$

12. $\sum_{n=1}^{\infty} n \left(\frac{3}{4}\right)^n$

13. $\sum_{n=1}^{\infty} \frac{1 + 1/n}{e^n}$

14. $\sum_{n=1}^{\infty} \frac{2 \cdot 4 \cdot \dots \cdot (2n)}{3 \cdot 5 \cdot \dots \cdot (2n+1)} \left(\frac{1}{n^2}\right)$

15. $\sum_{n=1}^{\infty} \frac{n^n}{(n+1)^{n+1}}$

16. $\sum_{n=1}^{\infty} \frac{(n+1)^n}{n^{n+1}}$

17. $\sum_{n=1}^{\infty} \frac{n^4 + 3}{5^{n/2}}$

18. $\sum_{n=1}^{\infty} \frac{2^n + n^2 3^n}{4^n}$

19. $\sum_{n=1}^{\infty} \frac{n^2 2^n - n}{n^3 + 1}$

20. $\sum_{n=1}^{\infty} \frac{(2n)!}{(3n)!} 5^{2n}$

21. For what integer values of a is the series

$$\sum_{n=1}^{\infty} \frac{(n!)^2}{(an)!}$$

convergent?

SECTION 11.6

Series with Positive and Negative Terms

The convergence tests of Sections 11.4 and 11.5 are applicable to nonnegative series, series whose terms are all nonnegative. Series with infinitely many positive and negative terms are more complicated. It is fortunate, however, that all our test are still useful in discussing convergence of series with positive and negative terms. What makes this possible is the following definition and Theorem 11.11.

Definition 11.7

A series $\sum_{n=1}^{\infty} c_n$ is said to be **absolutely convergent** if the series of absolute values $\sum_{n=1}^{\infty} |c_n|$ converges.

At first glance it might seem that absolute convergence is a strange concept indeed. What possible good could it do to consider the series of absolute values, which is quite different from the original series? The fact is that when the series of absolute values converges, it automatically follows from the next theorem that the original series converges also. And since the series of absolute values has all nonnegative terms, we can use the comparison, limit comparison, limit ratio, limit root, or integral test to consider its convergence.

Theorem 11.11

If a series is absolutely convergent, then it is convergent.

Proof Let $\{S_n\}$ be the sequence of partial sums of the absolutely convergent series $\sum c_n$. Define sequences $\{P_n\}$ and $\{N_n\}$, where P_n is the sum of all positive terms in

S_n, and N_n is the sum of the absolute values of all negative terms in S_n. Then

$$S_n = P_n - N_n.$$

The sequence of partial sums for the series of absolute values $\sum |c_n|$ is

$$\{P_n + N_n\},$$

and this sequence must be nondecreasing and bounded. Since each of the sequences $\{P_n\}$ and $\{N_n\}$ is nondecreasing and a part of $\{P_n + N_n\}$, it follows that each is bounded, and therefore has a limit, say, P and N, respectively. As a result, sequence $\{P_n - N_n\} = \{S_n\}$ has limit $P - N$, and series $\sum c_n$ converges to $P - N$.

EXAMPLE 11.11

Show that the following series are absolutely convergent:

(a) $\sum_{n=1}^{\infty} \dfrac{(-1)^n n}{(n+1)2^n}$

(b) $1 - \dfrac{1}{2^2} - \dfrac{1}{3^3} + \dfrac{1}{4^4} + \dfrac{1}{5^5} + \dfrac{1}{6^6} - \dfrac{1}{7^7} - \dfrac{1}{8^8} - \dfrac{1}{9^9} - \dfrac{1}{10^{10}} + \dfrac{1}{11^{11}} + \cdots + \dfrac{1}{15^{15}} - \cdots$.

SOLUTION (a) The series of absolute values is $\sum n/[(n+1)2^n]$. We use the limit comparison test to show that it converges. Since

$$l = \lim_{n \to \infty} \dfrac{\dfrac{n}{(n+1)2^n}}{\dfrac{1}{2^n}} = 1,$$

and $\sum (1/2)^n$ is convergent (a geometric series with $r = 1/2$), it follows that the series of absolute values converges. The given series therefore converges absolutely.

(b) The series of absolute values is $\sum 1/n^n$. We use the comparison test to verify that this series converges. Since

$$\dfrac{1}{n^n} \le \dfrac{1}{n^2}$$

and $\sum 1/n^2$ converges, so does $\sum 1/n^n$. The given series therefore converges absolutely. ∎

In Example 11.11, absolute convergence of the given series implies convergence of the series, but it is customary to omit such a statement. It is important to realize, however, that it is the given series that is being analyzed, and its convergence is guaranteed by Theorem 11.11.

We now ask whether series can converge without converging absolutely. If there are such series, and indeed there are, we must devise new convergence tests. We describe these series as follows.

Definition 11.8

A series that converges but does not converge absolutely is said to **converge conditionally**.

The most important type of series with both positive and negative terms is an alternating series. As the name suggests, an **alternating series** has terms that are alternately positive and negative. For example,

$$\sum_{n=1}^{\infty} \frac{(-1)^{n+1}}{n} = 1 - \frac{1}{2} + \frac{1}{3} - \frac{1}{4} + \frac{1}{5} - \cdots$$

is an alternating series called the **alternating harmonic series**.

Given an alternating series to examine for convergence, we first test for absolute convergence as in Example 11.11(a). Should this fail, we check for conditional convergence with the following test.

Theorem 11.12 (Alternating Series Test) *An alternating series $\sum_{n=1}^{\infty} c_n$ converges if the sequence of absolute values of the terms $\{|c_n|\}$ is nonincreasing and has limit zero.*

Proof Suppose $c_1 > 0$, in which case all odd terms are positive and all even terms are negative. If $\{S_n\}$ is the sequence of partial sums of $\sum c_n$, then the even partial sums S_{2n} can be expressed in two forms:

$$S_{2n} = (c_1 + c_2) + (c_3 + c_4) + \cdots + (c_{2n-1} + c_{2n}),$$
$$S_{2n} = c_1 + (c_2 + c_3) + (c_4 + c_5) + \cdots + (c_{2n-2} + c_{2n-1}) + c_{2n}.$$

Since $\{|c_n|\}$ is nonincreasing ($|c_n| \geq |c_{n+1}|$), each term in the parentheses of the first expression is nonnegative. Consequently, the subsequence $\{S_{2n}\}$ of even partial sums of $\{S_n\}$ is nondecreasing. Since each term in the parentheses in the second expression is nonpositive, as is c_{2n}, it follows that $S_{2n} \leq c_1$ for all n. By Theorem 11.1 then, the sequence $\{S_{2n}\}$ has a limit, say, S.

The subsequence of odd partial sums $\{S_{2n-1}\}$ is such that $S_{2n-1} = S_{2n} - c_{2n}$. Since $\{S_{2n}\}$ has limit S and $\{c_{2n}\}$ has limit zero, it follows that $\{S_{2n-1}\}$ has limit S also. Consequently, $\{S_n\}$ has limit S, and series $\sum c_n$ converges.

A similar proof holds when $c_1 < 0$.

EXAMPLE 11.12 Determine whether the following series converge absolutely, converge conditionally, or diverge:

(a) $\sum_{n=1}^{\infty} \frac{(-1)^{n+1}}{n}$ (b) $\sum_{n=1}^{\infty} (-1)^n \frac{\sqrt{n^2 + 5n}}{n^{3/2}}$ (c) $\sum_{n=1}^{\infty} (-1)^{n+1} \frac{4^n}{n^5 3^n}$

SOLUTION (a) The alternating harmonic series $\sum (-1)^{n+1}/n$ is not absolutely convergent because the series of absolute values $\sum 1/n$ diverges. Since the sequence of absolute values of the terms $\{1/n\}$ is decreasing with limit zero, series $\sum (-1)^{n+1}/n$ converges conditionally.

(b) For this alternating series, we first consider the series of absolute values

$$\sum_{n=1}^{\infty} \frac{\sqrt{n^2 + 5n}}{n^{3/2}}.$$

We use the limit comparison test on this series. Since

$$l = \lim_{n \to \infty} \frac{\frac{\sqrt{n^2 + 5n}}{n^{3/2}}}{\frac{1}{n^{1/2}}} = \lim_{n \to \infty} \frac{n\sqrt{1 + \frac{5}{n}}}{n^{3/2}} \cdot n^{1/2} = 1,$$

and $\sum 1/n^{1/2}$ diverges, so does $\sum \sqrt{n^2 + 5n}/n^{3/2}$. The original series does not therefore converge absolutely. We now resort to the alternating series test. The sequence $\{\sqrt{n^2 + 5n}/n^{3/2}\}$ of absolute values of the terms of the series is nonincreasing if

$$\frac{\sqrt{(n+1)^2 + 5(n+1)}}{(n+1)^{3/2}} \leq \frac{\sqrt{n^2 + 5n}}{n^{3/2}}.$$

When we square and cross multiply, the inequality becomes

$$n^3(n^2 + 7n + 6) \leq (n^2 + 5n)(n+1)^3$$
$$= n^5 + 8n^4 + 18n^3 + 16n^2 + 5n;$$

that is,

$$n^4 + 12n^3 + 16n^2 + 5n \geq 0,$$

which is obviously valid because $n \geq 1$. Since $\lim_{n \to \infty} \sqrt{n^2 + 5n}/n^{3/2} = 0$, we conclude that the alternating series $\sum (-1)^n (\sqrt{n^2 + 5n}/n^{3/2})$ converges conditionally.

(c) If we apply the limit ratio test to the series of absolute values $\sum 4^n/(n^5 3^n)$, we have

$$L = \lim_{n \to \infty} \frac{\frac{4^{n+1}}{(n+1)^5 3^{n+1}}}{\frac{4^n}{n^5 3^n}} = \lim_{n \to \infty} \frac{4}{3} \left(\frac{n}{n+1}\right)^5 = \frac{4}{3}.$$

Since $L > 1$, the series $\sum 4^n/(n^5 3^n)$ diverges. The original alternating series does not therefore converge absolutely. But $L = \frac{4}{3}$ implies that for large n, each term in the series of absolute values is approximately $\frac{4}{3}$ times the term that precedes it, and therefore

$$\lim_{n \to \infty} \frac{4^n}{n^5 3^n} = \infty.$$

Consequently,

$$\lim_{n \to \infty} (-1)^{n+1} \frac{4^n}{n^5 3^n}$$

cannot possibly exist, and the given series diverges by the nth term test. ∎

We have noted several times that the essential question for convergence of a nonnegative series is, "Do the terms approach zero quickly enough to guarantee convergence of the series?" With a series that has infinitely many positive and negative terms, this question is inappropriate. Such a series may converge because of a partial cancelling effect; for example, a negative term may offset the effect of a large positive term. This kind of process may produce a convergent series even though the series would be divergent if all terms were replaced by their absolute values. A specific example is the alternating harmonic series which converges (conditionally) because of this cancelling effect, whereas the harmonic series itself, which has no cancellations, diverges.

In Sections 11.4–6, we have obtained a number of tests for determining whether series converge or diverge. To test a series for convergence, we suggest the following procedure:

1. Try the nth term test for divergence.
2. If $\{c_n\}$ has limit zero and the series is nonnegative, try the comparison, limit comparison, limit ratio, limit root, or integral test.
3. If $\{c_n\}$ has limit zero and the series contains both positive and negative terms, test for absolute convergence using the tests in 2. If this fails and the series is alternating, test for conditional convergence with the alternating series test.

Each of the comparison, limit comparison, limit ratio, limit root, integral, and alternating series tests requires conditions to be satisfied for all terms of the series. Specifically, the comparison, limit comparison, and limit root tests require $c_n \geq 0$ for all n; the limit ratio test requires $c_n > 0$; the integral test requires $f(n)$ to be positive, continuous, and decreasing; and the alternating series test requires $\{|c_n|\}$ to be nonincreasing and $\{c_n\}$ to be alternately positive and negative. None of these requirements is essential for all n; in fact, so long as they are satisfied for all terms in the series beyond some point, say for n greater than or equal to some integer N, the particular test may be used on the series $\sum_{n=N}^{\infty} c_n$. The original series $\sum_{n=1}^{\infty} c_n$ then converges if and only if $\sum_{n=N}^{\infty} c_n$ converges.

EXERCISES 11.6

In Exercises 1–14 determine whether the series converges absolutely, converges conditionally, or diverges.

1. $\sum_{n=1}^{\infty} (-1)^n \dfrac{n}{n^3 + 1}$

2. $\sum_{n=1}^{\infty} (-1)^n \dfrac{n}{n^2 + 1}$

3. $\sum_{n=1}^{\infty} \dfrac{\cos(n\pi/2)}{2n^2}$

4. $\sum_{n=1}^{\infty} (-1)^n \dfrac{n^3}{3^n}$

5. $\sum_{n=1}^{\infty} \dfrac{(-1)^{n+1}}{\sqrt{n}}$

6. $\sum_{n=1}^{\infty} (-1)^n \dfrac{3^n}{n^3}$

7. $\sum_{n=1}^{\infty} (-1)^n \dfrac{n}{n^2 + n + 1}$

8. $\sum_{n=1}^{\infty} \dfrac{n \sin(n\pi/4)}{2^n}$

9. $\sum_{n=1}^{\infty} (-1)^{n+1} \left(\dfrac{n}{n+1} \right)$

10. $\sum_{n=1}^{\infty} (-1)^{n+1} \dfrac{\sqrt{3n-2}}{n}$

11. $\sum_{n=1}^{\infty} (-1)^n \left(\dfrac{n}{n+1} \right)^n$

12. $\sum_{n=1}^{\infty} (-1)^n \dfrac{\sqrt{n^2 + 3}}{n^2 + 5}$

13. $\sum_{n=2}^{\infty} (-1)^{n-1} \dfrac{\ln n}{n}$

14. $\sum_{n=1}^{\infty} \dfrac{\cos(n\pi/10) \operatorname{Cot}^{-1} n}{n^3 + 5n}$

15. Discuss convergence of the series

$$\sum_{n=1}^{\infty} \dfrac{\sin(nx)}{n^2}.$$

16. Prove that if $\sum c_n$ converges absolutely, then $\sum c_n^p$ converges absolutely for all integers $p > 1$.

17. Discuss convergence of the series

$$\sum_{n=1}^{\infty} (-1)^n \dfrac{n^n}{(n+1)^{n+1}}.$$

SECTION 11.7

Approximating the Sum of a Series

We have considered only half of the convergence problem for infinite series of constants. The comparison, limit comparison, limit ratio, limit root, integral, and alternating series tests may determine whether a series converges or diverges, but they do not determine the sum of the series in the case of a convergent series. This part of the problem, as suggested before, can sometimes be more complicated.

If the convergent series is a geometric series, no problem exists; we can use formula 11.13b to find its sum. It may also happen that the nth partial sum S_n of the series can be calculated in a simple form, in which case the sum of the series is $\lim_{n \to \infty} S_n$. Cases of this latter type are very rare. These two methods exhaust our present capabilities for obtaining the sum of a convergent series, but we will find additional methods in Section 11.11.

In many applications of infinite series we only need a reasonable approximation to the sum of a series, and therefore turn our attention to the problem of estimating the sum of a convergent series. The easiest method for estimating the sum S of a series $\sum c_n$ is simply to choose the partial sum S_N for some N as an approximation; that is, truncate the series after N terms and choose

$$S \approx S_N = c_1 + c_2 + \cdots + c_N.$$

But an approximation is of value only if we can make some definitive statement about its accuracy. In truncating the series, we have neglected the infinity of terms $\sum_{n=N+1}^{\infty} c_n$, and the accuracy of the approximation is therefore determined by the size of $\sum_{n=N+1}^{\infty} c_n$; the smaller it is, the better the approximation. The problem is that we do not know the exact value of $\sum_{n=N+1}^{\infty} c_n$; if we did, there would be no need to approximate the sum of the original series in the first place. What we must do is estimate the sum $\sum_{n=N+1}^{\infty} c_n$.

When the integral test or the alternating series test are used to prove that a series converges, simple formulas give accuracy estimates on the truncated series. Let us illustrate these first.

Truncating an Alternating Series

It is very simple to obtain an estimate of the accuracy of a truncated alternating series $\sum c_n$ provided the sequence $\{|c_n|\}$ is nonincreasing with limit zero. In the proof of Theorem 11.12, we showed that when $c_1 > 0$, the subsequence $\{S_{2n}\}$ of even partial sums is nondecreasing and therefore approaches the sum of the series $\sum c_n$ from below (Figure 11.18). In a similar way, we can show that the subsequence $\{S_{2n-1}\}$ of odd partial sums is nonincreasing and approaches the sum of the series from above. It follows that the sum $\sum c_n$ must be between any two terms of the subsequences $\{S_{2n}\}$ and $\{S_{2n-1}\}$. In particular, it must be between any two successive partial sums. Thus, if $\sum c_n$ is approximated by S_N, the maximum possible error is $S_{N+1} - S_N = c_{N+1}$, the next term. *When an alternating series is truncated, the maximum possible error is the next term.*

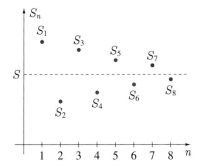

FIGURE 11.18

EXAMPLE 11.13

Use the first twenty terms of the series $\sum_{n=1}^{\infty} (-1)^{n+1}/n^3$ to estimate its sum. Obtain an error estimate.

SOLUTION The sum of the first twenty terms of the series is $0.901\,485$. Since the series is alternating, the maximum possible error in this estimate is the twenty-first term, $1/21^3 < 0.000\,108$. Thus,

$$0.901\,485 < \sum_{n=1}^{\infty} \frac{(-1)^{n+1}}{n^3} < 0.901\,593.$$

∎

In practical situations, we often have to decide how many terms of a series to take in order to guarantee a certain degree of accuracy. Once again this is easy for alternating series.

EXAMPLE 11.14

How many terms in the series $\sum_{n=2}^{\infty} (-1)^{n+1}/(n^3 \ln n)$ ensure a truncation error of less than 10^{-5}?

SOLUTION The maximum error in truncating this alternating series when $n = N$ is

$$\frac{(-1)^{N+2}}{(N+1)^3 \ln(N+1)}.$$

The absolute value of this error is less than 10^{-5} when

$$\frac{1}{(N+1)^3 \ln(N+1)} < 10^{-5} \quad \text{or}$$

$$(N+1)^3 \ln(N+1) > 10^5.$$

A calculator quickly reveals that the smallest integer for which this is valid is $N = 30$. Thus, the truncated series has the required accuracy after the 29th term. Note that the first term corresponds to $n = 2$ not $n = 1$.

∎

Truncating a Series Whose Convergence was Established with the Integral Test

Suppose now that a series $\sum_{n=1}^{\infty} c_n$ has been shown to converge with the integral test; that the integral

$$\int_1^{\infty} f(x)\,dx$$

converges where $f(n) = c_n$. If the series is truncated after the Nth term, the error $c_{N+1} + c_{N+2} \cdots$ is shown as the sum of the areas of the rectangles in Figure 11.19. Clearly the sum of these areas is less than the area under $y = f(x)$ to the right of $x = N$. In other words, the error in truncating the series with the Nth term must be less than

$$\int_N^{\infty} f(x)\,dx. \tag{11.21}$$

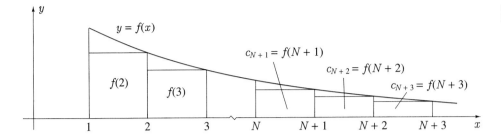

FIGURE 11.19

EXAMPLE 11.15

What is a maximum possible error when the series

$$\sum_{n=2}^{\infty} \frac{1}{n(\ln n)^4}$$

is truncated after 100 terms?

SOLUTION The error cannot be larger than

$$\int_{100}^{\infty} \frac{1}{x(\ln x)^4}\, dx = \left\{ \frac{-1}{3(\ln x)^3} \right\}_{100}^{\infty} = \frac{1}{3(\ln 100)^3} < 0.0035.$$

∎

When convergence of a series is established by the comparison, limit comparison, limit ratio, or limit root tests, we often estimate the truncation error $\sum_{n=N+1}^{\infty} c_n$ by comparing it to something that is summable. We illustrate this in the following two examples.

EXAMPLE 11.16

Estimate the error in truncating the series $\sum_{n=0}^{\infty} 1/n!$ after 8 terms. The first term of this series contains $0!$. This is defined as 1.

SOLUTION The sum of the first 8 terms of this series is $S_8 = 2.718\,254\,0$. The truncation error in using this approximation is

$$\sum_{n=8}^{\infty} \frac{1}{n!} = \frac{1}{8!} + \frac{1}{9!} + \frac{1}{10!} + \frac{1}{11!} + \cdots$$

$$= \frac{1}{8!}\left(1 + \frac{1}{9} + \frac{1}{9\cdot 10} + \frac{1}{9\cdot 10\cdot 11} + \cdots\right)$$

$$< \frac{1}{8!}\left(1 + \frac{1}{9} + \frac{1}{9^2} + \frac{1}{9^3} + \cdots\right)$$

$$= \frac{1}{8!}\frac{1}{1 - \frac{1}{9}} \qquad \text{(using equation 11.13b)}$$

$$= \frac{9}{8\cdot 8!}$$

$$< 0.000\,028\,0.$$

SECTION 11.7 APPROXIMATING THE SUM OF A SERIES

We may write, therefore, that

$$2.718\,254\,0 < \sum_{n=0}^{\infty} \frac{1}{n!} < 2.718\,282\,0.$$

You might recognize that these same digits appeared in the approximation of the number e as the limit of the sequence $\{(1 + 1/n)^n\}$ in Section 11.2. In Section 11.11, we will show that $\sum_{n=0}^{\infty} 1/n!$ converges to e,

$$e = \sum_{n=0}^{\infty} \frac{1}{n!}. \qquad (11.22)$$

The advantage of the series definition is obvious; it is very easy to obtain an accuracy statement wherever the series is truncated.

EXAMPLE 11.17

How many terms in the convergent series $\sum_{n=1}^{\infty} n/[(n+1)3^n]$ ensure a truncation error of less than 10^{-5}?

SOLUTION If this series is truncated after the Nth term, the error is

$$\sum_{n=N+1}^{\infty} \frac{n}{(n+1)3^n} = \frac{N+1}{(N+2)3^{N+1}} + \frac{N+2}{(N+3)3^{N+2}} + \cdots$$

$$< \frac{1}{3^{N+1}} + \frac{1}{3^{N+2}} + \cdots$$

$$= \frac{\frac{1}{3^{N+1}}}{1 - \frac{1}{3}} \qquad \text{(using equation 11.13b)}$$

$$= \frac{1}{2 \cdot 3^N}.$$

Consequently, the error is guaranteed to be less than 10^{-5} if N satisfies the inequality

$$\frac{1}{2 \cdot 3^N} < 10^{-5} \qquad \text{or}$$

$$2 \cdot 3^N > 10^5.$$

With the help of a calculator, we find that this is true if $N \geq 10$, and therefore 10 or more terms yield the required accuracy.

EXERCISES 11.7

In Exercises 1–8 use the number of terms indicated to find an approximation to the sum of the series. In each case, obtain an estimate of the truncation error.

1. $\sum_{n=1}^{\infty} \dfrac{1}{n 2^n}$ (10 terms)

2. $\sum_{n=1}^{\infty} \dfrac{(-1)^n}{n^4}$ (20 terms)

3. $\sum_{n=1}^{\infty} \dfrac{1}{n^n}$ (5 terms)

4. $\sum_{n=2}^{\infty} \dfrac{(-1)^{n+1}}{n^3 3^n}$ (3 terms)

5. $\sum_{n=1}^{\infty} \dfrac{1}{2^n} \sin\left(\dfrac{\pi}{n}\right)$ (15 terms)

6. $\sum_{n=2}^{\infty} \dfrac{2^n - 1}{3^n + n}$ (20 terms)

7. $\sum_{n=2}^{\infty} \dfrac{2^n + 1}{3^n + n}$ (20 terms)

8. $\sum_{n=1}^{\infty} \dfrac{(-1)^n}{n}$ (100 terms)

In Exercises 9–11 how many terms in the series guarantee an approximation to the sum with a truncation error of less than 10^{-4}?

9. $\sum_{n=1}^{\infty} \dfrac{(-1)^n}{n^2}$

10. $\sum_{n=1}^{\infty} \dfrac{1}{n^2 4^n}$

11. $\sum_{n=1}^{\infty} \dfrac{2^n}{n!}$

12. This exercise shows that we must be very careful in predicting the accuracy of a result. Consider the series

$$S = 3.125\,100\,1 - 0.000\,090\,18\left(1 + \dfrac{1}{10} + \dfrac{1}{10^2} + \dfrac{1}{10^3} + \cdots\right).$$

(a) Show that the sum of this series is exactly $S = 3.124\,999\,9$. To two decimals, then, the value of S is 3.12.

(b) Verify that the first four partial sums of the series are

$S_1 = 3.125\,100\,1,$
$S_2 = 3.125\,009\,92,$
$S_3 = 3.125\,000\,902,$
$S_4 = 3.125\,000\,000\,2.$

(c) If $E_n = S_n - S$ are the differences between the sum of the series and its first four partial sums, show that

$E_1 = 0.000\,100\,2,$
$E_2 = 0.000\,010\,02,$
$E_3 = 0.000\,001\,002,$
$E_4 = 0.000\,000\,100\,2.$

What can you say about the accuracy of $S_1, S_2, S_3,$ and S_4 as approximations to S?

(d) If S is approximated by any of $S_1, S_2, S_3,$ or S_4 to two decimals, the result is 3.13, not 3.12 as in (a). Thus, in spite of the accuracy predicted in (c), S_1, $S_2, S_3,$ and S_4 do not predict S correctly to two decimals. Do they predict S correctly to three or four decimals?

SECTION 11.8

Power Series and Intervals of Convergence

Our work on sequences and series of constants in the first seven sections of this chapter has prepared the way for discussions on sequences and series of functions. An infinite sequence of functions is the assignment of functions to integers,

$$\{f_n(x)\}_0^{\infty} = f_0(x), f_1(x), \ldots, \quad (11.23)$$

where we have initiated our assignment with the integer $n = 0$ rather than $n = 1$. The reason for this will become apparent shortly. From such a sequence we may form an infinite series of functions:

$$\sum_{n=0}^{\infty} f_n(x) = f_0(x) + f_1(x) + f_2(x) + \cdots. \quad (11.24)$$

When $f_n(x)$ is a constant multiplied by x^n,

$$f_n(x) = a_n x^n, \qquad a_n = \text{a constant},$$

series 11.24 takes the form

$$\sum_{n=0}^{\infty} a_n x^n = a_0 + a_1 x + a_2 x^2 + \cdots + a_n x^n + \cdots, \qquad (11.25)$$

and is called a **power series** in x.

Unlike series of constants, the variable of summation n in 11.25 is one step out of phase with the counting of terms: $n = 0$ identifies the first term, $n = 1$ the second term, etc. Instead, the value of n now identifies the power of x. In the remainder of the chapter we often write $\sum a_n x^n$ in place of $\sum_{n=0}^{\infty} a_n x^n$ whenever the lower limit is $n = 0$.

For each value of x that we substitute into a power series, we obtain a series of constants. For example, if into the power series

$$\sum_{n=0}^{\infty} x^n = 1 + x + x^2 + x^3 + \cdots$$

we substitute $x = 1/2$, we obtain the series of constants

$$\sum_{n=0}^{\infty} \left(\frac{1}{2}\right)^n = 1 + \frac{1}{2} + \frac{1}{2^2} + \frac{1}{2^3} + \cdots,$$

a geometric series that converges to 2. When we substitute $x = -3$, we obtain the divergent geometric series

$$\sum_{n=0}^{\infty} (-3)^n = 1 - 3 + 9 - 27 + \cdots.$$

Definition 11.9

The totality of values of x for which a power series converges is called its **interval of convergence**.

Obviously the interval of convergence for a power series $\sum a_n x^n$ always includes the value $x = 0$, but what other possibilities are there? "Interval of convergence" suggests that the values of x for which a power series converges form some kind of interval. This is indeed true, as we shall soon see.

EXAMPLE 11.18

Find the interval of convergence for the power series $\sum_{n=0}^{\infty} x^n$.

SOLUTION Since the series is a geometric series with common ratio x, its interval of convergence is $-1 < x < 1$. ∎

EXAMPLE 11.19

Find intervals of convergence for the following power series:

(a) $\sum_{n=1}^{\infty} \dfrac{x^n}{n^n}$ (b) $\sum_{n=1}^{\infty} n!\, x^n$ (c) $\sum_{n=1}^{\infty} \dfrac{(-1)^{n+1}}{n} x^n$

SOLUTION (a) Consider the series of absolute values $\sum |x|^n / n^n$. We use the limit root test on this series, and calculate

$$R = \lim_{n \to \infty} \left(\dfrac{|x|^n}{n^n} \right)^{1/n} = \lim_{n \to \infty} \dfrac{|x|}{n} = 0,$$

and this is true for any x whatsoever. The power series therefore converges absolutely for all x.

(b) Consider again the series of absolute values $\sum n!\, |x|^n$. We use the limit ratio test and obtain

$$L = \lim_{n \to \infty} \dfrac{(n+1)!\, |x|^{n+1}}{n!\, |x|^n} = |x| \lim_{n \to \infty} (n+1) = \infty.$$

Since $L = \infty$ for all $x \neq 0$, the original power series does not converge absolutely for any $x \neq 0$. But this result implies much more. It shows that the terms in the sequence $\{n!\, |x|^n\}$ become indefinitely large as $n \to \infty$. But then $\lim_{n \to \infty} n!\, x^n$ cannot possibly exist, and the series diverges by the nth term test. The power series therefore converges only for $x = 0$.

(c) Once again for the series of absolute values, we obtain

$$L = \lim_{n \to \infty} \dfrac{\dfrac{|x|^{n+1}}{n+1}}{\dfrac{|x|^n}{n}} = \lim_{n \to \infty} |x| \dfrac{n}{n+1} = |x|.$$

Since $L = |x|$, it follows that the series of absolute values converges if $|x| < 1$ and diverges if $|x| > 1$. The original power series therefore converges absolutely for $-1 < x < 1$. Furthermore, when $|x| > 1$, so too is L, and it follows that the terms in the sequence $\{|x|^n / n\}$ must become indefinitely large. But then

$$\lim_{n \to \infty} \dfrac{(-1)^{n+1}}{n} x^n$$

cannot possibly exist, and the series diverges by the nth term test. If $x < -1$, terms in the sequence $\{(-1)^{n+1} x^n / n\}$ are all negative and "approach negative infinity"; if $x > 1$, the terms are alternately positive and negative with increasing absolute values. When $x = 1$, the power series reduces to the alternating harmonic series

$$\sum_{n=1}^{\infty} \dfrac{(-1)^{n+1}}{n} = 1 - \dfrac{1}{2} + \dfrac{1}{3} - \dfrac{1}{4} + \cdots,$$

which converges conditionally; and when $x = -1$, it reduces to

$$\sum_{n=1}^{\infty} \dfrac{(-1)^{n+1}}{n} (-1)^n = -1 - \dfrac{1}{2} - \dfrac{1}{3} - \dfrac{1}{4} - \cdots,$$

which diverges. The interval of convergence is therefore $-1 < x \leq 1$. ∎

Example 11.19 has illustrated that there are at least three possible types of intervals of convergence for power series $\sum a_n x^n$:

1. The power series converges only for $x = 0$;
2. The power series converges absolutely for all x;
3. There exists a number $R > 0$ such that the power series converges absolutely for $|x| < R$, diverges for $|x| > R$, and may or may not converge for $x = \pm R$

These are in fact the only possibilities for an interval of convergence, and this is proved in Exercises 32 and 33. In (3) we call R the **radius of convergence** of the power series. It is half the length of the interval of convergence, or, the distance we may proceed in either direction along the x-axis from $x = 0$ and expect absolute convergence of the power series, with the possible exceptions of $x = \pm R$. In order to have a radius of convergence associated with every power series, we say in (1) and (2) that $R = 0$ and $R = \infty$ respectively.

Every power series $\sum a_n x^n$ now has a radius of convergence R. If $R = 0$, the power series converges only for $x = 0$; if $R = \infty$, the power series converges absolutely for all x; and if $0 < R < \infty$, the power series converges absolutely for $|x| < R$, diverges for $|x| > R$, and may or may not converge for $x = \pm R$. For many power series the radius of convergence can be calculated according to the following theorem.

Theorem 11.13

The radius of convergence of a power series $\sum_{n=0}^{\infty} a_n x^n$ is given by

$$R = \lim_{n \to \infty} \left| \frac{a_n}{a_{n+1}} \right| \quad \text{or,} \tag{11.26a}$$

$$R = \lim_{n \to \infty} \frac{1}{\sqrt[n]{|a_n|}}, \tag{11.26b}$$

provided either limit exists or is equal to infinity.

Proof We verify 11.26a using the limit ratio test; verification of 11.26b is similar, using the limit root test.

If the limit ratio test is applied to the series of absolute values $\sum |a_n x^n|$,

$$L = \lim_{n \to \infty} \frac{|a_{n+1} x^{n+1}|}{|a_n x^n|} = |x| \lim_{n \to \infty} \left| \frac{a_{n+1}}{a_n} \right|.$$

Assuming that limit 11.26a exists or is equal to infinity, there are three possibilities:

(i) If $\lim_{n \to \infty} |a_n/a_{n+1}| = 0$, then $\lim_{n \to \infty} |a_{n+1}/a_n| = \infty$. Therefore $L = \infty$, and the power series diverges for all $x \neq 0$. In other words,

$$R = 0 = \lim_{n \to \infty} \left| \frac{a_n}{a_{n+1}} \right|.$$

(ii) If $\lim_{n \to \infty} |a_n/a_{n+1}| = \infty$, then $\lim_{n \to \infty} |a_{n+1}/a_n| = 0$. Therefore $L = 0$, and the power series converges absolutely for all x. Consequently,

$$R = \infty = \lim_{n \to \infty} \left| \frac{a_n}{a_{n+1}} \right|.$$

(iii) If $\lim_{n\to\infty} |a_n/a_{n+1}| = R$, then $\lim_{n\to\infty} |a_{n+1}/a_n| = 1/R$. In this case, $L = |x|/R$. Since the power series converges absolutely for $L < 1$ and diverges for $L > 1$, it follows that absolute convergence occurs for $|x| < R$ and divergence for $|x| > R$. This implies that R is the radius of convergence of the power series.

With this theorem the limit ratio test and limit root test can be avoided in determining intervals of convergence for many power series. For the series in Example 11.19 we proceed as follows:

(a) For the power series $\sum_{n=1}^{\infty} x^n/n^n$, we use 11.26b to obtain

$$R = \lim_{n\to\infty} \frac{1}{\sqrt[n]{|a_n|}} = \lim_{n\to\infty} \frac{1}{(1/n^n)^{1/n}} = \lim_{n\to\infty} n = \infty.$$

The power series therefore converges absolutely for all x.

(b) For the series $\sum_{n=1}^{\infty} n!\, x^n$, we use 11.26a,

$$R = \lim_{n\to\infty} \left|\frac{a_n}{a_{n+1}}\right| = \lim_{n\to\infty} \frac{n!}{(n+1)!} = \lim_{n\to\infty} \frac{1}{n+1} = 0,$$

and the power series converges only for $x = 0$.

(c) For the power series $\sum_{n=1}^{\infty} [(-1)^{n+1}/n]x^n$, 11.26a gives

$$R = \lim_{n\to\infty} \left|\frac{a_n}{a_{n+1}}\right| = \lim_{n\to\infty} \left|\frac{\frac{(-1)^{n+1}}{n}}{\frac{(-1)^{n+2}}{n+1}}\right| = \lim_{n\to\infty} \left(\frac{n+1}{n}\right) = 1.$$

When $x = 1$, the power series reduces to the alternating harmonic series

$$\sum_{n=1}^{\infty} \frac{(-1)^{n+1}}{n} = 1 - \frac{1}{2} + \frac{1}{3} - \frac{1}{4} + \cdots,$$

which converges conditionally. When $x = -1$, it reduces to

$$\sum_{n=1}^{\infty} \frac{(-1)^{n+1}}{n}(-1)^n = -1 - \frac{1}{2} - \frac{1}{3} - \frac{1}{4} - \cdots,$$

which diverges. The interval of convergence is therefore $-1 < x \leq 1$.

EXAMPLE 11.20

Find the interval of convergence for the power series $\sum_{n=1}^{\infty} nx^n/5^{2n}$.

SOLUTION Since

$$R = \lim_{n\to\infty} \left|\frac{a_n}{a_{n+1}}\right| = \lim_{n\to\infty} \frac{\frac{n}{5^{2n}}}{\frac{n+1}{5^{2n+2}}} = 25,$$

the power series converges absolutely for $-25 < x < 25$. At $x = 25$, the series reduces to $\sum_{n=1}^{\infty} n$, which diverges; and at $x = -25$, it reduces to $\sum_{n=1}^{\infty} n(-1)^n$, which also diverges. The interval of convergence is therefore $-25 < x < 25$. ∎

EXAMPLE 11.21

Find the interval of convergence for the power series $\sum_{n=1}^{\infty} x^{2n+1}/(n^2 2^n)$.

SOLUTION Since coefficients of even powers of x are 0 in this power series, the sequence $\{a_n/a_{n+1}\}$ is not defined. We cannot therefore find its radius of convergence directly using Theorem 11.13. Instead we write

$$\sum_{n=1}^{\infty} \frac{1}{n^2 2^n} x^{2n+1} = x \sum_{n=1}^{\infty} \frac{1}{n^2 2^n} (x^2)^n$$

and set $y = x^2$ in the series:

$$\sum_{n=1}^{\infty} \frac{1}{n^2 2^n} (x^2)^n = \sum_{n=1}^{\infty} \frac{1}{n^2 2^n} y^n.$$

According to equation 11.26a the radius of convergence of this series in y is

$$\lim_{n \to \infty} \frac{\frac{1}{n^2 2^n}}{\frac{1}{(n+1)^2 2^{n+1}}} = \lim_{n \to \infty} 2\left(\frac{n+1}{n}\right)^2 = 2.$$

Since $x = \pm\sqrt{y}$, it follows that the radius of convergence of the power series in x is $R = \sqrt{2}$. When $x = \sqrt{2}$, the series reduces to $\sum_{n=1}^{\infty} \sqrt{2}/n^2 = \sqrt{2} \sum_{n=1}^{\infty} 1/n^2$, which converges; and when $x = -\sqrt{2}$, it reduces to the negative of this series. The interval of convergence is therefore $-\sqrt{2} \leq x \leq \sqrt{2}$. ∎

We describe what it means for a function to be the sum of a power series in the following definition.

Definition 11.10

A function $f(x)$ is said to be the **sum** of a power series $\sum_{n=0}^{\infty} a_n x^n$ if at each point x in the interval of convergence of the power series, the sum of the series is the same as the value of $f(x)$ at that x.

We write

$$f(x) = \sum_{n=0}^{\infty} a_n x^n \qquad (11.27)$$

to signify that $f(x)$ is the sum of the power series. For example, the sum of the geometric series

$$\sum_{n=0}^{\infty} x^n = 1 + x + x^2 + x^3 + \cdots$$

is $1/(1-x)$ for $|x| < 1$, and we write

$$\frac{1}{1-x} = 1 + x + x^2 + x^3 + \cdots, \qquad |x| < 1.$$

We affix to the equation the interval of convergence of the power series, thus identifying for which values of x the function $1/(1-x)$ is the sum of the series. The function

$1/(1-x)$ is defined for all x, except $x = 1$, but it represents the sum of the series only in the interval $-1 < x < 1$.

In Section 11.3 we defined the sum of an infinite series of constants as the limit of its sequence of partial sums. In the case of power series 11.27, the sequence of partial sums is a sequence of polynomials:

$$S_1(x) = a_0,$$
$$S_2(x) = a_0 + a_1 x,$$
$$S_3(x) = a_0 + a_1 x + a_2 x^2,$$
$$\vdots \qquad \vdots$$
$$S_n(x) = a_0 + a_1 x + a_2 x^2 + \cdots + a_{n-1} x^{n-1},$$
$$\vdots \qquad \vdots$$

By equation 11.27 we are saying that for each x in the interval of convergence of the power series,

$$f(x) = \lim_{n \to \infty} S_n(x).$$

For example, the nth partial sum of the geometric series $\sum_{n=0}^{\infty} x^n$ is

$$S_n(x) = 1 + x + x^2 + \cdots + x^{n-1} = \begin{cases} n & x = 1 \\ \dfrac{1 - x^n}{1 - x} & x \neq 1 \end{cases}$$

(see equation 11.13a). Algebraically, it is clear that for large n, $S_n(x)$ approaches $1/(1-x)$, provided $|x| < 1$, and we write therefore

$$\frac{1}{1-x} = 1 + x + x^2 + x^3 + \cdots, \qquad |x| < 1.$$

In Figure 11.20 we show the first five partial sums of this geometric series together with the sum, just to illustrate how the partial sums of the series approximate the sum more closely for increasing n. Note how these figures illustrate the following facts about the geometric series and its partial sums:

1. The partial sums $S_n(x)$ are defined for all x and the sum $1/(1-x)$ is defined for all $x \neq 1$, but the curves $y = S_n(x)$ approach $y = 1/(1-x)$ only for $|x| < 1$.

2. For $x > 0$, the partial sums $S_n(x)$ approach $1/(1-x)$ from below; for $x < 0$, the $S_n(x)$ oscillate about $1/(1-x)$, but gradually approach $1/(1-x)$.

(b)   (c)   **FIGURE 11.20**

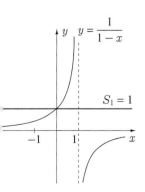
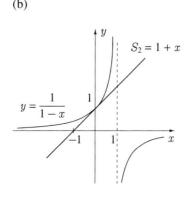
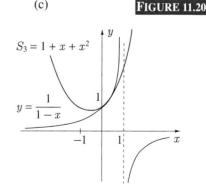

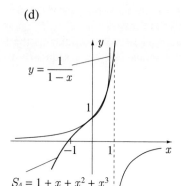

(d) $y = \dfrac{1}{1-x}$, $S_4 = 1 + x + x^2 + x^3$

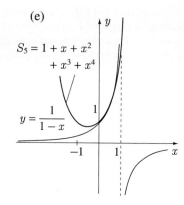

(e) $S_5 = 1 + x + x^2 + x^3 + x^4$, $y = \dfrac{1}{1-x}$

EXAMPLE 11.22 Find the sum of the series

$$\sum_{n=0}^{\infty} \frac{(-1)^n}{2^n} x^{(2n+1)/2}.$$

SOLUTION We write

$$\sum_{n=0}^{\infty} \frac{(-1)^n}{2^n} x^{(2n+1)/2} = x^{1/2} \sum_{n=0}^{\infty} \frac{(-1)^n}{2^n} x^n = x^{1/2} \sum_{n=0}^{\infty} \left(-\frac{x}{2}\right)^n$$

and note that the power series on the right is a geometric series with sum

$$\sum_{n=0}^{\infty} \left(-\frac{x}{2}\right)^n = \frac{1}{1 + \dfrac{x}{2}}, \quad \left|-\frac{x}{2}\right| < 1.$$

Consequently,

$$\sum_{n=0}^{\infty} \frac{(-1)^n}{2^n} x^{(2n+1)/2} = \sqrt{x}\, \frac{1}{1 + \dfrac{x}{2}} = \frac{2\sqrt{x}}{x+2}, \quad 0 \le x < 2.$$

For many power series there is no known function that is equal to the sum of the power series over its interval of convergence. In such cases we continue to write equation 11.27, but say that the power series defines the value of $f(x)$ at each x in the interval of convergence. For instance, a very important series in engineering and physics is

$$\sum_{n=0}^{\infty} \frac{(-1)^n}{2^{2n}(n!)^2} x^{2n},$$

which converges absolutely for all x. It arises so often in applications that it is given a special name, the *Bessel function of the first kind of order zero*, and is denoted by $J_0(x)$. In other words, we write

$$J_0(x) = \sum_{n=0}^{\infty} \frac{(-1)^n}{2^{2n}(n!)^2} x^{2n},$$

and say that the power series defines the value of $J_0(x)$ for each x.

We have called 11.25 a power series in x. It is also said to be a *power series about* 0 (meaning the point 0 on the x-axis), where we note that for any interval of convergence whatsoever, 0 is always at its centre. This suggests that power series about other points might be considered, and this is indeed the case. The general power series about a point c on the x-axis is

$$\sum_{n=0}^{\infty} a_n(x-c)^n = a_0 + a_1(x-c) + a_2(x-c)^2 + \cdots \qquad (11.28)$$

and is said to be a power series in $x - c$.

A power series in $x - c$ has an interval of convergence and a radius of convergence analogous to a power series in x. In particular, every power series in $x - c$ has a radius of convergence R such that if $R = 0$, the power series converges only for $x = c$; if $R = \infty$, the power series converges absolutely for all x; and if $0 < R < \infty$, the series converges absolutely for $|x - c| < R$, diverges for $|x - c| > R$, and may or may not converge for $x = c \pm R$. The radius of convergence is again given by equations 11.26, provided the limits exist or are equal to infinity. For power series in $x - c$, then, the point c is the centre of the interval of convergence.

EXAMPLE 11.23

Find the interval of convergence of the power series

$$2^2(x+2) + 2^3(x+2)^2 + 2^4(x+2)^3 + \cdots + 2^{n+1}(x+2)^n + \cdots .$$

SOLUTION By writing the power series in sigma notation,

$$\sum_{n=1}^{\infty} 2^{n+1}(x+2)^n,$$

we can use equation 11.26a (or 11.26b) to calculate its radius of convergence,

$$R = \lim_{n \to \infty} \left| \frac{a_n}{a_{n+1}} \right| = \lim_{n \to \infty} \frac{2^{n+1}}{2^{n+2}} = \frac{1}{2}.$$

Since this is a power series about -2, the ends of the interval of convergence are $x = -5/2$ and $x = -3/2$. At $x = -5/2$, the power series reduces to $\sum_{n=1}^{\infty} 2(-1)^n$, which diverges; and at $x = -3/2$, it reduces to $\sum_{n=1}^{\infty} 2$, which also diverges. The interval of convergence is therefore $-5/2 < x < -3/2$. ∎

EXERCISES 11.8

In Exercises 1–22 find the interval of convergence for the power series.

1. $\sum_{n=1}^{\infty} \frac{1}{n} x^n$

2. $\sum_{n=1}^{\infty} n^2 x^n$

3. $\sum_{n=0}^{\infty} \frac{1}{(n+1)^3} x^n$

4. $\sum_{n=0}^{\infty} n^2 3^n x^n$

5. $\sum_{n=0}^{\infty} \frac{1}{2^n} (x-1)^n$

6. $\sum_{n=4}^{\infty} (-1)^n n^3 (x+3)^n$

7. $\sum_{n=1}^{\infty} \frac{1}{\sqrt{n}} (x+2)^n$

8. $\sum_{n=2}^{\infty} 2^n \left(\frac{n-1}{n+2} \right)^2 (x-4)^n$

9. $\sum_{n=1}^{\infty} \frac{1}{n^2} x^{2n}$

10. $\sum_{n=0}^{\infty} (-1)^n x^{3n}$

11. $\sum_{n=0}^{\infty} \frac{n-1}{n+1} (2x)^n$

12. $\sum_{n=0}^{\infty} \frac{1}{\sqrt{n+1}} x^{3n+1}$

13. $\sum_{n=0}^{\infty} \frac{(-1)^n}{3^n} x^{2n+1}$ 14. $\sum_{n=1}^{\infty} \frac{(-e)^n}{n^2} x^n$

15. $\frac{x}{9} + \frac{4}{3^4} x^2 + \frac{9}{3^6} x^3 + \cdots + \frac{n^2}{3^{2n}} x^n + \cdots$

16. $x + 2^2 x^2 + 3^3 x^3 + \cdots + n^n x^n + \cdots$

17. $\frac{1}{36}(x+10)^6 + \frac{1}{49}(x+10)^7 + \frac{1}{64}(x+10)^8 + \cdots + \frac{1}{n^2}(x+10)^n + \cdots$

18. $3x + 8(3x)^2 + 27(3x)^3 + \cdots + n^3(3x)^n + \cdots$

19. $\frac{3}{4} x^2 + x^4 + \frac{27}{16} x^6 + \cdots + \frac{3^n}{(n+1)^2} x^{2n} + \cdots$

20. $1 + \frac{1}{5} x^3 + \frac{1}{25} x^6 + \cdots + \frac{1}{5^n} x^{3n} + \cdots$

21. $\sum_{n=2}^{\infty} \frac{1}{\ln n} x^n$ 22. $\sum_{n=2}^{\infty} \frac{1}{n^2 \ln n} x^n$

In Exercises 23–25 find the radius of convergence of the power series.

23. $\sum_{n=1}^{\infty} \frac{(n!)^3}{(3n)!} x^n$

24. $\sum_{n=1}^{\infty} \frac{2 \cdot 4 \cdot 6 \cdot \cdots \cdot (2n)}{3 \cdot 5 \cdot 7 \cdot \cdots \cdot (2n+1)} x^n$

25. $\sum_{n=1}^{\infty} \frac{[1 \cdot 3 \cdot 5 \cdot \cdots \cdot (2n+1)]^2}{2^{2n}(2n)!} x^n$

In Exercises 26–29 find the sum of the power series.

26. $\sum_{n=0}^{\infty} \frac{1}{4^n} x^{3n}$ 27. $\sum_{n=1}^{\infty} (-e)^n x^n$

28. $\sum_{n=1}^{\infty} \frac{1}{3^{2n}} (x-1)^n$ 29. $\sum_{n=2}^{\infty} (x+5)^{2n}$

30. If m is a nonnegative integer, the Bessel function of order m of the first kind is defined by the power series

$$J_m(x) = \sum_{n=0}^{\infty} \frac{(-1)^n}{2^{2n+m} n! (n+m)!} x^{2n+m}.$$

(a) Write out the first five terms of $J_0(x)$, $J_1(x)$, and $J_m(x)$.

(b) Find the interval of convergence for each $J_m(x)$.

31. The hypergeometric series is

$$1 + \frac{\alpha \beta}{\gamma} + \frac{\alpha(\alpha+1)\beta(\beta+1)}{2! \gamma(\gamma+1)} x^2 + \frac{\alpha(\alpha+1)(\alpha+2)\beta(\beta+1)(\beta+2)}{3! \gamma(\gamma+1)(\gamma+2)} x^3 + \cdots,$$

where α, β, and γ are all constants.

(a) Write this series in sigma notation.

(b) What is the radius of convergence of the hypergeometric series if γ is not zero or a negative integer?

32. Show that if a power series $\sum a_n x^n$ converges for $x_1 \neq 0$, then it converges absolutely for all x in the interval $|x| < |x_1|$.

33. Prove that the interval of convergence of a power series $\sum a_n x^n$ must be one of the three possibilities listed on page 554. Use Exercise 32 and the *completeness property* for real numbers, which states: If a nonempty set S of real numbers has an upper bound, then it has a least upper bound; that is, there exists a number l such that all numbers in S are less than or equal to l and there is no upper bound smaller than l.

SECTION 11.9

Taylor's Remainder Formula and Taylor Series

If $f(x)$ is the sum of a power series $\sum_{n=0}^{\infty} a_n(x-c)^n$, we write

$$f(x) = \sum_{n=0}^{\infty} a_n(x-c)^n. \qquad (11.29)$$

In Section 11.8 we considered this equation in situations where the power series is given and the sum is to be determined. An alternative view is to consider the function $f(x)$ as given and the power series as an expansion of the function about the point c. For example, given the function $f(x) = 1/(2x+3)$, it is quite simple to use properties of geometric series to find a power series expansion in x for $f(x)$. If we write

$$f(x) = \frac{1}{2x+3} = \frac{1}{3\left(1 + \dfrac{2x}{3}\right)}$$

and interpret the right-hand side as one-third the sum of a geometric series with first term equal to 1 and common ratio $-2x/3$, then

$$f(x) = \frac{1}{3} \sum_{n=0}^{\infty} \left(-\frac{2x}{3}\right)^n, \qquad \left|-\frac{2x}{3}\right| < 1$$

$$= \sum_{n=0}^{\infty} \frac{(-1)^n 2^n}{3^{n+1}} x^n, \qquad |x| < \frac{3}{2}.$$

Naturally it is not clear at this time why we should want to expand a function in the form of an infinite series, but in Section 11.12, when we consider applications of series, we will find such expansions extremely useful.

In this section we establish two results. First we give conditions that gurarantee that a function $f(x)$ has a power series expansion about a point c and that this power series converges to $f(x)$. Second, we show that if a function has a power series expansion about a point, then it has only one such series. The first result is an immediate consequence of Theorem 11.14.

Theorem 11.14

(Taylor's Remainder Formula) If $f(x)$ and its first $n-1$ derivatives are continuous on the closed interval between c and x, and if $f(x)$ has an nth derivative on the open interval between c and x, then there exists a point z_n between c and x such that

$$f(x) = f(c) + f'(c)(x-c) + \frac{f''(c)}{2!}(x-c)^2 + \frac{f'''(c)}{3!}(x-c)^3 + \cdots$$
$$+ \frac{f^{(n-1)}(c)}{(n-1)!}(x-c)^{n-1} + \frac{f^{(n)}(z_n)}{n!}(x-c)^n. \qquad (11.30)$$

The notation $f^{(n)}(z_n)$ represents the nth derivative of $f(x)$ evaluated at $x = z_n$. This result can be proved by applying Rolle's theorem (Theorem 3.13) to the function

$$F(y) = f(x) - f(y) - f'(y)(x-y) - \frac{f''(y)}{2!}(x-y)^2 - \cdots$$
$$- \frac{f^{(n-1)}(y)}{(n-1)!}(x-y)^{n-1} - \left(\frac{x-y}{x-c}\right)^n \Big[f(x) - f(c) - f'(c)(x-c)$$
$$- \frac{f''(c)}{2!}(x-c)^2 - \cdots - \frac{f^{(n-1)}(c)}{(n-1)!}(x-c)^{n-1}\Big]$$

defined on the interval $c \le y \le x$, or $x \le y \le c$, depending on whether x is larger or smaller than c.

Note that when $n = 1$, Taylor's remainder formula reduces to the mean value theorem of Section 3.9. In other words, the mean value theorem is a special case of Taylor's remainder formula.

Consider now a function $f(x)$ that has derivatives of all orders on the closed interval between c and x. Taylor's remainder formula can therefore be written down for all values of n. Let us write it out for $n = 1, 2, 3,$ and 4:

$$f(x) = f(c) + f'(z_1)(x - c), \tag{11.31a}$$

$$f(x) = f(c) + f'(c)(x - c) + \frac{f''(z_2)}{2!}(x - c)^2, \tag{11.31b}$$

$$f(x) = f(c) + f'(c)(x - c) + \frac{f''(c)}{2!}(x - c)^2 + \frac{f'''(z_3)}{3!}(x - c)^3, \tag{11.31c}$$

$$f(x) = f(c) + f'(c)(x - c) + \frac{f''(c)}{2!}(x - c)^2$$
$$+ \frac{f'''(c)}{3!}(x - c)^3 + \frac{f''''(z_4)}{4!}(x - c)^4. \tag{11.31d}$$

The last terms in these equations are called **remainders**, and when they are denoted by $R_1, R_2, R_3,$ and R_4 respectively,

$$f(x) = f(c) + R_1, \tag{11.32a}$$

$$f(x) = f(c) + f'(c)(x - c) + R_2, \tag{11.32b}$$

$$f(x) = f(c) + f'(c)(x - c) + \frac{f''(c)}{2!}(x - c)^2 + R_3, \tag{11.32c}$$

$$f(x) = f(c) + f'(c)(x - c) + \frac{f''(c)}{2!}(x - c)^2$$
$$+ \frac{f'''(c)}{3!}(x - c)^3 + R_4, \tag{11.32d}$$

and in general,

$$f(x) = f(c) + f'(c)(x - c) + \frac{f''(c)}{2!}(x - c)^2 + \cdots$$
$$+ \frac{f^{(n-1)}(c)}{(n-1)!}(x - c)^{n-1} + R_n. \tag{11.32e}$$

We call R_n the remainder for the simple reason that if we were to drop the remainders from equations (11.32), then what remains is a sequence of polynomial approximations to $f(x)$:

$$f(x) \approx f(c), \tag{11.33a}$$
$$f(x) \approx f(c) + f'(c)(x - c), \tag{11.33b}$$
$$f(x) \approx f(c) + f'(c)(x - c) + \frac{f''(c)}{2!}(x - c)^2, \tag{11.33c}$$
$$f(x) \approx f(c) + f'(c)(x - c) + \frac{f''(c)}{2!}(x - c)^2 + \frac{f'''(c)}{3!}(x - c)^3. \tag{11.33d}$$

We give a detailed discussion on the accuracy of these approximations in Section 11.12, but for now we can at least appreciate why Theorem 11.14 is called Taylor's remainder formula.

The nth remainder is

$$R_n = R_n(c,x) = \frac{f^{(n)}(z_n)}{n!}(x-c)^n, \qquad (11.34)$$

where z_n (which, as the notation suggests, varies with n) is always between c and x. We have written $R_n(c,x)$ to emphasize the fact that the remainders depend on the point of expansion c and the value of x being considered.

Suppose the sequence of remainders has limit zero,

$$\lim_{n\to\infty} R_n = 0. \qquad (11.35)$$

What this means is that the sequence of polynomials in 11.33 approximates $f(x)$ more and more closely as n gets larger and larger. In fact, these polynomials are the partial sums for the following infinite series,

$$f(c) + f'(c)(x-c) + \frac{f''(c)}{2!}(x-c)^2 + \frac{f'''(c)}{3!}(x-c)^3 + \cdots . \qquad (11.36)$$

Thus, if the sequence of remainders $\{R_n\}$ approaches zero, then power series 11.36 converges to $f(x)$. We may write therefore that

$$f(x) = f(c) + f'(c)(x-c) + \frac{f''(c)}{2!}(x-c)^2 + \frac{f'''(c)}{3!}(x-c)^3 + \cdots , \qquad (11.37)$$

and this equation is valid for all those x's for which $\lim_{n\to\infty} R_n(c,x) = 0$. This power series in $x - c$ is called the **Taylor series** of $f(x)$ about the point c. We have shown that *if the sequence of remainders $\{R_n\}$ for a function $f(x)$ exists and has limit zero, the Taylor series converges to $f(x)$ for that x.*

When $c = 0$, the Taylor series takes the form

$$f(x) = f(0) + f'(0)x + \frac{f''(0)}{2!}x^2 + \frac{f'''(0)}{3!}x^3 + \cdots , \qquad (11.38)$$

a power series in x called the **Maclaurin series** for $f(x)$.

To obtain the Taylor or Maclaurin series for a function $f(x)$ is quite simple; evaluate all derivatives of $f(x)$ at the point of expansion and use formula 11.37 or 11.38. To show that the series converges to $f(x)$ is a different matter. We must show that the sequence $\{R_n\}$ has limit zero, and this can be very difficult because R_n is defined in terms of an unknown point z_n. This problem is usually circumvented by finding a maximum value for R_n and showing that this maximum value approaches zero. We illustrate this procedure in the next three examples. To do so, however, we require the result of the following theorem. Apart from the fact that Theorem 11.15 is needed for Examples 11.24 and 11.25, its proof is interesting in that it interchanges the usual role of sequences and series. Heretofore, we have used sequences to determine convergence of series. This theorem uses series to verify convergence of a certain sequence.

Theorem 11.15

The sequence $\{|x|^n/n!\}$ has limit 0 *for any x whatsoever,*

$$\lim_{n\to\infty} \frac{|x|^n}{n!} = 0. \qquad (11.39)$$

Proof Consider the series $\sum_{n=1}^{\infty} |x|^n/n!$. Since the ratio test gives

$$L = \lim_{n \to \infty} \frac{\frac{|x|^{n+1}}{(n+1)!}}{\frac{|x|^n}{n!}} = \lim_{n \to \infty} \frac{|x|}{n+1} = 0,$$

the series converges for any x. But then the nth term test implies that its nth term must have limit 0.

EXAMPLE 11.24

Find the Maclaurin series for $\sin x$ and show that it converges to $\sin x$ for all x.

SOLUTION If $f(x) = \sin x$, then:

$$f(0) = \sin x_{|x=0} = 0;$$
$$f'(0) = \cos x_{|x=0} = 1;$$
$$f''(0) = -\sin x_{|x=0} = 0;$$
$$f'''(0) = -\cos x_{|x=0} = -1;$$
$$f''''(0) = \sin x_{|x=0} = 0; \quad \text{etc.}$$

Taylor's remainder formula for $\sin x$ and $c = 0$ yields

$$\sin x = x - \frac{x^3}{3!} + \frac{x^5}{5!} - \frac{x^7}{7!} + \cdots + (n^{\text{th}} \text{ term}) + R_n,$$

where

$$R_n = \frac{d^n}{dx^n}(\sin x)_{|x=z_n} \frac{x^n}{n!}.$$

But the nth derivative of $\sin x$ is $\pm \sin x$ or $\pm \cos x$, so that

$$\left| \frac{d^n}{dx^n}(\sin x)_{|x=z_n} \right| \leq 1.$$

Hence,

$$|R_n| \leq \frac{|x|^n}{n!}.$$

But according to equation 11.39, $\lim_{n \to \infty} |x|^n/n! = 0$ for any x whatsoever, and therefore

$$\lim_{n \to \infty} R_n = 0.$$

The Maclaurin series for $\sin x$ therefore converges to $\sin x$ for all x, and we may write

$$\sin x = x - \frac{x^3}{3!} + \frac{x^5}{5!} - \frac{x^7}{7!} + \cdots$$
$$= \sum_{n=0}^{\infty} \frac{(-1)^n}{(2n+1)!} x^{2n+1}, \quad -\infty < x < \infty.$$

Often it is necessary to consider separately points on either side of the point of expansion. This is illustrated in the next example.

EXAMPLE 11.25

Find the Maclaurin series for e^x and show that it converges to e^x for all x.

SOLUTION Since

$$\frac{d^n}{dx^n}(e^x)|_{x=0} = e^x|_{x=0} = 1,$$

Taylor's remainder formula for e^x and $c=0$ gives

$$e^x = 1 + x + \frac{x^2}{2!} + \frac{x^3}{3!} + \cdots + \frac{x^{n-1}}{(n-1)!} + R_n,$$

where

$$R_n = \frac{d^n}{dx^n}(e^x)|_{x=z_n}\frac{x^n}{n!} = e^{z_n}\frac{x^n}{n!}.$$

Now, if $x < 0$, then $x < z_n < 0$, and

$$|R_n| < e^0 \frac{|x|^n}{n!},$$

which approaches zero as n becomes infinite (see equation 11.39). If $x > 0$, then $0 < z_n < x$, and

$$|R_n| < e^x \frac{|x|^n}{n!},$$

which again has limit zero as n approaches infinity. Thus, for any x whatsoever, $\lim_{n\to\infty} R_n = 0$, and the Maclaurin series for e^x converges to e^x:

$$e^x = 1 + x + \frac{x^2}{2!} + \frac{x^3}{3!} + \cdots$$
$$= \sum_{n=0}^{\infty} \frac{1}{n!} x^n, \qquad -\infty < x < \infty.$$

∎

The following example is more difficult. The remainders do not apporach zero for all x so that the Taylor series does not converge to the function for all x. The problem would have been even more difficult had we not suggested the values of x to consider.

EXAMPLE 11.26

Find the Taylor series for $\ln x$ about the point 1 and show that it converges to $\ln x$ for $1/2 \leq x \leq 2$.

SOLUTION If $f(x) = \ln x$, then:

$$f(1) = \ln x|_{x=1} = 0;$$
$$f'(1) = \frac{1}{x}\bigg|_{x=1} = 1;$$
$$f''(1) = \frac{-1}{x^2}\bigg|_{x=1} = -1;$$

$$f'''(1) = \frac{2}{x^3}\Big|_{x=1} = 2;$$

$$f''''(1) = \frac{-3!}{x^4}\Big|_{x=1} = -3!; \quad \text{etc.}$$

Taylor's remainder formula for $\ln x$ and $c = 1$ yields

$$\ln x = (x-1) - \frac{1}{2!}(x-1)^2 + \frac{2!}{3!}(x-1)^3 + \cdots + \frac{(-1)^n(n-2)!}{(n-1)!}(x-1)^{n-1} + R_n,$$

where

$$R_n = \frac{d^n}{dx^n}(\ln x)\Big|_{x=z_n}\frac{(x-1)^n}{n!} = \frac{(-1)^{n+1}(n-1)!}{(z_n)^n}\frac{(x-1)^n}{n!} = \frac{(-1)^{n+1}}{n}\left(\frac{x-1}{z_n}\right)^n,$$

and z_n is between 1 and x.

If $1 < x \leq 2$, then the largest value of $x-1$ is 1. Furthermore, z_n must be larger than 1. It follows that

$$|R_n| < \frac{1}{n}\left(\frac{1}{1}\right)^n = \frac{1}{n}$$

and therefore

$$\lim_{n\to\infty} R_n = 0.$$

If $1/2 \leq x < 1$, then $-1/2 \leq x-1 < 0$. Combine this with $x < z_n < 1$, and we can state that $-1 < (x-1)/z_n < 0$. Then,

$$|R_n| < \frac{1}{n} \quad \text{and} \quad \lim_{n\to\infty} R_n = 0.$$

Thus, for $1/2 \leq x \leq 2$, the sequence of remainders $\{R_n\}$ approaches zero, and the Taylor series converges to $\ln x$ for those values of x:

$$\ln x = (x-1) - \frac{1}{2}(x-1)^2 + \frac{1}{3}(x-1)^3 + \cdots$$

$$= \sum_{n=1}^{\infty} \frac{(-1)^{n+1}}{n}(x-1)^n, \quad \frac{1}{2} \leq x \leq 2.$$

This series actually converges to $\ln x$ on the larger interval $0 < x \leq 2$. We will prove this in Example 11.30. ∎

We have shown that if the sequence of Taylor's remainders $\{R_n\}$ for a function $f(x)$ and a point c has limit zero, then $f(x)$ has at least one power series expansion about c, namely its Taylor series,

$$f(x) = \sum_{n=0}^{\infty} \frac{f^{(n)}(c)}{n!}(x-c)^n, \quad f^{(0)}(c) = f(c), \tag{11.40}$$

and this series converges to $f(x)$. We now show that this is the *only* power series expansion of $f(x)$ about c. To do this we require the following theorem.

> **Theorem 11.16**
>
> If a function $f(x)$ has a power series expansion $f(x) = \sum_{n=0}^{\infty} a_n(x-c)^n$ with positive radius of convergence R, then each of the following series has radius of convergence R:
>
> $$f'(x) = \sum_{n=0}^{\infty} na_n(x-c)^{n-1}, \qquad (11.41)$$
>
> $$\int f(x)\,dx = \sum_{n=0}^{\infty} \frac{a_n}{n+1}(x-c)^{n+1} + C. \qquad (11.42)$$
>
> In other words, power series expansions for $f'(x)$ and $\int f(x)\,dx$ can be obtained by term-by-term differentiation and integration of the power series of $f(x)$, and all three series have the same radius of convergence.

Due to the difficulty in proving this theorem, and in order to preserve the continuity of our discussion, we omit a proof. Note that the theorem is stated in terms of radii of convergence rather than intervals of convergence. This is due to the fact that in differentiating a power series we may lose the end points of the original interval of convergence, and in integrating we may pick them up.

The next theorem implies that there is at most one power series expansion of a function about a given point.

> **Theorem 11.17**
>
> If a power series expansion of a function has a positive radius of convergence, then it is the Taylor series.

Proof Suppose the power series expansion

$$f(x) = \sum_{n=0}^{\infty} a_n(x-c)^n = a_0 + a_1(x-c) + a_2(x-c)^2 + \cdots$$

has a positive radius of convergence. Substitution of $x = c$ gives

$$f(c) = a_0.$$

If we differentiate the power series according to Theorem 11.16, we obtain

$$f'(x) = a_1 + 2a_2(x-c) + 3a_3(x-c)^2 + \cdots,$$

and if we substitute $x = c$, the result is

$$f'(c) = a_1.$$

If we differentiate the power series for $f'(x)$, we obtain

$$f''(x) = 2a_2 + 3 \cdot 2a_3(x-c) + 4 \cdot 3a_4(x-c)^2 + \cdots,$$

and substitute $x = c$,

$$f''(c) = 2a_2 \quad \text{or} \quad a_2 = \frac{f''(c)}{2!}.$$

Continued differentiation and substitution leads to the result that for all n,

$$a_n = \frac{f^{(n)}(c)}{n!}.$$

The power series $f(x) = \sum_{n=0}^{\infty} a_n(x-c)^n$ is therefore the Taylor series.

The following corollary is an immediate consequence of this theorem.

Corollary *If two power series $\sum_{n=0}^{\infty} a_n(x-c)^n$ and $\sum_{n=0}^{\infty} b_n(x-c)^n$ with positive radii of convergence have identical sums,*

$$\sum_{n=0}^{\infty} a_n(x-c)^n = \sum_{n=0}^{\infty} b_n(x-c)^n,$$

then $a_n = b_n$ for all n.

Our theory of power series expansions of functions is now complete. Given a function $f(x)$ and a point c, there can only be one power series expansion for $f(x)$ about c, its Taylor series. To show that the Taylor series does indeed have sum $f(x)$, it is sufficient to show that the sequence of Taylor's remainders $\{R_n\}$ has limit zero. When this is the case we write

$$f(x) = \sum_{n=0}^{\infty} \frac{f^{(n)}(c)}{n!}(x-c)^n. \qquad (11.43)$$

Equation 11.43 says, then, that the right-hand side is the Taylor series of $f(x)$ and that the Taylor series converges to $f(x)$. This point is most important because there do exist functions $f(x)$ whose Taylor series exist but do not converge to $f(x)$ (see Exercise 12). We therefore write 11.43 only when the Taylor series of $f(x)$ exists and that it converges to $f(x)$.

Only one point of consideration remains. Although we have a formula for calculation of the Taylor series of a function $f(x)$, determination of $f^{(n)}(c)$ could be very complicated. In addition, if a reasonable formula for $f^{(n)}(c)$ cannot be found, how will we ever prove that the sequence of Taylor's remainders has limit zero? Fortunately, in more common situations, special devices exist that enable us to find Taylor series without actually calculating $f^{(n)}(c)$. Indeed, if by any method, we can obtain a power series representation of $f(x)$, then we can be assured by Theorem 11.17 that it is the Taylor series. In Section 11.10 we illustrate some of these techniques.

EXERCISES 11.9

In Exercises 1–6 use Taylor's remainder formula to find Taylor series for the function $f(x)$ about the point indicated. In each case, find the interval on which the Taylor series converges to $f(x)$.

1. $f(x) = \cos x$ about $x = 0$
2. $f(x) = e^{5x}$ about $x = 0$
3. $f(x) = \sin(10x)$ about $x = 0$
4. $f(x) = \sin x$ about $x = \pi/4$
5. $f(x) = e^{2x}$ about $x = 1$
6. $f(x) = 1/(3x+2)$ about $x = 2$

7. Prove the corollary to Theorem 11.17.

8. Prove that if a power series with positive radius of convergence has sum zero, $\sum_{n=0}^{\infty} a_n(x-c)^n = 0$, then $a_n = 0$ for all n.

9. In Section 4.4 we stated the second-derivative test for determining whether a critical point x_0 at which $f'(x_0) = 0$ yields a relative maximum or a relative minimum. Use Taylor's remainder formula to verify this result when $f'(x)$ and $f''(x)$ are continuous on an open interval containing x_0.

10. Extend the result of Exercise 9 to verify the extrema test of Exercise 34 in Section 4.4.

11. There is an integral form for the remainder $R_n(c, x)$ in Taylor's remainder formula that is sometimes more useful than the derivative form in Theorem 11.15: If $f(x)$ and its first n derivatives are continuous on the closed interval between c and x, then

$$f(x) = f(c) + f'(c)(x-c) + \frac{f''(c)}{2!}(x-c)^2 + \cdots$$
$$+ \frac{f^{(n-1)}(c)}{(n-1)!}(x-c)^{n-1} + R_n(c, x),$$

where

$$R_n(c, x) = \frac{1}{(n-1)!} \int_c^x (x-t)^{n-1} f^{(n)}(t)\, dt.$$

Use the following outline of steps to prove this result.

(a) Show that

$$f(x) = f(c) + \int_c^x f'(t)\, dt.$$

(b) Use integration by parts with $u = f'(t)$, $du = f''(t)\, dt$, $dv = dt$, and $v = t - x$ on the integral in (a) to obtain

$$f(x) = f(c) + f'(c)(x-c) + \int_c^x (x-t) f''(t)\, dt.$$

(c) Use integration by parts with $u = f''(t)$, $du = f'''(t)\, dt$, $dv = (x-t)\, dt$, and $v = -(1/2)(x-t)^2$ to obtain

$$f(x) = f(c) + f'(c)(x-c) + \frac{f''(c)}{2!}(x-c)^2$$
$$+ \frac{1}{2!} \int_c^x (x-t)^2 f'''(t)\, dt.$$

(d) Continue this process to obtain the integral form for Taylor's remainder formula.

12. (a) Sketch a graph of the function

$$f(x) = \begin{cases} e^{-1/x^2} & x \neq 0 \\ 0 & x = 0. \end{cases}$$

(b) Use L'Hôpital's rule to show that for every positive integer n,

$$\lim_{x \to 0} \frac{e^{-1/x^2}}{x^n} = 0.$$

(c) Prove, by mathematical induction, that $f^{(n)}(0) = 0$ for $n \geq 1$.

(d) What is the Maclaurin series for $f(x)$?

(e) For what values of x does the Maclaurin series of $f(x)$ converge to $f(x)$?

SECTION 11.10

Taylor Series Expansions of Functions

In Section 11.9 we pointed out that if by any means we can find a power series representation for a function $f(x)$, then we can be assured by Theorem 11.17 that it is the Taylor series for $f(x)$. In this section we illustrate various techniques for producing power series. This has the advantage of avoiding Taylor remainders in establishing convergence of the series to $f(x)$. Our first example uses geometric series.

EXAMPLE 11.27

Find (a) the Maclaurin series for $1/(4 + 5x)$; (b) the Taylor series about 5 for $1/(13 - 2x)$.

SOLUTION (a) If we write

$$\frac{1}{4+5x} = \frac{1}{4\left(1 + \dfrac{5x}{4}\right)}$$

and interpret the right-hand side as one-quarter of the sum of a geometric series with first term equal to 1 and common ratio $-5x/4$, then

$$\frac{1}{4+5x} = \frac{1}{4}\left[1 + \left(-\frac{5x}{4}\right) + \left(-\frac{5x}{4}\right)^2 + \cdots\right], \qquad \left|-\frac{5x}{4}\right| < 1,$$

$$= \sum_{n=0}^{\infty} \frac{(-1)^n 5^n}{4^{n+1}} x^n, \qquad |x| < \frac{4}{5}.$$

(b) By a similar procedure, we have

$$\frac{1}{13-2x} = \frac{1}{3-2(x-5)}$$

$$= \frac{1}{3\left[1 - \frac{2}{3}(x-5)\right]}$$

$$= \frac{1}{3}\left[1 + \frac{2}{3}(x-5) + \left(\frac{2}{3}\right)^2 (x-5)^2 + \cdots\right], \qquad \left|\frac{2}{3}(x-5)\right| < 1,$$

$$= \sum_{n=0}^{\infty} \frac{2^n}{3^{n+1}}(x-5)^n, \qquad |x-5| < \frac{3}{2}.$$

In both examples, properties of geometric series gave not only the required series, but also their intervals of convergence. To appreciate the simplicity of these solutions, we suggest using Taylor remainders in an attempt to obtain the series with the same intervals of convergence. You will quickly abort.

Addition and Subtraction of Power Series

In Section 11.3 we noted that convergent series of constants can be added or subtracted to give convergent series. It follows, therefore, that if $f(x) = \sum a_n(x-c)^n$ and $g(x) = \sum b_n(x-c)^n$, then

$$f(x) \pm g(x) = \sum_{n=0}^{\infty} (a_n \pm b_n)(x-c)^n, \qquad (11.44)$$

and these results are valid for every x that is common to the intervals of convergence of the two series. We use this result in the following example.

EXAMPLE 11.28 Find the Maclaurin series for $f(x) = 5x/(x^2 - 3x - 4)$.

SOLUTION We decompose $f(x)$ into its partial fractions,

$$f(x) = \frac{5x}{x^2 - 3x - 4} = \frac{4}{x-4} + \frac{1}{x+1},$$

and expand each of these terms in a Maclaurin series,

$$\frac{4}{x-4} = \frac{-1}{1-\frac{x}{4}} = -\left(1 + \frac{x}{4} + \frac{x^2}{4^2} + \cdots\right), \qquad |x| < 4, \qquad \text{and}$$

$$\frac{1}{1+x} = 1 - x + x^2 - x^3 + \cdots, \qquad |x| < 1.$$

Addition of these series within their common interval of convergence gives the Maclaurin series for $f(x)$:

$$\frac{5x}{x^2 - 3x - 4} = \left(-1 - \frac{x}{4} - \frac{x^2}{4^2} - \frac{x^3}{4^3} - \cdots\right) + \left(1 - x + x^2 - x^3 + \cdots\right)$$

$$= \left(-1 - \frac{1}{4}\right)x + \left(1 - \frac{1}{4^2}\right)x^2 + \left(-1 - \frac{1}{4^3}\right)x^3 + \cdots$$

$$= \sum_{n=1}^{\infty} \left[(-1)^n - \frac{1}{4^n}\right]x^n, \qquad |x| < 1.$$

∎

Differentiation and Integration of Power Series

Perhaps the most useful technique for generating power series expansions is to differentiate or integrate known expansions according to Theorem 11.16.

EXAMPLE 11.29

Find Maclaurin series for the following functions:

(a) $\cos x$ (b) $\dfrac{1}{(1-x)^3}$

SOLUTION (a) Recall from Example 11.24 that

$$\sin x = x - \frac{x^3}{3!} + \frac{x^5}{5!} - \frac{x^7}{7!} + \cdots, \qquad -\infty < x < \infty.$$

Term-by-term differentiation gives

$$\cos x = 1 - \frac{x^2}{2!} + \frac{x^4}{4!} - \frac{x^6}{6!} + \cdots$$

$$= \sum_{n=0}^{\infty} \frac{(-1)^n}{(2n)!} x^{2n}, \qquad -\infty < x < \infty.$$

(b) Term-by-term differentiation of the series

$$\frac{1}{1-x} = 1 + x + x^2 + x^3 + \cdots + x^n + \cdots, \qquad |x| < 1$$

gives

$$\frac{1}{(1-x)^2} = 1 + 2x + 3x^2 + \cdots + nx^{n-1} + \cdots, \qquad |x| < 1.$$

Another differentiation yields

$$\frac{2}{(1-x)^3} = 2 + 3 \cdot 2x + 4 \cdot 3x^2 + \cdots + n(n-1)x^{n-2} + \cdots, \qquad |x| < 1.$$

Division by 2 now gives

$$\frac{1}{(1-x)^3} = 1 + 3x + 6x^2 + \cdots + \frac{n(n-1)}{2}x^{n-2} + \cdots$$

$$= \sum_{n=2}^{\infty} \frac{n(n-1)}{2} x^{n-2}$$

$$= \frac{1}{2} \sum_{n=0}^{\infty} (n+2)(n+1) x^n, \qquad |x| < 1.$$

∎

EXAMPLE 11.30 Find the Taylor series about 1 for $\ln x$.

SOLUTION Noting that $\ln x$ is an antiderivative of $1/x$, we first expand $1/x$ in a Taylor series about 1:

$$\frac{1}{x} = \frac{1}{(x-1)+1} = 1 - (x-1) + (x-1)^2 - (x-1)^3 + \cdots, \qquad |x-1| < 1.$$

If we integrate this series term-by-term, we have

$$\ln|x| = \left[x - \frac{1}{2}(x-1)^2 + \frac{1}{3}(x-1)^3 - \frac{1}{4}(x-1)^4 + \cdots\right] + C.$$

Substitution of $x = 1$ implies that $0 = 1 + C$; that is, $C = -1$, and hence,

$$\ln|x| = (x-1) - \frac{1}{2}(x-1)^2 + \frac{1}{3}(x-1)^3 - \cdots$$

$$= \sum_{n=1}^{\infty} \frac{(-1)^{n+1}}{n} (x-1)^n.$$

According to Theorem 11.16, the radius of convergence of this series is also $R = 1$. So the series certainly converges to $\ln|x|$ for $0 < x < 2$. But when $x = 2$, the series reduces to the alternating harmonic series, which converges conditionally. Consequently, we can extend the interval of convergence to include $x = 2$ and write

$$\ln x = \sum_{n=1}^{\infty} \frac{(-1)^{n+1}}{n} (x-1)^n, \qquad 0 < x \le 2.$$

Convergence of this series at $x = 2$ does not, by itself, imply convergence to $\ln 2$, as this equation suggests. It is, however, true, and this is a direct application of the following theorem.

∎

> **Theorem 11.18**
>
> If the Taylor series $\sum_{n=0}^{\infty} a_n(x-c)^n$ of a function $f(x)$ converges at the endpoint $x = c + R$ of its interval of convergence, and if $f(x)$ is continuous at $x = c + R$, then the Taylor series evaluated at $c + R$ converges to $f(c + R)$. The same result is valid at the other endpoint $x = c - R$.

Comparison of the solutions in Examples 11.26 and 11.30 indicates once again the advantage of avoiding the use of Taylor's remainder formula.

Multiplication of Power Series

In Example 11.28 we added the Maclaurin series for $4/(x-4)$ and $1/(1+x)$ to obtain the Maclaurin series for $5x/(x^2 - 3x - 4)$. An alternative procedure might be to multiply the two series since

$$\frac{5x}{x^2 - 3x - 4} = 5x \left(\frac{1}{x-4}\right)\left(\frac{1}{x+1}\right)$$

$$= \frac{5x}{-4}\left(1 + \frac{x}{4} + \frac{x^2}{4^2} + \frac{x^3}{4^3} + \cdots\right)(1 - x + x^2 - x^3 + \cdots).$$

The rules of algebra would demand that we multiply every term of the first series by every term of the second. If we do this and group all products with like powers of x, we obtain

$$\frac{5x}{x^2 - 3x - 4} = \frac{5x}{-4}\left[1 + \left(-1 + \frac{1}{4}\right)x + \left(1 - \frac{1}{4} + \frac{1}{4^2}\right)x^2 + \left(-1 + \frac{1}{4} - \frac{1}{4^2} + \frac{1}{4^3}\right)x^3 + \cdots\right].$$

It is clear that the coefficient of x^n is a finite geometric series to which we can apply formula 11.13a:

$$(-1)^n\left[1 - \frac{1}{4} + \frac{1}{4^2} - \cdots + \frac{(-1)^n}{4^n}\right] = (-1)^n\left[\frac{1 - \left(-\frac{1}{4}\right)^{n+1}}{1 + \frac{1}{4}}\right]$$

$$= (-1)^n \frac{4}{5}\left[1 - \left(-\frac{1}{4}\right)^{n+1}\right].$$

Consequently,

$$\frac{5x}{x^2 - 3x - 4} = \frac{5x}{-4} \sum_{n=0}^{\infty} (-1)^n \frac{4}{5}\left[1 - \left(-\frac{1}{4}\right)^{n+1}\right] x^n$$

$$= \sum_{n=0}^{\infty} (-1)^{n+1}\left[1 - \frac{(-1)^{n+1}}{4^{n+1}}\right] x^{n+1}$$

$$= \sum_{n=0}^{\infty} \left[(-1)^{n+1} - \frac{1}{4^{n+1}}\right] x^{n+1}$$

$$= \sum_{n=1}^{\infty} \left[(-1)^n - \frac{1}{4^n}\right] x^n.$$

For this example, then, multiplication as well as addition of power series leads to the Maclaurin series. Clearly, addition of power series is much simpler for this example, but we have at least demonstrated that power series can be multiplied together. That this is generally possible is stated in the following theorem.

Theorem 11.19

If $f(x) = \sum_{n=0}^{\infty} a_n(x-c)^n$ and $g(x) = \sum_{n=0}^{\infty} b_n(x-c)^n$, then

$$f(x)g(x) = \sum_{n=0}^{\infty} d_n(x-c)^n, \qquad (11.45\text{a})$$

where

$$d_n = \sum_{i=0}^{n} a_i b_{n-i} = a_0 b_n + a_1 b_{n-1} + \cdots + a_{n-1} b_1 + a_n b_0, \qquad (11.45\text{b})$$

and this series converges absolutely at every point at which the series for $f(x)$ and $g(x)$ converge absolutely.

EXAMPLE 11.31

Find the Maclaurin series for $f(x) = [1/(x-1)]\ln(1-x)$.

SOLUTION If we integrate the Maclaurin series

$$\frac{1}{1-x} = 1 + x + x^2 + x^3 + \cdots, \qquad |x| < 1,$$

we find

$$-\ln|1-x| = \left(x + \frac{x^2}{2} + \frac{x^3}{3} + \frac{x^4}{4} + \cdots\right) + C.$$

By setting $x = 0$, we obtain $C = 0$, and

$$\ln(1-x) = -x - \frac{x^2}{2} - \frac{x^3}{3} - \frac{x^4}{4} - \cdots.$$

Note that we have dropped the absolute value signs since the radius of convergence of the series is 1. We now form the Maclaurin series for $f(x)$:

$$\frac{1}{x-1}\ln(1-x) = \frac{-1}{1-x}\ln(1-x)$$

$$= (1 + x + x^2 + x^3 + \cdots)\left(x + \frac{x^2}{2} + \frac{x^3}{3} + \cdots\right)$$

$$= x + \left(1 + \frac{1}{2}\right)x^2 + \left(1 + \frac{1}{2} + \frac{1}{3}\right)x^3 + \cdots$$

$$= \sum_{n=1}^{\infty} \left(1 + \frac{1}{2} + \frac{1}{3} + \cdots + \frac{1}{n}\right)x^n.$$

Since both of the multiplied series converge absolutely on the interval $|x| < 1$, the Maclaurin series for $(x-1)^{-1}\ln(1-x)$ is also valid on this interval. ∎

EXAMPLE 11.32

Find the first three nonzero terms in the Maclaurin series for $\tan x$.

SOLUTION If $\tan x = \sum a_n x^n$, then by setting $\tan x = \sin x / \cos x$, we have

$$\sin x = \cos x \sum_{n=0}^{\infty} a_n x^n.$$

We now substitute the Maclaurin series for $\sin x$ and $\cos x$:

$$x - \frac{x^3}{3!} + \frac{x^5}{5!} - \frac{x^7}{7!} + \cdots = \left(1 - \frac{x^2}{2!} + \frac{x^4}{4!} - \frac{x^6}{6!} + \cdots\right)(a_0 + a_1 x + a_2 x^2 + \cdots).$$

According to the corollary to Theorem 11.17, two power series can be identical only if their corresponding coefficients are equal. We therefore multiply the right-hand side and equate coefficients of like powers of x:

$x^0:\quad 0 = a_0;$

$x:\quad 1 = a_1;$

$x^2:\quad 0 = a_2 - \dfrac{a_0}{2!},\quad$ which implies $a_2 = 0;$

$x^3:\quad -\dfrac{1}{3!} = a_3 - \dfrac{a_1}{2!},\quad$ which implies $a_3 = \dfrac{1}{2!} - \dfrac{1}{3!} = \dfrac{1}{3};$

$x^4:\quad 0 = a_4 - \dfrac{a_2}{2!} + \dfrac{a_0}{4!},\quad$ from which $a_4 = 0;$

$x^5:\quad \dfrac{1}{5!} = a_5 - \dfrac{a_3}{2!} + \dfrac{a_1}{4!},\quad$ from which $a_5 = \dfrac{1}{5!} + \dfrac{1}{6} - \dfrac{1}{4!} = \dfrac{2}{15}.$

The first three nonzero terms in the Maclaurin series for $\tan x$ are therefore

$$\tan x = x + \frac{1}{3}x^3 + \frac{2}{15}x^5 + \cdots.$$

Binomial Expansion

One of the most widely used power series is the binomial expansion. We are well acquainted with the binomial theorem, which predicts the product $(a + b)^m$ for any positive integer m:

$$(a + b)^m = \sum_{n=0}^{m} \binom{m}{n} a^n b^{m-n}. \tag{11.46}$$

With the usual definition of the binomial coefficients,

$$\binom{m}{n} = \frac{m!}{(m-n)!\,n!} = \frac{m(m-1)(m-2)\cdots(m-n+1)}{n!},$$

the binomial theorem becomes

$$(a+b)^m = a^m + m a^{m-1} b + \frac{m(m-1)}{2!} a^{m-2} b^2 + \cdots + m a b^{m-1} + b^m. \tag{11.47}$$

Even when m is not a positive integer, this form for the binomial theorem remains almost intact. To show this, we consider the power series

$$1 + \sum_{n=1}^{\infty} \frac{m(m-1)(m-2)\cdots(m-n+1)}{n!} x^n$$

for any real number m except a nonnegative integer. The radius of convergence of this power series is

$$R = \lim_{n \to \infty} \left| \frac{m(m-1)(m-2)\cdots(m-n+1)}{n!} \cdot \frac{(n+1)!}{m(m-1)(m-2)\cdots(m-n)} \right|$$

$$= \lim_{n \to \infty} \left| \frac{n+1}{m-n} \right| = 1.$$

The power series therefore converges absolutely for $|x| < 1$. Whether the series converges at the end points $x = \pm 1$ depends on the value of m. For the time being, we will work on the interval $|x| < 1$, and at the end of the discussion, we will state the complete result. Let us denote the sum of the series by

$$f(x) = 1 + \sum_{n=1}^{\infty} \frac{m(m-1)(m-2)\cdots(m-n+1)}{n!} x^n, \qquad |x| < 1.$$

If we differentiate this series term-by-term according to Theorem 11.16,

$$f'(x) = \sum_{n=1}^{\infty} \frac{m(m-1)\cdots(m-n+1)}{(n-1)!} x^{n-1}, \qquad |x| < 1,$$

and then multiply both sides by x, we have

$$xf'(x) = \sum_{n=1}^{\infty} \frac{m(m-1)\cdots(m-n+1)}{(n-1)!} x^n, \qquad |x| < 1.$$

If we add these results, we obtain

$$f'(x) + xf'(x) = \sum_{n=1}^{\infty} \frac{m(m-1)\cdots(m-n+1)}{(n-1)!} x^{n-1} + \sum_{n=1}^{\infty} \frac{m(m-1)\cdots(m-n+1)}{(n-1)!} x^n.$$

We now change the variable of summation in the first sum,

$$(1+x)f'(x) = \sum_{n=0}^{\infty} \frac{m(m-1)\cdots(m-n)}{n!} x^n + \sum_{n=1}^{\infty} \frac{m(m-1)\cdots(m-n+1)}{(n-1)!} x^n.$$

When these summations are added over their common range, beginning at $n = 1$, and the $n = 0$ term in the first summation is written out separately, the result is

$$(1+x)f'(x) = m + \sum_{n=1}^{\infty} \frac{m(m-1)\cdots(m-n+1)}{(n-1)!} \left(\frac{m-n}{n} + 1 \right) x^n$$

$$= m \left[1 + \sum_{n=1}^{\infty} \frac{m(m-1)\cdots(m-n+1)}{n!} x^n \right]$$

$$= mf(x).$$

Consequently, the function $f(x)$ must satisfy the differential equation

$$\frac{f'(x)}{f(x)} = \frac{m}{1+x}.$$

Integration immediately gives

$$\ln|f(x)| = m\ln|1+x| + C, \quad \text{or}$$

$$f(x) = D(1+x)^m.$$

To evaluate D, we note that from the original definition of $f(x)$ as the sum of the power series, $f(0) = 1$, and this implies that $D = 1$. Thus,

$$f(x) = (1+x)^m,$$

and we may write finally that

$$(1+x)^m = 1 + \sum_{n=1}^{\infty} \frac{m(m-1)(m-2)\cdots(m-n+1)}{n!} x^n, \quad |x| < 1 \quad (11.48\,\text{a})$$

$$= 1 + mx + \frac{m(m-1)}{2!}x^2 + \frac{m(m-1)(m-2)}{3!}x^3 + \cdots, \quad |x| < 1. \quad (11.48\,\text{b})$$

This is called the binomial expansion of $(1+x)^m$. We have verified the binomial expansion for m any real number except a nonnegative integer, but in the case of a nonnegative integer, the series terminates after $m+1$ terms and is therefore valid for these values of m also. We mentioned earlier that the binomial expansion may also converge at the end points $x = \pm 1$ depending on the value of m. The complete result states that 11.48 is valid for

$$\begin{array}{ll} -\infty < x < \infty & \text{if } m \text{ is a nonnegative integer,} \\ -1 < x < 1 & \text{if } m \leq -1, \\ -1 < x \leq 1 & \text{if } -1 < m < 0, \\ -1 \leq x \leq 1 & \text{if } m > 0 \text{ but not an integer.} \end{array}$$

It is not difficult to generalize this result to expand $(a+b)^m$ for real m. If $|b| < |a|$, we write

$$(a+b)^m = a^m\left(1 + \frac{b}{a}\right)^m$$

and now expand the bracketed term by means of 11.48:

$$(a+b)^m = a^m\left[1 + m\left(\frac{b}{a}\right) + \frac{m(m-1)}{2!}\left(\frac{b}{a}\right)^2 + \cdots\right] \quad |b| < |a|,$$

$$= a^m + ma^{m-1}b + \frac{m(m-1)}{2!}a^{m-2}b^2 + \cdots \quad |b| < |a|, \quad (11.49)$$

which, as we predicted, is equation 11.47 except that the series does not terminate.

EXAMPLE 11.33 Use the binomial expansion to find the Maclaurin series for $1/(1-x)^3$.

SOLUTION By 11.48b, we have

$$\frac{1}{(1-x)^3} = (1-x)^{-3}$$

$$= 1 + (-3)(-x) + \frac{(-3)(-4)}{2!}(-x)^2 + \frac{(-3)(-4)(-5)}{3!}(-x)^3 + \cdots$$

$$= 1 + 3x + \frac{3 \cdot 4}{2}x^2 + \frac{4 \cdot 5}{2}x^3 + \frac{5 \cdot 6}{2}x^4 + \cdots$$

$$= \frac{1}{2}\sum_{n=0}^{\infty}(n+1)(n+2)x^n, \quad |x| < 1.$$

This result was also obtained in Example 11.29 by differentiation of the Maclaurin series for $1/(1-x)$.

EXAMPLE 11.34 Find the Maclaurin series for $\text{Sin}^{-1}x$.

SOLUTION By the binomial expansion, we have

$$\frac{1}{\sqrt{1-x^2}} = 1 + \left(-\frac{1}{2}\right)(-x^2) + \frac{\left(-\frac{1}{2}\right)\left(-\frac{3}{2}\right)}{2!}(-x^2)^2 + \frac{\left(-\frac{1}{2}\right)\left(-\frac{3}{2}\right)\left(-\frac{5}{2}\right)}{3!}(-x^2)^3 + \cdots$$

$$= 1 + \frac{1}{2}x^2 + \frac{3}{2^2 2!}x^4 + \frac{3 \cdot 5}{2^3 3!}x^6 + \frac{3 \cdot 5 \cdot 7}{2^4 4!}x^8 + \cdots, \quad |x| < 1.$$

Integration of this series gives

$$\text{Sin}^{-1}x = \left(x + \frac{1}{2 \cdot 3}x^3 + \frac{3}{2^2 2! 5}x^5 + \frac{3 \cdot 5}{2^3 3! 7}x^7 + \frac{3 \cdot 5 \cdot 7}{2^4 4! 9}x^9 + \cdots\right) + C.$$

Evaluation of both sides of this equation at $x = 0$ gives us $C = 0$. According to Theorem 11.16, the radius of convergence of this series must be 1, and we can write

$$\text{Sin}^{-1}x = x + \sum_{n=1}^{\infty} \frac{1 \cdot 3 \cdot 5 \cdot \cdots \cdot (2n-1)}{2^n n!(2n+1)}x^{2n+1}$$

$$= x + \sum_{n=1}^{\infty} \frac{1 \cdot 2 \cdot 3 \cdot 4 \cdot 5 \cdot \cdots \cdot (2n-2)(2n-1)(2n)}{2 \cdot 4 \cdot \cdots \cdot (2n)2^n n!(2n+1)}x^{2n+1}$$

$$= x + \sum_{n=1}^{\infty} \frac{(2n)!}{(2n+1)2^{2n}(n!)^2}x^{2n+1}, \quad |x| < 1.$$

In this case, integration of the series for $(1-x^2)^{-1/2}$ does not pick up the end points $x = \pm 1$, although this is difficult to prove.

EXERCISES 11.10

In Exercises 1–20 find the power series expansion of the function about the indicated point.

1. $f(x) = \dfrac{1}{3x+2}$ about $x=0$

2. $f(x) = \dfrac{1}{4+x^2}$ about $x=0$

3. $f(x) = \dfrac{1}{x+3}$ about $x=2$

4. $f(x) = \cos(x^2)$ about $x=0$

5. $f(x) = \dfrac{1}{\sqrt{1+x}}$ about $x=0$

6. $f(x) = e^{5x}$ about $x=0$

7. $f(x) = \cosh x$ about $x=0$

8. $f(x) = \sinh x$ about $x=0$

9. $f(x) = \ln(1+2x)$ about $x=0$

10. $f(x) = (1+3x)^{3/2}$ about $x=0$

11. $f(x) = 1/x$ about $x=4$

12. $f(x) = x^4 + 3x^2 - 2x + 1$ about $x=0$

13. $f(x) = \dfrac{1}{(x+2)^3}$ about $x=0$

14. $f(x) = x^4 + 3x^2 - 2x + 1$ about $x=-2$

15. $f(x) = \dfrac{1}{x^2+8x+15}$ about $x=0$

16. $f(x) = e^x$ about $x=3$

17. $f(x) = \text{Tan}^{-1} x$ about $x=0$

18. $f(x) = \sqrt{x+3}$ about $x=0$

19. $f(x) = \dfrac{x^2}{(1+x^2)^2}$ about $x=0$

20. $f(x) = x(1-x)^{1/3}$ about $x=0$

In Exercises 21–23 find the first four nonzero terms in the Maclaurin series for the function.

21. $f(x) = \tan 2x$

22. $f(x) = \sec x$

23. $f(x) = e^x \sin x$

24. Find the Maclaurin series for $\cos^2 x$.

In Exercises 25–28 find the Maclaurin series for the function.

25. $f(x) = \dfrac{1}{x^6 - 3x^3 - 4}$

26. $f(x) = \text{Sin}^{-1}(x^2)$

27. $f(x) = \dfrac{2x^2+4}{x^2+4x+3}$

28. $f(x) = \ln\left[\dfrac{1+x/\sqrt{2}}{1-x/\sqrt{2}}\right]$

29. If during a working day, one person drinks from a fountain every 30 seconds (on the average), then the probability that exactly n people drink in a time interval of length t seconds is given by the *Poisson distribution*:

$$P_n(t) = \dfrac{1}{n!}\left(\dfrac{t}{30}\right)^n e^{-t/30}.$$

Calculate $\sum_{n=0}^{\infty} P_n(t)$ and interpret the result.

30. A certain experiment is to be performed until it is successful. The probability that it will be successful in any given attempt is p ($0 < p < 1$), and therefore the probability that it will fail is $q = 1 - p$. The expected number of times that the experiment must be performed in order to be successful can be shown to be represented by the infinite series

$$\sum_{n=1}^{\infty} npq^{n-1} = \sum_{n=1}^{\infty} np(1-p)^{n-1}.$$

(a) What is the sum of this series?

(b) If p is the probability that a single die will come up 6, is the answer in (a) what you would expect?

31. Find the Maclaurin series for the *error function* $\text{erf}(x)$ defined by

$$\text{erf}(x) = \dfrac{2}{\sqrt{\pi}} \int_0^x e^{-t^2} dt.$$

32. Find Maclaurin series for the *Fresnel integrals* $C(x)$ and $S(x)$ defined by

$$C(x) = \int_0^x \cos(\pi t^2/2)\, dt, \quad S(x) = \int_0^x \sin(\pi t^2/2)\, dt.$$

Show that Bessel functions of the first kind (defined in Exercise 30 of Section 11.8) satisfy the properties in Exercises 33 and 34.

33. $2m J_m(x) - x J_{m-1}(x) = x J_{m+1}(x)$

34. $J_{m-1}(x) - J_{m+1}(x) = 2 J'_m(x)$

35. If the function $(1 - 2\mu x + x^2)^{-1/2}$ is expanded in a Maclaurin series in x,

$$\frac{1}{\sqrt{1-2\mu x + x^2}} = \sum_{n=0}^{\infty} P_n(\mu) x^n,$$

the coefficients $P_n(\mu)$ are called the *Legendre polynomials*. Find $P_0(\mu)$, $P_1(\mu)$, $P_2(\mu)$, and $P_3(\mu)$.

36. (a) If we define $f(x) = x/(e^x - 1)$ at $x = 0$ as $f(0) = 1$, it turns out that $f(x)$ has a Maclaurin series expansion with positive radius of convergence. When this expansion is expressed in the form

$$\frac{x}{e^x - 1} = 1 + B_1 x + \frac{B_2}{2!} x^2 + \frac{B_3}{3!} x^3 + \cdots,$$

the coefficients $B_1, B_2, B_3, \ldots$ are called the *Bernoulli numbers*. Write this equation in the form

$$x = (e^x - 1)\left(1 + B_1 x + \frac{B_2}{2!} x^2 + \cdots\right),$$

and substitute the Maclaurin series for e^x to find the first five Bernoulli numbers.

(b) Show that the odd Bernoulli numbers all vanish for $n \geq 3$.

37. By substituting power series expansions for $e^{xt/2}$ and $e^{-x/(2t)}$ in terms of powers of t and $1/t$ respectively, show that

$$e^{x(t-1/t)/2} = \sum_{n=0}^{\infty} J_n(x) t^n.$$

For a definition of $J_n(x)$, see Exercises 30 in Section 11.8.

SECTION 11.11

Sums of Power Series and Series of Constants

Theorem 11.16 provides an important technique for finding sums of power series. If an unknown series can be reduced to a known series by differentiations or integrations, then its sum can be obtained when these operations are reversed.

EXAMPLE 11.35

Find the sum of the series $\sum_{n=0}^{\infty} (n+1) x^n$.

SOLUTION The series converges absolutely for $|x| < 1$. If we denote the sum of the series by $S(x)$,

$$S(x) = \sum_{n=0}^{\infty} (n+1) x^n,$$

and use Theorem 11.16 to integrate the series term-by-term,

$$\int S(x)\, dx = \sum_{n=0}^{\infty} x^{n+1} + C.$$

But the series on the right is a geometric series with sum $x/(1-x)$, provided $|x| < 1$, and we may write therefore

$$\int S(x)\, dx = \frac{x}{1-x} + C, \qquad |x| < 1.$$

If we now differentiate this equation, we obtain

$$S(x) = \frac{(1-x)(1) - x(-1)}{(1-x)^2} = \frac{1}{(1-x)^2},$$

and therefore

$$\sum_{n=0}^{\infty} (n+1) x^n = \frac{1}{(1-x)^2}, \qquad |x| < 1.$$

EXAMPLE 11.36

Find the sum of the series $\sum_{n=1}^{\infty} x^n/n$.

SOLUTION The series converges for $-1 \leq x < 1$. If we set

$$S(x) = \sum_{n=1}^{\infty} \frac{1}{n} x^n,$$

and differentiate the series term-by-term,

$$S'(x) = \sum_{n=1}^{\infty} x^{n-1}.$$

This is a geometric series with sum $1/(1-x)$, so that

$$S'(x) = \frac{1}{1-x}.$$

Antidifferentiation now gives

$$S(x) = -\ln(1-x) + C.$$

Since $S(0) = 0$, it follows that

$$0 = -\ln(1) + C.$$

Hence, $C = 0$, and $S(x) = -\ln(1-x)$. We have shown therefore that

$$\sum_{n=1}^{\infty} \frac{1}{n} x^n = -\ln(1-x), \qquad -1 \leq x < 1.$$

∎

By substituting values of x into known power series we obtain formulas for sums of series of constants. For instance, in Example 11.25, we verified that the Maclaurin series for e^x is

$$e^x = \sum_{n=0}^{\infty} \frac{1}{n!} x^n, \qquad -\infty < x < \infty.$$

By substituting $x = 1$, we obtain a series which converges to e,

$$e = \sum_{n=0}^{\infty} \frac{1}{n!} = 1 + \frac{1}{1!} + \frac{1}{2!} + \frac{1}{3!} + \cdots.$$

This was the series that we truncated in Example 11.16.

The following example is another illustration of this idea.

EXAMPLE 11.37

Find the sum of the series

$$\sum_{n=0}^{\infty} \frac{(-1)^n}{(2n+1)2^n}.$$

SOLUTION There are many power series which reduce to the given series upon substitution of a specific value of x. For instance, substitution of $-1/2$, 1, and $1/\sqrt{2}$ into the following power series, respectively, lead to the given series:

$$\sum_{n=0}^{\infty} \frac{1}{2n+1} x^n, \qquad \sum_{n=0}^{\infty} \frac{(-1)^n}{(2n+1)2^n} x^n, \qquad \sum_{n=0}^{\infty} \frac{\sqrt{2}(-1)^n}{2n+1} x^{2n+1}$$

Which should we consider? Although it is not the simplest, the third series looks most promising; the fact that the power on x corresponds to the coefficient in the denominator suggests that we can find the sum of this series. We therefore set

$$S(x) = \sum_{n=0}^{\infty} \frac{\sqrt{2}(-1)^n}{2n+1} x^{2n+1},$$

and this series converges for $-1 \leq x \leq 1$. If we differentiate the series with respect to x, then

$$S'(x) = \sum_{n=0}^{\infty} \sqrt{2}(-1)^n x^{2n} = \sum_{n=0}^{\infty} \sqrt{2}(-x^2)^n = \frac{\sqrt{2}}{1-(-x^2)} = \frac{\sqrt{2}}{1+x^2}.$$

Antidifferentiation now gives

$$S(x) = \int \frac{\sqrt{2}}{1+x^2} \, dx = \sqrt{2} \operatorname{Tan}^{-1} x + C.$$

Since $S(0) = 0$, it follows that $C = 0$, and

$$\sum_{n=0}^{\infty} \frac{\sqrt{2}(-1)^n}{2n+1} x^{2n+1} = \sqrt{2} \operatorname{Tan}^{-1} x, \qquad |x| \leq 1.$$

If we now set $x = 1/\sqrt{2}$,

$$\sum_{n=0}^{\infty} \frac{\sqrt{2}(-1)^n}{2n+1} \left(\frac{1}{\sqrt{2}}\right)^{2n+1} = \sqrt{2} \operatorname{Tan}^{-1}\left(\frac{1}{\sqrt{2}}\right).$$

Consequently,

$$\sum_{n=0}^{\infty} \frac{(-1)^n}{(2n+1)2^n} = \sqrt{2} \operatorname{Tan}^{-1}\left(\frac{1}{\sqrt{2}}\right).$$

EXERCISES 11.11

In Exercises 1–10 find the sum of the power series.

1. $\displaystyle\sum_{n=1}^{\infty} nx^{n-1}$

2. $\displaystyle\sum_{n=2}^{\infty} n(n-1)x^{n-2}$

3. $\displaystyle\sum_{n=1}^{\infty} (n+1)x^{n-1}$

4. $\displaystyle\sum_{n=1}^{\infty} n^2 x^{n-1}$

5. $\displaystyle\sum_{n=1}^{\infty} (n^2 + 2n)x^n$

6. $\displaystyle\sum_{n=0}^{\infty} \frac{1}{n+1} x^n$

7. $\displaystyle\sum_{n=0}^{\infty} \frac{(-1)^n}{2n+1} x^{2n+1}$

8. $\displaystyle\sum_{n=1}^{\infty} \frac{(-1)^n}{n} x^{2n}$

9. $\displaystyle\sum_{n=2}^{\infty} n3^n x^{2n}$

10. $\displaystyle\sum_{n=0}^{\infty} \left(\frac{n+1}{n+2}\right) x^n$

In Exercises 11–20 verify that the sum of the series is as indicated.

11. $\displaystyle\sum_{n=0}^{\infty} \frac{2^n}{n!} = e^2$

12. $\displaystyle\sum_{n=0}^{\infty} \frac{(-1)^n}{(2n+1)!} = \sin 1$

13. $\displaystyle\sum_{n=0}^{\infty} \frac{(-1)^n 3^{2n}}{(2n)!} = \cos 3$

14. $\displaystyle\sum_{n=1}^{\infty} \frac{(-1)^n}{n!} = \frac{1}{e} - 1$

15. $\sum_{n=1}^{\infty} \frac{(-1)^n}{3^{2n}(2n+1)!} = 3\sin\left(\frac{1}{3}\right) - 1$

16. $\sum_{n=2}^{\infty} \frac{(-1)^{n+1} 2^{2n+3}}{(2n)!} = -8(1 + \cos 2)$

17. $\sum_{n=1}^{\infty} \frac{2^n}{n 3^n} = \ln 3$

18. $\sum_{n=1}^{\infty} \frac{1}{n 2^n} = \ln 2$

19. $\sum_{n=1}^{\infty} \frac{(-1)^n}{2^{2n}} = -\frac{1}{5}$

20. $\sum_{n=1}^{\infty} \frac{n}{2^n} = 2$

21. Find the Maclaurin series for $\text{Tan}^{-1} x$ and use it to evaluate
$$\sum_{n=1}^{\infty} \frac{(-1)^n}{2n+1}.$$

22. Find the Maclaurin series for $x/(1+x^2)^2$ and use it to evaluate
$$\sum_{n=1}^{\infty} \frac{n(-1)^n}{3^{2n}}.$$

SECTION 11.12
Applications of Taylor Series and Taylor's Remainder Formula

If Taylor's remainder $R_n(c, x)$ is truncated from remainder formula 11.32e, a polynomial approximation to $f(x)$ is obtained

$$f(x) \approx f(c) + f'(c)(x-c) + \frac{f''(c)}{2!}(x-c)^2 + \cdots + \frac{f^{(n-1)}(c)}{(n-1)!}(x-c)^{n-1}.$$

The remainder

$$R_n(c, x) = \frac{f^{(n)}(z_n)}{n!}(x-c)^n,$$

where z_n is between c and x represents the error of the approximation; the smaller R_n, the better the approximation. In this section we use Taylor series and Taylor's remainder formula in a number of situations that require approximations.

Approximations of Functions

Consider using the first three terms of the Maclaurin series for e^x to approximate e^x on the interval $0 \le x \le 1/2$. Taylor's remainder formula, with $c = 0$, states that

$$e^x = 1 + x + \frac{x^2}{2} + R_3$$

where

$$R_3 = \frac{d^3}{dx^3}(e^x)\big|_{x=z} \frac{x^3}{3!} = \frac{e^z x^3}{6},$$

and z is between 0 and x. Although z is unknown — except that it is between 0 and x, we can say that because only the values $0 \le x \le \frac{1}{2}$ are under consideration, z must be less than $\frac{1}{2}$. It follows that

$$R_3 < \frac{e^{1/2} x^3}{6} \le \frac{\sqrt{e}(1/2)^3}{6} < 0.035.$$

Thus the quadratic function $1 + x + x^2/2$ approximates e^x on the interval $0 \le x \le \frac{1}{2}$ with an error no greater than 0.035.

In the following example, we determine the number of terms of a Maclaurin series required to guarantee a certain accuracy.

EXAMPLE 11.38

How many terms in the Maclaurin series of $\ln(1+x)$ guarantee a truncation error of less than 10^{-6} for any x in the interval $0 \leq x \leq 1/2$?

SOLUTION The nth derivative of $\ln(1+x)$ is

$$\frac{d^n}{dx^n}\ln(1+x) = \frac{(-1)^{n+1}(n-1)!}{(x+1)^n}, \quad n \geq 1,$$

and therefore Taylor's remainder formula with $c = 0$ states that

$$\ln(1+x) = x - \frac{x^2}{2} + \frac{x^3}{3} - \frac{x^4}{4} + \cdots + \frac{(-1)^n}{n-1}x^{n-1} + R_n.$$

Now that we see terms in the Maclaurin series, for $0 \leq x \leq \frac{1}{2}$, the series is alternating. We could therefore discuss accuracy from an alternating series point of view and avoid Taylor remainders totally. Recall from Section 11.7 that when an alternating series is truncated, the maximum possible error is the next term. Consequently, if the Maclaurin series for $\ln(1+x)$ is truncated after the term $(-1)^n x^{n-1}/(n-1)$, the maximum error is

$$\frac{(-1)^{n+1}}{n}x^n.$$

Considering values of x in the interval $0 \leq x \leq \frac{1}{2}$, this error is a maximum at $x = \frac{1}{2}$. Thus, the error in truncating the series after $(-1)^n x^{n-1}/(n-1)$ is no greater in absolute value than

$$\frac{(1/2)^n}{n}.$$

This is less than 10^{-6} if

$$\frac{(1/2)^n}{n} < 10^{-6} \quad \text{or}$$

$$n2^n > 10^6.$$

A calculator quickly indicates that the smallest value of n for which this is true is $n = 16$. Consequently, if $\ln(1+x)$ is approximated by the 15th degree polynomial

$$\ln(1+x) \approx x - \frac{x^2}{2} + \frac{x^3}{3} - \frac{x^4}{4} + \cdots + \frac{x^{15}}{15}$$

on the interval $0 \leq x \leq 1/2$, the truncation error is less than 10^{-6}.

It is interesting to note that Taylor remainders do no better. To see this we calculate that

$$R_n = \frac{d^n}{dx^n}\ln(1+x)\Big|_{x=z}\frac{x^n}{n!} = \frac{(-1)^{n+1}(n-1)!}{(z+1)^n}\frac{x^n}{n!} = \frac{(-1)^{n+1}}{n(z+1)^n}x^n.$$

Since z is between 0 and x and $0 \leq x \leq \frac{1}{2}$, we can state that x must be less than or equal to $\frac{1}{2}$, and z must be greater than 0. Hence,

$$|R_n| < \frac{1}{n(1)^n}(1/2)^n = \frac{1}{n2^n},$$

and this is the same error expression previously obtained when analysis was performed from an alternating series point of view. ∎

Evaluation of Definite Integrals

In Section 9.9 we developed three numerical techniques for approximating definite integrals of functions $f(x)$ that have no obvious antiderivatives: the rectangular rule, the trapezoidal rule, and Simpson's rule. Each method divides the interval of integration into a number of subintervals and approximates $f(x)$ by a more elementary function on each subinterval. The rectangular rule replaces $f(x)$ by a step function, the trapezoidal rule replaces $f(x)$ by a succession of linear functions, and Simpson's rule uses a sequence of quadratic functions.

Another possibility is to replace $f(x)$ by a truncated power series (a polynomial) over the entire interval of integration. For instance, consider the definite integral

$$\int_0^{1/2} \frac{\sin x}{x} \, dx,$$

where $(\sin x)/x$ is defined as 1 at $x = 0$. Suppose we expand the integrand in a Maclaurin series:

$$\frac{1}{x} \sin x = \frac{1}{x}\left(x - \frac{x^3}{3!} + \frac{x^5}{5!} - \cdots\right) = 1 - \frac{x^2}{3!} + \frac{x^4}{5!} - \cdots, \qquad -\infty < x < \infty.$$

According to Theorem 11.16,

$$\int \frac{\sin x}{x} \, dx = \left(x - \frac{x^3}{3 \cdot 3!} + \frac{x^5}{5 \cdot 5!} - \cdots\right) + C, \qquad -\infty < x < \infty,$$

and therefore

$$\int_0^{1/2} \frac{\sin x}{x} \, dx = \frac{1}{2} - \frac{\left(\frac{1}{2}\right)^3}{3 \cdot 3!} + \frac{\left(\frac{1}{2}\right)^5}{5 \cdot 5!} - \cdots.$$

If we approximate this alternating series with its first three terms, the truncation error is the fourth term. Since

$$\frac{1}{2} - \frac{\left(\frac{1}{2}\right)^3}{3 \cdot 3!} + \frac{\left(\frac{1}{2}\right)^5}{5 \cdot 5!} = 0.493\,107\,6 \qquad \text{and}$$

$$\frac{\left(\frac{1}{2}\right)^7}{7 \cdot 7!} < 0.000\,000\,222,$$

it follows that

$$0.493\,107\,4 < \int_0^{1/2} \frac{\sin x}{x} \, dx < 0.493\,107\,6.$$

Consequently, using only three terms of the Maclaurin series for $(\sin x)/x$, we can say that to five decimals

$$\int_0^{1/2} \frac{\sin x}{x} \, dx = 0.49311.$$

EXAMPLE 11.39

A very important function in engineering and physics is the error function erf(x) defined by

$$\text{erf}(x) = \frac{2}{\sqrt{\pi}} \int_0^x e^{-t^2}\, dt.$$

Calculate erf(1) correct to three decimals.

SOLUTION If we replace the integrand by its Maclaurin series, we have

$$\frac{\sqrt{\pi}}{2}\text{erf}(1) = \int_0^1 \left[1 - t^2 + \frac{(-t^2)^2}{2!} + \frac{(-t^2)^3}{3!} + \cdots \right] dt$$

$$= \left\{ t - \frac{t^3}{3} + \frac{t^5}{5 \cdot 2!} - \frac{t^7}{7 \cdot 3!} + \cdots \right\}_0^1$$

$$= 1 - \frac{1}{3} + \frac{1}{5 \cdot 2!} - \frac{1}{7 \cdot 3!} + \frac{1}{9 \cdot 4!} - \cdots .$$

Since

$$\frac{2}{\sqrt{\pi}} \left(1 - \frac{1}{3} + \frac{1}{5 \cdot 2!} - \frac{1}{7 \cdot 3!} + \frac{1}{9 \cdot 4!} - \frac{1}{11 \cdot 5!} \right) = 0.842\,593\,67$$

and the seventh nonzero term is

$$\frac{2}{\sqrt{\pi}} \frac{1}{13 \cdot 6!} < 0.000\,120\,56,$$

it follows that

$$0.842\,593\,67 < \text{erf}(1) < 0.842\,714\,23,$$

and to three decimals

$$\text{erf}(1) = 0.843.$$

Limits

We have customarily used L'Hôpital's rule to evaluate limits of the indeterminate form $0/0$. Maclaurin and Taylor series can sometimes be used to advantage. Consider

$$\lim_{x \to 0} \frac{x - \sin x}{x^3}.$$

Three applications of L'Hôpital's rule give a limit of $\frac{1}{6}$. Alternatively, if we substitute the Maclaurin series for $\sin x$,

$$\lim_{x \to 0} \frac{x - \sin x}{x^3} = \lim_{x \to 0} \frac{1}{x^3} \left[x - \left(x - \frac{x^3}{3!} + \frac{x^5}{5!} - \cdots \right) \right]$$

$$= \lim_{x \to 0} \left[\frac{1}{6} - \frac{x^2}{5!} + \cdots \right]$$

$$= \frac{1}{6}.$$

Here is another example.

EXAMPLE 11.40

Evaluate
$$\lim_{\lambda \to 0^+} \frac{\lambda^{-5}}{e^{c/\lambda} - 1}$$
where $c > 0$ is a constant (see also Exercise 58 in Section 4.9).

SOLUTION We begin by making the change of variable $v = 1/\lambda$ in the limit,
$$\lim_{\lambda \to 0^+} \frac{\lambda^{-5}}{e^{c/\lambda} - 1} = \lim_{v \to \infty} \frac{v^5}{e^{cv} - 1}.$$

We now expand e^{cv} into its Maclaurin series,
$$\lim_{\lambda \to 0^+} \frac{\lambda^{-5}}{e^{c/\lambda} - 1} = \lim_{v \to \infty} \frac{v^5}{\left(1 + cv + \frac{c^2 v^2}{2!} + \cdots\right) - 1}$$
$$= \lim_{v \to \infty} \frac{v^5}{cv + \frac{c^2 v^2}{2!} + \cdots}.$$

If we now divide numerator and denominator by v^5,
$$\lim_{\lambda \to 0^+} \frac{\lambda^{-5}}{e^{c/\lambda} - 1} = \lim_{v \to \infty} \frac{1}{\frac{c}{v^4} + \frac{c^2}{2 v^3} + \frac{c^3}{3! v^2} + \frac{c^4}{4! v} + \frac{c^5}{5!} + \frac{c^6 v}{6!} + \cdots} = 0. \quad\blacksquare$$

Differential Equations

Many differential equations arising in physics and engineering have solutions that can be expressed only in terms of infinite series. One such equation is Bessel's differential equation of order zero for a function $y = f(x)$:
$$xy'' + y' + xy = 0.$$

Before considering this rather difficult differential equation, we introduce the ideas through an easier example.

EXAMPLE 11.41

Determine whether the differential equation
$$\frac{dy}{dx} - 2y = x$$
has a solution which can be expressed as a power series $y = \sum_{n=0}^{\infty} a_n x^n$ with positive radius of convergence.

SOLUTION If
$$y = f(x) = \sum_{n=0}^{\infty} a_n x^n = a_0 + a_1 x + a_2 x^2 + \cdots$$

is to be a solution of the differential equation, we may substitute the power series into the differential equation,

$$[a_1 + 2a_2x + 3a_3x^2 + 4a_4x^3 + \cdots] - 2[a_0 + a_1x + a_2x^2 + \cdots] = x.$$

We now gather together like terms in the various powers of x,

$$0 = (a_1 - 2a_0) + (2a_2 - 2a_1 - 1)x + (3a_3 - 2a_2)x^2 + (4a_4 - 2a_3)x^3 + \cdots.$$

Since the power series on the right has sum zero, its coefficients must all vanish (see Exercise 8 in Section 11.9), and therefore we must set

$$a_1 - 2a_0 = 0,$$
$$2a_2 - 2a_1 - 1 = 0,$$
$$3a_3 - 2a_2 = 0,$$
$$4a_4 - 2a_3 = 0,$$

and so on. These equations imply that

$$a_1 = 2a_0;$$
$$a_2 = \frac{1}{2}(1 + 2a_1) = \frac{1}{2}(1 + 4a_0);$$
$$a_3 = \frac{2}{3}a_2 = \frac{2}{3!}(1 + 4a_0);$$
$$a_4 = \frac{2}{4}a_3 = \frac{2^2}{4!}(1 + 4a_0).$$

The pattern emerging is

$$a_n = \frac{2^{n-2}}{n!}(1 + 4a_0), \qquad n \geq 2.$$

Thus,

$$f(x) = a_0 + 2a_0x + \frac{1}{2}(1 + 4a_0)x^2 + \cdots + \frac{2^{n-2}}{n!}(1 + 4a_0)x^n + \cdots$$
$$= a_0 + 2a_0x + \frac{1}{4}(1 + 4a_0)\left(\frac{2^2}{2!}x^2 + \frac{2^3}{3!}x^3 + \cdots + \frac{2^n}{n!}x^n + \cdots\right).$$

We can find the sum of the series in parentheses by noting that the Maclaurin series for e^{2x} is

$$e^{2x} = 1 + (2x) + \frac{(2x)^2}{2!} + \frac{(2x)^3}{3!} + \cdots.$$

Therefore the solution of the differential equation is

$$y = f(x) = a_0 + 2a_0x + \frac{1}{4}(1 + 4a_0)\left[e^{2x} - 1 - 2x\right]$$
$$= -\frac{1}{4} - \frac{x}{2} + \frac{1}{4}(1 + 4a_0)e^{2x}$$
$$= Ce^{2x} - \frac{1}{4} - \frac{x}{2}. \qquad \blacksquare$$

Using power series to solve the differential equation in Example 11.41 is certainly not the most expedient method. You will learn better methods if and when you undertake an indepth study of differential equations. But the example clearly illustrated the procedure by which power series are used to solve differential equations. We now apply the procedure to Bessel's differential equation of order zero.

EXAMPLE 11.42

Find a power series solution $y = \sum_{n=0}^{\infty} a_n x^n$, with positive radius of convergence, for Bessel's differential equation of order zero,

$$xy'' + y' + xy = 0.$$

SOLUTION In this example we abandon the $\cdots$ notation used in Example 11.41, and maintain sigma notation throughout. When we substitute $y = \sum_{n=0}^{\infty} a_n x^n$ into the differential equation, we obtain

$$0 = x \sum_{n=2}^{\infty} n(n-1) a_n x^{n-2} + \sum_{n=1}^{\infty} n a_n x^{n-1} + x \sum_{n=0}^{\infty} a_n x^n$$

$$= \sum_{n=2}^{\infty} n(n-1) a_n x^{n-1} + \sum_{n=1}^{\infty} n a_n x^{n-1} + \sum_{n=0}^{\infty} a_n x^{n+1}.$$

In order to bring these three summations together as one, and combine terms in like powers of x, we lower the index of summation in the last term by 2,

$$0 = \sum_{n=2}^{\infty} n(n-1) a_n x^{n-1} + \sum_{n=1}^{\infty} n a_n x^{n-1} + \sum_{n=2}^{\infty} a_{n-2} x^{n-1}.$$

We now combine the three summations over their common interval, beginning at $n = 2$, and write separately the $n = 1$ term in the second summation,

$$0 = a_1 + \sum_{n=2}^{\infty} [n(n-1) a_n + n a_n + a_{n-2}] x^{n-1}.$$

But the only way a power series can be equal to zero is for all of its coefficients to be equal to zero; that is,

$$a_1 = 0; \qquad n(n-1) a_n + n a_n + a_{n-2} = 0, \quad n \geq 2.$$

Thus,

$$a_n = -\frac{a_{n-2}}{n^2}, \qquad n \geq 2,$$

a recursive relation defining the unknown coefficient a_n of x^n in terms of the coefficient a_{n-2} of x^{n-2}. Since $a_1 = 0$, it follows that

$$0 = a_1 = a_3 = a_5 = \cdots.$$

For $n = 2$,

$$a_2 = -\frac{a_0}{2^2}.$$

For $n = 4$,
$$a_4 = -\frac{a_2}{4^2} = \frac{a_0}{2^2 4^2} = \frac{a_0}{2^4(2!)^2}.$$

For $n = 6$,
$$a_6 = -\frac{a_4}{6^2} = \frac{-a_0}{2^4(2!)^2 6^2} = -\frac{a_0}{2^6(3!)^2}.$$

The solution is therefore
$$y = a_0 - \frac{a_0}{2^2}x^2 + \frac{a_0}{2^4(2!)^2}x^4 - \frac{a_0}{2^6(3!)^2}x^6 + \cdots$$
$$= a_0 \sum_{n=0}^{\infty} \frac{(-1)^n}{2^{2n}(n!)^2} x^{2n}.$$

The function defined by the infinite series
$$J_0(x) = \sum_{n=0}^{\infty} \frac{(-1)^n}{2^{2n}(n!)^2} x^{2n}, \qquad -\infty < x < \infty$$

is called the zero-order Bessel function of the first kind. ∎

EXERCISES 11.12

In Exercises 1–10 find a maximum possible error in using the given terms of the Taylor series to approximate the function on the interval specified.

1. $e^x \approx 1 + x + \dfrac{x^2}{2} + \dfrac{x^3}{6}$ for $0 \le x \le 0.01$

2. $e^x \approx 1 + x + \dfrac{x^2}{2} + \dfrac{x^3}{6}$ for $0 \le x < 0.01$

3. $e^x \approx 1 + x + \dfrac{x^2}{2} + \dfrac{x^3}{6}$ for $-0.01 \le x \le 0$

4. $e^x \approx 1 + x + \dfrac{x^2}{2} + \dfrac{x^3}{6}$ for $|x| \le 0.01$

5. $\sin x \approx x - \dfrac{x^3}{3!}$ for $0 \le x \le 1$

6. $\cos x \approx 1 - \dfrac{x^2}{2!} + \dfrac{x^4}{4!}$ for $|x| \le 0.1$

7. $\ln(1-x) \approx -x - \dfrac{x^2}{2} - \dfrac{x^3}{3}$ $0 \le x \le 0.01$

8. $\dfrac{1}{(1-x)^3} \approx 1 + 3x + 6x^2 + 10x^3$ for $|x| < 0.2$

9. $\operatorname{Tan}^{-1} x \approx x - \dfrac{x^3}{3} + \dfrac{x^5}{5} - \cdots - \dfrac{x^{99}}{99}$ for $|x| < 1$

10. $\ln x \approx (x-1) - \dfrac{1}{2}(x-1)^2 + \dfrac{1}{3}(x-1)^3 - \dfrac{1}{4}(x-1)^4$ for $|x-1| \le 1/2$

In Exercises 11–18 evaluate the integral correct to three decimals.

11. $\displaystyle\int_0^1 \frac{\sin x}{x} \, dx$

12. $\displaystyle\int_0^{1/2} \cos(x^2) \, dx$

13. $\displaystyle\int_0^{2/3} \frac{1}{x^4 + 1} \, dx$

14. $\displaystyle\int_{-1}^1 x^{11} \sin x \, dx$

15. $\displaystyle\int_0^{1/2} \frac{1}{\sqrt{1+x^3}} \, dx$

16. $\displaystyle\int_0^{0.3} e^{-x^2} \, dx$

17. $\displaystyle\int_{-0.1}^0 \frac{1}{x-1} \ln(1-x) \, dx$

18. $\displaystyle\int_0^{1/2} \frac{1}{x^6 - 3x^3 - 4} \, dx$

In Exercises 19–24 use series to evaluate the limit.

19. $\displaystyle\lim_{x \to 0} \frac{\tan x}{x}$

20. $\displaystyle\lim_{x \to 0} \frac{1 - \cos x}{x^2}$

21. $\displaystyle\lim_{x \to 0} \frac{(1 - \cos x)^2}{3x^4}$

22. $\displaystyle\lim_{x \to 0} \frac{\sqrt{1+x} - 1}{x}$

23. $\displaystyle\lim_{x \to \infty} x \sin\left(\frac{1}{x}\right)$

24. $\displaystyle\lim_{x \to 0} \left(\frac{e^x + e^{-x}}{e^x - e^{-x}} - \frac{1}{x}\right)$

In Exercises 25–28 determine where the Maclaurin series for the function may be truncated in order to guarantee the accuracy indicated.

25. $\sin(x/3)$ on $|x| \leq 4$ with error less than 10^{-3}

26. $1/\sqrt{1+x^3}$ on $0 < x < 1/2$ with error less than 10^{-4}

27. $\ln(1-x)$ on $|x| < 1/3$ with error less than 10^{-2}

28. $\cos^2 x$ on $|x| < 0.1$ with error less than 10^{-3}

In Exercises 29–34 find a series solution in powers of x for the differential equation.

29. $y' + 3y = 4$
30. $y'' + y' = 0$
31. $xy' - 4y = 3x$
32. $4xy'' + 2y' + y = 0$
33. $y'' + y = 0$
34. $xy'' + y = 0$

35. Find the natural logarithm of $0.999\,999\,999\,9$ accurate to 15 decimal places.

36. In special relativity theory, the kinetic energy K of an object moving with speed v is defined by

$$K = c^2(m - m_0),$$

where c is a constant (the speed of light), m_0 is the rest mass of the object, and m is its mass when moving with speed v. The masses m and m_0 are related by

$$m = \frac{m_0}{\sqrt{1 - v^2/c^2}}.$$

Use the binomial expansion to show that

$$K = \frac{1}{2}m_0 v^2 + m_0 c^2 \left(\frac{3}{8}\frac{v^4}{c^4} + \frac{5}{16}\frac{v^6}{c^6} + \cdots\right),$$

and hence, to a first approximation, kinetic energy is defined by the classical expression $m_0 v^2/2$.

37. The ellipse $b^2 x^2 + a^2 y^2 = a^2 b^2$ can be represented parametrically by

$$x = a\cos t, \quad y = b\sin t, \quad 0 \leq t < 2\pi.$$

(a) Show that the length of the circumference of the ellipse is defined by the definite integral

$$L = 4b \int_0^{\pi/2} \sqrt{1 - k^2 \sin^2 t}\, dt, \quad k^2 = 1 - \frac{a^2}{b^2}.$$

(b) Use the binomial expansion to show that

$$L = 2\pi b \left(1 - \frac{k^2}{4} - \frac{3k^4}{64} - \cdots\right)$$

so that to a first approximation, L is the circumference of a circle of radius b.

38. In determining the radiated power from a half-wave antenna, it is necessary to evaluate

$$\int_0^{2\pi} \frac{1 - \cos\theta}{\theta}\, d\theta.$$

Find a two-decimal approximation for this integral.

39. Planck's law for the energy density Ψ of blackbody radiation of wavelength λ states that

$$\Psi(\lambda) = \frac{8\pi ch \lambda^{-5}}{e^{ch/(\lambda kT)} - 1},$$

where $h > 0$ is Planck's constant, c is the (constant) speed of light, and T is temperature, also assumed constant. Show that for long wavelengths, Planck's law reduces to the Rayleigh-Jeans law:

$$\Psi(\lambda) = \frac{8\pi kT}{\lambda^4}.$$

SUMMARY

An infinite sequence of constants is the assignment of numbers to positive integers. In most applications of sequences, the prime consideration is whether the sequence has a limit. If the sequence has its terms defined explicitly, then our ability to take limits of continuous functions ("limits at infinity" in Chapter 2 and L'Hôpital's rule in Chapter 4) can be very helpful. If the sequence is defined recursively, existence of the limit can sometimes be established by showing that the sequence is monotonic and bounded.

An expression of the form

$$\sum_{n=1}^{\infty} c_n = c_1 + c_2 + \cdots + c_n + \cdots$$

is called an infinite series. We define the sum of this series as the limit of its sequence of partial sums $\{S_n\}$, provided the sequence has a limit. Unfortunately, for most series we cannot find a simple formula for S_n, and therefore analysis of the limit of the sequence $\{S_n\}$ is impossible. To remedy this, we developed various convergence tests that avoided the sequence $\{S_n\}$: n^{th} term, comparison, limit comparison, limit ratio, limit root, integral, and alternating series tests. Note the sequences that are associated with a series $\sum c_n$:

$\{S_n\}$	sequence of partial sums for the definition of a sum;
$\{c_n\}$	sequence of terms for the nth term test;
$\{c_n/b_n\}$	sequence for limit comparison test;
$\{c_{n+1}/c_n\}$	sequence for the limit ratio test;
$\{\sqrt[n]{c_n}\}$	sequence for the limit root test;
$\{|c_n|\}$	sequence for the alternating series test.

Depending on the limits of these sequences — if they exist, we may be able to infer something about convergence of the series.

From infinite sequences and series of constants we proceeded to infinite sequences and series of functions — in particular, power series. We first considered situations where a power series was given, and the sum was to be determined. We saw that every power series $\sum a_n(x-c)^n$ has a radius of convergence R and an associated interval of convergence. If $R = 0$, the interval of convergence consists of only one point $x = c$; if $R = \infty$, the power series converges absolutely for all x; and if $0 < R < \infty$, the interval of convergence must be one of four possibilities: $c - R < x < c + R$, $c - R \leq x < c + R$, $c - R < x \leq c + R$, or $c - R \leq x \leq c + R$. The radius of convergence is given by $\lim_{n \to \infty} |a_n/a_{n+1}|$ or $\lim_{n \to \infty} |a_n|^{-1/n}$ provided the limits exist or are equal to infinity. If at each point in the interval of convergence of the power series the value of a function $f(x)$ is the same as the sum of the series, we write $f(x) = \sum a_n(x-c)^n$ and call $f(x)$ the sum of the series.

We also considered situations where a function $f(x)$ and a point c are given, and ask whether $f(x)$ has a power series expansion about c. We saw that there can be at most one power series expansion of $f(x)$ about c with a positive radius of convergence, and this series must be its Taylor series. One way to verify that $f(x)$ does indeed have a Taylor series about c and that this series converges to $f(x)$ is to show that the sequence of Taylor's remainders $\{R_n(c,x)\}$ exists and has limit zero. Often, however, it is much easier to find Taylor series by adding, multiplying, differentiating, and integrating known series.

When a Taylor series is truncated, Taylor's remainder $R_n(c,x)$ represents the truncation error and, in spite of the fact that R_n is expressed in terms of some unknown point z_n, it is often possible to calculate a maximum value for the error. Sometimes $R_n(c,x)$ can be avoided altogether. For instance, if the Taylor series is an alternating series, then the maximum possible truncation error is the value of the next term.

Power series are often used in situations that require approximations. Taylor series provide polynomial approximations to complicated functions, and they offer an alternative to the numerical techniques of Section 9.9 in the evaluation of definite integrals. Power series are also useful in situations that do not require approximations. They are sometimes helpful in evaluating limits, and they are the only way to solve many differential equations.

Key Terms and Formulas

In reviewing this chapter, you should be able to define or discuss the following key terms:

Sequence of constants
Explicit sequence
Recursive sequence
Limit of a sequence
Convergent sequence
Divergent sequence
Increasing sequence
Decreasing sequence
Nonincreasing sequence
Nondecreasing sequence
Monotonic sequence
Upper bound
Lower bound
Successive substitutions
Series of constants
Geometric series
Harmonic series
Sequence of partial sums
Convergent series
Divergent series

n^{th} term test
Nonnegative series
Integral test
p-series
Comparison test
Limit comparison test
Limit ratio test
Limit root test
Absolutely convergent series
Conditionally convergent series
Alternating harmonic series
Truncation error
Power series
Interval of convergence
Radius of convergence
Sum of a power series
Taylor remainder formula
Taylor series
Maclaurin series
Binomial expansion

REVIEW EXERCISES

In Exercises 1–4 discuss, with all necessary proofs, whether the sequence is monotonic and has an upper bound, a lower bound, and a limit.

1. $\left\{\dfrac{n^2 - 5n + 3}{n^2 + 5n + 4}\right\}$

2. $c_1 = 1$, $c_{n+1} = (1/2)\sqrt{c_n^2 + 1}$, $n \geq 1$

3. $\left\{\dfrac{\text{Tan}^{-1}(1/n)}{n^2 + 1}\right\}$

4. $c_1 = 7$, $c_{n+1} = 15 + \sqrt{c_n - 2}$, $n \geq 1$

5. Use Newton's iterative procedure and the method of successive substitutions to approximate the root of the equation
$$x = \left(\dfrac{x + 5}{x + 4}\right)^2$$
between $x = 1$ and $x = 2$.

6. For what values of k does the sequence
$$c_1 = k, \quad c_{n+1} = c_n^2, \quad n \geq 1$$
converge?

7. Find an explicit definition for the sequence
$$c_1 = 1, \quad c_{n+1} = \sqrt{1 + c_n^2}, \quad n \geq 1.$$

8. Use the derivative of the function $f(x) = (\ln x)/x$ to prove that the sequence $\{\ln n/n\}$ is decreasing for $n \geq 3$.

In Exercises 9–28 determine whether the series converges or diverges. In the case of a convergent series that has both positive and negative terms, indicate whether it converges absolutely or conditionally.

9. $\displaystyle\sum_{n=1}^{\infty} \dfrac{n^2 - 3n + 2}{n^3 + 4n}$

10. $\displaystyle\sum_{n=1}^{\infty} \dfrac{n^2 + 5n + 3}{n^4 - 2n + 5}$

11. $\displaystyle\sum_{n=1}^{\infty} \dfrac{5^{2n}}{n!}$

12. $\displaystyle\sum_{n=1}^{\infty} \dfrac{n^2 + 3}{n3^n}$

13. $\displaystyle\sum_{n=1}^{\infty} \dfrac{(\ln n)^2}{\sqrt{n}}$

14. $\displaystyle\sum_{n=1}^{\infty} (-1)^n \left(\dfrac{n+1}{n^2}\right)$

15. $\displaystyle\sum_{n=1}^{\infty} (-1)^n \left(\dfrac{n+1}{n^3}\right)$

16. $\displaystyle\sum_{n=1}^{\infty} \text{Cos}^{-1}\left(\dfrac{1}{n}\right)$

17. $\displaystyle\sum_{n=1}^{\infty} \dfrac{1}{n}\text{Cos}^{-1}\left(\dfrac{1}{n}\right)$

18. $\displaystyle\sum_{n=1}^{\infty} \dfrac{1}{n^2}\text{Cos}^{-1}\left(\dfrac{1}{n}\right)$

19. $\displaystyle\sum_{n=1}^{\infty} \dfrac{2 \cdot 4 \cdot 6 \cdot \ldots \cdot (2n)}{n!}$

20. $\displaystyle\sum_{n=1}^{\infty} \dfrac{3 \cdot 6 \cdot 9 \cdot \ldots \cdot (3n)}{(2n)!}$

21. $\displaystyle\sum_{n=1}^{\infty} \sqrt{\frac{n^2+1}{n^2+5}}$

22. $\displaystyle\sum_{n=1}^{\infty} (-1)^{n+1} \left(1+\frac{1}{n}\right)^3$

23. $\displaystyle\sum_{n=1}^{\infty} \frac{1}{n^2} \sin n$

24. $\displaystyle\sum_{n=1}^{\infty} \frac{10^n}{5^{3n+2}}$

25. $\displaystyle\sum_{n=1}^{\infty} (-1)^n \frac{\ln n}{n}$

26. $\displaystyle\sum_{n=1}^{\infty} \frac{1}{e^{n\pi}}$

27. $\displaystyle\sum_{n=1}^{\infty} \frac{2^n + 2^{-n}}{3^n}$

28. $\displaystyle\sum_{n=1}^{\infty} \frac{1}{\sqrt{n}} \cos(n\pi)$

In Exercises 29–36 find the interval of convergence for the power series.

29. $\displaystyle\sum_{n=0}^{\infty} \frac{n+1}{n^2+1} x^n$

30. $\displaystyle\sum_{n=1}^{\infty} \frac{1}{n^2 2^n} x^n$

31. $\displaystyle\sum_{n=0}^{\infty} (n+1)^3 x^n$

32. $\displaystyle\sum_{n=1}^{\infty} \frac{1}{n^n} x^n$

33. $\displaystyle\sum_{n=0}^{\infty} \frac{1}{4^n}(x-2)^n$

34. $\displaystyle\sum_{n=2}^{\infty} \sqrt{\frac{n+1}{n-1}}(x+3)^n$

35. $\displaystyle\sum_{n=1}^{\infty} n 3^n x^{2n}$

36. $\displaystyle\sum_{n=1}^{\infty} \frac{2^n}{n} x^{3n}$

In Exercises 37–45 find the power series expansion of the function about the indicated point.

37. $f(x) = \sqrt{1+x^2}$, about $x = 0$

38. $f(x) = e^{x+5}$, about $x = 0$

39. $f(x) = \cos(x + \pi/4)$, about $x = 0$

40. $f(x) = x \ln(2x+1)$, about $x = 0$

41. $f(x) = \sin x$, about $x = \pi/4$

42. $f(x) = x/(x^2 + 4x + 3)$, about $x = 0$

43. $f(x) = e^x$, about $x = 3$

44. $f(x) = (x+1) \ln(x+1)$, about $x = 0$

45. $f(x) = x^3 e^{x^2}$, about $x = 0$

46. How many terms in the Maclaurin series for $f(x) = e^{-x^2}$ guarantee a truncation error of less than 10^{-5} for all x in the interval $0 \leq x \leq 2$?

47. Find a power series solution in powers of x for the differential equation

$$y'' - 4y = 0.$$

48. Find the Maclaurin series for $f(x) = \sqrt{1 + \sin x}$ valid for $-\pi/2 \leq x \leq \pi/2$ by first showing that $f(x)$ can be written in the form

$$f(x) = \sin(x/2) + \cos(x/2).$$

Why is the restriction $-\pi/2 \leq x \leq \pi/2$ necessary?

49. On a calculator take the cosine of 1 (radian). Take the cosine of this result, and again, and again, and again, ... What happens? Interpret what is going on.

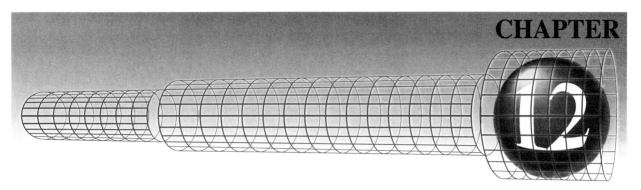

CHAPTER 12

Vectors and Three-dimensional Analytic Geometry

Chapters 1–11 dealt with single-variable calculus–differentiation and integration of functions $f(x)$ of one variable. In Chapters 12–15 we study multivariable calculus. Discussions of three-dimensional analytic geometry and vectors in Sections 1–5 of this chapter prepare the way. In Sections 6–10, we differentiate and integrate vector functions, and apply the results to the geometry of curves in space and the motion of objects.

SECTION 12.1

Rectangular Coordinates In Space

The coordinate of a point on the real line is defined as its directed distance from a fixed point called the origin. The Cartesian coordinates of a point in a plane are its directed distances from two fixed lines called the coordinate axes. In space, Cartesian coordinates are directed distances from three fixed planes called the coordinate planes. In particular, we draw through a point O, called the **origin**, three mutually perpendicular lines called the x-, y-, and z-axes (Figure 12.1). Each of the axes is coordinatized with some unit distance (which need not be the same for all three axes). These three coordinate axes determine three planes called **coordinate planes**: The xy-coordinate plane is that plane containing the x- and y-axes, the yz-coordinate plane contains the y- and z-axes, and the xz-coordinate plane contains the x- and z-axes.

FIGURE 12.1

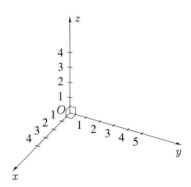

If P is any point in space, we draw lines from P perpendicular to the three coordinate planes (Figure 12.2). The directed distance from the yz-coordinate plane to P is parallel to the x-axis, and is called the x-coordinate of P. Similarly, y- and z-coordinates are defined as directed distances from the xz- and xy-coordinate planes to P. These three coordinates of P, written (x, y, z), are called the **Cartesian** or **rectangular coordinates** of P. Note that if we draw lines through P that are perpendicular to the axes, then the directed distances from O to the points of intersection of these perpendiculars with the axes are also the Cartesian coordinates of P. (Figure 12.3).

FIGURE 12.2 **FIGURE 12.3**

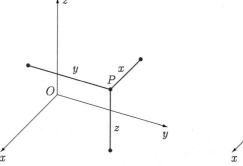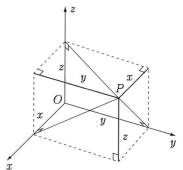

By either definition, each point in space has a unique ordered set of Cartesian coordinates (x, y, z); conversely, every ordered triple of real numbers (x, y, z) is the set of coordinates for one and only one point in space. For example, points with coordinates $(1, 1, 1)$, $(2, -3, 4)$, $(3, 4, -1)$, and $(-2, 5, 3)$ are shown in Figure 12.4.

FIGURE 12.4

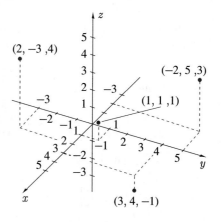

The coordinate systems in Figures 12.1–12.4 are called **right-handed coordinate systems**, because if we curl the fingers on our right hand from the positive x-direction toward the positive y-direction, then the thumb points in the positive z-direction (Figure 12.5). The coordinate system in Figure 12.6, on the other hand, is a **left-handed coordinate system**, since the thumb of the left hand points in the positive z-direction when the fingers of this hand are curled from the positive x-direction to the positive y-direction. We always use right-handed systems in this book.

FIGURE 12.5 **FIGURE 12.6**

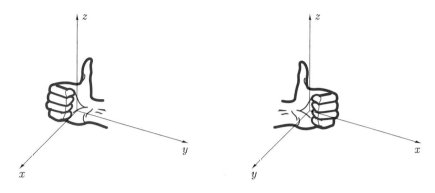

Suppose we construct for any two points P_1 and P_2 with coordinates (x_1, y_1, z_1) and (x_2, y_2, z_2), respectively, a box with sides parallel to the coordinate planes and with line segment $P_1 P_2$ as diagonal (Figure 12.7). Because triangles $P_1 AB$ and $P_1 BP_2$ are right-angled, we can write

$$\begin{aligned} ||P_1 P_2||^2 &= ||P_1 B||^2 + ||BP_2||^2 \\ &= ||P_1 A||^2 + ||AB||^2 + ||BP_2||^2 \\ &= (x_2 - x_1)^2 + (y_2 - y_1)^2 + (z_2 - z_1)^2. \end{aligned}$$

In other words, the length of the line segment joining two points $P_1(x_1, y_1, z_1)$ and $P_2(x_2, y_2, z_2)$ is

$$||P_1 P_2|| = \sqrt{(x_2 - x_1)^2 + (y_2 - y_1)^2 + (z_2 - z_1)^2}. \qquad (12.1)$$

This is the analogue of formula 1.7 for the length of a line segment joining two points in the xy–plane.

FIGURE 12.7

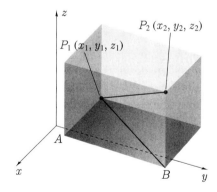

Just as the x- and y-axes divide the xy-plane into four regions called quadrants, the xy-, yz- and xz-coordinate planes divide xyz-space into eight regions called **octants**. The region where x-, y-, and z-coordinates are all positive is called the first octant. There is no commonly accepted way to number the remaining seven octants.

EXERCISES 12.1

1. Draw a Cartesian coordinate system and show the points $(1,2,1)$, $(-1,3,2)$, $(1,-2,4)$, $(3,4,-5)$, $(-1,-2,-3)$, $(-2,-5,4)$, $(8,-3,-6)$, and $(-4,3,-5)$.

2. Find the length of the line segment joining the points $(1,-2,5)$ and $(-3,2,4)$.

3. Prove that the triangle with vertices $(2,0,4\sqrt{2})$, $(3,-1,5\sqrt{2})$, and $(4,-2,4\sqrt{2})$ is right-angled and isosceles.

4. A cube has sides of length 2 units. What are coordinates of its corners if one corner is at the origin, three of its faces lie in the coordinate planes, and one corner has all three coordinates positive?

5. Show that the (undirected, perpendicular) distances from a point (x,y,z) to the x-, y-, and z-axes are, respectively, $\sqrt{y^2+z^2}$, $\sqrt{x^2+z^2}$, and $\sqrt{x^2+y^2}$.

In Exercises 6–9, find the (undirected) distances from the point to (a) the origin; (b) the x-axis; (c) the y-axis; (d) the z-axis.

6. $(2,3,-4)$

7. $(1,-5,-6)$

8. $(4,3,0)$

9. $(-2,1,-3)$

10. Prove that the three points $(1,3,5)$, $(-2,0,3)$ and $(7,9,9)$ are collinear.

11. Find that point in the xy-plane that is equidistant from the points $(1,3,2)$ and $(2,4,5)$ and has a y-coordinate equal to three times its x-coordinate.

12. Find an equation describing all points that are equidistant from the points $(-3,0,4)$ and $(2,1,5)$. What does this equation describe geometrically?

13. (a) If $(\sqrt{3}-3, 2+2\sqrt{3}, 2\sqrt{3}-1)$ and $(2\sqrt{3},4,\sqrt{3}-2)$ are two vertices of an equilateral triangle, and if the third vertex lies on the z-axis, find the third vertex.
    (b) Can you find a third vertex on the x-axis?

14. A birdhouse is built from a box $1/2$ m on each side with a roof as shown in Figure 12.8. If the distance from each corner of the roof to the peak is $3/4$ m, find the coordinates of the nine corners of the house. (The sides of the box are parallel to the coordinate planes.)

FIGURE 12.8

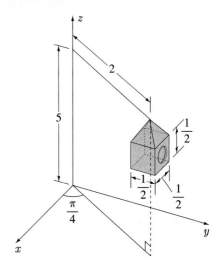

15. If P and Q (Figure 12.9) have coordinates (x_1, y_1, z_1) and (x_2, y_2, z_2), show that coordinates of the point R midway between P and Q are $\left(\dfrac{x_1+x_2}{2}, \dfrac{y_1+y_2}{2}, \dfrac{z_1+z_2}{2} \right)$.

FIGURE 12.9

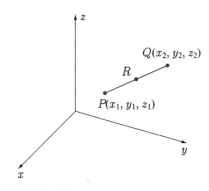

16. (a) Find the midpoint of the line segment joining the points $P(1,-1,-3)$ and $Q(3,2,-4)$.
    (b) If the line segment joining P and Q is extended its own length beyond Q to a point R, find the coordinates of R.

17. The four sided object in Figure 12.10 is a *tetrahedron*. If the four vertices of the tetrahedron are as shown, prove that the three lines joining the midpoints of opposite edges (one of which is PQ) meet at a point that bisects each of them.

FIGURE 12.10

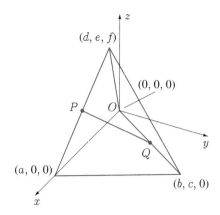

18. Let A, B, C, and D be the vertices of a quadrilateral in space (not necessarily planar). Show that the line segments joining midpoints of opposite sides of the quadrilateral intersect in a point which bisects each.

19. Generalize the result of Exercise 15 to prove that if a point R divides the length PQ so that $\dfrac{\|PR\|}{\|RQ\|} = \dfrac{r_1}{r_2}$, where r_1 and r_2 are positive integers, then the coordinates of R are

$$x = \frac{r_1 x_2 + r_2 x_1}{r_1 + r_2}, \quad y = \frac{r_1 y_2 + r_2 y_1}{r_1 + r_2}, \quad z = \frac{r_1 z_2 + r_2 z_1}{r_1 + r_2}.$$

20. A man 2 m tall walks along the edge of a straight road 10 m wide (Figure 12.11). On the other edge of the road stands a streetlight 8 m high. A building runs parallel to the road and 1 m from it. If Cartesian coordinates are set up as shown (with the x- and y-axes in the plane of the road), find the coordinates of the tip of the man's shadow when he is at the position shown.

FIGURE 12.11

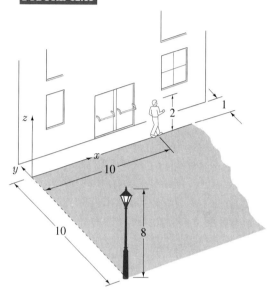

SECTION 12.2

Curves and Surfaces

An equation involving the x- and y-coordinates of points in the xy-plane usually specifies a curve. For example, the equation $x^2 + y^2 = 4$ describes a circle of radius 2 centred at the origin (Figure 12.12). We now ask what is defined by an equation involving the Cartesian coordinates (x, y, z) of points in space. For example, the equation $z = 0$ describes all points in the xy-plane since all such points have a z-coordinate equal to zero. Similarly, $y = 2$ describes all points in the plane parallel to and two units to the right of the xz-plane. What does the equation $x^2 + y^2 = 4$ describe? In other words, regarded as a restriction on the x-, y-, and z-coordinates of points in space, rather than a restriction on the x- and y-coordinates of points in the xy-plane, what does it represent? Because the equation says nothing about z, there is no restriction whatsoever on z. In other words, the z-coordinate can take on all possible values, but the x- and y-coordinates must be restricted by $x^2 + y^2 = 4$. If we consider those points in the xy-plane ($z = 0$) that satisfy $x^2 + y^2 = 4$, we obtain the circle in Figure 12.12. In space, each of these points has coordinates $(x, y, 0)$, where x and y still satisfy $x^2 + y^2 = 4$ (Figure 12.13). If we now take any point Q that is either directly above or directly below a point $P(x, y, 0)$ on this circle, it has exactly the same x- and y-coordinates as P; only its z-coordinate differs. Thus the x- and y-coordinates of Q also satisfy $x^2 + y^2 = 4$. Since we can do this for any point P on the circle, it follows that $x^2 + y^2 = 4$ describes the right-circular cylinder of radius 2 and infinite extent in Figure 12.13.

FIGURE 12.12

FIGURE 12.13

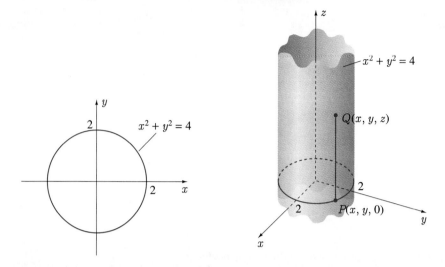

By reasoning similar to that used above, we can show that the equation $2x + y = 2$ describes the plane in Figure 12.14 parallel to the z-axis and standing on the straight line $2x + y = 2$, $z = 0$ in the xy-plane.

FIGURE 12.14

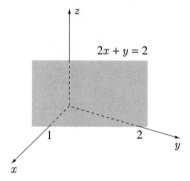

Finally, consider the equation $x^2 + y^2 + z^2 = 9$. Since $\sqrt{x^2 + y^2 + z^2}$ is the distance from the origin to a point with coordinates (x, y, z), this equation describes all points that are three units away from the origin. In other words, $x^2 + y^2 + z^2 = 9$ describes the points on a sphere of radius 3 centred at the origin.

It appears that one equation in the coordinates (x, y, z) of points in space specifies a surface. The shape of the surface is determined by the form of the equation. If one equation in the coordinates (x, y, z) specifies a surface, it is easy to see what two simultaneous equations specify. For instance, suppose we ask for all points in space whose coordinates satisfy both of the equations

$$x^2 + y^2 = 4, \quad z = 1.$$

By itself, $x^2 + y^2 = 4$ describes the cylinder in Figure 12.13. The equation $z = 1$ describes all points in a plane parallel to the xy-plane and one unit above it. To ask for all points that satisfy $x^2 + y^2 = 4$ and $z = 1$ simultaneously is to ask for all points that lie on both surfaces. Consequently, the equations $x^2 + y^2 = 4$, $z = 1$ describe the curve of intersection of the two surfaces—the circle in Figure 12.15.

FIGURE 12.15

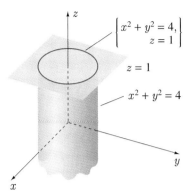

The equation $x = 0$ describes the yz-plane; the equation $y = 0$ describes the xz-plane. If we put the two equations together, $x = 0$ and $y = 0$, we obtain all points that lie on both the yz-plane and the xz-plane; i.e., the z-axis. In other words, equations for the z-axis are $x = 0$, $y = 0$.

Finally, $x^2 + y^2 + z^2 = 9$ is the equation of a sphere of radius 3 centred at the origin, and $y = 2$ is the equation of a plane parallel to the xz-plane and two units to the right. Together, the equations $x^2 + y^2 + z^2 = 9$, $y = 2$ describe the curve of intersection of the two surfaces—the circle in Figure 12.16. Note that by substituting $y = 2$ into the equation of the sphere, we can write alternatively that $x^2 + z^2 = 5$, $y = 2$. This pair of equations is equivalent to the original pair because all points that satisfy $x^2 + y^2 + z^2 = 9$, $y = 2$ also satisfy $x^2 + z^2 = 5$, $y = 2$, and vice versa. This new pair of equations provides an alternative way of visualizing the curve. Again $y = 2$ is the plane of Figure 12.16, but $x^2 + z^2 = 5$ describes a right-circular cylinder of radius $\sqrt{5}$ and infinite extent around the y-axis (Figure 12.17). Our discussion has shown that the cylinder and plane intersect in the same curve as the sphere and plane.

FIGURE 12.16

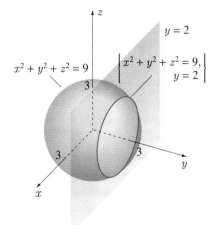

FIGURE 12.17

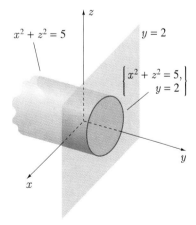

In summary, we have illustrated that one equation in the coordinates (x, y, z) of a point specifies a surface; two simultaneous equations specify a curve, the curve of intersection of the two surfaces (provided, of course, that the surfaces do intersect).

In Chapters 4–10 we learned to appreciate the value of curve sketching in the xy-plane. Sometimes a sketch serves as a device by which we can interpret algebraic statements geometrically (such as the mean value theorem or the interpretation of a critical point of a function as a point where the tangent line to the graph of the function is horizontal, vertical, or does not exist). Sometimes it plays an integral part in the solution of a problem (such as when the definite integral is used to find areas, volumes, etc.). Sometimes a sketch is a complete solution to a problem (such as to determine whether a given function has an inverse). We will find that sketching surfaces can be just as useful for multivariable calculus in Chapters 13–15. Unfortunately, surface sketching is more difficult than curve sketching, principally because we are drawing a space diagram on a page. Our ability to sketch curves, as we will see, can be of immense help.

One of the most helpful techniques for sketching a surface is to imagine the intersection of the surface with various planes—in particular, the coordinate planes. From these cross-sections of the surface, it is sometimes possible to visualize the entire surface. For example, if we intersect the surface $z = x^2 + y^2$ with the yz-plane, we obtain the parabola $z = y^2$, $x = 0$. Similarly, the parabola $z = x^2$, $y = 0$ is the intersection curve with the xz-plane. These curves, shown in Figure 12.18(a), would lead us to suspect that the surface $z = x^2 + y^2$ might be shaped as shown in Figure 12.18(b). To verify this, we intersect the surface with a plane $z = k$ (k a constant), giving the curve

$$z = x^2 + y^2, \quad \text{or} \quad x^2 + y^2 = k,$$
$$z = k, \qquad\qquad\qquad z = k.$$

These latter equations indicate that cross-sections of $z = x^2 + y^2$ with planes $z = k$ are circles centred on the z-axis with radii $\sqrt{k}$ that increase as k increases. This certainly confirms the sketch in Figure 12.18(b).

FIGURE 12.18

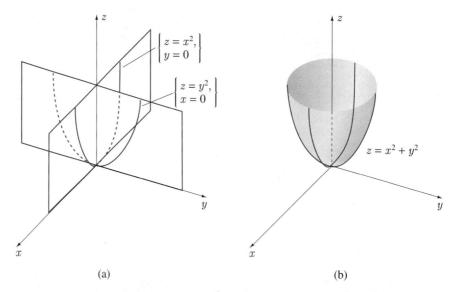

(a) (b)

Intersections of the surface $y = z + x^2$ with the xy-, xz-, and yz-coordinate planes give two parabolas and a straight line, shown in Figure 12.19(a). These really do not seem to help us visualize the surface. If, however, we intersect the surface with planes $z = k$, we obtain the parabolas

$$y = z + x^2, \quad \text{or} \quad y = x^2 + k,$$
$$z = k, \qquad\qquad\qquad z = k.$$

These parabolas, shown in Figure 12.19(b), indicate that the surface $y = z + x^2$ should be drawn as in Figure 12.19(c).

FIGURE 12.19

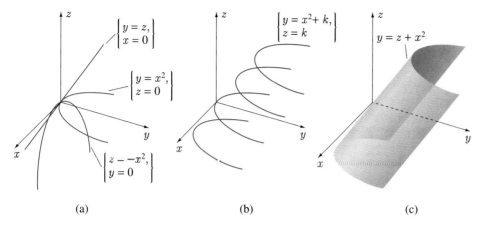

(a)   (b)   (c)

We can sometimes "build" surfaces in much the same way that we "built" curves in Chapter 1. For the surface $z = 1 - x^2 - y^2$, we first draw the surface $z = x^2 + y^2$ in Figure 12.18(b). To sketch $z = -(x^2 + y^2)$, we turn $z = x^2 + y^2$ upside down (Figure 12.20a), and finally we see that $z = 1 - x^2 - y^2$ is $z = -(x^2 + y^2)$ shifted upward one unit (Figure 12.20b).

FIGURE 12.20

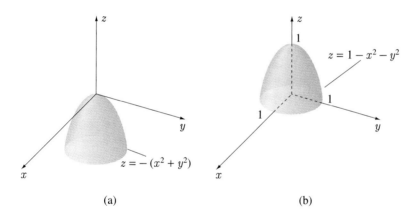

(a)   (b)

EXAMPLE 12.1

Sketch the surface defined by each of the following equations:

(a) $z = \sqrt{4x + 2y - x^2 - y^2 - 4}$

(b) $y = 1 + \sqrt{x^2 + z^2}$

SOLUTION

(a) If we square the equation, and at the same time complete the squares on $-x^2 + 4x$ and $-y^2 + 2y$, we have

$$z^2 = -(x-2)^2 - (y-1)^2 + 1,$$

or

$$(x-2)^2 + (y-1)^2 + z^2 = 1.$$

SECTION 12.2   CURVES AND SURFACES   **603**

Because $\sqrt{(x-2)^2 + (y-1)^2 + z^2}$ is the distance from a point (x,y,z) to $(2,1,0)$, this equation states that (x,y,z) must always be a unit distance from $(2,1,0)$; i.e., the equation $(x-2)^2+(y-1)^2+z^2 = 1$ defines a sphere of radius 1 centred at $(2,1,0)$ (Figure 12.21a). Because the original equation requires z to be nonnegative, the required surface is the upper half of this sphere—the hemisphere in Figure 12.21(b).

FIGURE 12.21

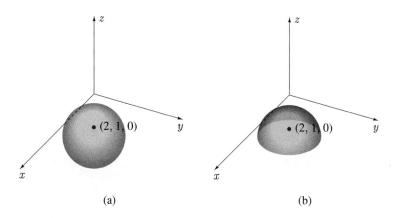

(a)   (b)

(b) If we intersect the surface $y = \sqrt{x^2 + z^2}$ with the xy-plane, we obtain the broken straight line $y = |x|$, $z = 0$ in Figure 12.22(a). Intersections of the surface with planes $y = k$ (k a constant) give

$$y = \sqrt{x^2 + z^2}, \qquad \text{or} \qquad x^2 + z^2 = k^2,$$
$$y = k, \qquad\qquad\qquad\qquad\quad y = k.$$

These define circles of radii k in the planes $y = k$ (Figure 12.22b). Consequently, $y = \sqrt{x^2 + z^2}$ defines the right-circular cone in Figure 12.22(c). The surface $y = 1 + \sqrt{x^2 + z^2}$ can now be obtained by shifting the cone one unit in the y-direction (Figure 12.22d).

FIGURE 12.22

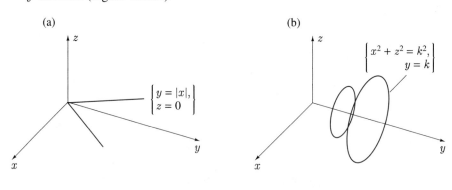

(a)   (b)

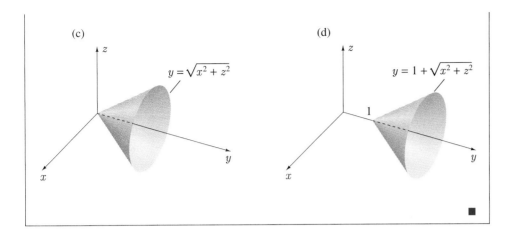

Cylinders

When one of the coordinates is missing from the equation of a surface, cross-sections of the surface with planes perpendicular to the axis of the missing variable are all identical. Such a surface is said to be a **cylinder**. For example, z is missing from the equation $x^2 + y^2 = 4$, and we saw in Figure 12.13 that this is the equation of a right-circular cylinder around the z-axis. Every cross-section of this surface with a plane $z = k$ is a circle of radius 2 centred on the z-axis. The equation $z = x^2$ is free of y. Each cross-section of this surface with a plane $y = k$ is the parabola $z = x^2$ in the plane $y = k$. Consequently, $z = x^2$ is the equation for the parabolic cylinder in Figure 12.23.

FIGURE 12.23

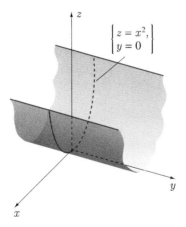

Quadric Surfaces

A **quadric surface** is a surface whose equation is quadratic in x, y, and z, the most general such equation being

$$Ax^2 + By^2 + Cz^2 + Dxy + Eyz + Fxz + Gx + Hy + Iz + J = 0. \qquad (12.2)$$

For the most part, we encounter quadric surfaces whose equations are of the forms

$$Ax^2 + By^2 + Cz^2 = J \quad \text{or} \quad Ax^2 + By^2 = Iz,$$

or these equations with x, y, and z interchanged. Surfaces with these equations fall into nine major classes depending on whether the constants are positive, negative, or zero. They are illustrated in Figures 12.24–12.32.

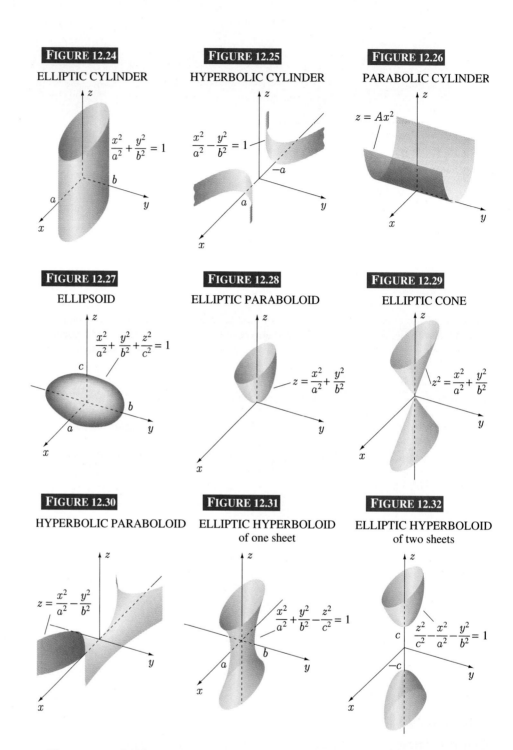

FIGURE 12.24 ELLIPTIC CYLINDER
$\dfrac{x^2}{a^2} + \dfrac{y^2}{b^2} = 1$

FIGURE 12.25 HYPERBOLIC CYLINDER
$\dfrac{x^2}{a^2} - \dfrac{y^2}{b^2} = 1$

FIGURE 12.26 PARABOLIC CYLINDER
$z = Ax^2$

FIGURE 12.27 ELLIPSOID
$\dfrac{x^2}{a^2} + \dfrac{y^2}{b^2} + \dfrac{z^2}{c^2} = 1$

FIGURE 12.28 ELLIPTIC PARABOLOID
$z = \dfrac{x^2}{a^2} + \dfrac{y^2}{b^2}$

FIGURE 12.29 ELLIPTIC CONE
$z^2 = \dfrac{x^2}{a^2} + \dfrac{y^2}{b^2}$

FIGURE 12.30 HYPERBOLIC PARABOLOID
$z = \dfrac{x^2}{a^2} - \dfrac{y^2}{b^2}$

FIGURE 12.31 ELLIPTIC HYPERBOLOID of one sheet
$\dfrac{x^2}{a^2} + \dfrac{y^2}{b^2} - \dfrac{z^2}{c^2} = 1$

FIGURE 12.32 ELLIPTIC HYPERBOLOID of two sheets
$\dfrac{z^2}{c^2} - \dfrac{x^2}{a^2} - \dfrac{y^2}{b^2} = 1$

The names of these surfaces are derived from the fact that their cross-sections are ellipses, hyperbolas, or parabolas. For example, cross-sections of the hyperbolic paraboloid with planes $z = k$ are hyperbolas $x^2/a^2 - y^2/b^2 = k$. Cross-sections with planes $x = k$ are parabolas $z = k^2/a^2 - y^2/b^2$, as are cross-sections with planes $y = k$.

In applications of multiple integrals in Chapter 14, it is often necessary to project a space curve into one of the coordinate planes and find the equations of the projection. To illustrate, consider the curve of intersection of the cylinder $x^2 + z^2 = 4$ and the plane $2y + z = 4$ (the first octant part of which is shown in Figure 12.33). Since the curve of intersection lies on the cylinder $x^2 + z^2 = 4$, its projection in the xz-plane is the circle $x^2 + z^2 = 4$, $y = 0$. To find its projection in the xy-plane, we eliminate z between the equations $2y + z = 4$ and $x^2 + z^2 = 4$. The result is $x^2 + (4 - 2y)^2 = 4$, or, $x^2 + 4(y - 2)^2 = 4$. This shows that the curve of intersection lies on the elliptic cylinder $x^2 + 4(y - 2)^2 = 4$, and therefore it projects onto the ellipse $x^2 + 4(y - 2)^2 = 4$, $z = 0$ in the xy-plane. The projection of the curve in the yz-plane is that part of the line $2y + z = 4$, $x = 0$ between the points $(0, 1, 2)$ and $(0, 3, -2)$.

FIGURE 12.33

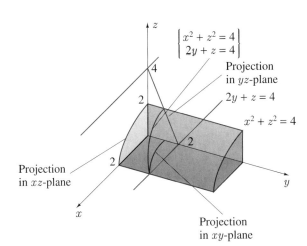

EXERCISES 12.2

In Exercises 1–35 sketch the surface defined by the equation.

1. $2y + 3z = 6$
2. $2x - 3y = 0$
3. $y = x^2 + 2$
4. $z = x^3$
5. $y^2 + z^2 = 1$
6. $x^2 + y^2 + z^2 = 4$
7. $x^2 + 4y^2 = 1$
8. $y^2 - z^2 = 4$
9. $z = 2(x^2 + y^2)$
10. $x = \sqrt{y^2 + z^2}$
11. $x = \sqrt{1 - y^2}$
12. $z = 2 - x$
13. $x^2 = y^2$
14. $x = z^2 + 2$
15. $z = y + 3$
16. $4z = 3\sqrt{x^2 + y^2}$
17. $x^2 - 2x + z^2 = 0$
18. $yz = 1$
19. $x^2 + y^2 + (z - 1)^2 = 3$
20. $z + 5 = 4(x^2 + y^2)$
21. $x^2 + z^2 = y^2$
22. $x^2 + z^2 = y^2 + 1$
23. $y^2 + z^2 = x^6$
24. $x^2 + y^2 + 4z^2 = 1$
25. $9z^2 = x^2 + y^2 + 1$
26. $(y^2 + z^2)^2 = x + 1$
27. $z^2 + 4y^2 = 1$
28. $y - z^2 = 0$
29. $x^2 - z^2 = 4$
30. $x^2 + y^2/4 + z^2/9 = 1$
31. $z = x^2/4 + y^2/25$
32. $x^2 = z^2 + 9y^2$
33. $z = y^2/16 - x^2/4$
34. $x^2 + y^2/4 - z^2/25 = 1$
35. $z^2 - 9x^2 - 16y^2 = 1$

In Exercises 36–45 sketch the curve defined by the equations.

36. $x^2 + y^2 = 2$, $z = 4$
37. $x + 2y = 6$, $y - 2z = 3$
38. $z = x^2 + y^2$, $x^2 + y^2 = 5$
39. $x^2 + y^2 = 2$, $x + z = 1$
40. $z = \sqrt{x^2 + y^2}$, $y = x$
41. $z + 2x^2 = 1$, $y = z$

42. $z = \sqrt{4 - x^2 - y^2}$, $x^2 + y^2 - 2y = 0$

43. $z = y$, $y = x^2$

44. $x^2 + z^2 = 1$, $y^2 + z^2 = 1$

45. $z = x^2$, $z = y^2$

In Exercises 46–55 find equations for the projections of the curve in the xy-, yz-, and xz-coordinate planes. In each case sketch the curve.

46. $x + y = 3$, $2y + 3z = 4$

47. $x + y + z = 4$, $2x - y + z = 6$

48. $x^2 + y^2 = 4$, $z = 4$

49. $x^2 + y^2 = 4$, $y = x$

50. $x^2 + y^2 = 4$, $x = z$

51. $x^2 + y^2 = 4$, $x + y + z = 2$

52. $y^2 + z^2 = 3$, $x^2 + z^2 = 3$

53. $z = x^2 + y^2$, $x + z = 1$

54. $z = \sqrt{x^2 + y^2}$, $z = 6 - x^2 - y^2$

55. $x^2 + y^2 + z^2 = 1$, $y = x$

In Exercises 56–61 find equations for the projection of the curve in the specified plane. Sketch each curve.

56. $z = x^2 - y^2$, $z = 2x + 4y$ in the xy-plane

57. $x^2 + y^2 - 4z^2 = 1$, $x + y = 2$ in the xz-plane

58. $y = z + x^2$, $y + z = 1$ in the xy-plane

59. $x = \sqrt{1 + 2y^2 + 4z^2}$, $x^2 + 9y^2 + 4z^2 = 36$ in the yz-plane

60. $z = x^2 + y^2$, $z = 4(x - 1)^2 + 4(y - 1)^2$ in the xy-plane

61. $x^2 + y^2 - 2y = 0$, $z^2 = x^2 + y^2$ in the xz-plane

In Exercises 62–71 sketch whatever is defined by the equation or equations.

62. $(x - 2)^2 + y^2 + z^2 = 0$

63. $x = 0$, $y = 5$

64. $\sqrt{x} + \sqrt{y} = 1$, $z = x$

65. $x + y = 15$, $y - x = 4$

66. $z = 1 - (x^2 + y^2)^{1/3}$

67. $z = |x|$

68. $z = x^2$, $y = z^2$

69. $x = \ln(y^2 + z^2)$

70. $z = |x - y|$

71. $x = 2$, $y = 4$, $z^2 - 1 = 0$

SECTION 12.3

Vectors

Physical quantities that have associated with them only a magnitude can be represented by real numbers. Some examples are temperature, density, area, moment of inertia, speed, and pressure. They are called **scalars**. There are many quantities, however, that have associated with them both magnitude and direction, and these quantities cannot be described by a single real number. Velocity, acceleration, and force are perhaps the most notable concepts in this category. To represent such quantities mathematically, we introduce **vectors**.

Definition 12.1

A **vector** is defined as a directed line segment.

To denote a vector we use a letter in boldface type, such as **v**. In Figures 12.34(a), (b), and (c) we show two vectors **u** and **v** along a line, three vectors **u**, **v**, and **w** in a plane, and three vectors **u**, **v**, and **w** in space, respectively. It is customary to place an arrowhead on a vector and call this end the **tip** of the vector. The other end is called the **tail** of the vector, and the direction of the vector is from tail to tip. A vector then has both *direction* and *length*.

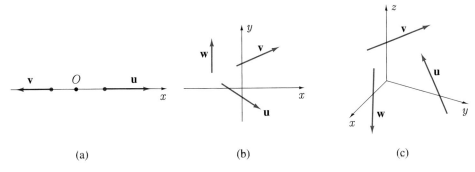

FIGURE 12.34

(a)   (b)   (c)

Definition 12.1 for a vector says nothing about its point of application; i.e., where its tail should be placed. This means that we may place the tail anywhere we wish. This suggests the following definition for equality of vectors.

> **Definition 12.2**
>
> Two vectors are equal if and only if they have the same length and direction. Their points of application are irrelevant.

For example, vectors **u** and **v** in Figure 12.35 have exactly the same length and direction, and are therefore one and the same. Although the vector **w** in the same figure is parallel to **u** and **v** and has the same length, it points in the opposite direction and is not, therefore, the same as **u** and **v**.

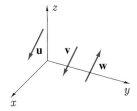

FIGURE 12.35

Components of Vectors

We realized in Chapter 1 that to solve geometric problems, it is often helpful to represent them algebraically. In fact, our entire development of single-variable calculus has hinged on our ability to represent a curve by an algebraic equation and also to draw the curve described by an equation. We now show that vectors can be represented algebraically. Suppose we denote by $\overrightarrow{PQ}$ the vector from P to Q in Figure 12.36. If P and Q have coordinates (x_1, y_1, z_1) and (x_2, y_2, z_2) in the coordinate system shown, then the length of $\overrightarrow{PQ}$ is

$$\text{length of } \overrightarrow{PQ} = \sqrt{(x_2 - x_1)^2 + (y_2 - y_1)^2 + (z_2 - z_1)^2}. \qquad (12.3)$$

SECTION 12.3 VECTORS

FIGURE 12.36

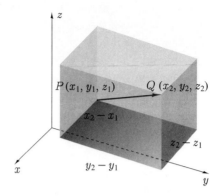

Note also that if we start at point P, proceed $x_2 - x_1$ units in the x-direction, then $y_2 - y_1$ units in the y-direction, and finally $z_2 - z_1$ units in the z-direction, then we arrive at Q. In other words, the three numbers $x_2 - x_1$, $y_2 - y_1$, and $z_2 - z_1$ characterize both the direction and the length of the vector joining P to Q. Because of this we make the following agreement.

Definition 12.3

If the tail of a vector **v** is at $P(x_1, y_1, z_1)$ and its tip is at $Q(x_2, y_2, z_2)$, then **v** shall be represented by the triple of numbers $x_2 - x_1$, $y_2 - y_1$, $z_2 - z_1$. In such a case we enclose the numbers in parentheses and write

$$\mathbf{v} = (x_2 - x_1, y_2 - y_1, z_2 - z_1). \tag{12.4}$$

The equal sign in 12.4 means "is represented by." The number $x_2 - x_1$ is called the x-**component** of **v**, $y_2 - y_1$ the y-**component**, and $z_2 - z_1$ the z-**component**. Vectors in the xy-plane have only an x- and a y-component:

$$\mathbf{v} = (x_2 - x_1, y_2 - y_1),$$

where (x_1, y_1) and (x_2, y_2) are the coordinates of the tail and tip of **v**. Vectors along the x-axis have only an x-component $x_2 - x_1$, where x_1 and x_2 are the coordinates of the tail and tip of **v**.

We now have an algebraic representation for vectors. Each vector has associated with it a set of components that can be found by subtracting the coordinates of its tail from the coordinates of its tip. Conversely, given a set of real numbers (a, b, c), there is one and only one vector with these numbers as components. We can visualize this vector by placing its tail at the origin and its tip at the point with coordinates (a, b, c) (Figure 12.37). Alternatively, we can place the tail of the vector at any point (x_1, y_1, z_1) and its tip at the point $(x_1 + a, y_1 + b, z_1 + c)$.

FIGURE 12.37

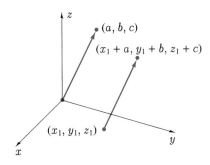

FIGURE 12.38

It is worth emphasizing once again that the same components of a vector are obtained for any point of application whatsoever. For example, the two vectors in Figure 12.38 are identical, and in both cases the components $(2,2)$ are obtained by subtracting the coordinates of the tail from the tip. What we are saying is that Definition 12.2 for equality of vectors can be stated algebraically as follows.

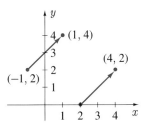

Theorem 12.1

Two vectors are equal if and only if they have the same components.

EXAMPLE 12.2

Find the components of a vector in the xy-plane that has length 5, its tail at the origin, and makes an angle of $\pi/6$ radians with the positive x-axis.

SOLUTION Figure 12.39 illustrates that there are two such vectors, **u** and **v**. From the triangles shown, it is clear that

$$||OP|| = 5\cos(\pi/6) = 5\sqrt{3}/2 \quad \text{and} \quad ||PQ|| = ||PR|| = 5\sin(\pi/6) = \frac{5}{2}.$$

Consequently, Q and R have coordinates $Q = (5\sqrt{3}/2, \frac{5}{2})$ and $R = (5\sqrt{3}/2, -\frac{5}{2})$, and

$$\mathbf{u} = \left(\frac{5\sqrt{3}}{2}, \frac{5}{2}\right), \quad \mathbf{v} = \left(\frac{5\sqrt{3}}{2}, -\frac{5}{2}\right).$$

FIGURE 12.39

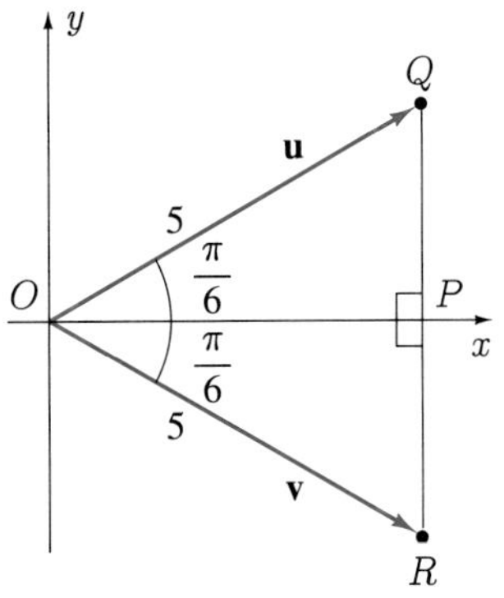

EXAMPLE 12.3

Find the components of the vector in the xy-plane that has its tail at the point $(4,5)$, has length 3, and points directly toward the point $(2,-3)$.

SOLUTION In Figure 12.40, $\|PQ\| = \sqrt{2^2 + 8^2} = 2\sqrt{17}$. Because of similar triangles, we can write that

$$\frac{\|ST\|}{\|PS\|} = \frac{\|QR\|}{\|PQ\|} \quad \text{or} \quad \|ST\| = \frac{3(2)}{2\sqrt{17}} = \frac{3}{\sqrt{17}}.$$

Similarly,

$$\|PT\| = \|PS\| \frac{\|PR\|}{\|PQ\|} = \frac{3(8)}{2\sqrt{17}} = \frac{12}{\sqrt{17}}.$$

Since $\|ST\|$ and $\|PT\|$ represent differences in the x- and y-coordinates of P and S (except for signs), the components of $\vec{PS}$ are $(-3/\sqrt{17}, -12/\sqrt{17})$.

FIGURE 12.40

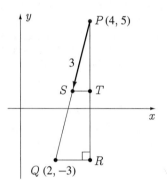

Unit Vectors and Scalar Multiplication

If the x-, y-, and z-components of a vector **v** are (v_x, v_y, v_z), often called the **Cartesian components** of **v**, then these components represent the differences in the coordinates of its tip and tail. But then according to equation 12.3, the length of the vector, which we denote by $|\mathbf{v}|$, is

$$|\mathbf{v}| = \sqrt{v_x^2 + v_y^2 + v_z^2}. \tag{12.5}$$

In words, the length of a vector is the square root of the sum of the squares of its components.

Definition 12.4

A vector **v** is said to be a **unit vector** if it has length equal to one unit; i.e., **v** is a unit vector if

$$v_x^2 + v_y^2 + v_z^2 = 1. \tag{12.6}$$

To indicate that a vector is a unit vector, we place a circumflex ˆ above it: $\hat{\mathbf{v}}$.

EXAMPLE 12.4

What is the length of the vector from $(1,-1,0)$ to $(2,-3,-5)$?

SOLUTION Since the components of the vector are $(1,-2-5)$, its length is

$$\sqrt{(1)^2 + (-2)^2 + (-5)^2} = \sqrt{30}.$$

∎

We now have vectors, which are directed line segments, and real numbers, which are scalars. We know that scalars can be added, subtracted, multiplied, and divided, but can we do the same with vectors, and can we combine vectors and scalars? In the remainder of this section we show how to add and subtract vectors and multiply vectors by scalars; in Sections 12.4 and 12.5 we define two ways to multiply vectors. Each of these operations can be approached either algebraically or geometrically. The geometric approach uses the geometric properties of vectors, namely, length and direction; the algebraic approach uses components of vectors. Neither method is suitable for all situations. Sometimes an idea is more easily introduced with a geometric approach; sometimes an algebraic approach is more suitable. We will choose whichever we feel expresses the idea more clearly. But, whenever we take a geometric approach, we will be careful to follow it up with the algebraic equivalent; conversely, when an algebraic approach is taken, we will always illustrate the geometric significance of the results.

To introduce multiplication of a vector by a scalar, consider the vectors **u** and **v** in Figure 12.41, both of which have their tails at the origin; **v** is in the same direction as **u** but is twice as long as **u**. In such a situation we would like to say that **v** is equal to $2\mathbf{u}$ and write $\mathbf{v} = 2\mathbf{u}$. Vector **w** is in the opposite direction to **r** and is three times as long as **r**, and we would like to denote this vector by $\mathbf{w} = -3\mathbf{r}$. Both of these situations are realized if we adopt the following definition for multiplication of a vector by a scalar.

FIGURE 12.41

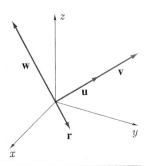

Definition 12.5

If $\lambda > 0$ is a scalar and **v** is a vector, then $\lambda\mathbf{v}$ is the vector that is in the same direction as **v** and λ times as long as **v**; if $\lambda < 0$, then $\lambda\mathbf{v}$ is the vector that is in the opposite direction to **v** and $|\lambda|$ times as long as **v**.

This is a geometric definition of scalar multiplication; it describes the length and direction of $\lambda\mathbf{v}$. We now show that the components of $\lambda\mathbf{v}$ are λ times the components of **v**. In Figure 12.42 we show a box with faces parallel to the coordinate planes and $\lambda\mathbf{v}$ as diagonal, and have given vector **v** components (v_x, v_y, v_z). From the pairs of similar triangles OAB and OCD, and OBE and ODF, we can write that

$$\frac{||OC||}{v_x} = \frac{||CD||}{v_y} = \frac{||OD||}{||OB||} = \frac{||DF||}{v_z} = \frac{|\lambda\mathbf{v}|}{|\mathbf{v}|} = \lambda.$$

Hence

$$||OC|| = \lambda v_x, \quad ||CD|| = \lambda v_y, \quad ||DF|| = \lambda v_z,$$

where $||OC||$, $||CD||$, and $||DF||$ are the components of $\lambda\mathbf{v}$. In other words, the components of $\lambda\mathbf{v}$ are λ times the components of **v**,

$$\lambda\mathbf{v} = \lambda(v_x, v_y, v_z) = (\lambda v_x, \lambda v_y, \lambda v_z). \tag{12.7}$$

FIGURE 12.42

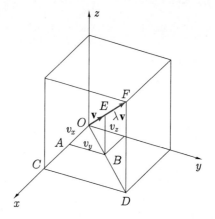

To multiply a vector by a scalar, then, we multiply each component by the scalar.

EXAMPLE 12.5

Find components for the unit vector in the same direction as $\mathbf{v} = (2, -2, 1)$.

SOLUTION The length of $\mathbf{v}$ is $|\mathbf{v}| = \sqrt{(2)^2 + (-2)^2 + 1^2} = 3$. According to our definition of multiplication of a vector by a scalar, the vector $\frac{1}{3}\mathbf{v}$ must have length 1 ($\frac{1}{3}$ that of $\mathbf{v}$) and the same direction as $\mathbf{v}$. Consequently, a unit vector in the same direction as $\mathbf{v}$ is

$$\hat{\mathbf{v}} = \frac{1}{3}\mathbf{v} = \frac{1}{3}(2, -2, 1) = \left(\frac{2}{3}, -\frac{2}{3}, \frac{1}{3}\right).$$

This example illustrates that a unit vector in the same direction as a given vector $\mathbf{v}$ is

$$\hat{\mathbf{v}} = \frac{\mathbf{v}}{|\mathbf{v}|}. \tag{12.8}$$

EXAMPLE 12.6

Find components for the vector of length 4 in the direction opposite that of $\mathbf{v} = (1, 2, -3)$.

SOLUTION Since $|\mathbf{v}| = \sqrt{1 + 4 + 9} = \sqrt{14}$, a unit vector in the same direction as $\mathbf{v}$ is

$$\hat{\mathbf{v}} = \frac{1}{\sqrt{14}}\mathbf{v}.$$

The vector of length 4 in the opposite direction to $\mathbf{v}$ must therefore be

$$(-4)\hat{\mathbf{v}} = \left(\frac{-4}{\sqrt{14}}\right)\mathbf{v} = \left(\frac{-4}{\sqrt{14}}\right)(1, 2, -3) = \left(-\frac{4}{\sqrt{14}}, -\frac{8}{\sqrt{14}}, \frac{12}{\sqrt{14}}\right).$$

Note that with the operation of scalar multiplication, we can simplify the solution of Example 12.3. The vector that points from $(4, 5)$ to $(2, -3)$ is $\mathbf{v} = (-2, -8)$, and therefore the unit vector in this direction is

$$\hat{\mathbf{v}} = \frac{1}{\sqrt{4+64}}(-2,-8) = \frac{1}{2\sqrt{17}}(-2,-8) = \frac{1}{\sqrt{17}}(-1,-4).$$

The required vector of length 3 is

$$3\hat{\mathbf{v}} = \frac{3}{\sqrt{17}}(-1,-4) = \left(-\frac{3}{\sqrt{17}}, -\frac{12}{\sqrt{17}}\right).$$

Addition and Subtraction of Vectors

In Figure 12.43 we show two parallel vectors $\mathbf{u}$ and $\mathbf{v}$ and have placed the tail of $\mathbf{v}$ on the tip of $\mathbf{u}$. It would seem natural to denote the vector that has its tail at the tail of $\mathbf{u}$ and its tip at the tip of $\mathbf{v}$ by $\mathbf{u} + \mathbf{v}$. For instance, if $\mathbf{u}$ and $\mathbf{v}$ were equal, then we would simply be saying that $\mathbf{u} + \mathbf{u} = 2\mathbf{u}$. We use this idea to define addition of vectors even when the vectors are not parallel.

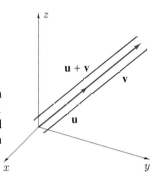

> **Definition 12.6**
>
> The sum of two vectors $\mathbf{u}$ and $\mathbf{v}$, denoted by $\mathbf{u} + \mathbf{v}$, is the vector from the tail of $\mathbf{u}$ to the tip of $\mathbf{v}$ when the tail of $\mathbf{v}$ is placed on the tip of $\mathbf{u}$.

Because the three vectors $\mathbf{u}$, $\mathbf{v}$ and $\mathbf{u} + \mathbf{v}$ then form a triangle (Figure 12.44), we call this **triangular addition** of vectors.

Note that were we to place tails of $\mathbf{u}$ and $\mathbf{v}$ both at the same point (Figure 12.45), and complete the parallelogram with $\mathbf{u}$ and $\mathbf{v}$ as sides, then the diagonal of this parallelogram would also represent the vector $\mathbf{u} + \mathbf{v}$. This is an equivalent method for geometrically finding $\mathbf{u} + \mathbf{v}$, and it is called **parallelogram addition** of vectors.

FIGURE 12.44

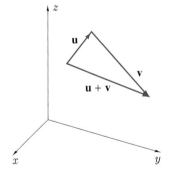

FIGURE 12.45

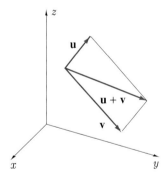

FIGURE 12.46

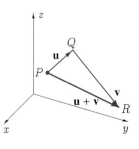

Algebraically, vectors are added component by component; i.e., if $\mathbf{u} = (u_x, u_y, u_z)$ and $\mathbf{v} = (v_x, v_y, v_z)$, then

$$\mathbf{u} + \mathbf{v} = (u_x + v_x, u_y + v_y, u_z + v_z). \qquad (12.9)$$

To verify this we simply note that differences in the coordinates of P and Q in Figure 12.46 are (u_x, u_y, u_z), and differences in those of Q and R are (v_x, v_y, v_z). Consequently, differences in the coordinates of P and R must be $(u_x+v_x, u_y+v_y, u_z+v_z)$.

It is not difficult to show (see Exercise 26) that vector addition and scalar multiplication obey the following rules:

$$\mathbf{u} + \mathbf{v} = \mathbf{v} + \mathbf{u}; \tag{12.10a}$$
$$(\mathbf{u} + \mathbf{v}) + \mathbf{w} = \mathbf{u} + (\mathbf{v} + \mathbf{w}); \tag{12.10b}$$
$$\lambda(\mathbf{u} + \mathbf{v}) = \lambda\mathbf{u} + \lambda\mathbf{v}; \tag{12.10c}$$
$$(\lambda + \mu)\mathbf{v} = \lambda\mathbf{v} + \mu\mathbf{v}. \tag{12.10d}$$

If we denote the vector $(-1)\mathbf{v}$ by $-\mathbf{v}$, then the components of $-\mathbf{v}$ are the negatives of those of $\mathbf{v}$:

$$-\mathbf{v} = (-1)\mathbf{v} = (-1)(v_x, v_y, v_z) = (-v_x, -v_y, -v_z). \tag{12.11}$$

This vector has the same length as $\mathbf{v}$, but is opposite in direction to $\mathbf{v}$ (Figure 12.47).

FIGURE 12.47

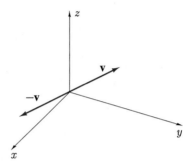

When $\mathbf{v}$ is added to $-\mathbf{v}$, the resultant vector has components that are all zero:

$$\mathbf{v} + (-\mathbf{v}) = (v_x, v_y, v_z) + (-v_x, -v_y, -v_z) = (0, 0, 0).$$

This vector, called the **zero vector**, is denoted by $\mathbf{0}$, and has the property that

$$\mathbf{v} + \mathbf{0} = \mathbf{0} + \mathbf{v} = \mathbf{v} \tag{12.12}$$

for any vector $\mathbf{v}$ whatsoever.

To subtract a vector $\mathbf{v}$ from $\mathbf{u}$, we add $-\mathbf{v}$ to $\mathbf{u}$.

Definition 12.7

The difference $\mathbf{u} - \mathbf{v}$ between two vectors $\mathbf{u}$ and $\mathbf{v}$ is the vector

$$\mathbf{u} - \mathbf{v} = \mathbf{u} + (-\mathbf{v}). \tag{12.13}$$

In Figure 12.48 $\mathbf{u} - \mathbf{v}$ is determined by a triangle, and in Figure 12.49 by a parallelogram. Alternatively, if we denote by $\mathbf{r}$ the vector joining the tip of $\mathbf{v}$ to the tip of $\mathbf{u}$ as in Figure 12.50, then by triangle addition, we have $\mathbf{v} + \mathbf{r} = \mathbf{u}$. Addition of $-\mathbf{v}$ to each side of this equation gives

$$-\mathbf{v} + \mathbf{v} + \mathbf{r} = -\mathbf{v} + \mathbf{u} \quad \text{or} \quad \mathbf{0} + \mathbf{r} = -\mathbf{v} + \mathbf{u}.$$

Thus $\mathbf{r} = \mathbf{u} - \mathbf{v}$, and $\mathbf{u} - \mathbf{v}$ is the vector joining the tip of $\mathbf{v}$ to the tip of $\mathbf{u}$. Definition 12.7 implies that vectors are subtracted component by component:

$$\mathbf{u} - \mathbf{v} = (u_x, u_y, u_z) - (v_x, v_y, v_z) = (u_x - v_x, u_y - v_y, u_z - v_z). \tag{12.14}$$

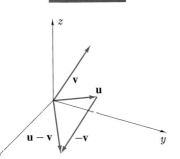

FIGURE 12.48

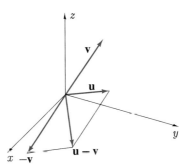

FIGURE 12.49

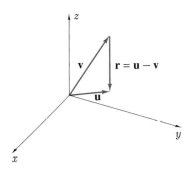

FIGURE 12.50

EXAMPLE 12.7

If $\mathbf{u} = (1,1,1)$, $\mathbf{v} = (-2,3,0)$, and $\mathbf{w} = (-10,10,-2)$, find:

(a) $3\mathbf{u} + 2\mathbf{v} - \mathbf{w}$; (b) $2\mathbf{u} - 4\mathbf{v} + \mathbf{w}$; (c) $|\mathbf{u}|\mathbf{v} + \dfrac{4}{|\mathbf{v}|}\mathbf{w}$.

SOLUTION

(a)
$$3\mathbf{u} + 2\mathbf{v} - \mathbf{w} = 3(1,1,1) + 2(-2,3,0) - (-10,10,-2)$$
$$= (3,3,3) + (-4,6,0) + (10,-10,2)$$
$$= (9,-1,5)$$

(b)
$$2\mathbf{u} - 4\mathbf{v} + \mathbf{w} = 2(1,1,1) - 4(-2,3,0) + (-10,10,-2)$$
$$= (0,0,0) = \mathbf{0}$$

(c) Since $|\mathbf{u}| = \sqrt{1^2 + 1^2 + 1^2} = \sqrt{3}$ and $|\mathbf{v}| = \sqrt{(-2)^2 + 3^2} = \sqrt{13}$,
$$|\mathbf{u}|\mathbf{v} + \frac{4}{|\mathbf{v}|}\mathbf{w} = \sqrt{3}(-2,3,0) + \frac{4}{\sqrt{13}}(-10,10,-2)$$
$$= \left(-2\sqrt{3} - \frac{40}{\sqrt{13}}, 3\sqrt{3} + \frac{40}{\sqrt{13}}, -\frac{8}{\sqrt{13}}\right).$$ ∎

Forces

We have already mentioned that quantities such as temperature, area, and density have associated with them only a magnitude and are therefore represented by scalars. There are many quantities, however, that have associated with them both magnitude and direction, and these are described by vectors. The most notable of this group are forces. When we speak of a force, we mean a push or pull of some size in some specific direction. For example, when the boy in Figure 12.51(a) pulls his wagon, he exerts a force in the direction indicated by the handle. Suppose that he pulls with a force of ten newtons (10 N) and that the angle between the handle and the horizontal is $\pi/4$ radians. To represent this force as a vector $\mathbf{F}_1$, we choose the coordinate system in Figure 12.51(b), and make the agreement that the *length of $\mathbf{F}_1$ be equal to the magnitude of the force*. Since $\mathbf{F}_1$ represents a force of 10 N, it follows that the length of $\mathbf{F}_1$ is 10 units. Furthermore, because $\mathbf{F}_1$ makes an angle of $\pi/4$ radians with the positive x- and y-

axes, the difference in the x-coordinates (and the y-coordinates) of its tip and tail must be $10\cos(\pi/4) = 5\sqrt{2}$. The components of $\mathbf{F}_1$ are therefore $\mathbf{F}_1 = (5\sqrt{2}, 5\sqrt{2})$. If the boy's young sister drags her feet on the ground, then she effectively exerts a force $\mathbf{F}_2$ in the negative x-direction. If the magnitude of this force is 3 N, then its vector representation is $\mathbf{F}_2 = (-3, 0)$. Finally, if the combined weight of the wagon and the girl is 200 N, then the force $\mathbf{F}_3$ of gravity on the wagon and its load is $\mathbf{F}_3 = (0, -200)$. In mechanics we replace the individual forces $\mathbf{F}_1$, $\mathbf{F}_2$ and $\mathbf{F}_3$ by a single force that has the same effect on the wagon as all three forces combined. This force, called the **resultant** of $\mathbf{F}_1$, $\mathbf{F}_2$, and $\mathbf{F}_3$, is represented by the vector $\mathbf{F}$, which is the sum of the vectors $\mathbf{F}_1$, $\mathbf{F}_2$, and $\mathbf{F}_3$:

$$\mathbf{F} = \mathbf{F}_1 + \mathbf{F}_2 + \mathbf{F}_3$$
$$= (5\sqrt{2}, 5\sqrt{2}) + (-3, 0) + (0, -200)$$
$$= (5\sqrt{2} - 3, 5\sqrt{2} - 200).$$

The magnitude of this force corresponds to the length of $\mathbf{F}$,

$$|\mathbf{F}| = \sqrt{(5\sqrt{2} - 3)^2 + (5\sqrt{2} - 200)^2} = 193.0,$$

and must therefore be 193.0 N. Its direction is shown in Figure 12.51(c), where

$$\theta = \operatorname{Tan}^{-1}\left(\frac{200 - 5\sqrt{2}}{5\sqrt{2} - 3}\right) = 1.55 \text{ radians.}$$

FIGURE 12.51

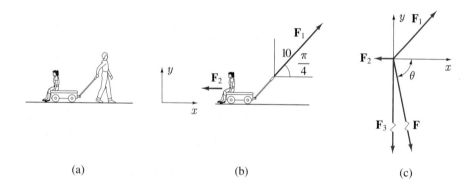

(a)   (b)   (c)

If two water skiers S_1 and S_2 exert forces of 500 N and 600 N, respectively, on the boat shown in Figure 12.52(a), what is the resultant force on the boat?

FIGURE 12.52

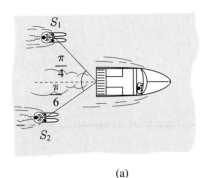
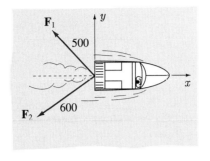

(a) (b)

SOLUTION If we introduce the coordinate system in Figure 12.52(b), then lengths of the force vectors $\mathbf{F}_1$ and $\mathbf{F}_2$ exerted by S_1 and S_2 are 500 and 600 units, respectively. The components of $\mathbf{F}_1$ and $\mathbf{F}_2$ are

$$\mathbf{F}_1 = \left(-\frac{500}{\sqrt{2}}, \frac{500}{\sqrt{2}}\right) \qquad \mathbf{F}_2 = \left(-\frac{600\sqrt{3}}{2}, -\frac{600}{2}\right)$$
$$= (-250\sqrt{2}, 250\sqrt{2}); \qquad = (-300\sqrt{3}, -300).$$

The resultant force $\mathbf{F}$ on the boat is then the sum of $\mathbf{F}_1$ and $\mathbf{F}_2$:

$$\mathbf{F} = \mathbf{F}_1 + \mathbf{F}_2 = (-250\sqrt{2} - 300\sqrt{3}, 250\sqrt{2} - 300)\,\text{N}.$$

By the x-, y-, and z-components (v_x, v_y, v_z) of a vector $\mathbf{v}$, we mean that if we start at a point P (Figure 12.53) and proceed v_x units in the x-direction, v_y units in the y-direction, and v_z units in the z-direction to a point Q, then $\mathbf{v}$ is the directed line segment joining P and Q. To phrase this another way, we introduce three special vectors parallel to the coordinate axes. We define $\hat{\mathbf{i}}$ as a unit vector in the positive x-direction, $\hat{\mathbf{j}}$ as a unit vector in the positive y-direction, and $\hat{\mathbf{k}}$ as a unit vector in the positive z-direction. We have shown these vectors with their tails at the origin in Figure 12.54, and it is clear that their components are

$$\hat{\mathbf{i}} = (1, 0, 0), \quad \hat{\mathbf{j}} = (0, 1, 0), \quad \hat{\mathbf{k}} = (0, 0, 1). \tag{12.15}$$

FIGURE 12.53

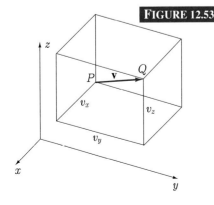

FIGURE 12.54

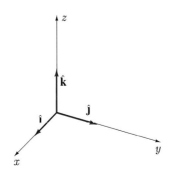

But note, then, that we can write the vector $\mathbf{v} = (v_x, v_y, v_z)$ in the form

$$\mathbf{v} = (v_x, 0, 0) + (0, v_y, 0) + (0, 0, v_z)$$
$$= v_x(1, 0, 0) + v_y(0, 1, 0) + v_z(0, 0, 1)$$
$$= v_x\hat{\mathbf{i}} + v_y\hat{\mathbf{j}} + v_z\hat{\mathbf{k}}.$$

In other words, every vector in space can be written as a linear combination of the three vectors $\hat{\mathbf{i}}$, $\hat{\mathbf{j}}$, and $\hat{\mathbf{k}}$ (i.e., as a constant times $\hat{\mathbf{i}}$ plus a constant times $\hat{\mathbf{j}}$ plus a constant times $\hat{\mathbf{k}}$). Furthermore, the constants multiplying $\hat{\mathbf{i}}$, $\hat{\mathbf{j}}$, and $\hat{\mathbf{k}}$ are precisely the Cartesian components of the vector. This result is equally clear geometrically. In Figure 12.53, we have shown the vector $\mathbf{v}$ from $\mathbf{P}$ to $\mathbf{Q}$. If we define points $\mathbf{A}$ and $\mathbf{B}$ as shown in Figure 12.55, then

$$\mathbf{v} = \overrightarrow{PQ} = \overrightarrow{PB} + \overrightarrow{BQ} = \overrightarrow{PA} + \overrightarrow{AB} + \overrightarrow{BQ}.$$

But because $\overrightarrow{PA}$ is a vector in the positive x-direction and has length v_x, it follows that $\overrightarrow{PA} = v_x\hat{\mathbf{i}}$. Similarly, $\overrightarrow{AB} = v_y\hat{\mathbf{j}}$ and $\overrightarrow{BQ} = v_z\hat{\mathbf{k}}$, and therefore

$$\mathbf{v} = v_x\hat{\mathbf{i}} + v_y\hat{\mathbf{j}} + v_z\hat{\mathbf{k}}. \qquad (12.16)$$

To say then that v_x, v_y and v_z are the x-, y-, and z-components of a vector $\mathbf{v}$ is equivalent to saying that $\mathbf{v}$ can be written in the form 12.16.

Some authors refer to v_x, v_y, and v_z as the **scalar components** of the vector $\mathbf{v}$, and the vectors $v_x\hat{\mathbf{i}}$, $v_y\hat{\mathbf{j}}$, and $v_z\hat{\mathbf{k}}$ as the **vector components** of $\mathbf{v}$. By "component," we always mean "scalar component."

Vectors in the xy-plane have only an x- and a y-component, and can therefore be written in terms of $\hat{\mathbf{i}}$ and $\hat{\mathbf{j}}$. If $\mathbf{v} = (v_x, v_y)$, then we write equivalently that $\mathbf{v} = v_x\hat{\mathbf{i}} + v_y\hat{\mathbf{j}}$ (Figure 12.56.). Vectors along the x-axis have only an x-component and can therefore be written in the form $\mathbf{v} = v_x\hat{\mathbf{i}}$.

FIGURE 12.55

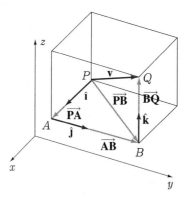

FIGURE 12.56

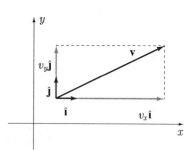

We use this new notation in the following example.

EXAMPLE 12.9

The force **F** exerted on a point charge q_1 coulombs by a charge q_2 coulombs is defined by Coulomb's law as

$$\mathbf{F} = \frac{q_1 q_2}{4\pi\epsilon_0 r^2}\hat{\mathbf{r}} \ \ \text{N},$$

where ϵ_0 is a positive constant, r is the distance in metres between the charges, and $\hat{\mathbf{r}}$ is a unit vector in the direction from q_2 to q_1 (Figure 12.57). When q_1 and q_2 are both positive charges or both negative charges, then **F** is repulsive, and when one is positive and the other is negative, **F** is attractive. In particular, suppose that charges of 2 C and -2 C are placed at $(0,0,0)$ and $(3,0,0)$, respectively, and a third charge of 1 C is placed at $(1,1,1)$. According to Coulomb's law, the 2-C charge will exert a repulsive force on the 1-C charge, and the -2-C charge will exert an attractive force on the 1-C charge. Find the resultant of these two forces on the 1-C charge.

SOLUTION If $\mathbf{F}_1$ is the force exerted on the 1-C charge by the -2-C charge (Figure 12.58), then

$$\mathbf{F}_1 = \frac{(1)(-2)}{4\pi\epsilon_0 r^2}\hat{\mathbf{r}},$$

where the distance between the charges is $r = \sqrt{(-2)^2 + 1^2 + 1^2} = \sqrt{6}$. The vector from $(3,0,0)$ to $(1,1,1)$ is $(-2,1,1)$, and therefore

$$\hat{\mathbf{r}} = \frac{1}{\sqrt{6}}(-2\hat{\mathbf{i}} + \hat{\mathbf{j}} + \hat{\mathbf{k}}).$$

Consequently,

$$\mathbf{F}_1 = \frac{-2}{4\pi\epsilon_0(6)} \cdot \frac{1}{\sqrt{6}}(-2\hat{\mathbf{i}} + \hat{\mathbf{j}} + \hat{\mathbf{k}}) = \frac{1}{12\sqrt{6}\pi\epsilon_0}(2\hat{\mathbf{i}} - \hat{\mathbf{j}} - \hat{\mathbf{k}}).$$

Similarly, the force $\mathbf{F}_2$ exerted on the 1-C charge by the charge at the origin is

$$\mathbf{F}_2 = \frac{(1)(2)}{4\pi\epsilon_0(3)} \cdot \frac{1}{\sqrt{3}}(\hat{\mathbf{i}} + \hat{\mathbf{j}} + \hat{\mathbf{k}}) = \frac{1}{6\sqrt{3}\pi\epsilon_0}(\hat{\mathbf{i}} + \hat{\mathbf{j}} + \hat{\mathbf{k}}).$$

The resultant of these forces is

$$\mathbf{F} = \mathbf{F}_1 + \mathbf{F}_2 = \frac{1}{12\sqrt{6}\pi\epsilon_0}\{(2 + 2\sqrt{2})\hat{\mathbf{i}} + (2\sqrt{2} - 1)\hat{\mathbf{j}} + (2\sqrt{2} - 1)\hat{\mathbf{k}}\} \ \text{N}.$$

FIGURE 12.57

FIGURE 12.58

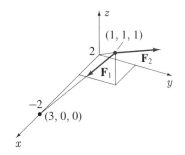

EXERCISES 12.3

If $\mathbf{u} = (1,3,6)$, $\mathbf{v} = (-2,0,4)$, and $\mathbf{w} = (4,3,-2)$, find components for the vector in Exercises 1–10.

1. $3\mathbf{u} - 2\mathbf{v}$
2. $2\mathbf{w} + 3\mathbf{v}$
3. $\mathbf{w} - 3\mathbf{u} - 3\mathbf{v}$
4. $\hat{\mathbf{v}}$
5. $2\hat{\mathbf{w}} - 3\mathbf{v}$
6. $|\mathbf{v}|\mathbf{v} - 2|\hat{\mathbf{v}}|\mathbf{w}$
7. $(15 - 2|\mathbf{w}|)(\mathbf{u} + \mathbf{v})$
8. $|3\mathbf{u}|\mathbf{v} - |-2\mathbf{v}|\mathbf{u}$
9. $|2\mathbf{u} + 3\mathbf{v} - \mathbf{w}|\hat{\mathbf{w}}$
10. $\dfrac{\mathbf{v} - \mathbf{w}}{|\mathbf{v} + \mathbf{w}|}$

If $\mathbf{u} = 2\hat{\mathbf{i}} + \hat{\mathbf{j}}$ and $\mathbf{v} = -\hat{\mathbf{i}} + 3\hat{\mathbf{j}}$, find components of the vector in Exercises 11–14 and illustrate the vector geometrically.

11. $\mathbf{u} + \mathbf{v}$
12. $\mathbf{u} - \mathbf{v}$
13. $2\hat{\mathbf{u}}$
14. $\hat{\mathbf{v}} + \hat{\mathbf{u}}$

In Exercises 15–24 find the Cartesian components for the spatial vector described. In each case, draw the vector.

15. From $(1,3,2)$ to $(-1,4,5)$
16. With length 5 in the positive x-direction
17. With length 2 in the negative z-direction
18. With tail at $(1,1,1)$, length 3, and pointing toward the point $(1,3,5)$
19. With positive y-component, length 1, and parallel to the line through $(1,3,6)$ and $(-2,1,4)$
20. In the same direction as the vector from $(1,0,-1)$ to $(3,2,-4)$ but only half as long
21. With positive and equal x- and y-components, length 10, and z-component equal to 4
22. Has its tail at the origin, makes angles of $\pi/3$ and $\pi/4$ radians with the positive x- and y-axes respectively, and has length $5/2$
23. From $(1,3,-2)$ to the midpoint of the line segment joining $(2,4,-3)$ and $(1,5,6)$
24. Has its tail at the origin, makes equal angles with the positive coordinate axes, has all positive components, and has length 2
25. If P, Q, and R are the points with coordinates $(3,2,-1)$, $(0,1,4)$, and $(6,5,-2)$ respectively, find the coordinates of a point S in order that $\overrightarrow{PQ} = \overrightarrow{RS}$.
26. Prove that vector addition and scalar multiplication have the properties in equations 12.10.
27. Draw all spatial vectors of length 1 that have equal x- and y-components and tails at the origin.
28. Draw all spatial vectors of length 2 that have their tails at the origin and make an angle of $\pi/4$ with the positive z-axis.
29. If $\mathbf{u} = 3\hat{\mathbf{i}} + 2\hat{\mathbf{j}} - 4\hat{\mathbf{k}}$, and $\mathbf{v} = \hat{\mathbf{i}} + 6\hat{\mathbf{j}} + 5\hat{\mathbf{k}}$, find scalars λ and ρ so that the vector $\mathbf{w} = 5\hat{\mathbf{i}} - 18\hat{\mathbf{j}} - 32\hat{\mathbf{k}}$ can be written in the form $\mathbf{w} = \lambda\mathbf{u} + \rho\mathbf{v}$.
30. Find a vector $\mathbf{T}$ of length 3 along the tangent line to the curve $y = x^2$ at the point $(2,4)$ (Figure 12.59).

FIGURE 12.59

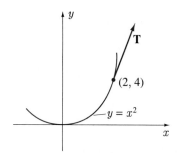

31. Use Coulomb's law (see Example 12.9) to find the force on a charge of 2 coulombs at the origin due to equal charges of 3 coulombs at the points $(1,1,2)$ and $(2,-1,-2)$.
32. Newton's universal law of gravitation states that the force of attraction $\mathbf{F}$, in newtons, exerted on a mass of m kilograms by a mass of M kilograms is

$$\mathbf{F} = \frac{GmM}{r^2}\hat{\mathbf{r}},$$

where $G = 6.67 \times 10^{-11}$ is a constant, r is the distance in metres between the masses, and $\hat{\mathbf{r}}$ is a unit vector in the direction from m to M. If point masses, each of 5 kilograms, are situated at $(5,1,3)$ and $(-1,2,1)$, what is the resultant force on a mass of 10 kilograms at $(2,2,2)$?

33. Illustrate geometrically the triangle inequality for vectors $|\mathbf{u} + \mathbf{v}| \leq |\mathbf{u}| + |\mathbf{v}|$. Prove the result algebraically. Vectors $\mathbf{u}$, $\mathbf{v}$, and $\mathbf{w}$ are said to be linearly dependent if there exist three scalars a, b, and c, not all zero, such that $a\mathbf{u} + b\mathbf{v} + c\mathbf{w} = \mathbf{0}$. If this equation can only be satisfied with $a = b = c = 0$, the vectors are said to be linearly independent. In Exercises 34–37 determine whether the vectors are linearly dependent or linearly independent.

34. $\mathbf{u} = (1, 1, 1)$, $\mathbf{v} = (2, 1, 3)$, $\mathbf{w} = (4, 2, 6)$

35. $\mathbf{u} = (1, 1, 1)$, $\mathbf{v} = (2, 1, 3)$, $\mathbf{w} = (1, 6, 4)$

36. $\mathbf{u} = (-1, 3, -5)$, $\mathbf{v} = (2, 4, -1)$, $\mathbf{w} = (3, 11, -7)$

37. $\mathbf{u} = (4, 2, 6)$, $\mathbf{v} = (1, 3, -2)$, $\mathbf{w} = (7, 1, 4)$

38. Use vectors to show that the line segment joining the midpoints of two sides of a triangle is parallel to the third side and its length is one-half the length of the third side (Figure 12.60).

FIGURE 12.60

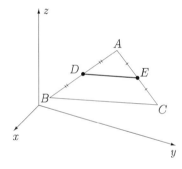

39. Use vectors to show that the medians of a triangle (Figure 12.61) all meet in a point with coordinates
$$\left(\frac{x_1 + x_2 + x_3}{3}, \frac{y_1 + y_2 + y_3}{3}, \frac{z_1 + z_2 + z_3}{3} \right).$$

FIGURE 12.61

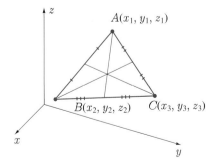

40. If n point masses m_i are located at points (x_i, y_i) in the xy-plane, equations 7.29 and 7.30 define the centre of mass $(\overline{x}, \overline{y})$ of the system. Show that these two scalar equations are represented by the one vector equation
$$M(\overline{x}, \overline{y}) = \sum_{i=1}^{n} m_i \mathbf{r}_i$$
where $\mathbf{r}_i$ is the vector joining the origin to the point (x_i, y_i).

41. If $\mathbf{u} = u_x \hat{\mathbf{i}} + u_y \hat{\mathbf{j}}$ and $\mathbf{v} = v_x \hat{\mathbf{i}} + v_y \hat{\mathbf{j}}$, show that every vector $\mathbf{w}$ in the xy-plane can be written in the form $\mathbf{w} = \lambda \mathbf{u} + \rho \mathbf{v}$ provided $u_x v_y - u_y v_x \neq 0$.

42. Two springs (with constants k_1 and k_2 and unstretched lengths l) are fixed at points A and B (Figure 12.62). They are joined to a sleeve that slides along the x-axis. Find the resultant force of the springs on the sleeve at any point between O and C.

FIGURE 12.62

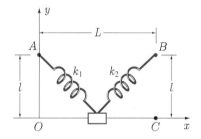

43. In Figure 12.63(a), the xy-plane is the interface between two materials that both transmit light. If a ray of light strikes the surface in a direction defined by the unit vector $\hat{\mathbf{u}} = (u_x, u_y, u_z)$, then some of the light is reflected along a vector $\hat{\mathbf{v}}$, and some is refracted along $\hat{\mathbf{w}}$. The three vectors $\hat{\mathbf{u}}$, $\hat{\mathbf{v}}$, and $\hat{\mathbf{w}}$ all lie in a plane that is perpendicular to the xy-plane.

(a) If the angle of incidence i (Figure 12.63b) is equal to the angle of reflection ϕ, find the components for $\hat{\mathbf{v}}$ in terms of those of $\hat{\mathbf{u}}$.

(b) If the angle of refraction θ is related to the angle of incidence by $n_1 \sin i = n_2 \sin \theta$, where n_1 and n_2 are the indices of refraction of the two materials, find the components of $\hat{\mathbf{w}}$.

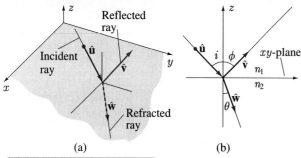

FIGURE 12.63

(a)

(b)

44. Vectors $\mathbf{v}_1, \mathbf{v}_2, \ldots, \mathbf{v}_n$ are drawn from the centre of a regular n-sided polygon in the plane to each of its vertices. Show that the sum of these vectors is the zero vector.

SECTION 12.4

The Scalar Product

In Section 12.3 we learned how to multiply a vector by a scalar and how to add and subtract vectors. The next question naturally is, "Can vectors be multiplied?" The answer is yes, and in fact we define two products for vectors, one of which yields a scalar and the other a vector. The first is defined algebraically as follows.

Definition 12.8

The **scalar product** (**dot product** or **inner product**) of two vectors $\mathbf{u}$ and $\mathbf{v}$ with Cartesian components (u_x, u_y, u_z) and (v_x, v_y, v_z) is defined as

$$\mathbf{u} \cdot \mathbf{v} = u_x v_x + u_y v_y + u_z v_z. \tag{12.17}$$

If $\mathbf{u}$ and $\mathbf{v}$ have only x- and y-components, Definition 12.8 reduces to

$$\mathbf{u} \cdot \mathbf{v} = u_x v_x + u_y v_y, \tag{12.18}$$

and if they have only x-components, it becomes

$$\mathbf{u} \cdot \mathbf{v} = u_x v_x. \tag{12.19}$$

It is straightforward to check that

$$\hat{\mathbf{i}} \cdot \hat{\mathbf{i}} = \hat{\mathbf{j}} \cdot \hat{\mathbf{j}} = \hat{\mathbf{k}} \cdot \hat{\mathbf{k}} = 1, \tag{12.20a}$$

$$\hat{\mathbf{i}} \cdot \hat{\mathbf{j}} = \hat{\mathbf{j}} \cdot \hat{\mathbf{k}} = \hat{\mathbf{i}} \cdot \hat{\mathbf{k}} = 0, \tag{12.20b}$$

and that the scalar product is commutative and distributive:

$$\mathbf{u} \cdot \mathbf{v} = \mathbf{v} \cdot \mathbf{u}, \tag{12.21a}$$

$$\mathbf{u} \cdot (\lambda \mathbf{v} + \rho \mathbf{w}) = \lambda (\mathbf{u} \cdot \mathbf{v}) + \rho (\mathbf{u} \cdot \mathbf{w}). \tag{12.21b}$$

EXAMPLE 12.10

If $\mathbf{u} = (-2, 1, 3)$ and $\mathbf{v} = (3, -2, -1)$, evaluate each of the following:

(a) $\mathbf{u} \cdot \mathbf{v}$; (b) $3\mathbf{u} \cdot (2\mathbf{u} - 4\mathbf{v})$.

SOLUTION

(a) $\qquad \mathbf{u} \cdot \mathbf{v} = (-2)(3) + (1)(-2) + (3)(-1) = -11$

(b) $\qquad 3\mathbf{u} \cdot (2\mathbf{u} - 4\mathbf{v}) = (-6, 3, 9) \cdot (-16, 10, 10)$
$$= (-6)(-16) + (3)(10) + (9)(10)$$
$$= 216.$$

By taking the scalar product of a vector $\mathbf{v} = (v_x, v_y, v_z)$ with itself, we obtain

$$\mathbf{v} \cdot \mathbf{v} = v_x^2 + v_y^2 + v_z^2 = |\mathbf{v}|^2. \qquad (12.22)$$

Because Definition 12.8 for the scalar product of two vectors $\mathbf{u}$ and $\mathbf{v}$ is phrased in terms of the components of $\mathbf{u}$ and $\mathbf{v}$, and these components depend on the coordinate system used, it follows that this definition also depends on the fact that we have used Cartesian coordinates. Were we to use a different set of coordinates (such as polar coordinates), then the definition of $\mathbf{u} \cdot \mathbf{v}$ in terms of components in that coordinate system would be different. For this reason we now find a geometric definition for the scalar product (which is therefore independent of coordinate systems).

Theorem 12.2

If two nonzero vectors $\mathbf{u}$ and $\mathbf{v}$ are placed tail to tail, and θ is the angle between these vectors ($0 \leq \theta \leq \pi$), then

$$\mathbf{u} \cdot \mathbf{v} = |\mathbf{u}||\mathbf{v}| \cos \theta. \qquad (12.23)$$

FIGURE 12.64

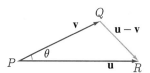

Proof The cosine law applied to the triangle in Figure 12.64 gives

$$|\overrightarrow{QR}|^2 = |\overrightarrow{PQ}|^2 + |\overrightarrow{PR}|^2 - 2|\overrightarrow{PQ}||\overrightarrow{PR}| \cos \theta,$$

or

$$|\mathbf{u} - \mathbf{v}|^2 = |\mathbf{v}|^2 + |\mathbf{u}|^2 - 2|\mathbf{v}||\mathbf{u}| \cos \theta.$$

Consequently,

$$|\mathbf{u}||\mathbf{v}| \cos \theta = \frac{1}{2}\{|\mathbf{u}|^2 + |\mathbf{v}|^2 - |\mathbf{u} - \mathbf{v}|^2\},$$

and if (u_x, u_y, u_z) and (v_x, v_y, v_z) are the Cartesian components of $\mathbf{u}$ and $\mathbf{v}$, then

$$|\mathbf{u}||\mathbf{v}| \cos \theta = \frac{1}{2}\{(u_x^2 + u_y^2 + u_z^2) + (v_x^2 + v_y^2 + v_z^2)$$
$$- [(u_x - v_x)^2 + (u_y - v_y)^2 + (u_z - v_z)^2]\}$$
$$= u_x v_x + u_y v_y + u_z v_z$$
$$= \mathbf{u} \cdot \mathbf{v}.$$

An immediate consequence of this result is the following.

Corollary Two nonzero vectors **u** and **v** are perpendicular if and only if

$$\mathbf{u} \cdot \mathbf{v} = 0. \qquad (12.24)$$

For example, the vectors $\mathbf{u} = (1,2,-1)$ and $\mathbf{v} = (4,2,8)$ are perpendicular since $\mathbf{u} \cdot \mathbf{v} = (1)(4) + (2)(2) + (-1)(8) = 0$.

Expression 12.23 doesn't just tell us whether or not the angle between two vectors is $\pi/2$ radians; it can also be used to determine the exact angle between any two nonzero vectors **u** and **v**, simply by solving for θ:

$$\cos\theta = \frac{\mathbf{u} \cdot \mathbf{v}}{|\mathbf{u}||\mathbf{v}|}. \qquad (12.25)$$

Since principal values of the inverse cosine function lie between 0 and π, precisely the range for θ, we can write that

$$\theta = \mathrm{Cos}^{-1}\left(\frac{\mathbf{u} \cdot \mathbf{v}}{|\mathbf{u}||\mathbf{v}|}\right). \qquad (12.26)$$

EXAMPLE 12.11

Find the angle between the vectors $\mathbf{u} = (2,-3,1)$ and $\mathbf{v} = (5,2,4)$.

SOLUTION According to formula 12.26,

$$\theta = \mathrm{Cos}^{-1}\left(\frac{\mathbf{u} \cdot \mathbf{v}}{|\mathbf{u}||\mathbf{v}|}\right) = \mathrm{Cos}^{-1}\left(\frac{10-6+4}{\sqrt{4+9+1}\sqrt{25+4+16}}\right) = \mathrm{Cos}^{-1}\left(\frac{8}{3\sqrt{70}}\right) = 1.25.$$

EXAMPLE 12.12

Find the angle between the lines $x + 2y = 3$ and $4x - 3y = 5$ in the xy-plane.

SOLUTION Since the slope of $x + 2y = 3$ is $-\frac{1}{2}$, a vector along this line is $\mathbf{u} = (-2,1)$. Similarly, a vector along $4x - 3y = 5$ is $\mathbf{v} = (3,4)$. If θ is the angle between these vectors, and therefore between the lines (Figure 12.65), then

$$\theta = \mathrm{Cos}^{-1}\left(\frac{\mathbf{u} \cdot \mathbf{v}}{|\mathbf{u}||\mathbf{v}|}\right) = \mathrm{Cos}^{-1}\left(\frac{-6+4}{\sqrt{5}\sqrt{25}}\right) = \mathrm{Cos}^{-1}\left(\frac{-2}{5\sqrt{5}}\right) = 1.75$$

radians. The acute angle between the lines is $\pi - 1.75 = 1.39$ radians. Formula 1.14 gives the same result.

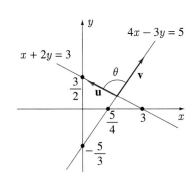

FIGURE 12.65

Planes

Scalar products can be used to find equations for planes in space. A plane can be characterized in various ways: by two intersecting lines, by a line and a point not on the line, or by three noncollinear points. For our present purposes, we use the fact that given a point $P(x_1, y_1, z_1)$ and a vector (A, B, C) (Figure 12.66), there is one and only one plane through P that is perpendicular to (A, B, C).

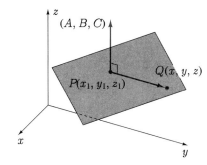

FIGURE 12.66

To find the equation of this plane we note that if $Q(x, y, z)$ is any other point in the plane, then the vector $\overrightarrow{PQ} = (x - x_1, y - y_1, z - z_1)$ lies in the plane. But $\overrightarrow{PQ}$ must then be perpendicular to (A, B, C); hence by the corollary to Theorem 12.2,

$$(A, B, C) \cdot (x - x_1, y - y_1, z - z_1) = 0.$$

Because this equation must be satisfied by every point (x, y, z) in the plane (and at the same time is not satisfied by any point not in the plane), it must be the equation of the plane through P and perpendicular to (A, B, C). If we expand the scalar product, we obtain the equation of the plane in the form

$$A(x - x_1) + B(y - y_1) + C(z - z_1) = 0.$$

This result is worth stating as a theorem.

> **Theorem 12.3**
>
> An equation for the plane through (x_1, y_1, z_1) perpendicular to the vector (A, B, C) is
> $$A(x - x_1) + B(y - y_1) + C(z - z_1) = 0. \quad (12.27)$$

Equation 12.27 can also be written in the form

$$Ax + By + Cz + D = 0, \qquad (12.28)$$

where $D = -(Ax_1 + By_1 + Cz_1)$, and this equation is said to be **linear** in x, y, and z. We have shown then that every plane has a linear equation, and the coefficients A, B, and C of x, y, and z in the equation are the components of a vector (A, B, C) that is perpendicular to the plane. Conversely, every linear equation in the form of 12.28 describes a plane. Instead of saying that (A, B, C) is perpendicular to the plane $Ax + By + Cz + D = 0$, we often say that (A, B, C) is **normal** to the plane or that (A, B, C) is a **normal vector** to the plane.

EXAMPLE 12.13

Find an equation for the plane through the point $(4, -3, 5)$ and normal to the vector $(4, -8, 3)$.

SOLUTION According to 12.27, the equation of the plane is

$$4(x - 4) - 8(y + 3) + 3(z - 5) = 0, \quad \text{or,} \quad 4x - 8y + 3z = 55.$$

∎

EXAMPLE 12.14

Determine whether the planes $x + 2y - 4z = 10$ and $2x + 4y - 8z = 11$ are parallel.

SOLUTION Normal vectors to these planes are $\mathbf{N}_1 = (1, 2, -4)$ and $\mathbf{N}_2 = (2, 4, -8)$. Since $\mathbf{N}_2 = 2\mathbf{N}_1$, the normal vectors are in the same direction, and therefore, the planes are parallel.

∎

Resolution of Vectors into Perpendicular Directions

In many applications of mathematics, we are called on to find distances between geometric objects. For example, we already have formulas 1.7 and 12.1 for the distances between points in a plane and points in space. In a plane we might want to find the distance from a point to a line or between two parallel lines. In space, there are many other distances that might prove useful: distance from a point to a plane or a line, between parallel planes, between nonintersecting lines, etc. Each of these distances can be calculated in a number of ways, but the easiest method is to use components of vectors.

We have defined what is meant by Cartesian components of a vector. In particular, Cartesian components (v_x, v_y, v_z) of a vector $\mathbf{v}$ are those scalars that multiply $\hat{\mathbf{i}}$, $\hat{\mathbf{j}}$ and $\hat{\mathbf{k}}$ so that $\mathbf{v} = v_x \hat{\mathbf{i}} + v_y \hat{\mathbf{j}} + v_z \hat{\mathbf{k}}$. They can be represented in terms of scalar products as

$$v_x = \mathbf{v} \cdot \hat{\mathbf{i}}, \quad v_y = \mathbf{v} \cdot \hat{\mathbf{j}}, \quad v_z = \mathbf{v} \cdot \hat{\mathbf{k}}. \qquad (12.29)$$

Geometrically they can be found by drawing $\mathbf{v}$, $\hat{\mathbf{i}}$, $\hat{\mathbf{j}}$, and $\hat{\mathbf{k}}$ all at the origin and dropping perpendiculars from the tip of $\mathbf{v}$ to the x-, y-, and z-axes (Figure 12.67). We now generalize our definition of a component along an axis to define the component of a vector in any direction whatsoever.

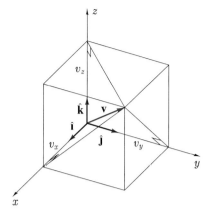

FIGURE 12.67

> To define the component of a vector **v** in a direction **u**, we place **v** and **u** tail to tail at a point P and draw the perpendicular from the tip of **v** to the line containing **u** (Figure 12.68). The directed distance PR is called the component of **v** in the direction **u**. If R is on the same side of P as the tip of **u**, PR is taken as positive; and if R is on that side of P opposite to the tip of **u**, PR is negative.

Definition 12.9

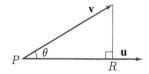

FIGURE 12.68

In Figure 12.68 the component of **v** in direction **u** is positive, and in Figure 12.69 the component of **v** in direction **u** is negative. Note that the length of **u** is irrelevant; it is only the direction of **u** that determines the component of **v** in the direction **u**. If θ is the angle between **u** and **v**, then

$$PR = |\mathbf{v}| \cos \theta.$$

FIGURE 12.69

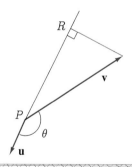

The right-hand side of this equation looks very much like the scalar product of **v** and a vector that makes an angle θ with **v**. It lacks only the length of this second vector. Clearly, **u** is a vector that makes an angle θ with **v**, but we *cannot* write $PR = |\mathbf{v}||\mathbf{u}| \cos \theta$, since the length of **u** need not be 1. If, however, $\hat{\mathbf{u}}$ is the unit vector in the same direction as **u**, then we can write

$$PR = |\mathbf{v}||\hat{\mathbf{u}}| \cos \theta = \mathbf{v} \cdot \hat{\mathbf{u}}. \qquad (12.30)$$

In other words, we have the following theorem.

> The component of a vector **v** in a direction **u** is $\mathbf{v} \cdot \hat{\mathbf{u}}$, where $\hat{\mathbf{u}}$ is the unit vector in direction **u**.

Theorem 12.4

This result agrees with equation 12.29 for the x-, y-, and z- components of **v**.

But what good does it do to have a component of a vector in every direction? For one thing, we can represent a vector in ways other than equation 12.16. For instance, consider the two perpendicular unit vectors $\hat{\mathbf{e}}_1$ and $\hat{\mathbf{e}}_2$ in Figure 12.70. The components of vector $\mathbf{v} = 3\hat{\mathbf{i}} + 4\hat{\mathbf{j}}$ in the directions of $\hat{\mathbf{e}}_1$ and $\hat{\mathbf{e}}_2$ are

$$\mathbf{v} \cdot \hat{\mathbf{e}}_1 = \frac{3}{\sqrt{2}} + \frac{4}{\sqrt{2}} = \frac{7}{\sqrt{2}}$$

and
$$\mathbf{v} \cdot \hat{\mathbf{e}}_2 = -\frac{3}{\sqrt{2}} + \frac{4}{\sqrt{2}} = \frac{1}{\sqrt{2}}.$$

Geometrically it is clear that we can express $\mathbf{v}$ as the sum
$$\mathbf{v} = \frac{7}{\sqrt{2}}\hat{\mathbf{e}}_1 + \frac{1}{\sqrt{2}}\hat{\mathbf{e}}_2.$$

In other words, $\mathbf{v}$, and every other vector in the plane, can be expressed as a linear combination of the perpendicular vectors $\hat{\mathbf{e}}_1$ and $\hat{\mathbf{e}}_2$ (instead of $\hat{\mathbf{i}}$ and $\hat{\mathbf{j}}$). We say that $\mathbf{v}$ has been **resolved** into components along $\hat{\mathbf{e}}_1$ and $\hat{\mathbf{e}}_2$. We will find this idea of resolving vectors in terms of perpendicular vectors other than $\hat{\mathbf{i}}$, $\hat{\mathbf{j}}$, and $\hat{\mathbf{k}}$ very important in Section 12.10.

FIGURE 12.70

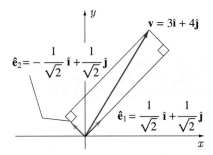

Distances Between Points, Lines, and Planes

We stated earlier that components of vectors in arbitrary directions can be used to calculate distances between various geometric objects. The first problem that we consider is the distance from a point $P(x,y)$ to a line l: $Ax+By+C=0$ in the xy-plane (Figure 12.71). There are many ways to calculate this distance, and by this distance we mean the perpendicular (or, equivalently, the shortest) distance d from P to l. One way is to use the formula $|Ax+By+C|/\sqrt{A^2+B^2}$ developed in Exercise 45 in Section 1.3. A second way is to find the function that describes the distance from P to any point on l, and then determine the absolute minimum for the function. Another method is to use components of vectors. If Q is any point on l, then d is the component of $\overrightarrow{PQ}$ in direction $\overrightarrow{PR}$; i.e.,

$$d = \overrightarrow{PQ} \cdot \widehat{\overrightarrow{PR}},$$

where $\widehat{\overrightarrow{PR}}$ is the unit vector in the direction of $\overrightarrow{PR}$. To illustrate, suppose that P has coordinates $(3,5)$ and l is the line $2y = x + 2$ (Figure 12.72). Since $Q(0,1)$ is a point on l, the vector $\overrightarrow{PQ}$ has Cartesian components $(-3,-4)$. The distance from P to l is the component of $\overrightarrow{PQ}$ in the direction of $\overrightarrow{PR}$. Now we do not know the coordinates of R (otherwise the problem would be finished), but we really do not need them. What we want is the unit vector along $\overrightarrow{PR}$. Since the slope of l is $\frac{1}{2}$, a vector along l is $(2,1)$. A vector perpendicular to l is therefore $(1,-2)$ (and note that the scalar product of $(2,1)$ and $(1,-2)$ vanishes). Consequently, the unit vector along $\overrightarrow{PR}$ is

$$\widehat{\overrightarrow{PR}} = \frac{1}{\sqrt{5}}(1,-2) \quad \text{and} \quad d = (-3,-4) \cdot \frac{1}{\sqrt{5}}(1,-2) = \frac{5}{\sqrt{5}} = \sqrt{5}.$$

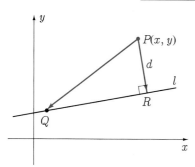

FIGURE 12.71

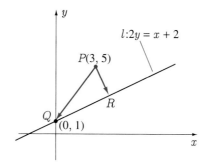

FIGURE 12.72

The following examples illustrate that the same technique may be applied to other distances.

EXAMPLE 12.15

Find the distance between the lines $x - 3y = 4$ and $x - 3y = -5$.

SOLUTION The required distance can be calculated as the distance from $P(4,0)$ (a point on $x - 3y = 4$) to $x - 3y = -5$ (Figure 12.73). For $Q(-5,0)$ on $x - 3y = -5$, $\overrightarrow{PQ} = (-9, 0)$. The required distance is the component of $\overrightarrow{PQ}$ in the direction $\overrightarrow{PR}$. Since $(3,1)$ is a vector along either of the lines (each has slope $\frac{1}{3}$), a vector perpendicular to the lines is $(-1, 3)$. Consequently,

$$d = \overrightarrow{PQ} \cdot \widehat{PR} = (-9, 0) \cdot \frac{1}{\sqrt{10}}(-1, 3) = \frac{9}{\sqrt{10}}.$$

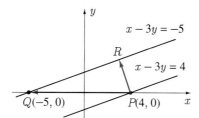

FIGURE 12.73

EXAMPLE 12.16

Find the distance from the point $(1, 2, 5)$ to the plane $x + y + 2z = 4$.

SOLUTION If Q is any point in the plane, say $(4, 0, 0)$, then the distance d from $P(1, 2, 5)$ to the plane is the component of $\overrightarrow{PQ}$ in the direction $\overrightarrow{PR}$ normal to the plane; i.e., $d = \overrightarrow{PQ} \cdot \widehat{PR}$ (Figure 12.74). Since $(1, 1, 2)$ is a vector normal to the plane,

$$\widehat{PR} = \frac{1}{\sqrt{6}}(-1, -1, -2)$$

and

$$d = (3, -2, -5) \cdot \frac{1}{\sqrt{6}}(-1, -1, -2) = \frac{9}{\sqrt{6}}.$$

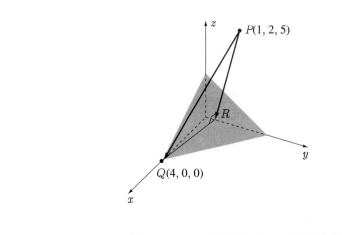

FIGURE 12.74

We will find distances between other geometric objects in Section 12.5 when we have defined a "vector product" of vectors. But the approach continues to be through components of vectors in arbitrary directions.

Work

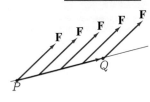

FIGURE 12.75

In Section 7.4 we described work as the product of force and distance. Now that we have represented forces by vectors, we can be more precise with our definition. In particular, if a particle moves along the line in Figure 12.75 from P to Q, then vector $\overrightarrow{PQ}$ represents its displacement. If a constant force $\mathbf{F}$ acts on the particle during this motion, then the work done by $\mathbf{F}$ is defined as

$$W = \mathbf{F} \cdot \overrightarrow{PQ}. \tag{12.31}$$

It is important to keep in mind exactly when this definition of work can be used: for constant forces acting along straight lines, and by constant $\mathbf{F}$, we mean that $\mathbf{F}$ is constant in both magnitude and direction. Note that when $\mathbf{F}$ and $\overrightarrow{PQ}$ are both in the same direction, then the angle between the vectors is zero. In this case,

$$W = |\mathbf{F}||\overrightarrow{PQ}|\cos(0) = |\mathbf{F}||\overrightarrow{PQ}|,$$

and this is essentially the equation dealt with in Section 7.4. When $\mathbf{F}$ and $\overrightarrow{PQ}$ are not in the same direction, then

$$W = |\mathbf{F}||\overrightarrow{PQ}|\cos\theta.$$

Since $|\mathbf{F}|\cos\theta$ is the component of $\mathbf{F}$ along $\overrightarrow{PQ}$, this equation simply states that when $\mathbf{F}$ and $\overrightarrow{PQ}$ are not in the same direction, $|\overrightarrow{PQ}|$ should be multiplied by the component of $\mathbf{F}$ in the direction of $\overrightarrow{PQ}$.

EXAMPLE 12.17

If the boy in Figure 12.76 pulls the wagon handle with a force of 10 N at an angle of $\pi/4$ radians with the horizontal, how much work does he do in walking a distance of 20 m in a straight line?

SOLUTION The force $\mathbf{F}$ exerted by the boy has magnitude $|\mathbf{F}| = 10$, and points in a direction that makes an angle of $\pi/4$ radians with the displacement vector. If $\mathbf{d}$ is the displacement vector, then

$$W = \mathbf{F} \cdot \mathbf{d} = |\mathbf{F}||\mathbf{d}|\cos\theta = 10(20)\cos\left(\frac{\pi}{4}\right) = \frac{200}{\sqrt{2}} = 100\sqrt{2} \text{ J}.$$

FIGURE 12.76

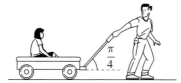

When motion is along a straight line, but $\mathbf{F}$ is *not* constant in direction or magnitude or both, we must use integration. The following example illustrates such a situation.

EXAMPLE 12.18

The spring in Figure 12.77 is fixed at A and moves the sleeve frictionlessly along the rod from B to C. If the spring is unstretched when the sleeve is at C, find the work done by the spring.

FIGURE 12.77

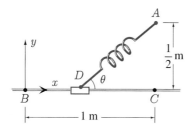

SOLUTION If we set up a coordinate system as shown, then at position D, the force $\mathbf{F}$ exerted by the spring on the sleeve has magnitude

$$|\mathbf{F}| = k\left[\sqrt{(1-x)^2 + \frac{1}{4}} - \frac{1}{2}\right],$$

where k is the spring constant. Since the spring is always stretched during the motion, the direction of $\mathbf{F}$ is along the vector $\overrightarrow{DA}$. Clearly, then, $\mathbf{F}$ changes in both magnitude and direction as the sleeve moves from B to C. For a small displacement dx at position D, the amount of work done by $\mathbf{F}$ is (approximately)

$$\mathbf{F} \cdot (dx\hat{\mathbf{i}}) = |\mathbf{F}|dx(\cos\theta) = k\left[\sqrt{(1-x)^2 + \frac{1}{4}} - \frac{1}{2}\right]dx \frac{1-x}{\sqrt{(1-x)^2 + \frac{1}{4}}}$$

$$= k(1-x)\left\{1 - \frac{\frac{1}{2}}{\sqrt{(1-x)^2 + \frac{1}{4}}}\right\}dx.$$

The total work done by the spring as the sleeve moves from B to C must therefore be

$$W = \int_0^1 k(1-x)\left\{1 - \frac{1}{2\sqrt{(1-x)^2 + \frac{1}{4}}}\right\}dx$$

$$= k\int_0^1 \left\{1 - x - \frac{1-x}{2\sqrt{(1-x)^2 + \frac{1}{4}}}\right\}dx$$

$$= k\left\{x - \frac{x^2}{2} + \frac{1}{2}\sqrt{(1-x)^2 + \frac{1}{4}}\right\}_0^1$$

$$= \frac{k}{4}(3 - \sqrt{5}) \text{ J.}$$

EXERCISES 12.4

If $\mathbf{u} = 2\hat{\mathbf{i}} - 3\hat{\mathbf{j}} + \hat{\mathbf{k}}$, $\mathbf{v} = \hat{\mathbf{j}} - \hat{\mathbf{k}}$, and $\mathbf{w} = 6\hat{\mathbf{i}} - 2\hat{\mathbf{j}} + 3\hat{\mathbf{k}}$, evaluate the scalar or find the components of the vector in Exercises 1–10.

1. $\mathbf{u} \cdot \mathbf{v}$
2. $(\mathbf{v} \cdot \mathbf{w})\mathbf{u}$
3. $(2\mathbf{u} - 3\mathbf{v}) \cdot \mathbf{w}$
4. $2\hat{\mathbf{i}} \cdot \hat{\mathbf{u}}$
5. $|2\mathbf{u}|\mathbf{v} \cdot \mathbf{w}$
6. $(3\mathbf{u} - 4\mathbf{w}) \cdot (2\hat{\mathbf{i}} + 3\mathbf{u} - 2\mathbf{v})$
7. $\mathbf{w} \cdot \hat{\mathbf{w}}$
8. $\dfrac{(105\mathbf{u} + 240\mathbf{v}) \cdot (105\mathbf{u} + 240\mathbf{v})}{|105\mathbf{u} + 240\mathbf{v}|^2}$
9. $|\mathbf{u} - \mathbf{v} + \hat{\mathbf{k}}|(\hat{\mathbf{j}} + \mathbf{w}) \cdot \hat{\mathbf{k}}$
10. $\mathbf{u} \cdot \mathbf{v} + \mathbf{v} \cdot \mathbf{w} - (\mathbf{u} + \mathbf{w}) \cdot \mathbf{v}$

In Exercises 11–14 determine whether the vectors are perpendicular.

11. $(1, 2), (3, 5)$
12. $(2, 4), (-8, 4)$
13. $(1, 3, 6), (-2, 1, -4)$
14. $(2, 3, -6), (-6, 6, 1)$

In Exercises 15–20 find the angle between the vectors.

15. $(3, 4), (2, -5)$
16. $(1, 6), (-4, 7)$
17. $(4, 2, 3), (1, 5, 6)$
18. $(3, 1, -1), (-2, 1, 4)$
19. $(2, 0, 5), (0, 3, 0)$
20. $(1, 3, -2), (-2, -6, 4)$

Find the equation for the plane in Exercises 21 and 22.

21. Through the point $(1, -1, 3)$ and normal to the vector $(4, 3, -2)$
22. Through the point $(2, 1, 5)$ and normal to the vector joining $(2, 1, 5)$ and $(4, 2, 3)$

In Exercises 23–28 find the indicated distance.

23. From the point $(1, 2)$ to the line $x - y = 1$

24. From the point $(3, 2)$ to the line $2x + 3y = 18$

25. From the point $(-1, 4)$ to the line $4x + 3y = 8$

26. From the point $(1, 3, 4)$ to the plane $x + y - 2z = 0$

27. Between the lines $x + 4 = y$ and $3x + 7 = 3y$

28. Between the planes $2x + 3y - z = 15$ and $4x + 6y - 2z = 7$

29. Verify the results in equations 12.21.

30. Show that a vector perpendicular to the line $Ax + By + C = 0$ in the xy-plane is (A, B).

31. The acute angle between two intersecting planes is defined as the acute angle between their normals. Find the acute angle between the planes $x - 2y + 4z = 6$ and $2x + y = z + 4$.

32. Use equation 12.22 to prove the parallelogram law:
$$|\mathbf{u} + \mathbf{v}|^2 + |\mathbf{u} - \mathbf{v}|^2 = 2|\mathbf{u}|^2 + 2|\mathbf{v}|^2.$$
Why is this called the parallelogram law?

33. The angles between a vector $\mathbf{v} = (v_x, v_y, v_z)$ and the vectors $\hat{\mathbf{i}}, \hat{\mathbf{j}}$, and $\hat{\mathbf{k}}$ are called the direction angles $\alpha, \beta,$ and γ of $\mathbf{v}$. Show that
$$\alpha = \operatorname{Cos}^{-1}\left(\frac{\mathbf{v} \cdot \hat{\mathbf{i}}}{|\mathbf{v}|}\right) = \operatorname{Cos}^{-1}\left(\frac{v_x}{|\mathbf{v}|}\right);$$
$$\beta = \operatorname{Cos}^{-1}\left(\frac{\mathbf{v} \cdot \hat{\mathbf{j}}}{|\mathbf{v}|}\right) = \operatorname{Cos}^{-1}\left(\frac{v_y}{|\mathbf{v}|}\right);$$
$$\gamma = \operatorname{Cos}^{-1}\left(\frac{\mathbf{v} \cdot \hat{\mathbf{k}}}{|\mathbf{v}|}\right) = \operatorname{Cos}^{-1}\left(\frac{v_z}{|\mathbf{v}|}\right).$$

34. Find direction angles for the vectors (a) $(1, 2, -3)$; (b) $(0, 1, -3)$; (c) $(-1, -2, 6)$.

35. If $\mathbf{F}$ is a constant force, show that the work done by $\mathbf{F}$ on an object moving around any closed polygon is zero.

In Exercises 36 and 37 verify that $\hat{\mathbf{v}}$ and $\hat{\mathbf{w}}$ are perpendicular, and then resolve $\mathbf{u}$ into components along $\hat{\mathbf{v}}$ and $\hat{\mathbf{w}}$; that is, find λ and ρ so that $\mathbf{u} = \lambda\hat{\mathbf{v}} + \rho\hat{\mathbf{w}}$.

36. $\mathbf{u} = (2, 1); \hat{\mathbf{v}} = (1/\sqrt{2}, 1/\sqrt{2}), \hat{\mathbf{w}} = (1/\sqrt{2}, -1/\sqrt{2})$

37. $\mathbf{u} = 3\hat{\mathbf{i}} - 2\hat{\mathbf{j}}; \hat{\mathbf{v}} = (\hat{\mathbf{i}} - 2\hat{\mathbf{j}})/\sqrt{5}, \hat{\mathbf{w}} = (2\hat{\mathbf{i}} + \hat{\mathbf{j}})/\sqrt{5}$

In Exercises 38 and 39 verify that $\hat{\mathbf{u}}, \hat{\mathbf{v}}$, and $\hat{\mathbf{w}}$ are mutually perpendicular, and then resolve $\mathbf{r}$ into components along $\hat{\mathbf{u}}, \hat{\mathbf{v}}$, and $\hat{\mathbf{w}}$.

38. $\mathbf{r} = (1, 3, -4); \hat{\mathbf{u}} = (2, 1, 0)/\sqrt{5},$
$\hat{\mathbf{v}} = (-1, 2, 3)/\sqrt{14}, \hat{\mathbf{w}} = (3, -6, 5)/\sqrt{70}$

39. $\mathbf{r} = 2\hat{\mathbf{i}} - \hat{\mathbf{k}}; \hat{\mathbf{u}} = (\hat{\mathbf{i}} + \hat{\mathbf{j}} + \hat{\mathbf{k}})/\sqrt{3},$
$\hat{\mathbf{v}} = (\hat{\mathbf{i}} + \hat{\mathbf{j}} - 2\hat{\mathbf{k}})/\sqrt{6}, \hat{\mathbf{w}} = (\hat{\mathbf{i}} - \hat{\mathbf{j}})/\sqrt{2}$

40. If $\mathbf{u} = (3, 2), \mathbf{v} = (1, -3)$, and $\mathbf{w} = (6, 2)$, verify that $\mathbf{v}$ and $\mathbf{w}$ are perpendicular, and find scalars λ and ρ so that $\mathbf{u} = \lambda\mathbf{v} + \rho\mathbf{w}$.

41. If $\mathbf{u} = (1, 0, 1), \mathbf{v} = (1, 1, -1), \mathbf{w} = (-1, 2, 1),$ and $\mathbf{r} = (-2, -3, 4)$, verify that $\mathbf{u}, \mathbf{v}$, and $\mathbf{w}$ are mutually perpendicular, and find scalars $\lambda, \rho,$ and μ so that $\mathbf{r} = \lambda\mathbf{u} + \rho\mathbf{v} + \mu\mathbf{w}$.

42. Find the equation of a plane normal to $\hat{\mathbf{i}} - 2\hat{\mathbf{j}} + 3\hat{\mathbf{k}}$ and two units from the point $(1, 2, 3)$.

43. If a, b, and c (all positive constants) are the intercepts of a plane with the x-, y-, and z-axes, and p is the length of the perpendicular from the origin to the plane, show that
$$\frac{1}{p^2} = \frac{1}{a^2} + \frac{1}{b^2} + \frac{1}{c^2}.$$

44. Repeat Example 12.18 if the spring has an unstretched length l that is less than the length of AC.

45. Two positive charges q_1 and q_2 are placed at positions $(5, 5)$ and $(-2, 3)$ in the xy-plane respectively. A third positive charge q_3 is moved along the x-axis from $x = 1$ to $x = -1$. Find the total work done by the electrostatic forces of q_1 and q_2 on q_3.

46. When the rocket in Figure 12.78 passes close to the spherical asteroid, it is attracted to the asteroid by a gravitational force with magnitude GmM/r^2 where m and M are the masses of the rocket and asteroid, r is the distance from the rocket to the centre of the asteroid, and G is Newton's gravitational constant. Determine the work the rocket must do against this force in order to follow the straight-line path from A to B.

FIGURE 12.78

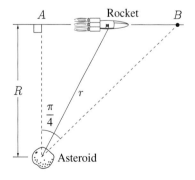

47. Prove that the (undirected) distance from a point $P(x_1, y_1, z_1)$ to the plane $Ax + By + Cz + D = 0$

is
$$\frac{|Ax_1 + By_1 + Cz_1 + D|}{\sqrt{A^2 + B^2 + C^2}}.$$

48. Prove that the (undirected) distance between two parallel planes $Ax + By + Cz + D_1 = 0$ and $Ax + By + Cz + D_2 = 0$ is
$$\frac{|D_1 - D_2|}{\sqrt{A^2 + B^2 + C^2}}.$$

49. Show that the vector $(|\mathbf{v}|\mathbf{u} + |\mathbf{u}|\mathbf{v})/||\mathbf{u}|\mathbf{v} + |\mathbf{v}|\mathbf{u}|$ is a unit vector which bisects the angle between $\mathbf{u}$ and $\mathbf{v}$.

SECTION 12.5

The Vector Product

In many applications of vectors we need a vector perpendicular to two given vectors. For instance, suppose we must find the equation for the plane passing through the three given points P_1, P_2 and P_3 in Figure 12.79. Clearly, $\overrightarrow{P_1P_2}$ and $\overrightarrow{P_1P_3}$ are vectors that lie in the plane, and to find a normal to the plane we therefore require a vector perpendicular to both $\overrightarrow{P_1P_2}$ and $\overrightarrow{P_1P_3}$.

FIGURE 12.79

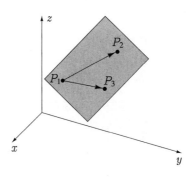

Consider, then, finding a vector $\mathbf{r} = (a, b, c)$ perpendicular to two given vectors $\mathbf{u} = (u_x, u_y, u_z)$ and $\mathbf{v} = (v_x, v_y, v_z)$. The corollary to Theorem 12.2 requires a, b, and c to satisfy
$$0 = \mathbf{r} \cdot \mathbf{u} = au_x + bu_y + cu_z,$$
$$0 = \mathbf{r} \cdot \mathbf{v} = av_x + bv_y + cv_z.$$

When we solve these equations, we find that there is an infinite number of solutions all represented by
$$a = s(u_yv_z - u_zv_y), \qquad b = s(u_zv_x - u_xv_z), \qquad c = s(u_xv_y - u_yv_x),$$

where s is any real number. In other words, any vector of the form
$$\mathbf{r} = s(u_yv_z - u_zv_y, u_zv_x - u_xv_z, u_xv_y - u_yv_x)$$

is perpendicular to the vectors $\mathbf{u} = (u_x, u_y, u_z)$ and $\mathbf{v} = (v_x, v_y, v_z)$. When we choose $s = 1$, the resulting vector is called the vector product of $\mathbf{u}$ and $\mathbf{v}$.

> **Definition 12.10**
>
> The **vector product (cross product** or **outer product)** of two vectors **u** and **v** with Cartesian components (u_x, u_y, u_z) and (v_x, v_y, v_z) is defined as
>
> $$\mathbf{u} \times \mathbf{v} = (u_y v_z - u_z v_y)\hat{\mathbf{i}} + (u_z v_x - u_x v_z)\hat{\mathbf{j}} + (u_x v_y - u_y v_x)\hat{\mathbf{k}}. \quad (12.32)$$

To eliminate the need for memorizing the exact placing of the six components of **u** and **v** in this definition, we borrow the notation for determinants from linear algebra. A brief discussion of determinants and their properties is given in Appendix F. We set up a 3×3 determinant with $\hat{\mathbf{i}}, \hat{\mathbf{j}}$, and $\hat{\mathbf{k}}$ across the top row and the components of **u** and **v** across the second and third rows:

$$\begin{vmatrix} \hat{\mathbf{i}} & \hat{\mathbf{j}} & \hat{\mathbf{k}} \\ u_x & u_y & u_z \\ v_x & v_y & v_z \end{vmatrix}.$$

In actual fact, this is not a determinant, since three of the entries are vectors and six are scalars. If we ignore this fact, and apply the rules for the expansion of a 3×3 determinant along its first row (namely, $\hat{\mathbf{i}}$ times the 2×2 determinant obtained by deleting the row and column containing $\hat{\mathbf{i}}$, minus $\hat{\mathbf{j}}$ times the 2×2 determinant obtained by deleting the row and column containing $\hat{\mathbf{j}}$, plus $\hat{\mathbf{k}}$ times the 2×2 determinant obtained by deleting the row and column containing $\hat{\mathbf{k}}$), we obtain

$$\begin{vmatrix} \hat{\mathbf{i}} & \hat{\mathbf{j}} & \hat{\mathbf{k}} \\ u_x & u_y & u_z \\ v_x & v_y & v_z \end{vmatrix} = \begin{vmatrix} u_y & u_z \\ v_y & v_z \end{vmatrix} \hat{\mathbf{i}} - \begin{vmatrix} u_x & u_z \\ v_x & v_z \end{vmatrix} \hat{\mathbf{j}} + \begin{vmatrix} u_x & u_y \\ v_x & v_y \end{vmatrix} \hat{\mathbf{k}}.$$

But the definition for the value of a 2×2 determinant is

$$\begin{vmatrix} a & b \\ c & d \end{vmatrix} = ad - bc.$$

Consequently,

$$\begin{vmatrix} \hat{\mathbf{i}} & \hat{\mathbf{j}} & \hat{\mathbf{k}} \\ u_x & u_y & u_z \\ v_x & v_y & v_z \end{vmatrix} = (u_y v_z - u_z v_y)\hat{\mathbf{i}} - (u_x v_z - u_z v_x)\hat{\mathbf{j}} + (u_x v_y - u_y v_x)\hat{\mathbf{k}},$$

and this is the same as the right-hand side of 12.32. We may therefore write, as a memory-saving device, that

$$\mathbf{u} \times \mathbf{v} = \begin{vmatrix} \hat{\mathbf{i}} & \hat{\mathbf{j}} & \hat{\mathbf{k}} \\ u_x & u_y & u_z \\ v_x & v_y & v_z \end{vmatrix}, \quad (12.33)$$

so long as we evaluate the right-hand side using the general rules for the expansion of a determinant along its first row.

For example, if $\mathbf{u} = (1, -1, 2)$ and $\mathbf{v} = (2, 3, -5)$, then

$$\mathbf{u} \times \mathbf{v} = \begin{vmatrix} \hat{\mathbf{i}} & \hat{\mathbf{j}} & \hat{\mathbf{k}} \\ 1 & -1 & 2 \\ 2 & 3 & -5 \end{vmatrix} = (5 - 6)\hat{\mathbf{i}} - (-5 - 4)\hat{\mathbf{j}} + (3 + 2)\hat{\mathbf{k}} = -\hat{\mathbf{i}} + 9\hat{\mathbf{j}} + 5\hat{\mathbf{k}}.$$

It is straightforward to verify that

$$\hat{\mathbf{i}} \times \hat{\mathbf{j}} = \hat{\mathbf{k}}, \quad \hat{\mathbf{j}} \times \hat{\mathbf{k}} = \hat{\mathbf{i}}, \quad \hat{\mathbf{k}} \times \hat{\mathbf{i}} = \hat{\mathbf{j}}, \qquad (12.34)$$

and that the cross-product is anticommutative and distributive:

$$\mathbf{u} \times \mathbf{v} = -\mathbf{v} \times \mathbf{u}, \qquad (12.35\text{a})$$

$$\mathbf{u} \times (\lambda \mathbf{v} + \rho \mathbf{w}) = \lambda(\mathbf{u} \times \mathbf{v}) + \rho(\mathbf{u} \times \mathbf{w}). \qquad (12.35\text{b})$$

Our preliminary analysis indicated that $\mathbf{u} \times \mathbf{v}$ is perpendicular to both $\mathbf{u}$ and $\mathbf{v}$. The following theorem relates the length of $\mathbf{u} \times \mathbf{v}$ to the lengths of $\mathbf{u}$ and $\mathbf{v}$.

Theorem 12.5

If θ is the angle between two vectors $\mathbf{u}$ and $\mathbf{v}$, then

$$|\mathbf{u} \times \mathbf{v}| = |\mathbf{u}||\mathbf{v}| \sin \theta. \qquad (12.36)$$

Proof Since θ is an angle between 0 and π, $\sin \theta$ must be positive, and we can write from equation 12.25 that

$$\sin \theta = \sqrt{1 - \cos^2 \theta} = \sqrt{1 - \left(\frac{\mathbf{u} \cdot \mathbf{v}}{|\mathbf{u}||\mathbf{v}|} \right)^2} = \frac{1}{|\mathbf{u}||\mathbf{v}|} \sqrt{|\mathbf{u}|^2 |\mathbf{v}|^2 - (\mathbf{u} \cdot \mathbf{v})^2}.$$

Consequently,

$$|\mathbf{u}||\mathbf{v}| \sin \theta = \sqrt{|\mathbf{u}|^2 |\mathbf{v}|^2 - (\mathbf{u} \cdot \mathbf{v})^2}.$$

If $\mathbf{u} = (u_x, u_y, u_z)$ and $\mathbf{v} = (v_x, v_y, v_z)$, then

$$\begin{aligned}
|\mathbf{u}|^2 |\mathbf{v}|^2 \sin^2 \theta &= (u_x^2 + u_y^2 + u_z^2)(v_x^2 + v_y^2 + v_z^2) - (u_x v_x + u_y v_y + u_z v_z)^2 \\
&= u_x^2(v_x^2 + v_y^2 + v_z^2) + u_y^2(v_x^2 + v_y^2 + v_z^2) \\
&\quad + u_z^2(v_x^2 + v_y^2 + v_z^2) - (u_x^2 v_x^2 + u_y^2 v_y^2 + u_z^2 v_z^2 \\
&\quad + 2 u_x v_x u_y v_y + 2 u_x v_x u_z v_z + 2 u_y v_y u_z v_z) \\
&= (u_y^2 v_z^2 - 2 u_y v_y u_z v_z + u_z^2 v_y^2) + (u_x^2 v_z^2 - 2 u_x v_x u_z v_z + u_z^2 v_x^2) \\
&\quad + (u_x^2 v_y^2 - 2 u_x v_x u_y v_y + u_y^2 v_x^2) \\
&= (u_y v_z - u_z v_y)^2 + (u_x v_z - u_z v_x)^2 + (u_x v_y - u_y v_x)^2 \\
&= |\mathbf{u} \times \mathbf{v}|^2,
\end{aligned}$$

or

$$|\mathbf{u} \times \mathbf{v}| = |\mathbf{u}||\mathbf{v}| \sin \theta.$$

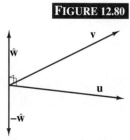

FIGURE 12.80

We now know that $\mathbf{u} \times \mathbf{v}$ is perpendicular to $\mathbf{u}$ and $\mathbf{v}$, and has length $|\mathbf{u}||\mathbf{v}| \sin \theta$, where θ is the angle between $\mathbf{u}$ and $\mathbf{v}$. Figure 12.80 illustrates that there are only two directions that are perpendicular to $\mathbf{u}$ and $\mathbf{v}$, and one is the negative of the other. Let us denote by $\hat{\mathbf{w}}$ the unit vector along that direction which is perpendicular to $\mathbf{u}$ and $\mathbf{v}$ and is determined by the right-hand rule (curl the fingers of the right hand from $\mathbf{u}$ toward $\mathbf{v}$ and the thumb points in the direction $\hat{\mathbf{w}}$).

We now show that $\mathbf{u} \times \mathbf{v}$ always points in the direction determined by the right-hand rule (i.e., in the direction $\hat{\mathbf{w}}$ rather than $-\hat{\mathbf{w}}$). To see this we place $\mathbf{u}$ and $\mathbf{v}$ tail to tail and establish a coordinate system with this common point as origin and the positive x-axis along $\mathbf{u}$ (Figure 12.81a). Let the plane determined by $\mathbf{u}$ and $\mathbf{v}$ be the xy-plane. In this coordinate system, $\mathbf{u}$ has only an x-component, $\mathbf{u} = u_x\hat{\mathbf{i}}$ ($u_x > 0$), and $\mathbf{v}$ has only x- and y- components, $\mathbf{v} = v_x\hat{\mathbf{i}} + v_y\hat{\mathbf{j}}$. The cross-product of $\mathbf{u}$ and $\mathbf{v}$ is therefore

$$\mathbf{u} \times \mathbf{v} = \begin{vmatrix} \hat{\mathbf{i}} & \hat{\mathbf{j}} & \hat{\mathbf{k}} \\ u_x & 0 & 0 \\ v_x & v_y & 0 \end{vmatrix} = u_x v_y \hat{\mathbf{k}}.$$

For $\mathbf{v}$ in Figure 12.81(a), v_y is clearly positive and therefore $u_x v_y$, the component of $\mathbf{u} \times \mathbf{v}$, is also positive. But then $\mathbf{u} \times \mathbf{v}$ is indeed determined by the right-hand rule since $\hat{\mathbf{w}} = \hat{\mathbf{k}}$.

FIGURE 12.81

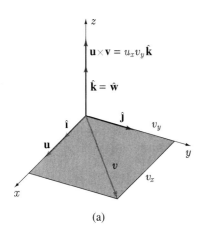

(a)

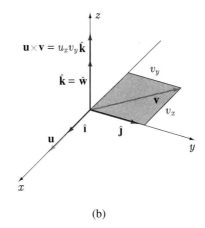

(b)

When the tip of $\mathbf{v}$ lies in the second quadrant of the xy-plane (Figure 12.81b), then $\mathbf{u} \times \mathbf{v}$ is once again in the positive z-direction, the direction of $\hat{\mathbf{w}}$. When the tip of $\mathbf{v}$ is in the third or fourth quadrant (Figure 12.82), $v_y < 0$, and $\mathbf{u} \times \mathbf{v}$ therefore has a negative z-component. But in this case $\hat{\mathbf{w}} = -\hat{\mathbf{k}}$, and the direction of $\mathbf{u} \times \mathbf{v}$ is once again determined by the right-hand rule.

What we have now established is the following coordinate-free definition for the vector product of two vectors $\mathbf{u}$ and $\mathbf{v}$:

FIGURE 12.82

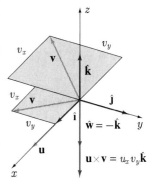

$$\mathbf{u} \times \mathbf{v} = (|\mathbf{u}||\mathbf{v}|\sin\theta)\hat{\mathbf{w}}. \qquad (12.37)$$

The unit vector $\hat{\mathbf{w}}$ defines the direction of $\mathbf{u} \times \mathbf{v}$ and the factor $|\mathbf{u}||\mathbf{v}|\sin\theta$ is its length.

The fact that the vector product $\mathbf{u} \times \mathbf{v}$ is perpendicular to both $\mathbf{u}$ and $\mathbf{v}$ makes it a powerful tool in many applications.

EXAMPLE 12.19

Find the equation of the plane through the points $(1,2,3)$, $(-2,0,4)$, and $(5,2,-1)$.

SOLUTION If the plane passes through $P(1,2,3)$, $Q(-2,0,4)$, and $R(5,2,-1)$, then $\overrightarrow{PQ} = (-3,-2,1)$ and $\overrightarrow{PR} = (4,0,-4)$ are vectors that lie in the plane. It follows that a vector normal to the plane is

$$\overrightarrow{PQ} \times \overrightarrow{PR} = \begin{vmatrix} \hat{\mathbf{i}} & \hat{\mathbf{j}} & \hat{\mathbf{k}} \\ -3 & -2 & 1 \\ 4 & 0 & -4 \end{vmatrix} = (8-0)\hat{\mathbf{i}} - (12-4)\hat{\mathbf{j}} + (0+8)\hat{\mathbf{k}} = 8(\hat{\mathbf{i}} - \hat{\mathbf{j}} + \hat{\mathbf{k}}).$$

Since $\hat{\mathbf{i}} - \hat{\mathbf{j}} + \hat{\mathbf{k}}$ is therefore normal to the plane, the equation of the plane is

$$(1, -1, 1) \cdot (x - 1, y - 2, z - 3) = 0, \quad \text{or}, \quad x - y + z = 2.$$

∎

Lines

In Section 12.2 we indicated that space curves can be described by two simultaneous equations in x, y, and z, and that such a representation describes the curve as the intersection of two surfaces. In this section we use vectors to discuss straight lines. A straight line in space is uniquely characterized by a point on it and a vector parallel to it. For instance, there is one and only one line l through the point $P_0(x_0, y_0, z_0)$ and in the direction $\mathbf{v} = (v_x, v_y, v_z)$ in Figure 12.83.

FIGURE 12.83

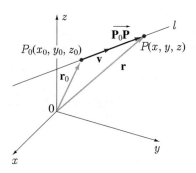

If $P(x, y, z)$ is any point on this line, then the vector $\mathbf{r}$ joining the origin to P has components $\mathbf{r} = (x, y, z)$. Now this vector can be expressed as the sum of $\mathbf{r}_0$, the vector from O to P_0 and $\overrightarrow{P_0P}$:

$$\mathbf{r} = \mathbf{r}_0 + \overrightarrow{P_0P}.$$

But $\overrightarrow{P_0P}$ is in the same direction as $\mathbf{v}$; therefore it must be some scalar multiple of $\mathbf{v}$, $\overrightarrow{P_0P} = t\mathbf{v}$. Consequently, the vector joining O to any point on l can be written in the form

$$\mathbf{r} = \mathbf{r}_0 + t\mathbf{v}, \tag{12.38}$$

for an appropriate value of t. Because the components of $\mathbf{r} = (x, y, z)$ are also the coordinates of the point (x, y, z) on l, this equation is called the **vector equation** of the line l through (x_0, y_0, z_0) in the direction $\mathbf{v}$. If we substitute components for $\mathbf{r}$, $\mathbf{r}_0$, and $\mathbf{v}$, then

$$(x, y, z) = (x_0, y_0, z_0) + t(v_x, v_y, v_z)$$
$$= (x_0 + tv_x, y_0 + tv_y, z_0 + tv_z).$$

Since two vectors are equal if and only if corresponding components are identical, we can write that

$$x = x_0 + v_x t, \qquad (12.39\text{a})$$
$$y = y_0 + v_y t, \qquad (12.39\text{b})$$
$$z = z_0 + v_z t. \qquad (12.39\text{c})$$

These three scalar equations are equivalent to vector equation 12.38; they are called **parametric equations** for line l. They illustrate once again that a line in space is characterized by a point (x_0, y_0, z_0) on it and a vector (v_x, v_y, v_z) along it. Each value of t substituted into 12.39 yields a point (x, y, z) on l, and conversely, every point on l is represented by some value of t. For instance, $t = 0$ yields P_0, and $t = 1$ gives the point at the tip of **v** in Figure 12.83.

If none of v_x, v_y, and v_z is equal to zero, we can solve equations 12.39 for t and equate the three expressions to obtain

$$\frac{x - x_0}{v_x} = \frac{y - y_0}{v_y} = \frac{z - z_0}{v_z}. \qquad (12.40)$$

These are called **symmetric equations** for the line l through (x_0, y_0, z_0) parallel to $\mathbf{v} = (v_x, v_y, v_z)$. There are only two independent equations in 12.40, which therefore substantiates our previous result that a curve (in this case, a line) can be described by two equations in x, y, and z. We could, for instance, write

$$v_x(y - y_0) = v_y(x - x_0) \quad \text{and} \quad v_z(y - y_0) = v_y(z - z_0),$$

or

$$v_y x - v_x y = v_y x_0 - v_x y_0, \qquad v_y z - v_z y = v_y z_0 - v_z y_0.$$

Since the first of these is linear in x and y and the second is linear in y and z, each describes a plane. This means that the line has been described as the curve of intersection of two planes.

EXAMPLE 12.20

Find, if possible, vector, parametric, and symmetric equations for the line through the points $(-1, 2, 1)$ and $(3, -2, 1)$.

SOLUTION A vector along the line is $(3, -2, 1) - (-1, 2, 1) = (4, -4, 0)$, and so too is $(1, -1, 0)$. A vector equation for the line is

$$\mathbf{r} = (-1, 2, 1) + t(1, -1, 0) = (t - 1, -t + 2, 1).$$

Parametric equations are therefore

$$x = t - 1, \quad y = -t + 2, \quad z = 1.$$

Because the z-component of every vector along the line is zero, we cannot write full symmetric equations for the line. By eliminating t between the x- and y-equations, however, we can write

$$x + 1 = \frac{y - 2}{-1}, \quad z = 1.$$

If we set $x + 1 = 2 - y$, $z = 1$, or $x + y = 1$, $z = 1$, we represent the line as the intersection of the planes $x + y = 1$ and $z = 1$ (Figure 12.84).

FIGURE 12.84

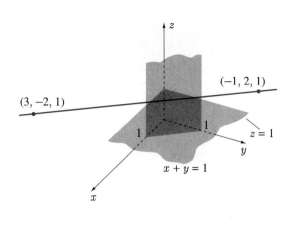

EXAMPLE 12.21

Find the equation of the plane containing the origin and the line $2x + y - z = 4$, $x + z = 5$.

SOLUTION We can easily find two more points on the plane. For instance, if $x = 0$, then the equations of the line require $z = 5$ and $y = 9$; and if $z = 0$, then $x = 5$ and $y = -6$. Thus $P(0,9,5)$ and $Q(5,-6,0)$, as well as $O(0,0,0)$, are points on the plane. It follows then that $\overrightarrow{OP} = (0,9,5)$ and $\overrightarrow{OQ} = (5,-6,0)$ are vectors in the plane, and a vector normal to the plane is

$$\overrightarrow{OP} \times \overrightarrow{OQ} = \begin{vmatrix} \hat{\mathbf{i}} & \hat{\mathbf{j}} & \hat{\mathbf{k}} \\ 0 & 9 & 5 \\ 5 & -6 & 0 \end{vmatrix} = (30, 25, -45).$$

The vector $(6, 5, -9)$ is also normal to the plane, and the equation of the plane is

$$0 = (6, 5, -9) \cdot (x - 0, y - 0, z - 0) = 6x + 5y - 9z.$$

EXAMPLE 12.22

Find symmetric equations for the line $x + y - 2z = 6$, $2x - 3y + 4z = 10$.

SOLUTION To find symmetric equations, we require a vector parallel to the line and a point on it. By setting $x = 0$ and solving $y - 2z = 6$, $-3y + 4z = 10$, we obtain $y = -22$, $z = -14$. Consequently, $(0, -22, -14)$ is a point on the line. To find a vector along the line, we could find another point on the line, say, by setting $z = 0$ and solving $x + y = 6$, $2x - 3y = 10$ for $x = \frac{28}{5}$, $y = \frac{2}{5}$. A vector along the line is therefore $(\frac{28}{5}, \frac{2}{5}, 0) - (0, -22, -14) = (\frac{28}{5}, \frac{112}{5}, 14)$, and so too is $(\frac{5}{14})(\frac{28}{5}, \frac{112}{5}, 14) = (2, 8, 5)$.

Alternatively, we know that $(1, 1, -2)$ and $(2, -3, 4)$ are vectors that are normal to the planes $x + y - 2z = 6$ and $2x - 3y + 4z = 10$, and a vector along the line of intersection of the planes must be perpendicular to both of these vectors (Figure

12.85). Consequently, a vector along the line of intersection is

$$\begin{vmatrix} \hat{\imath} & \hat{\jmath} & \hat{k} \\ 1 & 1 & -2 \\ 2 & -3 & 4 \end{vmatrix} = (-2, -8, -5).$$

Symmetric equations for the line are therefore

$$\frac{x}{2} = \frac{y+22}{8} = \frac{z+14}{5}.$$

FIGURE 12.85

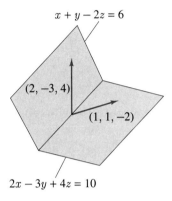

EXAMPLE 12.23

Find the distance from the point $(1, 3, 6)$ to the line l:

$$x - 1 = \frac{y-2}{2} = \frac{z-4}{3}.$$

SOLUTION Clearly $Q(1, 2, 4)$ is a point on the line, and therefore the required distance d is the component of $\overrightarrow{PQ}$ in the direction $\overrightarrow{PR}$ (Figure 12.86). By equation 12.30, then, $d = \overrightarrow{PQ} \cdot \widehat{PR}$. To find $\widehat{PR}$ we need a vector in the direction $\overrightarrow{PR}$. This is not nearly as simple as finding $\overrightarrow{PR}$ for a line in the xy-plane. Since $S(0, 0, 1)$ is also a point on l, $\overrightarrow{QS} = (-1, -2, -3)$ is a vector along l. But then

$$\overrightarrow{QS} \times \overrightarrow{PQ} = \begin{vmatrix} \hat{\imath} & \hat{\jmath} & \hat{k} \\ -1 & -2 & -3 \\ 0 & -1 & -2 \end{vmatrix} = \hat{\imath} - 2\hat{\jmath} + \hat{k}$$

is perpendicular to both $\overrightarrow{PQ}$ and $\overrightarrow{QS}$. It now follows that a vector in the direction $\overrightarrow{PR}$ must be

$$(\overrightarrow{QS} \times \overrightarrow{PQ}) \times \overrightarrow{QS} = \begin{vmatrix} \hat{\imath} & \hat{\jmath} & \hat{k} \\ 1 & -2 & 1 \\ -1 & -2 & -3 \end{vmatrix} = 8\hat{\imath} + 2\hat{\jmath} - 4\hat{k}.$$

Finally, then,
$$\widehat{PR} = \frac{(8,2,-4)}{\sqrt{64+4+16}} = \frac{(4,1,-2)}{\sqrt{21}}$$
and
$$d = (0,-1,-2) \cdot \frac{(4,1,-2)}{\sqrt{21}} = \frac{\sqrt{21}}{7}.$$

FIGURE 12.86

FIGURE 12.87

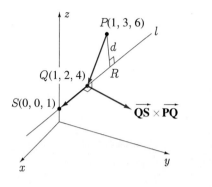
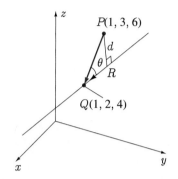

An alternative procedure is to note in Figure 12.87 that $d = |\overrightarrow{PQ}|\sin\theta$. But the right-hand side of this equation looks like the length of the cross-product of $\overrightarrow{PQ}$ and a unit vector that makes an angle θ with $\overrightarrow{PQ}$. Since $\widehat{RQ}$ makes an angle θ with $\overrightarrow{PQ}$, we could write that
$$d = |\overrightarrow{PQ}||\widehat{RQ}|\sin\theta = |\overrightarrow{PQ} \times \widehat{RQ}|.$$

Now $\overrightarrow{PQ} = (0,-1,-2)$ and a vector along l is $(1,2,3)$. Thus $\widehat{RQ} = (-1,-2,-3)/\sqrt{14}$ and
$$\overrightarrow{PQ} \times \widehat{RQ} = \begin{vmatrix} \hat{\mathbf{i}} & \hat{\mathbf{j}} & \hat{\mathbf{k}} \\ 0 & -1 & -2 \\ -1/\sqrt{14} & -2/\sqrt{14} & -3/\sqrt{14} \end{vmatrix} = \frac{1}{\sqrt{14}}(-1,2,-1).$$

Finally,
$$d = \frac{1}{\sqrt{14}}\sqrt{1+4+1} = \frac{\sqrt{21}}{7}.$$

■

This example suggests the following formula for the area of a triangle.

Theorem 12.6 If A, B, and C are the vertices of a triangle, then
$$\text{Area of } \triangle ABC = \frac{1}{2}|\overrightarrow{AB} \times \overrightarrow{AC}|. \tag{12.41}$$

Proof The area of triangle ABC in Figure 12.88 is

$$\frac{1}{2}|\overrightarrow{AC}||\overrightarrow{BD}| = \frac{1}{2}|\overrightarrow{AC}||\overrightarrow{AB}|\sin\theta = \frac{1}{2}|\overrightarrow{AB}\times\overrightarrow{AC}|.$$

FIGURE 12.88

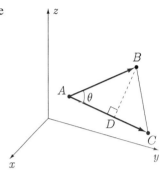

For the triangle with vertices $A(1,1,1)$, $B(2,-3,2)$, and $C(4,1,5)$ in Figure 12.89,

$$\overrightarrow{AB}\times\overrightarrow{AC} = \begin{vmatrix} \hat{i} & \hat{j} & \hat{k} \\ 1 & -4 & 1 \\ 3 & 0 & 4 \end{vmatrix} = (-16,-1,12).$$

The area of the triangle is therefore

$$\frac{1}{2}|(-16,-1,12)| = \frac{1}{2}\sqrt{256+1+144} = \frac{1}{2}\sqrt{401}.$$

The following corollary is a direct consequence.

Corollary The area of a parallelogram with coterminal sides $\overrightarrow{AB}$ and $\overrightarrow{AC}$ (Figure 12.90) is

$$\text{Area} = |\overrightarrow{AB}\times\overrightarrow{AC}|. \qquad (12.42)$$

FIGURE 12.89

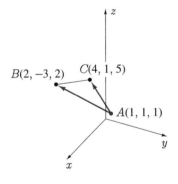

FIGURE 12.90

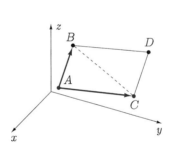

EXERCISES 12.5

If $\mathbf{u} = (3,1,4)$, $\mathbf{v} = (-1,2,0)$, and $\mathbf{w} = (-2,-3,5)$, evaluate the scalar or find the components of the vector in Exercises 1–10.

1. $\mathbf{v}\times\mathbf{w}$

2. $(-3\mathbf{u})\times(2\mathbf{v})$

3. $\mathbf{u}\cdot(\mathbf{v}\times\mathbf{w})$

4. $\hat{\mathbf{u}}\times\hat{\mathbf{w}}$

5. $(3\mathbf{u})\times\mathbf{w} + \mathbf{u}\times\mathbf{v}$

6. $\mathbf{u}\times(3\mathbf{v}-\mathbf{w})$

7. $\dfrac{\mathbf{w}\times\mathbf{u}}{|\mathbf{u}\times\mathbf{v}|}$

8. $\mathbf{u}\times\mathbf{w} - \mathbf{u}\times\mathbf{v} + \mathbf{u}\times(2\mathbf{u}+\mathbf{v})$

9. $(\mathbf{u}\times\mathbf{v})\times\mathbf{w}$

10. $\mathbf{u}\times(\mathbf{v}\times\mathbf{w})$

In Exercises 11–14 find components for the vector.

11. Perpendicular to the vectors $(1,3,5)$ and $(-2,1,4)$

12. Perpendicular to the y-axis and the vector joining the points $(2,4,-3)$ and $(1,5,6)$

13. Perpendicular to the plane containing the lines $x = 1+3t$, $y = 2-4t$, $z = -1+6t$, and $x = 1+5t$, $y = 2-t$, $z = -1$

14. Perpendicular to the triangle with vertices $(-1,0,3)$,

(5,1,2) and (−6,2,4)

15. What is the area of the triangle in Exercise 14?

16. Verify the results in equations 12.35.

17. Find the area of the parallelogram with vertices $(1,2,3)$, $(4,3,7)$, $(-1,3,6)$ and $(2,4,10)$.

In Exercises 18–27 find vector, parametric, and symmetric (if possible) equations for the straight line.

18. Through the point $(1,-1,3)$ and parallel to the vector $(2,4,-3)$

19. Through the point $(-1,3,6)$ and parallel to the vector $(2,-3,0)$

20. Through the points $(2,-3,4)$ and $(5,2,-1)$

21. Through the points $(-2,3,3)$ and $(-2,-3,-3)$

22. Through the points $(1,3,4)$ and $(1,3,5)$

23. Through the point $(1,-3,5)$ and parallel to the line

$$\frac{x}{5} = \frac{y-2}{3} = \frac{z+4}{-2}$$

24. Through the point $(2,0,3)$ and parallel to the line $x = 4+t, y = 2, z = 6 - 2t$

25. Through the point of intersection of the lines

$$\frac{x-1}{2} = \frac{y+4}{-3} = \frac{z-2}{5}; \quad \frac{x-1}{6} = \frac{y+4}{3} = \frac{z-2}{4};$$

and parallel to the line joining the points $(1,3,-2)$ and $(2,-2,1)$

26. $2x - y = 5, 3x + 4y + z = 10$

27. Through the point $(-2,3,1)$ and parallel to the line $x + y = 3, 2x - y + z = -2$

In Exercises 28–35 find an equation for the plane.

28. Containing the points $(1,3,2)$, $(-2,0,-2)$, $(1,4,3)$

29. Containing the point $(2,-4,3)$ and the line $(x-1)/3 = (y+5)/4 = z+2$

30. Containing the lines $x = 2y = (z+1)/4$ and $x = t, y = 2t, z = 6t - 1$

31. Containing the lines $(x-1)/6 = y/8 = (z+2)/2$ and $(x+1)/3 = (y-2)/4 = z + 5$

32. Containing the line $x - y + 2z = 4, 2x + y + 3z = 6$ and the point $(1,-2,4)$

33. Containing the two lines $x + 2y + 4z = 21, x - y + 6z = 13$ and $x = 2 + 3t, y = 4, z = -3 + 5t$

34. Containing the two lines $3x + 4y = -6$, $x + 2y + z = 2$ and $2y + 3z = 19$, $3x - 2y - 9z = -58$

35. Containing the line $x + y - 4z = 6, 2x + 3y + 5z = 10$ and

(a) perpendicular to the xy-plane
(b) perpendicular to the xz-plane
(c) perpendicular to the yz-plane

36. Does the line $(x-3)/2 = y - 2 = (z+1)/4$ lie in the plane $x - y + 2z = -1$?

37. Show that the lines joining the midpoints of the sides of any quadrilateral form a parallelogram.

38. Show that if a plane has nonzero intercepts a, b, and c on the x-, y-, and z-axes, then its equation is $x/a + y/b + z/c = 1$.

39. Find equations for the four faces of the tetrahedron in Figure 12.10.

40. Find equations for the nine planes forming the sides, bottom and roof of the birdhouse in Figure 12.8.

41. Show that the cross-product is not associative; that is, in general $\mathbf{u} \times (\mathbf{v} \times \mathbf{w}) \neq (\mathbf{u} \times \mathbf{v}) \times \mathbf{w}$.

42. Show that the three lines $(x-4)/3 = (y-8)/4 = (z+7)/(-4)$; $(x+5)/3 = (y+2)/2 = (z+1)/2$; and $x = 1, y = 5 + t, z = -6 - 3t$ form a triangle and find its area.

In Exercises 43–46 find the distance.

43. From the point $(1,2,-3)$ to the line $x = 2(y+1) = (z-4)/2$

44. From the point $(3,-2,0)$ to the line $x = t, y = 3 - 2t, z = 4 + t$

45. From the line $x - 1 = 3(y + 4) = -z - 1$ to the plane $2x - 3y + z = 4$

46. From the line $x = 1 - 6t, y = 4t + 2, z = -t$ to the plane $x + y - 2z = 1$

If a force $\mathbf{F}$ acts at a point Q (Figure 12.91), then the moment of $\mathbf{F}$ about a point P is defined as the vector $\mathbf{M} = \mathbf{r} \times \mathbf{F}$, where $\mathbf{r} = \overrightarrow{PQ}$. In Exercises 47–51 calculate the moment of the force about the given point.

FIGURE 12.91

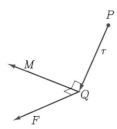

47. $\mathbf{F} = 2\hat{\mathbf{i}} + 3\hat{\mathbf{j}} - 4\hat{\mathbf{k}}$ at $(1,3,2)$ about $(-1,4,2)$

48. $\mathbf{F} = \hat{\mathbf{i}} + 2\hat{\mathbf{j}}$ at $(1,1,0)$ about $(2,1,-5)$

49. $\mathbf{F} = -\hat{\mathbf{i}} + 3\hat{\mathbf{k}}$ at $(0,0,0)$ about $(-1,3,0)$

50. $\mathbf{F} = 3\hat{\mathbf{i}} - \hat{\mathbf{j}} + 4\hat{\mathbf{k}}$ at $(1,1,1)$ about $(2,2,2)$

51. $\mathbf{F} = 6\hat{\mathbf{i}}$ at $(0,1,3)$ about $(2,0,0)$

52. (a) If $\mathbf{u} \neq \mathbf{0}$, show that if both the conditions $\mathbf{u} \cdot \mathbf{v} = \mathbf{u} \cdot \mathbf{w}$ and $\mathbf{u} \times \mathbf{v} = \mathbf{u} \times \mathbf{w}$ are satisfied, then $\mathbf{v} = \mathbf{w}$.
(b) Show that if one of the conditions in (a) is satisfied, but the other is not, then $\mathbf{v}$ cannot be equal to $\mathbf{w}$.

53. The scalar $\mathbf{u} \cdot (\mathbf{v} \times \mathbf{w})$ is called the *scalar triple product* of $\mathbf{u}$, $\mathbf{v}$, and $\mathbf{w}$.
(a) Find $\mathbf{u} \cdot (\mathbf{v} \times \mathbf{w})$ if $\mathbf{u} = (6,-1,0)$, $\mathbf{v} = (1,3,4)$, and $\mathbf{w} = (-2,-1,4)$.
(b) Prove that $\mathbf{u} \cdot (\mathbf{v} \times \mathbf{w}) = (\mathbf{u} \times \mathbf{v}) \cdot \mathbf{w}$.
(c) Show that $|\mathbf{u} \cdot (\mathbf{v} \times \mathbf{w})|$ can be interpreted as the volume of the parallelepiped with $\mathbf{u}$, $\mathbf{v}$, and $\mathbf{w}$ as coterminal sides (Figure 12.92).
(d) Verify that three nonzero vectors $\mathbf{u}$, $\mathbf{v}$, and $\mathbf{w}$ all lie in the same plane if and only if $\mathbf{u} \cdot (\mathbf{v} \times \mathbf{w}) = 0$.

FIGURE 12.92

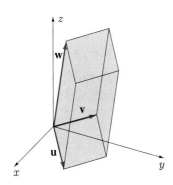

In Exercises 54 and 55 prove the identity.

54. $(\mathbf{u} \times \mathbf{v}) \cdot (\mathbf{w} \times \mathbf{r}) = (\mathbf{u} \cdot \mathbf{w})(\mathbf{v} \cdot \mathbf{r}) - (\mathbf{u} \cdot \mathbf{r})(\mathbf{v} \cdot \mathbf{w})$

55. $\mathbf{u} \times (\mathbf{v} \times \mathbf{w}) = (\mathbf{u} \cdot \mathbf{w})\mathbf{v} - (\mathbf{u} \cdot \mathbf{v})\mathbf{w}$

56. Verify that the equation of the plane passing through the three points $P_1(x_1,y_1,z_1)$, $P_2(x_2,y_2,z_2)$, and $P_3(x_3,y_3,z_3)$ can be written in the form $\overrightarrow{P_1P} \cdot (\overrightarrow{P_1P_2} \times \overrightarrow{P_1P_3}) = 0$, where $P(x,y,z)$ is any point in the plane.

57. Use vectors to prove the sine law
$$\frac{\sin A}{a} = \frac{\sin B}{b} = \frac{\sin C}{c}$$
for the triangle in Figure 12.93.
(Hint: Note that $\overrightarrow{PQ} + \overrightarrow{QR} + \overrightarrow{RP} = \mathbf{0}$. Cross this equation with $\overrightarrow{PQ}$ and take lengths.)

FIGURE 12.93

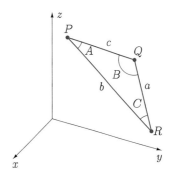

58. The region S_{xy} in the xy-plane bounded by the straight lines $x = 0$, $x = 1$, $y = 0$ and $y = 1$ is a rectangle with unit area.
(a) Show that the region in the plane $y = z$ that projects onto S_{xy} is also a rectangle, but with area $\sqrt{2}$.
(b) Show that the region in the plane $x + y - 2z = 0$ that projects onto S_{xy} is a parallelogram with area $\sqrt{6}/2$.
(c) Generalize the results of (a) and (b) to show that if S is the area in a plane $Ax + By + Cz + D = 0$ ($C \neq 0$) that projects onto S_{xy}, then the area of S is $\sec \gamma$ where γ is the acute angle between $\hat{\mathbf{k}}$ and the normal to the given plane.

59. Find the shortest distance between the lines
$$\frac{x-1}{2} = \frac{y+3}{3} = 4-z; \quad x = -1+t, \; y = 2t, \; z = 3-2t.$$

SECTION 12.6

Differentiation and Integration of Vectors

In Sections 12.4 and 12.5, vectors were used to represent forces and moments; they can also be used to describe many other physical quantities, such as position, velocity, acceleration, electric and magnetic fields, and fluid flow. In applications such as these, vectors seldom have constant components; instead they have components that are either functions of position, or functions of some parameter, such as time, or both. For instance, the spring force in Example 12.18 varies in both magnitude and direction as the sleeve moves from B to C. Consequently, components of the vector $\mathbf{F}$ representing this force are functions of position x between B and C:

$$\mathbf{F} = F_x\hat{\mathbf{i}} + F_y\hat{\mathbf{j}} = F_x(x)\hat{\mathbf{i}} + F_y(x)\hat{\mathbf{j}}.$$

When a particle moves along a curve C in the xy-plane defined parametrically by

$$C: \quad x = x(t), \quad y = y(t), \quad \alpha \leq t \leq \beta,$$

its position (x, y) relative to the origin is represented by the vector $\mathbf{r} = x\hat{\mathbf{i}} + y\hat{\mathbf{j}}$ (Figure 12.94). This vector is called the **position vector** or **displacement vector** of the particle relative to the origin. We will have more to say about it in Section 12.10. Note, however, that if we substitute from the parametric equations for C, we have

$$\mathbf{r} = x(t)\hat{\mathbf{i}} + y(t)\hat{\mathbf{j}},$$

which indicates that the displacement vector has components that are functions of the parameter t.

FIGURE 12.94

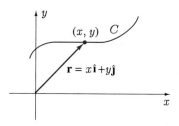

In this section we consider the general situation in which the components v_x, v_y, and v_z of a vector $\mathbf{v}$ are functions of some parameter t,

$$\mathbf{v} = v_x(t)\hat{\mathbf{i}} + v_y(t)\hat{\mathbf{j}} + v_z(t)\hat{\mathbf{k}}, \qquad (12.43)$$

and show how the operations of differentiation and integration can be applied to such a vector. In Sections 12.8 and 12.9 we will use these results to discuss the geometry of curves, and this will pave the way for an analysis of the motion of particles in Section 12.10.

Because $\mathbf{v}$ in 12.43 has components that are functions of t, we say that $\mathbf{v}$ itself is a function of t, and write $\mathbf{v} = \mathbf{v}(t)$. Each of the component functions $v_x(t)$, $v_y(t)$, and $v_z(t)$ has a domain, and their common domain is called the domain of the vector function $\mathbf{v}(t)$. Given that this domain is some interval $\alpha \leq t \leq \beta$, we express 12.43 more fully in the form

$$\mathbf{v} = \mathbf{v}(t) = v_x(t)\hat{\mathbf{i}} + v_y(t)\hat{\mathbf{j}} + v_z(t)\hat{\mathbf{k}}, \quad \alpha \leq t \leq \beta. \qquad (12.44)$$

To differentiate and integrate vector functions, we first require the concept of a limit.

Definition 12.11

If $\mathbf{v}(t) = v_x(t)\hat{\mathbf{i}} + v_y(t)\hat{\mathbf{j}} + v_z(t)\hat{\mathbf{k}}$, then

$$\lim_{t \to t_0} \mathbf{v}(t) = \left(\lim_{t \to t_0} v_x(t)\right)\hat{\mathbf{i}} + \left(\lim_{t \to t_0} v_y(t)\right)\hat{\mathbf{j}} + \left(\lim_{t \to t_0} v_z(t)\right)\hat{\mathbf{k}}, \qquad (12.45)$$

provided each of the limits on the right exists.

This definition states that to take the limit of a vector function, we take the limit of each component separately. As an illustration, consider the following example.

EXAMPLE 12.24

If $\mathbf{v} = \mathbf{v}(t) = (t^2 + 1)\hat{\mathbf{i}} + 3t\hat{\mathbf{j}} - (\sin t)\hat{\mathbf{k}}$, calculate $\lim_{t \to 5} \mathbf{v}(t)$.

SOLUTION According to Definition 12.11,

$$\lim_{t \to 5} \mathbf{v}(t) = \left(\lim_{t \to 5}(t^2 + 1)\right)\hat{\mathbf{i}} + \left(\lim_{t \to 5}(3t)\right)\hat{\mathbf{j}} - \left(\lim_{t \to 5} \sin t\right)\hat{\mathbf{k}}$$
$$= 26\hat{\mathbf{i}} + 15\hat{\mathbf{j}} - (\sin 5)\hat{\mathbf{k}}.$$

In the remainder of this section we define continuity, derivatives, and antiderivatives for vector functions. Each definition is an exact duplicate of the corresponding definition for a scalar function $y(t)$, except that $y(t)$ is replaced by $\mathbf{v}(t)$. We then show that the vector definition can be rephrased in terms of components of the vector. We begin with continuity in the following definition.

Definition 12.12

A vector function $\mathbf{v}(t)$ is said to be **continuous** at a point t_0 if

$$\mathbf{v}(t_0) = \lim_{t \to t_0} \mathbf{v}(t). \qquad (12.46)$$

It is a simple matter to prove the next theorem.

Theorem 12.7

A vector function is continuous at a point if and only if its components are continuous at that point.

Proof If $\mathbf{v}(t) = v_x(t)\hat{\mathbf{i}} + v_y(t)\hat{\mathbf{j}} + v_z(t)\hat{\mathbf{k}}$, then according to 12.46, $\mathbf{v}(t)$ is continuous at t_0 if and only if

$$v_x(t_0)\hat{\mathbf{i}} + v_y(t_0)\hat{\mathbf{j}} + v_z(t_0)\hat{\mathbf{k}} = \lim_{t \to t_0}\{v_x(t)\hat{\mathbf{i}} + v_y(t)\hat{\mathbf{j}} + v_z(t)\hat{\mathbf{k}}\}.$$

Definition 12.11 implies that we can write this condition in the form

$$v_x(t_0)\hat{\mathbf{i}} + v_y(t_0)\hat{\mathbf{j}} + v_z(t_0)\hat{\mathbf{k}} = \left(\lim_{t \to t_0} v_x(t)\right)\hat{\mathbf{i}} + \left(\lim_{t \to t_0} v_y(t)\right)\hat{\mathbf{j}} + \left(\lim_{t \to t_0} v_z(t)\right)\hat{\mathbf{k}}.$$

But because two vectors are equal if and only if their components are equal, we can say that $\mathbf{v}(t)$ is continuous at t_0 if and only if

$$v_x(t_0) = \lim_{t \to t_0} v_x(t); \quad v_y(t_0) = \lim_{t \to t_0} v_y(t); \quad v_z(t_0) = \lim_{t \to t_0} v_z(t);$$

i.e., $\mathbf{v}(t)$ is continuous at t_0 if and only if its components are continuous at t_0.

For example, the vector function

$$\mathbf{v}(t) = (t-1)\hat{\mathbf{i}} + (1/t)\hat{\mathbf{j}} + (t^2 - 1)^{-1}\hat{\mathbf{k}}$$

is discontinuous for $t = 0$ (since $v_y(0)$ is not defined) and for $t = \pm 1$ (since $v_z(\pm 1)$ is not defined).

The derivative of a scalar function $y(t)$ is its instantaneous rate of change

$$\frac{dy}{dt} = \lim_{h \to 0} \frac{y(t+h) - y(t)}{h}.$$

The derivative of a vector function is also a rate of change defined by a similar limit.

Definition 12.13

The derivative of a vector function $\mathbf{v}(t)$ is defined as

$$\frac{d\mathbf{v}}{dt} = \lim_{h \to 0} \frac{\mathbf{v}(t+h) - \mathbf{v}(t)}{h}, \tag{12.47}$$

provided the limit exists.

In practice, we seldom use the definition of a derivative to calculate dy/dt for a scalar function $y(t)$; formulas such as the power, product, quotient, and chain rules are more convenient. It would be helpful to have corresponding formulas for derivatives of vector functions. The following theorem shows that to differentiate a vector function we simply differentiate its Cartesian components.

Theorem 12.8

If $\mathbf{v}(t) = v_x(t)\hat{\mathbf{i}} + v_y(t)\hat{\mathbf{j}} + v_z(t)\hat{\mathbf{k}}$, and $d\mathbf{v}/dt$ exists, then

$$\frac{d\mathbf{v}}{dt} = \frac{dv_x}{dt}\hat{\mathbf{i}} + \frac{dv_y}{dt}\hat{\mathbf{j}} + \frac{dv_z}{dt}\hat{\mathbf{k}}. \tag{12.48}$$

Proof If we substitute the components of $\mathbf{v}(t+h)$ and $\mathbf{v}(t)$ into Definition 12.13, then we have

$$\frac{d\mathbf{v}}{dt} = \lim_{h \to 0} \left\{ \frac{[v_x(t+h)\hat{\mathbf{i}} + v_y(t+h)\hat{\mathbf{j}} + v_z(t+h)\hat{\mathbf{k}}] - [v_x(t)\hat{\mathbf{i}} + v_y(t)\hat{\mathbf{j}} + v_z(t)\hat{\mathbf{k}}]}{h} \right\}$$

$$= \lim_{h \to 0} \left\{ \left[\frac{v_x(t+h) - v_x(t)}{h}\right]\hat{\mathbf{i}} + \left[\frac{v_y(t+h) - v_y(t)}{h}\right]\hat{\mathbf{j}} + \left[\frac{v_z(t+h) - v_z(t)}{h}\right]\hat{\mathbf{k}} \right\}$$

(according to equation 12.14), and

$$\frac{d\mathbf{v}}{dt} = \left(\lim_{h \to 0} \frac{v_x(t+h) - v_x(t)}{h}\right)\hat{\mathbf{i}} + \left(\lim_{h \to 0} \frac{v_y(t+h) - v_y(t)}{h}\right)\hat{\mathbf{j}}$$
$$+ \left(\lim_{h \to 0} \frac{v_z(t+h) - v_z(t)}{h}\right)\hat{\mathbf{k}}$$

(according to equation 12.45). If $d\mathbf{v}/dt$ exists, then each of the three limits on the right must exist, and we can write that

$$\frac{d\mathbf{v}}{dt} = \frac{dv_x}{dt}\hat{\mathbf{i}} + \frac{dv_y}{dt}\hat{\mathbf{j}} + \frac{dv_z}{dt}\hat{\mathbf{k}}.$$

This proof is also reversible, giving the following corollary to Theorem 12.8.

Corollary *If the derivatives dv_x/dt, dv_y/dt, and dv_z/dt of the components of a vector function $\mathbf{v}(t) = v_x(t)\hat{\mathbf{i}} + v_y(t)\hat{\mathbf{j}} + v_z(t)\hat{\mathbf{k}}$ exist at a point, then so does $d\mathbf{v}/dt$.*

Theorem 12.8 gives us a working rule for differentiating vector functions: To differentiate a vector function, we differentiate its Cartesian components.

EXAMPLE 12.25

If $\mathbf{v}(t) = t^2\hat{\mathbf{i}} + (3t^3 - 2t)\hat{\mathbf{j}} + 5\hat{\mathbf{k}}$, find $\mathbf{v}'(3)$.

SOLUTION According to 12.48

$$\frac{d\mathbf{v}}{dt} = 2t\hat{\mathbf{i}} + (9t^2 - 2)\hat{\mathbf{j}}.$$

Consequently,

$$\mathbf{v}'(3) = 6\hat{\mathbf{i}} + 79\hat{\mathbf{j}}.$$

The sum rule 3.8 for differentiation of scalar functions has its counterpart in the sum rule for vector functions,

$$\frac{d}{dt}(\mathbf{u} + \mathbf{v}) = \frac{d\mathbf{u}}{dt} + \frac{d\mathbf{v}}{dt} \qquad (12.49)$$

(see Exercise 22). There are three types of products associated with vectors: the product of a scalar and a vector, the dot product of two vectors, and the cross product of two vectors. Corresponding to each, we have a product rule for differentiation, but all resemble the product rule for scalar functions.

Theorem 12.9

If $f(t)$ is a differentiable function and $\mathbf{u}(t)$ and $\mathbf{v}(t)$ are differentiable vector functions, then:

$$\frac{d}{dt}(f\mathbf{v}) = \frac{df}{dt}\mathbf{v} + f\frac{d\mathbf{v}}{dt}, \qquad (12.50)$$

$$\frac{d}{dt}(\mathbf{u} \cdot \mathbf{v}) = \mathbf{u} \cdot \frac{d\mathbf{v}}{dt} + \frac{d\mathbf{u}}{dt} \cdot \mathbf{v}, \qquad (12.51)$$

$$\frac{d}{dt}(\mathbf{u} \times \mathbf{v}) = \mathbf{u} \times \frac{d\mathbf{v}}{dt} + \frac{d\mathbf{u}}{dt} \times \mathbf{v}. \qquad (12.52)$$

For a proof of these results, see Exercise 23.

EXAMPLE 12.26

If $f(t) = t^2 + 2t + 3$, $\mathbf{u}(t) = t\hat{\mathbf{i}} + t^2\hat{\mathbf{j}} - 3\hat{\mathbf{k}}$, and $\mathbf{v}(t) = t(\hat{\mathbf{i}} + \hat{\mathbf{j}} + \hat{\mathbf{k}})$, use (12.50)–(12.52) to evaluate:

$$\text{(a) } \frac{d}{dt}(f\mathbf{u}); \quad \text{(b) } \frac{d}{dt}(\mathbf{u}\cdot\mathbf{v}); \quad \text{(c) } \frac{d}{dt}(\mathbf{u}\times\mathbf{v}).$$

SOLUTION

(a) With 12.50,

$$\frac{d}{dt}(f\mathbf{u}) = \frac{df}{dt}\mathbf{u} + f\frac{d\mathbf{u}}{dt} = (2t+2)(t\hat{\mathbf{i}} + t^2\hat{\mathbf{j}} - 3\hat{\mathbf{k}}) + (t^2 + 2t + 3)(\hat{\mathbf{i}} + 2t\hat{\mathbf{j}})$$

$$= (3t^2 + 4t + 3)\hat{\mathbf{i}} + (4t^3 + 6t^2 + 6t)\hat{\mathbf{j}} - 6(t+1)\hat{\mathbf{k}}.$$

(b) With 12.51,

$$\frac{d}{dt}(\mathbf{u}\cdot\mathbf{v}) = \mathbf{u}\cdot\frac{d\mathbf{v}}{dt} + \frac{d\mathbf{u}}{dt}\cdot\mathbf{v}$$

$$= (t\hat{\mathbf{i}} + t^2\hat{\mathbf{j}} - 3\hat{\mathbf{k}})\cdot(\hat{\mathbf{i}} + \hat{\mathbf{j}} + \hat{\mathbf{k}}) + (\hat{\mathbf{i}} + 2t\hat{\mathbf{j}})\cdot(t\hat{\mathbf{i}} + t\hat{\mathbf{j}} + t\hat{\mathbf{k}})$$

$$= (t + t^2 - 3) + (t + 2t^2) = 3t^2 + 2t - 3.$$

(c) With 12.52,

$$\frac{d}{dt}(\mathbf{u}\times\mathbf{v}) = \mathbf{u}\times\frac{d\mathbf{v}}{dt} + \frac{d\mathbf{u}}{dt}\times\mathbf{v} = (t\hat{\mathbf{i}} + t^2\hat{\mathbf{j}} - 3\hat{\mathbf{k}})\times(\hat{\mathbf{i}} + \hat{\mathbf{j}} + \hat{\mathbf{k}})$$

$$+ (\hat{\mathbf{i}} + 2t\hat{\mathbf{j}})\times(t\hat{\mathbf{i}} + t\hat{\mathbf{j}} + t\hat{\mathbf{k}})$$

$$= \begin{vmatrix} \hat{\mathbf{i}} & \hat{\mathbf{j}} & \hat{\mathbf{k}} \\ t & t^2 & -3 \\ 1 & 1 & 1 \end{vmatrix} + \begin{vmatrix} \hat{\mathbf{i}} & \hat{\mathbf{j}} & \hat{\mathbf{k}} \\ 1 & 2t & 0 \\ t & t & t \end{vmatrix}$$

$$= [(t^2 + 3)\hat{\mathbf{i}} - (3+t)\hat{\mathbf{j}} + (t - t^2)\hat{\mathbf{k}}]$$

$$+ [2t^2\hat{\mathbf{i}} - t\hat{\mathbf{j}} + (t - 2t^2)\hat{\mathbf{k}}]$$

$$= (3t^2 + 3)\hat{\mathbf{i}} - (3 + 2t)\hat{\mathbf{j}} + (2t - 3t^2)\hat{\mathbf{k}}.$$

∎

If vector functions can be differentiated, then they can be antidifferentiated. Formally we make the following statement.

Definition 12.14

A vector function $\mathbf{V}(t)$ is said to be an **antiderivative** or **indefinite integral** of $\mathbf{v}(t)$ on the interval $\alpha < t < \beta$ if

$$\frac{d\mathbf{V}}{dt} = \mathbf{v}(t) \quad \text{for} \quad \alpha < t < \beta. \tag{12.53}$$

For example, an antiderivative of $\mathbf{v}(t) = 2t\hat{\mathbf{i}} - \hat{\mathbf{j}} + 3t^2\hat{\mathbf{k}}$ is

$$\mathbf{V}(t) = t^2\hat{\mathbf{i}} - t\hat{\mathbf{j}} + t^3\hat{\mathbf{k}}.$$

If we add to $\mathbf{V}(t)$ in 12.53 any vector with constant components, denoted by $\mathbf{C}$, then $\mathbf{V}(t) + \mathbf{C}$ is also an antiderivative of $\mathbf{v}(t)$. We call this vector *the* antiderivative or *the* indefinite integral of $\mathbf{v}(t)$, and write

$$\int \mathbf{v}(t)\,dt = \mathbf{V}(t) + \mathbf{C}. \qquad (12.54)$$

For our example, then,

$$\int (2t\hat{\mathbf{i}} - \hat{\mathbf{j}} + 3t^2\hat{\mathbf{k}})\,dt = t^2\hat{\mathbf{i}} - t\hat{\mathbf{j}} + t^3\hat{\mathbf{k}} + \mathbf{C}.$$

Because vectors can be differentiated component by component, it follows that they may also be integrated component by component; i.e., if $\mathbf{v}(t) = v_x(t)\hat{\mathbf{i}} + v_y(t)\hat{\mathbf{j}} + v_z(t)\hat{\mathbf{k}}$, then

$$\int \mathbf{v}(t)\,dt = \left(\int v_x(t)\,dt\right)\hat{\mathbf{i}} + \left(\int v_y(t)\,dt\right)\hat{\mathbf{j}} + \left(\int v_z(t)\,dt\right)\hat{\mathbf{k}}. \qquad (12.55)$$

EXAMPLE 12.27

Find the antiderivative of $\mathbf{v}(t) = \sqrt{t-1}\,\hat{\mathbf{i}} + e^t\hat{\mathbf{j}} + 6t^2\hat{\mathbf{k}}$.

SOLUTION According to equation 12.55,

$$\int \mathbf{v}(t)\,dt = \left(\frac{2}{3}(t-1)^{3/2} + C_1\right)\hat{\mathbf{i}} + (e^t + C_2)\hat{\mathbf{j}} + (2t^3 + C_3)\hat{\mathbf{k}}$$

$$= \frac{2}{3}(t-1)^{3/2}\hat{\mathbf{i}} + e^t\hat{\mathbf{j}} + 2t^3\hat{\mathbf{k}} + \mathbf{C},$$

where $\mathbf{C} = C_1\hat{\mathbf{i}} + C_2\hat{\mathbf{j}} + C_3\hat{\mathbf{k}}$ is a constant vector.

EXERCISES 12.6

In Exercises 1–5 find the largest possible domain for the vector function.

1. $\mathbf{v}(t) = t^2\hat{\mathbf{i}} + \sqrt{t-1}\,\hat{\mathbf{j}} + \hat{\mathbf{k}}$
2. $\mathbf{v}(t) = (\sin t)\hat{\mathbf{i}} + (\cos t)\hat{\mathbf{j}} - t^3\hat{\mathbf{k}}$
3. $\mathbf{v}(t) = (\mathrm{Sin}^{-1} t)\hat{\mathbf{i}} - t^2\hat{\mathbf{j}} + (t+1)\hat{\mathbf{k}}$
4. $\mathbf{v}(t) = \ln(t+4)(\hat{\mathbf{i}} + \hat{\mathbf{j}})$
5. $\mathbf{v}(t) = e^t\hat{\mathbf{i}} + (\cos^2 t)\hat{\mathbf{j}} - (e^t \cos t)\hat{\mathbf{k}}$

If $f(t) = t^2 + 3$, $g(t) = 2t^3 - 3t$, $\mathbf{u}(t) = t\hat{\mathbf{i}} - t^2\hat{\mathbf{j}} + 2t\hat{\mathbf{k}}$, and $\mathbf{v}(t) = \hat{\mathbf{i}} - 2t\hat{\mathbf{j}} + 3t^2\hat{\mathbf{k}}$, find the scalar or the components of the vector in Exercises 6–21.

6. $\dfrac{d\mathbf{u}}{dt}$
7. $\dfrac{d}{dt}[f(t)\mathbf{v}(t)]$
8. $\dfrac{d}{dt}[g(t)\mathbf{u}(t)]$
9. $\dfrac{d}{dt}(\mathbf{u} \times \mathbf{v})$
10. $\dfrac{d}{dt}(\mathbf{u} \times t\mathbf{v})$
11. $\dfrac{d}{dt}(2\mathbf{u} \cdot \mathbf{v})$
12. $\dfrac{d}{dt}(3\mathbf{u} + 4\mathbf{v})$
13. $\displaystyle\int \mathbf{u}(t)\,dt$
14. $\dfrac{d}{dt}[f(t)\mathbf{u} + g(t)\mathbf{v}]$
15. $\displaystyle\int 4\mathbf{v}(t)\,dt$
16. $\dfrac{d}{dt}[t(\mathbf{u} \times \mathbf{v})]$

17. $\int [f(t)\mathbf{u}(t)]\, dt$

18. $\int [3g(t)\mathbf{v}(t) + \mathbf{u}(t)]\, dt$

19. $\int [f(t)\mathbf{u} \cdot \mathbf{v}]\, dt$

20. $\mathbf{u} \times \dfrac{d\mathbf{v}}{dt} - f(t)\mathbf{u} \cdot \dfrac{d\mathbf{v}}{dt}\mathbf{v}$

21. $\mathbf{u} \cdot \dfrac{d\mathbf{v}}{dt} - \mathbf{v} \cdot \int \mathbf{u}(t)\, dt$

22. Prove equation 12.49.

23. Verify the results in equations 12.50–12.52.

24. Prove that for differentiable functions $\mathbf{u}(t)$, $\mathbf{v}(t)$, and $\mathbf{w}(t)$,

$$\dfrac{d}{dt}(\mathbf{u} \cdot \mathbf{v} \times \mathbf{w}) = \dfrac{d\mathbf{u}}{dt} \cdot \mathbf{v} \times \mathbf{w} + \mathbf{u} \cdot \dfrac{d\mathbf{v}}{dt} \times \mathbf{w} + \mathbf{u} \cdot \mathbf{v} \times \dfrac{d\mathbf{w}}{dt}.$$

25. Prove that if a differentiable function $\mathbf{v}(t)$ has constant length, then at any point at which $d\mathbf{v}/dt \neq \mathbf{0}$, the vector $d\mathbf{v}/dt$ is perpendicular to $\mathbf{v}$.

26. If $\mathbf{v} = \mathbf{v}(s)$ is a differentiable vector function and $s = s(t)$ is a differentiable scalar function, prove that

$$\dfrac{d\mathbf{v}}{dt} = \dfrac{d\mathbf{v}}{ds}\dfrac{ds}{dt}.$$

This result is called the *chain rule* for differentiation of vector functions.

27. Show that the following definition for the limit of a vector function is equivalent to Definition 12.11: A vector function $\mathbf{v}(t)$ is said to have limit $\mathbf{V}$ as t approaches t_0 if given any $\epsilon > 0$, there exists a $\delta > 0$ such that $|\mathbf{v}(t) - \mathbf{V}| < \epsilon$ whenever $0 < |t - t_0| < \delta$.

SECTION 12.7

Parametric and Vector Representations of Curves

In Section 12.2 we presented curves in space as the intersection of two surfaces. For example, each of the equations

$$x^2 + y^2 + z^2 = 9, \quad y = 2$$

describes a surface (the first a sphere and the second a plane), and together they describe the curve of intersection of the surfaces—the circle in Figure 12.95.

FIGURE 12.95

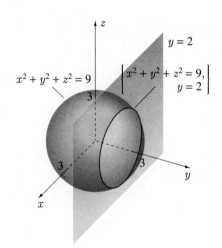

Parametric Representation of Curves

In many applications it is more convenient to have a curve defined "parametrically."

> **Definition 12.15**
>
> A curve in space is defined parametrically by three functions:
>
> $$C: \quad x = x(t), \quad y = y(t), \quad z = z(t), \quad \alpha \leq t \leq \beta. \qquad (12.56)$$

Each value of t in the interval $\alpha \leq t \leq \beta$ is substituted into the three functions, and the triple $(x, y, z) = (x(t), y(t), z(t))$ specifies a point on the curve. Definition 12.15 clearly corresponds to parametric Definition 10.1 for a plane curve.

It is customary to assign a direction to a curve by calling that point on C corresponding to $t = \alpha$ the initial point and that point corresponding to $t = \beta$ the final point, and the direction of C is that direction along C from initial point to final point (Figure 12.96). Note in particular that the direction of a curve always corresponds to the direction in which the parameter t *increases* along the curve. Because of this, whenever we describe a curve in nonparametric form but with a specified direction, we must be careful in setting up parametric equations to ensure that the parameter increases in the appropriate direction.

FIGURE 12.96

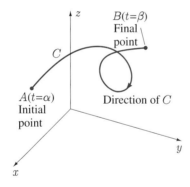

When a curve is described as the curve of intersection of two surfaces, we can obtain parametric equations for the curve by specifying one of x, y, or z as a function of t, and then solving the equations for the other two as functions of t. Considerable ingenuity is sometimes required in arriving at an initial function of t. We illustrate this in the following example.

> **EXAMPLE 12.28**
>
> Find parametric equations for each of the following curves:
> (a) $z - 1 = x^2 + y^2$, $x - y = 0$ directed so that z increases when x and y are positive;
> (b) $x + 2y + z = 4$, $2x + y + 3z = 6$ directed so that y increases along the curve;
> (c) $x^2 + (y - 1)^2 = 4$, $z = x$ directed so that y increases when x is positive.
>
> **SOLUTION**

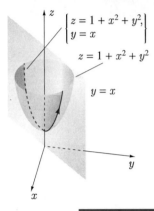

FIGURE 12.97

(a) The curve of intersection of the circular paraboloid $z = 1 + x^2 + y^2$ and the plane $y = x$ is shown in Figure 12.97. If we choose x as the parameter along the curve by setting $x = t$, then

$$x = t, \quad y = t, \quad z = 1 + 2t^2$$

(and note that z increases when x and y are positive).

(b) The straight-line intersection of the two planes is shown in Figure 12.98. If we choose y as the parameter by setting $y = t$ (thus forcing y to increase as t increases), then

$$x + z = 4 - 2t, \quad 2x + 3z = 6 - t.$$

The solution of these equations for x and z in terms of t gives the parametric equations

$$x = 6 - 5t, \quad y = t, \quad z = -2 + 3t.$$

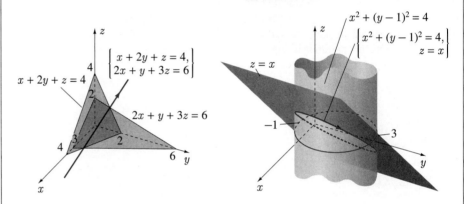

FIGURE 12.98

FIGURE 12.99

(c) The curve of intersection of the right-circular cylinder $x^2 + (y - 1)^2 = 4$ and the plane $z = x$ is shown in Figure 12.99. If we set $x = 2 \cos t$, then $y = 1 \pm 2 \sin t$. A set of parametric equations for the curve is therefore

$$x = 2 \cos t, \quad y = 1 + 2 \sin t, \quad z = 2 \cos t, \quad 0 \leq t < 2\pi.$$

Any range of values of t of length 2π traces the curve exactly once. Note that had we chosen the equations

$$x = 2 \cos t, \quad y = 1 - 2 \sin t, \quad z = 2 \cos t, \quad 0 \leq t < 2\pi,$$

we would have generated the same set of points traced in the opposite direction. ∎

Definition 12.16

A curve $C: x = x(t), y = y(t), z = z(t), \alpha \leq t \leq \beta$, is said to be **continuous** if each of the functions $x(t)$, $y(t)$, and $z(t)$ is continuous for $\alpha \leq t \leq \beta$.

Geometrically, this implies that the curve is at no point separated. Each of the curves in Example 12.28 is therefore continuous.

A curve is said to be **closed** if its initial and final points are the same. Circles and ellipses are closed curves. Straight line segments, parabolas and hyperbolas are not closed.

Vector Representation of Curves

The position vector or displacement vector of a point $P(x, y, z)$ in space is the vector

$$\mathbf{r} = (x, y, z) = x\hat{\mathbf{i}} + y\hat{\mathbf{j}} + z\hat{\mathbf{k}}.$$

If we consider only points that lie on a curve defined parametrically by 12.56, then for these points we can write that

$$\mathbf{r} = \mathbf{r}(t) = x(t)\hat{\mathbf{i}} + y(t)\hat{\mathbf{j}} + z(t)\hat{\mathbf{k}}, \quad \alpha \leq t \leq \beta. \qquad (12.57)$$

As t varies from $t = \alpha$ to $t = \beta$, the tip of this vector traces the curve C from initial point to final point (Figure 12.100). We call 12.57 the **vector representation** of a curve.

FIGURE 12.100

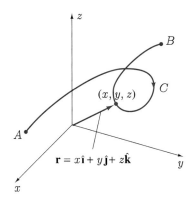

EXAMPLE 12.29

Sketch the curve with position vector

$$\mathbf{r} = \mathbf{r}(t) = (2\cos t)\hat{\mathbf{i}} + (3\sin t)\hat{\mathbf{j}}, \quad 0 \leq t \leq \pi.$$

SOLUTION When we set $x = 2\cos t$ and $y = 3\sin t$, it is clear that $x^2/4 + y^2/9 = 1$. The position vector $\mathbf{r} = \mathbf{r}(t)$ therefore describes points on an ellipse in the xy-plane. As t varies from 0 to π, x varies from 2 to -2, and y from 0 to 3 to 0 again. This means that only the top half of the ellipse is defined by $\mathbf{r} = \mathbf{r}(t)$ (Figure 12.101).

FIGURE 12.101

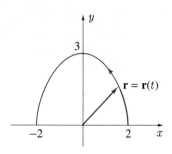

EXAMPLE 12.30 Sketch the curve with position vector

$$\mathbf{r} = \mathbf{r}(t) = t\hat{\mathbf{i}} + t^2\hat{\mathbf{j}} + t\hat{\mathbf{k}}, \quad t \geq 0.$$

SOLUTION When we set $x = t$, $y = t^2$, and $z = t$, then $z = x$ and $y = x^2$. These imply that $\mathbf{r} = \mathbf{r}(t)$ describes points on the curve of intersection of the surfaces $y = x^2$ and $z = x$ (Figure 12.102). Because $t \geq 0$, only that half of the curve of intersection in the first octant is defined by $\mathbf{r} = \mathbf{r}(t)$.

FIGURE 12.102

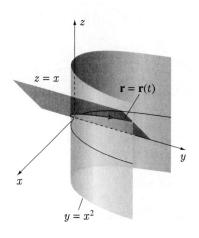

EXERCISES 12.7

In Exercises 1–10 find parametric and vector representations for the curve. Sketch each curve.

1. $x + 2y + 3z = 6$, $y - 2z = 3$ directed so that z increases along the curve

2. $x^2 + y^2 = 2$, $z = 4$ directed so that y increases in the first octant

3. $x^2 + y^2 = 2$, $x + y + z = 1$ directed so that y decreases when x is positive

4. $z = x^2 + y^2$, $x^2 + y^2 = 5$ directed clockwise as viewed from the origin

5. $z + 2x^2 = 1$, $y = z$ directed so that x decreases along the curve

6. $z = \sqrt{x^2 + y^2}$, $y = x$ directed so that y increases when x is positive

7. $z = x + y$, $y = x^2$ directed so that x increases along the curve

8. $z = \sqrt{4 - x^2 - y^2}$, $x^2 + y^2 - 2y = 0$ directed so that z decreases when x is positive

9. $x = \sqrt{z}$, $z = y^2$ directed away from the origin in the first octant

10. $z = \sqrt{x^2 + y^2}$, $y = x^2$ directed so that y decreases in the first octant

In Exercises 11–15 sketch the curve with the given position vector.

11. $\mathbf{r}(t) = t\hat{\mathbf{i}} + t\hat{\mathbf{j}} + t^2 \hat{\mathbf{k}}$, $t \geq 0$

12. $\mathbf{r}(t) = (2\cos t)\hat{\mathbf{i}} + (2\sin t)\hat{\mathbf{j}} + 3t\hat{\mathbf{k}}$, $0 \leq t \leq 4\pi$

13. $\mathbf{r}(t) = (t - 2)\hat{\mathbf{i}} + (2 - 3t)\hat{\mathbf{j}} + 5t\hat{\mathbf{k}}$

14. $\mathbf{r}(t) = (t^2 - t)\hat{\mathbf{i}} + t\hat{\mathbf{j}} + 5\hat{\mathbf{k}}$

15. $\mathbf{r}(t) = (\cos t)\hat{\mathbf{i}} + (\sin t)\hat{\mathbf{j}} + (\cos t)\hat{\mathbf{k}}$, $0 \leq t \leq \pi$

SECTION 12.8

Tangent Vectors and Lengths of Curves

If C is a curve in the xy-plane (Figure 12.103), then the tangent line to C at P is defined as the limiting position of the line PQ as Q moves along C toward P (see Section 3.1). We take the same approach in defining tangent vectors to curves in an arbitrary plane or in space. On curve C defined by 12.56, we let P and Q be the points corresponding to the parameter values t and $t + h$. Position vectors of P and Q are then

$$\mathbf{r}(t) = x(t)\hat{\mathbf{i}} + y(t)\hat{\mathbf{j}} + z(t)\hat{\mathbf{k}}$$

and

$$\mathbf{r}(t + h) = x(t + h)\hat{\mathbf{i}} + y(t + h)\hat{\mathbf{j}} + z(t + h)\hat{\mathbf{k}}$$

(Figure 12.104), and the vector $\overrightarrow{PQ}$ joining P to Q is clearly equal to

$$\overrightarrow{PQ} = \mathbf{r}(t + h) - \mathbf{r}(t).$$

FIGURE 12.103

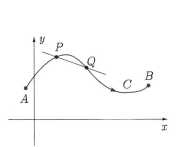

FIGURE 12.104

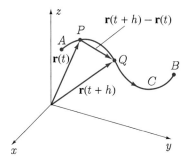

If we let h approach zero, then the point Q moves along C toward P, and the direction of $\overrightarrow{PQ}$ becomes closer to what seems to be a reasonable definition of the tangent direction to C at P. Perhaps then we should define $\lim_{h\to 0}[\mathbf{r}(t+h) - \mathbf{r}(t)]$ as a tangent vector to C at P. Unfortunately, the limit vector has length zero, and therefore

$$\lim_{h\to 0}[\mathbf{r}(t+h) - \mathbf{r}(t)] = \mathbf{0}.$$

If, however, we divide $\mathbf{r}(t+h) - \mathbf{r}(t)$ by h, then the resulting vector

$$\frac{\mathbf{r}(t+h) - \mathbf{r}(t)}{h}$$

is not equal to $\overrightarrow{PQ}$, but it does have the same direction as $\overrightarrow{PQ}$. Consider, then, taking the limit of this vector as h approaches zero:

$$\lim_{h\to 0} \frac{\mathbf{r}(t+h) - \mathbf{r}(t)}{h}.$$

If the limit vector exists, then it will be tangent to C at P. But according to equation 12.47, this limit defines the derivative $d\mathbf{r}/dt$,

$$\frac{d\mathbf{r}}{dt} = \lim_{h\to 0} \frac{\mathbf{r}(t+h) - \mathbf{r}(t)}{h} = \frac{dx}{dt}\hat{\mathbf{i}} + \frac{dy}{dt}\hat{\mathbf{j}} + \frac{dz}{dt}\hat{\mathbf{k}}, \qquad (12.58)$$

provided each of the derivatives dx/dt, dy/dt, and dz/dt exists. We have just established the following result.

Theorem 12.10

If $\mathbf{r} = \mathbf{r}(t) = x(t)\hat{\mathbf{i}} + y(t)\hat{\mathbf{j}} + z(t)\hat{\mathbf{k}}$, $\alpha \leq t \leq \beta$, is the vector representation of a curve C, then at any point on C at which $x'(t)$, $y'(t)$, and $z'(t)$ all exist and do not simultaneously vanish,

$$\mathbf{T} = \frac{d\mathbf{r}}{dt} = \frac{dx}{dt}\hat{\mathbf{i}} + \frac{dy}{dt}\hat{\mathbf{j}} + \frac{dz}{dt}\hat{\mathbf{k}} \qquad (12.59)$$

is a tangent vector to C (Figure 12.105).

There are two tangent directions at any point on a curve. One of these has been shown to be $d\mathbf{r}/dt$; the other must be defined by $-d\mathbf{r}/dt$ (Figure 12.106). How can we tell which one is $d\mathbf{r}/dt$? A closer analysis of the limit in 12.58 indicates the following (see Exercise 17).

Corollary The tangent vector $d\mathbf{r}/dt$ to a curve C: $x = x(t)$, $y = y(t)$, $z = z(t)$, $\alpha \leq t \leq \beta$, always points in the direction in which the parameter t increases along C.

FIGURE 12.105

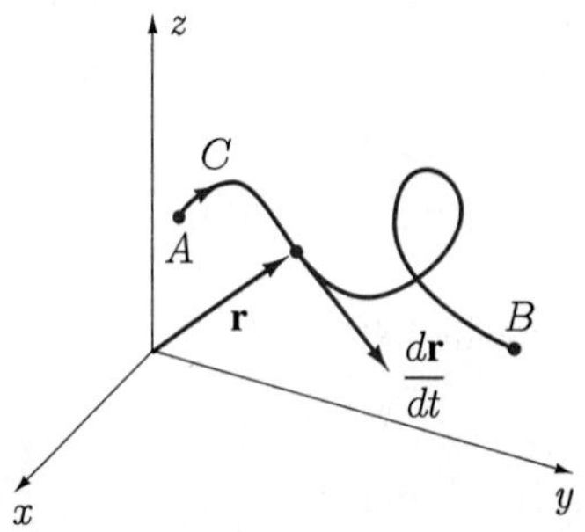

FIGURE 12.106

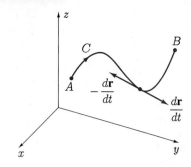

Definition 12.17

A curve C: $x = x(t)$, $y = y(t)$, $z = z(t)$, $\alpha \leq t \leq \beta$, is said to be **smooth** if the derivatives $x'(t)$, $y'(t)$, and $z'(t)$ are all continuous for $\alpha < t < \beta$ and do not simultaneously vanish for $\alpha < t < \beta$.

Since $x'(t)$, $y'(t)$, and $z'(t)$ are the components of a tangent vector to C, this definition implies that along a smooth curve, small changes in t produce small changes in the direction of the tangent vector. In other words, the tangent vector turns gradually, or "smoothly." The curve in Figure 12.107 is smooth; that in Figure 12.108 is not because abrupt changes in the direction of the curve occur at P and Q. According to the following definition, this curve is piecewise smooth.

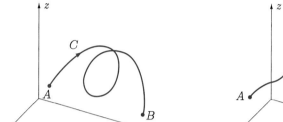

FIGURE 12.107

FIGURE 12.108

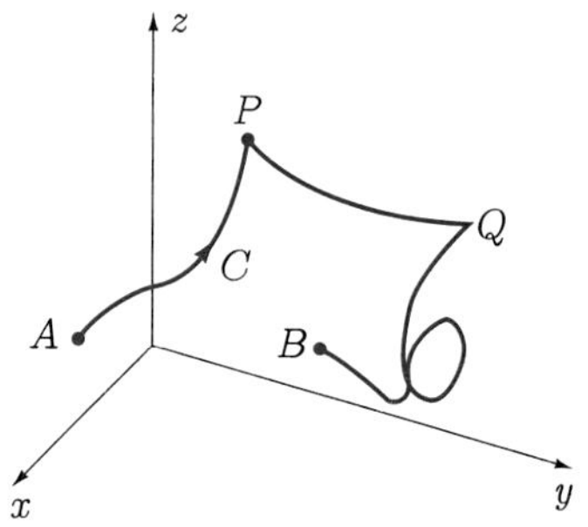

Definition 12.18

A curve is said to be **piecewise smooth** if it is continuous and can be divided into a finite number of smooth subcurves.

EXAMPLE 12.31

For the curve in Example 30, find a tangent vector at the point $(3, 9, 3)$.

SOLUTION A tangent vector to this curve at any point on the curve is

$$\frac{d\mathbf{r}}{dt} = \frac{dx}{dt}\hat{\mathbf{i}} + \frac{dy}{dt}\hat{\mathbf{j}} + \frac{dz}{dt}\hat{\mathbf{k}} = \hat{\mathbf{i}} + 2t\hat{\mathbf{j}} + \hat{\mathbf{k}}.$$

Since $t = 3$ yields the point $(3, 9, 3)$, a tangent vector at this point is $\mathbf{r}'(3) = \hat{\mathbf{i}} + 6\hat{\mathbf{j}} + \hat{\mathbf{k}}$.

EXAMPLE 12.32

Find a tangent vector at the point $(2, 0, 3)$ to the helix

$$x = 2\cos t, \quad y = 2\sin t, \quad z = \frac{3t}{2\pi}, \quad t \geq 0.$$

Is the helix smooth?

SOLUTION

A tangent vector to the helix at any point is

$$\frac{d\mathbf{r}}{dt} = \frac{dx}{dt}\hat{\mathbf{i}} + \frac{dy}{dt}\hat{\mathbf{j}} + \frac{dz}{dt}\hat{\mathbf{k}} = (-2\sin t)\hat{\mathbf{i}} + (2\cos t)\hat{\mathbf{j}} + \left(\frac{3}{2\pi}\right)\hat{\mathbf{k}}.$$

Since $t = 2\pi$ yields the point $(2, 0, 3)$, a tangent vector at this point is $\mathbf{r}'(2\pi) = 2\hat{\mathbf{j}} + (3/(2\pi))\hat{\mathbf{k}}$ (see Figure 12.109). Since $x'(t)$, $y'(t)$, and $z'(t)$ are all continuous functions, and they are not simultaneously zero, the helix is indeed smooth. ∎

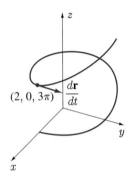

FIGURE 12.109

Unit Tangent Vectors

When a curve in the xy-plane is defined parametrically by

$$C: \quad x = x(t), \quad y = y(t), \quad \alpha \leq t \leq \beta, \tag{12.60}$$

a tangent vector to C is

$$\mathbf{T} = \frac{d\mathbf{r}}{dt} = \frac{dx}{dt}\hat{\mathbf{i}} + \frac{dy}{dt}\hat{\mathbf{j}}, \tag{12.61}$$

and this tangent vector points in the direction in which t increases along C. To produce a unit tangent vector to C at any point, we divide $\mathbf{T}$ by its length:

$$\hat{\mathbf{T}} = \frac{\mathbf{T}}{|\mathbf{T}|} = \frac{d\mathbf{r}/dt}{|d\mathbf{r}/dt|}. \tag{12.62}$$

We now show that if length along C is used as the parameter by which to specify its points, then division by $|\mathbf{T}|$ is unnecessary.

In Section 7.3 we showed that small lengths along a plane curve C can be approximated by straight-line lengths along tangent lines to the curve, and that the total length of a smooth curve from A to B (Figure 12.110) is

$$L = \int_A^B \sqrt{(dx)^2 + (dy)^2}.$$

FIGURE 12.110

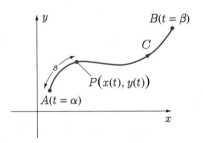

With parametric equations 12.60 we can write this formula as a definite integral with respect to t (see also equation 10.3):

$$L = \int_\alpha^\beta \sqrt{\left[\left(\frac{dx}{dt}\right)^2 + \left(\frac{dy}{dt}\right)^2\right](dt)^2} = \int_\alpha^\beta \sqrt{\left(\frac{dx}{dt}\right)^2 + \left(\frac{dy}{dt}\right)^2}\, dt. \quad (12.63)$$

Furthermore, if we denote by $s = s(t)$ the length of that part of C from its initial point A (where $t = \alpha$) to any point $P(x(t), y(t))$ on C (Figure 12.110), then $s(t)$ is defined by the integral

$$s(t) = \int_\alpha^t \sqrt{\left(\frac{dx}{dt}\right)^2 + \left(\frac{dy}{dt}\right)^2}\, dt. \quad (12.64)$$

It follows, then, that the derivative of $s(t)$ is

$$\frac{ds}{dt} = \sqrt{\left(\frac{dx}{dt}\right)^2 + \left(\frac{dy}{dt}\right)^2}. \quad (12.65)$$

But according to 12.61, $d\mathbf{r}/dt$ is a tangent vector to C with the same length:

$$\left|\frac{d\mathbf{r}}{dt}\right| = \sqrt{\left(\frac{dx}{dt}\right)^2 + \left(\frac{dy}{dt}\right)^2} = \frac{ds}{dt}. \quad (12.66)$$

When this equation is multiplied by dt, it gives

$$|d\mathbf{r}| = \sqrt{(dx)^2 + (dy)^2} = ds. \quad (12.67)$$

This equation states that ds is the length of the tangent vector $d\mathbf{r} = dx\hat{\mathbf{i}} + dy\hat{\mathbf{j}}$, and therefore ds is a measure of length along the tangent line to C. In spite of this we often think of ds as a measure of small lengths along C itself (Figure 12.111), and that ds is approximated by the tangential straight-line length

$$|d\mathbf{r}| = \sqrt{(dx)^2 + (dy)^2}.$$

Note too that if we use length s along C as a parameter, then the chain rule applied to $\mathbf{r} = \mathbf{r}(s)$, $s = s(t)$ gives

$$\frac{d\mathbf{r}}{dt} = \frac{d\mathbf{r}}{ds}\frac{ds}{dt}. \quad (12.68)$$

FIGURE 12.111

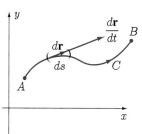

(The chain rule is proved in Exercise 26 of Section 12.6.) Consequently, equation 12.66 implies that

$$\frac{d\mathbf{r}}{ds} = \frac{d\mathbf{r}/dt}{ds/dt} = \frac{d\mathbf{r}/dt}{|d\mathbf{r}/dt|}. \quad (12.69)$$

What we have shown, then, is that if we choose length along a curve C as the parameter by which to specify points on the curve (C: $x = x(s)$, $y = y(s)$), then the vector

$$\hat{\mathbf{T}} = \frac{d\mathbf{r}}{ds} = \frac{dx}{ds}\hat{\mathbf{i}} + \frac{dy}{ds}\hat{\mathbf{j}} \quad (12.70)$$

is a unit tangent vector to C. In addition, the corollary to Theorem 12.10 implies that $d\mathbf{r}/ds$ points in the direction in which s increases along C. This suggests perhaps that we should always set up parametric equations for a curve with length along the curve as parameter. Theoretically this is quite acceptable, but practically it is impossible. For most curves we have enough difficulty just finding a set of parametric equations, let alone finding that set with length along the curve as parameter. If we then use a parameter t other than length along the curve, a unit tangent vector is calculated according to 12.62.

These results can be extended to space curves as well. When a smooth curve C has parametric equations 12.56, equation 12.62 still defines a unit tangent vector to C, but because C is a space curve, $d\mathbf{r}/dt$ is calculated according to 12.59.

Corresponding to formula 12.64 for length along a curve in the xy-plane, length along a smooth curve in space is defined by the definite integral

$$s(t) = \int_\alpha^t \sqrt{\left(\frac{dx}{dt}\right)^2 + \left(\frac{dy}{dt}\right)^2 + \left(\frac{dz}{dt}\right)^2}\, dt. \qquad (12.71)$$

These two results imply that

$$\left|\frac{d\mathbf{r}}{dt}\right| = \sqrt{\left(\frac{dx}{dt}\right)^2 + \left(\frac{dy}{dt}\right)^2 + \left(\frac{dz}{dt}\right)^2} = \frac{ds}{dt}, \qquad (12.72)$$

the three-space analogue of 12.66. Once again we are led to the fact that when s is used as parameter along C, then

$$\hat{\mathbf{T}} = \frac{d\mathbf{r}}{ds} = \frac{dx}{ds}\hat{\mathbf{i}} + \frac{dy}{ds}\hat{\mathbf{j}} + \frac{dz}{ds}\hat{\mathbf{k}} \qquad (12.73)$$

is a unit tangent vector to C. In addition, multiplication of 12.72 by dt yields

$$|d\mathbf{r}| = ds = \sqrt{(dx)^2 + (dy)^2 + (dz)^2}, \qquad (12.74)$$

indicating that small lengths ds along C (Figure 12.112) are defined in terms of small lengths $|d\mathbf{r}|$ along the tangent line to C.

FIGURE 12.112

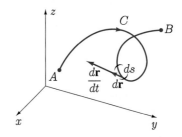

EXAMPLE 12.33

Find a unit tangent vector to the curve

$$C: \quad x = \sin t, \quad y = 2\cos t, \quad z = 2t/\pi, \quad t \geq 0$$

at the point $(0, -2, 2)$.

SOLUTION A tangent vector to C at any point is

$$\frac{d\mathbf{r}}{dt} = \frac{dx}{dt}\hat{\mathbf{i}} + \frac{dy}{dt}\hat{\mathbf{j}} + \frac{dz}{dt}\hat{\mathbf{k}} = \cos t\,\hat{\mathbf{i}} - 2\sin t\,\hat{\mathbf{j}} + \left(\frac{2}{\pi}\right)\hat{\mathbf{k}}.$$

Since $t = \pi$ yields the point $(0, -2, 2)$, a tangent vector at this point is $\mathbf{r}'(\pi) = -\hat{\mathbf{i}} + (2/\pi)\hat{\mathbf{k}}$. A unit tangent vector is then

$$\hat{\mathbf{T}} = \frac{-\hat{\mathbf{i}} + (2/\pi)\hat{\mathbf{k}}}{\sqrt{1 + 4/\pi^2}} = \frac{-\pi\hat{\mathbf{i}} + 2\hat{\mathbf{k}}}{\sqrt{4 + \pi^2}}.$$

∎

EXAMPLE 12.34

Find the length of that part of the curve $x = y^{2/3}$, $x = z^{2/3}$ between the points $(0, 0, 0)$ and $(4, 8, 8)$.

SOLUTION If we use $x = t$, $y = t^{3/2}$, $z = t^{3/2}$, $0 \leq t \leq 4$, as parametric equations for the curve, then

$$L = \int_0^4 \sqrt{\left(\frac{dx}{dt}\right)^2 + \left(\frac{dy}{dt}\right)^2 + \left(\frac{dz}{dt}\right)^2}\, dt = \int_0^4 \sqrt{1 + \left(\frac{3}{2}\sqrt{t}\right)^2 + \left(\frac{3}{2}\sqrt{t}\right)^2}\, dt$$

$$= \int_0^4 \sqrt{1 + \frac{9t}{2}}\, dt = \left\{\frac{4}{27}\left(1 + \frac{9t}{2}\right)^{3/2}\right\}_0^4 = \frac{4}{27}(19\sqrt{19} - 1).$$

∎

EXERCISES 12.8

In Exercises 1–5 express the curve in vector form and find the unit tangent vector $\hat{\mathbf{T}}$ at each point on the curve.

1. $x = \sin t$, $y = \cos t$, $z = t$, $-\infty < t < \infty$
2. $x = t$, $y = t^2$, $z = t^3$, $t \geq 1$
3. $x = (t-1)^2$, $y = (t+1)^2$, $z = -t$, $-3 \leq t \leq 4$
4. $x + y = 5$, $x^2 - y = z$ from $(5, 0, 25)$ to $(0, 5, -5)$
5. $x + y + z = 4$, $x^2 + y^2 = 4$, $y \geq 0$ from $(2, 0, 2)$ to $(-2, 0, 6)$

In Exercises 6–10, find $\hat{\mathbf{T}}$ at the point.

6. $x = 4\cos t$, $y = 6\sin t$, $z = 2\sin t$, $-\infty < t < \infty$; $(2\sqrt{2}, 3\sqrt{2}, \sqrt{2})$
7. $x = 2 - 5t$, $y = 1 + t$, $z = 6 + 4t$, $-\infty < t < \infty$; $(7, 0, 2)$
8. $x^2 + y^2 + z^2 = 4$, $z = \sqrt{x^2 + y^2}$, directed so that x increases when y is positive; $(1, 1, \sqrt{2})$
9. $x = y^2 + 1$, $z = x + 5$, directed so that y increases along the curve; $(5, 2, 10)$
10. $x^2 + (y - 1)^2 = 4$, $z = x$, directed so that z decreases when y is negative; $(2, 1, 2)$

In Exercises 11–14 find the length of the curve. Sketch each curve.

11. $x = 2\cos t$, $y = 2\sin t$, $z = 3t$, $0 \leq t \leq 2\pi$
12. $x = 2 - 5t$, $y = 1 + t$, $z = 6 + 4t$, $-1 \leq t \leq 0$
13. $x = t^3$, $y = t^2$, $z = t^3$, $0 \leq t \leq 1$
14. $x = t$, $y = t^{3/2}$, $z = 4t^{3/2}$, $1 \leq t \leq 4$

15. In Definition 12.17 why are the derivatives assumed not to vanish simultaneously? Hint: Consider the curve $x = t^3$, $y = t^2$, $z = 0$.

16. Find a unit tangent vector to the curve $\mathbf{r} = (\cos t + t\sin t)\hat{\mathbf{i}} + (\sin t - t\cos t)\hat{\mathbf{j}}$, called an involute of a circle.

17. Show that the tangent vector $d\mathbf{r}/dt$ to a curve described by equation 12.57 always points in the direction in which t increases along the curve.

18. (a) What happens when equation 12.59 is used to determine a tangent vector to the curve $x = t^2$, $y = t^3$, $z = t^2$, $-\infty < t < \infty$, at the origin?
    (b) Can you devise a way in which to obtain a tangent vector?

SECTION 12.9

Normal Vectors, Curvature, and Radius of Curvature

In discussing curves we distinguish between two types of properties: intrinsic and not intrinsic. An intrinsic property is one that is independent of the parameter used to specify the curve; a property that is not intrinsic is parameter-dependent. To illustrate, the tangent vector $\mathbf{T} = d\mathbf{r}/dt$ in 12.59 is not intrinsic; a change of parameter results in a change in the length of $\mathbf{T}$. The unit tangent vector $\hat{\mathbf{T}}$, on the other hand, is intrinsic; there is only one unit tangent vector in the direction of the curve. Length of a curve from its initial point to an arbitrary point is an intrinsic property; a change of parameter along the curve does not affect length between points.

Because length along a curve is an intrinsic property, it is customary in theoretical discussions to use it as the parameter by which to specify points on the curve. When C is a smooth curve in the xy-plane, parametric equations for C in terms of length s along C take the form

$$C: \quad x = x(s), \quad y = y(s), \quad 0 \le s \le L. \tag{12.75}$$

Normal Vectors to Curves

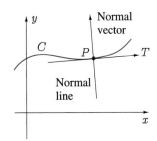

FIGURE 12.113

The normal line at a point P on a smooth curve C in the xy-plane is that line which is perpendicular to the tangent line to C at P (Figure 12.113). Any vector along this normal line is said to be a normal vector to the curve at P. Since the unit tangent vector to C at P is

$$\hat{\mathbf{T}} = \frac{d\mathbf{r}}{ds} = \frac{dx}{ds}\hat{\mathbf{i}} + \frac{dy}{ds}\hat{\mathbf{j}},$$

it follows that

$$\hat{\mathbf{N}} = -\frac{dy}{ds}\hat{\mathbf{i}} + \frac{dx}{ds}\hat{\mathbf{j}} \tag{12.76}$$

is a unit normal vector to C at P (note that $\hat{\mathbf{T}} \cdot \hat{\mathbf{N}} = 0$). Because there is only one direction normal to C at P, every normal vector to C at P must be some multiple $\lambda \hat{\mathbf{N}}$ of $\hat{\mathbf{N}}$.

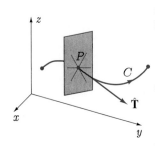

FIGURE 12.114

The situation is quite different for space curves (Figure 12.114). If $\hat{\mathbf{T}}$ is the unit tangent vector to a smooth curve C, then there is an entire plane of normal vectors to C at P. In the following discussion, we single out two normal vectors called the **principal normal** and the **binormal**. Suppose that C is defined parametrically by

$$C: \quad x = x(s), \quad y = y(s), \quad z = z(s), \quad 0 \le s \le L, \tag{12.77}$$

and that $\hat{\mathbf{T}}$ is the unit tangent vector to C defined by 12.73. Because $\hat{\mathbf{T}}$ has unit length,

$$1 = \hat{\mathbf{T}} \cdot \hat{\mathbf{T}}.$$

If we use equation 12.51 to differentiate this equation with respect to s, we have

$$0 = \frac{d\hat{\mathbf{T}}}{ds} \cdot \hat{\mathbf{T}} + \hat{\mathbf{T}} \cdot \frac{d\hat{\mathbf{T}}}{ds} = 2\left(\hat{\mathbf{T}} \cdot \frac{d\hat{\mathbf{T}}}{ds}\right).$$

But if neither of the vectors $\hat{\mathbf{T}}$ nor $d\hat{\mathbf{T}}/ds$ is equal to zero, then the fact that their scalar product is equal to zero implies that they are perpendicular. In other words,

$$\mathbf{N} = \frac{d\hat{\mathbf{T}}}{ds} \qquad (12.78)$$

is a normal vector to C at any point. The unit normal vector in this direction

$$\hat{\mathbf{N}} = \frac{\mathbf{N}}{|\mathbf{N}|} = \frac{d\hat{\mathbf{T}}/ds}{|d\hat{\mathbf{T}}/ds|} \qquad (12.79)$$

is called the **principal normal** (vector) to C (Figure 12.115).

FIGURE 12.115

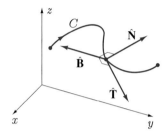

Because $\hat{\mathbf{N}}$ is defined in terms of intrinsic properties $\hat{\mathbf{T}}$ and s for a curve, it must also be an intrinsic property. It follows, then, that no matter what parameter is used to specify points on a curve, $\hat{\mathbf{N}}$ is always the same. But how do we find $\hat{\mathbf{N}}$ when a curve C is specified in terms of a parameter other than length along C, say, in the form

$$C: \quad x = x(t), \quad y = y(t), \quad z = z(t), \quad \alpha \leq t \leq \beta? \qquad (12.80)$$

If $s = s(t)$ is length along C (measured from $t = \alpha$), then by the chain rule

$$\frac{d\hat{\mathbf{T}}}{ds} = \frac{d\hat{\mathbf{T}}}{dt}\frac{dt}{ds},$$

where dt/ds must be positive since both s and t increase along C. Consequently,

$$\hat{\mathbf{N}} = \frac{d\hat{\mathbf{T}}/ds}{|d\hat{\mathbf{T}}/ds|} = \frac{(d\hat{\mathbf{T}}/dt)(dt/ds)}{|(d\hat{\mathbf{T}}/dt)(dt/ds)|} = \frac{(d\hat{\mathbf{T}}/dt)(dt/ds)}{|d\hat{\mathbf{T}}/dt|dt/ds} = \frac{d\hat{\mathbf{T}}/dt}{|d\hat{\mathbf{T}}/dt|}. \qquad (12.81)$$

In other words, for any parametrization of C whatsoever, the vector $d\hat{\mathbf{T}}/dt$ always points in the direction of the principal normal, and to find $\hat{\mathbf{N}}$, we simply find the unit vector in the direction of $d\hat{\mathbf{T}}/dt$.

In the study of space curves, a second normal vector to C, called the **binormal**, is defined by

$$\hat{\mathbf{B}} = \hat{\mathbf{T}} \times \hat{\mathbf{N}}. \qquad (12.82)$$

Since the cross product of two vectors is always perpendicular to each of the vectors, it follows that the binormal is perpendicular to both $\hat{\mathbf{T}}$ and $\hat{\mathbf{N}}$, and must therefore indeed be a normal vector to C (Figure 12.115).

We have singled out three vectors at each point P on a curve C: a unit tangent vector $\hat{\mathbf{T}}$ and two unit normal vectors $\hat{\mathbf{N}}$ and $\hat{\mathbf{B}}$. As P moves along C, these vectors constantly change direction but always have unit length.

EXAMPLE 12.35

Find $\hat{\mathbf{T}}$, $\hat{\mathbf{N}}$, and $\hat{\mathbf{B}}$ for the curve $x = t$, $y = t^2$, $z = t^2$, $t \geq 0$.

SOLUTION The unit tangent vector $\hat{\mathbf{T}}$ is defined by

$$\hat{\mathbf{T}} = \frac{d\mathbf{r}/dt}{|d\mathbf{r}/dt|} = \frac{(1, 2t, 2t)}{\sqrt{1 + 4t^2 + 4t^2}} = \frac{(1, 2t, 2t)}{\sqrt{1 + 8t^2}}.$$

The principal normal $\hat{\mathbf{N}}$ lies along the vector $\mathbf{N} = d\hat{\mathbf{T}}/dt$, and according to equation 12.50, we can write that

$$\begin{aligned}
\mathbf{N} = \frac{d\hat{\mathbf{T}}}{dt} &= \frac{d}{dt}\left(\frac{1}{\sqrt{1 + 8t^2}}\right)(1, 2t, 2t) + \frac{1}{\sqrt{1 + 8t^2}}\frac{d}{dt}(1, 2t, 2t) \\
&= \frac{-8t}{(1 + 8t^2)^{3/2}}(1, 2t, 2t) + \frac{1}{\sqrt{1 + 8t^2}}(0, 2, 2) \\
&= \frac{1}{(1 + 8t^2)^{3/2}}\{-8t(1, 2t, 2t) + (1 + 8t^2)(0, 2, 2)\} \\
&= \frac{(-8t, 2, 2)}{(1 + 8t^2)^{3/2}}.
\end{aligned}$$

The principal normal is therefore

$$\hat{\mathbf{N}} = \frac{\mathbf{N}}{|\mathbf{N}|} = \frac{(-8t, 2, 2)}{\sqrt{64t^2 + 4 + 4}} = \frac{(-4t, 1, 1)}{\sqrt{2 + 16t^2}}.$$

The binormal is

$$\begin{aligned}
\hat{\mathbf{B}} = \hat{\mathbf{T}} \times \hat{\mathbf{N}} &= \frac{(1, 2t, 2t)}{\sqrt{1 + 8t^2}} \times \frac{(-4t, 1, 1)}{\sqrt{2 + 16t^2}} \\
&= \frac{1}{\sqrt{1 + 8t^2}\sqrt{2 + 16t^2}} \begin{vmatrix} \hat{\mathbf{i}} & \hat{\mathbf{j}} & \hat{\mathbf{k}} \\ 1 & 2t & 2t \\ -4t & 1 & 1 \end{vmatrix} \\
&= \frac{1}{\sqrt{2}\sqrt{1 + 8t^2}\sqrt{1 + 8t^2}}(0, -1 - 8t^2, 1 + 8t^2) \\
&= \frac{1 + 8t^2}{\sqrt{2}(1 + 8t^2)}(0, -1, 1) \\
&= \frac{(0, -1, 1)}{\sqrt{2}}.
\end{aligned}$$

The significance of the fact that the binormal has constant direction can be seen from a sketch of the curve. Because the parametric equations imply that $y = x^2$ and $z = y$, the curve is the curve of intersection of these two surfaces (Figure 12.116). Since the curve lies in the plane $-y + z = 0$, and a normal vector to this plane is $(0, -1, 1)$, it follows that $(0, -1, 1)$ is always normal to the curve. But this is precisely the direction of $\hat{\mathbf{B}}$. In other words, constant $\hat{\mathbf{B}}$ implies that the curve lies in a plane that has $\hat{\mathbf{B}}$ as normal (see Exercise 31).

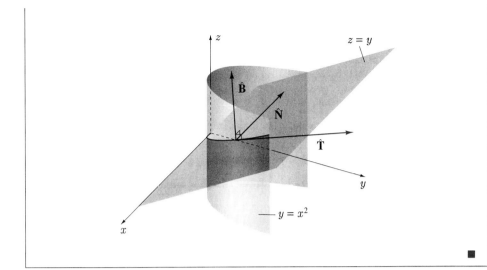

FIGURE 12.116

EXAMPLE 12.36

Show that for a smooth curve C: $x = x(t)$, $y = y(t)$, $\alpha \leq t \leq \beta$ in the xy-plane, the principal normal is

$$\hat{\mathbf{N}} = \text{sgn}\left(\frac{dy}{dt}\frac{d^2x}{dt^2} - \frac{dx}{dt}\frac{d^2y}{dt^2}\right) \frac{(dy/dt, -dx/dt)}{\sqrt{(dx/dt)^2 + (dy/dt)^2}},$$

where the signum function $\text{sgn}(u)$ is defined by

$$\text{sgn}(u) = \begin{cases} 1, & \text{if } u > 0, \\ 0, & \text{if } u = 0, \\ -1, & \text{if } u < 0. \end{cases}$$

SOLUTION The unit tangent vector to C is

$$\hat{\mathbf{T}} = \frac{(dx/dt, dy/dt)}{\sqrt{(dx/dt)^2 + (dy/dt)^2}}.$$

For simplicity in notation, we use a dot "·" above a variable to indicate that the variable is differentiated with respect to t. For example, $\dot{x} = dx/dt$ and $\ddot{x} = d^2x/dt^2$. With this notation,

$$\hat{\mathbf{T}} = \frac{(\dot{x}, \dot{y})}{\sqrt{\dot{x}^2 + \dot{y}^2}}.$$

By equation 12.8, $\hat{\mathbf{N}} = (d\hat{\mathbf{T}}/dt)/|d\hat{\mathbf{T}}/dt|$, where

$$\frac{d\hat{\mathbf{T}}}{dt} = \frac{d}{dt}\frac{(\dot{x}, \dot{y})}{\sqrt{\dot{x}^2 + \dot{y}^2}} = \frac{d}{dt}\left(\frac{1}{\sqrt{\dot{x}^2 + \dot{y}^2}}\right)(\dot{x}\hat{\mathbf{i}} + \dot{y}\hat{\mathbf{j}}) + \frac{1}{\sqrt{\dot{x}^2 + \dot{y}^2}}(\ddot{x}\hat{\mathbf{i}} + \ddot{y}\hat{\mathbf{j}})$$

$$= \left(\frac{-\dot{x}\ddot{x} - \dot{y}\ddot{y}}{(\dot{x}^2 + \dot{y}^2)^{3/2}}\right)(\dot{x}\hat{\mathbf{i}} + \dot{y}\hat{\mathbf{j}}) + \frac{1}{\sqrt{\dot{x}^2 + \dot{y}^2}}(\ddot{x}\hat{\mathbf{i}} + \ddot{y}\hat{\mathbf{j}})$$

$$= \frac{1}{(\dot{x}^2+\dot{y}^2)^{3/2}}\{-(\dot{x}\ddot{x}+\dot{y}\ddot{y})(\dot{x}\hat{\mathbf{i}}+\dot{y}\hat{\mathbf{j}})+(\dot{x}^2+\dot{y}^2)(\ddot{x}\hat{\mathbf{i}}+\ddot{y}\hat{\mathbf{j}})\}$$

$$= \frac{1}{(\dot{x}^2+\dot{y}^2)^{3/2}}\{(-\dot{x}^2\ddot{x}-\dot{x}\dot{y}\ddot{y}+\dot{x}^2\ddot{x}+\dot{y}^2\ddot{x})\hat{\mathbf{i}}$$
$$+ (-\dot{x}\dot{y}\ddot{x}-\dot{y}^2\ddot{y}+\dot{x}^2\ddot{y}+\dot{y}^2\ddot{y})\hat{\mathbf{j}}\}$$

$$= \frac{1}{(\dot{x}^2+\dot{y}^2)^{3/2}}\{\dot{y}(\dot{y}\ddot{x}-\dot{x}\ddot{y})\hat{\mathbf{i}}+\dot{x}(\dot{x}\ddot{y}-\dot{y}\ddot{x})\hat{\mathbf{j}}\}$$

$$= \frac{\dot{y}\ddot{x}-\dot{x}\ddot{y}}{(\dot{x}^2+\dot{y}^2)^{3/2}}(\dot{y}\hat{\mathbf{i}}-\dot{x}\hat{\mathbf{j}}).$$

If $\dot{y}\ddot{x}-\dot{x}\ddot{y}$ is positive, then

$$\hat{\mathbf{N}} = \frac{\dot{y}\hat{\mathbf{i}}-\dot{x}\hat{\mathbf{j}}}{\sqrt{\dot{x}^2+\dot{y}^2}};$$

whereas if $\dot{y}\ddot{x}-\dot{x}\ddot{y}$ is negative, then

$$\hat{\mathbf{N}} = \frac{-\dot{y}\hat{\mathbf{i}}+\dot{x}\hat{\mathbf{j}}}{\sqrt{\dot{x}^2+\dot{y}^2}}.$$

In other words,

$$\hat{\mathbf{N}} = \operatorname{sgn}(\dot{y}\ddot{x}-\dot{x}\ddot{y})\left(\frac{\dot{y}\hat{\mathbf{i}}-\dot{x}\hat{\mathbf{j}}}{\sqrt{\dot{x}^2+\dot{y}^2}}\right).$$

∎

Curvature and Radius of Curvature

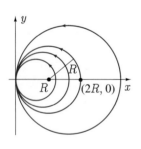

FIGURE 12.117

When length s along a smooth curve C is used as the parameter by which to identify points on the curve, the vector $\hat{\mathbf{T}} = d\mathbf{r}/ds$ is a unit tangent vector to C. Suppose we differentiate $\hat{\mathbf{T}}$ with respect to s to form $d\hat{\mathbf{T}}/ds = d^2\mathbf{r}/ds^2$. Since $\hat{\mathbf{T}}$ has constant unit length, only its direction can change; therefore, the derivative $d\hat{\mathbf{T}}/ds$ must be a measure of the rate of change of the direction of $\hat{\mathbf{T}}$. Since $\hat{\mathbf{T}}$ is really our way of specifying the direction of the curve itself, we can also say that $d\hat{\mathbf{T}}/ds$ is a measure of how fast the direction of C changes. But exactly how does a vector $d\hat{\mathbf{T}}/ds$ which has both magnitude and direction measure the rate of change of the direction of C? We illustrate by example that it cannot be the direction of $d\hat{\mathbf{T}}/ds$; it must be its magnitude that measures the rate of change of the direction of C. In Figure 12.117 we show a number of circles in the xy-plane, all of which are tangent to the y-axis at the origin. Parametric equations for the circle with centre $(R,0)$ and radius R in terms of length s along the circle (as measured from $(2R,0)$) are

$$x = R + R\cos(s/R), \quad y = R\sin(s/R), \quad 0 \le s < 2\pi R.$$

Consequently,

$$\hat{\mathbf{T}} = \frac{d\mathbf{r}}{ds} = -\sin\left(\frac{s}{R}\right)\hat{\mathbf{i}} + \cos\left(\frac{s}{R}\right)\hat{\mathbf{j}}$$

and

$$\frac{d\hat{\mathbf{T}}}{ds} = -\frac{1}{R}\cos\left(\frac{s}{R}\right)\hat{\mathbf{i}} - \frac{1}{R}\sin\left(\frac{s}{R}\right)\hat{\mathbf{j}}.$$

At the origin, $s = \pi R$, and

$$\left.\frac{d\hat{\mathbf{T}}}{ds}\right|_{s=\pi R} = \frac{1}{R}\hat{\mathbf{i}}.$$

Thus for each of the circles in Figure 12.117, the vector $d\hat{\mathbf{T}}/ds$ has exactly the same direction. Yet the rate of change of the direction of $\hat{\mathbf{T}}$ is not the same for each circle; the direction changes more rapidly as the radius of the circle decreases. We must conclude, therefore, that it cannot be the direction of $d\hat{\mathbf{T}}/ds$ that measures the rate of change of $\hat{\mathbf{T}}$. Since a vector has only length and direction, it must be the length of $d\hat{\mathbf{T}}/ds$ that measures this rate of change. The circles in Figure 12.117 certainly support this claim; the length of $d\hat{\mathbf{T}}/ds$ is $1/R$, and this quantity increases as the radii of the circles decrease. This agrees with the fact that the rate at which $\hat{\mathbf{T}}$ turns increases as R decreases. According to the following definition, we call $|d\hat{\mathbf{T}}/ds|$ **curvature** and $1/|d\hat{\mathbf{T}}/ds|$ **radius of curvature**.

> **Definition 12.19**
>
> If $x = x(s)$, $y = y(s)$, $z = z(s)$, $0 \leq s \leq L$, are parametric equations for a smooth curve in terms of length s along the curve, we define the **curvature** of the curve at a point as
>
> $$\kappa(s) = \left|\frac{d\hat{\mathbf{T}}}{ds}\right|, \qquad (12.83)$$
>
> its **radius of curvature** as
>
> $$\rho(s) = \frac{1}{\kappa(s)}, \qquad (12.84)$$
>
> and its **circle of curvature** as that circle in the plane of $\hat{\mathbf{T}}$ and $\hat{\mathbf{N}}$ with centre at $\mathbf{r}(s) + \rho(s)\hat{\mathbf{N}}$ and radius $\rho(s)$.

The circle of curvature is illustrated in Figure 12.118.

FIGURE 12.118

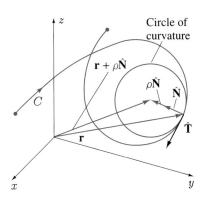

For the circles in Figure 12.117 we have already shown that $d\hat{\mathbf{T}}/ds = R^{-1}\hat{\mathbf{i}}$. Consequently, for these circles, the curvature is always R^{-1}, and the radius of curvature is R, the radius of the circle. In other words, for a circle, the circle of curvature is the circle itself, its radius of curvature is its radius, and its curvature is the inverse of its radius. For the case when a curve is not a circle, we show in Exercise 27 that at any point on the curve the circle of curvature is in some sense the "best-fitting" circle to the curve at that point.

Because curvature and radius of curvature have been defined in terms of intrinsic properties $\hat{\mathbf{T}}$ and s for a curve, they must also be intrinsic properties. It follows, then, that no matter what parameter is used to specify points on a curve, curvature and radius of curvature are always the same. The following theorem shows how to calculate κ and ρ when the curve is specified in terms of a parameter other than length along the curve.

Theorem 12.11

When a smooth curve is defined parametrically by

$$C: \quad x = x(t), \quad y = y(t), \quad z = z(t), \quad \alpha \leq t \leq \beta,$$

its curvature $\kappa(t)$ is given by

$$\kappa(t) = \frac{|\dot{\mathbf{r}} \times \ddot{\mathbf{r}}|}{|\dot{\mathbf{r}}|^3}, \tag{12.85}$$

where $\dot{\mathbf{r}} = d\mathbf{r}/dt$ and $\ddot{\mathbf{r}} = d^2\mathbf{r}/dt^2$.

Proof If $s(t)$ is length along C (measured from $t = \alpha$), then by the chain rule

$$\kappa = \left|\frac{d\hat{\mathbf{T}}}{ds}\right| = \left|\frac{d\hat{\mathbf{T}}}{dt}\frac{dt}{ds}\right| = \frac{|d\hat{\mathbf{T}}/dt|}{|ds/dt|} = \frac{|d\hat{\mathbf{T}}/dt|}{ds/dt} = \frac{|d\hat{\mathbf{T}}/dt|}{|\dot{\mathbf{r}}|}$$

(see equation 12.72). Now, we can write $\dot{\mathbf{r}}$ in the form $\dot{\mathbf{r}} = |\dot{\mathbf{r}}|\hat{\mathbf{T}}$; therefore, using 12.50, we have

$$\ddot{\mathbf{r}} = \left(\frac{d}{dt}|\dot{\mathbf{r}}|\right)\hat{\mathbf{T}} + |\dot{\mathbf{r}}|\frac{d\hat{\mathbf{T}}}{dt}$$

$$= \left(\frac{d}{dt}|\dot{\mathbf{r}}|\right)\hat{\mathbf{T}} + (|\dot{\mathbf{r}}||d\hat{\mathbf{T}}/dt|)\frac{d\hat{\mathbf{T}}/dt}{|d\hat{\mathbf{T}}/dt|}$$

$$= \left(\frac{d}{dt}|\dot{\mathbf{r}}|\right)\hat{\mathbf{T}} + \left(|\dot{\mathbf{r}}|\left|\frac{d\hat{\mathbf{T}}}{dt}\right|\right)\hat{\mathbf{N}}.$$

If we take the cross product of this vector with $\dot{\mathbf{r}}$, we get

$$\dot{\mathbf{r}} \times \ddot{\mathbf{r}} = \left(\frac{d}{dt}|\dot{\mathbf{r}}|\right)\dot{\mathbf{r}} \times \hat{\mathbf{T}} + \left(|\dot{\mathbf{r}}|\left|\frac{d\hat{\mathbf{T}}}{dt}\right|\right)\dot{\mathbf{r}} \times \hat{\mathbf{N}}$$

$$= \left(|\dot{\mathbf{r}}|\left|\frac{d\hat{\mathbf{T}}}{dt}\right|\right)\dot{\mathbf{r}} \times \hat{\mathbf{N}}. \quad \text{(Since $\dot{\mathbf{r}}$ is parallel to $\hat{\mathbf{T}}$)}$$

Because $\dot{\mathbf{r}}$ is perpendicular to $\hat{\mathbf{N}}$, it follows that $|\dot{\mathbf{r}} \times \hat{\mathbf{N}}| = |\dot{\mathbf{r}}||\hat{\mathbf{N}}|\sin(\pi/2) = |\dot{\mathbf{r}}|$, and therefore

$$|\dot{\mathbf{r}} \times \ddot{\mathbf{r}}| = \left(|\dot{\mathbf{r}}|\left|\frac{d\hat{\mathbf{T}}}{dt}\right|\right)|\dot{\mathbf{r}}|.$$

Consequently,

$$\left|\frac{d\hat{\mathbf{T}}}{dt}\right| = \frac{|\dot{\mathbf{r}} \times \ddot{\mathbf{r}}|}{|\dot{\mathbf{r}}|^2}$$

and

$$\kappa = \kappa(t) = \frac{|\dot{\mathbf{r}} \times \ddot{\mathbf{r}}|}{|\dot{\mathbf{r}}|^3}.$$

For the radius of curvature, we have the following.

Corollary When a smooth curve is defined in terms of an arbitrary parameter t,

$$\rho(t) = \frac{|\dot{\mathbf{r}}|^3}{|\dot{\mathbf{r}} \times \ddot{\mathbf{r}}|}. \tag{12.86}$$

EXAMPLE 12.37

Find the curvature and radius of curvature for the curve in Example 35.

SOLUTION According to 12.85,

$$\kappa(t) = \frac{|\dot{\mathbf{r}} \times \ddot{\mathbf{r}}|}{|\dot{\mathbf{r}}|^3} = \frac{|(1, 2t, 2t) \times (0, 2, 2)|}{|(1, 2t, 2t)|^3}$$

$$= \frac{1}{(1 + 4t^2 + 4t^2)^{3/2}} \begin{vmatrix} \hat{\mathbf{i}} & \hat{\mathbf{j}} & \hat{\mathbf{k}} \\ 1 & 2t & 2t \\ 0 & 2 & 2 \end{vmatrix} = \frac{1}{(1 + 8t^2)^{3/2}} |(0, -2, 2)|$$

$$= \frac{2\sqrt{2}}{(1 + 8t^2)^{3/2}};$$

$$\rho(t) = \frac{1}{\kappa(t)} = \frac{(1 + 8t^2)^{3/2}}{2\sqrt{2}}.$$

Note in particular that as t increases, so does ρ, a fact that is certainly supported by Figure 12.116.

EXAMPLE 12.38

Show that for a smooth curve $y = y(x)$ in the xy-plane,

$$\kappa(x) = \frac{|y''|}{[1 + (y')^2]^{3/2}}.$$

SOLUTION When we use x as parameter along the curve $y = y(x)$, parametric equations are $x = x$, $y = y(x)$. Then,

$$\dot{\mathbf{r}} = (1, y'(x)), \quad \ddot{\mathbf{r}} = (0, y''(x)),$$

and

$$\dot{\mathbf{r}} \times \ddot{\mathbf{r}} = \begin{vmatrix} \hat{\mathbf{i}} & \hat{\mathbf{j}} & \hat{\mathbf{k}} \\ 1 & y' & 0 \\ 0 & y'' & 0 \end{vmatrix} = y''\hat{\mathbf{k}}.$$

Thus

$$\kappa(x) = \frac{|\dot{\mathbf{r}} \times \ddot{\mathbf{r}}|}{|\dot{\mathbf{r}}|^3} = \frac{|y''|}{|(1, y')|^3} = \frac{|y''|}{[1 + (y')^2]^{3/2}}.$$

EXERCISES 12.9

In Exercises 1–5 find $\hat{\mathbf{N}}$ and $\hat{\mathbf{B}}$ at each point on the curve.

1. $x = \sin t$, $y = \cos t$, $z = t$, $-\infty < t < \infty$

2. $x = t$, $y = t^2$, $z = t^3$, $t \geq 1$

3. $x = (t-1)^2$, $y = (t+1)^2$, $z = -t$, $-3 \leq t \leq 4$

4. $x + y = 5$, $x^2 - y = z$, from $(5, 0, 25)$ to $(0, 5, -5)$

5. $z = x$, $x^2 + y^2 = 4$, $y \geq 0$, from $(2, 0, 2)$ to $(-2, 0, -2)$

In Exercises 6–10 find $\hat{\mathbf{N}}$ and $\hat{\mathbf{B}}$ at the point.

6. $x = 4\cos t$, $y = 6\sin t$, $z = 2\sin t$, $-\infty < t < \infty$; $(2\sqrt{2}, 3\sqrt{2}, \sqrt{2})$

7. $x = 2 - 5t$, $y = 1 + t$, $z = 6 + 4t^3$, $-\infty < t < \infty$; $(7, 0, 2)$

8. $x^2 + y^2 + z^2 = 4$, $z = \sqrt{x^2 + y^2}$, directed so that x

increases when y is positive; $(1, 1, \sqrt{2})$

9. $x = y^2 + 1$, $z = x + 5$, directed so that y increases along the curve; $(5, 2, 10)$

10. $x^2 + (y-1)^2 = 4$, $x = z$, directed so that z decreases when y is negative; $(2, 1, 2)$

In Exercises 11–18 find the curvature and the radius of curvature of the curve (if they exist). Sketch each curve.

11. $(x-h)^2 + (y-k)^2 = R^2$, $z = 0$, directed counterclockwise

12. $x = x_0 + at$, $y = y_0 + bt$, $z = z_0 + ct$, $-\infty < t < \infty$ (x_0, y_0, z_0, a, b, c all constants)

13. $x = t$, $y = t^2$, $z = 0$, $t \geq 0$

14. $x = e^t \cos t$, $y = e^t \sin t$, $z = t$, $-\infty < t < \infty$

15. $x = t$, $y = t^3$, $z = t^2$, $t \geq 0$

16. $x = 2\cos t$, $y = 2\sin t$, $z = 2\sin t$, $0 \leq t < 2\pi$

17. $x = t + 1$, $y = t^2 - 1$, $z = t + 1$, $-\infty < t < \infty$

18. $x = t^2$, $y = t^4$, $z = 2t$, $-1 \leq t \leq 5$

19. At which points on the ellipse $b^2 x^2 + a^2 y^2 = a^2 b^2$ ($a > b$) is the curvature a maximum and at which points is the curvature a minimum?

20. Show that curvature for a smooth curve $x = x(t)$, $y = y(t)$, $\alpha \leq t \leq \beta$, in the xy-plane can be expressed in the form

$$\kappa(t) = \frac{\left| \dfrac{dy}{dt} \dfrac{d^2 x}{dt^2} - \dfrac{dx}{dt} \dfrac{d^2 y}{dt^2} \right|}{\left[\left(\dfrac{dx}{dt} \right)^2 + \left(\dfrac{dy}{dt} \right)^2 \right]^{3/2}}.$$

21. Show that the only curves for which curvature is identically equal to zero are straight lines.

22. What happens to curvature at a point of inflection on the graph of a function $y = f(x)$?

23. Let C be the curve $x = t$, $y = t^2$ in the xy-plane.
    (a) At each point on C calculate the unit tangent vector $\hat{\mathbf{T}}$ and the principal normal $\hat{\mathbf{N}}$. What is $\hat{\mathbf{B}}$? (See Example 36 for $\hat{\mathbf{N}}$.)
    (b) $\mathbf{F} = t^2 \hat{\mathbf{i}} + t^4 \hat{\mathbf{j}}$ is a vector that is defined at each point P on C. Denote by F_T and F_N the components of $\mathbf{F}$ in the directions $\hat{\mathbf{T}}$ and $\hat{\mathbf{N}}$ at P. Find F_T and F_N as functions of t.
    (c) Express $\mathbf{F}$ in terms of $\hat{\mathbf{T}}$ and $\hat{\mathbf{N}}$.

24. Repeat Exercise 23 for the curve C : $x = 2\cos t$, $y = 2\sin t$, and the vector $\mathbf{F} = x^2 \hat{\mathbf{i}} + y^2 \hat{\mathbf{j}}$.

25. The vectors $\hat{\mathbf{T}}$, $\hat{\mathbf{N}}$, and $\hat{\mathbf{B}}$ were calculated at each point on the curve $x = t$, $y = t^2$, $z = t^2$ in Example 12.35. If $\mathbf{F} = t^2 \hat{\mathbf{i}} + 2t \hat{\mathbf{j}} - 3 \hat{\mathbf{k}}$ is a vector defined along C, find the components of $\mathbf{F}$ in the directions $\hat{\mathbf{T}}$, $\hat{\mathbf{N}}$, and $\hat{\mathbf{B}}$. Express $\mathbf{F}$ in terms of $\hat{\mathbf{T}}$, $\hat{\mathbf{N}}$, and $\hat{\mathbf{B}}$.

26. Calculate $\hat{\mathbf{T}}$, $\hat{\mathbf{N}}$, and $\hat{\mathbf{B}}$ for the curve $x = \cos t$, $y = \sin t$, $z = t$. Express the vector $\mathbf{F} = x\hat{\mathbf{i}} + xy^2 \hat{\mathbf{j}} + \hat{\mathbf{k}}$ in terms of $\hat{\mathbf{T}}$, $\hat{\mathbf{N}}$, and $\hat{\mathbf{B}}$.

27. In this exercise we discuss our claim that the circle of curvature is the "best-fitting" circle to the curve at a point.
    (a) First, is it true that the circle of curvature at a point on a curve passes through that point?
    (b) Second, show that the circle of curvature and curve share the same tangent line at their common point.
    (c) Finally, verify that the circle of curvature and curve have the same curvature at their common point.

28. If ϕ is the angle between $\hat{\mathbf{i}}$ and $\hat{\mathbf{T}}$ for a curve in the xy-plane (Figure 12.119), show that

$$\kappa(s) = \left| \frac{d\phi}{ds} \right|.$$

FIGURE 12.119

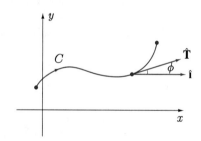

29. In differential geometry the rates of change $d\hat{\mathbf{T}}/ds$, $d\hat{\mathbf{N}}/ds$, and $d\hat{\mathbf{B}}/ds$ with respect to length along a curve are represented by the Frenet-Serret formulas. We develop these equations in this exercise.
    (a) The first Frenet-Serret formula is

$$\frac{d\hat{\mathbf{T}}}{ds} = \kappa \hat{\mathbf{N}}.$$

    Verify this result.
    (b) The second Frenet-Serret formula is

$$\frac{d\hat{\mathbf{B}}}{ds} = -\tau \hat{\mathbf{N}},$$

where τ is called the torsion of the curve. To verify this result, differentiate the equations $\hat{\mathbf{B}} = \hat{\mathbf{T}} \times \hat{\mathbf{N}}$ and $\hat{\mathbf{B}} \cdot \hat{\mathbf{B}} = 1$ to show that $d\hat{\mathbf{B}}/ds$ is perpendicular to $\hat{\mathbf{T}}$ and $\hat{\mathbf{B}}$ and therefore parallel to $\hat{\mathbf{N}}$.

(c) The third formula is

$$\frac{d\hat{\mathbf{N}}}{ds} = \tau\hat{\mathbf{B}} - \kappa\hat{\mathbf{T}}.$$

Verify this result by showing that $\hat{\mathbf{N}} = \hat{\mathbf{B}} \times \hat{\mathbf{T}}$ and then calculating $d\hat{\mathbf{N}}/ds$.

30. Show that the torsion in Exercise 29 can be calculated according to

$$\tau(t) = \frac{(\dot{\mathbf{r}} \times \ddot{\mathbf{r}}) \cdot \dddot{\mathbf{r}}}{|\dot{\mathbf{r}} \times \ddot{\mathbf{r}}|^2}.$$

31. Show that a curve lies in a plane if and only if its torsion vanishes.

SECTION 12.10

Displacement, Velocity, and Acceleration

In Sections 4.7 and 5.2 we introduced the concepts of displacement, velocity, and acceleration for moving objects, but indicated that our terminology at that time was somewhat loose. In particular, we stated that if $x = x(t)$ represents the position of a particle moving along the x-axis, then the instantaneous velocity of the particle is

$$v = \frac{dx}{dt}, \qquad (12.87)$$

provided, of course, that t is time, and the acceleration of the particle is

$$a = \frac{dv}{dt} = \frac{d^2 x}{dt^2}. \qquad (12.88)$$

We illustrated by examples that given any one of $x(t)$, $v(t)$, or $a(t)$ and sufficient initial conditions, it is always possible to find the other two. There was nothing wrong with the calculations in the examples—they were correct—but our terminology was not quite correct. We now rectify this situation and give precise definitions of velocity and acceleration.

Suppose a particle moves along some curve C in space (under perhaps the influence of various forces), and that C is defined as a function of time t by the parametric equations

$$C: \quad x = x(t), \quad y = y(t), \quad z = z(t), \quad t \geq 0. \qquad (12.89)$$

The position of the particle can then be described as a function of time by its position or displacement vector:

$$\mathbf{r} = \mathbf{r}(t) = x(t)\hat{\mathbf{i}} + y(t)\hat{\mathbf{j}} + z(t)\hat{\mathbf{k}}, \quad t \geq 0. \qquad (12.90)$$

The **velocity** $\mathbf{v}$ of the particle at any time t is defined as the time rate of change of its displacement vector:

$$\mathbf{v} = \frac{d\mathbf{r}}{dt}. \qquad (12.91)$$

Velocity, then, is a vector, and because of Theorem 12.8, the components of velocity are the derivatives of the components of displacement:

$$\mathbf{v} = \frac{d\mathbf{r}}{dt} = \frac{dx}{dt}\hat{\mathbf{i}} + \frac{dy}{dt}\hat{\mathbf{j}} + \frac{dz}{dt}\hat{\mathbf{k}}. \qquad (12.92)$$

But according to Theorem 12.10, the vector $d\mathbf{r}/dt$ is tangent to the curve C (Figure 12.120). In other words, if a particle is at position P, and we draw its velocity vector with tail at P, then $\mathbf{v}$ is tangent to the trajectory.

FIGURE 12.120

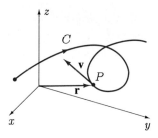

In some applications it is the length or magnitude of velocity that is important, not its direction. This quantity, called **speed**, is therefore defined by

$$|\mathbf{v}| = \sqrt{\left(\frac{dx}{dt}\right)^2 + \left(\frac{dy}{dt}\right)^2 + \left(\frac{dz}{dt}\right)^2}. \tag{12.93}$$

Equation 12.72 implies that if $s(t)$ is length along the trajectory C (where $s(0) = 0$), then $|\mathbf{v}| = ds/dt$. In other words, speed is the time rate of change of distance traveled along C.

It is important to understand this difference between velocity and speed. Velocity is the time derivative of displacement; speed is the time derivative of distance traveled. Velocity is a vector; speed is a scalar—the magnitude of velocity.

The **acceleration** of the particle as it moves along the curve C in equations 12.89 is defined as the rate of change of velocity with respect to time:

$$\mathbf{a} = \frac{d\mathbf{v}}{dt} = \frac{d^2\mathbf{r}}{dt^2} = \frac{d^2x}{dt^2}\hat{\mathbf{i}} + \frac{d^2y}{dt^2}\hat{\mathbf{j}} + \frac{d^2z}{dt^2}\hat{\mathbf{k}}. \tag{12.94}$$

Acceleration, then, is also a vector; it is the derivative of velocity, and therefore its components are the derivatives of the components of the velocity vector. Alternatively, it is the second derivative of displacement and has components that are the second derivatives of the displacement vector.

In the special case in which C is a curve in the xy-plane, the definitions of displacement, velocity, speed, and acceleration become, respectively,

$$\mathbf{r} = x(t)\hat{\mathbf{i}} + y(t)\hat{\mathbf{j}}, \tag{12.95}$$

$$\mathbf{v} = \frac{d\mathbf{r}}{dt} = \frac{dx}{dt}\hat{\mathbf{i}} + \frac{dy}{dt}\hat{\mathbf{j}}, \tag{12.96}$$

$$|\mathbf{v}| = \sqrt{\left(\frac{dx}{dt}\right)^2 + \left(\frac{dy}{dt}\right)^2}, \tag{12.97}$$

$$\mathbf{a} = \frac{d\mathbf{v}}{dt} = \frac{d^2\mathbf{r}}{dt^2} = \frac{d^2x}{dt^2}\hat{\mathbf{i}} + \frac{d^2y}{dt^2}\hat{\mathbf{j}}. \tag{12.98}$$

For motion along the x-axis,

$$\mathbf{r} = x(t)\hat{\mathbf{i}}, \tag{12.99}$$

$$\mathbf{v} = \frac{d\mathbf{r}}{dt} = \frac{dx}{dt}\hat{\mathbf{i}}, \tag{12.100}$$

$$|\mathbf{v}| = \left|\frac{dx}{dt}\right|, \tag{12.101}$$

$$\mathbf{a} = \frac{d\mathbf{v}}{dt} = \frac{d^2\mathbf{r}}{dt^2} = \frac{d^2x}{dt^2}\hat{\mathbf{i}}. \tag{12.102}$$

If we compare equations 12.100 and 12.102 with equations 12.87 and 12.88, we see that for motion along the x-axis, $x(t)$, $v(t)$, and $a(t)$ are the components of the displacement, velocity, and acceleration vectors, respectively. Because these are the only components of $\mathbf{r}(t)$, $\mathbf{v}(t)$, and $\mathbf{a}(t)$, it follows that consideration of the components of the vectors is equivalent to consideration of the vectors themselves. For one-dimensional motion, then, we can drop the vector notation and work with components (and this is precisely the procedure that we followed in Sections 4.7 and 5.2).

Newton's second law describes the effects of forces on the motion of objects. It states that if an object of mass m is subjected to a force $\mathbf{F}$, then the time rate of change of its momentum ($m\mathbf{v}$) is equal to $\mathbf{F}$:

$$\mathbf{F} = \frac{d}{dt}(m\mathbf{v}). \qquad (12.103)$$

In most cases, the mass of the object is constant, and this equation then yields its acceleration:

$$\mathbf{F} = m\frac{d\mathbf{v}}{dt} = m\mathbf{a}. \qquad (12.104)$$

If $\mathbf{F}$ is known as a function of time t, $\mathbf{F} = \mathbf{F}(t)$, then 12.104 defines the acceleration of the object as a function of time,

$$\mathbf{a}(t) = \frac{1}{m}\mathbf{F}(t),$$

and integration of this equation leads to expressions for the velocity $\mathbf{v}(t)$ and position $\mathbf{r}(t)$ as functions of time.

EXAMPLE 12.39

Find the velocity and acceleration of a particle and describe its motion, if its position as a function of time is given by

$$x = 2t + \sqrt{2}, \quad y = \sqrt{4t^2 + 4\sqrt{2}\,t + 1}, \quad z = 0, \quad t \geq 0.$$

SOLUTION The velocity of the particle is

$$\mathbf{v} = \frac{dx}{dt}\hat{\mathbf{i}} + \frac{dy}{dt}\hat{\mathbf{j}} = 2\hat{\mathbf{i}} + \left(\frac{4t + 2\sqrt{2}}{\sqrt{4t^2 + 4\sqrt{2}\,t + 1}}\right)\hat{\mathbf{j}},$$

and its acceleration is

$$\mathbf{a} = \frac{d^2x}{dt^2}\hat{\mathbf{i}} + \frac{d^2y}{dt^2}\hat{\mathbf{j}} = \frac{-4}{(4t^2 + 4\sqrt{2}\,t + 1)^{3/2}}\hat{\mathbf{j}}.$$

Because $t = (x - \sqrt{2})/2$, an explicit definition of the path is

$$y = \sqrt{4\left(\frac{x - \sqrt{2}}{2}\right)^2 + 4\sqrt{2}\left(\frac{x - \sqrt{2}}{2}\right) + 1} = \sqrt{x^2 - 1}.$$

Since $t \geq 0$, it follows that $x \geq \sqrt{2}$, and the path is that part of the hyperbola $x^2 - y^2 = 1$ in Figure 12.121. Note that $dx/dt = 2$, so that the x-component of the velocity is always equal to 2. This is also reflected in the fact that the acceleration has no x-component. Since the y-component of acceleration is always negative, the y-component of velocity is decreasing in time.

FIGURE 12.121

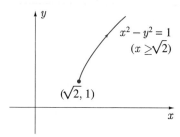

EXAMPLE 12.40

A particle starts at time $t = 0$ from position $(1, 1)$ with speed 2 m/s in the negative y-direction. It is subjected to an acceleration that is given as a function of time by

$$\mathbf{a}(t) = \frac{1}{\sqrt{t+1}}\hat{\mathbf{i}} + 6t\hat{\mathbf{j}} \text{ m/s}^2.$$

Find its velocity and position as functions of time.

SOLUTION If $\mathbf{a} = d\mathbf{v}/dt = (1/\sqrt{t+1})\hat{\mathbf{i}} + 6t\hat{\mathbf{j}}$, then

$$\mathbf{v} = 2\sqrt{t+1}\,\hat{\mathbf{i}} + 3t^2\hat{\mathbf{j}} + \mathbf{C},$$

where $\mathbf{C}$ is some constant vector. Because the initial velocity of the particle is 2 m/s in the negative y-direction, $\mathbf{v}(0) = -2\hat{\mathbf{j}}$. Consequently,

$$-2\hat{\mathbf{j}} = 2\hat{\mathbf{i}} + \mathbf{C},$$

which implies that $\mathbf{C} = -2\hat{\mathbf{i}} - 2\hat{\mathbf{j}}$. The velocity, then, of the particle at any time $t \geq 0$ is

$$\mathbf{v}(t) = (2\sqrt{t+1} - 2)\hat{\mathbf{i}} + (3t^2 - 2)\hat{\mathbf{j}} \text{ m/s}.$$

Because $\mathbf{v} = d\mathbf{r}/dt$, integration gives

$$\mathbf{r} = \left[\frac{4}{3}(t+1)^{3/2} - 2t\right]\hat{\mathbf{i}} + (t^3 - 2t)\hat{\mathbf{j}} + \mathbf{D}.$$

Since the particle starts from position $(1, 1)$, $\mathbf{r}(0) = \hat{\mathbf{i}} + \hat{\mathbf{j}}$, and

$$\hat{\mathbf{i}} + \hat{\mathbf{j}} = \frac{4}{3}\hat{\mathbf{i}} + \mathbf{D}, \quad \text{or,} \quad \mathbf{D} = -\frac{1}{3}\hat{\mathbf{i}} + \hat{\mathbf{j}}.$$

The displacement of the particle is therefore

$$\mathbf{r}(t) = \left[\frac{4}{3}(t+1)^{3/2} - 2t - \frac{1}{3}\right]\hat{\mathbf{i}} + (t^3 - 2t + 1)\hat{\mathbf{j}} \text{ m}.$$

EXAMPLE 12.41

A shell is fired from an artillery gun with speed v_0 m/s at an angle θ with the horizontal (Figure 12.122). If gravity is assumed to be the only force acting on the shell:

(a) Find the position of the projectile as a function of time t.
(b) Find the range of the shell; i.e., find the horizontal distance traveled by the shell.
(c) Find the maximum height attained by the shell.
(d) Show that the path of the shell is a parabola.

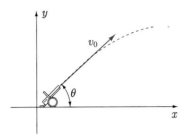

FIGURE 12.122

SOLUTION

(a) If m is the mass of the shell, then the force of gravity on the shell is

$$\mathbf{F} = -9.81m\hat{\mathbf{j}}.$$

According to Newton's second law, the acceleration of the shell is defined by

$$-9.81m\hat{\mathbf{j}} = m\mathbf{a}, \quad \text{or,} \quad \mathbf{a} = -9.81\hat{\mathbf{j}}.$$

Since $\mathbf{a} = d\mathbf{v}/dt$, we can write that

$$\frac{d\mathbf{v}}{dt} = -9.81\hat{\mathbf{j}},$$

and integration of this equation gives

$$\mathbf{v}(t) = -9.81t\hat{\mathbf{j}} + \mathbf{C}.$$

If we choose time $t = 0$ at the instant the shell is fired, then $\mathbf{v}(0) = v_0\cos\theta\hat{\mathbf{i}} + v_0\sin\theta\hat{\mathbf{j}}$, and hence

$$v_0\cos\theta\hat{\mathbf{i}} + v_0\sin\theta\hat{\mathbf{j}} = \mathbf{C}.$$

Because $\mathbf{v} = d\mathbf{r}/dt$, the position vector of the shell is

$$\mathbf{r}(t) = -4.905t^2\hat{\mathbf{j}} + \mathbf{C}t + \mathbf{D}.$$

Since $\mathbf{r}(0) = \mathbf{0}$, we must set $\mathbf{D} = \mathbf{0}$, and

$$\mathbf{r}(t) = -4.905t^2\hat{\mathbf{j}} + v_0\cos\theta t\hat{\mathbf{i}} + v_0\sin\theta t\hat{\mathbf{j}}$$
$$= (tv_0\cos\theta)\hat{\mathbf{i}} + (-4.905t^2 + tv_0\sin\theta)\hat{\mathbf{j}}.$$

(b) The shell strikes the ground when the y-component of $\mathbf{r}$ vanishes; i.e., when

$$0 = -4.905t^2 + tv_0 \sin\theta$$
$$= t(-4.905t + v_0 \sin\theta).$$

Clearly, $t = (v_0 \sin\theta)/4.905$, and at this time, the x-component of $\mathbf{r}$ is the range

$$v_0 \cos\theta \left(\frac{v_0 \sin\theta}{4.905}\right) = \frac{v_0^2 \sin\theta \cos\theta}{4.905} = \frac{v_0^2 \sin 2\theta}{9.81} \text{ m}.$$

(c) The shell is at its maximum height when its y-component of velocity is equal to zero; i.e., when

$$0 = -9.81t + v_0 \sin\theta, \text{ or, } t = \frac{v_0 \sin\theta}{9.81}.$$

At this time the y-component of the position vector is

$$-4.905\left(\frac{v_0 \sin\theta}{9.81}\right)^2 + v_0 \sin\theta\left(\frac{v_0 \sin\theta}{9.81}\right) = \frac{v_0^2 \sin^2\theta}{19.62} \text{ m}.$$

(d) Since $x(t) = tv_0 \cos\theta$ and $y(t) = -4.905t^2 + tv_0 \sin\theta$, it follows that

$$y = -4.905\left(\frac{x}{v_0 \cos\theta}\right)^2 + v_0 \sin\theta\left(\frac{x}{v_0 \cos\theta}\right)$$
$$= \frac{-4.905}{v_0^2 \cos^2\theta}x^2 + x\tan\theta,$$

a parabola. ∎

Tangential and Normal Components of Velocity and Acceleration

For some types of motion it is inconvenient to express velocity and acceleration in terms of Cartesian components; sometimes it is an advantage to resolve these vectors into tangential and normal components. When the trajectory C of a particle is specified as a function of time t by 12.89, its velocity $\mathbf{v} = d\mathbf{r}/dt$ is tangent to C, and we can therefore write

$$\mathbf{v} = |\mathbf{v}|\hat{\mathbf{T}}. \qquad (12.105)$$

In other words, the tangential component of velocity is speed, and $\mathbf{v}$ has no normal component. Differentiation of this equation gives the particle's acceleration:

$$\mathbf{a} = \frac{d\mathbf{v}}{dt} = \left(\frac{d}{dt}|\mathbf{v}|\right)\hat{\mathbf{T}} + |\mathbf{v}|\frac{d\hat{\mathbf{T}}}{dt}$$
$$= \left(\frac{d}{dt}|\mathbf{v}|\right)\hat{\mathbf{T}} + \left(|\mathbf{v}|\left|\frac{d\hat{\mathbf{T}}}{dt}\right|\right)\frac{d\hat{\mathbf{T}}/dt}{|d\hat{\mathbf{T}}/dt|}$$
$$= \left(\frac{d}{dt}|\mathbf{v}|\right)\hat{\mathbf{T}} + \left(|\mathbf{v}|\left|\frac{d\hat{\mathbf{T}}}{dt}\right|\right)\hat{\mathbf{N}}. \qquad (12.106)$$

We have therefore expressed **a** in terms of the unit tangent vector $\hat{\mathbf{T}}$ to C and the principal normal $\hat{\mathbf{N}}$ (Figure 12.123). We call $d(|\mathbf{v}|)/dt$ and $|\mathbf{v}||d\hat{\mathbf{T}}/dt|$ the tangential and normal components of acceleration, respectively. If a_T and a_N denote these components, we can write that

$$\mathbf{a} = a_T \hat{\mathbf{T}} + a_N \hat{\mathbf{N}}, \quad (12.107\text{a})$$

where

$$a_T = \mathbf{a} \cdot \hat{\mathbf{T}} = \frac{d}{dt}|\mathbf{v}|, \quad a_N = \mathbf{a} \cdot \hat{\mathbf{N}} = |\mathbf{v}|\left|\frac{d\hat{\mathbf{T}}}{dt}\right|. \quad (12.107\text{b})$$

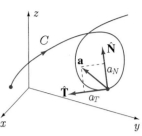

FIGURE 12.123

Note that the tangential component of acceleration is the time rate of change of speed. Since acceleration is the rate of change of velocity, the normal component of acceleration must determine the rate of change of the direction of **v**.

To calculate a_N using 12.107b is often quite complicated. A far easier formula results if we take the scalar product of **a** as defined by 12.107a with itself,

$$\begin{aligned}\mathbf{a} \cdot \mathbf{a} &= (a_T \hat{\mathbf{T}} + a_N \hat{\mathbf{N}}) \cdot (a_T \hat{\mathbf{T}} + a_N \hat{\mathbf{N}}) \\ &= a_T^2 \hat{\mathbf{T}} \cdot \hat{\mathbf{T}} + 2a_T a_N \hat{\mathbf{T}} \cdot \hat{\mathbf{N}} + a_N^2 \hat{\mathbf{N}} \cdot \hat{\mathbf{N}} \\ &= a_T^2 + a_N^2,\end{aligned}$$

since $\hat{\mathbf{T}} \cdot \hat{\mathbf{N}} = 0$ and $\hat{\mathbf{T}} \cdot \hat{\mathbf{T}} = \hat{\mathbf{N}} \cdot \hat{\mathbf{N}} = 1$. Consequently,

$$a_N^2 = \mathbf{a} \cdot \mathbf{a} - a_T^2 = |\mathbf{a}|^2 - a_T^2,$$

and because a_N is always positive (see equation 12.107b),

$$a_N = \sqrt{|\mathbf{a}|^2 - a_T^2}. \quad (12.108)$$

EXAMPLE 12.42

If the trajectory of a particle is defined by

$$x = t^2 + 1, \quad y = 2t^2 - 1, \quad z = t^2 + 5t, \quad t \geq 0,$$

where t is time, find the tangential and normal components of the particle's velocity and acceleration.

SOLUTION The velocity and acceleration of the particle have Cartesian components

$$\mathbf{v} = 2t\hat{\mathbf{i}} + 4t\hat{\mathbf{j}} + (2t+5)\hat{\mathbf{k}}, \quad \mathbf{a} = 2\hat{\mathbf{i}} + 4\hat{\mathbf{j}} + 2\hat{\mathbf{k}}.$$

The tangential component of the particle's velocity is its speed:

$$|\mathbf{v}| = \sqrt{(2t)^2 + (4t)^2 + (2t+5)^2} = \sqrt{24t^2 + 20t + 25}.$$

According to 12.107b, the tangential component of the acceleration is

$$a_T = \frac{d}{dt}|\mathbf{v}| = \frac{24t + 10}{\sqrt{24t^2 + 20t + 25}}.$$

With 12.108, the normal component of acceleration is

$$a_N = \sqrt{|\mathbf{a}|^2 - a_T^2} = \left\{(4+16+4) - \frac{(24t+10)^2}{24t^2+20t+25}\right\}^{1/2}$$

$$= \left\{\frac{24(24t^2+20t+25) - (24t+10)^2}{24t^2+20t+25}\right\}^{1/2}$$

$$= \frac{10\sqrt{5}}{\sqrt{24t^2+20t+25}}.$$

EXAMPLE 12.43

Prove that the tangential component of the acceleration of a particle is equal to zero if and only if its speed is constant.

SOLUTION The tangential component of the acceleration of a particle is $a_T = d(|\mathbf{v}|)/dt$ and is equal to zero, therefore, if and only if

$$\frac{d}{dt}|\mathbf{v}| = 0.$$

But this is valid if and only if

$$|\mathbf{v}| = \text{a constant}.$$

What this example shows is that when a particle moves with constant speed, its acceleration always points along the principal normal to the curve. In the special case in which the trajectory is a circle (Figure 12.124), the principal normal always points directly toward the centre of the circle; therefore, the acceleration always points toward the centre of the circle also. Understand, however, that this is true only when speed is constant. If speed is not constant for circular motion, then a_T is not equal to zero; therefore, acceleration is not directed toward the centre of the circle.

FIGURE 12.124

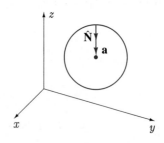

EXERCISES 12.10

In Exercises 1–5 find the velocity, speed, and acceleration of a particle if the given equations represent its position as a function of time.

1. $x(t) = \sqrt{t^2 + 1}$, $y(t) = t\sqrt{t^2 + 1}$, $t \geq 0$

2. $x(t) = t + 1/t$, $y(t) = t - 1/t$, $t \geq 1$

3. $x(t) = \sin t$, $y(t) = 3\cos t$, $z(t) = \sin t$, $0 \leq t \leq 10\pi$

4. $x(t) = t^2 + 1$, $y(t) = 2te^t$, $z(t) = 1/t^2$, $1 \leq t \leq 5$

5. $x(t) = e^{-t^2}$, $y(t) = t \ln t$, $z(t) = 5$, $t \geq 1$

In Exercises 6 and 7 a particle at $(1, 2, -1)$ starts from rest at time $t = 0$. Find its position as a function of time if the given function defines its acceleration.

6. $\mathbf{a}(t) = 3t^2 \hat{\mathbf{i}} + (t+1)\hat{\mathbf{j}} - 4t^3 \hat{\mathbf{k}}$, $t \geq 0$

7. $\mathbf{a}(t) = 3\hat{\mathbf{i}} + \hat{\mathbf{j}}/(t+1)^3$, $t \geq 0$

In Exercises 8 and 9 find the tangential and normal components of acceleration for a particle moving with position defined by the given functions (where t is time).

8. $x(t) = t$, $y(t) = t^2 + 1$, $t \geq 0$

9. $x(t) = \cos t$, $y(t) = \sin t$, $z = t$, $t \geq 0$

10. Show that the normal component of acceleration of a particle can be expressed in the form $a_N = |\mathbf{v}|^2/\rho = \kappa |\mathbf{v}|^2$.

11. The kinetic energy of an object of mass m moving with velocity $\mathbf{v}$ is defined as $K = m|\mathbf{v}|^2/2$. Find the kinetic energy for the particle in Exercises 1–5 if its mass is 2 g. Assume that x, y, and z are measured in metres and t in seconds.

12. A particle starts at the origin and moves along the curve $4y = x^2$ to the point $(4, 4)$.
    (a) If the y-component of its acceleration is always equal to 2 and the y-component of its velocity is initially zero, find the x-component of its acceleration.
    (b) If the x-component of its acceleration is equal to $24t^2$ (t being time) and the x-component of its velocity is initially zero, find the y-component of its acceleration.

13. A particle moves along the curve $x(t) = t$, $y(t) = t^3 - 3t^2 + 2t$, $0 \leq t \leq 5$ in the xy-plane (where t is time). Is there any point at which its velocity is parallel to its displacement?

14. A particle moves along the curve $y = x^3 - 2x + 3$ so that its x-component of velocity is always equal to 5. Find its acceleration.

15. If a particle starts at time $t = 0$ from rest at position $(3, 4)$ and experiences an acceleration $\mathbf{a} = -5t^4 \hat{\mathbf{i}} - (2t^3 + 1)\hat{\mathbf{j}}$, find its speed at $t = 2$.

16. A particle travels counterclockwise around the circle $(x-h)^2 + (y-k)^2 = R^2$ in Figure 12.125. Show that the speed of the particle at any time is $|\mathbf{v}| = \omega R$, where $\omega = d\theta/dt$ is called the *angular speed* of the particle.

FIGURE 12.125

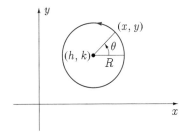

17. A particle travels around the circle $x^2 + y^2 = 4$ counterclockwise at constant speed, making 2 revolutions each second. If x and y are measured in metres, what is the velocity of the particle when it is at the point $(1, -\sqrt{3})$?

18. (a) Show that if an object moves with constant speed in a circular path of radius R, the magnitude of its acceleration is $|\mathbf{a}| = |\mathbf{v}|^2/R$.
    (b) If a satellite moves with constant speed in a circular orbit 200 km above the earth's surface, what is its speed? (Hint: Use Newton's universal law of gravitation (see Exercise 32 in Section 12.3) to determine the acceleration $\mathbf{a}$ of the satellite. Assume that the earth is a sphere with radius 6370 km and density 5.52×10^3 kg/m^3.)

19. Two particles move along curves C_1 and C_2 (Figure 12.126). If at some instant of time the particles are at positions P_1 and P_2, then the vector $\overrightarrow{P_1 P_2}$ is the displacement of P_2 with respect to P_1. Clearly, $\overrightarrow{OP_1} + \overrightarrow{P_1 P_2} = \overrightarrow{OP_2}$. Show that when this equation is differentiated with respect to time, we have

$$\mathbf{v}_{P_1/O} + \mathbf{v}_{P_2/P_1} = \mathbf{v}_{P_2/O},$$

where $\mathbf{v}_{P_1/O}$ and $\mathbf{v}_{P_2/O}$ are velocities of P_1 and P_2 with respect to the origin, and $\mathbf{v}_{P_2/P_1}$ is the velocity of P_2 with respect to P_1. Can this equation be rewritten in the form

$$\mathbf{v}_{P_1/O} + \mathbf{v}_{O/P_2} = \mathbf{v}_{P_1/P_2}?$$

FIGURE 12.126

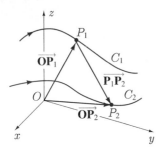

20. A plane flies on a course N30°E with air speed 650 km/h (i.e., the speed of the plane relative to the air is 650). If the air is moving at 40 km/h due east, find the ground velocity and speed of the plane. (Hint: Use Exercise 19.)

21. A plane flies with speed 600 km/h in still air. The plane is to fly in a straight line from city A to city B where B is 1000 km northwest of A. What should be its bearing if the wind is blowing from the west at 50 km/h? How long will the trip take?

22. A straight river is 200 m wide and the water flows at 3 km/h. If you can paddle your canoe at 4 km/h in still water, in what direction should you paddle if you wish the canoe to go straight across the river? How long will it take to cross?

23. (a) In Figure 12.127 a cannon is fired up an inclined plane. If the speed at which the ball is ejected from the cannon is S, show that the range R of the ball is given by

$$R = \frac{2S^2 \cos\theta \sin(\theta - \alpha)}{g \cos^2 \alpha},$$

where g is the acceleration due to gravity.
(b) What angle θ maximizes R?

FIGURE 12.127

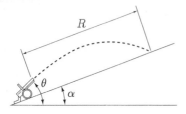

24. What constant acceleration must a particle experience if it is to travel from $(1,2,3)$ to $(4,5,7)$ along the straight line joining the points, starting from rest, and covering the distance in 2 units of time?

25. Calculate the normal component a_N of the acceleration of the particle in Example 12.42 using equation 12.107.

26. A particle moves along the curve $x(t) = 2 + \sqrt{1-t^2}$, $y(t) = t$, $0 \leq t \leq 1/2$, where t is time. Is there a time at which its acceleration is perpendicular to its velocity?

27. (a) Show that motion along a straight line results in both of the following situations:
    (i) The initial velocity is zero, and the acceleration is constant.
    (ii) The initial velocity is nonzero, and the acceleration is constant and parallel to the initial velocity.
    (b) Can we generalize the results of (a) and state that constant acceleration produces straight-line motion? Illustrate.

28. A particle starts from position $\mathbf{r}_0 = (x_0, y_0, z_0)$ at time $t = t_0$ with velocity $\mathbf{v}_0$. If it experiences constant acceleration $\mathbf{a}$, show that

$$\mathbf{r} = \mathbf{r}_0 + \mathbf{v}_0(t - t_0) + \frac{1}{2}\mathbf{a}(t - t_0)^2.$$

29. A ladder 8 m long has its upper end against a vertical wall and its lower end on a horizontal floor. Suppose that the lower end slips away from the wall at constant speed 1 m/s.
    (a) Find the velocity and acceleration of the middle point of the ladder when the foot of the ladder is 3 m from the wall.
    (b) How fast does the middle point of the ladder strike the floor?

30. Water issues from the nozzle of a fire hose at speed S (Figure 12.128). Show that the maximum height attainable by the water on the building is given by $(S^4 - g^2 d^2)/(2gS^2)$, where g is the acceleration due to gravity.

FIGURE 12.128

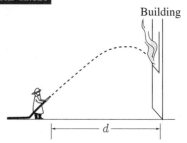

31. A boy stands on a cliff 50 m high that overlooks a river 85 m wide (Figure 12.129). If he can throw a stone at 25 m/s, can he throw it across the river?

FIGURE 12.129

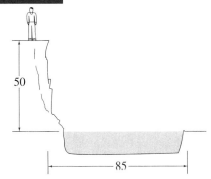

32. A golfer can drive a maximum of 300 m in the air on a level fairway. From the tee in Figure 12.130 can he expect to clear the stream?

FIGURE 12.130

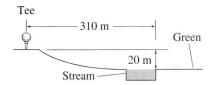

33. A particle is confined to move in a circular path of radius R in the xy-plane if and only if its position vector is $\mathbf{r} = x\hat{\mathbf{i}} + y\hat{\mathbf{j}}$, where $x = h + R\cos\omega(t)$, $y = k + R\sin\omega(t)$ and $\omega(t)$ is some function of time t. Show that the acceleration of the particle is directed radially toward the centre of the circular path if and only if $\omega(t) = At + B$, where A and B are constant. Furthermore, in this case verify that the speed of the particle is constant.

34. If $\mathbf{r}$ is the position vector of a particle with mass m moving under the action of a force $\mathbf{F}$, the torque (or moment) of $\mathbf{F}$ about the origin is $\mathbf{M} = \mathbf{r} \times \mathbf{F}$. The angular momentum of m about O is defined as $\mathbf{H} = \mathbf{r} \times m\mathbf{v}$. Use Newton's second law in the form 12.103 to show that $\mathbf{M} = d\mathbf{H}/dt$.

35. A stone is embedded in the tread of a tire and the tire rolls (without slipping) along the x-axis (Figure 12.131). If the stone starts at the origin, the path that it traces is called a cycloid.
(a) Show that if θ is the angle through which the stone has turned, then parametric equations for the cycloid are $x = R(\theta - \sin\theta)$, $y = R(1 - \cos\theta)$.
(b) Verify that if the centre of the tire moves at constant speed S, with $t = 0$ when the stone is at the origin, then $\theta = St/R$.
(c) Find the velocity, speed, and acceleration of the stone at any time.
(d) What are the normal and tangential components of the stone's acceleration?

FIGURE 12.131

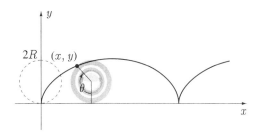

36. If the stone in Exercise 35 is embedded in the side of the tire, its path is called a trochoid (Figure 12.132).
(a) Show that if the distance from the centre of the tire to the stone is b, then parametric equations for the trochoid are $x = R\theta - b\sin\theta$, $y = R - b\cos\theta$.
(b) Find the velocity, speed, and acceleration of the stone if the tire rolls so that its centre has constant speed S.
(c) What are normal and tangential components of the stone's acceleration?

FIGURE 12.132

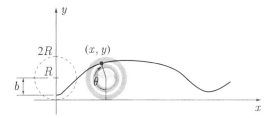

37. Circles C_1 and C_2 in Figure 12.133 represent cross-sections of two cylinders. The left cylinder remains stationary while the right one rolls (without slipping) around the left one, and the cylinders always remain in contact. If the right cylinder picks up a speck of dirt at point $(R, 0)$, the path that the dirt traces out during one revolution is a cardioid.
(a) Show that parametric equations for the cardioid are $x = R(2\cos\theta - \cos 2\theta)$, $y = R(2\sin\theta - \sin 2\theta)$.
(b) Verify that if the point of contact moves at constant speed S, with $t = 0$ when the speck of dirt is picked up, then $\theta = St/R$.
(c) Find the velocity, speed, and acceleration of the speck of dirt.
(d) What are normal and tangential components of the dirt's acceleration?

FIGURE 12.133

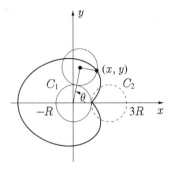

38. Show that the path of a particle lies on a sphere if its displacement and velocity are always perpendicular during its motion.

39. Suppose in Exercise 22 that, due to an injured elbow, you can only paddle at 2 km/h. What should be your heading to travel straight to a point L km downstream on the opposite shore? Are there any restrictions on L?

40. Suppose position vectors of a system of n masses m_i are denoted by $\mathbf{r}_i$ and forces acting on these masses are $\mathbf{F}_i$. Show that if $\mathbf{F} = \sum_{i=1}^{n} \mathbf{F}_i$, then the acceleration $\mathbf{a}$ of the centre of mass of the system (see Section 7.6) is given by $\mathbf{F} = M\mathbf{a}$, where $M = \sum_{i=1}^{n} m_i$.

41. If the force acting on a particle is always tangent to the particle's trajectory, what can you conclude about the trajectory?

SUMMARY

We have now established the groundwork for multivariable calculus. We discussed curves and surfaces in space and introduced vectors. Our approach to three-dimensional analytic geometry paralleled that to plane analytic geometry in Chapter 1. We described points by Cartesian coordinates (x, y, z) and then illustrated that an equation $F(x, y, z) = 0$ in these coordinates usually defines a surface. A second equation $G(x, y, z) = 0$ also defines a surface, and the pair of simultaneous equations $F(x, y, z) = 0$, $G(x, y, z) = 0$ describes the curve of intersection of the two surfaces (if the two surfaces do intersect). It is often more useful to have parametric equations for a curve, and these can be obtained by specifying one of x, y, or z as a function of a parameter t and solving the given equations for the other two in terms of t: $x = x(t)$, $y = y(t)$, $z = z(t)$.

The most common surfaces that we encountered were planes and quadric surfaces. Every plane has an equation of the form $Ax + By + Cz + D = 0$; conversely, every such equation describes a plane (provided A, B, and C are not all zero). A plane is uniquely defined by a vector that is perpendicular to it [and (A, B, C) is one such vector] and a point on it. Quadric surfaces are surfaces whose equations are quadratic in x, y, and z, the most important of which were sketched in Figures 12.24–12.33.

Every straight line in space is characterized by a vector along it and a point on it. (Contrast this with the above characterization of a plane.) If (a, b, c) are the components of a vector along a line and (x_0, y_0, z_0) are the coordinates of a point on it, then the vector, symmetric, and parametric equations for the line are, respectively,

$$(x, y, z) = (x_0, y_0, z_0) + t(a, b, c);$$
$$\frac{x - x_0}{a} = \frac{y - y_0}{b} = \frac{z - z_0}{c};$$
$$x = x_0 + at,$$
$$y = y_0 + bt,$$
$$z = z_0 + ct.$$

Geometrically, vectors are defined as directed line segments; algebraically, they are represented by ordered sets of real numbers (v_x, v_y, v_z), called their Cartesian components. Vectors can be added or subtracted geometrically using triangles or parallelograms; algebraically, they are added and subtracted component by component. Vectors can also be multiplied by scalars to give parallel vectors of different lengths.

We defined two products of vectors: the scalar product and the vector product. The scalar product of two vectors $\mathbf{u} = (u_x, u_y, u_z)$ and $\mathbf{v} = (v_x, v_y, v_z)$ is defined as

$$\mathbf{u} \cdot \mathbf{v} = u_x v_x + u_y v_y + u_z v_z = |\mathbf{u}||\mathbf{v}|\cos\theta,$$

where $|\mathbf{u}| = \sqrt{u_x^2 + u_y^2 + u_z^2}$ is the length of $\mathbf{u}$ and θ is the angle between $\mathbf{u}$ and $\mathbf{v}$. If the components of $\mathbf{u}$ and $\mathbf{v}$ are known, then this definition defines the angle θ between the vectors. The scalar product has many uses: finding components of vectors in arbitrary directions, calculating distances between geometric objects, and finding mechanical work, among others.

The vector product of two vectors $\mathbf{u}$ and $\mathbf{v}$ is

$$\mathbf{u} \times \mathbf{v} = \begin{vmatrix} \hat{\mathbf{i}} & \hat{\mathbf{j}} & \hat{\mathbf{k}} \\ u_x & u_y & u_z \\ v_x & v_y & v_z \end{vmatrix} = |\mathbf{u}||\mathbf{v}|\sin\theta\,\hat{\mathbf{w}},$$

where $\hat{\mathbf{w}}$ is the unit vector perpendicular to $\mathbf{u}$ and $\mathbf{v}$ determined by the right-hand rule. Because of the perpendicularity property, the vector product is indispensable in finding vectors perpendicular to other vectors. We used this fact when finding a vector along the line of intersection of two planes, a vector perpendicular to the plane containing three given points, and distances between geometric objects. It can also be used to find areas of triangles and parallelograms.

If a curve is represented vectorially in the form $\mathbf{r}(t) = x(t)\hat{\mathbf{i}} + y(t)\hat{\mathbf{j}} + z(t)\hat{\mathbf{k}}$, then a unit vector tangent to the curve at any point is

$$\hat{\mathbf{T}} = \frac{d\mathbf{r}/dt}{|d\mathbf{r}/dt|}.$$

Two unit vectors normal to the curve are the principal normal $\hat{\mathbf{N}}$ and the binormal $\hat{\mathbf{B}}$:

$$\hat{\mathbf{N}} = \frac{d\hat{\mathbf{T}}/dt}{|d\hat{\mathbf{T}}/dt|}; \quad \hat{\mathbf{B}} = \hat{\mathbf{T}} \times \hat{\mathbf{N}}.$$

These three vectors form a moving triad of mutually perpendicular unit vectors along the curve. The curvature of a curve, defined by $\kappa(t) = |\dot{\mathbf{r}} \times \ddot{\mathbf{r}}|/|\dot{\mathbf{r}}|^3$, measures the rate at which the curve changes direction: The larger κ is, the faster the curve turns. The reciprocal of curvature $\rho = \kappa^{-1}$ is called radius of curvature. It is the radius of that circle which best approximates the curve at any point.

If parametric equations for a curve represent the position of a particle and t is time, then the velocity and acceleration of the particle are, respectively,

$$\mathbf{v} = \frac{d\mathbf{r}}{dt}; \quad \mathbf{a} = \frac{d\mathbf{v}}{dt} = \frac{d^2\mathbf{r}}{dt^2};$$

and its speed is the magnitude of velocity, $|\mathbf{v}|$.

Tangential and normal components of velocity and acceleration of the particle are defined by

$$\mathbf{v} = |\mathbf{v}|\hat{\mathbf{T}}; \quad \mathbf{a} = a_T\hat{\mathbf{T}} + a_N\hat{\mathbf{N}};$$

where

$$a_T = \frac{d}{dt}|\mathbf{v}| \quad \text{and} \quad a_N = |\mathbf{v}|\left|\frac{d\hat{\mathbf{T}}}{dt}\right| = \sqrt{|\mathbf{a}|^2 - a_T^2}.$$

What these results say is that velocity is always tangent to the trajectory of the particle, and its acceleration always lies in the plane of the velocity vector and the principal normal.

Key Terms and Formulas

In reviewing this chapter, you should be able to define or discuss the following key terms:

Coordinate plane
Right (left) handed coordinate system
Quadric Surface
Vector
Tip of a vector
Tail of a vector
Components of a vector
Equality of vectors
Unit vector
Scalar multiplication
Sum of two vectors
Triangular addition of vectors
Parallelogram addition of vectors
Zero vector
Resultant of forces
Scalar product
Equation of a plane
Vector product
Vector equation of a line
Parametric equations of a line
Symmetric equations of a line
Scalar triple product

Continuous vector function
Derivative of a vector function
Antiderivative of a vector function
Parametric definition of a curve
Continuous curve
Closed curve
Displacement vector
Vector representation of a curve
Tangent vector to a curve
Smooth curve
Piecewise smooth curve
Unit tangent vector to a curve
Length of a curve
Principal normal vector
Binormal vector
Curvature
Radius of curvature
Circle of curvature
Velocity
Acceleration
Tangential component of acceleration
Normal component of acceleration

REVIEW EXERCISES

In Exercises 1–10, find the value of the scalar or the components of the vector if $\mathbf{u} = (1, 3, -2)$, $\mathbf{v} = (2, 4, -1)$, $\mathbf{w} = (0, 2, 1)$, and $\mathbf{r} = (2, 0, -1)$.

1. $2\mathbf{u} - 3\mathbf{w} + \mathbf{r}$
2. $\mathbf{u} \cdot (\mathbf{v} \times \mathbf{w})$
3. $(3\mathbf{u} \times 4\mathbf{v}) - \mathbf{w}$
4. $3\mathbf{u} \times (4\mathbf{v} - \mathbf{w})$
5. $|\mathbf{u}|\mathbf{v} - |\mathbf{v}|\mathbf{r}$
6. $(\mathbf{u} + \mathbf{v}) \cdot (\mathbf{r} - \mathbf{w})$
7. $(\mathbf{u} + \mathbf{v}) \times (\mathbf{r} - \mathbf{w})$
8. $(\mathbf{u} \times \mathbf{v}) \times (\mathbf{r} \times \mathbf{w})$
9. $(\mathbf{u} \cdot \mathbf{v})\mathbf{r} - 3(\mathbf{v} \cdot \mathbf{w})\mathbf{u}$
10. $\dfrac{2\mathbf{r}}{\mathbf{v} \cdot \mathbf{w}} + 3(\mathbf{v} + \mathbf{u})$

In Exercises 11–26 sketch whatever the equation, or equations, describe in space.

11. $x - y + 2z = 6$
12. $x^2 + z^2 = 1$
13. $x = \sqrt{y^2 + z^2}$
14. $x - y = 5$, $2x + y = 6$
15. $x^2 + y^2 + z^2 = 6z + 10$
16. $x^2 + y^2 + z^2 = 6z - 10$
17. $x + y = 5$, $2x - 3y + 6z = 1$, $y = z$
18. $x = t^2$, $y = t$, $z = t^3$
19. $x = t$, $y = t^3 + 1$
20. $\dfrac{x-1}{3} = \dfrac{y+5}{2} = z$

21. $z = 4 - x^2 - 2y^2$

22. $y^2 + z^2 = 1$, $y = z$

23. $y^2 + z^2 = 1$, $x = z$

24. $x^2 + y^2 = z^2 + 1$

25. $x = y^2$, $x = z^2$

26. $z^2 = x^2 - y^2$

In Exercises 27–30 find equations for the line.

27. Through the points $(-2, 3, 0)$ and $(1, -2, 4)$

28. Through $(6, 6, 2)$ and perpendicular to the plane $5x - 2y + z = 4$

29. Parallel to the line $x - y = 5$, $2x + 3y + 6z = 4$ and through the origin

30. Perpendicular to the line $x = t + 2$, $y = 3 - 2t$, $z = 4 + t$, intersecting this line, and through the point $(1, 3, 2)$

In Exercises 31–34 find the equation for the plane.

31. Through the points $(1, 3, 2)$, $(2, -1, 0)$ and $(6, 1, 3)$

32. Through the point $(1, 2, -1)$ and perpendicular to the line $y = z$, $x + y = 4$

33. Containing the line $x - y + z = 3$, $3x + 4y = 6$ and the point $(2, 2, 2)$

34. Containing the lines $x = 3t$, $y = 1 + 2t$, $z = 4 - t$ and $x = y = z$

In Exercises 35–39 find the distance.

35. Between the points $(1, 3, -2)$ and $(6, 4, 1)$

36. From the point $(6, 2, 1)$ to the plane $6x + 2y - z = 4$

37. From the line $x - y + z = 2$, $2x + y + z = 4$ to the plane $x - y = 5$

38. From the line $x - y + z = 2$, $2x + y + z = 4$ to the plane $3x + 6y = 4$

39. From the point $(6, 2, 3)$ to the line $x - y + z = 6$, $2x + y + 4z = 1$

40. Find the area of the triangle with vertices $(1, 1, 1)$, $(-2, 1, 0)$ and $(6, 3, -2)$.

41. If the points in Exercise 40 are three vertices of a parallelogram, what are the possibilities for the fourth vertex? What are the areas of these parallelograms?

In Exercises 42 and 43 find the unit tangent vector $\hat{\mathbf{T}}$, the principal normal vector $\hat{\mathbf{N}}$, and the binormal vector $\hat{\mathbf{B}}$ for the curve.

42. $x = 2\sin t$, $y = 2\cos t$, $z = t$

43. $x = t^3$, $y = 2t^2$, $z = t + 4$

44. If a particle has a trajectory defined by $x = t$, $y = t^2$, $z = t^2$, where t is time, find its velocity, speed, and acceleration at any time. What are normal and tangential components of its velocity and acceleration?

45. A force $\mathbf{F} = x^{-2}(2\hat{\mathbf{i}} + 3\hat{\mathbf{j}})$ acts on a particle moving from $x = 1$ to $x = 4$ along the x-axis. How much work does it do?

46. A ball rolls off a table 1 m high with speed 0.5 m/s (Figure 12.134).
    (a) With what speed does it strike the floor?
    (b) What is its displacement vector relative to the point where it left the table when it strikes the floor?
    (c) If it rebounds in the direction shown but loses 20% of its speed in the bounce, find the position of its second bounce.

FIGURE 12.134

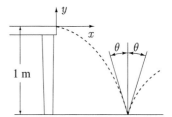

47. In Figure 12.135 a spring (with constant k) is fixed at A and attached to a sleeve at C. The sleeve is free to slide without friction on a vertical rod, and when the spring is horizontal (at B), it is unstretched. If the sleeve is slowly lowered, there is a position at which the vertical component of the spring force on C is balanced by the force of gravity on the sleeve (ignoring the weight of the spring itself). If the mass of the sleeve is m, find an equation determining s in terms of d, m, k and the acceleration g due to gravity.

FIGURE 12.135

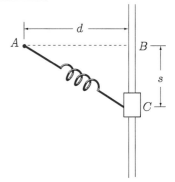

48. Find the Cartesian components of the spring force $\mathbf{F}$ on the sleeve in Example 12.18.

49. If a toy train travels around the oval track in Figure 12.136 with constant speed, show that its acceleration at A (the point at which the circular end meets the straight section) is discontinuous.

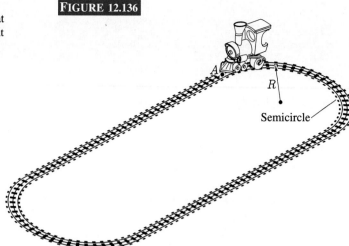

FIGURE 12.136

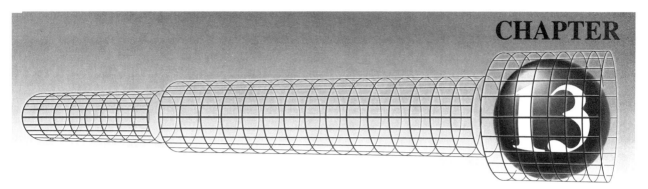

CHAPTER 13

Differential Calculus of Multivariable Functions

Few quantities in real life depend on only one variable; most depend on a multitude of interrelated variables. In order to understand such complicated relationships, we initiate discussions in this chapter with derivatives of functions of more than one variable. Much of the theory and many of our examples involve functions of two or three variables, because in these cases we can give geometric as well as analytic explanations. If the situation is completely analogous for functions of more variables, then it is likely that no mention of this fact will be made; on the other hand, if the situation is different for a higher number of variables, we will be careful to point out these differences.

SECTION 13.1

Multivariable Functions

If a variable T depends on other variables x, y, z, and t, we write $T = f(x,y,z,t)$ and speak of T as a function of x, y, z, and t. For example, T might be temperature, x, y, and z might be the coordinates of points in some region of space, and t might be time. The stopping distance D of a car depends on many factors: the initial speed s, the reaction time t of the driver to move from the accelerator to the brake, the texture T of the road, the moisture level M on the road, etc. We write $D = f(s,t,T,M,...)$ to represent this functional dependence. The function $P = f(I,R) = I^2 R$ represents the power necessary to maintain a current I through a wire with resistance R.

More precisely, a variable z is said to be a function of two independent variables x and y if x and y are not related and each pair of values of x and y determines a unique value of z. We write $z = f(x,y)$ to indicate that z is a function of x and y. Each possible pair of values x and y of the independent variables can be represented geometrically as a point (x,y) in the xy-plane. The totality of all points for which $f(x,y)$ is defined forms a region in the xy-plane called the **domain** of the function. Figure 13.1, for example, illustrates a rectangular domain. If for each point (x,y) in the domain we plot a point $f(x,y)$ units above the xy-plane, we obtain a surface, such as the one in Figure 13.1. Each point on this surface has coordinates (x,y,z) that satisfy the equation

$$z = f(x,y), \qquad (13.1)$$

FIGURE 13.1

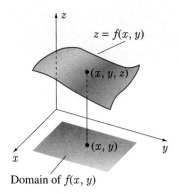

Domain of $f(x, y)$

and therefore 13.1 is the equation of the surface. This surface is a pictorial representation of the function.

It is clear that functions of more than two independent variables cannot be represented pictorially as surfaces. For example, if $u = f(x, y, z)$ is a function of three independent variables, values (x, y, z) of these independent variables can be represented geometrically as points in space. To graph $u = f(x, y, z)$ as above would require a u-axis perpendicular to the x-, y-, and z-axes, a somewhat difficult task geometrically. We can certainly think of $u = f(x, y, z)$ as defining a surface in four-dimensional $xyzu$-space, but visually we are stymied.

Although every function $f(x, y)$ of two independent variables can be represented geometrically as a surface, not every surface represents a function $f(x, y)$. A given surface does represent a function $f(x, y)$ if and only if every vertical line (in the z-direction) that intersects the surface does so in exactly one point.

EXAMPLE 13.1

Sketch the surface defined by the function $f(x, y) = x^2 + 4y^2$.

SOLUTION To sketch the surface $z = x^2 + 4y^2$, we note that if the surface is intersected with a plane $z = k > 0$, then the ellipse $x^2 + 4y^2 = k$, $z = k$ is obtained. As k increases, the ellipse becomes larger. In other words, cross-sections of this surface are ellipses that expand with increasing z. If we intersect the surface with the yz-plane ($x = 0$), we obtain the parabola $z = 4y^2$, $x = 0$. Similarly, intersection of the surface with the xz-plane gives the parabola $z = x^2$, $y = 0$. These facts lead to Figure 13.2.

FIGURE 13.2

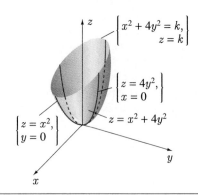

EXAMPLE 13.2

The ends of a taut string are fixed at $x = 0$ and $x = 2$ on the x-axis. At time $t = 0$, the string is given a displacement in the y-direction of $y = \sin(\pi x/2)$ (Figure 13.3a). If the string is then released, its displacement thereafter is given by

$$y = f(x, t) = \sin(\pi x/2) \cos(8\pi t).$$

Physically, this function need only be considered for $t \geq 0$ and $0 \leq x \leq 2$, and is pictured graphically as the surface in Figure 13.3b. Interpret physically the intersections of this surface with planes $t = t_0$ (t_0 = a constant) and $x = x_0$ (x_0 = a constant).

SOLUTION If we intersect the surface $y = f(x,t)$ with a plane $t = t_0$ ($t_0 \geq 0$) (Figure 13.3c), the curve of intersection is a picture of the position of the string at time t_0.

If we intersect the surface with a plane $x = x_0$ ($0 \leq x_0 \leq 2$)(Figure 13.3d), we obtain a graphical history of the displacement of the particle at position x_0.

FIGURE 13.3

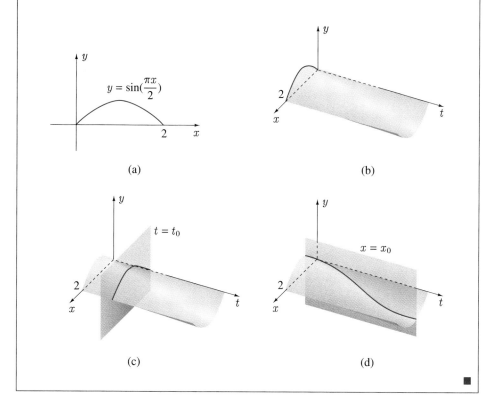

Many software packages draw surfaces $z = f(x, y)$ corresponding to functions $f(x, y)$ of two independent variables. They use cross-sections of the surface with planes parallel to the xz- and yz-planes to give excellent visualizations of the surface. The examples in Figure 13.4 were produced with MathCAD.

FIGURE 13.4

(a)

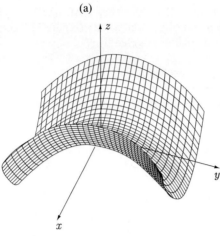

$z = x^2 - y^2$ $\quad -1 \leq x \leq 1$
$\quad -1 \leq y \leq 1$

(b)

$z = 3\sqrt{x^2 + y^2} - (x^2 + y^2)$ $\quad -2 \leq x \leq 2$
$\quad -2 \leq y \leq 2$

(c)

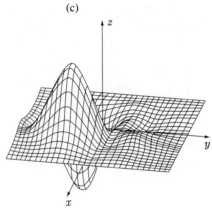

$z = x(y-1)^2 e^{-(x^2 + y^2)/4}$ $\quad -5 \leq x \leq 5$
$\quad -5 \leq y \leq 5$

(d)

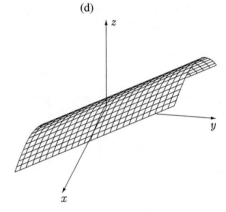

$z = 4y - x^2$ $\quad -2 \leq x \leq 1$
$\quad -1 \leq y \leq 1$

(e)

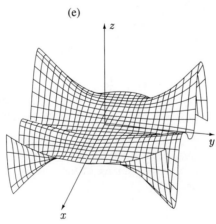

$z = xy^2 \cos x$ $\quad -5 \leq x \leq 5$
$\quad -10 \leq y \leq 10$

(f)

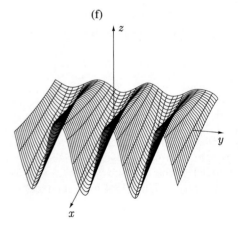

$z = e^x \sin y$ $\quad -1 \leq x \leq 1$
$\quad -10 \leq y \leq 10$

FIGURE 13.4

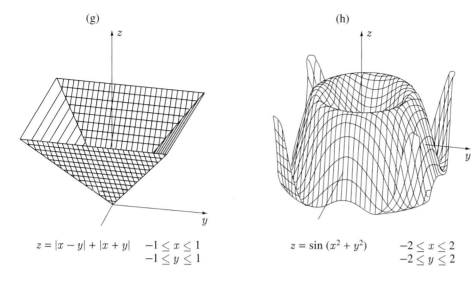

(g) $z = |x - y| + |x + y|$ $-1 \leq x \leq 1$, $-1 \leq y \leq 1$

(h) $z = \sin(x^2 + y^2)$ $-2 \leq x \leq 2$, $-2 \leq y \leq 2$

Another way to visualize a function $f(x,y)$ of two independent variables is through level curves. Curves $f(x,y) = C$ are drawn in the xy-plane for various values of C. Effectively, the surface $z = f(x,y)$ is sliced with a plane $z = C$, and the curve of intersection is projected into the xy-plane. Each curve joins all points for which $f(x,y)$ has the same value; or, it joins all points which have the same height on the surface $z = f(x,y)$. A few level curves for the surface in Figure 13.2 are shown in Figure 13.5; they are ellipses.

FIGURE 13.5

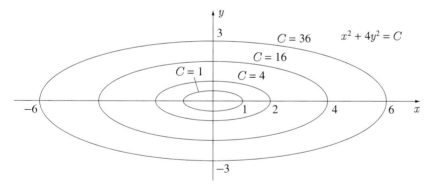

Level curves for the function in Figure 13.6(a) are shown in Figure 13.6(b). This technique is used on topographical maps to indicate land elevation, on marine charts to indicate water depth, and on climatic maps to indicate curves of constant temperature (isotherms) and curves of constant barometric pressure (isobars).

FIGURE 13.6

(a)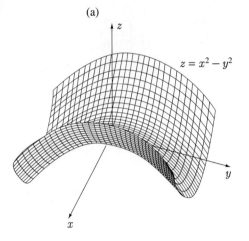

(b)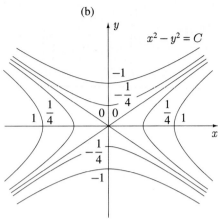

EXERCISES 13.1

1. If $f(x,y) = x^3 y + x \sin y$, evaluate
   (a) $f(1,2)$
   (b) $f(-2,-2)$
   (c) $f(x^2 + y, x - y^2)$
   (d) $f(x + h, y) - f(x, y)$

2. If $f(x,y,z) = x^2 y^2 - x^4 + 4zx^2$, show that $f(a + b, a - b, ab) = 0$.

In Exercises 3–6 find and illustrate geometrically the largest possible domain for the function.

3. $f(x,y) = \sqrt{4 - x^2 - y^2}$

4. $f(x,y) = \ln(1 - x^2 + y^2)$

5. $f(x,y) = \text{Sin}^{-1}(x^2 y + 1)$

6. $f(x,y,z) = 1/(x^2 + y^2 + z^2)$

7. For what values of x and y is the function
$$f(x,y) = \frac{12xy - x^2 y^2}{2(x+y)}$$
equal to zero? Illustrate these values as points in the xy-plane. What is the largest domain of the function?

In Exercises 8–21 sketch the surface defined by the function.

8. $f(x,y) = y^2$

9. $f(x,y) = 4 - x - 2y$

10. $f(x,y) = x^2 + y^2$

11. $f(x,y) = \sqrt{x^2 + y^2}$

12. $f(x,y) = y + x$

13. $f(x,y) = 1 - x^3$

14. $f(x,y) = 2(x^2 + y^2)$

15. $f(x,y) = 1 - x^2 - 4y^2$

16. $f(x,y) = xy$

17. $f(x,y) = y - x^2$

18. $f(x,y) = e^{-x^2 - y^2}$

19. $f(x,y) = |x - y|$

20. $f(x,y) = x^2 - y^2$

21. $f(x,y) = \sqrt{1 + x^2 - y^2}$

In Exercises 22–25 draw level curves $f(x,y) = C$ corresponding to the values $C = -2, -1, 0, 1, 2$.

22. $f(x,y) = 4 - \sqrt{4x^2 + y^2}$

23. $f(x,y) = y - x^2$

24. $f(x,y) = \ln(x^2 + y^2)$

25. $f(x,y) = x^2 - y^2$

26. A closed box is to have total surface area 30 m². Find a formula for the volume of the box in terms of its length l and width w.

27. (a) A company wishes to construct a storage tank in the form of a rectangular box. If material for sides and top costs $1.25 per square metre and material for the bottom costs $4.75 per square metre, find the cost of building the tank as a function of its length l, width w, and height h.
   (b) If the tank must hold 1000 m³, find the construction cost in terms of l and w.
   (c) Repeat (a) and (b) if the 12 edges of the tank must be welded at a cost of $7.50 per metre of weld.

28. A rectangular box is inscribed inside the ellipsoid $x^2/a^2 + y^2/b^2 + z^2/c^2 = 1$ with sides parallel to the coordinate planes and corners on the ellipsoid. Find a formula for the volume of the box in terms of x and y.

29. (a) A silo is to be built in the shape of a right-circular cylinder surmounted by a right-circular cone. If the radius of each is 6 m, find a formula for the volume V in the silo as a function of the heights H and h of the cylinder and cone.

    (b) If the total surface area of the silo must be 200 m² (not including the base), find V as a function of h. (The area of the curved surface of a cone is $\pi r \sqrt{r^2 + h^2}$.)

30. A long piece of metal 1 m wide is bent in two places A and B (Figure 13.7) to form a channel with three straight sides. Find a formula for the cross-sectional area of the channel in terms of x, θ, and ϕ.

FIGURE 13.7

31. A uniform circular rod has flat ends at $x = 0$ and $x = \pi$ on the x-axis. The round side of the rod is insulated and the faces at $x = 0$ and $x = \pi$ are both kept at temperature $0\,^\circ\mathrm{C}$ for time $t > 0$. If the initial temperature (at time $t = 0$) of the rod is given by $100\sin x$, $0 \leq x \leq \pi$, then the temperature thereafter is

$$T = f(x,t) = 100\,e^{-kt}\sin x \quad (k > 0 \text{ constant}).$$

(a) Sketch the surface $T = f(x,t)$.

(b) Interpret physically the curves of intersection of this surface with planes $x = x_0$ and $t = t_0$.

32. A cow's daily diet consists of three foods: hay, grain, and supplements. The animal is always given 11 kg of hay per day, 50% of which is digestive material and 12% of which is protein. Grain is 74% digestive and 8.8% protein. Supplements are 62% digestive material and 34% protein. Hay costs $27.50 for 1000 kg, whereas grain and supplements cost $110 and $175 respectively for 1000 kg. A healthy cow's daily diet must contain between 9.5 kg and 11.5 kg of digestive material and between 1.9 kg and 2.0 kg of protein. Find a formula for the cost per day C of feeding a cow in terms of the number of kilograms of grain G and supplements S fed to the cow daily. What is the domain of this function?

SECTION 13.2

Limits and Continuity

The concepts of limit and continuity for multivariable functions are exactly the same as for functions of one variable; on the other hand, the work involved with the application of these concepts is much more complicated for multivariable functions.

Intuitively, a function $f(x, y)$ is said to have limit L as x and y approach x_0 and y_0, if $f(x, y)$ gets arbitrarily close to L, and stays close to L, as x and y get arbitrarily close to x_0 and y_0. To give a mathematical definition of limits, it is convenient to represent pairs of independent variables as points (x, y) in the xy-plane. We then have the following definition.

Definition 13.1

A function $f(x, y)$ has limit L as (x, y) approaches (x_0, y_0), written

$$\lim_{(x,y) \to (x_0,y_0)} f(x, y) = L, \qquad (13.2)$$

if given any $\epsilon > 0$, we can find a $\delta > 0$ such that

$$|f(x, y) - L| < \epsilon$$

whenever $0 < \sqrt{(x - x_0)^2 + (y - y_0)^2} < \delta$ and (x, y) is in the domain of $f(x, y)$.

In other words, $f(x,y)$ has limit L as (x,y) approaches (x_0,y_0) if $f(x,y)$ can be made arbitrarily close to L (within ϵ) by choosing points (x,y) sufficiently close to (x_0,y_0) (within a circle of radius δ). Note the similarity of this definition to that for the limit of a function $f(x)$ of one variable in Section 2.5.

It is clear that
$$\lim_{(x,y)\to(2,1)} (x^2 + 2xy - 5) = 3,$$
but the limit
$$\lim_{(x,y)\to(0,0)} \frac{x^2 - y^2}{x^2 + y^2}$$
presents a problem, since both numerator and denominator approach zero as x and y approach zero.

To conclude that $\lim_{x\to a} f(x) = L$, the limit must be L no matter how x approaches a—be it through numbers larger than a, through numbers smaller than a, or through any other approach. For limit 13.2, the limit of $f(x,y)$ must also be L for all possible ways of approaching (x_0,y_0). But in this case there might be a multitude of ways of approaching (x_0,y_0). We might be able to approach (x_0,y_0) along straight lines with various slopes, along parabolas, along cubics, etc. Definition 13.1 implies, then, that the limit exists only if it is independent of the manner of approach. It is assumed, however, that we approach (x_0,y_0) only through points (x,y) that lie in the domain of definition of the function.

For the second example above, suppose we approach the origin along the straight line $y = mx$. Then, along this line,

$$\lim_{(x,y)\to(0,0)} \frac{x^2 - y^2}{x^2 + y^2} = \lim_{x\to 0} \frac{x^2 - m^2 x^2}{x^2 + m^2 x^2} = \frac{1 - m^2}{1 + m^2}.$$

Because this result depends on m, we have shown that as (x,y) approaches $(0,0)$ along various straight lines, the function $(x^2 - y^2)/(x^2 + y^2)$ approaches different numbers. We conclude, therefore, that the function does not have a limit as (x,y) approaches $(0,0)$. In Figure 13.8 we show a portion of this function to illustrate our conclusion.

FIGURE 13.8

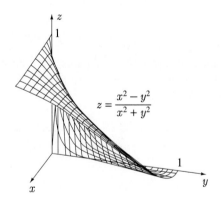

EXAMPLE 13.3

Evaluate

$$\lim_{(x,y)\to(0,0)} \frac{x^2 - y^2}{x + y}$$

if it exists.

SOLUTION Because points on the line $y = -x$ are not within the domain of definition of the function, we can write that

$$\lim_{(x,y)\to(0,0)} \frac{x^2 - y^2}{x + y} = \lim_{(x,y)\to(0,0)} \frac{(x - y)(x + y)}{x + y} = \lim_{(x,y)\to(0,0)} (x - y) = 0.$$

The concept of continuity for multivariable functions is contained in the following definition.

Definition 13.2

A function $f(x, y)$ is said to be **continuous** at a point (x_0, y_0) if

$$\lim_{(x,y)\to(x_0,y_0)} f(x, y) = f(x_0, y_0). \tag{13.3}$$

Just as Definition 2.1 for continuity of a function of only one variable contains three conditions, so too does Definition 13.2. It demands that:

(1) $f(x, y)$ be defined at (x_0, y_0);
(2) $\lim_{(x,y)\to(x_0,y_0)} f(x, y)$ exist;
(3) the numbers in (1) and (2) be the same.

Geometrically, a function $f(x, y)$ is continuous at a point (x_0, y_0) if the surface $z = f(x, y)$ does not have a separation at the point $(x_0, y_0, f(x_0, y_0))$. The function $f(x, y) = (x^2 - y^2)/(x^2 + y^2)$ in Figure 13.8 is discontinuous at $(0, 0)$.

The function $f(x, y) = 1 - e^{-1/(x^2+y^2)}$ is discontinuous at $(0, 0)$ since it is undefined for $x = 0$ and $y = 0$. The surface has a hole at $(0, 0)$ (Figure 13.9). The function $f(x, y) = \text{sgn}[(x - 1)^2 + (y - 2)^2]$ (see Exercise 36 in Section 2.4) is discontinuous at $(1, 2)$; it has value 0 at $x = 1$ and $y = 2$, and value 1 for all other values of x and y (Figure 13.10).

FIGURE 13.9

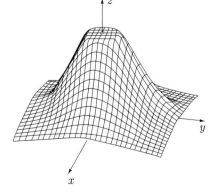

FIGURE 13.10

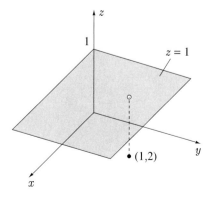

EXERCISES 13.2

In Exercises 1–20 evaluate the limit, if it exists.

1. $\lim\limits_{(x,y)\to(2,-3)} \dfrac{x^2-1}{x+y}$

2. $\lim\limits_{(x,y)\to(1,1)} \dfrac{x^3+2y^3}{x^3+4y^3}$

3. $\lim\limits_{(x,y)\to(3,2)} \dfrac{2x-3y}{x+y}$

4. $\lim\limits_{(x,y,z)\to(2,3,-1)} \dfrac{xyz}{x^2+y^2+z^2}$

5. $\lim\limits_{(x,y)\to(1,0)} \dfrac{x}{y}$

6. $\lim\limits_{(x,y,z)\to(0,\pi/2,1)} \mathrm{Tan}^{-1}[x/(yz)]$

7. $\lim\limits_{(x,y,z)\to(0,\pi/2,1)} \mathrm{Tan}^{-1}(yz/x)$

8. $\lim\limits_{(x,y,z)\to(0,\pi/2,1)} \mathrm{Tan}^{-1}|yz/x|$

9. $\lim\limits_{(x,y)\to(3,4)} \dfrac{|x^2-y^2|}{x^2-y^2}$

10. $\lim\limits_{(x,y)\to(3,4)} \dfrac{|x^2+y^2|}{x^2+y^2}$

11. $\lim\limits_{(x,y)\to(2,1)} \dfrac{x^2-y^2}{x-y}$

12. $\lim\limits_{(x,y)\to(2,2)} \dfrac{x^2-y^2}{x-y}$

13. $\lim\limits_{(x,y)\to(0,0)} \dfrac{x^2-y^2}{x-y}$

14. $\lim\limits_{(x,y)\to(0,0)} \dfrac{x-y}{x+y}$

15. $\lim\limits_{(x,y,z)\to(0,0,0)} \dfrac{x^2-y^2}{y^2+z^2+1}$

16. $\lim\limits_{(x,y)\to(2,1)} \dfrac{(x-2)^2(y+1)}{x-2}$

17. $\lim\limits_{(x,y,z)\to(1,1,1)} |2x-y-z|$

18. $\lim\limits_{(x,y)\to(0,0)} \dfrac{3x^3-y^3}{2x^3+4y^3}$

19. $\lim\limits_{(x,y)\to(0,0)} \mathrm{Sec}^{-1}\left(\dfrac{-1}{x^2+y^2}\right)$

20. $\lim\limits_{(x,y)\to(0,0)} \mathrm{Sec}^{-1}(x^2+y^2)$

In Exercises 21–26 find all points of discontinuity for the function.

21. $f(x,y) = \dfrac{x^2-1}{x+y}$

22. $f(x,y) = \dfrac{xy}{x^2+y^2}$

23. $f(x,y) = \dfrac{1}{1-x^2-y^2}$

24. $f(x,y,z) = \dfrac{1}{xyz}$

25. $f(x,y) = |x-y|$

26. $f(x,y) = \dfrac{x+y}{x^2y+xy^2}$

27. Evaluate $\lim\limits_{(x,y)\to(a,a)}\left[\cos(x+y)-\sqrt{1-\sin^2(x+y)}\right]$ where $0 \le a \le \pi/2$.

In Exercises 28–31 evaluate the limit, if it exists.

28. $\lim\limits_{(x,y)\to(0,0)} \dfrac{x^4+y^2}{x^4-y^2}$ Hint: Approach $(0,0)$ along parabolas.

29. $\lim\limits_{(x,y)\to(1,1)} \dfrac{x^2-2x+y^2+2y-2}{x^2-y^2-2x+2y}$ Hint: Approach $(1,1)$ along straight lines.

30. $\lim\limits_{(x,y)\to(1,0)} \dfrac{\sqrt{x+y}-\sqrt{x-y}}{y}$

31. $\lim\limits_{(x,y)\to(0,0)} \dfrac{\sin(x^2+y^2)}{x^2+y^2}$

32. (a) Does the limit $\lim\limits_{(x,y)\to(1,1)} \dfrac{\sin(x-y)}{x-y}$ exist? Explain. Is the function continuous at $(1,1)$?

    (b) If we define the function everywhere by giving it the value 1 along the line $y=x$, does the limit of the function exist at $(1,1)$? Is the function continuous at $(1,1)$?

33. Give a mathematical definition for
$$\lim_{(x,y,z)\to(x_0,y_0,z_0)} f(x,y,z) = L.$$

34. Is the following statement true or false? If a function $f(x,y)$ is undefined at every point on a curve C, then for any point (x_0,y_0) on C, $\lim\limits_{(x,y)\to(x_0,y_0)} f(x,y)$ does not exist. Explain. Give an example.

35. Let $f(x,y) = \begin{cases} \dfrac{x^2 y^2}{x^4+y^4} & \text{if } (x,y)\ne(0,0) \\ 0 & \text{if } (x,y)=(0,0). \end{cases}$

    (a) Show that $f(x,y)$ is continuous in each variable separately at $(0,0)$. In other words, show that

$f(x,0)$ and $f(0,y)$ are continuous at $x = 0$ and $y = 0$.

(b) Show that $f(x,y)$ is not continuous at $(0,0)$.

36. Prove that:

(a) $\lim_{(x,y) \to (0,0)} (xy + 5) = 5$

(b) $\lim_{(x,y) \to (1,1)} (x^2 + 2xy + 5) = 8$

SECTION 13.3

Partial Derivatives

We now define partial derivatives of multivariable functions and interpret these derivatives algebraically and geometrically.

Definition 13.3

The **partial derivative** of a function $f(x,y)$ with respect to x is

$$\frac{\partial f}{\partial x} = \lim_{\Delta x \to 0} \frac{f(x + \Delta x, y) - f(x, y)}{\Delta x}, \qquad (13.4)$$

and the partial derivative with respect to y is

$$\frac{\partial f}{\partial y} = \lim_{\Delta y \to 0} \frac{f(x, y + \Delta y) - f(x, y)}{\Delta y}. \qquad (13.5)$$

It is evident from 13.4 that the partial derivative of $f(x,y)$ with respect to x is simply the ordinary derivative of $f(x,y)$ with respect to x, where y is considered a constant. Similarly, $\partial f/\partial y$ is the ordinary derivative of $f(x,y)$ with respect to y, holding x constant.

For the partial derivative of a function of more than two independent variables, we again permit one variable to vary, but hold all others constant. For example, the partial derivative of $f(x,y,z,t,...)$ with respect to z is

$$\frac{\partial f}{\partial z} = \lim_{\Delta z \to 0} \frac{f(x,y,z+\Delta z,t,...) - f(x,y,z,t,...)}{\Delta z}. \qquad (13.6)$$

Hence, we differentiate with respect to z while treating $x, y, t, ...$ as constants.

Other notations for the partial derivative are common. In particular, for $\partial f/\partial x$ when $z = f(x,y)$, there are also

$$\frac{\partial z}{\partial x}, f_x, z_x, \left(\frac{\partial f}{\partial x}\right)_y \quad \text{and} \quad \left(\frac{\partial z}{\partial x}\right)_y,$$

the last two indicating that the variable y is held constant when differentiation with respect to x is performed.

EXAMPLE 13.4

Find $\partial z/\partial x$ and $\partial z/\partial y$ if $z = f(x,y) = x^2 y^3 + e^{xy}$.

SOLUTION   For this function,

$$\frac{\partial z}{\partial x} = 2xy^3 + ye^{xy} \quad \text{and} \quad \frac{\partial z}{\partial y} = 3x^2 y^2 + xe^{xy}.$$

EXAMPLE 13.5

Find $\partial f/\partial x$ at the point $(1,2,3)$ if $f(x,y,z) = x^2/y^4 + 3xz + 4$.

SOLUTION   Since $\partial f/\partial x = 2x/y^4 + 3z$,

$$\left.\frac{\partial f}{\partial x}\right|_{(1,2,3)} = 2(1)/2^4 + 3(3) = \frac{73}{8}.$$

For a function $y = f(x)$ of one variable, we defined differentials dx and dy in such a way that the derivative dy/dx could be thought of as a quotient. This is *not* done for functions of more than one variable. Although we write the partial derivative $\frac{\partial f}{\partial x}$ in the form $\partial f/\partial x$ (for typographical reasons), we *never* consider it as a quotient.

Algebraically, the partial derivative $\partial f/\partial x$ represents the rate of change of $f(x,y,...)$ with respect to x when all other variables in $f(x,y,...)$ are held constant. For instance, $V = \pi r^2 h/3$ represents the volume of a right-circular cone with height h and radius r, and therefore $\partial V/\partial r = 2\pi rh/3$ represents the rate of change of the volume of the cone as the base radius changes and the height remains fixed.

We can interpret the partial derivative of a function geometrically when the function can be interpreted geometrically, namely, when there are only two independent variables. Consider, then, a function $f(x,y)$ that is represented geometrically as a surface $z = f(x,y)$ in Figure 13.11. If we intersect this surface with a plane $y = y_0 = $ a constant, we obtain a curve with equations

$$y = y_0, \qquad z = f(x, y_0). \tag{13.7}$$

Because this curve lies in the plane $y = y_0$, we can talk about its tangent line at the point (x_0, y_0, z_0), where $z_0 = f(x_0, y_0)$. The slope of this tangent is the derivative of z with respect to x, but because y is being held constant at y_0, it must be the partial derivative of z with respect to x. In other words, the slope of the tangent line to the curve in Figure 13.11 at the point (x_0, y_0, z_0) is $\partial f/\partial x|_{(x_0, y_0)}$. Similarly, the partial derivative $\partial f/\partial y$ evaluated at (x_0, y_0) represents the slope of the tangent line to the curve of intersection of $z = f(x,y)$ and the plane $x = x_0$ at the point (x_0, y_0, z_0).

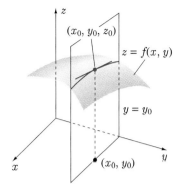

FIGURE 13.11

EXERCISES 13.3

In Exercises 1–20 evaluate $\partial f/\partial x$ and $\partial f/\partial y$.

1. $f(x,y) = x^3 y^2 + 2xy$
2. $f(x,y) = 3xy - 4x^4 y^4$
3. $f(x,y) = x^4/y^3$
4. $f(x,y) = x/(x+y) - x/y$
5. $f(x,y) = x/(2x^2 + y)$
6. $f(x,y) = \sin(xy)$
7. $f(x,y) = x\cos(x+y)$
8. $f(x,y) = \sqrt{x^2 + y^2}$
9. $f(x,y) = x\sqrt{x^2 - y^2}$
10. $f(x,y) = \tan(2x^2 + y^2)$
11. $f(x,y) = e^{x+y}$
12. $f(x,y) = e^{xy}$
13. $f(x,y) = xye^{xy}$
14. $f(x,y) = \ln(x^2 + y^2)$
15. $f(x,y) = (x+1)\ln(xy)$
16. $f(x,y) = \sin(ye^x)$
17. $f(x,y) = \mathrm{Tan}^{-1}(x/y)$
18. $f(x,y) = \sqrt[3]{1 - \cos^3(x^2 y)}$
19. $f(x,y) = \dfrac{\sin x}{\cos y}$
20. $f(x,y) = \ln\left(\sec\sqrt{x+y}\right)$

In Exercises 21–30 evaluate the indicated derivative.

21. $\partial f/\partial x$ if $f(x,y,z) = xyze^{x^2+y^2}$
22. $\partial f/\partial z$ if $f(x,z) = \mathrm{Tan}^{-1}[1/(x^2 + z^2)]$
23. $\partial f/\partial y$ at $(1,1,0)$ if $f(x,y,z) = xy(x^2 + y^2 + z^2)^{1/3}$
24. $\partial f/\partial x$ at $(1,-1,1,-1)$ if $f(x,y,z,t) = zt/(x^2 + y^2 - t^2)$
25. $\partial f/\partial t$ if $f(x,y,t) = x\sqrt{t^2 - y^2}/t^2 + (x/y)\mathrm{Sec}^{-1}(t/3)$
26. $\partial f/\partial x$ if $f(x,y,z) = \mathrm{Cot}^{-1}(1 + x + y + z)$
27. $\partial f/\partial y$ at $(1,2,3)$ if $f(x,y,t) = \mathrm{Sin}^{-1}(xyt)/\mathrm{Cos}^{-1}(xyt)$
28. $\partial f/\partial x$ if $f(x,y,z) = x^3/y + x\sin(yz/x)$
29. $\partial f/\partial t$ if $f(x,y,z,t) = xyz\ln(x^2 + y^2 + z^2)$
30. $\partial f/\partial z$ if $f(x,z) = (z^2/2)\mathrm{Sin}^{-1}(x/z) + (x/2)\sqrt{z^2 - x^2}$

31. If $f(x,y) = x^3 y/(x-y)$, show that
$$x\frac{\partial f}{\partial x} + y\frac{\partial f}{\partial y} = 3f(x,y).$$

32. If $f(x,y,z) = (x^4 + y^4 + z^4)/(xyz)$, show that
$$x\frac{\partial f}{\partial x} + y\frac{\partial f}{\partial y} + z\frac{\partial f}{\partial z} = f(x,y,z).$$

33. If $f(x,y,z) = (x^2 + y^2)\cos[(y+z)/x]$, show that
$$x\frac{\partial f}{\partial x} + y\frac{\partial f}{\partial y} + z\frac{\partial f}{\partial z} = 2f(x,y,z).$$

34. To evaluate $\partial f/\partial x$ for $f(x,y)$ at the point $(1,2)$ which of the following are acceptable:
   (a) differentiate $f(x,y)$ with respect to x holding y constant, and then set $x=1$ and $y=2$
   (b) set $x=1$ and $y=2$, and then differentiate with respect to x

(c) set $y = 2$, differentiate with respect to x, and set $x = 1$

(d) set $x = 1$, differentiate with respect to x, and set $y = 2$

35. Temperature at points (x, y) in a semicircular plate defined by $x^2 + y^2 \leq 4$, $y \geq 0$ is given by $T(x, y) = 16x^2 - 24xy + 40y^2$. Find, if possible: **(a)** $T_x(1, 1)$ **(b)** $T_y(1, 1)$ **(c)** $T_x(1, 0)$ **(d)** $T_y(1, 0)$ **(e)** $T_x(0, 2)$ **(f)** $T_y(0, 2)$

36. Can you find a function $f(x, y)$ so that $f_x(x, y) = 2x - 3y$ and $f_y(x, y) = 3x + 4y$?

37. Suppose a, b, and c are the lengths of the sides of a triangle and A, B, and C are the opposite angles. Find:
(a) $a_A(b, c, A)$ **(b)** $A_a(a, b, c)$ **(c)** $a_b(b, c, A)$
(d) $A_b(a, b, c)$

38. The equation of continuity for three-dimensional unsteady flow of a compressible fluid is

$$\frac{\partial \rho}{\partial t} + \frac{\partial}{\partial x}(\rho u) + \frac{\partial}{\partial y}(\rho v) + \frac{\partial}{\partial z}(\rho w) = 0,$$

where $\rho(x, y, z, t)$ is the density of the fluid, and $u\hat{\mathbf{i}} + v\hat{\mathbf{j}} + w\hat{\mathbf{k}}$ is the velocity of the fluid at position (x, y, z) and time t. Determine whether the continuity equation is satisfied if:

(a) $\rho = $ constant, $u = (2x^2 - xy + z^2)t$, $v = (x^2 - 4xy + y^2)t$, $w = (-2xy - yz + y^2)t$

(b) $\rho = xy + zt$, $u = x^2y + t$, $v = y^2z - 2t^2$, $w = 5x + 2z$

39. In complex variable theory, two functions $u(x, y)$ and $v(x, y)$ are said to be *harmonic conjugates* in a region R if in R they satisfy the Cauchy-Riemann equations

$$\frac{\partial u}{\partial x} = \frac{\partial v}{\partial y}, \qquad \frac{\partial v}{\partial x} = -\frac{\partial u}{\partial y}.$$

Show that the following pairs of functions are harmonic conjugates:
(a) $u(x, y) = -3xy^2 + y + x^3$, $v(x, y) = 3x^2y - y^3 - x + 5$
(b) $u(x, y) = (x^2 + x + y^2)/(x^2 + y^2)$, $v(x, y) = -y/(x^2 + y^2)$
(c) $u(x, y) = e^x(x \cos y - y \sin y)$, $v(x, y) = e^x(x \sin y + y \cos y)$

40. If r and θ are polar coordinates, then the Cauchy-Riemann equations in Exercise 39 for functions $u(r, \theta)$ and $v(r, \theta)$ take the form

$$\frac{\partial u}{\partial r} = \frac{1}{r}\frac{\partial v}{\partial \theta}, \qquad \frac{1}{r}\frac{\partial u}{\partial \theta} = -\frac{\partial v}{\partial r}, \qquad r \neq 0.$$

Show that the following pairs of functions satisfy these equations:
(a) $u(r, \theta) = (r^2 + r \cos \theta)/(1 + r^2 + 2r \cos \theta)$, $v(r, \theta) = r \sin \theta / (1 + r^2 + 2r \cos \theta)$
(b) $u(r, \theta) = \sqrt{r} \cos(\theta/2)$, $v(r, \theta) = \sqrt{r} \sin(\theta/2)$
(c) $u(r, \theta) = \ln r$, $v(r, \theta) = \theta$

SECTION 13.4

Gradients

Suppose a function $f(x, y, z)$ is defined at each point in some region of space, and that at each point of the region all three partial derivatives

$$\frac{\partial f}{\partial x}, \quad \frac{\partial f}{\partial y}, \quad \frac{\partial f}{\partial z}$$

exist. For example, if $f(x, y, z)$ represents the present temperature at each point in the room in which you are working, then these derivatives represent the rates of change of temperature in directions parallel to the x-, y-, and z-axes, respectively. There is a particular combination of these derivatives that will prove very useful in later work. This combination is contained in the following definition.

Definition 13.4

If a function $f(x, y, z)$ has partial derivatives $\partial f/\partial x$, $\partial f/\partial y$, and $\partial f/\partial z$ at each point in some region D of space, then at each point in D we define a vector called the **gradient** of $f(x, y, z)$, written grad f or ∇f, by

$$\operatorname{grad} f = \nabla f = \frac{\partial f}{\partial x}\hat{\mathbf{i}} + \frac{\partial f}{\partial y}\hat{\mathbf{j}} + \frac{\partial f}{\partial z}\hat{\mathbf{k}}. \qquad (13.8)$$

For a function $f(x, y)$ of only two independent variables, we have

$$\nabla f = \frac{\partial f}{\partial x}\hat{\mathbf{i}} + \frac{\partial f}{\partial y}\hat{\mathbf{j}}. \qquad (13.9)$$

EXAMPLE 13.6

If $f(x, y, z) = x^2 yz - 2x/y$, find ∇f at $(1, -1, 3)$.

SOLUTION Since

$$\nabla f = (2xyz - 2/y)\hat{\mathbf{i}} + (x^2 z + 2x/y^2)\hat{\mathbf{j}} + (x^2 y)\hat{\mathbf{k}},$$

then

$$\nabla f|_{(1,-1,3)} = -4\hat{\mathbf{i}} + 5\hat{\mathbf{j}} - \hat{\mathbf{k}}.$$

EXAMPLE 13.7

If $f(x, y, z) = \operatorname{Tan}^{-1}(xy/z)$, what is ∇f?

SOLUTION

$$\nabla f = \left\{\frac{1}{1+(xy/z)^2}\left(\frac{y}{z}\right)\right\}\hat{\mathbf{i}} + \left\{\frac{1}{1+(xy/z)^2}\left(\frac{x}{z}\right)\right\}\hat{\mathbf{j}} + \left\{\frac{1}{1+(xy/z)^2}\left(\frac{-xy}{z^2}\right)\right\}\hat{\mathbf{k}}$$

$$= \frac{yz}{z^2 + x^2 y^2}\hat{\mathbf{i}} + \frac{xz}{z^2 + x^2 y^2}\hat{\mathbf{j}} - \frac{xy}{z^2 + x^2 y^2}\hat{\mathbf{k}}$$

$$= (yz\hat{\mathbf{i}} + xz\hat{\mathbf{j}} - xy\hat{\mathbf{k}})/(z^2 + x^2 y^2).$$

Gradients arise in a multitude of applications in applied mathematics—heat conduction, electromagnetic theory, and fluid flow, to name a few— and two of the properties that make them so indispensable are discussed in detail in Sections 13.8 and 13.9. Examples 13.8 and 13.9 suggest these properties, but we make no attempt at a complete discussion here. For the moment we simply want you to be familiar with the definition of gradients and be able to calculate them.

EXAMPLE 13.8

The equation $F(x,y,z) = 0$, where $F(x,y,z) = x^2 + y^2 + z^2 - 4$, defines a sphere of radius 2 centred at the origin. Show that the gradient vector ∇F at any point on the sphere is perpendicular to the sphere.

SOLUTION If $P(x,y,z)$ is any point on the sphere (Figure 13.12), then the position vector $\mathbf{r} = x\hat{\mathbf{i}} + y\hat{\mathbf{j}} + z\hat{\mathbf{k}}$ from the origin to P is clearly perpendicular to the sphere. On the other hand,

$$\nabla F = 2x\hat{\mathbf{i}} + 2y\hat{\mathbf{j}} + 2z\hat{\mathbf{k}} = 2\mathbf{r}.$$

Consequently, at any point P on the sphere, ∇F is also perpendicular to the sphere.

FIGURE 13.12

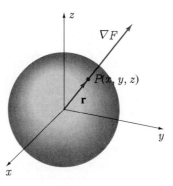

This example suggests that gradients may be useful in finding perpendiculars to surfaces (and, as we will see, perpendiculars to curves).

EXAMPLE 13.9

The function $f(x,y) = 2x^2 - 4x + 3y^2 + 2y + 6$ is defined at every point in the xy-plane. If we start at the origin (0, 0) and move along the positive x-axis, the rate of change of the function is $f_x(0,0) = -4$; if we move along the y-axis, the rate of change is $f_y(0,0) = 2$. Calculate the rate of change of $f(x,y)$ at (0, 0) if we move toward the point (1, 1) along the line $y = x$, and show that it is equal to the component of $\nabla f|_{(0,0)}$ in the direction $\mathbf{v} = (1,1)$.

SOLUTION The difference in the values of $f(x,y)$ at any point (x,y) and (0, 0) is $f(x,y) - f(0,0) = (2x^2 - 4x + 3y^2 + 2y + 6) - (6) = 2x^2 - 4x + 3y^2 + 2y$. If we divide this by the length of the line joining (0, 0) and (x,y), we obtain

$$\frac{f(x,y) - f(0,0)}{\sqrt{x^2 + y^2}} = \frac{2x^2 - 4x + 3y^2 + 2y}{\sqrt{x^2 + y^2}}.$$

The limit of this quotient as (x,y) approaches (0, 0) along the line $y = x$ should yield the required rate of change. We therefore set $y = x$ and take the limit as x approaches zero through positive numbers:

$$\lim_{x \to 0^+} \frac{2x^2 - 4x + 3x^2 + 2x}{\sqrt{x^2 + x^2}} = \lim_{x \to 0^+} \frac{x(5x - 2)}{\sqrt{2}\,x} = -\sqrt{2}.$$

This quantity, then, is the rate of change of $f(x,y)$ at $(0, 0)$ along the line $y = x$ toward the point $(1, 1)$.

On the other hand,

$$\nabla f_{|(0,0)} = (4x - 4, 6y + 2)_{|(0,0)} = (-4, 2),$$

and the component of this vector in the direction $\mathbf{v} = (1, 1)$ is

$$\nabla f_{|(0,0)} \cdot \hat{\mathbf{v}} = (-4, 2) \cdot \frac{(1,1)}{\sqrt{2}} = \frac{-4+2}{\sqrt{2}} = -\sqrt{2}.$$

This example indicates that gradients may be useful in calculating rates of change of functions in directions other than those parallel to the coordinate axes.

EXAMPLE 13.10

The electrostatic potential at a point (x, y, z) in space due to a charge q fixed at the origin is given by

$$V = \frac{q}{4\pi\epsilon_0 r},$$

where $r = \sqrt{x^2 + y^2 + z^2}$. If a second charge Q is placed at (x, y, z), it experiences a force $\mathbf{F}$, where

$$\mathbf{F} = \frac{qQ}{4\pi\epsilon_0 r^3}\mathbf{r},$$

where $\mathbf{r} = x\hat{\mathbf{i}} + y\hat{\mathbf{j}} + z\hat{\mathbf{k}}$. Show that $\mathbf{F} = -Q\nabla V$.

SOLUTION

$$\nabla V = \nabla\left(\frac{q}{4\pi\epsilon_0 r}\right) = \frac{q}{4\pi\epsilon_0}\nabla\left(\frac{1}{\sqrt{x^2+y^2+z^2}}\right)$$

$$= \frac{q}{4\pi\epsilon_0}\left\{\frac{-x}{(x^2+y^2+z^2)^{3/2}}\hat{\mathbf{i}} + \frac{-y}{(x^2+y^2+z^2)^{3/2}}\hat{\mathbf{j}} + \frac{-z}{(x^2+y^2+z^2)^{3/2}}\hat{\mathbf{k}}\right\}$$

$$= \frac{-q}{4\pi\epsilon_0 r^3}(x\hat{\mathbf{i}} + y\hat{\mathbf{j}} + z\hat{\mathbf{k}}) = \frac{-q}{4\pi\epsilon_0 r^3}\mathbf{r} = -\frac{\mathbf{F}}{Q}$$

EXERCISES 13.4

In Exercises 1–10 find the gradient of the function.

1. $f(x,y,z) = x^2 y + xz + yz^2$
2. $f(x,y,z) = x^2 yz$
3. $f(x,y,z) = x^2 y/z - 2xz^6$
4. $f(x,y) = x^2 y + xy^2$
5. $f(x,y) = \sin(x+y)$
6. $f(x,y,z) = \text{Tan}^{-1}(xyz)$
7. $f(x,y) = \text{Tan}^{-1}(y/x)$
8. $f(x,y,z) = e^{x+y+z}$
9. $f(x,y) = 1/(x^2 + y^2)$

10. $f(x, y, z) = 1/\sqrt{x^2 + y^2 + z^2}$

In Exercises 11–15 find the gradient of the function at the point.

11. $f(x, y) = xy + x + y$ at $(1, 3)$
12. $f(x, y, z) = \cos(x + y + z)$ at $(-1, 1, 1)$
13. $f(x, y, z) = (x^2 + y^2 + z^2)^2$ at $(0, 3, 6)$
14. $f(x, y) = e^{-x^2 - y^2}$ at $(2, 2)$
15. $f(x, y, z) = xy \ln(x + y)$ at $(4, -2)$
16. The equation $F(x, y, z) = Ax + By + Cz + D = 0$ defines a plane in space. Show that at any point on the plane, the vector ∇F is perpendicular to the plane.
17. Use the result of Exercise 16 to illustrate that a vector along the line
$$F(x, y, z) = 2x + 3y - 2z + 4 = 0,$$
$$G(x, y, z) = x - y + 3z + 6 = 0$$
is $\nabla F \times \nabla G$. Find parametric equations for the line.
18. Prove that if $f(x, y, z)$ and $g(x, y, z)$ both have gradients, then $\nabla(fg) = f \nabla g + g \nabla f$.
19. Repeat Example 13.9 for the functions (a) $f(x, y) = x^2 + y^2$ (b) $f(x, y) = 2x^3 - 3y$.
20. The equation $F(x, y) = x^3 + xy + y^4 - 5 = 0$ implicitly defines a curve in the xy-plane. Show that at any point on the curve, ∇F is a normal vector to the curve.

Sketch the surface defined by the equation in Exercises 21 and 22. At what points on the surface is ∇F not defined?

21. $F(x, y, z) = z - \sqrt{x^2 + y^2} = 0$
22. $F(x, y, z) = z - |x - y| = 0$
23. If $f(x, y) = 1 - x^2 - y^2$, find ∇f. Find the point (x, y) at which $\nabla f = \mathbf{0}$, and illustrate graphically the nature of the surface $z = f(x, y)$ at this point.
24. If the gradient of a function $f(x, y)$ is $\nabla f = (2xy - y)\hat{\mathbf{i}} + (x^2 - x)\hat{\mathbf{j}}$, what is $f(x, y)$?
25. Repeat Exercise 24 if $\nabla f = (2x/y + 1)\hat{\mathbf{i}} + (-x^2/y^2 + 2)\hat{\mathbf{j}}$.
26. If the gradient of a function $f(x, y, z)$ is $\nabla f = yz\hat{\mathbf{i}} + (xz + 2yz)\hat{\mathbf{j}} + (xy + y^2)\hat{\mathbf{k}}$, what is $f(x, y, z)$?
27. Repeat Exercise 26 if $\nabla f = (x\hat{\mathbf{i}} + y\hat{\mathbf{j}} + z\hat{\mathbf{k}})/\sqrt{x^2 + y^2 + z^2}$.
28. If $f(x, y)$ and $g(x, y)$ have first partial derivatives in a region R of the xy-plane, and if in R, $\nabla f = \nabla g$, how are $f(x, y)$ and $g(x, y)$ related?
29. If $\nabla f = \mathbf{0}$ for all points in some region R of space, what can we say about $f(x, y, z)$ in R?
30. Show that if the equation $F(x, y) = 0$ implicitly defines a curve C in the xy-plane, then at any point on C, the vector ∇F is perpendicular to C.

SECTION 13.5

Higher-Order Partial Derivatives

If $f(x, y) = x^3 y^2 + ye^x$, then

$$\frac{\partial f}{\partial x} = 3x^2 y^2 + ye^x$$

and

$$\frac{\partial f}{\partial y} = 2x^3 y + e^x.$$

Since each of these partial derivatives is a function of x and y, we can take further partial derivatives. The partial derivative of $\partial f / \partial x$ with respect to x is called the second partial derivative of $f(x, y)$ with respect to x, and is written

$$\frac{\partial}{\partial x}\left(\frac{\partial f}{\partial x}\right) = \frac{\partial^2 f}{\partial x^2} = 6xy^2 + ye^x.$$

Similarly, we have three more second partial derivatives:

$$\frac{\partial}{\partial y}\left(\frac{\partial f}{\partial x}\right) = \frac{\partial^2 f}{\partial y \partial x} = 6x^2 y + e^x,$$

$$\frac{\partial}{\partial x}\left(\frac{\partial f}{\partial y}\right) = \frac{\partial^2 f}{\partial x \partial y} = 6x^2 y + e^x,$$

$$\frac{\partial}{\partial y}\left(\frac{\partial f}{\partial y}\right) = \frac{\partial^2 f}{\partial y^2} = 2x^3.$$

Note that the second partial derivatives $\partial^2 f / \partial x \partial y$ and $\partial^2 f / \partial y \partial x$ are identical. This is not a peculiarity of this function, but according to the following theorem it is what we should expect normally.

Theorem 13.1

If $f(x,y)$, $\partial f / \partial x$, $\partial f / \partial y$, $\partial^2 f / \partial x \partial y$, and $\partial^2 f / \partial y \partial x$ are all defined inside a circle centred at a point P, and are continuous at P, then at P,

$$\frac{\partial^2 f}{\partial x \partial y} = \frac{\partial^2 f}{\partial y \partial x}. \tag{13.10}$$

For most functions with which we will be concerned, this theorem can be extended to say that a mixed partial derivative may be calculated in any order whatsoever. For example, if we require $\partial^{10} f / \partial x^3 \partial y^7$, where $f(x, y) = \ln(y^y) + x^2 y^{10}$, it is advantageous to reverse the order of differentiation,

$$\frac{\partial^{10} f}{\partial x^3 \partial y^7} = \frac{\partial^{10} f}{\partial y^7 \partial x^3} = \frac{\partial^7}{\partial y^7}\left(\frac{\partial^3 f}{\partial x^3}\right) = 0.$$

EXAMPLE 13.11

Show that $f(x,y) = \text{Tan}^{-1}(2xy/(x^2 - y^2))$ satisfies the equation

$$\frac{\partial^2 f}{\partial x^2} + \frac{\partial^2 f}{\partial y^2} = 0$$

at all points in the xy-plane that are not on the lines $y = \pm x$.

SOLUTION Since

$$\frac{\partial f}{\partial x} = \frac{1}{1 + \left(\dfrac{2xy}{x^2 - y^2}\right)^2}\left\{\frac{(x^2 - y^2)(2y) - 2xy(2x)}{(x^2 - y^2)^2}\right\}$$

$$= \frac{(x^2 - y^2)^2}{(x^2 - y^2)^2 + 4x^2 y^2}\left\{\frac{2y(x^2 - y^2 - 2x^2)}{(x^2 - y^2)^2}\right\}$$

$$= \frac{-2y(x^2 + y^2)}{x^4 - 2x^2 y^2 + y^4 + 4x^2 y^2} = \frac{-2y(x^2 + y^2)}{(x^2 + y^2)^2} = \frac{-2y}{x^2 + y^2},$$

the second derivative with respect to x is

$$\frac{\partial^2 f}{\partial x^2} = \frac{2y}{(x^2+y^2)^2}(2x) = \frac{4xy}{(x^2+y^2)^2}.$$

Similarly,

$$\frac{\partial f}{\partial y} = \frac{1}{1+\left(\frac{2xy}{x^2-y^2}\right)^2}\left\{\frac{(x^2-y^2)(2x)-2xy(-2y)}{(x^2-y^2)^2}\right\}$$

$$= \frac{(x^2-y^2)^2}{(x^2+y^2)^2}\left\{\frac{2x(x^2-y^2+2y^2)}{(x^2-y^2)^2}\right\} = \frac{2x}{x^2+y^2}.$$

Consequently,

$$\frac{\partial^2 f}{\partial y^2} = \frac{-2x}{(x^2+y^2)^2}(2y) = \frac{-4xy}{(x^2+y^2)^2},$$

and addition of these two expressions for $\partial^2 f/\partial x^2$ and $\partial^2 f/\partial y^2$ completes the proof. ∎

The equation

$$\frac{\partial^2 f}{\partial x^2} + \frac{\partial^2 f}{\partial y^2} = 0 \qquad (13.11)$$

for a function $f(x,y)$ is one of the most important equations in applied mathematics. It is called Laplace's equation in two variables (x and y). Laplace's equation for a function $f(x,y,z)$ of three variables is

$$\frac{\partial^2 f}{\partial x^2} + \frac{\partial^2 f}{\partial y^2} + \frac{\partial^2 f}{\partial z^2} = 0. \qquad (13.12)$$

A function is said to be **harmonic** in a region R if it satisfies Laplace's equation in R and has continuous second partial derivatives in R. In particular, the function $f(x,y)$ in Example 13.11 is harmonic in any region that does not contain points on the lines $y = \pm x$. In Exercises 29–31, we illustrate various areas of applied mathematics that make use of Laplace's equation.

EXERCISES 13.5

In Exercises 1–20 find the derivative.

1. $\partial^2 f/\partial x^2$ if $f(x,y) = x^2 y^2 - 2x^3 y$
2. $\partial^3 f/\partial y^3$ if $f(x,y) = 2x/y + 3x^3 y^4$
3. $\partial^2 f/\partial z^2$ if $f(x,y,z) = \sin(xyz)$
4. $\partial^2 f/\partial y \partial z$ if $f(x,y,z) = xyze^{x+y+z}$
5. $\partial^2 f/\partial y \partial x$ if $f(x,y) = \sqrt{x^2+y^2}$
6. $\partial^3 f/\partial x^2 \partial y$ if $f(x,y) = e^{x+y} - x^2/y^2$
7. $\partial^3 f/\partial y^3$ at $(1,3)$ if $f(x,y) = 3x^3 y^3 - 3x/y$
8. $\partial^3 f/\partial x \partial y \partial z$ at $(1,0,-1)$ if $f(x,y,z) = x^2 y^2 + x^2 z^2 + y^2 z^2$
9. $\partial^2 f/\partial x^2$ if $f(x,y) = \sqrt{1-x^2-y^2}$
10. $\partial^2 f/\partial z^2$ if $f(x,y,z) = \ln\sqrt{x^2+y^2+z^2}$
11. $\partial^3 f/\partial x^2 \partial y$ if $f(x,y) = x^2 e^y + y^2 e^x$
12. $\partial^2 f/\partial x^2$ if $f(x,y) = \text{Tan}^{-1}(y/x)$
13. $\partial^3 f/\partial x \partial y^2$ if $f(x,y,z) = \cot(x^2+y^2+z^2)$
14. $\partial^2 f/\partial x \partial y$ at $(-2,-2)$ if $f(x,y) = \text{Sin}^{-1}(x^2+y^2)^{-1}$

15. $\partial^{10} f/\partial x^7 \partial y^3$ if $f(x,y) = x^7 e^x y^2 + 1/y^6$

16. $\partial^8 f/\partial x^8$ if $f(x,y,z) = x^8 y^9 z^{10}$

17. $\partial^6 f/\partial x^2 \partial y^2 \partial z^2$ if $f(x,y,z) = 1/x^2 + 1/y^2 + 1/z^2$

18. $\partial^4 f/\partial x^3 \partial y$ if $f(x,y) = \cos(x + y^3)$

19. $\partial^4 f/\partial x \partial y \partial z \partial t$ if $f(x,y,z,t) = \sqrt{x^2 + y^2 + z^2 - t^2}$

20. $\partial^2 f/\partial x \partial y$ if $f(x,y) = \text{Sec}^{-1}(xy)$

21. If $z = x^2 + xy + y^2 \sin(x/y)$, show that

$$x\frac{\partial z}{\partial x} + y\frac{\partial z}{\partial y} = 2z = x^2\frac{\partial^2 z}{\partial x^2} + 2xy\frac{\partial^2 z}{\partial x \partial y} + y^2\frac{\partial^2 z}{\partial y^2}.$$

22. If $u = x + y + ze^{y/x}$, show that

$$x^2\frac{\partial^2 u}{\partial x^2} + y^2\frac{\partial^2 u}{\partial y^2} + z^2\frac{\partial^2 u}{\partial z^2} + 2xy\frac{\partial^2 u}{\partial x \partial y} + 2yz\frac{\partial^2 u}{\partial y \partial z}$$
$$+ 2xz\frac{\partial^2 u}{\partial x \partial z} = 0.$$

In Exercises 23–28 find a region (if possible) in which the function is harmonic.

23. $f(x,y) = x^2 - y^2 + 2xy + y$

24. $f(x,y) = \ln(x^2 + y^2)$

25. $f(x,y) = x^3 y^2 - 3xy$

26. $f(x,y,z) = 3x^2 yz - y^3 z + xy$

27. $f(x,y,z) = 1/\sqrt{x^2 + y^2 + z^2}$

28. $f(x,y,z) = x^3 y^3 z^3$

29. (a) Show that the electrostatic potential function $V(x,y,z) = q/(4\pi\epsilon_0 r)$ in Example 13.10 satisfies Laplace's equation 13.12.
(b) If $V(x,y,z)$ represents the electrostatic potential at a point (x,y,z) due to a system of n point charges at points (x_i, y_i, z_i), does $V(x,y,z)$ satisfy Laplace's equation?

30. The gravitational potential at a point (x,y,z) in space due to a uniform spherical mass distribution (mass M) at the origin is defined as $V = GM/r$ where G is a constant and $r = \sqrt{x^2 + y^2 + z^2}$. Show that $V(x,y,z)$ satisfies Laplace's equation 13.12.

31. Figure 13.13 shows a plate bounded by the lines $x = 0$, $y = 0$, $x = 1$, and $y = 1$. Temperature along the first three sides is kept at $0\,°C$, while that along $y = 1$ varies according to $f(x) = \sin(3\pi x) - 2\sin(4\pi x)$, $0 \le x \le 1$. The temperature at any point interior to the plate is then

$$T(x,y) = C(e^{3\pi y} - e^{-3\pi y})\sin(3\pi x)$$
$$+ D(e^{4\pi y} - e^{-4\pi y})\sin(4\pi x),$$

where $C = (e^{3\pi} - e^{-3\pi})^{-1}$ and $D = -2(e^{4\pi} - e^{-4\pi})^{-1}$. Show that $T(x,y)$ is harmonic in the region $0 < x < 1$, $0 < y < 1$, and that it also satisfies the boundary conditions $T(0,y) = 0$, $T(1,y) = 0$, $T(x,0) = 0$, and $T(x,1) = f(x)$.

FIGURE 13.13

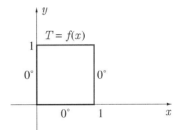

32. If $\partial^2 z/\partial x^2$ is positive for $x_1 \le x \le x_2$ and $y = y_0$, what can you say geometrically about the surface $z = f(x,y)$?

33. Two functions $u(x,y)$ and $v(x,y)$ are said to be harmonic conjugates if they satisfy the Cauchy-Riemann equations of Exercise 39 in Section 13.3. Show that if $u(x,y)$ and $v(x,y)$ are harmonic conjugates and have continuous second partial derivatives in a region R, then each is harmonic in R.

34. If the ends of a taut string are fastened at $x = 0$ and $x = L$ on the x-axis and if the string vibrates in the y-direction, then its displacement $y(x,t)$ must satisfy the one-dimensional wave equation

$$\frac{\partial^2 y}{\partial t^2} = \frac{T}{\rho}\frac{\partial^2 y}{\partial x^2},$$

where T is the constant tension in the string and ρ is its constant density. Determine whether the wave equation is satisfied by:

(a) $y(x,t) = \sin\left(\frac{5\pi x}{L}\right) \cos\left(\frac{5\pi}{L}\sqrt{\frac{T}{\rho}}t\right)$

(b) $y(x,t) = \sum_{n=1}^{50} a_n \sin\left(\frac{n\pi x}{L}\right) \cos\left(\frac{n\pi}{L}\sqrt{\frac{T}{\rho}}t\right)$,
where a_n are constants

35. A uniform circular rod has flat ends at $x = 0$ and $x = L$ on the x-axis. If the round side of the rod is insulated, then heat flows in only the x-direction. If, further, the face at $x = L$ is insulated and the face at $x = 0$ is kept at temperature $100\,°C$, the temperature $T(x,t)$ at any point in the rod must satisfy the one-dimensional heat conduction problem:

$$\frac{\partial T}{\partial t} = k \frac{\partial^2 T}{\partial x^2}, \quad k = \text{a constant},$$
$$T(0,t) = 100,$$
$$T_x(L,t) = 0.$$

Show that for any positive integer n,

$$T(x,t) = 100 - \frac{2L}{(2n-1)\pi} e^{-k[(2n-1)\pi/(2L)]^2 t} \times$$
$$\sin\left[\left(\frac{2n-1}{2}\right)\frac{\pi x}{L}\right]$$

satisfies the heat conduction problem.

36. (a) Show that the function $u(x,y) = x^2 - y^2$ is harmonic in the entire xy-plane.

(b) Use the Cauchy-Riemann equations in Exercise 39 of Section 13.3 to find a function $v(x,y)$ so that u and v are harmonic conjugates.

37. Repeat Exercise 36 if $u(x,y) = e^x \cos y + x$.

38. For what values of n does the function $(x^2 + y^2 + z^2)^n$ satisfy equation 13.12? In what regions are the functions harmonic?

SECTION 13.6

Chain Rules for Partial Derivatives

If $y = f(u)$ and $u = g(x)$, then the chain rule for the derivative dy/dx of the composite function $f[g(x)]$ is

$$\frac{dy}{dx} = \frac{dy}{du}\frac{du}{dx}. \tag{13.13}$$

Equation 13.13 can be extended in terms of more intermediate variables, say $y = f(u)$, $u = g(s)$, $s = h(x)$, in which case

$$\frac{dy}{dx} = \frac{dy}{du}\frac{du}{ds}\frac{ds}{dx}. \tag{13.14}$$

For multivariable functions, variations in chain rules are countless. We discuss two examples in considerable detail, and then show schematically how to obtain the correct chain rule for every possible situation.

Suppose z is a function of u and v and each of u and v is a function of x and y,

$$z = f(u,v), \quad u = g(x,y), \quad v = h(x,y). \tag{13.15}$$

By the substitutions

$$z = f[g(x,y), h(x,y)], \tag{13.16}$$

we could express z as a function of x and y, and could ask for the partial derivative $\partial z/\partial x$. However, if the functions in 13.15 are at all complicated, you can imagine how difficult the composite function in 13.16 might be to differentiate. As a result, we search for an alternative procedure for calculating $\partial z/\partial x$, namely, the appropriate chain rule. It is contained in the following theorem.

Theorem 13.2

Let $u = g(x,y)$ and $v = h(x,y)$ be continuous and have first partial derivatives with respect to x at a point (x,y), and let $z = f(u,v)$ have continuous first partial derivatives inside a circle centred at the point $(u,v) = (g(x,y), h(x,y))$. Then

$$\frac{\partial z}{\partial x} = \frac{\partial z}{\partial u}\frac{\partial u}{\partial x} + \frac{\partial z}{\partial v}\frac{\partial v}{\partial x}. \tag{13.17}$$

Proof This result can be proved in much the same way as chain rule 3.16 was proved in Section 3.6. By Definition 13.3,

$$\frac{\partial z}{\partial x} = \lim_{\Delta x \to 0} \frac{f[g(x+\Delta x, y), h(x+\Delta x, y)] - f[g(x,y), h(x,y)]}{\Delta x}.$$

Now the increment Δx in x produces changes in u and v, which we denote by

$$\Delta u = g(x+\Delta x, y) - g(x,y), \quad \Delta v = h(x+\Delta x, y) - h(x,y).$$

If we write u and v whenever $g(x,y)$ and $h(x,y)$ are evaluated at (x,y), and substitute for $g(x+\Delta x, y)$ and $h(x+\Delta x, y)$ in the definition for $\partial z/\partial x$, then

$$\begin{aligned}\frac{\partial z}{\partial x} &= \lim_{\Delta x \to 0} \frac{f(u+\Delta u, v+\Delta v) - f(u,v)}{\Delta x} \\ &= \lim_{\Delta x \to 0} \frac{[f(u+\Delta u, v+\Delta v) - f(u, v+\Delta v)] + [f(u, v+\Delta v) - f(u,v)]}{\Delta x} \\ &= \lim_{\Delta x \to 0} \left\{ \frac{f(u+\Delta u, v+\Delta v) - f(u, v+\Delta v)}{\Delta x} + \frac{f(u, v+\Delta v) - f(u,v)}{\Delta x} \right\}.\end{aligned}$$

We assumed that the derivative

$$\frac{\partial z}{\partial v} = \lim_{\Delta v \to 0} \frac{f(u, v+\Delta v) - f(u,v)}{\Delta v}$$

exists at (u,v), and an equivalent way to express the fact that this limit exists is to say that

$$\frac{f(u, v+\Delta v) - f(u,v)}{\Delta v} = \frac{\partial z}{\partial v} + \epsilon_1,$$

where ϵ_1 must satisfy the condition that $\lim_{\Delta v \to 0} \epsilon_1 = 0$. We can write, therefore, that

$$f(u, v+\Delta v) - f(u,v) = [z_v(u,v) + \epsilon_1]\Delta v.$$

Similarly, we can write that

$$f(u+\Delta u, v+\Delta v) - f(u, v+\Delta v) = [z_u(u, v+\Delta v) + \epsilon_2]\Delta u,$$

where $\lim_{\Delta u \to 0} \epsilon_2 = 0$ (provided Δv is sufficiently small). When these expressions are substituted into the limit for $\partial z/\partial x$, we have

$$\frac{\partial z}{\partial x} = \lim_{\Delta x \to 0} \left\{ [z_u(u, v+\Delta v) + \epsilon_2]\frac{\Delta u}{\Delta x} + [z_v(u,v) + \epsilon_1]\frac{\Delta v}{\Delta x} \right\}.$$

We now examine each part of this limit. Clearly,

$$\lim_{\Delta x \to 0} \frac{\Delta u}{\Delta x} = \frac{\partial u}{\partial x} \quad \text{and} \quad \lim_{\Delta x \to 0} \frac{\Delta v}{\Delta x} = \frac{\partial v}{\partial x}.$$

In addition, because $g(x,y)$ and $h(x,y)$ are continuous, $\Delta u \to 0$ and $\Delta v \to 0$ as $\Delta x \to 0$. Consequently,

$$\lim_{\Delta x \to 0} \epsilon_1 = \lim_{\Delta v \to 0} \epsilon_1 = 0 \quad \text{and} \quad \lim_{\Delta x \to 0} \epsilon_2 = \lim_{\Delta v \to 0} \epsilon_2 = 0.$$

Finally, because $\partial z/\partial u$ is continuous,

$$\lim_{\Delta x \to 0} z_u(u, v+\Delta v) = \lim_{\Delta v \to 0} z_u(u, v+\Delta v) = z_u(u,v).$$

When all these results are taken into account, we have

$$\frac{\partial z}{\partial x} = z_u(u,v)\frac{\partial u}{\partial x} + z_v(u,v)\frac{\partial v}{\partial x} = \frac{\partial z}{\partial u}\frac{\partial u}{\partial x} + \frac{\partial z}{\partial v}\frac{\partial v}{\partial x},$$

which completes the proof.

Chain rule 13.17 defines $\partial z/\partial x$ in terms of derivatives of the given functions in 13.15. We could be more explicit by indicating which variable is being held constant in each of the five derivatives:

$$\left.\frac{\partial z}{\partial x}\right)_y = \left.\frac{\partial z}{\partial u}\right)_v \left.\frac{\partial u}{\partial x}\right)_y + \left.\frac{\partial z}{\partial v}\right)_u \left.\frac{\partial v}{\partial x}\right)_y. \qquad (13.18)$$

From the point of view of rates of change, this result seems quite reasonable. The left side is the rate of change of z with respect to x holding y constant. The first term $(\partial z/\partial u)(\partial u/\partial x)$ accounts for the rate of change of z with respect to those x's that affect z through u. The second term $(\partial z/\partial v)(\partial v/\partial x)$ accounts for the rate of change of z with respect to those x's that affect z through v. The total rate of change is then the sum of the two parts.

Consider now the functional situation

$$z = f(u,v), \quad u = g(x,y,s), \quad v = h(x,y,s), \quad x = p(t), \quad y = q(t), \quad s = r(t). \qquad (13.19)$$

By the substitutions

$$z = f[g(p(t),q(t),r(t)), h(p(t),q(t),r(t))], \qquad (13.20)$$

we can express z as a function of t alone, and can therefore pose the problem of calculating dz/dt. If we reason as in the last paragraph, the appropriate chain rule for dz/dt must account for all t's affecting z through u and v. We obtain, then,

$$\frac{dz}{dt} = \frac{\partial z}{\partial u}\frac{du}{dt} + \frac{\partial z}{\partial v}\frac{dv}{dt},$$

where we have written du/dt and dv/dt because u and v can be expressed entirely in terms of t:

$$u = g[p(t),q(t),r(t)], \quad v = h[p(t),q(t),r(t)].$$

Chain rules for each of du/dt and dv/dt (similar to 13.17) yield

$$\frac{du}{dt} = \frac{\partial u}{\partial x}\frac{dx}{dt} + \frac{\partial u}{\partial y}\frac{dy}{dt} + \frac{\partial u}{\partial s}\frac{ds}{dt}, \quad \frac{dv}{dt} = \frac{\partial v}{\partial x}\frac{dx}{dt} + \frac{\partial v}{\partial y}\frac{dy}{dt} + \frac{\partial v}{\partial s}\frac{ds}{dt}.$$

Finally, then,

$$\frac{dz}{dt} = \frac{\partial z}{\partial u}\left\{\frac{\partial u}{\partial x}\frac{dx}{dt} + \frac{\partial u}{\partial y}\frac{dy}{dt} + \frac{\partial u}{\partial s}\frac{ds}{dt}\right\} + \frac{\partial z}{\partial v}\left\{\frac{\partial v}{\partial x}\frac{dx}{dt} + \frac{\partial v}{\partial y}\frac{dy}{dt} + \frac{\partial v}{\partial s}\frac{ds}{dt}\right\}, \qquad (13.21)$$

which expresses dz/dt in terms of derivatives of the given functions in 13.19.

These two examples illustrate that finding chain rules for complicated composite functions can be very difficult. Fortunately, there is an amazingly simple method that gives the correct chain rule in every situation. The method is not designed to help you understand the chain rule, but to find it quickly. We suggest that you test your understanding by developing a few chain rules in the exercises with a discussion such as in the second example above, and then check your result by the quicker method.

In the first example we represent the functional situation described in 13.15 by the schematic diagram below.

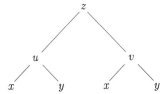

At the top of the diagram is the dependent variable z, which we wish to differentiate. In the line below z are the variables u and v in terms of which z is initially defined. In the line below u and v are x's and y's illustrating that each of u and v is defined in terms of x and y. To obtain the partial derivative $\partial z/\partial x$, we take all possible paths in this diagram from z to x. The two paths are from z through u to x, and from z through v to x. For each straight-line segment in a given path, we differentiate the upper variable with respect to the lower variable and multiply together all such derivatives in that path. The products are then added together to form the complete chain rule. In particular, for the path through u we form the product

$$\frac{\partial z}{\partial u}\frac{\partial u}{\partial x},$$

and for the path through v,

$$\frac{\partial z}{\partial v}\frac{\partial v}{\partial x}.$$

The complete chain rule is then the sum of these products,

$$\frac{\partial z}{\partial x} = \frac{\partial z}{\partial u}\frac{\partial u}{\partial x} + \frac{\partial z}{\partial v}\frac{\partial v}{\partial x},$$

and this result agrees with 13.17. The schematic diagram also indicates which variables are to be held constant in the derivatives on the right (as in 13.18). All variables on the same level are held constant.

For the second example in equations 13.19 the schematic diagram is:

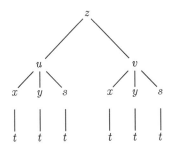

There are six possible paths from z to t so that the chain rule for dz/dt must have six terms. We find

$$\frac{dz}{dt} = \frac{\partial z}{\partial u}\frac{\partial u}{\partial x}\frac{dx}{dt} + \frac{\partial z}{\partial u}\frac{\partial u}{\partial y}\frac{dy}{dt} + \frac{\partial z}{\partial u}\frac{\partial u}{\partial s}\frac{ds}{dt} + \frac{\partial z}{\partial v}\frac{\partial v}{\partial x}\frac{dx}{dt} + \frac{\partial z}{\partial v}\frac{\partial v}{\partial y}\frac{dy}{dt} + \frac{\partial z}{\partial v}\frac{\partial v}{\partial s}\frac{ds}{dt},$$

and this agrees with 13.21. Note too that if when forming a derivative from the schematic diagram, there are two or more lines emanating from a variable, then we obtain a partial derivative; if there is only one line, then we have an ordinary derivative.

EXAMPLE 13.12

Find chain rules for

$$\left(\frac{\partial z}{\partial x}\right)_y \quad \text{and} \quad \left(\frac{\partial z}{\partial y}\right)_x$$

if

$$z = f(r,s,x), \quad r = g(x,y), \quad s = h(x,y).$$

SOLUTION From the schematic diagram below,

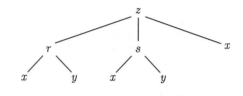

$$\left(\frac{\partial z}{\partial y}\right)_x = \left(\frac{\partial z}{\partial r}\right)_{s,x}\left(\frac{\partial r}{\partial y}\right)_x + \left(\frac{\partial z}{\partial s}\right)_{r,x}\left(\frac{\partial s}{\partial y}\right)_x,$$

$$\left(\frac{\partial z}{\partial x}\right)_y = \left(\frac{\partial z}{\partial r}\right)_{s,x}\left(\frac{\partial r}{\partial x}\right)_y + \left(\frac{\partial z}{\partial s}\right)_{r,x}\left(\frac{\partial s}{\partial x}\right)_y + \left(\frac{\partial z}{\partial x}\right)_{r,s}.$$

 In the previous example it is essential that we indicate which variables to hold constant in the partial derivatives. If we were to omit these designations, then in the second result we would have a term $\partial z/\partial x$ on both sides of the equation and might be tempted to cancel them. This cannot be done, however, since they are entirely different derivatives. The term $\partial z/\partial x)_y$ indicates the derivative of z with respect to x holding y constant if z were expressed entirely in terms of x and y; the term $\partial z/\partial x)_{r,s}$ indicates the derivative of the given function $f(r,s,x)$ with respect to x holding r and s constant.

EXAMPLE 13.13

Find dz/dt if

$$z = x^3 y^2 + x\sin y + tx, \quad x = 2t + \frac{1}{t}, \quad y = t^2 e^t.$$

SOLUTION From the schematic diagram below,

$$\frac{dz}{dt} = \frac{\partial z}{\partial x}\frac{dx}{dt} + \frac{\partial z}{\partial y}\frac{dy}{dt} + \frac{\partial z}{\partial t}$$

$$= (3x^2 y^2 + \sin y + t)(2 - 1/t^2) + (2x^3 y + x\cos y)(2te^t + t^2 e^t) + x.$$

When a chain rule is used to calculate a derivative, the result usually involves all intermediate variables. For instance, the derivative dz/dt in Example 13.13 involves not only t, but the intermediate variables x and y as well. Were dz/dt required at $t = 1$, values of x and y for $t = 1$ would be calculated — $x(1) = 3$ and $y(1) = e$ — and all three values substituted to obtain

$$\frac{dz}{dt}\bigg|_{t=1} = [3(3)^2(e)^2 + \sin(e) + 1](2-1)$$
$$+ [2(3)^3 e + (3)\cos(e)](2e + e) + 3$$
$$= 1378.6.$$

EXAMPLE 13.14

Find $\partial^2 z/\partial x^2$ if

$$z = s^2 t + 2\sin t, \quad s = xy - y, \quad t = x^2 + \frac{y}{x}.$$

SOLUTION From the schematic diagram below,

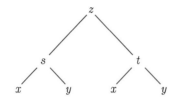

$$\frac{\partial z}{\partial x} = \frac{\partial z}{\partial s}\frac{\partial s}{\partial x} + \frac{\partial z}{\partial t}\frac{\partial t}{\partial x}$$
$$= (2st)(y) + (s^2 + 2\cos t)(2x - y/x^2).$$

Now $\partial z/\partial x$ is a function of s, t, x, and y, and therefore in order to find

$$\frac{\partial^2 z}{\partial x^2} = \frac{\partial}{\partial x}\left(\frac{\partial z}{\partial x}\right),$$

we form a schematic diagram for $\partial z/\partial x$:

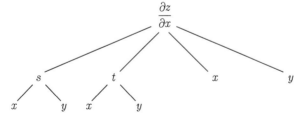

From this schematic diagram, we obtain

$$\frac{\partial^2 z}{\partial x^2} = \frac{\partial}{\partial s}\left(\frac{\partial z}{\partial x}\right)\frac{\partial s}{\partial x} + \frac{\partial}{\partial t}\left(\frac{\partial z}{\partial x}\right)\frac{\partial t}{\partial x} + \frac{\partial}{\partial x}\left(\frac{\partial z}{\partial x}\right)_{s,t,y}$$
$$= [2ty + 2s(2x - y/x^2)](y) + [2sy - 2\sin t(2x - y/x^2)](2x - y/x^2)$$
$$+ (s^2 + 2\cos t)(2 + 2y/x^3).$$ ∎

EXAMPLE 13.15

The temperature T at points in the atmosphere depends on both position (x, y, z) and time t: $T = T(x, y, z, t)$. When a weather balloon is released to take temperature readings, it is not free to take readings at just any point, but only at those points along the path that the winds force the balloon to follow. This path is a curve in space that can be defined parametrically by

$$C: \quad x = x(t), \quad y = y(t), \quad z = z(t), \quad t \geq 0,$$

t again being time. If we substitute from the equations for C into the temperature function, then T becomes a function of t alone,

$$T = T[x(t), y(t), z(t), t],$$

and this function of time describes the temperature at points along the path of the balloon. For the derivative of this function with respect to t, the schematic diagram

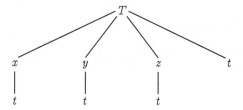

yields

$$\frac{dT}{dt} = \frac{\partial T}{\partial x}\frac{dx}{dt} + \frac{\partial T}{\partial y}\frac{dy}{dt} + \frac{\partial T}{\partial z}\frac{dz}{dt} + \frac{\partial T}{\partial t}.$$

The question we pose is "What is the physical difference between dT/dt and $\partial T/\partial t$?"

SOLUTION The temperature at a point in space is independent of the observer measuring it; hence $T[x(t), y(t), z(t), t]$ is the temperature at points on C as measured by both the balloon and any observer fixed in the xyz-reference system. If, however, these two observers calculate the rate of change of temperature with respect to time at some point (x, y, z) on C, they calculate different results. The observer fixed in the xyz-reference system (not restricted to move along C) calculates the rate of change of T with respect to t as the derivative of the function $T(x, y, z, t)$ partially with respect to t holding x, y, and z constant; i.e., the fixed observer calculates $\partial T/\partial t$ as the rate of change of temperature in time. The balloon, on the other hand, has no alternative but to take temperature readings as it moves along C; thus its measurement of T as a function of t is

$$T[x(t), y(t), z(t), t].$$

Therefore, when the balloon calculates the time variation of temperature, it is calculating dT/dt. It follows, then, that the terms

$$\frac{\partial T}{\partial x}\frac{dx}{dt} + \frac{\partial T}{\partial y}\frac{dy}{dt} + \frac{\partial T}{\partial z}\frac{dz}{dt}$$

describe that part of dT/dt caused by the motion of the balloon through space. ∎

Many important applications of the chain rule occur in the field of partial differential equations. To illustrate, we introduce the one-dimensional wave equation in the following example.

EXAMPLE 13.16

The one-dimensional wave equation

$$\frac{\partial^2 y}{\partial t^2} = c^2 \frac{\partial^2 y}{\partial x^2}, \quad c = \text{constant}$$

for functions $y(x,t)$ describes transverse vibrations of taut strings, and longitudinal and rotational vibrations of metal rods. Show that if $f(u)$ and $g(v)$ are twice differentiable functions of u and v, then $y(x,t) = f(x+ct) + g(x-ct)$ satisfies the wave equation.

SOLUTION The schematic diagram below describes the functional situation

$$y = f(u) + g(v)$$

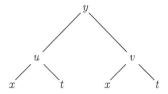

where $u = x + ct$ and $v = x - ct$. The chain rule for $\partial y/\partial t$ is

$$\frac{\partial y}{\partial t} = \frac{\partial y}{\partial u}\frac{\partial u}{\partial t} + \frac{\partial y}{\partial v}\frac{\partial v}{\partial t} = cf'(u) - cg'(v).$$

The schematic diagram for $\partial y/\partial t$ below leads to

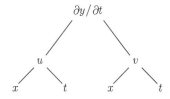

$$\frac{\partial^2 y}{\partial t^2} = \frac{\partial}{\partial u}\left(\frac{\partial y}{\partial t}\right)\frac{\partial u}{\partial t} + \frac{\partial}{\partial v}\left(\frac{\partial y}{\partial t}\right)\frac{\partial v}{\partial t}$$
$$= [cf''(u)]c + [-cg''(v)](-c)$$
$$= c^2[f''(u) + g''(v)].$$

A similar calculation gives $\dfrac{\partial^2 y}{\partial x^2} = f''(u) + g''(v)$. Hence $y(x,t)$ does indeed satisfy the wave equation. ∎

Homogeneous Functions

Homogeneous functions arise in numerous areas of applied mathematics. A function

$f(x,y,z)$ is said to be **positively homogeneous** of degree n if for every $t > 0$,

$$f(tx, ty, tz) = t^n f(x, y, z). \qquad (13.22)$$

For example, the function $f(x,y,z) = x^2 + y^2 + z^2$ is homogeneous of degree 2; the function $f(x,y) = x^3 \cos(y/x) + x^2 y + xy^2$ is homogeneous of degree 3; and $f(x,y,z,t) = \sqrt{x^2 + z^2}(x^2 y + yt^2)$ is homogeneous of degree 4. Partial derivatives of homogeneous functions satisfy many identities. In particular, their first derivatives satisfy Euler's theorem.

Theorem 13.3

(Euler's Theorem) If $f(x,y,z)$ is positively homogeneous of degree n, and has continuous first partial derivatives, then

$$x \frac{\partial f}{\partial x} + y \frac{\partial f}{\partial y} + z \frac{\partial f}{\partial z} = nf(x, y, z). \qquad (13.23)$$

Proof To verify 13.23 we differentiate 13.22 with respect to t holding x, y, and z constant. For the derivative of the left side we introduce variables $u = tx$, $v = ty$, and $w = tz$, and use the schematic diagram below. The result is

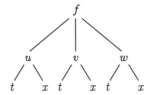

$$\frac{\partial f}{\partial u} \frac{\partial u}{\partial t} + \frac{\partial f}{\partial v} \frac{\partial v}{\partial t} + \frac{\partial f}{\partial w} \frac{\partial w}{\partial t} = nt^{n-1} f(x, y, z),$$

or,

$$x \frac{\partial f}{\partial u} + y \frac{\partial f}{\partial v} + z \frac{\partial f}{\partial w} = nt^{n-1} f(x, y, z).$$

When we set $t = 1$, we obtain $u = x$, $v = y$, $w = z$, and the above equation becomes 13.23.

The results of Exercises 31–33 in Section 13.3 are special cases of 13.23.

EXERCISES 13.6

In Exercises 1–10 we have defined a general functional situation and a specific example. Find the chain rule for the indicated derivative in the general situation, and then use that result to calculate the same derivative in the specific example.

1. dz/dt if $z = f(x,t)$, $x = g(t)$; $z = xt^2/(x+t)$, $x = e^{3t}$

2. $\left(\dfrac{\partial z}{\partial t}\right)_s$ if $z = f(x,y)$, $x = g(s,t)$, $y = h(s,t)$; $z = x^2 e^y + y \ln x$, $x = s^2 \cos t$, $y = 4 \operatorname{Sec}^{-1}(t^2 + 2s)$

3. $\left(\dfrac{\partial u}{\partial s}\right)_t$ if $u = f(x,y,z)$, $x = g(s,t)$, $y = h(s,t)$, $z = k(s,t)$; $u = \sqrt{x^2 + y^2 + z^2}$, $,x = 2st$, $y = s^2 + t^2$, $z = st$

4. dz/du if $z = f(x, y, v)$, $x = g(u)$, $y = h(u)$, $v = k(u)$; $z = x^2yv^3$, $x = u^3 + 2u$, $y = \ln(u^2 + 1)$, $v = ue^u$

5. $\left(\dfrac{\partial u}{\partial r}\right)_t$ if $u = f(x, y, s)$, $x = g(t)$, $y = h(r)$, $s = k(r, t)$; $u = \sqrt{x^2 + y^2 s}$, $x = t/(t+5)$, $y = \operatorname{Sin}^{-1}(r^2 + 5)$, $s = \tan(rt)$

6. $\left(\dfrac{\partial z}{\partial t}\right)_r$ if $z = f(x)$, $x = g(y)$, $y = h(r, t)$; $z = 3^{x+2}$, $x = y^2 + 5$, $y = \csc(r^2 + t)$

7. $\left(\dfrac{\partial u}{\partial x}\right)_y$ if $u = f(x, y, z)$, $z = g(x, y)$; $u = y/\sqrt{x^2 + y^2 + z^2}$, $z = x/y$

8. $\left(\dfrac{\partial x}{\partial y}\right)_z$ if $x = f(r, s, t)$, $r = g(y)$, $s = h(y, z)$, $t = k(y, z)$; $x = s^2r^2t^2$, $r = y^{-5}$, $s = 1/(y^2 + z^2)$, $t = 1/y^2 + 1/z^2$

9. $\left(\dfrac{\partial z}{\partial t}\right)_s$ if $z = f(x, y)$, $x = g(r)$, $y = h(r)$, $r = k(s, t)$; $z = e^{x+y}$, $x = 2r + 5$, $y = 2r - 5$, $r = t\ln(s^2 + t^2)$

10. dz/dt if $z = f(x, y, u)$, $x = g(v)$, $u = h(x, y)$, $v = k(t)$, $y = p(t)$; $z = x^2 + y^2 + u^2$, $x = v^3 - 3v^2$, $u = 1/(x^2 - y^2)$, $v = e^t$, $y = e^{4t}$

In Exercises 11–15 find the derivative.

11. $\left(\dfrac{\partial^2 z}{\partial t^2}\right)_s$ if $z = x^2y^2 + xe^y$, $x = s + t^2$, $y = s - t^2$

12. d^2x/dt^2 if $x = y^2 + yt - t^2$, $y = t^2e^t$

13. $\left(\dfrac{\partial^2 u}{\partial s^2}\right)_t$ if $u = x^2 + y^2 + z^2 + xyz$, $x = s^2 + t^2$, $y = s^2 - t^2$, $z = st$

14. d^2z/dv^2 if $z = \sin(xy)$, $x = 3\cos v$, $y = 4\sin v$

15. $\partial^2 u/\partial x\partial y$ if $u = y/\sqrt{x^2 + y^2 + z^2}$, $z = x/y$

16. Suppose that u is a differentiable function of r and $r = \sqrt{x^2 + y^2 + z^2}$. Show that

$$\left(\dfrac{\partial u}{\partial x}\right)^2 + \left(\dfrac{\partial u}{\partial y}\right)^2 + \left(\dfrac{\partial u}{\partial z}\right)^2 = \left(\dfrac{du}{dr}\right)^2.$$

17. Consider a gas that is moving through some region D of space. If we follow a particular particle of the gas, it traces out some curved path (Figure 13.14)

$$C: \quad x = x(t), \ y = y(t), \ z = z(t), \quad t \geq 0.$$

Suppose the density of the gas at any point in the region D at time t is denoted by $\rho(x, y, z, t)$. We can write that along C, $\rho = \rho[x(t), y(t), z(t), t]$.

(a) Obtain the chain rule defining $d\rho/dt$ in terms of $\partial \rho/\partial t$ and derivatives of x, y, and z with respect to t.

(b) Explain the physical difference between $d\rho/dt$ and $\partial \rho/\partial t$.

FIGURE 13.14

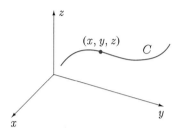

18. The radius and height of a right-circular cone are 10 and 20 cm respectively. If the radius is increasing at 1 cm/min and the height is decreasing at 2 cm/min, how fast is the volume changing? Do you need multivariable calculus to solve this problem?

19. If two sides of a triangle have lengths x and y and the angle between them is θ, then the area of the triangle is $A = (1/2)xy\sin\theta$. How fast is the area changing when x is 1 m, y is 2 m, and θ is 1/3 radian, if x and y are each increasing at $1/2$ m/s and θ is decreasing at $1/10$ radian per second?

20. When a rocket rises from the earth's surface, its mass decreases because fuel is being consumed at the rate of 50 kg/s. Use Newton's universal law of gravitation (see Example 7.33 in Section 7.9) to determine how fast the force of gravity of the earth on the rocket is changing when the rocket is 100 km above the earth's surface and travelling at 2 km/s. Assume that the mass of the rocket at this height is 12×10^6 kg.

21. If $z = f(u, v)$, $u = g(x, y)$, $v = h(x, y)$, find the chain rule for the second derivative $\partial^2 z/\partial x^2$.

22. Which of the following functions are positively homogeneous:

(a) $f(x, y) = x^2 + xy + 3y^2$
(b) $f(x, y) = x^2y + xy - 2xy^2$
(c) $f(x, y, z) = x^2\sin(y/z) + y^2 + y^3/z$
(d) $f(x, y, z) = xe^{y/z} - xyz$
(e) $f(x, y, z, t) = x^4 + y^4 + z^4 + t^4 - xyzt$
(f) $f(x, y, z, t) = e^{x^2+y^2}(z^2 + t^2)$
(g) $f(x, y, z) = \cos(xy)\sin(yz)$
(h) $f(x, y) = \sqrt{x^2 + xy + y^2}\,e^{y/x}(2x^2 - 3y^2)$

23. If $f(s)$ and $g(t)$ are differentiable functions, show that $\nabla f(x^2 - y^2) \cdot \nabla g(xy) = 0$.

24. If $f(s)$ is a differentiable function, show that $f(x - y)$ satisfies the equation
$$\frac{\partial f}{\partial y} = -\frac{\partial f}{\partial x}.$$

25. If $f(s)$ is a differentiable function, show that $u(x,y) = f(4x - 3y) + 5(y - x)$ satisfies the equation
$$3\frac{\partial u}{\partial x} + 4\frac{\partial u}{\partial y} = 5.$$

26. If $f(s)$ and $g(t)$ are twice differentiable, show that the function $u(x,y) = xf(x+y) + yg(x+y)$ satisfies
$$\frac{\partial^2 u}{\partial x^2} - 2\frac{\partial^2 u}{\partial x \partial y} + \frac{\partial^2 u}{\partial y^2} = 0.$$

27. If $f(s)$ and $g(t)$ are twice differentiable, show that $f(x-y) + g(x+y)$ satisfies
$$\frac{\partial^2 u}{\partial x^2} - \frac{\partial^2 u}{\partial y^2} = 0.$$

28. Show that if $f(v)$ is differentiable, then $u(x,y) = x^2 f(y/x)$ satisfies
$$x\frac{\partial u}{\partial x} + y\frac{\partial u}{\partial y} = 2u.$$

In Exercises 29–31 suppose that $f(x,y)$ satisfies the first partial differential equation. Show that with the change of independent variables, the function $F(u,v) = f[x(u,v), y(u,v)]$ must satisfy the second partial differential equation.

29. $\left(\frac{\partial f}{\partial x}\right)^2 + \left(\frac{\partial f}{\partial y}\right)^2 = 0$; $u = (x+y)/2$, $v = (x-y)/2$;
$\left(\frac{\partial F}{\partial u}\right)^2 + \left(\frac{\partial F}{\partial v}\right)^2 = 0$

30. $\frac{\partial^2 f}{\partial x^2} - \frac{\partial^2 f}{\partial y^2} = 0$; $u = (x+y)/2$, $v = (x-y)/2$;
$\frac{\partial^2 F}{\partial u \partial v} = 0$

31. $\left(\frac{\partial f}{\partial x}\right)^2 + \left(\frac{\partial f}{\partial y}\right)^2 = 0$; $x = u\cos v$, $y = u\sin v$;
$\left(\frac{\partial F}{\partial u}\right)^2 + \frac{1}{u^2}\left(\frac{\partial F}{\partial v}\right)^2 = 0$

32. An observer travels along the curve $x = t^2$, $y = 3t^3 + 1$, $z = 2t + 5$, where x, y, and z are in metres and $t \geq 0$ is in seconds. If the density ρ of a gas (in kg/m³) is given by $\rho = (3x^2 + y^2)/(z^2 + 5)$, find the time rate of change of the density of the gas as measured by the observer when $t = 2$ s.

33. If $f(r)$ is a differentiable function and $r = \sqrt{x^2 + y^2 + z^2}$, show that
$$\nabla f = \frac{f'(r)}{r}(x\hat{\mathbf{i}} + y\hat{\mathbf{j}} + z\hat{\mathbf{k}}).$$

34. If $f(x,y) = 0$ defines y as a function of x, show that
$$\frac{d^2 y}{dx^2} = -\frac{f_{xx}f_y^2 - 2f_{xy}f_x f_y + f_{yy}f_x^2}{f_y^3}.$$

35. If $f(x,y)$ is a harmonic function, show that the function $F(x,y) = f(x^2 - y^2, 2xy)$ is also harmonic.

36. Show that in polar coordinates, the two-dimensional Laplace equation 13.11 takes the form
$$\frac{\partial^2 f}{\partial r^2} + \frac{1}{r}\frac{\partial f}{\partial r} + \frac{1}{r^2}\frac{\partial^2 f}{\partial \theta^2} = 0.$$

37. Find an identity satisfied by the second partial derivatives of a function $f(x,y,z)$ that is positively homogeneous of degree n.

38. It is postulated in one of the theories of traffic flow that the average speed u at a point x on a straight highway (along the x-axis) is related to the concentration k of traffic by the differential equation
$$u\frac{\partial u}{\partial x} + \frac{\partial u}{\partial t} = -c^2 k^n \frac{\partial k}{\partial x},$$
where t is time, and $c > 0$ and n are constants.

(a) Use chain rules for $\partial u/\partial x$ and $\partial u/\partial t$ in the functional situation $u = f(k)$ and $k = g(x,t)$ to show that
$$\frac{du}{dk}\left(u\frac{\partial k}{\partial x} + \frac{\partial k}{\partial t}\right) + c^2 k^n \frac{\partial k}{\partial x} = 0.$$

(b) The equation of continuity for traffic flow states that
$$\frac{\partial k}{\partial t} + \frac{\partial(ku)}{\partial x} = 0.$$
Use these last two equations to obtain the differential equation relating speed and concentration:

$$\frac{du}{dk} = -ck^{(n-1)/2}.$$

(c) Solve the differential equation in (b) for $u = f(k)$.

39. A bead slides from rest at the origin on a frictionless wire in a vertical plane to the point (x_0, y_0) under the influence of gravity (Figure 13.15). It can be shown that the time for the bead to traverse the path is

$$t = \frac{1}{\sqrt{2g}} \int_0^{x_0} \sqrt{\frac{1+(y')^2}{y}}\, dx,$$

where g is the acceleration due to gravity and $y' = dy/dx$. The problem of finding that shape of wire which makes t as small as possible is called the *brachistochrone problem*. It is shown in the calculus of variations that $y = f(x)$ must satisfy the equation

$$\frac{d}{dx}\left(\frac{\partial F}{\partial y'}\right) - \frac{\partial F}{\partial y} = 0,$$

where

$$F(y, y') = \sqrt{\frac{1+(y')^2}{y}}.$$

FIGURE 13.15

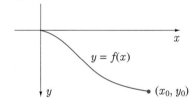

(a) Show that $f(x)$ must satisfy the differential equation $1 + (y')^2 + 2yy'' = 0$.
(b) Show that the curve that satisfies the equation in (a) is the cycloid defined parametrically by

$$x = a(\theta - \sin\theta), \quad y = a(1 - \cos\theta),$$

where a is a constant.

40. Two equal masses m are connected by springs having equal spring constant k so that the masses are free to slide on a frictionless table (Figure 13.16). The walls A and B are fixed.

(a) Use Newton's second law to show that the differential equations for the motions of the masses are

$$m\ddot{x}_1 = k(x_2 - 2x_1), \quad m\ddot{x}_2 = k(x_1 - 2x_2),$$

where x_1 and x_2 are the displacements of the masses from their equilibrium positions, $\ddot{x}_1 = d^2 x_1/dt^2$ and $\ddot{x}_2 = d^2 x_2/dt^2$.

(b) The Euler-Lagrange equations from theoretical mechanics for this system are

$$\frac{d}{dt}\left(\frac{\partial L}{\partial \dot{x}_1}\right) - \frac{\partial L}{\partial x_1} = 0, \quad \frac{d}{dt}\left(\frac{\partial L}{\partial \dot{x}_2}\right) - \frac{\partial L}{\partial x_2} = 0,$$

where L is defined as the kinetic energy of the two masses less the energy stored in the springs. Show that

$$L(x_1, x_2, \dot{x}_1, \dot{x}_2) = \frac{m}{2}(\dot{x}_1^2 + \dot{x}_2^2) - k(x_1^2 + x_2^2 - x_1 x_2).$$

(c) Obtain the equations in (a) from the Euler-Lagrange equations in (b).

FIGURE 13.16

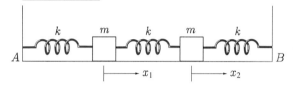

41. Suppose the second-order partial differential equation

$$p\frac{\partial^2 z}{\partial x^2} + q\frac{\partial^2 z}{\partial x \partial y} + r\frac{\partial^2 z}{\partial y^2} = F\left(x, y, z, \frac{\partial z}{\partial x}, \frac{\partial z}{\partial y}\right)$$

(p, q, and r are constants) is subjected to the change of variables

$$s = ax + by, \quad t = cx + dy,$$

where a, b, c, and d are constants. Show that the partial differential equation in s and t is

$$P\frac{\partial^2 z}{\partial s^2} + Q\frac{\partial^2 z}{\partial s \partial t} + R\frac{\partial^2 z}{\partial t^2} = G\left(s, t, z, \frac{\partial z}{\partial s}, \frac{\partial z}{\partial t}\right),$$

where $Q^2 - 4PR = (q^2 - 4pr)(ad - bc)^2$.

42. Show that if a solution $u = f(x, y, z)$ of the three-dimensional Laplace equation 13.12 can be expressed in the form $u = g(r)$, where $r = \sqrt{x^2 + y^2 + z^2}$, then $f(x, y, z)$ must be of the form

$$f(x, y, z) = \frac{C}{\sqrt{x^2 + y^2 + z^2}} + D,$$

where C and D are constants.

SECTION 13.7

Implicit Differentiation

In Section 3.7 we introduced the technique of implicit differentiation in order to obtain the derivative of a function $y = f(x)$ defined implicitly by an equation

$$F(x,y) = 0. \qquad (13.24)$$

Essentially, the technique is to differentiate all terms in the equation with respect to x, considering all the while that y is a function of x. For example, if y is defined implicitly by

$$x^2 y^3 + 3xy = 3x + 2,$$

implicit differentiation gives

$$2xy^3 + 3x^2 y^2 \frac{dy}{dx} + 3y + 3x\frac{dy}{dx} = 3.$$

We can now solve to obtain

$$\frac{dy}{dx} = \frac{3 - 2xy^3 - 3y}{3x^2 y^2 + 3x}.$$

With the chain rule we can actually present a formula for dy/dx. Since equation 13.24, when written in the form

$$F[x, f(x)] = 0,$$

must be valid for all x in the domain of the function $f(x)$, we can differentiate it with respect to x. From the schematic diagram below

the derivative of the left side of the equation is

$$\frac{dF}{dx} = \frac{\partial F}{\partial x} + \frac{\partial F}{\partial y}\frac{dy}{dx}.$$

If we equate this to the derivative of the right side of the equation, we find

$$F_x + F_y \frac{dy}{dx} = 0,$$

or

$$\frac{dy}{dx} = -\frac{F_x}{F_y}. \qquad (13.25)$$

For the function defined implicitly above by $x^2 y^3 + 3xy - 3x - 2 = 0$, equation 13.25 gives

$$\frac{dy}{dx} = -\frac{2xy^3 + 3y - 3}{3x^2 y^2 + 3x},$$

and this result is identical to that obtained by implicit differentiation.

Similarly, if the equation

$$F(x, y, z) = 0 \qquad (13.26)$$

defines z implicitly as a function of x and y, the schematic diagram below

immediately yields

$$\frac{\partial F}{\partial x} + \frac{\partial F}{\partial z}\frac{\partial z}{\partial x} = 0, \quad \frac{\partial F}{\partial y} + \frac{\partial F}{\partial z}\frac{\partial z}{\partial y} = 0.$$

From these we obtain the results

$$\frac{\partial z}{\partial x} = -\frac{F_x}{F_z}, \quad \frac{\partial z}{\partial y} = -\frac{F_y}{F_z}. \tag{13.27}$$

We do not suggest that formulas 13.25 and 13.27 be memorized. On the contrary, we obtain results in this section that include 13.25 and 13.27 as special cases. To develop these results we work with three equations in five variables,

$$F(x, y, u, v, w) = 0,$$
$$G(x, y, u, v, w) = 0, \tag{13.28}$$
$$H(x, y, u, v, w) = 0.$$

We assume that these equations define u, v, and w as functions of x and y for some domain of values of x and y (and do so implicitly). It might even be possible to solve the system and obtain explicit definitions of the functions

$$u = f(x, y),$$
$$v = g(x, y), \tag{13.29}$$
$$w = h(x, y).$$

We pose the problem of finding the six first-order partial derivatives of u, v, and w with respect to x and y supposing that it is undesirable or even impossible to obtain the explicit form of the functions. To do this, we note that were results 13.29 known and substituted into 13.28, then

$$F[x, y, f(x, y), g(x, y), h(x, y)] = 0,$$
$$G[x, y, f(x, y), g(x, y), h(x, y)] = 0,$$
$$H[x, y, f(x, y), g(x, y), h(x, y)] = 0,$$

would be identities in x and y. As a result we could differentiate each equation with respect to x, obtaining from the schematic diagram

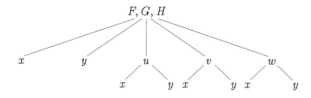

the following:
$$\frac{\partial F}{\partial x} + \frac{\partial F}{\partial u}\frac{\partial u}{\partial x} + \frac{\partial F}{\partial v}\frac{\partial v}{\partial x} + \frac{\partial F}{\partial w}\frac{\partial w}{\partial x} = 0,$$
$$\frac{\partial G}{\partial x} + \frac{\partial G}{\partial u}\frac{\partial u}{\partial x} + \frac{\partial G}{\partial v}\frac{\partial v}{\partial x} + \frac{\partial G}{\partial w}\frac{\partial w}{\partial x} = 0, \quad (13.30)$$
$$\frac{\partial H}{\partial x} + \frac{\partial H}{\partial u}\frac{\partial u}{\partial x} + \frac{\partial H}{\partial v}\frac{\partial v}{\partial x} + \frac{\partial H}{\partial w}\frac{\partial w}{\partial x} = 0,$$

or,
$$F_u \frac{\partial u}{\partial x} + F_v \frac{\partial v}{\partial x} + F_w \frac{\partial w}{\partial x} = -F_x,$$
$$G_u \frac{\partial u}{\partial x} + G_v \frac{\partial v}{\partial x} + G_w \frac{\partial w}{\partial x} = -G_x, \quad (13.31)$$
$$H_u \frac{\partial u}{\partial x} + H_v \frac{\partial v}{\partial x} + H_w \frac{\partial w}{\partial x} = -H_x.$$

We have in 13.31 three equations in the three unknowns $\partial u/\partial x$, $\partial v/\partial x$, and $\partial w/\partial x$, and because the equations are linear in the unknowns, solutions can be obtained using Cramer's rule.† In particular,

$$\frac{\partial u}{\partial x} = \frac{\begin{vmatrix} -F_x & F_v & F_w \\ -G_x & G_v & G_w \\ -H_x & H_v & H_w \end{vmatrix}}{\begin{vmatrix} F_u & F_v & F_w \\ G_u & G_v & G_w \\ H_u & H_v & H_w \end{vmatrix}} = -\frac{\begin{vmatrix} F_x & F_v & F_w \\ G_x & G_v & G_w \\ H_x & H_v & H_w \end{vmatrix}}{\begin{vmatrix} F_u & F_v & F_w \\ G_u & G_v & G_w \\ H_u & H_v & H_w \end{vmatrix}}. \quad (13.32)$$

The two determinants on the right of 13.32 involve only derivatives of the given functions F, G, and H, and we have therefore obtained a method for finding $\partial u/\partial x$ that avoids solving 13.28 for u, v, and w. We could list similar formulas for the remaining five derivatives, but first we introduce some simplifying notation.

Definition 13.5 The **Jacobian determinant** of functions F, G, and H with respect to variables u, v, and w is denoted by $\dfrac{\partial(F,G,H)}{\partial(u,v,w)}$ and is defined as the determinant

$$\frac{\partial(F,G,H)}{\partial(u,v,w)} = \begin{vmatrix} F_u & F_v & F_w \\ G_u & G_v & G_w \\ H_u & H_v & H_w \end{vmatrix} = \begin{vmatrix} \dfrac{\partial F}{\partial u} & \dfrac{\partial F}{\partial v} & \dfrac{\partial F}{\partial w} \\ \dfrac{\partial G}{\partial u} & \dfrac{\partial G}{\partial v} & \dfrac{\partial G}{\partial w} \\ \dfrac{\partial H}{\partial u} & \dfrac{\partial H}{\partial v} & \dfrac{\partial H}{\partial w} \end{vmatrix}. \quad (13.33)$$

With this notation we can write 13.32 in the form

$$\frac{\partial u}{\partial x} = -\frac{\dfrac{\partial(F,G,H)}{\partial(x,v,w)}}{\dfrac{\partial(F,G,H)}{\partial(u,v,w)}}. \quad (13.34)$$

---

† Cramer's rule is discussed in Appendix F.

The remaining derivatives of $v = g(x,y)$ and $w = h(x,y)$ with respect to x can also be obtained from equations 13.31 by Cramer's rule:

$$\frac{\partial v}{\partial x} = -\frac{\dfrac{\partial(F,G,H)}{\partial(u,x,w)}}{\dfrac{\partial(F,G,H)}{\partial(u,v,w)}}, \quad \frac{\partial w}{\partial x} = -\frac{\dfrac{\partial(F,G,H)}{\partial(u,v,x)}}{\dfrac{\partial(F,G,H)}{\partial(u,v,w)}}. \qquad (13.35)$$

A similar procedure yields

$$\frac{\partial u}{\partial y} = -\frac{\dfrac{\partial(F,G,H)}{\partial(y,v,w)}}{\dfrac{\partial(F,G,H)}{\partial(u,v,w)}}, \quad \frac{\partial v}{\partial y} = -\frac{\dfrac{\partial(F,G,H)}{\partial(u,y,w)}}{\dfrac{\partial(F,G,H)}{\partial(u,v,w)}}, \quad \frac{\partial w}{\partial y} = -\frac{\dfrac{\partial(F,G,H)}{\partial(u,v,y)}}{\dfrac{\partial(F,G,H)}{\partial(u,v,w)}}. \qquad (13.36)$$

Formulas 13.34–13.36 apply only to the situation in which equations 13.28 define u, v, and w as functions of x and y. It is, however, fairly evident how to construct formulas in other situations. Each formula has a Jacobian divided by a Jacobian (and do not forget the negative sign). In the denominator, it is the Jacobian of the functions defining the original equations with respect to the dependent variables. The only difference in the Jacobian in the numerator is that the dependent variable that is being differentiated is replaced by the independent variable with respect to which differentiation is being performed.

The results in equations 13.34–13.36 are valid provided, of course, that the Jacobian

$$\frac{\partial(F,G,H)}{\partial(u,v,w)} \neq 0.$$

In actual fact, it is this condition that guarantees that equations 13.28 do define u, v, and w as functions of x and y in the first place.

As a second example, if

$$F(x,y,s,t) = 0,$$
$$G(x,y,s,t) = 0,$$

define x and y as functions of s and t, then

$$\frac{\partial x}{\partial s} = -\frac{\dfrac{\partial(F,G)}{\partial(s,y)}}{\dfrac{\partial(F,G)}{\partial(x,y)}}, \quad \frac{\partial x}{\partial t} = -\frac{\dfrac{\partial(F,G)}{\partial(t,y)}}{\dfrac{\partial(F,G)}{\partial(x,y)}},$$

$$\frac{\partial y}{\partial s} = -\frac{\dfrac{\partial(F,G)}{\partial(x,s)}}{\dfrac{\partial(F,G)}{\partial(x,y)}}, \quad \frac{\partial y}{\partial t} = -\frac{\dfrac{\partial(F,G)}{\partial(x,t)}}{\dfrac{\partial(F,G)}{\partial(x,y)}}.$$

We now show that the results in equations 13.25 and 13.27 are special cases of these considerations. If equation 13.24 defines y implicitly as a function of x, then

$$\frac{dy}{dx} = -\frac{\dfrac{\partial(F)}{\partial(x)}}{\dfrac{\partial(F)}{\partial(y)}} = -\frac{F_x}{F_y}.$$

If equation 13.26 defines z implicitly as a function of x and y, then

$$\frac{\partial z}{\partial x} = -\frac{\dfrac{\partial(F)}{\partial(x)}}{\dfrac{\partial(F)}{\partial(z)}} = -\frac{F_x}{F_z}; \quad \frac{\partial z}{\partial y} = -\frac{\dfrac{\partial(F)}{\partial(y)}}{\dfrac{\partial(F)}{\partial(z)}} = -\frac{F_y}{F_z}.$$

EXAMPLE 13.17 If $x^2 y^2 z^3 + zx \sin y = 5$ defines z as a function of x and y, find $\partial z / \partial x$.

SOLUTION If we set $F(x,y,z) = x^2 y^2 z^3 + zx \sin y - 5 = 0$, then

$$\frac{\partial z}{\partial x} = -\frac{\dfrac{\partial(F)}{\partial(x)}}{\dfrac{\partial(F)}{\partial(z)}} = -\frac{F_x}{F_z} = -\frac{2xy^2 z^3 + z \sin y}{3x^2 y^2 z^2 + x \sin y}.$$

∎

EXAMPLE 13.18 The equations $x = r \cos \theta$, $y = r \sin \theta$ define Cartesian coordinates x and y in terms of polar coordinates r and θ. At the same time they implicitly define r and θ as functions of x and y. Find $\partial \theta / \partial x$.

SOLUTION If we set

$$F(x,y,r,\theta) = x - r \cos \theta = 0,$$
$$G(x,y,r,\theta) = y - r \sin \theta = 0,$$

then

$$\frac{\partial \theta}{\partial x} = -\frac{\dfrac{\partial(F,G)}{\partial(r,x)}}{\dfrac{\partial(F,G)}{\partial(r,\theta)}} = -\frac{\begin{vmatrix} F_r & F_x \\ G_r & G_x \end{vmatrix}}{\begin{vmatrix} F_r & F_\theta \\ G_r & G_\theta \end{vmatrix}} = -\frac{\begin{vmatrix} -\cos \theta & 1 \\ -\sin \theta & 0 \end{vmatrix}}{\begin{vmatrix} -\cos \theta & r \sin \theta \\ -\sin \theta & -r \cos \theta \end{vmatrix}} = -\frac{\sin \theta}{r}.$$

∎

We have already mentioned that partial derivatives are not to be considered quotients. In particular, note that had we thought of $\partial \theta / \partial x$ in Example 18 as a quotient, we might have been tempted to write

$$\frac{\partial \theta}{\partial x} = \frac{1}{\partial x / \partial \theta} = \frac{1}{-r \sin \theta},$$

which certainly does not agree with the result obtained.

EXAMPLE 13.19 In the theory of thermodynamics, the variables pressure P, temperature T, volume V, and internal energy U are related by two equations of state:

$$F(P,T,V,U) = 0, \quad G(P,T,V,U) = 0.$$

The second law of thermodynamics implies that if U and P are regarded as functions of T and V, then the functions $U(T,V)$ and $P(T,V)$ must satisfy the equation

$$\frac{\partial U}{\partial V} - T \frac{\partial P}{\partial T} + P = 0.$$

Show that if U and V are chosen as independent variables rather than T and V, then the second law takes the form

$$\frac{\partial T}{\partial V} + T\frac{\partial P}{\partial U} - P\frac{\partial T}{\partial U} = 0.$$

SOLUTION When U and V are taken as independent variables,

$$\frac{\partial T}{\partial V} = -\frac{\dfrac{\partial(F,G)}{\partial(V,P)}}{\dfrac{\partial(F,G)}{\partial(T,P)}},$$

$$\frac{\partial P}{\partial U} = -\frac{\dfrac{\partial(F,G)}{\partial(T,U)}}{\dfrac{\partial(F,G)}{\partial(T,P)}},$$

$$\frac{\partial T}{\partial U} = -\frac{\dfrac{\partial(F,G)}{\partial(U,P)}}{\dfrac{\partial(F,G)}{\partial(T,P)}}.$$

Consequently,

$$\frac{\partial T}{\partial V} + T\frac{\partial P}{\partial U} - P\frac{\partial T}{\partial U} = -\frac{1}{\dfrac{\partial(F,G)}{\partial(T,P)}}\left\{\frac{\partial(F,G)}{\partial(V,P)} + T\frac{\partial(F,G)}{\partial(T,U)} - P\frac{\partial(F,G)}{\partial(U,P)}\right\}.$$

To show that this expression must be equal to zero, we note that when T and V are independent variables, the second law implies that

$$0 = -\frac{\dfrac{\partial(F,G)}{\partial(V,P)}}{\dfrac{\partial(F,G)}{\partial(U,P)}} + T\frac{\dfrac{\partial(F,G)}{\partial(U,T)}}{\dfrac{\partial(F,G)}{\partial(U,P)}} + P$$

$$= -\frac{1}{\dfrac{\partial(F,G)}{\partial(U,P)}}\left\{\frac{\partial(F,G)}{\partial(V,P)} - T\frac{\partial(F,G)}{\partial(U,T)} - P\frac{\partial(F,G)}{\partial(U,P)}\right\}$$

$$= -\frac{1}{\dfrac{\partial(F,G)}{\partial(U,P)}}\left\{\frac{\partial(F,G)}{\partial(V,P)} + T\frac{\partial(F,G)}{\partial(T,U)} - P\frac{\partial(F,G)}{\partial(U,P)}\right\}.$$

Hence

$$\frac{\partial T}{\partial V} + T\frac{\partial P}{\partial U} - P\frac{\partial T}{\partial U} = 0. \qquad\blacksquare$$

EXERCISES 13.7

In Exercises 1–4 y is defined implicitly as a function of x. Find dy/dx.

1. $x^3 y^2 - 2xy + 5 = 0$

2. $(x + y)^2 = 2x$

3. $x(x - y) - 4y^3 = 2e^{xy} + 6$

4. $\sin(x + y) + y^2 = 12x^2 + y$

In Exercises 5–8 z is defined implicitly as a function of x and y. Find $\partial z/\partial x$ and $\partial z/\partial y$.

5. $x^2 \sin z - ye^z = 2x$

6. $x^2 z^2 + yz + 3x = 4$

7. $z \sin^2 y + y \sin^2 x = z^3$

8. $\text{Tan}^{-1}(yz) = xz$

In Exercises 9–13 find the required derivative. Assume that the system of equations does define the function(s) indicated.

9. $\partial u/\partial x$ and $\partial v/\partial y$ if $x^2 - y^2 + u^2 + 2v^2 = 1$, $x^2 + y^2 = 2 + u^2 + v^2$

10. $\partial x/\partial t$ if $\sin(x + t) - \sin(x - t) = z$

11. $\partial \phi/\partial x)_{y,z}$ if $x = r \sin \phi \cos \theta$, $y = r \sin \phi \sin \theta$, $z = r \cos \phi$

12. dz/dx if $x^2 + y^2 - z^2 + 2xy = 1$, $x^3 + y^3 - 5y = 4$

13. $\partial u/\partial y)_x$ if $xyu + vw = 4$, $y^2 + u^2 - u^2 v = y$, $yw + xu + v + 4 = 0$

14. Given that the equations
$$x^2 - y \cos(uv) + z^2 = 0,$$
$$x^2 + y^2 - \sin(uv) + 2z^2 = 2,$$
$$xy - \sin u \cos v + z = 0$$
define x, y, and z as functions of u and v, find $\partial x/\partial u)_v$ at the values $x = 1$, $y = 1$, $u = \pi/2$, $v = 0$, and $z = 0$.

15. If the equation $F(x, y, z) = 0$ defines each of x, y, and z as a function of the other two, show that
$$\left(\frac{\partial z}{\partial x}\right)_y \left(\frac{\partial x}{\partial y}\right)_z \left(\frac{\partial y}{\partial z}\right)_x = -1.$$

16. If $z = e^x \cos y$, where x and y are functions of t defined by
$$x^3 + e^x - t^2 - t = 1, \quad yt^2 + y^2 t - t + y = 0,$$
find dz/dt.

17. Find $\partial s/\partial u)_v$ if $s = x^2 + y^2$, and x and y are functions of u and v defined by
$$u = x^2 - y^2, \quad v = x^2 - y.$$

18. Find $\partial z/\partial y)_x$ if $z = u^3 v + \sin(uv)$, and u and v are functions of x and y defined by
$$x = e^u \cos v, \quad y = e^u \sin v.$$

19. Given that $z^3 - xz - y = 0$ defines z as a function of x and y, show that
$$\frac{\partial^2 z}{\partial x \partial y} = -\frac{3z^2 + x}{(3z^2 - x)^3}.$$

20. If the equations $x = u^2 - v^2$, $y = 2uv$, define u and v as functions of x and y, find $\partial^2 u/\partial x^2$.

21. (a) Given that the equation $z^4 x + y^3 z + 9x^3 = 2$ defines z as a function of x and y, and x as a function of y and z, are $\partial z/\partial x$ and $\partial x/\partial z$ reciprocals?
    (b) Given that the equations $z^4 x + y^3 z + 9x^3 = 2$, $x^2 y + xz = 1$, define z as a function of x, and x as a function of z, are dz/dx and dx/dz reciprocals?
    (c) Given that the equations $u^2 - v = 3x + y$, $u - 2v^2 = x - 2y$, define u and v as functions of x and y, and also define x and y as functions of u and v, are $\partial u/\partial x$ and $\partial x/\partial u$ reciprocals?

22. Given that the equations $x^2 - 2y^2 s^2 t - 2st^2 = 1$, $x^2 + 2y^2 s^2 t + 5st^2 = 1$, define s and t as functions of x and y, find $\partial^2 t/\partial y^2$.

23. Show that if V and P are chosen as independent variables in Example 13.19, then the second law of thermodynamics takes the form
$$T - P\frac{\partial T}{\partial P} + \frac{\partial(T, U)}{\partial(V, P)} = 0.$$

24. (a) Suppose the equations $F(u, v, s, t) = 0$, $G(u, v, s, t) = 0$ define u and v as functions of s and t, and the equations $H(s, t, x, y) = 0$, $I(s, t, x, y) = 0$ define s and t as functions of x and y. Show that
$$\frac{\partial(u, v)}{\partial(s, t)} \frac{\partial(s, t)}{\partial(x, y)} = \frac{\partial(u, v)}{\partial(x, y)}.$$

(b) If the equations $F(u, v, x, y) = 0$, $G(u, v, x, y) = 0$ define u and v as functions of x and y, and also define x and y as functions of u and v, show that

$$\frac{\partial(u,v)}{\partial(x,y)} = \frac{1}{\frac{\partial(x,y)}{\partial(u,v)}}.$$

25. Suppose the system of m linear equations in n unknowns ($n > m$)

$$\sum_{j=1}^{n} a_{ij} x_j = c_i, \quad i = 1, \ldots, m,$$

defines $x_1, x_2, \ldots, x_m$ as functions of $x_{m+1}, x_{m+2}, \ldots, x_n$. Show that if $1 \leq i \leq m$ and $m+1 \leq j \leq n$, then

$$\frac{\partial x_i}{\partial x_j} = -\frac{D_{ij}}{D},$$

where $D = |a_{ij}|_{m \times m}$, and D_{ij} is the same as determinant D except that its ith column is replaced by the jth column of $[a_{ij}]_{m \times n}$.

SECTION 13.8

Directional Derivatives

FIGURE 13.17

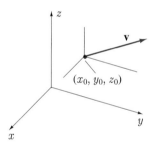

If a function $f(x, y, z)$ is defined throughout some region of space, then at any point (x_0, y_0, z_0) we can calculate its partial derivatives $\partial f / \partial x$, $\partial f / \partial y$, and $\partial f / \partial z$. These derivatives define rates of change of $f(x, y, z)$ at (x_0, y_0, z_0) in directions parallel to the x-, y-, and z-axes. But what if we want the rate of change of $f(x, y, z)$ at (x_0, y_0, z_0) in some arbitrary direction defined by a vector $\mathbf{v}$ (Figure 13.17)? By the rate of change of $f(x, y, z)$ in the direction $\mathbf{v}$, we mean the rate of change with respect to distance as measured along a line through (x_0, y_0, z_0) in the direction $\mathbf{v}$. Let us define s as a measure of directed distance along this line, taking $s = 0$ at (x_0, y_0, z_0) and positive s in the direction of $\mathbf{v}$. What we want, then, is the derivative of $f(x, y, z)$ with respect to s at $s = 0$. To express $f(x, y, z)$ in terms of s, we use parametric equations of the line through (x_0, y_0, z_0) along $\mathbf{v}$. If $\hat{\mathbf{v}} = (v_x, v_y, v_z)$ is a unit vector in the direction of $\mathbf{v}$, then parametric equations for this line (see equations 12.39) are

$$x = x_0 + v_x s, \quad y = y_0 + v_y s, \quad z = z_0 + v_z s. \quad (13.37)$$

From the schematic diagram below,

we obtain

$$\frac{df}{ds} = \frac{\partial f}{\partial x}\frac{dx}{ds} + \frac{\partial f}{\partial y}\frac{dy}{ds} + \frac{\partial f}{\partial z}\frac{dz}{ds}$$
$$= \frac{\partial f}{\partial x} v_x + \frac{\partial f}{\partial y} v_y + \frac{\partial f}{\partial z} v_z,$$

where all partial derivatives of $f(x, y, z)$ are to be evaluated at (x_0, y_0, z_0). This derivative df/ds is called the **directional derivative** of $f(x, y, z)$ in the direction $\mathbf{v}$ at the point (x_0, y_0, z_0), and is usually given the alternative notation

$$D_{\mathbf{v}} f = \frac{\partial f}{\partial x} v_x + \frac{\partial f}{\partial y} v_y + \frac{\partial f}{\partial z} v_z. \quad (13.38)$$

Now, v_x, v_y, and v_z are the components of the unit vector $\hat{\mathbf{v}}$ in the direction of $\mathbf{v}$, and $\partial f/\partial x$, $\partial f/\partial y$, and $\partial f/\partial z$ are the components of the gradient of $f(x, y, z)$. We can write, therefore, that

$$D_{\mathbf{v}} f = \nabla f \cdot \hat{\mathbf{v}}. \quad (13.39)$$

Consequently, the derivative (rate of change) of a function in any given direction is the scalar product of the gradient of the function and a unit vector in the required direction. We state this in the following theorem.

Theorem 13.4

> The directional derivative of a function in any direction is the component of the gradient of the function in that direction.

EXAMPLE 13.20

Find $D_{\mathbf{v}}f$ at $(4,0,16)$ if $f(x,y,z) = x^3 e^y + xz$ and $\mathbf{v}$ is the vector from $(4,0,16)$ to $(-2,1,4)$.

SOLUTION Since

$$\nabla f_{|(4,0,16)} = \{(3x^2 e^y + z)\hat{\mathbf{i}} + x^3 e^y \hat{\mathbf{j}} + x\hat{\mathbf{k}}\}_{|(4,0,16)}$$
$$= 64\hat{\mathbf{i}} + 64\hat{\mathbf{j}} + 4\hat{\mathbf{k}}$$

and

$$\hat{\mathbf{v}} = \frac{\mathbf{v}}{|\mathbf{v}|} = \frac{(-6,1,-12)}{\sqrt{36+1+144}} = \frac{-1}{\sqrt{181}}(6,-1,12),$$

we have

$$D_{\mathbf{v}}f = -(64,64,4) \cdot \frac{(6,-1,12)}{\sqrt{181}} = -\frac{368}{\sqrt{181}}.$$

The directional derivative gives us insight into some of the properties of the gradient vector. In particular, we have the next theorem.

Theorem 13.5

> The gradient ∇f of a function $f(x,y,z)$ defines the direction in which the function increases most rapidly, and the maximum rate of change is $|\nabla f|$.

Proof Theorem 13.4 states that the directional derivative of $f(x,y,z)$ in a direction $\mathbf{v}$ is the component of ∇f in that direction. Figure 13.18, which shows components of ∇f in various directions, makes it clear that $D_{\mathbf{v}}f$ is greatest when $\mathbf{v}$ is parallel to ∇f. Alternatively, if θ is the angle between $\mathbf{v}$ and ∇f, then

$$D_{\mathbf{v}}f = \nabla f \cdot \hat{\mathbf{v}} = |\nabla f||\hat{\mathbf{v}}|\cos\theta = |\nabla f|\cos\theta.$$

Obviously $D_{\mathbf{v}}f$ is a maximum when $\cos\theta$ is a maximum—i.e., when $\cos\theta = 1$ or $\theta = 0$—and this occurs when $\mathbf{v}$ is parallel to ∇f. Finally, when $\mathbf{v}$ is parallel to ∇f, we have $D_{\mathbf{v}}f = |\nabla f|$, and this completes the proof.

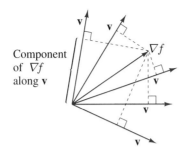

FIGURE 13.18

Note that for any function $f(x,y,z)$,

$$D_{\mathbf{i}}f = \frac{\partial f}{\partial x}, \quad D_{\mathbf{j}}f = \frac{\partial f}{\partial y}, \quad D_{\hat{\mathbf{k}}}f = \frac{\partial f}{\partial z}.$$

In other words, the partial derivatives of a function are its directional derivatives along the coordinate directions.

EXAMPLE 13.21

Find the direction at the point $(1,2,-3)$ in which the function $f(x,y,z) = x^2 y + xyz$ increases most rapidly.

SOLUTION According to Theorem 13.5, $f(x,y,z)$ increases most rapidly in the direction

$$\nabla f|_{(1,2,-3)} = (2xy + yz, x^2 + xz, xy)|_{(1,2,-3)} = (-2,-2,2).$$

You might feel that because the definition of the directional derivative $D_{\mathbf{v}}f$ does not involve a limit process, it is some strange new type of differentiation. To show that this is not the case, let us return to the calculation of the derivative of $f(x,y,z)$ at (x_0, y_0, z_0) in the direction $\mathbf{v}$ shown in Figure 13.17. With parametric equations 13.37 for the line through (x_0, y_0, z_0) along $\mathbf{v}$, the value of $f(x,y,z)$ at any point (x,y,z) along this line is $f(x_0 + v_x s, y_0 + v_y s, z_0 + v_z s)$. If we take the difference between this value and $f(x_0, y_0, z_0)$ and divide by the distance s between (x_0, y_0, z_0) and (x,y,z), then the limit of this expression as $s \to 0^+$ should define the derivative of $f(x,y,z)$ at (x_0, y_0, z_0) in the direction $\mathbf{v}$; that is,

$$D_{\mathbf{v}}f = \lim_{s \to 0^+} \frac{f(x_0 + v_x s, y_0 + v_y s, z_0 + v_z s) - f(x_0, y_0, z_0)}{s}. \tag{13.40}$$

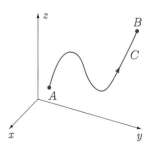

FIGURE 13.19

It can be shown that this limit (and this is perhaps the form we might have expected the derivative to take) also leads to the result contained in 13.39 (see Exercise 33).

Consider a curve C in space that is defined parametrically by

$$C: \quad x = x(t), \quad y = y(t), \quad z = z(t), \quad \alpha \le t \le \beta$$

(Figure 13.19). Imagine that C is the path traced out by some particle as it moves through space under the action of some system of forces, and suppose that $f(x,y,z)$ is a function defined along C. Perhaps the particle is a weather balloon and $f(x,y,z)$ is temperature at points along its trajectory C. In such applications we are frequently asked for the rate of change of $f(x,y,z)$ with respect to distance traveled along C. If we use s as a measure of distance along C (taking $s = 0$ at A), then the required rate of change is

df/ds. Since the coordinates of points (x,y,z) on C can be regarded as functions of s (although it might be difficult to find these functions explicitly), the chain rule gives

$$\frac{df}{ds} = \frac{\partial f}{\partial x}\frac{dx}{ds} + \frac{\partial f}{\partial y}\frac{dy}{ds} + \frac{\partial f}{\partial z}\frac{dz}{ds}$$
$$= \left(\frac{\partial f}{\partial x}, \frac{\partial f}{\partial y}, \frac{\partial f}{\partial z}\right) \cdot \left(\frac{dx}{ds}, \frac{dy}{ds}, \frac{dz}{ds}\right)$$
$$= \nabla f \cdot \frac{d\mathbf{r}}{ds}.$$

In Section 12.8 we saw that $d\mathbf{r}/ds$ is a unit tangent vector $\hat{\mathbf{T}}$ to C. Consequently,

$$\frac{df}{ds} = \nabla f \cdot \hat{\mathbf{T}}.$$

But this equation states that df/ds is the directional derivative of $f(x,y,z)$ along the tangent direction to the curve C. In other words, to calculate the rate of change of a function $f(x,y,z)$ with respect to distance as measured along a curve C, we calculate the directional derivative of $f(x,y,z)$ in the direction of the tangent vector to C.

EXAMPLE 13.22

Find the rate of change of the function $f(x,y,z) = x^2y - xz$ along the curve $y = x^2$, $z = x$ in the direction of decreasing x at the point $(2,4,2)$.

SOLUTION Since parametric equations for the curve are C: $x = -t$, $y = t^2$, $z = -t$, a tangent vector to C at any point is $\mathbf{T} = (-1, 2t, -1)$. At $(2,4,2)$, $t = -2$, and the tangent vector is $\mathbf{T} = (-1, -4, -1)$. A unit tangent vector to C at $(2,4,2)$ in the direction of decreasing x is therefore

$$\hat{\mathbf{T}} = \frac{(-1,-4,-1)}{\sqrt{18}} = \frac{-1}{3\sqrt{2}}(1,4,1).$$

The rate of change of $f(x,y,z)$ in this direction is

$$\nabla f \cdot \hat{\mathbf{T}} = (2xy - z, x^2, -x)|_{(2,4,2)} \cdot \frac{(1,4,1)}{-3\sqrt{2}}$$
$$= \frac{-1}{3\sqrt{2}}(14,4,-2) \cdot (1,4,1) = -\frac{28}{3\sqrt{2}}.$$

It is worthwhile at this point to discuss directional derivatives for a function $f(x,y)$ of two independent variables. Such a function can be represented graphically as a surface $z = f(x,y)$ (Figure 13.20).

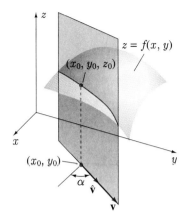

FIGURE 13.20

For a direction **v** at (x_0, y_0) in the xy-plane,

$$D_{\mathbf{v}} f = \nabla f \cdot \hat{\mathbf{v}},$$

where ∇f is evaluated at (x_0, y_0). Algebraically, this is the derivative (rate of change) of $f(x, y)$ in the direction **v**. Geometrically, this is the rate of change of the height z of the surface as we move along the curve of intersection of the surface and a vertical plane containing the vector **v**, or, the slope of this curve. Each direction **v** at (x_0, y_0) defines an angle α with a line through (x_0, y_0) parallel to the positive x-axis, and for this direction

$$\hat{\mathbf{v}} = \cos \alpha \, \hat{\mathbf{i}} + \sin \alpha \, \hat{\mathbf{j}}.$$

We can write, then,

$$D_{\mathbf{v}} f = \nabla f \cdot \hat{\mathbf{v}} = \left(\frac{\partial f}{\partial x} \hat{\mathbf{i}} + \frac{\partial f}{\partial y} \hat{\mathbf{j}} \right) \cdot (\cos \alpha \, \hat{\mathbf{i}} + \sin \alpha \, \hat{\mathbf{j}})$$

$$= \frac{\partial f}{\partial x} \cos \alpha + \frac{\partial f}{\partial y} \sin \alpha. \qquad (13.41)$$

If $D_{\mathbf{v}} f$ represents the slope of the curve of intersection of the surface and the vertical plane through **v**, then $D_{\mathbf{v}}(D_{\mathbf{v}} f)$ represents the rate of change of this slope. Now

$$D_{\mathbf{v}}(D_{\mathbf{v}} f) = \nabla(D_{\mathbf{v}} f) \cdot \hat{\mathbf{v}},$$

and

$$\nabla(D_{\mathbf{v}} f) = \left(\frac{\partial^2 f}{\partial x^2} \cos \alpha + \frac{\partial^2 f}{\partial x \partial y} \sin \alpha \right) \hat{\mathbf{i}} + \left(\frac{\partial^2 f}{\partial y \partial x} \cos \alpha + \frac{\partial^2 f}{\partial y^2} \sin \alpha \right) \hat{\mathbf{j}}.$$

Thus,

$$D_{\mathbf{v}}(D_{\mathbf{v}} f) = \left(\frac{\partial^2 f}{\partial x^2} \cos \alpha + \frac{\partial^2 f}{\partial x \partial y} \sin \alpha \right) \cos \alpha + \left(\frac{\partial^2 f}{\partial y \partial x} \cos \alpha + \frac{\partial^2 f}{\partial y^2} \sin \alpha \right) \sin \alpha$$

$$= \frac{\partial^2 f}{\partial x^2} \cos^2 \alpha + 2 \frac{\partial^2 f}{\partial x \partial y} \cos \alpha \sin \alpha + \frac{\partial^2 f}{\partial y^2} \sin^2 \alpha. \qquad (13.42)$$

We call $D_{\mathbf{v}}(D_{\mathbf{v}} f)$ the second directional derivative of $f(x, y)$ at (x_0, y_0) in the direction **v**. If it is positive, then the curve of intersection is concave upward, whereas if it is negative, the curve is concave downward. We will find these results useful in Section 13.10 when we discuss relative extrema of functions of two independent variables.

EXERCISES 13.8

In Exercises 1–8 calculate the directional derivative of the function at the point and in the direction indicated.

1. $f(x, y, z) = 2x^2 - y^2 + z^2$ at $(1, 2, 3)$ in the direction of the vector from $(1, 2, 3)$ to $(3, 5, 0)$.

2. $f(x, y, z) = x^2 y + xz$ at $(-1, 1, -1)$ in the direction of the vector that joins $(3, 2, 1)$ to $(3, 1, -1)$.

3. $f(x, y) = xe^y + y$ at $(3, 0)$ in the direction of the vector from $(3, 0)$ to $(-2, -4)$.

4. $f(x, y, z) = \ln(xy + yz + xz)$ at $(1, 1, 1)$ in the direction from $(1, 1, 1)$ toward the point $(-1, -2, 3)$.

5. $f(x, y) = \text{Tan}^{-1}(xy)$ at $(1, 2)$ along the line $y = 2x$ in the direction of increasing x.

6. $f(x, y) = \sin(x + y)$ at $(2, -2)$ along the line $3x + 4y = -2$ in the direction of decreasing y.

7. $f(x, y, z) = x^3 y \sin z$ at $(3, -1, -2)$ along the line $x = 3 + t$, $y = -1 + 4t$, $z = -2 + 2t$ in the direction of decreasing x.

8. $f(x, y, z) = x^2 y + y^2 z + z^2 x$ at $(1, -1, 0)$ along the line $x + 2y + 1 = 0$, $x - y + 2z = 2$ in the direction of decreasing z.

In Exercises 9–12 find the rate of change of the function with respect to distance travelled along the curve.

9. $f(x, y) = 2x - 3y$ at $(1, 1)$ along the curve $y = x^2$ in the direction of increasing x.

10. $f(x, y) = x^2 + y$ at $(-1, 3)$ along the curve $y = -3x^3$ in the direction of decreasing x.

11. $f(x, y, z) = xy + z^2$ at $(1, 0, -2)$ along the curve $y = x^2 - 1$, $z = -2x$ in the direction of increasing x.

12. $f(x, y, z) = x^2 y + xy^3 z$ at $(2, -1, 2)$ along the curve $x^2 - y^2 = 3$, $z = x$ in the direction of increasing x.

In Exercises 13–18 find the direction in which the function increases most rapidly at the point. What is the rate of change in that direction?

13. $f(x, y, z) = x^4 yz - xy^3 + z$ at $(1, 1, -3)$

14. $f(x, y) = 2xy + \ln(xy)$ at $(2, 1/2)$

15. $f(x, y, z) = 1/\sqrt{x^2 + y^2 + z^2}$ at $(1, -3, 2)$

16. $f(x, y, z) = -1/\sqrt{x^2 + y^2 + z^2}$ at $(1, -3, 2)$

17. $f(x, y, z) = \text{Tan}^{-1}(xyz)$ at $(3, 2, -4)$

18. $f(x, y) = xye^{xy}$ at $(1, 1)$

19. In what direction is the rate of change of $f(x, y, z) = xyz$ smallest at the point $(2, -1, 3)$?

20. In what directions (if any) is the rate of change of the function $f(x, y) = x^2 y + y^3$ at the point $(1, -1)$ equal to **(a)** 0? **(b)** 1? **(c)** 20?

21. In what directions (if any) is the rate of change of the function $f(x, y, z) = xy + z$ at the point $(0, 1, -2)$ equal to **(a)** 0? **(b)** 1? **(c)** -20?

22. Must there always be a direction in which the rate of change of a function at a point is equal to **(a)** 0? **(b)** 3?

23. In the derivation of 13.38, why was it necessary to use a unit vector $\hat{\mathbf{v}}$ to determine parametric equations for the line through (x_0, y_0, z_0) along $\mathbf{v}$? In other words, why could we not use the components of $\mathbf{v}$ itself to write parametric equations for the line?

24. How fast is the distance to the origin changing with respect to distance travelled along the curve $x = 2\cos t$, $y = 2\sin t$, $z = 3t$ at any point on the curve? What is the rate of change when $t = 0$? Would you expect this?

25. Find points on the curve $C: x = t$, $y = 1 - 2t$, $z = t$ at which the rate of change of $f(x, y, z) = x^2 + xyz$ with respect to distance travelled along the curve vanishes.

26. Repeat Exercise 25 for the curve $C: z = x$, $x = y^2$ and the function $f(x, y, z) = x^2 - y^2 + z^2$.

27. The path of a particle is defined parametrically by $x = (\cos t - t\sin t)\hat{\mathbf{i}} + (\sin t - t\cos t)\hat{\mathbf{j}}$ where t is time (the path is called an involute of a circle). Show that if the particle's speed is constant, then the rate of change of its distance from the origin with respect to distance travelled is also constant. Is the time rate of change of its distance from the origin also constant?

28. The rate of change of a function $f(x, y)$ at a point (x_0, y_0) in direction $\hat{\mathbf{i}} + 2\hat{\mathbf{j}}$ is 3 and the rate of change in direction $-2\hat{\mathbf{i}} - \hat{\mathbf{j}}$ is -1. Find its rate of change in direction $2\hat{\mathbf{i}} + 3\hat{\mathbf{j}}$.

29. Rates of change of a function $f(x, y, z)$ at a point (x_0, y_0, z_0) in directions $\hat{\mathbf{i}} + \hat{\mathbf{j}}$, $2\hat{\mathbf{i}} - \hat{\mathbf{k}}$, and $\hat{\mathbf{i}} - \hat{\mathbf{j}} + \hat{\mathbf{k}}$ are 1, 2, and -3 respectively. What is its partial derivative with respect to z at the point?

30. Find the second directional derivative of the function $f(x, y) = x^3 y^2$ at the point $(1, 1)$ in the direction of the vector $(1, -2)$.

31. Find the second directional derivative of the function $f(x,y,z) = x^2 + 2y^2 + 3z^2$ at the point $(-2,-1,3)$ in the direction $(1,1,-1)$.

32. The path followed by a stone embedded in the tread of a tire is a cycloid given parametrically by $x = R(\theta - \sin\theta)$, $y = R(1 - \cos\theta)$, $\theta \geq 0$ (see Exercise 46 in Section 10.1).
    (a) How fast is the distance from the origin changing with respect to distance travelled along the curve when $\theta = \pi/2$ and $\theta = \pi$?
    (b) How fast is the y-coordinate changing at these points?
    (c) How fast is the x-coordinate changing at these points?

33. Verify that expression 13.40 for $D_\mathbf{v} f$ leads to formula 13.39.

SECTION 13.9

Tangent Lines and Tangent Planes

Tangent Lines to Curves

One equation in the coordinates x, y, and z of points in space,

$$F(x,y,z) = 0, \qquad (13.43)$$

usually defines a surface. When each of the equations

$$F(x,y,z) = 0, \quad G(x,y,z) = 0 \qquad (13.44)$$

defines a surface, then together they define the curve of intersection of the two surfaces. Theoretically, we can find parametric equations for the curve by setting x equal to some function of a parameter t, say $x = x(t)$, and then solving equations 13.44 for y and z in terms of t: $y = y(t)$ and $z = z(t)$. The parametric definition, therefore, takes the form

$$x = x(t), \quad y = y(t), \quad z = z(t), \quad \alpha \leq t \leq \beta, \qquad (13.45)$$

where α and β specify the endpoints of the curve. Practical difficulties arise in choosing $x(t)$ and solving for $y(t)$ and $z(t)$. For some examples, it might be more convenient to specify $y(t)$ and solve for $x(t)$ and $z(t)$ or, alternatively, to specify $z(t)$ and solve for $x(t)$ and $y(t)$. We considered examples of such conversions in Section 12.7.

In Section 12.8 we indicated that if a curve C is defined parametrically by 13.45, then a tangent vector to C at any point P is

$$\frac{d\mathbf{r}}{dt} = \frac{dx}{dt}\hat{\mathbf{i}} + \frac{dy}{dt}\hat{\mathbf{j}} + \frac{dz}{dt}\hat{\mathbf{k}} \qquad (13.46)$$

(Figure 13.21). The tangent line to C at P is defined as the line through P having direction $d\mathbf{r}/dt$. If (x_0, y_0, z_0) are the coordinates of P and t_0 is the value of t yielding P, then the vector equation for the tangent line at P is

$$(x,y,z) = (x_0, y_0, z_0) + u\frac{d\mathbf{r}}{dt}\bigg|_{t=t_0} \qquad (13.47\text{a})$$

(see equation 12.38). Parametric equations for the tangent line are therefore

$$x = x_0 + x'(t_0)u, \quad y = y_0 + y'(t_0)u, \quad z = z_0 + z'(t_0)u, \qquad (13.47\text{b})$$

and in the case that none of $x'(t_0)$, $y'(t_0)$, and $z'(t_0)$ vanishes, we can also write symmetric equations for the tangent line:

$$\frac{x - x_0}{x'(t_0)} = \frac{y - y_0}{y'(t_0)} = \frac{z - z_0}{z'(t_0)}. \tag{13.47c}$$

FIGURE 13.21

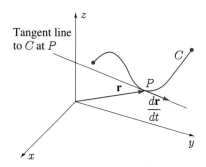

EXAMPLE 13.23

Find equations for the tangent line to the curve

$$C: \quad x = 2\cos t, \quad y = 4\sin t, \quad z = 2t/\pi$$

at $P(\sqrt{2}, 2\sqrt{2}, \frac{1}{2})$.

SOLUTION Since $t = \pi/4$ at P, a tangent vector to C at P is

$$\frac{d\mathbf{r}}{dt}\bigg|_{t=\pi/4} = (-2\sin t, 4\cos t, 2/\pi)|_{t=\pi/4} = (-\sqrt{2}, 2\sqrt{2}, 2/\pi).$$

Symmetric equations for the tangent line are therefore

$$\frac{x - \sqrt{2}}{-\sqrt{2}} = \frac{y - 2\sqrt{2}}{2\sqrt{2}} = \frac{z - \frac{1}{2}}{2/\pi}.$$

EXAMPLE 13.24

Find equations for the tangent line to the curve

$$C: \quad x = 2t^3 + t^2 + 4t, \quad y = t^2 - 3t + 5, \quad z = t^2 - 1$$

at the point $P(0, 5, -1)$.

SOLUTION Since $t = 0$ at P, a tangent vector to C at P is

$$\frac{d\mathbf{r}}{dt}\bigg|_{t=0} = (6t^2 + 2t + 4, 2t - 3, 2t)|_{t=0} = (4, -3, 0).$$

> Because the z-component of this vector is equal to zero, we cannot use the full symmetric equations 13.47c. We can, however, use them for the x- and y- coordinates of points on the line and write
>
> $$\frac{x}{4} = \frac{y-5}{-3}, \quad \text{or} \quad 3x + 4y = 20.$$
>
> Since the z-component of $d\mathbf{r}/dt$ vanishes, $d\mathbf{r}/dt$ must be parallel to the xy-plane, and so too must the tangent line to C at P. But then every point on the line has the same z-coordinate, and because the z-coordinate of P is -1, it follows that all points on the tangent line have z-coordinate equal to -1. Equations for the tangent line are therefore $3x + 4y = 20$, $z = -1$. ∎

Tangent Planes to Surfaces

We now consider the problem of finding the equation for the tangent plane at a point P on a surface S (Figure 13.22). We define the tangent plane as that plane which contains all tangent lines at P to curves in S through P (provided, of course, that such a plane exists). Suppose that the surface is defined by the equation

$$F(x, y, z) = 0, \qquad (13.48)$$

and that

$$C: \quad x = x(t), \quad y = y(t), \quad z = z(t), \quad \alpha \leq t \leq \beta,$$

is any curve in S through P. Since C is in S, the equation

$$F[x(t), y(t), z(t)] = 0$$

is valid for all t in $\alpha \leq t \leq \beta$. If $F(x, y, z)$ has continuous first partial derivatives, and $x(t)$, $y(t)$, and $z(t)$ are all differentiable, we may differentiate this equation using the chain rule:

$$\frac{\partial F}{\partial x}\frac{dx}{dt} + \frac{\partial F}{\partial y}\frac{dy}{dt} + \frac{\partial F}{\partial z}\frac{dz}{dt} = 0.$$

This equation, which holds at all points on C, and in particular at P, can be expressed vectorially as

$$0 = \left(\frac{\partial F}{\partial x}, \frac{\partial F}{\partial y}, \frac{\partial F}{\partial z}\right) \cdot \left(\frac{dx}{dt}, \frac{dy}{dt}, \frac{dz}{dt}\right) = \nabla F \cdot \frac{d\mathbf{r}}{dt}.$$

But if the scalar product of two vectors vanishes, the vectors are perpendicular (see equation 12.24). Consequently, ∇F is perpendicular to the tangent vector $d\mathbf{r}/dt$ to C at P. Since C is an arbitrary curve in S, it follows that ∇F at P is perpendicular to the tangent line to every curve C in S at P. In other words, ∇F at P must be perpendicular to the tangent plane to S at P (Figure 13.23). If the coordinates of P are (x_0, y_0, z_0), then the equation of the tangent plane to S at P is

$$\begin{aligned} 0 &= \nabla F|_P \cdot (x - x_0, y - y_0, z - z_0) \\ &= F_x(x_0, y_0, z_0)(x - x_0) + F_y(x_0, y_0, z_0)(y - y_0) \\ &\quad + F_z(x_0, y_0, z_0)(z - z_0) \end{aligned} \qquad (13.49)$$

(see equation 12.27).

FIGURE 13.22

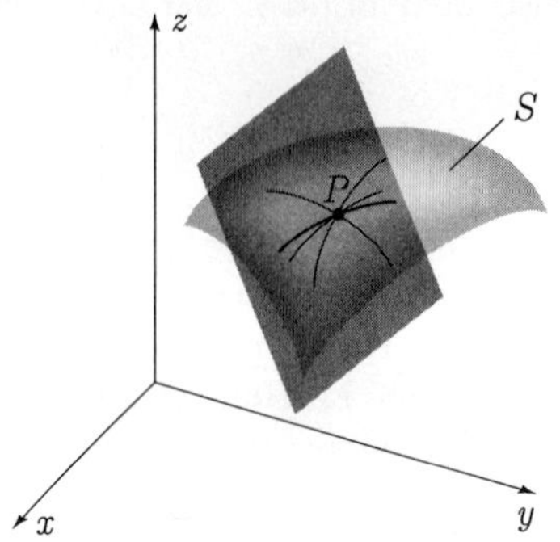

FIGURE 13.23

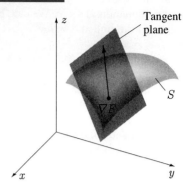

EXAMPLE 13.25

Find the equation of the tangent plane to the surface $xyz^3 + yz^2 = 4$ at the point $(1, 2, 1)$.

SOLUTION A vector perpendicular to the tangent plane is

$$\nabla(xyz^3 + yz^2 - 4)_{|(1,2,1)} = (yz^3, xz^3 + z^2, 3xyz^2 + 2yz)_{|(1,2,1)} = (2, 2, 10).$$

But then the vector $(1, 1, 5)$ must also be perpendicular to the tangent plane, and the equation of the plane is therefore

$$0 = (1, 1, 5) \cdot (x - 1, y - 2, z - 1) = x + y + 5z - 8.$$

We have shown in this section that if the equation $F(x, y, z) = 0$ defines a surface S, and if there is a tangent plane to S at a point P, then the vector $\nabla F_{|P}$ is normal to the tangent plane (Figure 13.23). It is customary to state in this situation that $\nabla F_{|P}$ is normal to the surface itself at P, rather than to the tangent plane to the surface. This fact proves to be another of the important properties of the gradient vector, and is worth stating as a theorem.

Theorem 13.6

If the equation $F(x, y, z) = 0$ defines a surface S, and $F(x, y, z)$ has continuous first partial derivatives, then at any point on S, the vector ∇F is perpendicular to S.

A geometric application of this fact is contained in the following example.

EXAMPLE 13.26

Find equations for the tangent line at the point $(1,2,2)$ to the curve C: $x^2+y^2+z^2 = 9$, $4(x^2+y^2) = 5z^2$.

SOLUTION Equation 13.46 indicates that to find a tangent vector to C we should first have parametric equations for C. These can be obtained by first solving each equation for $x^2 + y^2$ and equating the results:

$$9 - z^2 = 5z^2/4.$$

This equation implies that $z = \pm 2$, the positive result being required here. On C, then, $x^2 + y^2 = 5$, and parametric equations for C are

$$x = \sqrt{5}\cos t, \quad y = \sqrt{5}\sin t, \quad z = 2, \quad 0 \leq t < 2\pi.$$

According to 13.46, a tangent vector to C at $(1,2,2)$ is

$$\left(\frac{dx}{dt}, \frac{dy}{dt}, \frac{dz}{dt}\right)_{|(1,2,2)} = (-\sqrt{5}\sin t, \sqrt{5}\cos t, 0)_{|t=\text{Sin}^{-1}(2/\sqrt{5})} = (-2, 1, 0).$$

The tangent line therefore has equations

$$\frac{x-1}{-2} = \frac{y-2}{1}, \quad z = 2, \quad \text{or,} \quad x + 2y = 5, \quad z = 2.$$

The fact that gradients can be used to find normals to surfaces suggests an alternative solution. It is clear from Figure 13.24 that if we define $F(x,y,z) = x^2 + y^2 + z^2 - 9$, then ∇F evaluated at $(1,2,2)$ is perpendicular not only to the surface $F(x,y,z) = 0$, but also to the curve C. Similarly, if $G(x,y,z) = 4(x^2+y^2) - 5z^2$, then ∇G at $(1,2,2)$ is also perpendicular to C. Since a vector along the tangent line to C at $(1,2,2)$ is perpendicular to both of these vectors, it follows that a vector along the tangent line is

$$(\nabla F \times \nabla G)_{|(1,2,2)} = \{(2x, 2y, 2z) \times (8x, 8y, -10z)\}_{|(1,2,2)}$$
$$= (2,4,4) \times (8,16,-20)$$
$$= 8\begin{vmatrix} \hat{\mathbf{i}} & \hat{\mathbf{j}} & \hat{\mathbf{k}} \\ 1 & 2 & 2 \\ 2 & 4 & -5 \end{vmatrix}$$
$$= 8(-18, 9, 0)$$
$$= 72(-2, 1, 0).$$

Once again, we have obtained $(-2, 1, 0)$ as a tangent vector to the curve, and equations for the tangent line can be written down as before.

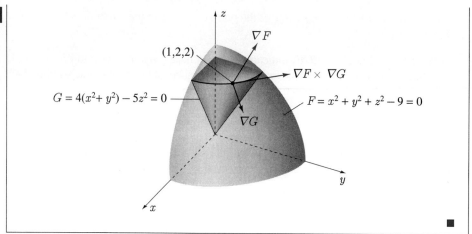

FIGURE 13.24

Example 13.26 illustrates that when a curve is defined as the intersection of two surfaces $F(x,y,z) = 0$, $G(x,y,z) = 0$ (Figure 13.25), then a vector tangent to the curve is

$$\mathbf{T} = \nabla F \times \nabla G. \qquad (13.50)$$

Thus to find a tangent vector to a curve we use 13.46 when the curve is defined parametrically. When the curve is defined as the intersection of two surfaces, we can either find parametric equations and use 13.46, or use 13.50. Note too that in order to find tangent lines to curves, it is not necessary to have a direction assigned to the curves.

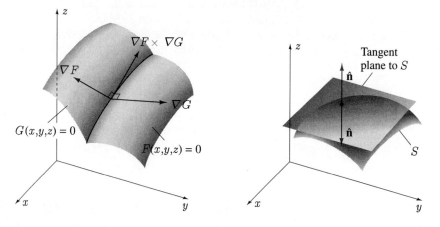

FIGURE 13.25

FIGURE 13.26

At each point on a surface S at which S has a tangent plane (Figure 13.26), we have defined a normal vector to S as a vector normal to the tangent plane to S. If we denote by $\hat{\mathbf{n}}$ a unit normal vector to S, then the direction of $\hat{\mathbf{n}}$ clearly varies as we move from point to point on S. We say that $\hat{\mathbf{n}}$ is a function of position (x, y, z) on S. Furthermore, at each point at which S has a unit normal vector, it has two such vectors, one in the opposite direction to the other. We say that a surface S is **smooth** if it can be assigned a unit normal $\hat{\mathbf{n}}$ that varies continuously on S. What this means geometrically is that for small changes in position, the unit normal $\hat{\mathbf{n}}$ will undergo small changes in direction. The sphere in Figure 13.27 is smooth, as is the paraboloid in Figure 13.28.

FIGURE 13.27

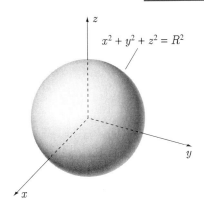

FIGURE 13.28

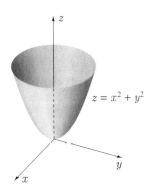

The surface bounding the cylindrical volume in Figure 13.29 is not smooth; a unit normal that varies continuously over the surface cannot be assigned at points on the circles $x^2 + y^2 = 1$, $z = \pm 1$. This surface can, however, be divided into a finite number of subsurfaces, each of which is smooth. In particular, we choose the three subsurfaces S_1: $z = 1$, $x^2 + y^2 \leq 1$; S_2: $z = -1$, $x^2 + y^2 \leq 1$; S_3: $x^2 + y^2 = 1$, $-1 < z < 1$. Such a surface is said to be **piecewise smooth**.

FIGURE 13.29

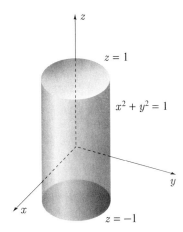

EXERCISES 13.9

In Exercises 1–20 find equations for the tangent line to the curve at the point.

1. $y = x^2$, $z = 0$ at $(-2, 4, 0)$
2. $x = t$, $y = t^2$, $z = t^3$ at $(1, 1, 1)$
3. $x = \cos t$, $y = \sin t$, $z = \cos t$ at $(1, 0, 1)$
4. $y = x^2$, $z = x$ at $(-2, 4, -2)$
5. $x^2 = y$, $z + x = y$ at $(1, 1, 0)$
6. $x = 2 - t^2$, $y = 3 + 2t$, $z = t$ at $(1, 5, 1)$
7. $x = 2\cos t$, $y = 3\sin t$, $z = 5$ at $(\sqrt{2}, -3/\sqrt{2}, 5)$
8. $x^2 y^3 + xy = 68$ at $(1, 4)$
9. $x + y + z = 4$, $x - y = 2$ at $(0, -2, 6)$
10. $x = e^{-t}\cos t$, $y = e^{-t}\sin t$, $z = t$ at $(1, 0, 0)$
11. $x = t^2 + 1$, $y = 2t - 4$, $z = t^3 + 3$ at $(2, -6, 2)$
12. $y^2 + z^2 = 6$, $x + z = 1$ at $(2, -\sqrt{5}, -1)$
13. $x^2 + y^2 + z^2 = 4$, $z^2 = x^2 + y^2$ at $(1, 1, -\sqrt{2})$
14. $x = t$, $y = 1$, $z = \sqrt{1 + t^2}$ at $(4, 1, \sqrt{17})$
15. $x = 1 + \cos t$, $y = 2 - \sin t$, $z = \sqrt{4 + t}$ at $(2, 2, 2)$

16. $x = z^2 + z^3$, $y = z - z^4$ at $(12, -14, 2)$

17. $x = y^2 + 3y^3 - 2y + 5$, $z = 0$ at $(7, 1, 0)$

18. $2x^2 + y^2 + 2y = 3$, $z = x + 1$ at $(0, 1, 1)$

19. $x = t^2$, $y = t$, $z = \sqrt{t + t^4}$ at $(1, 1, \sqrt{2})$

20. $x = t \sin t$, $y = t \cos t$, $z = 2t$ at $(0, 2\pi, 4\pi)$

In Exercises 21–26 find an equation for the tangent plane to the surface at the point.

21. $z = \sqrt{x^2 + y^2}$ at $(1, 1, \sqrt{2})$

22. $x = x^2 - y^3 z$ at $(2, -1, -2)$

23. $x^2 y + y^2 z + z^2 x + 3 = 0$ at $(2, -1, -1)$

24. $x + y + z = 4$ at $(1, 1, 2)$

25. $x = y \sin(\pi z/2)$ at $(-1, -1, 1)$

26. $x^2 + y^2 + 2y = 1$ at $(1, 0, 3)$

27. Show that the curve $x = 2(t^3 + 2)/3$, $y = 2t^2$, $z = 3t - 2$ intersects the surface $x^2 + 2y^2 + 3z^2 = 15$ at right angles at the point $(2, 2, 1)$.

28. Verify that the curve $x^2 - y^2 + z^2 = 1$, $xy + xz = 2$ is tangent to the surface $xyz - x^2 - 6y + 6 = 0$ at the point $(1, 1, 1)$.

29. Show that the equation of the tangent plane to a surface $S: z = f(x, y)$ at a point (x_0, y_0, z_0) on S can be written in the form

$$z - z_0 = (x - x_0) f_x(x_0, y_0) + (y - y_0) f_y(x_0, y_0).$$

In Exercises 30–32 find the indicated derivative for the function.

30. $f(x, y, z) = 2x^2 + y^2 z^2$ at $(3, 1, 0)$ with respect to distance along the curve $x + y + z = 4$, $x - y + z = 2$ in the direction of increasing x

31. $f(x, y, z) = xyz + xy + xz + yz$ at $(1, -2, 5)$ perpendicular to the surface $z = x^2 + y^2$

32. $f(x, y, z) = x^2 + y^2 - z^2$ at $(3, 4, 5)$ with respect to distance along the curve $x^2 + y^2 - z^2 = 0$, $2x^2 + 2y^2 - z^2 = 25$ in the direction of decreasing x

33. If $F(x, y) = 0$ defines a curve implicitly in the xy-plane, prove that at any point on the curve, ∇F is perpendicular to the curve.

34. Find the equation of the tangent plane to the ellipsoid $x^2/a^2 + y^2/b^2 + z^2/c^2 = 1$ at any point (x_0, y_0, z_0) on the surface.

35. Find all points on the surface $z = x^2/4 - y^2/9$ at which the tangent plane is parallel to the plane $x + y + z = 4$.

36. Find all points on the surface $z^2 = 4(x^2 + y^2)$ at which the tangent plane is parallel to the plane $x - y + 2z = 3$.

37. Suppose that the equations $F(x, y, z, t) = 0$, $G(x, y, z, t) = 0$, $H(x, y, z, t) = 0$ implicitly define parametric equations for a curve C (t being the parameter). If $P(x_0, y_0, z_0)$ is a point on C, show that equations for the tangent line to C at P can be written in the form

$$\frac{x - x_0}{\frac{\partial(F, G, H)}{\partial(t, y, z)}\bigg|_P} = \frac{y - y_0}{\frac{\partial(F, G, H)}{\partial(x, t, z)}\bigg|_P} = \frac{z - z_0}{\frac{\partial(F, G, H)}{\partial(x, y, t)}\bigg|_P},$$

provided none of the Jacobians vanish.

38. Find all points on the paraboloid $z = x^2 + y^2 - 1$ at which the normal to the surface coincides with the line joining the origin to the point.

39. Show that the sum of the intercepts on the x-, y-, and z-axes of the tangent plane to the surface $\sqrt{x} + \sqrt{y} + \sqrt{z} = \sqrt{a}$ at any point is a.

SECTION 13.10

Relative Maxima and Minima

We now study relative extrema of functions of more than one independent variable. Most of the discussion will be confined to functions $f(x, y)$ of two independent variables because we can discuss the concepts geometrically as well as algebraically. Unfortunately, not all the results are easily extended to functions of more than two independent variables, and we will therefore be careful to point out these limitations.

Before beginning the discussion, we briefly review maxima-minima results for functions $f(x)$ of one variable. We do this because maxima-minima theory for multivariable functions is essentially the same as that for single-variable functions. In fact, every definition that we make and every result that we discuss in this section has its counterpart in single-variable theory. Hence, a synopsis of single-variable results is in order. Unfortunately, proving results in the multivariable case is considerably more complicated than in the single-variable case, but if we can keep central ideas foremost in our minds and constantly make comparisons with single-variable calculus, we will find that discussions are not nearly as difficult as they might otherwise be.

Critical points of a function $f(x)$ are points at which $f'(x)$ is either equal to zero or does not exist. Geometrically, this means points at which the graph of $f(x)$ has a horizontal tangent line, a vertical tangent line, or no tangent line at all. Critical points for continuous functions are classified as yielding relative maxima, relative minima, horizontal points of inflection, vertical points of inflection, or just corners. There are two tests to determine whether a critical point x_0 yields a relative maximum or a relative minimum. The first-derivative test states that if $f'(x)$ changes from a positive quantity to a negative quantity as x increases through x_0, then x_0 gives a relative maximum; if $f'(x)$ changes from negative to positive, then a relative minimum is obtained. The second-derivative test indicates the nature of a critical point at which $f'(x_0) = 0$ whenever $f''(x_0) \neq 0$. If $f''(x_0) > 0$, then a relative minimum is obtained, and if $f''(x_0) < 0$, a relative maximum is found.

We begin our study of extrema theory for multivariable functions by defining critical points for functions of two independent variables.

Definition 13.6

A point (x_0, y_0) in the domain of a function $f(x, y)$ is said to be a **critical point** of $f(x, y)$ if

$$\frac{\partial f}{\partial x}\bigg|_{(x_0, y_0)} = 0, \quad \frac{\partial f}{\partial y}\bigg|_{(x_0, y_0)} = 0, \quad (13.51)$$

or if one (or both) of these partial derivatives does not exist at (x_0, y_0).

There are two ways to interpret critical points of $f(x, y)$ geometrically. In Section 13.3, we interpreted $\partial f/\partial x$ at (x_0, y_0) as the slope of the tangent line to the curve of intersection of the surface $z = f(x, y)$ and the plane $y = y_0$, and $\partial f/\partial y$ as the slope of the tangent line to the curve of intersection with $x = x_0$. It follows, then, that (x_0, y_0) is critical if both curves have horizontal tangent lines or if either curve has a vertical tangent line or no tangent line at all. Alternatively, recall that the equation of the tangent plane to the surface $z = f(x, y)$ at (x_0, y_0) is

$$z - z_0 = f_x(x_0, y_0)(x - x_0) + f_y(x_0, y_0)(y - y_0)$$

(see Exercise 29 in Section 13.9). If both partial derivatives vanish, then the tangent plane is horizontal with equation $z = z_0$. For example, at each of the critical points in Figures 13.30–13.33, $\partial f/\partial x = \partial f/\partial y = 0$ and the tangent plane is horizontal. The remaining functions in Figures 13.34–13.38 have critical points at which either $\partial f/\partial x$ or $\partial f/\partial y$ or both do not exist. In Figures 13.34–13.37, the surfaces do not have tangent planes at critical points, and in Figure 13.38, the tangent plane is vertical at each critical point. Consequently, (x_0, y_0) is a critical point of a function $f(x, y)$ if at (x_0, y_0) the surface $z = f(x, y)$ has a horizontal tangent plane, a vertical tangent plane, or no tangent plane at all.

FIGURE 13.30

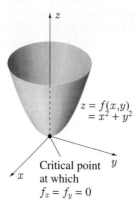

z
$z = f(x,y)$
$= x^2 + y^2$
Critical point
at which
$f_x = f_y = 0$
x
y

FIGURE 13.31

z
$z = f(x,y)$
$= \sqrt{1 - x^2 - (y-1)^2}$
(0,1)
2
x
y
Critical point
at which
$f_x = f_y = 0$

FIGURE 13.32

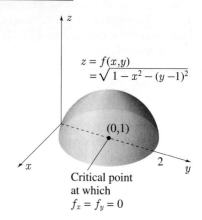

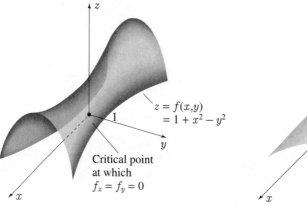

z
$z = f(x,y)$
$= 1 + x^2 - y^2$
1
y
Critical point
at which
$f_x = f_y = 0$
x

FIGURE 13.33

z
$z = f(x,y)$
$= y^3$
y
Critical
points
at which
$f_x = f_y = 0$
x

FIGURE 13.34

z
$z = f(x,y)$
$= \sqrt{(x-1)^2 + y^2}$
y
(1,0)
Critical point at which
f_x and f_y do not exist
x

FIGURE 13.35

z
$z = f(x,y)$
$= |x - y|$
Critical
points
at which
f_x and f_y
do not
exist
y
x

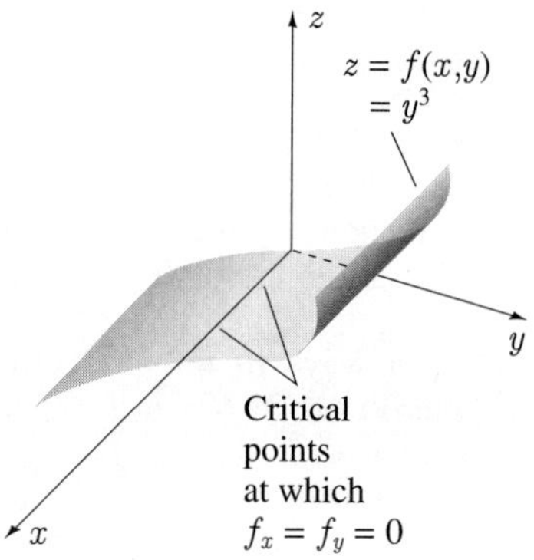

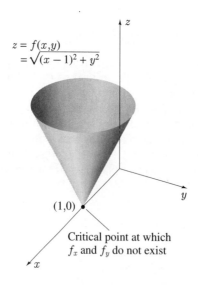

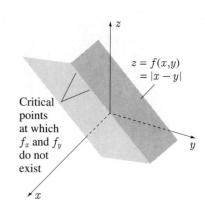

FIGURE 13.36

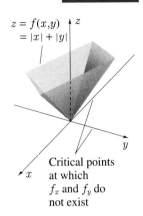

Critical points at which f_x and f_y do not exist

FIGURE 13.37

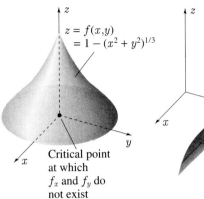

Critical point at which f_x and f_y do not exist

FIGURE 13.38

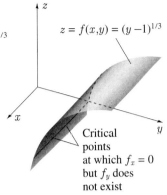

Critical points at which $f_x = 0$ but f_y does not exist

EXAMPLE 13.27

Find all critical points for the function

$$f(x,y) = x^2 y - 2xy^2 + 3xy + 4.$$

SOLUTION For critical points, we first solve

$$0 = \frac{\partial f}{\partial x} = 2xy - 2y^2 + 3y = y(2x - 2y + 3),$$

$$0 = \frac{\partial f}{\partial y} = x^2 - 4xy + 3x = x(x - 4y + 3).$$

To satisfy these two equations simultaneously, there are four possibilities:
1. $x = 0$, $y = 0$, which gives the critical point $(0,0)$;
2. $y = 0$, $x - 4y + 3 = 0$, which gives the critical point $(-3, 0)$;
3. $x = 0$, $2x - 2y + 3 = 0$, which gives the critical point $(0, \frac{3}{2})$;
4. $2x - 2y + 3 = 0$, $x - 4y + 3 = 0$, which gives the critical point $(-1, \frac{1}{2})$.

Since $\partial f/\partial x$ and $\partial f/\partial y$ are defined for all x and y, these are the only critical points. ∎

Critical points for functions of more than two independent variables can be defined algebraically, but because we have no geometric representation for such functions, there is no geometric interpretation for their critical points. For example, if $f(x, y, z, t)$ is a function of independent variables x, y, z, and t, then (x_0, y_0, z_0, t_0) is a critical point of $f(x, y, z, t)$ if all four of its first-order partial derivatives vanish at (x_0, y_0, z_0, t_0),

$$\frac{\partial f}{\partial x}\bigg|_{(x_0,y_0,z_0,t_0)} = \frac{\partial f}{\partial y}\bigg|_{(x_0,y_0,z_0,t_0)} = \frac{\partial f}{\partial z}\bigg|_{(x_0,y_0,z_0,t_0)} = \frac{\partial f}{\partial t}\bigg|_{(x_0,y_0,z_0,t_0)} = 0, \quad (13.52)$$

or if at least one of the partial derivatives does not exist at the point. Note that because the partial derivatives of a function are the components of the gradient of the function, we can say that a critical point of a function is a point at which its gradient is either equal to zero or undefined.

EXAMPLE 13.28

Find all critical points for the function

$$f(x,y,z) = xyz\sqrt{x^2 + y^2 + z^2}.$$

SOLUTION For critical points, we consider the equations

$$0 = \frac{\partial f}{\partial x} = yz\sqrt{x^2 + y^2 + z^2} + \frac{x^2yz}{\sqrt{x^2 + y^2 + z^2}}$$

$$= \frac{yz}{\sqrt{x^2 + y^2 + z^2}}(2x^2 + y^2 + z^2),$$

$$0 = \frac{\partial f}{\partial y} = xz\sqrt{x^2 + y^2 + z^2} + \frac{xy^2z}{\sqrt{x^2 + y^2 + z^2}}$$

$$= \frac{xz}{\sqrt{x^2 + y^2 + z^2}}(x^2 + 2y^2 + z^2),$$

$$0 = \frac{\partial f}{\partial z} = xy\sqrt{x^2 + y^2 + z^2} + \frac{xyz^2}{\sqrt{x^2 + y^2 + z^2}}$$

$$= \frac{xy}{\sqrt{x^2 + y^2 + z^2}}(x^2 + y^2 + 2z^2).$$

The partial derivatives are clearly undefined for $x = y = z = 0$, and therefore the origin $(0,0,0)$ is a critical point. If x, y, and z are not all zero, then the terms in parentheses cannot vanish, and we must set

$$yz = 0, \quad xz = 0, \quad xy = 0.$$

If any two of x, y, and z vanish, but the third does not, then these equations are satisfied. In other words, every point on the x-axis, every point on the y-axis, and every point on the z-axis is critical. ∎

We now turn our attention to the classification of critical points of a function $f(x,y)$ of two independent variables. Critical points $(0,1)$ in Figure 13.31 and $(0,0)$ in Figure 13.37 yield "high" points on the surfaces. We describe this property in the following definition.

Definition 13.7

A function $f(x,y)$ is said to have a **relative maximum** $f(x_0, y_0)$ at a point (x_0, y_0) if there exists a circle in the xy-plane centred at (x_0, y_0) such that for all points (x,y) inside this circle

$$f(x,y) \leq f(x_0, y_0). \tag{13.53}$$

The "low" points on the surfaces at $(0,0)$ in Figure 13.30 and $(1,0)$ in Figure 13.34 are relative minima according to the following.

Definition 13.8

A function $f(x,y)$ is said to have a **relative minimum** $f(x_0,y_0)$ at a point (x_0,y_0) if there exists a circle in the xy-plane centred at (x_0,y_0) such that for all points (x,y) inside this circle

$$f(x,y) \geq f(x_0,y_0). \tag{13.54}$$

Note that every critical point in Figure 13.35 yields a relative minimum of $f(x,x) = 0$, as does the critical point $(0,0)$ in Figure 13.36.

Definition 13.9

If a critical point of a function $f(x,y)$ at which $\partial f/\partial x = \partial f/\partial y = 0$ yields neither a relative maximum nor a relative minimum, it is said to yield a **saddle point**.

The critical point $(0,0)$ in Figure 13.32 therefore gives a saddle point, as does each of the critical points in Figure 13.33. Saddle points for surfaces $z = f(x,y)$ are clearly the analogues of horizontal points of inflection for curves $y = f(x)$. In both cases the derivative(s) of the function vanishes but there is neither a relative maximum nor a relative minimum.

The critical points in Figure 13.36 (except $(0,0)$) are the counterparts of corners for the graph of a function $f(x)$. They are points at which one or both of the partial derivatives of $f(x,y)$ do not exist, but like corners for $f(x)$, they do not necessarily yield relative extrema. Critical points in Figure 13.38 are the analogues of vertical points of inflection for a function $f(x)$.

Our discussion has made it clear that:

(a) At a relative maximum or minimum of $f(x,y)$ either $\partial f/\partial x$ and $\partial f/\partial y$ both vanish, or one or both of the partial derivatives do not exist;

(b) Saddle points may also occur where $\partial f/\partial x = \partial f/\partial y = 0$, and points where the derivatives do not exist may fail to yield relative extrema.

In other words, every relative extremum of $f(x,y)$ occurs at a critical point; but critical points do not always give relative extrema.

Given the problem of determining all relative maxima and minima of a function $f(x,y)$, we should first find its critical points. But how do we decide whether these critical points yield relative maxima, relative minima, saddle points, or none of these? We do not have a practical test that is equivalent to the first-derivative test for functions of one variable, but we do have a test that corresponds to the second-derivative test. For functions of two independent variables the situation is more complicated, however, since there are three second-order partial derivatives, but the idea of the test is essentially the same. It determines whether certain curves are concave upward or concave downward at the critical point. The complete result is contained in the following theorem.

Theorem 13.7

Suppose (x_0, y_0) is a critical point of $f(x, y)$ at which $\partial f/\partial x$ and $\partial f/\partial y$ both vanish. Suppose further that f_x, f_y, f_{xx}, f_{xy}, and f_{yy} are all continuous at (x_0, y_0). Define
$$A = f_{xx}(x_0, y_0), \quad B = f_{xy}(x_0, y_0), \quad C = f_{yy}(x_0, y_0).$$

If

(i) $B^2 - AC < 0$ and $A < 0$, then $f(x, y)$ has a relative maximum at (x_0, y_0);
(ii) $B^2 - AC < 0$ and $A > 0$, then $f(x, y)$ has a relative minimum at (x_0, y_0);
(iii) $B^2 - AC > 0$, then $f(x, y)$ has a saddle point at (x_0, y_0);
(iv) $B^2 - AC = 0$, then the test fails.

Proof

(i) Suppose we intersect the surface $z = f(x, y)$ with a plane parallel to the z-axis, through the point $(x_0, y_0, 0)$, and making an angle α with the line through $(x_0, y_0, 0)$ parallel to the positive x-axis (Figure 13.39). The slope of the curve of intersection of these surfaces at the point $(x_0, y_0, f(x_0, y_0))$ is given by the directional derivative

$$D_{\mathbf{v}}f_{|(x_0,y_0)} = \nabla f_{|(x_0,y_0)} \cdot \hat{\mathbf{v}} = \nabla f_{|(x_0,y_0)} \cdot (\cos\alpha, \sin\alpha)$$
$$= \frac{\partial f}{\partial x}\bigg|_{(x_0,y_0)} \cos\alpha + \frac{\partial f}{\partial y}\bigg|_{(x_0,y_0)} \sin\alpha$$

(see equation 13.41). Since (x_0, y_0) is a critical point at which $\nabla f = \mathbf{0}$, it follows that
$$D_{\mathbf{v}}f_{|(x_0,y_0)} = 0 \quad \text{for all} \quad \alpha.$$

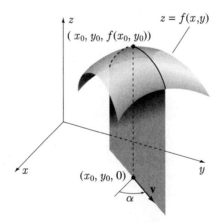

FIGURE 13.39

In Figure 13.39 we have illustrated the critical point as a relative maximum. But how do we verify that this is indeed the case? If we can show that each and every curve of intersection of the surface with a vertical plane through $(x_0, y_0, 0)$ is concave downward at $(x_0, y_0, 0)$, then (x_0, y_0) must give a relative maximum. But to discuss concavity of a curve we require the second derivative—in this case, the second directional derivative of $f(x, y)$. According to 13.42, the second directional derivative of $f(x, y)$ at (x_0, y_0) in the direction $\hat{\mathbf{v}} = (\cos\alpha, \sin\alpha)$ is

$$D_{\mathbf{v}}(D_{\mathbf{v}}f) = \frac{\partial^2 f}{\partial x^2}\bigg|_{(x_0,y_0)} \cos^2 \alpha + 2\frac{\partial^2 f}{\partial x \partial y}\bigg|_{(x_0,y_0)} \cos \alpha \sin \alpha + \frac{\partial^2 f}{\partial y^2}\bigg|_{(x_0,y_0)} \sin^2 \alpha$$
$$= A \cos^2 \alpha + 2B \cos \alpha \sin \alpha + C \sin^2 \alpha,$$

where we understand here that $D_{\mathbf{v}}(D_{\mathbf{v}}f)$ is implicitly suffixed by (x_0,y_0). In order, therefore, to verify that (x_0,y_0) gives a relative maximum, it is sufficient to show that $D_{\mathbf{v}}(D_{\mathbf{v}}f)$ is negative for each value of α in the interval $0 \le \alpha < 2\pi$. However, because $D_v(D_v f)$ is unchanged if α is replaced by $\alpha + \pi$, it is sufficient to verify that $D_{\mathbf{v}}(D_{\mathbf{v}}f)$ is negative for $0 \le \alpha < \pi$.

For any of these values of α except $\pi/2$, we can write

$$D_{\mathbf{v}}(D_{\mathbf{v}}f) = \cos^2 \alpha (A + 2B \tan \alpha + C \tan^2 \alpha),$$

and if we set $u = \tan \alpha$,

$$D_{\mathbf{v}}(D_{\mathbf{v}}f) = \cos^2 \alpha (A + 2Bu + Cu^2).$$

It is evident that $D_{\mathbf{v}}(D_{\mathbf{v}}f) < 0$ for all $\alpha \ne \pi/2$ if and only if

$$Q(u) = A + 2Bu + Cu^2 < 0 \quad \text{for } -\infty < u < \infty.$$

Were we to sketch a graph of the quadratic $Q(u)$, we would see that it crosses the u-axis where

$$u = \frac{-2B \pm \sqrt{4B^2 - 4AC}}{2C} = \frac{-B \pm \sqrt{B^2 - AC}}{C}.$$

But if we know, as stated in Theorem 13.7(i), that $B^2 - AC < 0$, then there are no real solutions of this equation, and therefore $Q(u)$ never crosses the u-axis. Since $Q(0) = A < 0$, it follows that $Q(u) < 0$ for all u. We have shown, then, that

$$D_{\mathbf{v}}(D_{\mathbf{v}}f) < 0 \quad \text{for all } \alpha \ne \pi/2.$$

When $\alpha = \pi/2$, $D_{\mathbf{v}}(D_{\mathbf{v}}f) = C$. Since $B^2 - AC < 0$ and $A < 0$, it follows that $C < 0$ also. Consequently, if $B^2 - AC < 0$ and $A < 0$, then $D_{\mathbf{v}}(D_{\mathbf{v}}f) < 0$ for all α, and (x_0, y_0) yields a relative maximum.

(ii) If $B^2 - AC < 0$ and $A > 0$, a similar argument leads to the conclusion that (x_0, y_0) yields a relative minimum; the only difference is that inequalities are reversed.

(iii) If $B^2 - AC > 0$, then $Q(u)$ has real distinct zeros, in which case $Q(u)$ is sometimes negative and sometimes positive. This means that the curve of intersection is sometimes concave upward and sometimes concave downward, and the point (x_0, y_0) therefore gives a saddle point.

(iv) If $B^2 - AC = 0$, the classification of the point determined by (x_0, y_0) depends on which of A, B, and C vanish, if any.

To illustrate that we can obtain a relative maximum, a relative minimum, or a saddle point for a critical point at which $B^2 - AC = 0$, consider the three functions $f(x,y) = -y^2$, $f(x,y) = y^2$, and $f(x,y) = y^3$ in Figures 13.40–13.42. The point $(0,0)$ is a critical point for each function, and at this point $B^2 - AC = 0$. Yet $(0,0)$ yields a relative maximum for $f(x,y) = -y^2$, a relative minimum for $f(x,y) = y^2$, and a saddle point for $f(x,y) = y^3$. In fact, every point on the x-axis is a relative maximum for $f(x,y) = -y^2$, a relative minimum for $f(x,y) = y^2$, and a saddle point for $f(x,y) = y^3$.

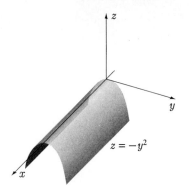

FIGURE 13.40

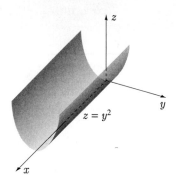

FIGURE 13.41

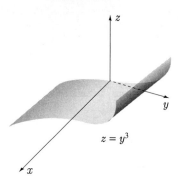

FIGURE 13.42

EXAMPLE 13.29 Find and classify critical points for each of the following functions as yielding relative maxima, relative minima, saddle points, or none of these:

(a) $f(x, y) = 4xy - x^4 - y^4$
(b) $f(x, y) = y^3 + x^2 - 6xy + 3x + 6y$
(c) $f(x, y) = \dfrac{12xy - x^2 y^2}{2(x + y)}$

SOLUTION

(a) Critical points of $f(x, y)$ are given by

$$0 = \frac{\partial f}{\partial x} = 4y - 4x^3, \quad 0 = \frac{\partial f}{\partial y} = 4x - 4y^3.$$

Solutions of these equations are $(0,0)$, $(1,1)$, and $(-1,-1)$. We now calculate

$$\frac{\partial^2 f}{\partial x^2} = -12x^2, \quad \frac{\partial^2 f}{\partial x \partial y} = 4, \quad \frac{\partial^2 f}{\partial y^2} = -12y^2.$$

At $(0,0)$, $B^2 - AC = 16 > 0$, and therefore $(0,0)$ yields a saddle point. At $(1,1)$, $B^2 - AC = -128$ and $A = -12$, and therefore $(1,1)$ gives a relative maximum. At $(-1,-1)$, $B^2 - AC = -128$ and $A = -12$, and $(-1,-1)$ also gives a relative maximum.

(b) Critical points for $f(x, y)$ are given by

$$0 = \frac{\partial f}{\partial x} = 2x - 6y + 3, \quad 0 = \frac{\partial f}{\partial y} = 3y^2 - 6x + 6.$$

Solutions of these equations are $(\frac{27}{2}, 5)$ and $(\frac{3}{2}, 1)$. The second derivatives of $f(x, y)$ are

$$\frac{\partial^2 f}{\partial x^2} = 2, \quad \frac{\partial^2 f}{\partial x \partial y} = -6, \quad \frac{\partial^2 f}{\partial y^2} = 6y.$$

At $(\frac{27}{2}, 5)$, $B^2 - AC = -24$ and $A = 2$, and therefore $(\frac{27}{2}, 5)$ yields a relative minimum. At $(\frac{3}{2}, 1)$, $B^2 - AC = 24$, and therefore $(\frac{3}{2}, 1)$ gives a saddle point.

(c) For critical points, we consider

$$0 = \frac{\partial f}{\partial x} = \frac{y^2(12 - x^2 - 2xy)}{2(x+y)^2}, \quad 0 = \frac{\partial f}{\partial y} = \frac{x^2(12 - y^2 - 2xy)}{2(x+y)^2},$$

or

$$0 = y^2(12 - x^2 - 2xy), \quad 0 = x^2(12 - y^2 - 2xy).$$

Solutions of these equations are $(2, 2)$ and $(-2, -2)$ (the solution $(0, 0)$ has been rejected since $f(x, y)$ is not defined there). To classify these critical points by means of the second-derivative test would lead to some messy calculations; thus we will show that the classification can be achieved without the test. First we note that $f(x, y) = 0$ whenever

$$0 = 12xy - x^2 y^2 = xy(12 - xy);$$

i.e., along the x-axis, the y-axis, and the hyperbola $xy = 12$ (except $(0, 0)$; see Figure 13.43). Further, between the axes and that part of the hyperbola in the first quadrant, the function is always positive. Finally, since $(2, 2)$ is the only critical point in this region, it follows that $(2, 2)$ could only yield a relative maximum. A similar argument in the third quadrant indicates that $(-2, -2)$ gives a relative minimum.

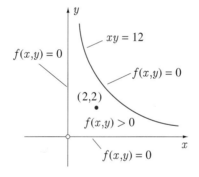

FIGURE 13.43

This completes our discussion of relative extrema of functions of two independent variables. Our next step should be extend the theory to functions of more than two variables. It is a simple matter to give definitions of relative maxima and minima for such functions; they are almost identical to Definitions 13.7 and 13.8 (see Exercise 19). On the other hand, to develop a theorem for functions of more than two independent variables that is analogous to Theorem 13.7 is beyond the scope of this book. We refer the interested reader to more advanced books.

EXERCISES 13.10

In Exercises 1–14 find all critical points for the function and classify each as yielding a relative maximum, a relative minimum, a saddle point, or none of these.

1. $f(x, y) = x^2 + 2xy + 2y^2 - 6y$

2. $f(x, y) = 3xy - x^3 - y^3$

3. $f(x, y) = x^3 - 3x + y^2 + 2y$

4. $f(x, y) = x^2 y^2 + 3x$

5. $f(x,y) = xy - x^2 + y^2$

6. $f(x,y) = x \sin y$

7. $f(x,y) = xye^{-(x^2+y^2)}$

8. $f(x,y) = x^2 - 2xy + y^2$

9. $f(x,y) = (x^2 + y^2)^{2/3}$

10. $f(x,y) = x^4 y^3$

11. $f(x,y) = 2xy^2 + 3xy + x^2 y^3$

12. $f(x,y) = |x| + y^2$

13. $f(x,y) = (1-x)(1-y)(x+y-1)$

14. $f(x,y) = x^4 + y^4 - x^2 - y^2 + 1$

In Exercises 15–18 find all critical points for the function.

15. $f(x,y,z) = x^2 + y^2 - z^2 + 3x - 2y + 5$

16. $f(x,y,z,t) = x^2 y^2 z^2 + t^2 x^2 + 3x$

17. $f(x,y,z) = xyz + x^2 yz - y$

18. $f(x,y,z) = xyze^{x^2+y^2+z^2}$

19. Give definitions for a relative maximum and a relative minimum for a function $f(x,y,z)$ at a point (x_0, y_0, z_0).

20. Suppose that $f(x,y)$ is harmonic in the region D: $x^2 + y^2 < 1$. Show that $f(x,y)$ cannot have a relative maximum or minimum at any point in D at which either f_{xx} or f_{xy} does not vanish.

21. Find and classify the critical points for the function $f(x,y) = y^2 - 4x^2 y + 3x^4$.

22. (a) Plot the ten points in the following table to show that they are reasonably close to being collinear.

x	1	2	3	4	5	6	7	8	9	10
y	6.05	8.32	10.74	13.43	15.90	18.38	20.93	23.32	24.91	28.36

(b) If $y = mx + b$ is the equation of a straight line that we might use to approximate the function $y = f(x)$ described by these points, then the following sum is a measure of how well the line fits the data:

$$S = S(m,b) = \sum_{i=1}^{10} (\bar{y}_i - mx_i - b)^2,$$

where $(x_i, \bar{y}_i)$ are the data points in the table. What does this sum represent geometrically?

(c) To find the best straight line to fit the data, the least-squares method suggests that m and b be chosen to minimize S. Show that S has only one critical point (m, b) which is defined by the linear equations

$$\left(\sum_{i=1}^{10} x_i^2\right) m + \left(\sum_{i=1}^{10} x_i\right) b = \sum_{i=1}^{10} x_i \bar{y}_i,$$

$$\left(\sum_{i=1}^{10} x_i\right) m + 10b = \sum_{i=1}^{10} \bar{y}_i.$$

(d) Solve these equations for m and b.

23. (a) Plot the sixteen points in the following table.

x	3.00	3.25	3.50	3.75	4.00	4.25	4.50	4.75
y	31.5	30.4	29.2	28.1	26.9	26.4	25.3	25.2
x	5.00	5.25	5.50	5.75	6.00	6.25	6.50	6.75
y	25.1	25.2	25.4	26.3	27.0	28.2	29.3	29.9

Do they seem to follow a parabolic path?

(b) If $y = ax^2 + bx + c$ is the equation of a parabola that is to approximate the function $y = f(x)$ described by these points, then the following sum is a measure of the accuracy of the fit:

$$S = S(a,b,c) = \sum_{i=1}^{16} (\bar{y}_i - ax_i^2 - bx_i - c)^2,$$

where $(x_i, \bar{y}_i)$ are the points in the table. To find the best possible fit, the least-squares method says to choose a, b, and c to minimize S. Show that S has only one critical point (a, b, c) which is defined by the linear equations

$$\left(\sum_{i=1}^{16} x_i^4\right) a + \left(\sum_{i=1}^{16} x_i^3\right) b + \left(\sum_{i=1}^{16} x_i^2\right) c = \sum_{i=1}^{16} x_i^2 \bar{y}_i,$$

$$\left(\sum_{i=1}^{16} x_i^3\right) a + \left(\sum_{i=1}^{16} x_i^2\right) b + \left(\sum_{i=1}^{16} x_i\right) c = \sum_{i=1}^{16} x_i \bar{y}_i,$$

$$\left(\sum_{i=1}^{16} x_i^2\right) a + \left(\sum_{i=1}^{16} x_i\right) b + 16c = \sum_{i=1}^{16} \bar{y}_i.$$

(c) Solve these equations for a, b, and c.

24. Find and classify the critical points of $f(x,y) = x^4 + 3xy^2 + y^2$ as yielding relative maxima, relative minima, or saddle points.

25. The equation $2x^2 + 3y^2 + z^2 - 12xy + 4xz = 35$ defines functions $z = f(x,y)$. Show that the point $x = 1$ and $y = 2$ is a critical point for any such function. Does it yield a relative extrema for the function?

26. (a) Show that the function $f(x,y,z) = x^2 + y^2 + z^2 - xyz$ has a critical point $(0,0,0)$. What are the other critical points?

    (b) Use the definition in Exercise 19 to show that $f(x,y,z)$ has a relative minimum at $(0,0,0)$.

SECTION 13.11

Absolute Maxima and Minima

So far as applications of maxima and minima are concerned, absolute maxima and minima are more important than relative maxima and minima. In this section and in Section 13.12 we discuss the theory of absolute extrema and consider a number of applications. Once again we begin with functions $f(x,y)$ of two independent variables and base our discussion on the theory of absolute extrema for functions of one variable.

We learned in Section 4.6 that a function $f(x)$ that is continuous on a finite interval $a \leq x \leq b$ must have an absolute maximum and an absolute minimum on that interval. Furthermore, these absolute extrema must occur at either critical points or at the ends $x = a$ and $x = b$ of the interval. Consequently, to find the absolute extrema of a function $f(x)$, we evaluate $f(x)$ at all critical points, at $x = a$, and at $x = b$; the largest of these numbers is the absolute maximum of $f(x)$ on $a \leq x \leq b$, and the smallest is the absolute minimum.

The procedure is much the same for a function $f(x,y)$ that is continuous on a region R that is finite and includes all the points on its boundary. First, however, we define exactly what we mean by absolute extrema of $f(x,y)$ and consider a number of simple examples. We will then be able to make general statements about the nature of all absolute extrema, and proceed to the important area of applications.

Definition 13.10

The **absolute maximum** of a function $f(x,y)$ on a region R is $f(x_0, y_0)$ if (x_0, y_0) is in R and

$$f(x,y) \leq f(x_0, y_0) \quad (13.55)$$

for all (x,y) in R. The **absolute minimum** of $f(x,y)$ on R is $f(x_0, y_0)$ if (x_0, y_0) is in R and

$$f(x,y) \geq f(x_0, y_0) \quad (13.56)$$

for all (x,y) in R.

In Figures 13.44–13.49, we have shown six functions defined on the circle R: $x^2 + y^2 \leq 1$. The absolute maxima and minima of these functions for this region are shown in Table 13.1.

FIGURE 13.44

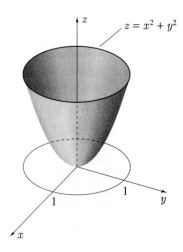

FIGURE 13.45

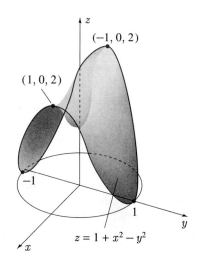

FIGURE 13.46

FIGURE 13.47

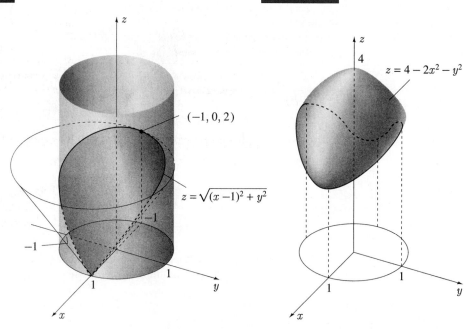

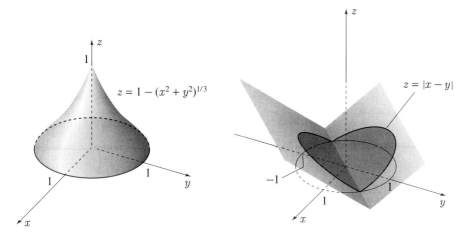

FIGURE 13.48

FIGURE 13.49

TABLE 13.1

Function $f(x, y)$	Position of absolute maximum	Value of absolute maximum	Position of absolute mimimum	Value of absolute mimimum
$x^2 + y^2$	Every point on $x^2 + y^2 = 1$	1	(0, 0)	0
$1 + x^2 - y^2$	$(\pm 1, 0)$	2	$(0, \pm 1)$	0
$\sqrt{(x-1)^2 + y^2}$	$(-1, 0)$	2	$(1, 0)$	0
$4 - 2x^2 - y^2$	$(0, 0)$	4	$(\pm 1, 0)$	2
$1 - (x^2 + y^2)^{1/3}$	$(0, 0)$	1	Every point on $x^2 + y^2 = 1$	0
$|x - y|$	$(\pm 1/\sqrt{2}, \mp 1/\sqrt{2})$	$\sqrt{2}$	Every point on $y = x$, $-1/\sqrt{2} \leq x \leq 1/\sqrt{2}$	0

For each of the functions in these figures, *absolute extrema occur at either a critical point or a point on the boundary of R*. This result is true for any *continuous function defined on a finite region that includes all the points on its boundary*. Although this result may seem fairly obvious geometrically, to prove it analytically is very difficult; we will be content to assume its validity and carry on from there.

It is very simple to look at the surface defined by a function $f(x, y)$ and pick off its absolute extrema, but in practice this just does not happen. Usually we must determine the absolute extrema algebraically from the function itself. Suppose, then, that a continuous function $f(x, y)$ is given and we are required to find its absolute extrema on a finite region R (which includes its boundary points). The previous discussion indicated that the extrema must occur either at critical points or on the boundary of R. Consequently, we should first determine all critical points of $f(x, y)$ in R, and evaluate

$f(x, y)$ at each of these points. These values should now be compared to the maximum and minimum values of $f(x, y)$ on the boundary of R. But how do we find the maximum and minimum values of $f(x, y)$ on the boundary? If the boundary of R is denoted by C (Figure 13.50), and if C has parametric equations $x = x(t)$, $y = y(t)$, $\alpha \leq t \leq \beta$, then on C we can express $f(x, y)$ in terms of t, and t alone:

$$f[x(t), y(t)], \quad \alpha \leq t \leq \beta.$$

To find the maximum and minimum values of $f(x, y)$ on C is now an absolute extrema problem for a function of one variable. The function $f[x(t), y(t)]$ should therefore be evaluated at each of its critical points and at $t = \alpha$ and $t = \beta$.

If the boundary of R consists of a number of curves (Figure 13.51), then this boundary procedure must be performed for each part. In other words, on each part of the boundary we express $f(x, y)$ as a function of one variable, and then evaluate this function at its critical points and at the ends of that part of the boundary to which it applies.

The absolute maximum of $f(x, y)$ on R is then the largest of all values of $f(x, y)$ evaluated at the critical points inside R, the critical points on the boundary of R, and the endpoints of each part of the boundary. The absolute minimum of $f(x, y)$ on R is the smallest of all these values.

FIGURE 13.50

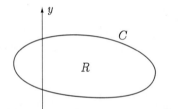

FIGURE 13.51

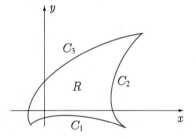

Recall that to find the absolute extrema of a function $f(x)$, continuous on $a \leq x \leq b$, we evaluate $f(x)$ at all critical points and at the boundary points $x = a$ and $x = b$. The procedure that we have established here for $f(x, y)$ is much the same—the difference is that for $f(x, y)$, the boundary consists not of two points, but of entire curves. Evaluation of $f(x, y)$ on the boundary therefore reduces to one or more extrema problems for functions of one variable. Note too that for $f(x, y)$ (or $f(x)$), it is not necessary to determine the nature of the critical points; it is necessary only to evaluate $f(x, y)$ at these points.

EXAMPLE 13.30

Find the maximum value of the function $z = f(x, y) = 4xy - x^4 - 2y^2$ on the region R: $-2 \leq x \leq 2$, $-2 \leq y \leq 2$.

SOLUTION
Critical points of $f(x, y)$ are given by

$$0 = \frac{\partial f}{\partial x} = 4y - 4x^3, \quad 0 = \frac{\partial f}{\partial y} = 4x - 4y.$$

Solutions of these equations are $(0, 0)$, $(1, 1)$, and $(-1, -1)$, and the values of $f(x, y)$ at these critical points are

$$f(0, 0) = 0, \quad f(1, 1) = 1, \quad f(-1, -1) = 1.$$

We denote the four parts of the boundary of R by C_1, C_2, C_3, and C_4 (Figure 13.52).

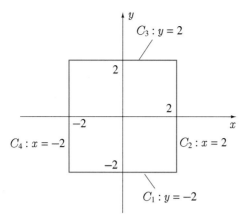

FIGURE 13.52

On C_1, $y = -2$, in which case

$$z = -8x - x^4 - 8, \quad -2 \leq x \leq 2.$$

For critical points of this function, we solve

$$0 = \frac{dz}{dx} = -8 - 4x^3.$$

The only solution is $x = -2^{1/3}$, at which the value of z is

$$z = 8 \cdot 2^{1/3} - 2^{4/3} - 8 = -0.44.$$

On C_2, $x = 2$, in which case

$$z = 8y - 16 - 2y^2, \quad -2 \leq y \leq 2.$$

Critical points are defined by

$$0 = \frac{dz}{dy} = 8 - 4y.$$

The only solution $y = 2$ defines one of the corners of the square, and at this point

$$z = -8.$$

On C_3, $y = 2$ and

$$z = 8x - x^4 - 8, \quad -2 \leq x \leq 2.$$

For critical points, we solve

$$0 = \frac{dz}{dx} = 8 - 4x^3.$$

At the single point $x = 2^{1/3}$,

$$z = 8 \cdot 2^{1/3} - 2^{4/3} - 8 = -0.44.$$

On the final curve C_4, $x = -2$ and
$$z = -8y - 16 - 2y^2, \quad -2 \le y \le 2.$$

Critical points are given by
$$0 = \frac{dz}{dy} = -8 - 4y.$$

The solution $y = -2$ defines another corner of the square at which
$$z = -8.$$

We have now evaluated $f(x, y)$ at all critical points inside R and at all critical points on the four parts of the boundary of R. It remains only to evaluate $f(x, y)$ at the corners of the square. Two corners have already been accounted for; the other two give
$$f(2, -2) = -40, \quad f(-2, 2) = -40.$$

The largest value of $f(x, y)$ produced is 1, and this is therefore the maximum value of $f(x, y)$ on R. ■

EXAMPLE 13.31

The temperature at each point (x, y) in a semicircular plate defined by $x^2 + y^2 \le 1$, $y \ge 0$ is given by
$$T(x, y) = 16x^2 - 24xy + 40y^2.$$

Find the hottest and coldest points in the plate.

SOLUTION
For critical points of $T(x, y)$, we solve
$$0 = \frac{\partial T}{\partial x} = 32x - 24y, \quad 0 = \frac{\partial T}{\partial y} = -24x + 80y.$$

The only solution of these equations, $(0, 0)$, is on the boundary. On the upper edge of the plate (Figure 13.53), we set $x = \cos t$, $y = \sin t$, $0 \le t \le \pi$, in which case
$$T = 16\cos^2 t - 24 \cos t \sin t + 40 \sin^2 t, \quad 0 \le t \le \pi.$$

For critical points of this function, we solve
$$0 = \frac{dT}{dt} = -32 \cos t \sin t - 24(-\sin^2 t + \cos^2 t) + 80 \sin t \cos t$$
$$= 24(\sin 2t - \cos 2t).$$

If we divide by $\cos 2t$ (since $\cos 2t = 0$ does not lead to a solution of this equation), we have
$$\tan 2t = 1.$$

The only solutions of this equation in the interval $0 \le t \le \pi$ are $t = \pi/8$ and $t = 5\pi/8$. When $t = \pi/8$, $T = 11.0$; and when $t = 5\pi/8$, $T = 45.0$. At the ends of this part of the boundary, $t = 0$ and $t = \pi$, and $T(1,0) = 16$ and $T(-1,0) = 16$. On the lower edge of the plate, $y = 0$, in which case

$$T = 16x^2, \quad -1 \leq x \leq 1.$$

The only critical point of this function is $x = 0$, at which $T = 0$. The hottest point in the plate is therefore $(\cos(5\pi/8), \sin(5\pi/8)) = (-0.38, 0.92)$, where the temperature is $45°$, and the coldest point is $(0,0)$ with temperature $0°$.

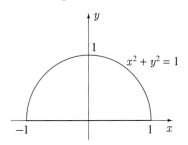

FIGURE 13.53

EXAMPLE 13.32

Find the point on the plane $2x - 3y - 4z = 25$ closest to the point $(3,2,1)$.

SOLUTION The distance D from $(3,2,1)$ to any point (x,y,z) in space is defined by

$$D^2 = (x-3)^2 + (y-2)^2 + (z-1)^2.$$

Because we want the minimum distance from $(3,2,1)$ to points in the plane $2x - 3y - 4z = 25$, we must minimize D but consider only those points (x,y,z) that satisfy the equation of the plane (Figure 13.54). At the moment, D^2 is a function of three variables x, y, and z, but they are not all independent because of the planar restriction. If we solve the equation of the plane for z in terms of x and y and substitute, then

$$D^2 = f(x,y) = (x-3)^2 + (y-2)^2 + \left(\frac{2x - 3y - 25}{4} - 1\right)^2$$

$$= (x-3)^2 + (y-2)^2 + \left(\frac{2x - 3y - 29}{4}\right)^2,$$

where x and y are independent variables. Now D is minimized when D^2 is minimized, and we will therefore find that point (x,y) which minimizes D^2. First we locate the critical points of D^2, by solving

$$0 = \frac{\partial f}{\partial x} = 2(x-3) + \left(\frac{2x - 3y - 29}{4}\right)$$

and

$$0 = \frac{\partial f}{\partial y} = 2(y-2) - \frac{3}{2}\left(\frac{2x - 3y - 29}{4}\right).$$

These reduce to

$$10x - 3y = 53, \quad -6x + 25y = -55.$$

The only solution is $x = 5$, $y = -1$. In this problem we do not have a finite region R in the xy-plane in which D^2 is defined. It is geometrically clear, however, that as x and y become very large, so too does D^2. It follows, therefore, that the one critical point obtained, namely $(5, -1)$, must yield the absolute minimum of D^2. Consequently, the point closest to $(3, 2, 1)$ in the plane $2x - 3y - 4z = 25$ is $(5, -1, -3)$.

FIGURE 13.54

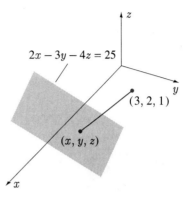

EXERCISES 13.11

In Exercises 1–8 find the maximum and minimum values of the function on the region.

1. $f(x, y) = x^2 + y^3$ on $R: x^2 + y^2 \leq 1$

2. $f(x, y) = x^2 + x + 3y^2 + y$ on the region R bounded by $y = x + 1$, $y = 1 - x$, $y = x - 1$, $y = -x - 1$

3. $f(x, y) = 3x + 4y$ on the region R bounded by the lines $x + y = 1$, $x + y = 4$, $y + 1 = x$, $y - 1 = x$

4. $f(x, y) = x^2 y + xy^2 + y$ on $R: -1 \leq x \leq 1$, $-1 \leq y \leq 1$

5. $f(x, y) = 3x^2 + 2xy - y^2 + 5$ on $R: 4x^2 + 9y^2 \leq 36$

6. $f(x, y) = x^3 - 3x + y^2 + 2y$ on the triangle bounded by $x = 0$, $y = 0$, $x + y = 1$

7. $f(x, y) = x^3 + y^3 - 3x - 12y + 2$ on the square $-3 \leq x \leq 3$, $-3 \leq y \leq 3$

8. $f(x, y) = x^3 + y^3 - 3x - 3y + 2$ on the circle $x^2 + y^2 \leq 1$

9. Find maximum and minimum values of the function $f(x, y, z) = xy^2 z^3$ on that part of the plane $x + y + z = 6$ for which **(a)** $x > 0$, $y > 0$, $z > 0$. **(b)** $x \geq 0$, $y \geq 0$, $z \geq 0$.

10. Find the point on the plane $x + y - 2z = 6$ closest to the origin.

11. Find the shortest distance from $(-1, 1, 2)$ to the plane $2x - 3y + 6z = 14$.

12. Find the point on the surface $z = x^2 + y^2$ closest to the point $(1, 1, 0)$.

13. Find the point on that part of the plane $x + y + 2z = 4$ in the first octant that is closest to the point $(3, 3, 1)$. For this question assume that the curves of intersection of the plane with the coordinate planes are part of the surface.

14. The electrostatic potential at each point in the region $0 \leq x \leq 1$, $0 \leq y \leq 1$ is given by $V(x, y) = 48xy - 32x^3 - 24y^2$. Find the maximum and minimum potentials in the region.

15. When a rectangular box is sent through the mail, the post office demands that the length of the box plus twice the sum of its height and width be no more than 250 cm. Find the dimensions of the box satisfying this requirement that encloses the largest possible volume.

16. An open tank in the form of a rectangular parallelepiped is to be built to hold 1000 L of acid. If the cost per unit area of lining the base of the tank is three times that of the sides, what dimensions minimize the cost of lining the tank?

17. Prove that the minimum distance from a point (x_1, y_1, z_1) to a plane $Ax + By + Cz + D = 0$ is $|Ax_1 + By_1 + Cz_1 + D|/\sqrt{A^2 + B^2 + C^2}$.

18. Prove that for triangles, the point that minimizes the sum of the squares of the distances to the vertices is the centroid.

19. Find the point on the curve $x^2 - xy + y^2 - z^2 = 1$, $x^2 + y^2 = 1$ closest to the origin.

20. Find the dimensions of the box with largest possible volume that can fit inside the ellipsoid $x^2/a^2 + y^2/b^2 + z^2/c^2 = 1$, assuming that its edges are parallel to the coordinate axes.

21. Find the maximum and minimum values of the function $f(x, y, z) = xyz$ on the sphere $x^2 + y^2 + z^2 = 1$.

22. If P is the perimeter of a triangle with sides of length x, y, and z, the area of the triangle is

$$A = \sqrt{\frac{P}{2}\left(\frac{P}{2} - x\right)\left(\frac{P}{2} - y\right)\left(\frac{P}{2} - z\right)},$$

where $P = x + y + z$. Show that A is maximized for fixed P when the triangle is equilateral.

23. Show that for any triangle with interior angles A, B, and C,

$$\sin(A/2) \sin(B/2) \sin(C/2) \leq 1/8.$$

Hint: Find the maximum value of the function $f(A, B, C) = \sin(A/2) \sin(B/2) \sin(C/2)$.

24. A silo is in the shape of a right-circular cylinder surmounted by a right-circular cone. If the radius of each is 6 m and the total surface area must be 200 m² (not including the base), what heights for the cone and cylinder yield maximum enclosed volume?

25. What values of x and y maximize the production function $P(x, y) = kx^\alpha y^\beta$ where k, α, and β are positive constants ($\alpha + \beta = 1$) when x and y must satisfy $Ax + By = C$, where A, B and C are positive constants.

26. A long piece of metal 1 m wide is bent at A and B, as shown in Figure 13.55, to form a channel with three straight sides. If the bends are equidistant from the ends, where should they be made in order to obtain maximum possible flow of fluid along the channel?

FIGURE 13.55

27. Find maximum and minimum values of the function $f(x, y, z) = xy + xz$ on the region $x^2 + y^2 + z^2 \leq 1$.

28. Find maximum and minimum values of the function $f(x, y, z) = x^2 yz$ on the region $x^2 + y^2 \leq 1$, $0 \leq z \leq 1$.

29. A cow's daily diet consists of three foods: hay, grain, and supplements. The cow is always given 11 kg of hay per day, 50% of which is digestive material and 12% of which is protein. Grain is 74% digestive material and 8.8% protein, whereas supplements are 62% digestive material and 34% protein. The cost of hay is $27.50 for 1000 kg, and grain and supplements cost $110 and $175 for 1000 kg. A healthy cow's diet must contain between 9.5 and 11.5 kg of digestive material and between 1.9 and 2.0 kg of protein. Determine the daily amounts of grain and supplements that the cow should be fed in order that total food costs be kept to a minimum.

30. Find the area of the largest triangle that has vertices on the circle $x^2 + y^2 = r^2$.

31. A rectangle is surmounted by an isosceles triangle as shown in Figure 13.56. Find x, y, and θ in order that the area of the figure be as large as possible under the restriction that its perimeter must be P.

FIGURE 13.56

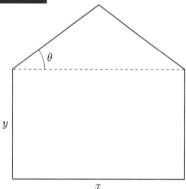

SECTION 13.12

Lagrange Multipliers

Many applied maxima and minima problems result in **constraint problems**. In particular, Examples 13.31 and 13.32 contain such problems. In Example 13.31, to find extreme values of T on the edge of the plate we maximized and minimized $T(x,y) = 16x^2 - 24x + 40y^2$ subject to first the constraint $x^2 + y^2 = 1$, and then the constraint $y = 0$. Our method there was to substitute from the constraint equation into $T(x,y)$ in order to obtain a function of one variable. In Example 13.32, to find the minimum distance from $(3,2,1)$ to the plane $2x - 3y - 4z = 25$, we minimized $D^2 = (x-3)^2 + (y-2)^2 + (z-1)^2$ subject to the constraint $2x - 3y - 4z = 25$. Again we substituted from the constraint to obtain D^2 as a function of two independent variables.

A natural question to ask is whether problems of this type can be solved without substituting from the constraint equation, for if the constraint equation is complicated, substitution may be very difficult or even impossible. To show that there is indeed an alternative, consider the situation in which a function $f(x, y, z)$ is to be maximized or minimized subject to two constraints:

$$F(x, y, z) = 0, \qquad (13.57a)$$
$$G(x, y, z) = 0. \qquad (13.57b)$$

Algebraically, we are to find extreme values of $f(x, y, z)$, considering only those values of x, y, and z that satisfy the two equations 13.57. Geometrically, we can interpret each of these conditions as specifying a surface, so that we are seeking extreme values of $f(x, y, z)$, considering only those points on the curve of intersection C of the surfaces $F(x, y, z) = 0$ and $G(x, y, z) = 0$ (Figure 13.57).

FIGURE 13.57

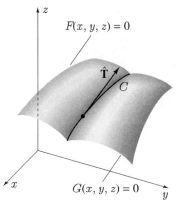

Extreme values of $f(x, y, z)$ along C will occur either at critical points of the function or at the ends of the curve. But what derivative or derivatives of $f(x, y, z)$ are we talking about when we say critical points? Since we are concerned only with values of $f(x, y, z)$ on C, we must mean the derivative of $f(x, y, z)$ along C; i.e., the directional derivative in the tangent direction to C. If $\mathbf{T}$, then, is a tangent vector to C, critical points of $f(x, y, z)$ along C are given by

$$0 = D_{\mathbf{T}} f = \nabla f \cdot \hat{\mathbf{T}},$$

or at points where the directional derivative is undefined. According to equation 13.50, a tangent vector to C is $\mathbf{T} = \nabla F \times \nabla G$, and hence a unit tangent vector is

$$\hat{\mathbf{T}} = \frac{\nabla F \times \nabla G}{|\nabla F \times \nabla G|}.$$

The directional derivative of $f(x, y, z)$ at points along C is therefore given by

$$D_{\mathbf{T}} f = \nabla f \cdot \frac{\nabla F \times \nabla G}{|\nabla F \times \nabla G|}.$$

It follows, then, that critical points of $f(x,y,z)$ are points (x,y,z) that satisfy the equation

$$\nabla f \cdot \nabla F \times \nabla G = 0,$$

or points at which the left side is not defined. Now vector $\nabla F \times \nabla G$ is perpendicular to both ∇F and ∇G. Since ∇f is perpendicular to $\nabla F \times \nabla G$ (their dot product is zero), it follows that ∇f must lie in the plane of ∇F and ∇G. Consequently, there exist scalars λ and μ such that

$$\nabla f = (-\lambda)\nabla F + (-\mu)\nabla G,$$

or

$$\nabla f + \lambda \nabla F + \mu \nabla G = \mathbf{0}. \quad (13.58)$$

This vector equation is equivalent to the three scalar equations

$$\frac{\partial f}{\partial x} + \lambda \frac{\partial F}{\partial x} + \mu \frac{\partial G}{\partial x} = 0, \quad (13.59\text{a})$$

$$\frac{\partial f}{\partial y} + \lambda \frac{\partial F}{\partial y} + \mu \frac{\partial G}{\partial y} = 0, \quad (13.59\text{b})$$

$$\frac{\partial f}{\partial z} + \lambda \frac{\partial F}{\partial z} + \mu \frac{\partial G}{\partial z} = 0, \quad (13.59\text{c})$$

and these equations must be satisfied at a critical point at which the directional derivative of $f(x,y,z)$ vanishes. Note too that at a point at which the directional derivative of $f(x,y,z)$ does not exist, one of the partial derivatives in these equations does not exist. In other words, we have shown that critical points of $f(x,y,z)$ are points that satisfy equations 13.59 or points at which the equations are undefined. These equations, however, contain five unknowns: x, y, z, λ, and μ. To complete the system we add equations 13.57 since they must also be satisfied by a critical point. Equations 13.57 and 13.59 therefore yield a system of five equations in the five unknowns x, y, z, λ, and μ; the first three unknowns (x,y,z) define a critical point of $f(x,y,z)$ along C. The advantage of this system of equations lies in the fact that the differentiations in 13.59 involve only the given functions (and no substitutions from the constraint equations are necessary). What we have sacrificed is a system of three equations in the three unknowns (x,y,z) for a system of five equations in the five unknowns (x,y,z,λ,μ).

Let us not forget that the original problem was to find extreme values for the function $f(x,y,z)$ subject to constraints 13.57. What we have shown so far is that critical points at which the directional derivative of $f(x,y,z)$ vanishes can be found by solving equations 13.57 and 13.59. In addition, critical points at which the directional derivative of $f(x,y,z)$ does not exist are points at which equations 13.59 are not defined. What remains is to evaluate $f(x,y,z)$ at all critical points and at the ends of C. *If C is a closed curve* (i.e., if C rejoins itself), then $f(x,y,z)$ *need be evaluated only at the critical points.*

Through the directional derivative and tangent vectors to curves, we have shown that equations 13.57 and 13.59 define critical points of a function $f(x,y,z)$ that is subject to two constraints: $F(x,y,z) = 0$ and $G(x,y,z) = 0$. But what about other situations? Let us say, for example, that we require extreme values of a function $f(x,y,z,t)$ subject to a single constraint $F(x,y,z,t) = 0$. How shall we find critical points of this function? Fortunately, as we now show, there is a very simple method that yields equations 13.57 and 13.59, and this method generalizes to other situations also.

To find critical points of $f(x,y,z)$ subject to constraints 13.57, we define a function

$$L(x,y,z,\lambda,\mu) = f(x,y,z) + \lambda F(x,y,z) + \mu G(x,y,z),$$

and regard it as a function of five independent variables x, y, z, λ, and μ. To find critical points of this function, we would first solve the equations obtained by setting each of the partial derivatives of L equal to zero:

$$0 = \frac{\partial L}{\partial x} = \frac{\partial f}{\partial x} + \lambda \frac{\partial F}{\partial x} + \mu \frac{\partial G}{\partial x},$$

$$0 = \frac{\partial L}{\partial y} = \frac{\partial f}{\partial y} + \lambda \frac{\partial F}{\partial y} + \mu \frac{\partial G}{\partial y},$$

$$0 = \frac{\partial L}{\partial z} = \frac{\partial f}{\partial z} + \lambda \frac{\partial F}{\partial z} + \mu \frac{\partial G}{\partial z},$$

$$0 = \frac{\partial L}{\partial \lambda} = F(x, y, z),$$

$$0 = \frac{\partial L}{\partial \mu} = G(x, y, z).$$

In addition, we would consider points at which the partial derivatives of L do not exist. Clearly, this means points (x, y, z) at which any of the partial derivatives of $f(x, y, z)$, $F(x, y, z)$, and $G(x, y, z)$ do not exist. But these are precisely equations 13.57 and 13.59. We have shown, then, that finding critical points (x, y, z) of $f(x, y, z)$ subject to $F(x, y, z) = 0$ and $G(x, y, z) = 0$ is equivalent to finding critical points (x, y, z, λ, μ) of $L(x, y, z, \lambda, \mu)$. The two unknowns λ and μ that accompany a critical point (x, y, z) of $f(x, y, z)$ are called **Lagrange multipliers**. They are not a part of the solution (x, y, z) to the original problem, but have been introduced as a convenience by which to arrive at that solution. The function $L(x, y, z, \lambda, \mu)$ is often called the **Lagrangian** of the problem.

The method for other constraint problems should now be evident. Given a function $f(x, y, z, t, \ldots)$ of n variables to maximize or minimize subject to m constraints

$$F_1(x, y, z, t, \ldots) = 0, \quad F_2(x, y, z, t, \ldots) = 0, \ldots, F_m(x, y, z, t, \ldots) = 0, \quad (13.60)$$

we introduce m Lagrange multipliers $\lambda_1, \lambda_2, \ldots, \lambda_m$ into a Lagrangian of $n + m$ independent variables $x, y, z, t, \ldots, \lambda_1, \lambda_2, \ldots, \lambda_m$:

$$L(x, y, z, t, \ldots \lambda_1, \lambda_2, \ldots, \lambda_m) = f(x, y, z, t, \ldots) + \lambda_1 F_1(x, y, z, t, \ldots) + \cdots$$

$$+ \lambda_m F_m(x, y, z, t, \ldots). \quad (13.61)$$

Critical points $(x, y, z, t, \ldots)$ of $f(x, y, z, t, \ldots)$ are then determined by the equations defining critical points of $L(x, y, z, t, \ldots, \lambda_1, \lambda_2, \ldots, \lambda_m)$, namely,

$$0 = \frac{\partial L}{\partial x} = \frac{\partial f}{\partial x} + \lambda_1 \frac{\partial F_1}{\partial x} + \cdots + \lambda_m \frac{\partial F_m}{\partial x}, \quad (13.62\text{a})$$

$$0 = \frac{\partial L}{\partial y} = \frac{\partial f}{\partial y} + \lambda_1 \frac{\partial F_1}{\partial y} + \cdots + \lambda_m \frac{\partial F_m}{\partial y}, \quad (13.62\text{b})$$

$$\vdots$$

$$0 = \frac{\partial L}{\partial \lambda_1} = F_1(x, y, z, t, \ldots), \quad (13.62\text{c})$$

$$0 = \frac{\partial L}{\partial \lambda_2} = F_2(x, y, z, t, \ldots), \quad (13.62\text{d})$$

$$\vdots$$

$$0 = \frac{\partial L}{\partial \lambda_m} = F_m(x, y, z, t, \ldots). \quad (13.62\text{e})$$

To use Lagrange multipliers in Example 13.32, we define the Lagrangian

$$L(x, y, z, \lambda) = D^2 + \lambda(2x - 3y - 4z - 25)$$
$$= (x - 3)^2 + (y - 2)^2 + (z - 1)^2 + \lambda(2x - 3y - 4z - 25).$$

Critical points of $L(x, y, z, \lambda)$ are defined by

$$0 = \frac{\partial L}{\partial x} = 2(x - 3) + 2\lambda,$$

$$0 = \frac{\partial L}{\partial y} = 2(y - 2) - 3\lambda,$$

$$0 = \frac{\partial L}{\partial z} = 2(z - 1) - 4\lambda,$$

$$0 = \frac{\partial L}{\partial \lambda} = 2x - 3y - 4z - 25.$$

The solution of this linear system is $(x, y, z, \lambda) = (5, -1, -3, -2)$, yielding as before the critical point $(5, -1, -3)$ of D^2.

EXAMPLE 13.33

Find the maximum value of the function

$$f(x, y, z) = 2x^2 y^2 + 2y^2 z^2 + 3z,$$

considering only those values of x, y, and z that satisfy the equations

$$z = x^2 + y^2, \quad x^2 + 3y^2 = 1.$$

SOLUTION For this example we illustrate the difference between a solution that utilizes Lagrange multipliers and one that does not. The solution that does not use Lagrange multipliers requires us first to express $f(x, y, z)$ in terms of a single independent variable. We can do this by solving the constraint equations for $x^2 = 1 - 3y^2$ and $y^2 = (1 - z)/2$ and substituting into $f(x, y, z)$:

$$f(x, y, z) = 2(1 - 3y^2)y^2 + 2y^2 z^2 + 3z$$

$$= 2\left(1 - \frac{3}{2} + \frac{3z}{2}\right)\left(\frac{1}{2} - \frac{z}{2}\right) + 2z^2\left(\frac{1}{2} - \frac{z}{2}\right) + 3z$$

$$= \frac{1}{2}(-2z^3 - z^2 + 10z - 1).$$

Since $z = x^2 + y^2 = (1 - 3y^2) + y^2 = 1 - 2y^2$, and y must be restricted to $|y| \leq 1/\sqrt{3}$, it follows that the only possible values for z are $\frac{1}{3} \leq z \leq 1$. We have shown, then, that maximizing $f(x, y, z)$ subject to the two constraints $z = x^2 + y^2$ and $x^2 + 3y^2 = 1$ is equivalent to maximizing

$$F(z) = \frac{1}{2}(-2z^3 - z^2 + 10z - 1), \quad \frac{1}{3} \leq z \leq 1.$$

For critical points of $F(z)$, we solve

$$0 = \frac{dF}{dz} = \frac{1}{2}(-6z^2 - 2z + 10),$$

from which we get

$$z = \frac{2 \pm \sqrt{4 + 240}}{-12} = \frac{-1 \pm \sqrt{61}}{6}.$$

These two critical points must be rejected as not lying within the interval $\frac{1}{3} \le z \le 1$, and the maximum value of $F(z)$ must therefore occur at either $z = \frac{1}{3}$ or $z = 1$. Since $F(\frac{1}{3}) = \frac{29}{27}$ and $F(1) = 3$, it follows that the maximum value of $F(z)$ and therefore of $f(x, y, z)$ is 3.

To maximize $f(x, y, z)$ using Lagrange multipliers, we define the Lagrangian

$$L(x, y, z, \lambda, \mu) = 2x^2 y^2 + 2y^2 z^2 + 3z + \lambda(x^2 + y^2 - z) + \mu(x^2 + 3y^2 - 1).$$

Critical points of L and therefore of $f(x, y, z)$ are defined by the equations

$$0 = \frac{\partial L}{\partial x} = 4xy^2 + 2\lambda x + 2\mu x = 2x(2y^2 + \lambda + \mu),$$

$$0 = \frac{\partial L}{\partial y} = 4x^2 y + 4yz^2 + 2\lambda y + 6\mu y = 2y(2x^2 + 2z^2 + \lambda + 3\mu),$$

$$0 = \frac{\partial L}{\partial z} = 4y^2 z + 3 - \lambda,$$

$$0 = \frac{\partial L}{\partial \lambda} = x^2 + y^2 - z,$$

$$0 = \frac{\partial L}{\partial \mu} = x^2 + 3y^2 - 1.$$

If we choose $x = 0$ to satisfy the first equation, then the remaining equations imply that $y = \pm 1/\sqrt{3}$, $z = \frac{1}{3}$ (and $\lambda = \frac{31}{9}$, $\mu = -\frac{11}{9}$). If we choose $y = 0$ to satisfy the second equation, then the remaining equations require $x = \pm 1$, $z = 1$ (and $\lambda = 3$, $\mu = -3$). The only other way to satisfy the first two equations is to set

$$2y^2 + \lambda + \mu = 0,$$
$$2x^2 + 2z^2 + \lambda + 3\mu = 0.$$

If we multiply the first of these by 3 and subtract the second, we have

$$\begin{aligned}
0 &= 6y^2 - 2x^2 - 2z^2 + 2\lambda \\
&= 6y^2 - 2x^2 - 2(x^2 + y^2)^2 + 2\lambda \\
&= 6y^2 - 2x^2 - 2(x^4 + 2x^2 y^2 + y^4) + 2\lambda \\
&= 6y^2 - 2(1 - 3y^2) - 2(1 - 3y^2)^2 \\
&\quad - 4y^2(1 - 3y^2) - 2y^4 + 2\lambda \\
&= 2(\lambda - 2 + 10y^2 - 4y^4).
\end{aligned}$$

But from the equation for $\partial L/\partial z$, we can also write

$$\begin{aligned}
0 &= 3 - \lambda + 4y^2(x^2 + y^2) \\
&= 3 - \lambda + 4y^2(1 - 3y^2 + y^2) \\
&= 3 - \lambda + 4y^2 - 8y^4.
\end{aligned}$$

These two equations in y and λ imply that

$$2 - 10y^2 + 4y^4 = 3 + 4y^2 - 8y^4, \quad \text{or} \quad 12y^4 - 14y^2 - 1 = 0.$$

Thus,
$$y^2 = \frac{14 \pm \sqrt{196 + 48}}{24} = \frac{7 \pm \sqrt{61}}{12},$$

where we reject the negative solution. But substitution of this result into the constraint $x^2 + 3y^2 = 1$ requires x^2 to be negative. Consequently, only four critical points (x, y, z) are obtained: $(0, \pm 1/\sqrt{3}, \frac{1}{3})$ and $(\pm 1, 0, 1)$. Since the curve defined by the constraints is closed, we need evaluate $f(x, y, z)$ only at these critical points:

$$f(0, \pm 1/\sqrt{3}, \frac{1}{3}) = 2(\frac{1}{3})(\frac{1}{9}) + 3(\frac{1}{3}) = \frac{29}{27},$$
$$f(\pm 1, 0, 1) = 3.$$

The maximum value of $f(x, y, z)$ is again 3.

Note that the endpoints in the first solution are critical points in the Lagrangian solution. This emphasizes the fact that, as we pointed out earlier, when the curve is closed and we use a Lagrangian, it is necessary to evaluate the function to be maximized or minimized only at its critical points. ∎

EXAMPLE 13.34

Find the maximum and minimum values of the function $f(x, y, z) = xyz$ on the sphere $x^2 + y^2 + z^2 = 1$.

SOLUTION If we define the Lagrangian

$$L(x, y, z, \lambda) = xyz + \lambda(x^2 + y^2 + z^2 - 1),$$

then critical points of L, and therefore of $f(x, y, z)$, are defined by the equations

$$0 = \frac{\partial L}{\partial x} = yz + 2\lambda x,$$
$$0 = \frac{\partial L}{\partial y} = xz + 2\lambda y,$$
$$0 = \frac{\partial L}{\partial z} = xy + 2\lambda z,$$
$$0 = \frac{\partial L}{\partial \lambda} = x^2 + y^2 + z^2 - 1.$$

If we multiply the first equation by y and the second by x, and equate the resulting expressions for $2\lambda xy$, we have

$$y^2 z = x^2 z.$$

Consequently, either $z = 0$ or $y = \pm x$.

Case I: $z = 0$. In this case the equations reduce to

$$\lambda x = 0, \quad \lambda y = 0, \quad xy = 0, \quad x^2 + y^2 = 1.$$

The first implies that either $x = 0$ or $\lambda = 0$. If $x = 0$, then $y = \pm 1$, and we have two critical points $(0, \pm 1, 0)$. If $\lambda = 0$, then the third equation requires $x = 0$ or $y = 0$. We therefore obtain two additional critical points $(\pm 1, 0, 0)$.

Case II: $y = x$. In this case the equations reduce to

$$xz + 2\lambda x = 0, \quad x^2 + 2\lambda z = 0, \quad 2x^2 + z^2 = 1.$$

The first implies that either $x = 0$ or $z = -2\lambda$. If $x = 0$, then $z = \pm 1$, and we have the two critical points $(0, 0, \pm 1)$. If $z = -2\lambda$, then the last two equations imply that $x = \pm 1/\sqrt{3}$, and we obtain the four critical points

$$(\pm 1/\sqrt{3}, \pm 1/\sqrt{3}, 1/\sqrt{3}) \quad \text{and} \quad (\pm 1/\sqrt{3}, \pm 1/\sqrt{3}, -1/\sqrt{3}).$$

Case III: $y = -x$. This case is similar to that for $y = x$, and leads to the additional four critical points

$$(\pm 1/\sqrt{3}, \mp 1/\sqrt{3}, 1/\sqrt{3}) \quad \text{and} \quad (\pm 1/\sqrt{3}, \mp 1/\sqrt{3}, -1/\sqrt{3}).$$

Because $x^2 + y^2 + z^2 = 1$ is a surface without a boundary, we complete the problem by evaluating $f(x, y, z)$ at each of the critical points:

$$f(\pm 1, 0, 0) = f(0, \pm 1, 0) = f(0, 0, \pm 1) = 0,$$
$$f(\pm 1/\sqrt{3}, \pm 1/\sqrt{3}, 1/\sqrt{3}) = f(\pm 1/\sqrt{3}, \mp 1/\sqrt{3}, -1/\sqrt{3}) = \sqrt{3}/9,$$
$$f(\pm 1/\sqrt{3}, \pm 1/\sqrt{3}, -1/\sqrt{3}) = f(\pm 1/\sqrt{3}, \mp 1/\sqrt{3}, 1/\sqrt{3}) = -\sqrt{3}/9.$$

The maximum and minimum values of $f(x, y, z)$ on $x^2 + y^2 + z^2 = 1$ are therefore $\sqrt{3}/9$ and $-\sqrt{3}/9$. ∎

To compare the Lagrangian solution in this example to that without a Lagrange multiplier, see Exercise 21 in Section 13.11.

EXAMPLE 13.35 When a thermonuclear reactor is built in the form of a right-circular cylinder, neutron diffusion theory requires its radius and height to satisfy the equation

$$\left(\frac{2.4048}{r}\right)^2 + \left(\frac{\pi}{h}\right)^2 = k,$$

where k is a constant. Find r and h in terms of k if the reactor is to occupy as small a volume as possible.

SOLUTION The volume of a right-circular cylinder is $V = \pi r^2 h$, and were there no constraints on r and h, this function would be considered for all points in the first quadrant of the rh-plane. However, r and h must satisfy a constraint that geometrically can be interpreted as a curve in the rh-plane. What we must do then is minimize $V = \pi r^2 h$, considering only those points (r, h) on the curve defined by the constraint. Clearly there is only one independent variable in the problem—either r or h, but not both. If we choose r as the independent variable, then we note from the constraint that as h becomes very large, r approaches $2.4048/\sqrt{k}$. Since there is no upper bound on r, we can state that the values of r to be considered in the minimization of V are $r > 2.4048/\sqrt{k}$.

To find critical points of V we introduce the Lagrangian

$$L(r, h, \lambda) = \pi r^2 h + \lambda \left\{ \left(\frac{2.4048}{r}\right)^2 + \left(\frac{\pi}{h}\right)^2 - k \right\},$$

and first solve the equations

$$0 = \frac{\partial L}{\partial r} = 2\pi r h + \lambda \left(\frac{-2(2.4048)^2}{r^3}\right),$$

$$0 = \frac{\partial L}{\partial h} = \pi r^2 + \lambda \left(\frac{-2\pi^2}{h^3}\right),$$

$$0 = \frac{\partial L}{\partial \lambda} = \left(\frac{2.4048}{r}\right)^2 + \left(\frac{\pi}{h}\right)^2 - k.$$

If we solve each of the first two equations for λ and equate the resulting expressions, we have

$$\frac{\pi r^4 h}{2.4048^2} = \frac{r^2 h^3}{2\pi}.$$

Since neither r nor h can be zero, we divide by $r^2 h$:

$$\frac{\pi r^2}{2.4048^2} = \frac{h^2}{2\pi},$$

from which

$$r = \frac{2.4048\, h}{\sqrt{2}\,\pi}.$$

Substitution of this result into the constraint equation gives

$$\left(\frac{\sqrt{2}\,\pi}{h}\right)^2 + \left(\frac{\pi}{h}\right)^2 = k,$$

and this equation can be solved for $h = \pi\sqrt{3/k}$. This gives

$$r = \frac{2.4048}{\sqrt{2}\,\pi} \frac{\pi\sqrt{3}}{\sqrt{k}} = 2.4048\sqrt{3/(2k)}.$$

We have obtained therefore only one critical point (r, h) at which the derivatives of L vanish. The only values of r and h at which the derivatives of L do not exist are $r = 0$ and $h = 0$, but these must be rejected since the constraint requires both r and h to be positive.

To finish the problem we note that

$$\lim_{r \to \infty} V = \infty, \quad \lim_{r \to 2.4048/\sqrt{k}^+} V = \lim_{h \to \infty} V = \infty.$$

It follows, therefore, that the single critical point at which $r = 2.4048\sqrt{3/(2k)}$ and $h = \pi\sqrt{3/k}$ must give the absolute minimum value of $V(r, h)$. ∎

EXERCISES 13.12

In Exercises 1–8 use Lagrange multipliers to find maximum and minimum values of the function subject to the constraints. In each case, interpret the constraints geometrically.

1. $f(x,y) = x^2 + y$ subject to $x^2 + y^2 = 4$
2. $f(x,y,z) = 5x - 2y + 3z + 4$ subject to $x^2 + 2y^2 + 4z^2 = 9$
3. $f(x,y) = x + y$ subject to $(x-1)^2 + y^2 = 1$
4. $f(x,y,z) = x^3 + y^3 + z^3$ subject to $x^2 + y^2 + z^2 = 9$
5. $f(x,y,z) = xyz$ subject to $x^2 + 2y^2 + 3z^2 = 12$
6. $f(x,y,z) = x^2y + z$ subject to $x^2 + y^2 = 1$, $z = y$
7. $f(x,y,z) = x^2 + y^2 + z^2$ subject to $x^2 + y^2 + z^2 = 2z$, $x + y + z = 1$
8. $f(x,y,z) = xyz - x^2z$ subject to $x^2 + y^2 = 1$, $z = \sqrt{x^2 + y^2}$

9. Use Lagrange multipliers to solve Exercise 11 in Section 13.11.
10. Use Lagrange multipliers to solve Exercise 12 in Section 13.11.
11. Use Lagrange multipliers to solve Exercise 20 in Section 13.11.
12. Use Lagrange multipliers to solve Exercise 22 in Section 13.11.
13. Use Lagrange multipliers to solve Exercise 23 in Section 13.11.
14. Use Lagrange multipliers to solve Exercise 24 in Section 13.11.
15. Use Lagrange multipliers to solve Exercise 17 in Section 13.11.
16. Suppose that $F(x,y) = 0$ and $G(x,y) = 0$ define two curves C_1 and C_2 in the xy-plane. Let $P(x_0, y_0)$ and $Q(X_0, Y_0)$ be the points on C_1 and C_2 that minimize the distance between C_1 and C_2. If C_1 and C_2 have tangent lines at P and Q, show that the line PQ is perpendicular to these tangent lines.
17. Find the points on the curve $x^2 + xy + y^2 = 1$ closest to and farthest from the origin.

In Exercises 18–20 use Lagrange multipliers to find maximum and minimum values of the function.

18. $f(x,y) = 3x^2 + 2xy - y^2 + 5$ for $4x^2 + 9y^2 \leq 36$
19. $f(x,y) = x^2y + xy^2 + y$ for $-1 \leq x \leq 1$, $-1 \leq y \leq 1$
20. $f(x,y,z) = xy + xz$ for $x^2 + y^2 + z^2 \leq 1$
21. Find the maximum value of $f(x,y,z) = x^2yz - xzy^2$ subject to the constraints $x^2 + y^2 = 1$, $z = \sqrt{x^2 + y^2}$.
22. The equation $3x^2 + 4xy + 6y^2 = 140$ describes an ellipse which has its centre at the origin, but major and minor axes are not along the x- and y-axes. Find coordinates of the ends of the major and minor axes.
23. Use Lagrange multipliers to find the point on the first octant part of the plane $Ax + By + Cz = D$, $(A, B, C, D$ all positive constants$)$ which maximizes the function $f(x,y,z) = x^p y^q z^r$ where p, q, and r are positive constants.
24. The folium of Descartes has parametric equations

$$x = \frac{3at}{1+t^3}, \quad y = \frac{3at^2}{1+t^3}, \quad (a > 0)$$

(see Exercise 51 in Section 3.7 and Exercise 49 in Section 10.1). Find the point in the first quadrant farthest from the origin in two ways:

(a) Express $D^2 = x^2 + y^2$ in terms of t and maximize this function of one variable.

(b) Show that an implicit equation for the curve is $x^3 + y^3 = 3axy$ and maximize $D^2 = x^2 + y^2$ subject to this constraint.

25. To find the point on the curve $x^2 - xy + y^2 - z^2 = 1$, $x^2 + y^2 = 1$ closest to the origin, we must minimize the function $f(x,y,z) = x^2 + y^2 + z^2$ subject to the constraints defined by the equations of the curve. Show that this can be done by

(a) using two Lagrange multipliers;

(b) expressing $f(x,y,z)$ in terms of x and y alone, $u = 1 - xy$, and minimizing this function subject to $x^2 + y^2 = 1$ (with one Lagrange multiplier);

(c) expressing $f(x,y,z)$ in terms of x alone, $u = 1 \pm x\sqrt{1-x^2}$, and minimizing these functions on appropriate intervals;

(d) writing $x = \cos t$, $y = \sin t$ along the curve, expressing $f(x,y,z)$ in terms of t, $u = 1 - \sin t \cos t$, and minimizing this function on appropriate intervals.

26. Find the smallest and largest distances from the origin to the curve $x^2 + y^2/4 + z^2/9 = 1$, $x + y + z = 0$.

27. Find the maximum value of $f(x,y,z) = (xy + x^2)/(z^2 + 1)$ subject to the constraint $x^2(4 - x^2) = y^2$.

SECTION 13.13

Differentials

If $y = f(x)$ is a function of one variable, the differential of y, defined by $dy = f'(x)\,dx$, was found to be an approximation to the increment $\Delta y = f(x + dx) - f(x)$ for small dx. In particular, dy is the change in y corresponding to the change dx in x if we follow the tangent line to the curve at (x, y) instead of the curve itself.

We take the same approach in defining differentials for multivariable functions. First consider a function $f(x, y)$ of two independent variables that can be represented geometrically as a surface with equation $z = f(x, y)$ (Figure 13.58). If we change the values of x and y by amounts $\Delta x = dx$ and $\Delta y = dy$, then the corresponding change in z is

$$\Delta z = f(x + dx, y + dy) - f(x, y).$$

Geometrically, this is the difference in the heights of the surface at the points $(x + dx, y + dy)$ and (x, y). If we draw the tangent plane to the surface at (x, y), then very near (x, y) the height of the tangent plane approximates the height of the surface (Figure 13.59). In particular, the height of the tangent plane at $(x + dx, y + dy)$ for small dx and dy approximates the height of the surface. We define the **differential** dz as the change in z corresponding to the changes dx and dy in x and y if we follow the tangent plane at (x, y) instead of the surface itself. To find dz in terms of dx and dy, we note that the vector joining the points (x, y, z) and $(x + dx, y + dy, z + dz)$ has components (dx, dy, dz), and this vector lies in the tangent plane. Since a normal vector to the tangent plane is

$$\nabla(z - f(x, y)) = (-f_x, -f_y, 1),$$

it follows that the vectors $(-f_x, -f_y, 1)$ and (dx, dy, dz) must be perpendicular. Consequently,

$$0 = (-f_x, -f_y, 1) \cdot (dx, dy, dz) = -\frac{\partial f}{\partial x}dx - \frac{\partial f}{\partial y}dy + dz,$$

and hence

$$dz = \frac{\partial f}{\partial x}dx + \frac{\partial f}{\partial y}dy. \tag{13.63}$$

Note that if y is held constant in the function $f(x, y)$, then $dy = 0$ and 13.63 for dz reduces to the definition of the differential of a function of one variable.

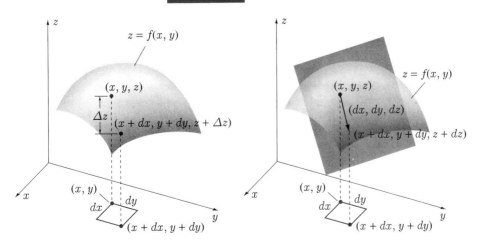

FIGURE 13.58

FIGURE 13.59

In Section 4.10 we indicated that we must be careful in using the differential dy as an approximation for the change Δy in a function $f(x)$. For the same reasons, we must be judicious in our use of dz as an approximation for Δz. Indeed, we have stated that dz is an approximation for Δz for small dx and dy, but the difficulty is deciding how small is small and how good is the approximation. In addition, note that if (x,y) is a critical point of the function $f(x,y)$, then either $dz = 0$ for all dx and dy, or dz is undefined. In other words, dz cannot be used to approximate Δz at a critical point.

EXAMPLE 13.36

If the radius of a right-circular cone is changed from 10 cm to 10.1 cm and the height is changed from 1 m to 0.99 m, use differentials to approximate the change in its volume.

SOLUTION The volume of a cone of radius r and height h is given by the formula $V = \pi r^2 h/3$. The differential of this function is

$$dV = \frac{\partial V}{\partial r} dr + \frac{\partial V}{\partial h} dh = \frac{2}{3}\pi rh\, dr + \frac{1}{3}\pi r^2\, dh.$$

If $r = 10$, $dr = 0.1$, $h = 100$, and $dh = -1$, then

$$dV = \frac{2}{3}\pi(10)(100)(0.1) + \frac{1}{3}\pi(10)^2(-1) = \frac{100\pi}{3} \text{ cm}^3.$$

Equation 13.63 suggests the following definition for the differential of a function of more than two independent variables.

Definition 13.11

If $u = f(x, y, z, t, \ldots, w)$, then the **differential** of $f(x, y, z, t, \ldots, w)$ is defined as

$$du = \frac{\partial f}{\partial x} dx + \frac{\partial f}{\partial y} dy + \frac{\partial f}{\partial z} dz + \frac{\partial f}{\partial t} dt + \cdots + \frac{\partial f}{\partial w} dw. \qquad (13.64)$$

EXAMPLE 13.37

The area of the triangle in Figure 13.60 is given by the formula $A = (\frac{1}{2})ab\sin\theta$. If when $\theta = \pi/3$, a and b are changed by $\frac{1}{3}\%$ and θ by $\frac{1}{2}\%$, use differentials to find the approximate percentage change in A.

FIGURE 13.60

SOLUTION

Since
$$dA = \frac{\partial A}{\partial a}da + \frac{\partial A}{\partial b}db + \frac{\partial A}{\partial \theta}d\theta$$
$$= \frac{1}{2}b\sin\theta\, da + \frac{1}{2}a\sin\theta\, db + \frac{1}{2}ab\cos\theta\, d\theta,$$

the approximate percentage change in A is

$$100\left(\frac{dA}{A}\right) = \frac{100}{A}\left\{\frac{1}{2}b\sin\theta\, da + \frac{1}{2}a\sin\theta\, db + \frac{1}{2}ab\cos\theta\, d\theta\right\}$$
$$= 100\left\{\frac{da}{a} + \frac{db}{b} + \cot\theta\, d\theta\right\}.$$

Since a and b are changed by $\frac{1}{3}\%$ and θ by $\frac{1}{2}\%$,

$$100\left(\frac{da}{a}\right) = 100\left(\frac{db}{b}\right) = \frac{1}{3} \quad \text{and} \quad 100\left(\frac{d\theta}{\theta}\right) = \frac{1}{2}.$$

Thus,
$$100\left(\frac{dA}{A}\right) = \frac{1}{3} + \frac{1}{3} + \frac{\theta}{2}\cot\theta = \frac{2}{3} + \frac{\theta}{2}\cot\theta,$$

and when $\theta = \pi/3$,

$$100\left(\frac{dA}{A}\right) = \frac{2}{3} + \frac{1}{2}\left(\frac{\pi}{3}\right)\left(\frac{1}{\sqrt{3}}\right) = 0.97.$$

The approximate percentage change in A is therefore 1%. ∎

EXERCISES 13.13

In Exercises 1–10 find the differential of the function.

1. $f(x, y) = x^2 y - \sin y$
2. $f(x, y) = \text{Tan}^{-1}(xy)$
3. $f(x, y, z) = xyz - x^3 e^z$
4. $f(x, y, z) = \sin(xyz) - x^2 y^2 z^2$
5. $f(x, y, z) = \ln(x^2 + y^2 + z^2)$
6. $f(x, y) = \text{Sin}^{-1}(xy)$
7. $f(x, y) = \text{Sin}^{-1}(x + y) + \text{Cos}^{-1}(x + y)$
8. $f(x, y, z, t) = xy + yz + zt + xt$
9. $f(x, y, z, w) = xy \tan(zw)$
10. $f(x, y, z, t) = e^{x^2 + y^2 + z^2 - t^2}$

11. A right-circular cone has radius 10 cm and height 20 cm. If its radius is increased by 0.1 cm and its height is decreased by 0.3 cm, use differentials to find the approximate change in its volume. Compare this with the actual change in volume.

12. When the ellipse $b^2 x^2 + a^2 y^2 = a^2 b^2$ is rotated about the x-axis, the volume V of the spheroid is $4\pi ab^2/3$. If a and b are each increased by 1%, use differentials to find the approximate percentage change in V.

13. When two resistors of resistances R_1 and R_2 are connected in parallel, their effective resistance is $R = R_1 R_2/(R_1 + R_2)$. Show that if R_1 and R_2 are both increased by a small percentage c, then the percentage increase of R is also c.

SECTION 13.14

Taylor Series for Multivariable Functions

Taylor series for functions of one variable can be used to generate Taylor series for multivariable functions. For simplicity, we once again work with functions of two independent variables. Extensions to functions of more than two independent variables will be clear. Suppose that a function $f(x, y)$ has continuous partial derivatives of all orders in some open circle centred at the point (c, d) (Figure 13.61).

FIGURE 13.61

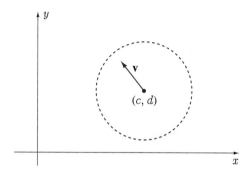

Parametric equations for the line through (c, d) in direction $\mathbf{v} = (v_x, v_y)$ are $x = c + v_x t$, $y = d + v_y t$. If we substitute these values into $f(x, y)$, we obtain a function $F(t)$ of one variable,

$$F(t) = f(c + v_x t, d + v_y t),$$

which represents the value of $f(x, y)$ at points along the line through (c, d) in direction $\mathbf{v}$. If we expand this function into its Maclaurin series we obtain

$$F(t) = F(0) + F'(0)t + \frac{F''(0)}{2!}t^2 + \cdots. \tag{13.65}$$

The schematic diagram to the right gives

$$F'(t) = \frac{\partial F}{\partial x}\frac{dx}{dt} + \frac{\partial F}{\partial y}\frac{dy}{dt} = f_x(x, y)v_x + f_y(x, y)v_y,$$

and therefore

$$F'(0) = f_x(c, d)v_x + f_y(c, d)v_y.$$

The second derivative of $F(t)$ is

$$\begin{aligned}F''(t) &= \frac{\partial}{\partial x}[F'(t)]\frac{dx}{dt} + \frac{\partial}{\partial y}[F'(t)]\frac{dy}{dt} \\ &= \frac{\partial}{\partial x}[f_x(x,y)v_x + f_y(x,y)v_y]v_x \\ &\quad + \frac{\partial}{\partial y}[f_x(x,y)v_x + f_y(x,y)v_y]v_y \\ &= f_{xx}(x,y)v_x^2 + 2f_{xy}(x,y)v_xv_y + f_{yy}(x,y)v_y^2,\end{aligned}$$

and therefore

$$F''(0) = f_{xx}(c,d)v_x^2 + 2f_{xy}(c,d)v_xv_y + f_{yy}(c,d)v_y^2.$$

A similar calculation gives

$$F'''(0) = f_{xxx}(c,d)v_x^3 + 3f_{xxy}(c,d)v_x^2 v_y + 3f_{xyy}(c,d)v_x v_y^2 + f_{yyy}(c,d)v_y^3,$$

and the pattern is emerging. When these results are substituted into 13.65,

$$F(t) = f(c + v_x t, d + v_y t)$$
$$= f(c,d) + [f_x(c,d)v_x + f_y(c,d)v_y]t$$
$$+ [f_{xx}(c,d)v_x^2 + 2f_x^y(c,d)v_x v_y + f_{yy}(c,d)v_y^2]\frac{t^2}{2!} + \cdots. \quad (13.66)$$

We now let **v** be the vector from (c,d) to point (x,y), so that $v_x = x - c$ and $v_y = y - d$, and at the same time set $t = 1$. Then $F(1) = f(c + x - c, d + y - d) = f(x,y)$, and 13.66 becomes

$$f(x,y) = f(c,d) + [f_x(c,d)(x-c) + f_y(c,d)(y-d)] + \frac{1}{2!}[f_{xx}(c,d)(x-c)^2$$
$$+ 2f_{xy}(c,d)(x-c)(y-d) + f_{yy}(c,d)(y-d)^2] + \cdots. \quad (13.67)$$

This is the Taylor series for $f(x,y)$ about the point (c,d). It gives the value of the function at the point (x,y) in terms of values of the function and its derivatives at the point (c,d).

EXAMPLE 13.38

Find the first six nonzero terms in the Taylor series for $f(x,y) = \sin(2x + 3y)$ about $(0,0)$.

SOLUTION We calculate that:

$$f(0,0) = 0$$
$$f_x(0,0) = 2\cos(2x+3y)|_{(0,0)} = 2$$
$$f_y(0,0) = 3\cos(2x+3y)|_{(0,0)} = 3$$
$$f_{xx}(0,0) = -4\sin(2x+3y)|_{(0,0)} = 0$$
$$f_{xy}(0,0) = -6\sin(2x+3y)|_{(0,0)} = 0$$
$$f_{yy}(0,0) = -9\sin(2x+3y)|_{(0,0)} = 0$$
$$f_{xxx}(0,0) = -8\cos(2x+3y)|_{(0,0)} = -8$$
$$f_{xxy}(0,0) = -12\cos(2x+3y)|_{(0,0)} = -12$$
$$f_{xyy}(0,0) = -18\cos(2x+3y)|_{(0,0)} = -18$$
$$f_{yyy}(0,0) = -27\cos(2x+3y)|_{(0,0)} = -27$$

Formula 13.67 then gives

$$\sin(2x+3y) = 0 + [2x+3y] + \frac{1}{2!}[0] + \frac{1}{3!}[-8x^3 - 36x^2y - 54xy^2 - 27y^3] + \ldots$$
$$= (2x+3y) - \frac{1}{3!}(2x+3y)^3 + \cdots.$$

This series could also have been obtained by substituting $2x + 3y$ for x in the Maclaurin series for $\sin x$. This is not always an alternative. ∎

EXERCISES 13.14

1. If $f(x,y) = F(x)G(y)$, is the Taylor series of $f(x,y)$ about $(0,0)$ the product of the Maclaurin series for $F(x)$ and $G(y)$?

2. What are the cubic terms in 13.67?

In Exercises 3–8 find the Taylor series of the function about the point by using Taylor series for functions of one variable.

3. $\cos(xy)$ about $(0,0)$

4. e^{2x-3y} about $(1,-1)$

5. $x^2 y\sqrt{1+x}$ about $(0,0)$

6. $\ln(1+x^2+y^2)$ about $(0,0)$

7. $\dfrac{1}{1+x+y}$ about $(3,-4)$

8. $\dfrac{xy^2}{1+y^2}$ about $(-1,0)$

In Exercises 9–14 find the Taylor series of the function up to and including quadratic terms.

9. $\dfrac{xy}{x^2+y^2}$ about $(-1,1)$

10. $\sqrt{1+xy}$ about $(2,1)$

11. $e^x \sin(3x-y)$ about $(-1,0)$

12. $(x+y)^2 \ln(x+y)$ about $(0,1)$

13. $\operatorname{Tan}^{-1}(3x+2y)$ about $(1,-1)$

14. $x^8 y^{10}$ about $(0,0)$

15. What are the terms in the Taylor series for a function $f(x,y,z)$ about the point (c,d,e) corresponding to those in equation 13.67?

16. Express 13.67 in sigma notation. Hint: Think about an operator
$$\left[(x-a)\dfrac{\partial}{\partial x} + (y-b)\dfrac{\partial}{\partial y}\right]^n$$
which is expanded as a binomial to operate on functions $f(x,y)$.

SUMMARY

We began the study of multivariable functions in this chapter, concentrating our attention on differentiation and its applications. We introduced two types of derivatives for a multivariable function: partial derivatives and directional derivatives. Partial derivatives are directional derivatives in directions parallel to the coordinate axes. The directional derivative of a function $f(x,y,z)$ in the direction $\mathbf{v}$ is given by the formula $D_\mathbf{v} f = \nabla f \cdot \hat{\mathbf{v}}$, where $\hat{\mathbf{v}}$ is the unit vector in the direction of $\mathbf{v}$, and the gradient ∇f is evaluated at the point at which $D_\mathbf{v} f$ is required. This formula leads to the fact that the gradient $\nabla f(x,y,z)$ points in the direction in which $f(x,y,z)$ increases most rapidly, and $|\nabla f|$ is the (maximum) rate of change of $f(x,y,z)$ in the direction ∇f. A second property of gradient vectors (which is related to the first) is that if $F(x,y,z) = 0$ is the equation of a surface, then at any point on the surface ∇F is perpendicular to the surface. This property, along with the fact that perpendicularity to a surface is synonymous with perpendicularity to its tangent plane, enables us to find equations for tangent planes to surfaces and tangent lines to curves.

We illustrated various ways to calculate partial derivatives of a multivariable function, depending on whether the function is defined explicitly, implicitly, or as a composite function. Since partial derivatives are ordinary derivatives with other variables held constant, there is no difficulty calculating partial derivatives when the function is defined explicitly; we simply use the rules from single-variable calculus. When the partial derivative of a composite function is required, we use a schematic diagram illustrating functional dependences to develop the appropriate chain rule. Partial derivatives for functions defined implicitly are calculated using Jacobians.

Critical points of a multivariable function are points at which all of its first partial derivatives vanish or at which one or more of these partial derivatives does not exist. Critical points can yield relative maxima, relative minima, saddle points, or none of these. For functions of two independent variables, a second-derivative test exists that may determine whether a critical point at which the partial derivatives vanish yields a relative maximum, a relative minimum, or a saddle point. This test is analogous to that for functions of one variable.

A continuous function of two independent variables defined on a region that includes its boundary always takes on a maximum value and a minimum value. To find these values we evaluate the function at each of its critical points and compare these numbers to the maximum and minimum values of the function on its boundary. Finding the extreme values on the boundary involves one or more extrema problems for a function of one variable, the number of such problems depending on the complexity of the boundary.

We suggest two methods for finding the extreme values of a function when the variables of the function are subject to constraints. We either solve the constraint equations for dependent variables and express the given function in terms of independent variables or we use Lagrange multipliers. Lagrange multipliers eliminate the necessity for solving the constraint equations, but they do, on the other hand, give a larger system of equations to solve for critical points.

Differentials of multivariable functions can be used to approximate changes in functions when small changes are made to its independent variables. Taylor series can also be used to approximate multivariable functions.

Key Terms and Formulas

In reviewing this chapter, you should be able to define or discuss the following key terms:

Domain
Limit
Continuous function
Partial derivative
Gradient
Harmonic function
Chain rule
Homogeneous function
Laplace's equation
Implicit differentiation
Jacobian determinant
Directional derivative

Tangent line to a curve
Tangent plane to a surface
Normal to a surface
Smooth surface
Piecewise smooth surface
Critical point
Relative maximum and minimum
Saddle point
Absolute maximum and minimum
Lagrange multiplier
Differential
Taylor series

REVIEW EXERCISES

In Exercises 1–20 find the derivative.

1. $\partial f/\partial x$ if $f(x,y) = x^2/y^3 - \operatorname{Sin}^{-1}(xy)$

2. $\partial^2 f/\partial y^2$ if $f(x,y,z) = \ln(x^2 + y^2 + z^2)$

3. $\partial^3 f/\partial x^2 \partial y$ if $f(x,y,z,t) = x^3 e^y - xzt^2 - \sin(x + y + z + t)$

4. $\partial z/\partial x$ if $z^2 x + \operatorname{Tan}^{-1} z + y = 3x$

5. $\partial u/\partial y$ if $u \cos y + y \cos(xu) + z^2 = 5x$

6. df/dt if $f(x,y) = x^2 + y^2 - e^{xy}$, $x = t^3 + 3t$, $y = t \ln t$

7. dy/dx if $x = y^3 + 3y^2 - 2y + 4$

8. $\partial u/\partial x)_y$ if $u^2 + v^2 - xy = 5$, $3u - 2v + x^2 u = 2v^3$

9. $\partial^2 f/\partial u \partial v$ if $f(u,v) = u^2/\sqrt{v} - v/\sqrt{u}$

10. df/dt if $f(x,y) = xy - x^2 - y^2$, $x = te^t$, $y = te^{-t}$

11. $\partial z/\partial t)_u$ if $z = x^2 - y^2$, $x = 2u - 3v^2 + 3uvt$, $y = u\cos(vt)$, $v = t^2 - 2t$

12. $\partial r/\partial x)_y$ if $x = r\cos\theta$, $y = r\sin\theta$

13. $\partial \theta/\partial x)_{y,z}$ if $x = r\sin\phi\cos\theta$, $y = r\sin\phi\sin\theta$, $z = r\cos\phi$

14. $\partial u/\partial r)_\theta$ if $u = x^2 - y^2 x^3$, $x = r\cos\theta$, $y = r\sin\theta$

15. $\partial^2 u/\partial r^2)_\theta$ if $u = x^2 - y^2 x^3$, $x = r\cos\theta$, $y = r\sin\theta$

16. $d^2 u/dt^2$ if $u = x/z^2 - z/x^2$, $x = t^3 - 3$, $z = 1/t^3$

17. dz/dt if $z = y - xy^2 + x$, and $x^2 - y^2 + xt = 2t$, $xy = 4t^2$

18. $\partial^2 z/\partial x^2$ if $xz - x^2 z^3 + y^2 = 3$

19. dy/dx if $yx - x^2 z^2 + 5x = 3$, $2xz - 3x^2 y^2 = 4z^4$

20. $\partial u/\partial t)_v$ if $u = xyt^2 - 3\operatorname{Sin}^{-1}(xy)$, $x = v^2 t^2 - 2t$, $y = v \tan t$

21. If $u = (x^2 + y^2)[1 + \sin(x/z)]$, show that
$$x\frac{\partial u}{\partial x} + y\frac{\partial u}{\partial y} + z\frac{\partial u}{\partial z} = 2u.$$

22. If $u = 2x^2 - 3y^2 + xy$, show that
$$x^2 \frac{\partial^2 u}{\partial x^2} + 2xy \frac{\partial^2 u}{\partial x \partial y} + y^2 \frac{\partial^2 u}{\partial y^2} = 2u.$$

23. If $f(s)$ is a differentiable function, show that $f(3x - 2y)$ satisfies $2\dfrac{\partial f}{\partial x} + 3\dfrac{\partial f}{\partial y} = 0$.

24. If $f(s,t)$ has continuous first partial derivatives, show that the function $f(x^2 - y^2, y^2 - x^2)$ satisfies
$$y\frac{\partial f}{\partial x} + x\frac{\partial f}{\partial y} = 0.$$

In Exercises 25–30 find the directional derivative.

25. $f(x,y) = x^2 \sin y$ at $(3,-1)$ in the direction $\mathbf{v} = (2,4)$

26. $f(x,y,z) = x^2 + y^2 + z^2$ at $(1,0,1)$ in the direction from $(1,0,1)$ to $(2,-1,3)$

27. $f(x,y,z) = z\operatorname{Tan}^{-1}(x+y)$ at $(-1,2,5)$ in the direction perpendicular to the surface $z = x^2 + y^2$ with positive z-component

28. $f(x,y,z) = x^2 + y - 2z$ at $(1,-1,2)$ along the line $x - y + z = 4$, $2x + 4y + 2 = 0$ in the direction of increasing x

29. $f(x,y) = \ln(x+y)$ at $(3,10)$ along the curve $y = x^2 + 1$ in the direction of decreasing y

30. $f(x,y,z) = 2xyz - x^2 - z^2$ at $(0,1,1)$ along the curve $x^2 + y^2 + z^2 = 2$, $y = z$ in the direction of increasing x

In Exercises 31–33 find the equation of the tangent plane to the surface.

31. $z = x^2 + y^2$ at $(1,3,10)$

32. $x^2 + z^3 = y^2$ at $(-1,3,2)$

33. $x^2 + y^2 = z^2 + 1$ at $(1,0,0)$

In Exercises 34–36 find equations for the tangent line to the curve.

34. $x = t^2 + 1$, $y = t^2 - 1$, $z = t^3 + 5t$ at $(2,0,6)$

35. $x + y + z = 0$, $2x - 3y - 6z = 11$ at $(1,1,-2)$

36. $z = xy$, $x^2 + y^2 = 2$ at $(1,1,1)$

In Exercises 37–40 find all critical points for the function and classify each as yielding a relative maximum, a relative minimum, or a saddle point.

37. $f(x,y) = x^3 + 3y^2 - 6x + 4$

38. $f(x,y) = ye^x$

39. $f(x,y) = x^2 - xy + y^2 + x - 4y$

40. $f(x,y) = (x^2 + y^2 - 1)^2$

41. If $f(x,y) = (x^2 + y^2) F(x,y)$ where $F(x,y) = x^3/y - y^3/x$, verify that
$$\frac{\partial^2 f}{\partial x^2} + \frac{\partial^2 f}{\partial y^2} = (x^2 + y^2)\left(\frac{\partial^2 F}{\partial x^2} + \frac{\partial^2 F}{\partial y^2}\right) + 12 F(x,y).$$

42. Find maximum and minimum values of the function $f(x,y) = xy$ on the circle $x^2 + y^2 \leq 1$.

43. Find maximum and minimum values of the function $f(x,y,z) = 2x + 3y - 4z$ on the sphere $x^2 + y^2 + z^2 \leq 2$.

44. Find the points on the curve $x^2 + x^4 + y^2 = 1$ closest to and farthest from the origin.

45. Find the point(s) on the surface $z^2 = 1 + xy$ closest to the origin.

46. Generalize Review Exercise 32 in Chapter 4 to incorporate a third crop, say sunflowers, with a yield of $r per hectare and a proportional loss cz.

47. If the equation $u = f(x - ut)$ defines u implicitly as a function of x and t, show that $\dfrac{\partial u}{\partial t} + u\dfrac{\partial u}{\partial x} = 0$.

48. Find the first six nonzero terms in the Taylor series for $x^3 \sin(x^2 y)$ about the point $(1, \pi/4)$.

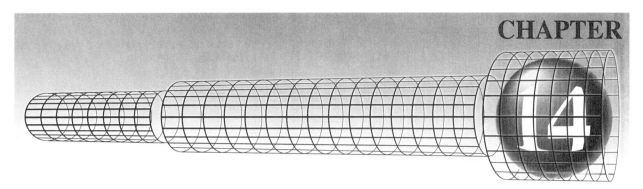

CHAPTER 14

Multiple Integrals

The definite integral of a function $f(x)$ of one variable is defined as the limit of a sum of the form

$$f(x_1^*)\Delta x_1 + f(x_2^*)\Delta x_2 + \cdots + f(x_n^*)\Delta x_n \quad (14.1)$$

where the norm of the partition approaches zero. We have seen that definite integrals can be used to calculate area, volume, work, fluid force, and moments. In spite of the fact that some of these are two- and three-dimensional concepts, we have been careful to emphasize that a definite integral with respect to x is an integration along the x-axis, and a definite integral with respect to y is an integration along the y-axis. In other words, independent of how we interpret the result of the integration, a definite integral is a "limit-summation" *along a line*.

Generalizations of these limiting sums to functions of two and three independent variables lead to definitions of double and triple integrals.

SECTION 14.1

Double Integrals and Double Iterated Integrals

Suppose a function $f(x, y)$ is defined in some region R of the xy-plane that has finite area (Figure 14.1). To define the double integral of $f(x, y)$ over R, we first divide R into n subregions of areas $\Delta A_1, \Delta A_2, \ldots, \Delta A_n$, in any manner whatsoever. In each subregion $\Delta A_i (i = 1, \ldots, n)$ we choose an arbitrary point (x_i^*, y_i^*) and form the sum

$$f(x_1^*, y_1^*)\Delta A_1 + f(x_2^*, y_2^*)\Delta A_2 + \cdots + f(x_n^*, y_n^*)\Delta A_n = \sum_{i=1}^{n} f(x_i^*, y_i^*)\Delta A_i. \quad (14.2)$$

The norm of the partition of R into sub–areas ΔA_i is the area of the largest of the sub–areas, denoted by $||\Delta A_i|| = \max_{i=1,\ldots,n} \Delta A_i$.

FIGURE 14.1

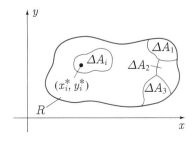

Suppose we increase the number of terms in 14.2 by increasing the number of sub-areas ΔA_i and decreasing the norm $||\Delta A_i||$. If the sum approaches a limit as the number of sub-areas becomes increasingly large and each sub-area shrinks to a point, we call the limit the **double integral** of $f(x,y)$ over the region R, and denote it by

$$\iint_R f(x,y)\,dA = \lim_{||\Delta A_i|| \to 0} \sum_{i=1}^n f(x_i^*, y_i^*)\Delta A_i. \qquad (14.3)$$

The notation $||\Delta A_i|| \to 0$ does not necessarily require that every ΔA_i shrink to a point. We implicitly assume, however, that this is always the case.

If limit 14.3 were dependent on the choice of subdivision ΔA_i or choice of star-points (x_i^*, y_i^*), double integrals would be of little use. We therefore demand that the limit of the sum be independent of the manner of subdivision of R and choice of star-points in the subregions. At first sight this requirement might seem rather severe since we must now check that all subdivisions and all choices of star-points lead to the same limit before concluding that the double integral exists. Fortunately, however, the following theorem indicates that for continuous functions this is unnecessary.

Theorem 14.1

Let C be a closed, piecewise-smooth curve that encloses a region R with finite area. If $f(x,y)$ is a continuous function inside and on C, then the double integral of $f(x,y)$ over R exists.

For a continuous function, then, the double integral exists and any choice of subdivision and star-points leads to the same value through limiting process 14.3. Note that continuity was also the condition that guaranteed existence of the definite integral in Theorem 6.2.

We cannot overemphasize the fact that a double integral is simply the limit of a sum. Moreover, any limit of form 14.3 may be interpreted as the double integral of a function $f(x,y)$ over a region defined by the ΔA_i.

The following properties of double integrals are easily proved using Definition 14.3:

(1) If the double integral of $f(x,y)$ over R exists and c is a constant, then

$$\iint_R cf(x,y)\,dA = c\iint_R f(x,y)\,dA. \qquad (14.4)$$

(2) If double integrals of $f(x,y)$ and $g(x,y)$ over R exist, then

$$\iint_R [f(x,y) + g(x,y)]\,dA = \iint_R f(x,y)\,dA + \iint_R g(x,y)\,dA. \qquad (14.5)$$

(3) If a region R is subdivided by a piecewise-smooth curve into two parts R_1 and R_2 that have at most boundary points in common (Figure 14.2), and the double integral of $f(x,y)$ over R exists, then

$$\iint_R f(x,y)\,dA = \iint_{R_1} f(x,y)\,dA + \iint_{R_2} f(x,y)\,dA. \qquad (14.6)$$

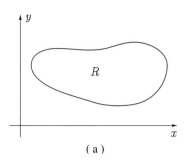
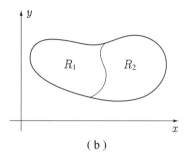

(a)   (b)

(4) The area of a region R is

$$\text{Area of } R = \iint_R dA. \tag{14.7}$$

In spite of the fact that double integrals are defined as "limits of sums," we do not evaluate them as such. Just as definite integrals are evaluated with indefinite integrals, we evaluate double integrals with double iterated integrals.

Double Iterated Integrals

We have already seen that a function $f(x, y)$ of two independent variables has two first-order partial derivatives, one with respect to x holding y constant, and one with respect to y holding x constant. We now reverse this process and define "partial" integration of $f(x, y)$ with respect to x and y. Quite naturally, we define *a* partial indefinite integral of $f(x, y)$ with respect to x as *an* antiderivative of $f(x, y)$ with respect to x holding y constant. For example, since

$$\frac{\partial}{\partial x}(x^3 + x^2 y) = 3x^2 + 2xy,$$

$x^3 + x^2 y$ is an antiderivative with respect to x of $3x^2 + 2xy$. But so too is $x^3 + x^2 y + y$. In fact for any differentiable function $C(y)$ of y, $x^3 + x^2 y + C(y)$ is an antiderivative of $3x^2 + 2xy$ with respect to x. Since this expression represents all antiderivatives of $3x^2 + 2xy$, we call it *the* indefinite integral of $3x^2 + 2xy$ with respect to x, and write

$$\int (3x^2 + 2xy)\,dx = x^3 + x^2 y + C(y).$$

Similarly, the indefinite integral of $3x^2 + 2xy$ with respect to y is

$$\int (3x^2 + 2xy)\,dy = 3x^2 y + xy^2 + D(x),$$

where $D(x)$ is an arbitrary differentiable function of x.

In this chapter we are concerned only with partial definite integrals. Limits on a partial definite integral with respect to x must not depend on x, but may depend on y. In general, then, a partial definite integral with respect to x is of the form

$$\int_{g(y)}^{h(y)} f(x, y)\,dx; \tag{14.8}$$

similarly, a partial definite integral with respect to y is of the form

$$\int_{g(x)}^{h(x)} f(x, y)\,dy. \tag{14.9}$$

Each of these partial definite integrals is evaluated by substituting the limits into a corresponding partial indefinite integral. For example,

$$\int_{x^2}^{x+2} (2y + xe^y)\,dy = \{y^2 + xe^y\}_{x^2}^{x+2}$$
$$= \{(x+2)^2 + xe^{x+2}\} - \{(x^2)^2 + xe^{x^2}\}$$
$$= (x+2)^2 - x^4 + x(e^{x+2} - e^{x^2}).$$

Once antidifferentiation in 14.8 is completed and the limits substituted, the result is a function of y alone. It is then possible to integrate this function with respect to y between any two limits, say from $y = c$ to $y = d$:

$$\int_c^d \left\{ \int_{g(y)}^{h(y)} f(x,y)\,dx \right\} dy.$$

In practice we omit the braces and simply write

$$\int_c^d \int_{g(y)}^{h(y)} f(x,y)\,dx\,dy, \qquad (14.10)$$

understanding that in the evaluation we proceed from the inner integral to the outer. This is called a **double iterated integral** first with respect to x and then with respect to y (or, more concisely, with respect to x and y). Double iterated integrals with respect to y and x take the form

$$\int_a^b \int_{g(x)}^{h(x)} f(x,y)\,dy\,dx. \qquad (14.11)$$

EXAMPLE 14.1

Evaluate each of the following double iterated integrals:

(a) $\int_0^1 \int_x^{x-1} (x^2 + e^y)\,dy\,dx$

(b) $\int_{-1}^1 \int_0^y xye^{x^2}\,dx\,dy$

SOLUTION

(a) $\int_0^1 \int_x^{x-1} (x^2 + e^y)\,dy\,dx = \int_0^1 \{x^2 y + e^y\}_x^{x-1}\,dx$

$$= \int_0^1 \{x^2(x-1) + e^{x-1} - x^3 - e^x\}\,dx$$

$$= \int_0^1 \{-x^2 + e^{x-1} - e^x\}\,dx$$

$$= \left\{ -\frac{x^3}{3} + e^{x-1} - e^x \right\}_0^1$$

$$= \left\{ -\frac{1}{3} + 1 - e \right\} - \{e^{-1} - 1\}$$

$$= \frac{5}{3} - e - e^{-1}.$$

(b) $\displaystyle\int_{-1}^{1}\int_{0}^{y} xye^{x^2}\,dx\,dy = \int_{-1}^{1}\left\{\frac{1}{2}ye^{x^2}\right\}_{0}^{y} dy = \frac{1}{2}\int_{-1}^{1}(ye^{y^2}-y)\,dy$

$\displaystyle = \frac{1}{2}\left\{\frac{1}{2}e^{y^2}-\frac{y^2}{2}\right\}_{-1}^{1} = \frac{1}{2}\left\{\frac{e}{2}-\frac{1}{2}-\frac{e}{2}+\frac{1}{2}\right\} = 0.$ ∎

EXERCISES 14.1

In Exercises 1–30 evaluate the double iterated integral.

1. $\displaystyle\int_{-1}^{2}\int_{y}^{y+2}(x^2-xy)\,dx\,dy$

2. $\displaystyle\int_{-3}^{3}\int_{-\sqrt{18-2y^2}}^{\sqrt{18-2y^2}} x\,dx\,dy$

3. $\displaystyle\int_{0}^{1}\int_{x^2}^{x}(2xy+3y^2)\,dy\,dx$

4. $\displaystyle\int_{-1}^{0}\int_{y}^{2}(1+y)^2\,dx\,dy$

5. $\displaystyle\int_{3}^{4}\int_{0}^{\pi/2} x\sin y\,dy\,dx$

6. $\displaystyle\int_{1}^{2}\int_{1}^{y} e^{x+y}\,dx\,dy$

7. $\displaystyle\int_{-1}^{1}\int_{-x}^{5}(x^2+y^2)\,dy\,dx$

8. $\displaystyle\int_{-1}^{1}\int_{x}^{2x}(xy+x^3y^3)\,dy\,dx$

9. $\displaystyle\int_{0}^{1}\int_{x}^{1}(x+y)^4\,dy\,dx$

10. $\displaystyle\int_{1}^{2}\int_{x}^{2x}\frac{1}{(x+y)^3}\,dy\,dx$

11. $\displaystyle\int_{0}^{1}\int_{0}^{3x}\sqrt{x+y}\,dy\,dx$

12. $\displaystyle\int_{-1}^{1}\int_{1}^{e}\frac{y}{x}\,dx\,dy$

13. $\displaystyle\int_{1}^{4}\int_{\sqrt{x}}^{x^2}(x^2+2xy-3y^2)\,dy\,dx$

14. $\displaystyle\int_{0}^{2}\int_{x^2}^{2x^2} x\cos y\,dy\,dx$

15. $\displaystyle\int_{0}^{1}\int_{1}^{\tan x}\frac{1}{1+y^2}\,dy\,dx$

16. $\displaystyle\int_{0}^{1}\int_{0}^{y^3}\frac{1}{1+y^2}\,dx\,dy$

17. $\displaystyle\int_{2}^{3}\int_{0}^{1}\frac{x}{\sqrt{1-y^2}}\,dy\,dx$

18. $\displaystyle\int_{0}^{2}\int_{-x}^{x}(8-2x^2)^{3/2}\,dy\,dx$

19. $\displaystyle\int_{0}^{1}\int_{0}^{x}\frac{1}{\sqrt{1-y^2}}\,dy\,dx$

20. $\displaystyle\int_{-9}^{0}\int_{0}^{x^2\sqrt{9+x}}\,dy\,dx$

21. $\displaystyle\int_{0}^{2}\int_{\sqrt{4-x^2}}^{2} y^2\,dy\,dx$

22. $\displaystyle\int_{-1}^{0}\int_{y}^{0} x\sqrt{x^2+y^2}\,dx\,dy$

23. $\displaystyle\int_{2}^{3}\int_{1}^{2x}\frac{1}{(xy+x^2)^2}\,dy\,dx$

24. $\displaystyle\int_{0}^{1}\int_{0}^{\mathrm{Cos}^{-1}x}(x\cos y)\,dy\,dx$

25. $\displaystyle\int_{0}^{1}\int_{\sqrt{y^2+y}}^{\sqrt{2y}} x^3\sqrt{x^2-y^2}\,dx\,dy$

26. $\displaystyle\int_{0}^{1}\int_{\sqrt{2y}}^{\sqrt{y^2+y}} x^3\sqrt{x^2-y^2}\,dx\,dy$

27. $\displaystyle\int_{-2}^{0}\int_{x^4}^{4x^2} \sqrt{y - x^4}\, dy\, dx$

28. $\displaystyle\int_{-2}^{0}\int_{y}^{0} \frac{x}{\sqrt{x^2 + y^2}}\, dx\, dy$

29. $\displaystyle\int_{-1}^{2}\int_{-1}^{y^3} \sqrt{1 + y}\, dx\, dy$

30. $\displaystyle\int_{0}^{1}\int_{0}^{x} \sqrt{x^2 + y^2}\, dy\, dx$

SECTION 14.2

Evaluation Of Double Integrals By Double Iterated Integrals

According to Theorem 14.1, if a function $f(x, y)$ is continuous on a finite region R with a piecewise-smooth boundary, then double integral 14.3 exists, and its evaluation by means of that limit is independent of both the manner of subdivision of R into areas ΔA_i and choice of star-points (x_i^*, y_i^*). We now show that if we make particular choices of ΔA_i, double integrals can be evaluated by means of double iterated integrals in x and y.

Consider first a rectangle R with edges parallel to the x- and y-axes as shown in Figure 14.3. We divide R into smaller rectangles by a network of $n + 1$ vertical lines and $m + 1$ horizontal lines identified by abscissae

$$a = x_0 < x_1 < x_2 < \cdots < x_{n-1} < x_n = b$$

and ordinates

$$c = y_0 < y_1 < y_2 < \cdots < y_{m-1} < y_m = d.$$

FIGURE 14.3

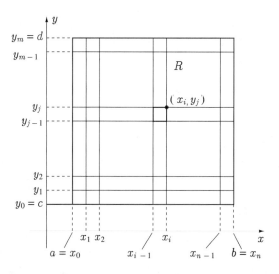

If the (i, j)th rectangle is that rectangle bounded by the lines $x = x_{i-1}$, $x = x_i$, $y = y_{j-1}$, and $y = y_j$, then its area is $\Delta x_i \Delta y_j$, where $\Delta x_i = x_i - x_{i-1}$ and $\Delta y_j = y_j - y_{j-1}$. We choose as star-point in the (i, j)th rectangle the upper right corner: $(x_i^*, y_j^*) = (x_i, y_j)$. With this rectangular subdivision of R and choice of star-points, Definition 14.3 for the double integral of $f(x, y)$ over R takes the form

$$\iint_R f(x, y)\, dA = \lim_{\|\Delta x_i \Delta y_j\| \to 0} \sum_{i=1}^{n} \sum_{j=1}^{m} f(x_i, y_j) \Delta x_i \Delta y_j. \qquad (14.12\text{a})$$

Since $\|\Delta x_i \Delta y_j\| \to 0$ if the norms $\|\Delta x_i\|$ and $\|\Delta y_j\|$ individually approach zero, we can write that

$$\iint_R f(x,y)\,dA = \lim_{\substack{\|\Delta x_i\| \to 0 \\ \|\Delta y_j\| \to 0}} \sum_{i=1}^{n} \sum_{j=1}^{m} f(x_i, y_j)\,\Delta x_i \Delta y_j. \qquad (14.12\text{b})$$

Suppose we choose to first perform the limit on y and then the limit on x, and write therefore

$$\iint_R f(x,y)\,dA = \lim_{\|\Delta x_i\| \to 0} \sum_{i=1}^{n} \left\{ \lim_{\|\Delta y_j\| \to 0} \sum_{j=1}^{m} f(x_i, y_j)\,\Delta y_j \right\} \Delta x_i.$$

Since x_i is constant in the limit with respect to y, the y-limit is precisely the definition of the definite integral of $f(x_i, y)$ with respect to y from $y = c$ to $y = d$; i.e.,

$$\lim_{\|\Delta y_j\| \to 0} \sum_{j=1}^{m} f(x_i, y_j)\,\Delta y_j = \int_c^d f(x_i, y)\,dy.$$

Consequently,

$$\iint_R f(x,y)\,dA = \lim_{\|\Delta x_i\| \to 0} \sum_{i=1}^{n} \left\{ \int_c^d f(x_i, y)\,dy \right\} \Delta x_i.$$

Because the term in braces is a function of x_i alone, we can interpret this limit as a definite integral with respect to x:

$$\iint_R f(x,y)\,dA = \int_a^b \left\{ \int_c^d f(x,y)\,dy \right\} dx = \int_a^b \int_c^d f(x,y)\,dy\,dx, \qquad (14.13)$$

a double iterated integral. By reversing the order of taking limits, we can show similarly that the double integral can be evaluated with a double iterated integral with respect to x and y:

$$\iint_R f(x,y)\,dA = \int_c^d \int_a^b f(x,y)\,dx\,dy. \qquad (14.14)$$

We have shown, then, that for the special case of a rectangle R with sides parallel to the axes, a double integral over R can be evaluated by using double iterated integrals. Conversely, every double iterated integral with constant limits represents a double integral over a rectangle. The double iterated integral simply indicates that a rectangular subdivision has been chosen to evaluate the double integral.

We have just stated that the choice of a double iterated integral to evaluate a double integral implies that the region of integration has been subdivided into small rectangles. We now show that the x- and y-integrations themselves can be interpreted geometrically. These interpretations will simplify the transition to more difficult regions of integration.

In the subdivision of R into rectangles, suppose we denote the dimensions of a representative rectangle at position (x, y) by dx and dy (Figure 14.4). In the inner integral

$$\int_c^d f(x,y)\,dy\,dx$$

of equation 14.13, x is held constant and integration is performed in the y-direction. This (partial) definite integral is therefore interpreted as summing over the rectangles in the vertical strip of width dx at position x. The limits $y = c$ and $y = d$ identify the initial and terminal positions of this vertical strip. It is important to note that we are not adding the areas of the rectangles of dimensions dx and dy in the strip. On the contrary, each rectangle of area $dy\,dx$ is multiplied by the value of $f(x,y)$ for that rectangle,

$$f(x,y)\,dy\,dx,$$

and it is these quantities that are added.

The x-integration in equation 14.13 is interpreted as adding over all strips starting at $x = a$ and ending at $x = b$. The limits on x, therefore, identify positions of the first and last strips.

FIGURE 14.4

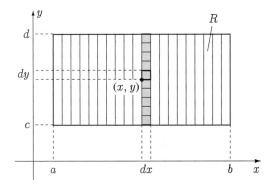

Although our diagram illustrates finite rectangles of dimensions dx and dy and finite strips of width dx, we must keep in mind that the integrations take limits as these dimensions approach zero.

Analogously, the double iterated integral in 14.14 is interpreted as adding over horizontal strips, as shown in Figure 14.5. Inner limits indicate where each strip starts and stops, and outer limits indicate the positions of first and last strips.

FIGURE 14.5

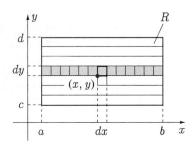

The transition now to general regions is quite straightforward. For the double integral of $f(x,y)$ over the region in Figure 14.6, we use a double iterated integral with respect to y and x. The y-integration adds the quantities $f(x,y)\,dy\,dx$ over rectangles in a vertical strip. We write

$$\int_{g(x)}^{h(x)} f(x,y)\,dy\,dx,$$

where $g(x)$ and $h(x)$ indicate that each vertical strip starts on the curve $y = g(x)$ and ends on the curve $y = h(x)$. The x-integration now adds over all strips, beginning at $x = a$ and ending at $x = b$:

$$\iint_R f(x,y)\,dA = \int_a^b \int_{g(x)}^{h(x)} f(x,y)\,dy\,dx. \qquad (14.15)$$

A double iterated integral in the reverse order is not convenient for this region because horizontal strips neither all start on the same curve nor all end on the same curve.

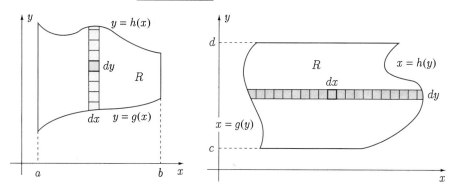

FIGURE 14.6

FIGURE 14.7

For the region in Figure 14.7, we obtain

$$\iint_R f(x,y)\,dA = \int_c^d \int_{g(y)}^{h(y)} f(x,y)\,dx\,dy. \qquad (14.16)$$

The limits on double iterated integrals have been interpreted schematically as follows:

$$\int_{\text{position of first horizontal strip}}^{\text{position of last horizontal strip}} \int_{\text{where each and every horizontal strip starts}}^{\text{where each and every horizontal strip stops}} f(x,y)\,dx\,dy;$$

$$\int_{\text{position of first vertical strip}}^{\text{position of last vertical strip}} \int_{\text{where each and every vertical strip starts}}^{\text{where each and every vertical strip stops}} f(x,y)\,dy\,dx.$$

With these interpretations on the limits, you can see how important it is to have a well-labeled diagram.

EXAMPLE 14.2

Evaluate the double integral of $f(x,y) = xy^2 + x^2$ over the region bounded by the curves $y = x^2$ and $x = y^2$.

SOLUTION If we use vertical strips (Figure 14.8), we have

$$\iint_R (xy^2 + x^2)\,dA = \int_0^1 \int_{x^2}^{\sqrt{x}} (xy^2 + x^2)\,dy\,dx = \int_0^1 \left\{ \frac{xy^3}{3} + x^2 y \right\}_{x^2}^{\sqrt{x}} dx$$

$$= \int_0^1 \left\{ \frac{1}{3} x^{5/2} + x^{5/2} - \frac{x^7}{3} - x^4 \right\} dx = \left\{ \frac{8 x^{7/2}}{21} - \frac{x^8}{24} - \frac{x^5}{5} \right\}_0^1$$

$$= \frac{8}{21} - \frac{1}{24} - \frac{1}{5} = \frac{39}{280}.$$

FIGURE 14.8

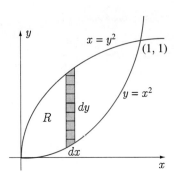

There are two distinct parts to every double integral: first, the function $f(x, y)$ being integrated, which is the integrand; second, the region R over which integration is being performed, and this region determines the limits on the corresponding double iterated integral. Note that we do not use $f(x, y)$ to determine limits on the double iterated integral; the region determines the limits. Conversely, if we are given a double iterated integral, then we know that it represents the double integral of its integrand over some region, and the region is completely defined by the limits on the iterated integral. This point is emphasized in the following example.

EXAMPLE 14.3

Evaluate the double iterated integral

$$\int_0^2 \int_y^2 e^{x^2} \, dx \, dy.$$

SOLUTION The function e^{x^2} does not have an elementary antiderivative with respect to x, and it is therefore impossible to evaluate the double iterated integral as it now stands. But the double iterated integral represents the double integral of e^{x^2} over some region R in the xy-plane. To find R we note that the inner integral indicates horizontal strips that all start on the line $x = y$ and stop on the line $x = 2$ (Figure 14.9a). The outer limits state that the first and last strips are at $y = 0$ and $y = 2$, respectively. This defines R as the triangle bounded by the straight lines $y = x$, $x = 2$, and $y = 0$ (Figure 14.9b). If we now reverse the order of integration and use vertical strips, we have

$$\int_0^2 \int_y^2 e^{x^2} \, dx \, dy = \iint_R e^{x^2} \, dA = \int_0^2 \int_0^x e^{x^2} \, dy \, dx = \int_0^2 \{ye^{x^2}\}_0^x \, dx$$

$$= \int_0^2 xe^{x^2} \, dx = \left\{\frac{1}{2}e^{x^2}\right\}_0^2 = \frac{e^4 - 1}{2}.$$

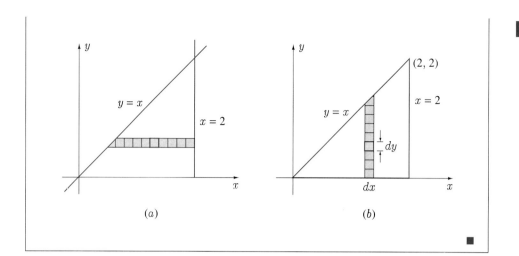

FIGURE 14.9

This example also points out that an iterated integral in one order may be much easier to evaluate than the corresponding iterated integral in the opposite order.

EXERCISES 14.2

In Exercises 1–12 evaluate the double integral over the region.

1. $\iint_R (x^2 + y^2)\, dA$ where R is bounded by $y = x^2$, $x = y^2$

2. $\iint_R (4 - x^2 - y)\, dA$ where R is bounded by $x = \sqrt{4 - y}$, $x = 0$, $y = 0$

3. $\iint_R (x + y)\, dA$ where R is bounded by $x = y^3 + 2$, $x = 1$, $y = 1$

4. $\iint_R xy^2\, dA$ where R is bounded by $x + y + 1 = 0$, $x + y^2 = 1$

5. $\iint_R xe^y\, dA$ where R is bounded by $y = x$, $y = 0$, $x = 1$

6. $\iint_R (x + y)\, dA$ where R is bounded by $x^2 + y^2 = 9$

7. $\iint_R x^2 y\, dA$ where R is bounded by $y = \sqrt{x + 4}$, $y = 0$, $x + y = 2$

8. $\iint_R (xy + y^2 - 3x^2)\, dA$ where R is bounded by $y = |x|$, $y = 1$, $y = 2$

9. $\iint_R (1 - x)^2\, dA$ where R is bounded by $x + y = 1$, $x + y = -1$, $x - y = 1$, $y - x = 1$

10. $\iint_R (x + y)\, dA$ where R is bounded by $x = y^2$, $x^2 - y^2 = 12$

11. $\iint_R x\, dA$ where R is bounded by $y = 3x$, $y = x$, $x + y = 4$

12. $\iint_R y^2\, dA$ where R is bounded by $x = 0$, $y = 1$, $y = 1/2$, $x = 1/\sqrt{y^4 + 12y^2}$

In Exercises 13–18 evaluate the double iterated integral by reversing the order of integration.

13. $\int_0^2 \int_0^{\sqrt{4-x^2}} (4 - y^2)^{3/2}\, dy\, dx$

14. $\int_0^1 \int_y^1 \sin(x^2)\, dx\, dy$

15. $\int_{-2}^0 \int_{-y}^2 y(x^2 + y^2)^8\, dx\, dy$

16. $\int_{-2}^0 \int_{-2}^x \dfrac{x}{\sqrt{x^2 + y^2}}\, dy\, dx$

17. $\int_0^2 \int_0^{x^2/2} \dfrac{x}{\sqrt{1 + x^2 + y^2}}\, dy\, dx$

18. $\displaystyle\int_0^2 \int_{-x^2/2}^0 \frac{x}{\sqrt{1+x^2+y^2}}\, dy\, dx$

19. Verify that if $m \le f(x,y) \le M$ for all (x,y) in R, then
$$m(\text{Area of } R) \le \iint_R f(x,y)\, dA \le M(\text{Area of } R).$$

20. Evaluate the double integral of $f(x,y) = 1/\sqrt{2x-x^2}$ over the region in the first quadrant bounded by $y^2 = 4 - 2x$.

In Exercises 21–28 either the integral has value 0 or it can be evaluated by doubling the double integral over half the region. By drawing the region and examining the integrand, determine which situation prevails. Do not evaluate the integral.

21. $\displaystyle\iint_R x^2 y^3\, dA$ where R is bounded by
$x = \sqrt{4-y^2},\ x = 0$

22. $\displaystyle\iint_R x^2 y^2\, dA$ where R is bounded by
$x = \sqrt{4-y^2},\ x = 0$

23. $\displaystyle\iint_R (x+y)\, dA$ where R is the square with vertices $(\pm 3, 0)$ and $(0, \pm 3)$

24. $\displaystyle\iint_R x^7 \cos(x^2)\, dA$ where R is bounded by
$y = 4 - |x|,\ y = x^2$

25. $\displaystyle\iint_R e^{x^2+y^2}\, dA$ where R is bounded by $y = 4 - 4x^2$, $y = x^2 - 1$

26. $\displaystyle\iint_R \cos(x^2 y)\, dA$ where R is bounded by $y = 0$, $y = x^3 - x$

27. $\displaystyle\iint_R \sin(x^2 y)\, dA$ where R is bounded by $y = 0$, $y = x^3 - x$

28. $\displaystyle\iint_R (x^2 y^3 + xy^2)\, dA$ where R is bounded by
$\sqrt{|x|} + \sqrt{|y|} = 1$

The average value of a function $f(x,y)$ over a region R with area A is defined as
$$\bar{f} = \frac{1}{A} \iint_R f(x,y)\, dA.$$

In Exercises 29–32 find the average value of the function over the region.

29. $f(x,y) = xy$ over the region in the first quadrant bounded by $x = 0$, $y = 0$, $y = \sqrt{1-x^2}$

30. $f(x,y) = x + y$ over the region bounded by $y = x$, $y = 0$, $y = \sqrt{2-x}$

31. $f(x,y) = x$ over the region between $y = \sin x$ and $y = 0$ for $0 \le x \le 2\pi$

32. $f(x,y) = e^{x+y}$ over the region bounded by $y = x+1$, $y = x-1$, $y = 1-x$, $y = -1-x$

In Exercises 33–39 evaluate the double integral over the region.

33. $\displaystyle\iint_R x^2\, dA$ where R is bounded by $x^2 + y^2 = 4$

34. $\displaystyle\iint_R (6 - x - 2y)\, dA$ where R is bounded by
$x^2 + y^2 = 4$

35. $\displaystyle\iint_R 6x^5\, dA$ where R is the region under $x + 5y = 16$, above $y = x - 4$, and bounded by $x = (y-2)^2$

36. $\displaystyle\iint_R ye^x\, dA$ where R is bounded by $y = x$,
$x + y = 2,\ y = 0$

37. $\displaystyle\iint_R \sqrt{1+y}\, dA$ where R is bounded by $x = -1$,
$y = 2,\ x = y^3$

38. $\displaystyle\iint_R y\sqrt{x^2+y^2}\, dA$ where R is bounded by $y = x$,
$x = -1,\ y = 0$

39. $\displaystyle\iint_R (x^2 + y^2)\, dA$ where R is bounded by $x^2 + y^2 = 9$

40. Evaluate the double iterated integral
$$\int_0^1 \int_0^1 |x - y|\, dy\, dx.$$

41. Evaluate the double integral $\displaystyle\iint_R |y - 2x^2 + 1|\, dA$ where R is the square bounded by $x = \pm 1$, $y = \pm 1$.

SECTION 14.3

Areas And Volumes Of Solids Of Revolution

Because equation 14.7 represents the area of a region R as a double integral, and double integrals are evaluated by means of double iterated integrals, it follows that areas can be calculated using double iterated integrals. In particular, to find the area of the region in Figure 14.10, we subdivide R into rectangles of dimensions dx and dy and therefore of area $dA = dy\,dx$. The areas of these rectangles are then added in the y-direction to give the area of a vertical strip

$$\int_{g(x)}^{h(x)} dy\,dx,$$

where the limits indicate that every vertical strip starts on the curve $y = g(x)$ and ends on the curve $y = h(x)$. Finally, the areas of the vertical strips are added together to give the total area:

$$\text{Area} = \int_a^b \int_{g(x)}^{h(x)} dy\,dx,$$

where a and b indicate the x-positions of the first and last strips.

FIGURE 14.10

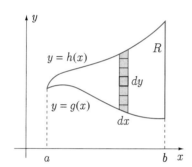

EXAMPLE 14.4

Find the area bounded by the curves $xy = 2$, $x = 2\sqrt{y}$, $y = 4$.

SOLUTION If we choose horizontal strips for this area (Figure 14.11), we have

$$\text{Area} = \int_1^4 \int_{2/y}^{2\sqrt{y}} dx\,dy = \int_1^4 \left\{ 2\sqrt{y} - \frac{2}{y} \right\} dy$$

$$= \left\{ \frac{4}{3} y^{3/2} - 2\ln|y| \right\}_1^4 = \frac{28}{3} - 2\ln 4.$$

For vertical strips (Figure 14.12), we require two iterated integrals because to the left of the line $x = 2$, strips begin on the hyperbola $xy = 2$, whereas to the right of $x = 2$, they begin on the parabola $x = 2\sqrt{y}$. We obtain

$$\text{Area} = \int_{1/2}^2 \int_{2/x}^4 dy\,dx + \int_2^4 \int_{x^2/4}^4 dy\,dx = \int_{1/2}^2 \left\{ 4 - \frac{2}{x} \right\} dx + \int_2^4 \left\{ 4 - \frac{x^2}{4} \right\} dx$$

$$= \{4x - 2\ln|x|\}_{1/2}^2 + \left\{ 4x - \frac{x^3}{12} \right\}_2^4 = \frac{28}{3} - 2\ln 4.$$

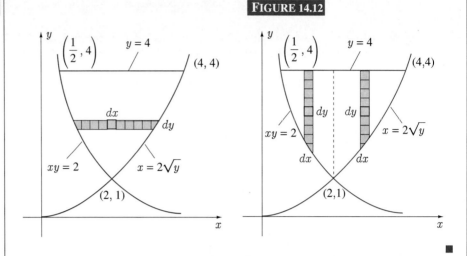

FIGURE 14.11

FIGURE 14.12

If we compare finding areas by definite integrals (Section 7.1) and finding the same areas by double integrals, it is clear that no great advantage is derived by using double integrals. In fact, it is probably more work because we must perform two, rather than one, integrations, although the first integration is trivial. The advantage of double integrals is therefore not in finding area; it is in finding volumes of solids of revolution, centres of mass, moments of inertia, and fluid forces, among other applications.

Volumes of Solids of Revolution

If the area in Figure 14.13 is rotated around the x-axis, the volume of the resulting solid of revolution can be evaluated by using the washer method introduced in Section 7.2:

$$\text{Volume} = \int_a^b \{\pi[h(x)]^2 - \pi[g(x)]^2\}\,dx. \tag{14.17}$$

If this area is rotated around the y-axis, the volume generated is calculated by using the cylindrical shell method:

$$\text{Volume} = \int_a^b 2\pi x[h(x) - g(x)]\,dx. \tag{14.18}$$

FIGURE 14.13

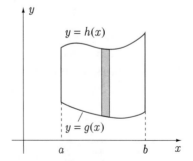

Thus once we have chosen to use vertical rectangles, the axis of revolution determines whether we use either the washer or cylindrical shell method. We now show that with double integrals one method works for both problems.

To rotate this area around the x-axis we subdivide R into small areas dA (Figure 14.14). If the area dA at a point (x, y) is rotated about the x-axis, it generates a "ring" with cross-sectional area dA. Since (x, y) travels a distance $2\pi y$ in traversing the ring, it follows that the volume in the ring is approximately $2\pi y\, dA$. To find the total volume obtained by rotating R about the x-axis, we add the volumes of all such rings and take the limit as the areas shrink to points. But this is precisely what we mean by the double integral of $2\pi y$ over the region R, and we therefore write

$$\text{Volume} = \iint_R 2\pi y\, dA. \tag{14.19}$$

FIGURE 14.14

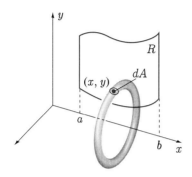

On the other hand, if dA is rotated about the y-axis, it again forms a ring, but with approximate volume $2\pi x\, dA$. The total volume, then, when R is rotated about the y-axis is

$$\text{Volume} = \iint_R 2\pi x\, dA. \tag{14.20}$$

Since double iterated integrals are used to evaluate double integrals, it follows that we can set up double iterated integrals to find the volumes represented by equations 14.19 and 14.20. The decision to use a double iterated integral with respect to y and x implies a subdivision of R into rectangles of dimensions dx and dy (Figure 14.15). The volume of the ring formed when this rectangle is rotated around the x-axis is $2\pi y\, dy\, dx$. If we choose to integrate first with respect to y, we are adding over all rectangles in a vertical strip

$$\int_{g(x)}^{h(x)} 2\pi y\, dy\, dx,$$

where the limits indicate that all vertical strips start on the curve $y = g(x)$ and end on the curve $y = h(x)$. This integral is the volume generated by rotating the vertical strip around the x-axis. Integration now with respect to x adds over all strips to give the required volume

$$\text{Volume} = \int_a^b \int_{g(x)}^{h(x)} 2\pi y\, dy\, dx. \tag{14.21}$$

Note that when we actually do perform the inner integration, we get

$$\text{Volume} = \int_a^b \{\pi y^2\}_{g(x)}^{h(x)} dx = \int_a^b \{\pi[h(x)]^2 - \pi[g(x)]^2\} dx,$$

and this is the result contained in equation 14.17.

FIGURE 14.15

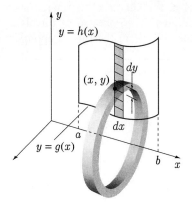

When R is rotated around the y-axis, the rectangular area $dy\,dx$ generates a ring of volume $2\pi x\,dy\,dx$. Addition over the rectangles in a vertical strip

$$\int_{g(x)}^{h(x)} 2\pi x\,dy\,dx$$

gives the volume generated by rotating the strip about the y-axis. Finally, integration with respect to x adds over all strips to give

$$\text{Volume} = \int_a^b \int_{g(x)}^{h(x)} 2\pi x\,dy\,dx. \qquad (14.22)$$

This time the inner integration leads to

$$\text{Volume} = \int_a^b \{2\pi xy\}_{g(x)}^{h(x)} dx = \int_a^b 2\pi x[h(x) - g(x)]\,dx,$$

the same result as in equation 14.18.

The advantage, then, in using double integrals to find volumes of solids of revolution is that it requires only one idea, that of rings. The first integration then leads to the concepts of washers or cylindrical shells.

EXAMPLE 14.5

Find the volumes of the solids of revolution if the area bounded by the curves $y = 2x - x^2$, $y = x^2 - 2x$ is rotated about:

(a) the y-axis;

(b) $x = -3$;

(c) $y = 2$.

SOLUTION

(a) If we use vertical strips (Figure 14.16), then

$$\text{Volume} = \int_0^2 \int_{x^2-2x}^{2x-x^2} 2\pi x \, dy \, dx = 2\pi \int_0^2 \{xy\}_{x^2-2x}^{2x-x^2} dx$$

$$= 2\pi \int_0^2 x\{(2x-x^2)-(x^2-2x)\} dx$$

$$= 4\pi \int_0^2 (2x^2 - x^3) dx = 4\pi \left\{ \frac{2x^3}{3} - \frac{x^4}{4} \right\}_0^2 = \frac{16\pi}{3}.$$

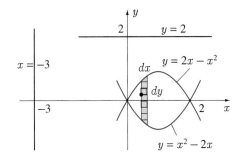

FIGURE 14.16

(b) In this case the radius of the ring formed by rotating the rectangle about $x = -3$ is $x + 3$, and therefore

$$\text{Volume} = \int_0^2 \int_{x^2-2x}^{2x-x^2} 2\pi(x+3) \, dy \, dx = 2\pi \int_0^2 \{y(x+3)\}_{x^2-2x}^{2x-x^2} dx$$

$$= 2\pi \int_0^2 (x+3)(4x-2x^2) dx = 4\pi \int_0^2 (6x - x^2 - x^3) dx$$

$$= 4\pi \left\{ 3x^2 - \frac{x^3}{3} - \frac{x^4}{4} \right\}_0^2 = \frac{64\pi}{3}.$$

(c) When the rectangle is rotated around $y = 2$, the radius of the ring is $2 - y$, and hence

$$\text{Volume} = \int_0^2 \int_{x^2-2x}^{2x-x^2} 2\pi(2-y) \, dy \, dx = 2\pi \int_0^2 \left\{ -\frac{1}{2}(2-y)^2 \right\}_{x^2-2x}^{2x-x^2} dx$$

$$= \pi \int_0^2 \{(2-x^2+2x)^2 - (2-2x+x^2)^2\} dx = 8\pi \int_0^2 (2x - x^2) dx$$

$$= 8\pi \left\{ x^2 - \frac{x^3}{3} \right\}_0^2 = \frac{32\pi}{3}.$$

∎

EXERCISES 14.3

In Exercises 1–10 use a double integral to find the area bounded by the curves.

1. $y = 4x^2$, $x = 4y^2$
2. $y = x^2$, $y = 5x + 6$
3. $x = y^2$, $x = 3y - 2$
4. $y = x^3 + 8$, $y = 4x + 8$
5. $y = 4/x^2$, $y = 5 - x^2$
6. $y = xe^{-x}$, $y = x$, $x = 2$
7. $x = 4y - 4y^2$, $y = x - 3$, $y = 1$, $y = 0$
8. $x = y(y - 2)$, $x + y = 12$
9. $y = x^3 - x^2 - 2x + 2$, $y = 2$
10. $x + y = 1$, $x + y = 5$, $y = 2x + 1$, $y = 2x + 6$

In Exercises 11–20 use a double integral to find the volume of the solid of revolution obtained by rotating the area bounded by the curves about the line.

11. $y = -\sqrt{4 - x}$, $x = 0$, $y = 0$ about $y = 0$
12. $4x^2 + 9y^2 = 36$ about $y = 0$
13. $y = (x - 1)^2$, $y = 1$ about $x = 0$
14. $y = x^2 + 4$, $y = 2x^2$ about $y = 0$
15. $x - 1 = y^2$, $x = 5$ about $x = 1$
16. $x = y^3$, $y = \sqrt{2 - x}$, $y = 0$ about $y = 1$
17. $y = 4x^2 - 4x$, $y = x^3$ about $y = -2$
18. $x = 3y - y^2$, $x = y^2 - 3y$ about $y = 4$
19. $x = 2y - y^2 - 2$, $x = -5$ about $x = 1$
20. $x + y = 4$, $y = 2\sqrt{x - 1}$, $y = 0$ about $y = -1$

In Exercises 21–30 use a double integral to find the area bounded by the curves.

21. $y = 2x^3$, $y = 4x + 8$, $y = 0$
22. $y = x/\sqrt{x + 3}$, $x = 1$, $x = 6$, $y = -x^2$
23. $x = y^2 + 2$, $x = -(y - 4)^2$, $y = 4 - x$, $y = 0$
24. $y = x^3 - x$, $x + y + 1 = 0$, $x = \sqrt{y + 1}$
25. $y^2 = x^2(4 - x^2)$
26. $x^2 + y^2 = 4$, $x^2 + y^2 = 4x$ (interior to both)
27. $x = 1/\sqrt{4 - y^2}$, $4x + y^2 = 0$, $y + 1 = 0$, $y - 1 = 0$
28. $y^2 = x^4(9 + x)$
29. $y = (x^2 + 1)/(x + 1)$, $x + 3y = 7$
30. $(x + 2)^2 y = 4 - x$, $x = 0$, $y = 0$ ($x \geq 0$, $y \geq 0$)

In Exercises 31–35 use a double integral to find the volume of the solid of revolution obtained by rotating the area bounded by the curves about the line.

31. $y = 4/(x^2 + 1)^2$, $y = 1$ about $x = 0$
32. $y = x^2 - 2$, $y = 0$ about $y = -1$
33. $y = |x^2 - 1|$, $x = -2$, $x = 2$, $y = -1$ about $y = -2$
34. $x = \sqrt{4 + 12y^2}$, $x - 20y = 24$, $y = 0$ about $y = 0$
35. $y = (x + 1)^{1/4}$, $y = -(x + 1)^2$, $x = 0$ about $x = 0$
36. Find the area common to the two circles $x^2 + y^2 = 4$ and $x^2 + y^2 = 6x$.

In Exercises 37–40 find the volume of the solid of revolution obtained by rotating the area bounded by the curves about the line. (Hint: Use the distance formula $|Ax_1 + By_1 + C|/\sqrt{A^2 + B^2}$ from a point (x_1, y_1) to a line $Ax + By + C = 0$ developed in Exercise 45 of Section 1.3.)

37. $x = 1$, $y = 1$, $x = 0$, $y = 0$ about $x + y = 2$
38. $y = x^2$, $y = 2x + 3$ about $y = 2x + 3$
39. $x = y^2$, $y = 0$, $x = 1$ about $y = 3x + 2$
40. $x = 2y$, $y = x - 1$, $y = 0$ about $x + y + 1 = 0$
41. Prove that the area above the line $y = h$ and under the circle $x^2 + y^2 = r^2$ ($r > h$) is given by

$$A = \pi r^2/2 - h\sqrt{r^2 - h^2} - r^2 \operatorname{Sin}^{-1}(h/r).$$

Fluid Pressure

In Section 7.5 we defined pressure at a point in a fluid as the magnitude of the force per unit area that would act on any surface placed at that point. We discovered that at a depth $d > 0$ below the surface of a fluid, pressure is given by

$$P = 9.81\rho d, \qquad (14.23)$$

where ρ is the density of the fluid. With these ideas and the definite integral, we were able to calculate fluid forces on flat surfaces in the fluid. In particular, the magnitude of the total force on each side of the vertical surface in Figure 14.17 is given by the definite integral

$$\text{Force} = \int_a^b -9.81\rho y[h(y) - g(y)]\,dy. \qquad (14.24)$$

Although horizontal rectangles are convenient for this problem, it is clear that they are not reasonable for the surface in Figure 14.18.

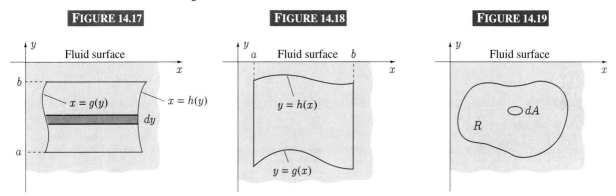

FIGURE 14.17   **FIGURE 14.18**   **FIGURE 14.19**

Double integrals, on the other hand, can be applied with equal ease to both surfaces. To see this we consider first the total force on each side of the surface in Figure 14.19. If the surface is divided into small areas dA, then the force on dA is $P\,dA$, where P is pressure at that depth. The total force on R is the sum of the forces on all such areas in R as the areas dA shrink to a point. But once again this is the concept of the double integral, and we therefore write

$$\text{Force} = \iint_R P\,dA. \qquad (14.25)$$

To set up a double iterated integral in order to evaluate this double integral, we use our interpretation of the double iterated integral as a limit of a sum in which the areas dA have been chosen as rectangles. In particular, for the surface in Figure 14.17 we draw rectangles of dimensions dx and dy, as shown in Figure 14.20. The force on this rectangle is its area $dx\,dy$ multiplied by pressure $-9.81\rho y$ at that depth, $-9.81\rho y\,dx\,dy$. Addition of these quantities over all rectangles in a horizontal strip gives the force on the strip,

$$\int_{g(y)}^{h(y)} -9.81\rho y\,dx\,dy,$$

where the limits indicate that all horizontal strips start on the curve $x = g(y)$ and end on the curve $x = h(y)$. Integration with respect to y now adds over all horizontal strips to give the total force on the surface:

$$\text{Force} = \int_a^b \int_{g(y)}^{h(y)} -9.81\rho y\, dx\, dy. \qquad (14.26)$$

When we perform the inner integration, we obtain

$$\text{Force} = \int_a^b \{-9.81\rho y x\}_{g(y)}^{h(y)}\, dy = \int_a^b -9.81\rho y[h(y) - g(y)]\, dy,$$

and this is the result contained in equation 14.24.

FIGURE 14.20   **FIGURE 14.21**

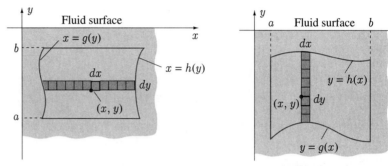

For the surface in Figure 14.18 we again draw rectangles of dimensions dx and dy and calculate the force on such a rectangle, $-9.81\rho y\, dy\, dx$. In this case it is more convenient to add over rectangles in a vertical strip to give the force on the strip (Figure 14.21):

$$\int_{g(x)}^{h(x)} -9.81\rho y\, dy\, dx.$$

The force on the entire surface can now be found by adding over all vertical strips:

$$\text{Force} = \int_a^b \int_{g(x)}^{h(x)} -9.81\rho y\, dy\, dx. \qquad (14.27)$$

EXAMPLE 14.6

The face of a dam is parabolic with breadth 100 m and height 50 m. Find the magnitude of the total force due to fluid pressure on the face.

SOLUTION If we use the coordinate system in Figure 14.22, then the edge of the dam has an equation of the form $y = kx^2$. Since $(50, 50)$ is a point on this curve, it follows that $k = \frac{1}{50}$. Because the force on the left half of the dam is the same as that on the right half, we can integrate for the right half and double the result; that is,

$$\text{Force} = 2\int_0^{50} \int_{x^2/50}^{50} 9.81(1000)(50 - y)\, dy\, dx$$

$$= 19\,620 \int_0^{50} \left\{50y - \frac{y^2}{2}\right\}_{x^2/50}^{50}\, dx = 19\,620 \int_0^{50} \left\{1250 - x^2 + \frac{x^4}{5000}\right\} dx$$

$$= 19\,620 \left\{1250x - \frac{x^3}{3} + \frac{x^5}{25\,000}\right\}_0^{50} = 6.54 \times 10^8\,\text{N}.$$

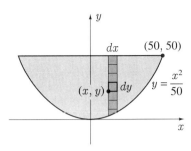

FIGURE 14.22

EXAMPLE 14.7

A tank in the form of a right-circular cylinder of radius $\frac{1}{2}$ m and length 10 m has its axis horizontal. If it is full of water, find the force due to water pressure on each end of the tank.

SOLUTION Since the force on that part of the end to the left of the y-axis (Figure 14.23) is identical to the force on that part to the right, we double the force on the right half,

$$\text{Force} = 2\int_{-1/2}^{1/2}\int_0^{\sqrt{1/4-y^2}} 9.81(1000)(\tfrac{1}{2}-y)\,dx\,dy$$

$$= 9810\int_{-1/2}^{1/2}\int_0^{\sqrt{1/4-y^2}} dx\,dy - 19\,620\int_{-1/2}^{1/2}\int_0^{\sqrt{1/4-y^2}} y\,dx\,dy.$$

The first double iterated integral represents the area of one-half the end of the tank. Consequently,

$$\text{Force} = 9810\{\tfrac{1}{2}\pi(\tfrac{1}{2})^2\} - 19\,620\int_{-1/2}^{1/2} y\sqrt{\tfrac{1}{4}-y^2}\,dy$$

$$= \frac{4905\pi}{4} - 19\,620\{-\tfrac{1}{3}(\tfrac{1}{4}-y^2)^{3/2}\}\Big|_{-1/2}^{1/2} = \frac{4905\pi}{4}\,\text{N}.$$

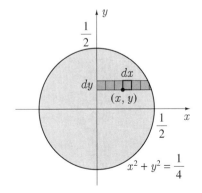

FIGURE 14.23

EXERCISES 14.4

In Exercises 1–8 the surface is submerged vertically in a fluid with density ρ. Find the force due to fluid pressure on one side of the surface.

1. An equilateral triangle of side length 2 with one side in the surface

2. A parabolic segment of base 12 and height 4 with the base in the surface

3. A square of side length 3 with one diagonal vertical and the uppermost vertex in the surface

4. A triangle of side lengths 5, 5, and 8, with the longest side uppermost, horizontal, and 3 units below the surface

5. A triangle of side lengths 5, 5, and 8, with the longest side below the the opposite vertex, horizontal, and 6 units below the surface

6. A trapezoid with vertical parallel sides of lengths 6 and 8, and a third side perpendicular to the parallel sides, of length 5, and in the surface

7. A triangle of side lengths 3, 3, and 4, with the longest side vertical, and the uppermost vertex 1 unit below the surface

8. A semicircle of radius 5 with the (diameter) base in the surface

9. The vertical end of a water trough is an isosceles triangle with width 2 m and depth 1 m. Find the force of the water on each end when the trough is one-half filled (by volume) with water.

10. A dam across a river has the shape of a parabola 36 m across the top and 9 m deep at the centre. Find the maximum force due to water pressure on the dam.

In Exercises 11–15 the surface is submerged vertically in a fluid with density ρ. Find the force due to fluid pressure on one side of the surface.

11. A circle of radius 2 with centre 3 units below the surface

12. A rectangle of side lengths 2 and 5, with one diagonal vertical and the uppermost vertex in the surface

13. An ellipse with major and minor axes of lengths 8 and 6, and with the major axis horizontal and 5 units below the surface

14. A parallelogram of side lengths 4 and 5, with one of the longer sides horizontal and in the surface, and two sides making an angle of $\pi/6$ radians with the surface

15. A triangle of side lengths 2, 3, and 4 with the longest side vertical, the side of length 2 above the side of length 3, and the uppermost vertex 1 unit below the surface

16. An oil can is in the form of a right-circular cylinder of radius r and height h. If the axis of the can is horizontal, and the can is full of oil with density ρ, find the force due to fluid pressure on each end.

17. Find the force due to water pressure on each side of the flat vertical plate in Figure 14.24.

FIGURE 14.24

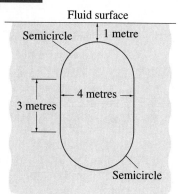

18. A square plate of side length 2 m has one side on the bottom of a swimming pool 3 m deep. The plate is inclined at an angle of $\pi/4$ radians with the bottom of the pool so that its horizontal upper edge is $3 - \sqrt{2}$ m below the surface. Find the force due to water pressure on each side of the plate.

19. A thin triangular piece of wood with sides of lengths 2 m, 2 m, and 3 m floats in a pond. A piece of rope is tied to the vertex opposite the longest side. A rock is then attached to the other end of the rope and lowered into the water. When the rock sits on the bottom (and the rope is taut), the longest side of the wood still floats in the surface of the water, but the opposite vertex is 1 m below the surface. Find the force due to water pressure on each side of the piece of wood.

SECTION 14.5

Centres Of Mass And Moments Of Inertia

We now show how double integrals can be used to replace definite integrals in calculating first moments, centres of mass, and moments of inertia of thin plates. Consider a thin plate with mass per unit area ρ such as that in Figure 14.25. Note that unlike our discussion in Section 7.6 where we assumed ρ constant, we have made no such assumption here. In other words, density could be a function of position, $\rho = \rho(x, y)$.

FIGURE 14.25

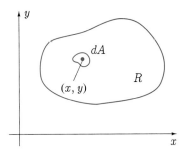

The centre of mass $(\overline{x}, \overline{y})$ of the plate is a point at which a particle of mass M (equal to the total mass of the plate) has the same first moments about the x-and y-axes as the plate itself. If we divide the plate into small areas dA, then the mass in dA is $\rho\, dA$. Addition over all such areas in R as each dA shrinks to a point gives the mass of the plate

$$M = \iint_R \rho\, dA. \tag{14.28}$$

Since the first moment of the mass in dA about the y-axis is $x\rho\, dA$, it follows that the first moment of the entire plate about the y-axis is

$$\iint_R x\rho\, dA.$$

But this must be equal to the first moment of the particle of mass M at $(\overline{x}, \overline{y})$ about the y-axis, and hence

$$M\overline{x} = \iint_R x\rho\, dA. \tag{14.29}$$

This equation can be solved for $\overline{x}$ once the double integral on the right and M have been calculated.

Similarly, $\overline{y}$ is determined by the equation

$$M\overline{y} = \iint_R y\rho\, dA, \tag{14.30}$$

where the double integral on the right is the first moment of the plate about the x-axis.

In any given problem, the double integrals in 14.28–14.30 are evaluated by means of double iterated integrals. For example, if we divide the plate in Figure 14.26 into rectangles of dimensions dx and dy and use vertical strips, then we have

$$M = \int_a^b \int_{g(x)}^{h(x)} \rho\, dy\, dx, \tag{14.31}$$

$$M\overline{x} = \int_a^b \int_{g(x)}^{h(x)} x\rho\, dy\, dx, \tag{14.32}$$

$$M\overline{y} = \int_a^b \int_{g(x)}^{h(x)} y\rho\, dy\, dx. \tag{14.33}$$

FIGURE 14.26

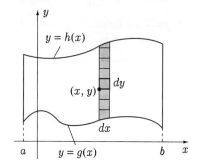

Once again we point out that equations 14.31–14.33 should not be memorized as formulas. Indeed, each can be derived as needed. For instance, to obtain equation 14.33, we reason that the first moment of the mass in a rectangle of dimensions dx and dy at position (x, y) about the x-axis is $y\rho\, dy\, dx$. Addition over the rectangles in a vertical strip gives the first moment of the strip about the x-axis:

$$\int_{g(x)}^{h(x)} y\rho\, dy\, dx,$$

and integration with respect to x now adds over all strips to give the first moment of the entire plate about the x-axis. Note that if ρ is constant and inner integrations are performed in each of equations 14.31–14.33, then

$$M = \int_a^b \{\rho y\}_{g(x)}^{h(x)}\, dx = \int_a^b \rho[h(x) - g(x)]\, dx,$$

$$M\overline{x} = \int_a^b \{\rho x y\}_{g(x)}^{h(x)}\, dx = \int_a^b \rho x[h(x) - g(x)]\, dx,$$

$$M\overline{y} = \int_a^b \left\{\rho \frac{y^2}{2}\right\}_{g(x)}^{h(x)}\, dx = \int_a^b \frac{\rho}{2}\{[h(x)]^2 - [g(x)]^2\}\, dx.$$

These are precisely the results in equations 7.31–7.33 (with different names for the curves), but the simplicity of the discussion leading to the double iterated integrals certainly demonstrates its advantage over use of the definite integral described in Section 7.6.

EXAMPLE 14.8

Find the centre of mass of a thin plate with constant mass per unit area ρ if its edges are defined by the curves $y = 2x - x^2$ and $y = x^2 - 4$.

SOLUTION For vertical strips as shown in Figure 14.27,

$$M = \int_{-1}^{2} \int_{x^2-4}^{2x-x^2} \rho \, dy \, dx$$

$$= \rho \int_{-1}^{2} \{(2x - x^2) - (x^2 - 4)\} dx$$

$$= \rho \left\{ x^2 - \frac{2x^3}{3} + 4x \right\}_{-1}^{2} = 9\rho.$$

If the centre of mass of the plate is $(\bar{x}, \bar{y})$, then

$$M\bar{x} = \int_{-1}^{2} \int_{x^2-4}^{2x-x^2} x\rho \, dy \, dx = \rho \int_{-1}^{2} x\{(2x - x^2) - (x^2 - 4)\} dx$$

$$= \rho \left\{ \frac{2x^3}{3} - \frac{x^4}{2} + 2x^2 \right\}_{-1}^{2} = \frac{9\rho}{2}.$$

Thus, $\bar{x} = \dfrac{9\rho}{2} \cdot \dfrac{1}{9\rho} = \dfrac{1}{2}$. Since

$$M\bar{y} = \int_{-1}^{2} \int_{x^2-4}^{2x-x^2} y\rho \, dy \, dx = \rho \int_{-1}^{2} \left\{ \frac{y^2}{2} \right\}_{x^2-4}^{2x-x^2} dx$$

$$= \frac{\rho}{2} \int_{-1}^{2} (-4x^3 + 12x^2 - 16) \, dx$$

$$= \frac{\rho}{2} \{-x^4 + 4x^3 - 16x\}_{-1}^{2} = -\frac{27\rho}{2},$$

we find $\bar{y} = -\dfrac{27\rho}{2} \cdot \dfrac{1}{9\rho} = -\dfrac{3}{2}$.

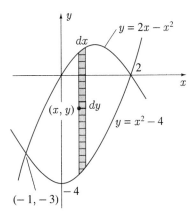

FIGURE 14.27

EXAMPLE 14.9

Find the first moment of area about the line $y = -2$ for the region bounded by the curves $x = |y|^3$ and $x = 2 - y^2$.

SOLUTION *Method 1* The first moment about $y = -2$ of a rectangle of dimensions

dx and dy at position (x, y) is $(y+2)\,dx\,dy$ (Figure 14.28). For the entire plate, then, the required first moment is

$$\int_{-1}^{0}\int_{-y^3}^{2-y^2}(y+2)\,dx\,dy + \int_{0}^{1}\int_{y^3}^{2-y^2}(y+2)\,dx\,dy$$

$$= \int_{-1}^{0}\{x(y+2)\}_{-y^3}^{2-y^2}\,dy + \int_{0}^{1}\{x(y+2)\}_{y^3}^{2-y^2}\,dy$$

$$= \int_{-1}^{0}(y^4 + y^3 - 2y^2 + 2y + 4)\,dy + \int_{0}^{1}(-y^4 - 3y^3 - 2y^2 + 2y + 4)\,dy$$

$$= \left\{\frac{y^5}{5} + \frac{y^4}{4} - \frac{2y^3}{3} + y^2 + 4y\right\}_{-1}^{0} + \left\{-\frac{y^5}{5} - \frac{3y^4}{4} - \frac{2y^3}{3} + y^2 + 4y\right\}_{0}^{1}$$

$$= \frac{17}{3}.$$

FIGURE 14.28

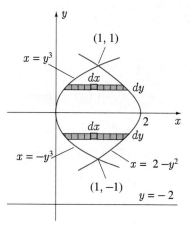

Method 2 By symmetry, the centroid of the region is somewhere along the x-axis. Hence, the required first moment is $2A$, where A is the area of the region and 2 is the distance from $y = -2$ to the centroid. Since the area of the region is equally distributed about the x-axis, we obtain the required first moment as

$$2(2)(\text{Area of plate above } x\text{-axis}) = 4\int_{0}^{1}\int_{y^3}^{2-y^2}dx\,dy = 4\int_{0}^{1}(2 - y^2 - y^3)\,dy$$

$$= 4\left\{2y - \frac{y^3}{3} - \frac{y^4}{4}\right\}_{0}^{1} = \frac{17}{3}.$$

∎

To calculate moments of inertia (second moments) of thin plates about lines parallel to the x- and y-axes is as easy as calculating first moments if we use double integrals. In particular, the mass in area dA in Figure 14.25 is $\rho\,dA$; thus its moments of inertia about the x- and y-axes are, respectively, $y^2\rho\,dA$ and $x^2\rho\,dA$. Moments of inertia of the entire plate about the x- and y-axes are therefore given by the double integrals

$$I_x = \iint_R y^2\rho\,dA \quad \text{and} \quad I_y = \iint_R x^2\rho\,dA. \tag{14.34}$$

For a plate such as that shown in Figure 14.29, we evaluate these double integrals by means of double iterated integrals with respect to x and y:

$$I_x = \int_a^b \int_{g(y)}^{h(y)} y^2 \rho \, dx \, dy, \qquad (14.35)$$

$$I_y = \int_a^b \int_{g(y)}^{h(y)} x^2 \rho \, dx \, dy. \qquad (14.36)$$

FIGURE 14.29

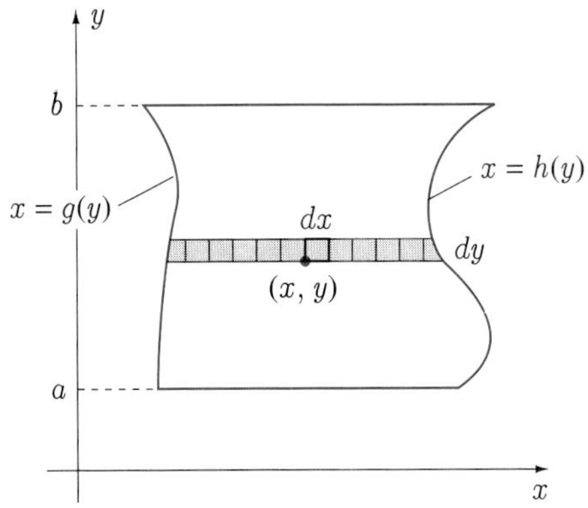

EXAMPLE 14.10

Find the moment of inertia about the x-axis of a thin plate with constant mass per unit area ρ if its edges are defined by the curves $y = x^3$, $y = \sqrt{2-x}$, and $x = 0$.

SOLUTION With vertical strips as shown in Figure 14.30, the moment of inertia about the x-axis is

$$\int_0^1 \int_{x^3}^{\sqrt{2-x}} y^2 \rho \, dy \, dx = \rho \int_0^1 \left\{ \frac{y^3}{3} \right\}_{x^3}^{\sqrt{2-x}} dx = \frac{\rho}{3} \int_0^1 \{(2-x)^{3/2} - x^9\} dx$$

$$= \frac{\rho}{3} \left\{ -\frac{2}{5}(2-x)^{5/2} - \frac{x^{10}}{10} \right\}_0^1 = \frac{16\sqrt{2} - 5}{30} \rho.$$

FIGURE 14.30

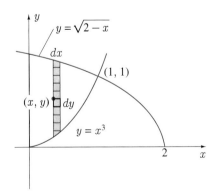

EXAMPLE 14.11

Find the second moment of area about the line $y = -1$ of the area bounded by the curves $x = y^2$ and $x = 2y$.

SOLUTION The second moment of area about the line $y = -1$ (Figure 14.31) is

$$\int_0^2 \int_{y^2}^{2y} (y+1)^2 \, dx \, dy = \int_0^2 \{x(y+1)^2\}_{y^2}^{2y} \, dy = \int_0^2 (y+1)^2 (2y - y^2) \, dy$$

$$= \int_0^2 (-y^4 + 3y^2 + 2y) \, dy = \left\{ -\frac{y^5}{5} + y^3 + y^2 \right\}_0^2$$

$$= \frac{28}{5}.$$

FIGURE 14.31

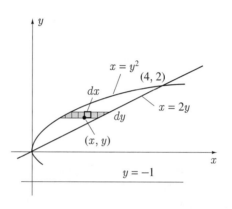

EXERCISES 14.5

In Exercises 1–10 find the centroid of the area bounded by the curves.

1. $x = y + 2$, $x = y^2$
2. $y = 8 - 2x^2$, $y + x^2 = 4$
3. $y = x^2 - 1$, $y + (x+1)^2 = 0$
4. $x + y = 5$, $xy = 4$
5. $y = e^x$, $y = 0$, $x = 0$, $x = 1$
6. $y = \sqrt{4 - x^2}$, $y = x$, $x = 0$
7. $y = 1/(x-1)$, $y = 1$, $y = 2$, $x = 0$
8. $x = 4y - 4y^2$, $x = y + 3$, $y = 1$, $y = 0$
9. $y = |x^2 - 1|$, $y = 2$
10. $y = x$, $y = 2x$, $2y = x + 3$

In Exercises 11–15 find the second moment of area of the region bounded by the curves about the line.

11. $y = x^2$, $y = x^3$ about the y-axis
12. $y = x$, $y = 2x + 4$, $y = 0$ about the x-axis
13. $y = x^2$, $2y = x^2 + 4$ about $y = 0$
14. $y = x^2 - 4$, $y = 2x - x^2$ about $x = -2$
15. $x = 1/\sqrt{y^4 + 12y^2}$, $x = 0$, $y = 1/2$, $y = 1$ about $y = 0$
16. Find the first moment about the line $y = -2$ of a thin plate of constant mass per unit area ρ if its edges are defined by the curves $y = 2 - 2x^2$ and $y = x^2 - 1$.

In Exercises 17–23 find the centroid of the area bounded by the curves.

17. $x = \sqrt{y + 2}$, $y = x$, $y = 0$
18. $y + x^2 = 0$, $x = y + 2$, $x + y + 2 = 0$, $y = 2$ (above $y + x^2 = 0$)
19. $y^2 = x^4(1 - x^2)$ (right loop)

20. $3x^2 + 4y^2 = 48$, $(x-2)^2 + y^2 = 1$

21. $y = \ln x$, $y + \sqrt{x-1} = 0$, $x = 2$

22. $y = \sqrt{2-x}$, $15y = x^2 - 4$

23. $y = x\sqrt{1-x^2}$, $x \geq 0$ and the x-axis

24. Find the moment of inertia of a uniform rectangular plate a units long and b units wide about a line through the centre of the plate and perpendicular to the plate.

In Exercises 25–27 find the second moment of area of the region bounded by the curves about the line.

25. $4x^2 + 9y^2 = 36$ about $y = -2$

26. $x = y^2$, $x + y = 2$ about $x = -1$

27. $y = \sqrt{a^2 - x^2}$, $y = a$, $x = a$ ($a > 0$) about the x-axis

28. Find the first moment of area about the line $x + y = 1$ for the area bounded by the curves $x = y^2 - 2$ and $y = x$.

29. Prove the *theorem of Pappus*: If a plane area is revolved about a coplanar axis not crossing the area, the volume generated is equal to the product of the area and the circumference of the circle described by the centroid of the area.

30. A thin flat plate of area A is immersed vertically in a fluid with density ρ. Show that the total force (due to fluid pressure) on each side of the plate is equal to the product of 9.81, A, ρ, and the depth of the centroid of the plate below the surface of the fluid. Use this result to find the forces in some of the problems in Exercises 14.4, say 1, 3, 4, 5, 7, 11, 12, 13, 16, and 17. For those problems involving triangles recall the result of Exercise 36 in Section 7.6 or Exercise 39 in Section 12.3.

31. Prove the *parallel axis theorem*: The moment of inertia of a thin plate (with constant mass per unit area) with respect to any coplanar line is equal to the moment of inertia with respect to the parallel line through the centre of mass plus the mass multiplied by the square of the distance between the lines.

SECTION 14.6

Surface Area

To find the length of a curve in Section 7.3 we approximated the curve by tangent line segments. To find the area of a surface we follow a similar procedure by approximating the surface with tangential planes. In particular, consider finding the area of a smooth surface S given that every vertical line that intersects the surface does so in exactly one point (Figure 14.32). If S_{xy} is the area in the xy-plane onto which S projects, we divide S_{xy} into n sub-areas ΔA_i in any fashion whatsoever, and choose a point (x_i, y_i) in each ΔA_i. At the point (x_i, y_i, z_i) on the surface S that projects onto (x_i, y_i), we draw the tangent plane to S. Suppose we now project ΔA_i upward onto S and onto the tangent plane at (x_i, y_i, z_i) and denote these projected areas by ΔS_i and ΔS_{Ti}, respectively.

FIGURE 14.32

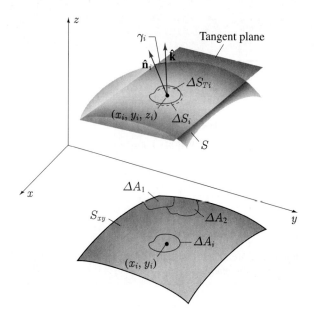

Now ΔS_{Ti} is an approximation to ΔS_i and, as long as ΔA_i is small, a reasonably good approximation. In fact, the smaller ΔA_i, the better the approximation. We therefore define the area of S as

$$\text{Area of } S = \lim_{n \to \infty} \sum_{i=1}^{n} \Delta S_{Ti}, \tag{14.37}$$

where in taking the limit we demand that each ΔA_i shrink to a point. We have therefore defined area on a curved surface in terms of flat areas on tangent planes to the surface. The advantage of this definition is that we can calculate ΔS_{Ti} in terms of ΔA_i. To see how, we denote by $\hat{\mathbf{n}}_i$ the unit normal vector to S at (x_i, y_i, z_i) with positive z-component and by γ_i the acute angle between $\hat{\mathbf{n}}_i$ and $\hat{\mathbf{k}}$. Now ΔS_{Ti} projects onto ΔA_i and γ_i is the acute angle between the planes containing ΔA_i and ΔS_{Ti} (Figure 14.33). It follows that ΔA_i and ΔS_{Ti} are related by

$$\Delta A_i = \cos \gamma_i \, \Delta S_{Ti} \tag{14.38}$$

(see Exercise 54 in Section 12.5). Note that if ΔS_{Ti} is horizontal, then $\gamma_i = 0$ and $\Delta S_{Ti} = \Delta A_i$; and if ΔS_{Ti} tends toward the vertical ($\gamma_i \to \pi/2$), then ΔS_{Ti} becomes very large for fixed ΔA_i.

FIGURE 14.33

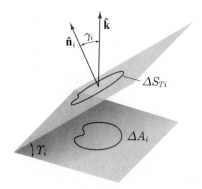

EXAMPLE 14.13

Find the area of the surface $z = x^{3/2}$ that projects onto the rectangle in the xy-plane bounded by the straight lines $x = 0$, $x = 2$, $y = 1$, and $y = 3$.

SOLUTION Since the surface projects one-to-one onto the rectangle (Figure 14.36), we find that

$$\text{Area} = \iint_{S_{xy}} \sqrt{1 + \left(\frac{\partial z}{\partial x}\right)^2 + \left(\frac{\partial z}{\partial y}\right)^2} \, dA = \iint_{S_{xy}} \sqrt{1 + \left(\frac{3}{2}x^{1/2}\right)^2} \, dA$$

$$= \frac{1}{2}\int_0^2 \int_1^3 \sqrt{4 + 9x} \, dy \, dx = \frac{1}{2}\int_0^2 \{y\sqrt{4 + 9x}\}_1^3 \, dx = \int_0^2 \sqrt{4 + 9x} \, dx$$

$$= \left\{\frac{2}{27}(4 + 9x)^{3/2}\right\}_0^2 = \frac{2}{27}(22\sqrt{22} - 8).$$

FIGURE 14.36

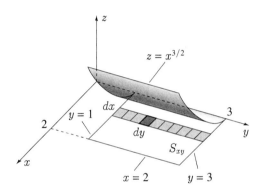

EXAMPLE 14.14

Find the area of the cone $y = \sqrt{x^2 + z^2}$ to the left of the plane $y = 1$.

SOLUTION *Method 1* The surface projects one-to-one onto the interior of the circle $x^2 + z^2 = 1$ in the xz-plane (Figure 14.37). Since the area of the surface is four times that in the first octant, if we let S_{xz} be the quarter-circle $x^2 + z^2 \leq 1$, $x \geq 0$, $z \geq 0$, then

$$\text{Area} = 4\iint_{S_{xz}} \sqrt{1 + \left(\frac{\partial y}{\partial x}\right)^2 + \left(\frac{\partial y}{\partial z}\right)^2} \, dA$$

$$= 4\iint_{S_{xz}} \sqrt{1 + \left(\frac{x}{\sqrt{x^2 + z^2}}\right)^2 + \left(\frac{z}{\sqrt{x^2 + z^2}}\right)^2} \, dA$$

$$= 4\iint_{S_{xz}} \sqrt{1 + \frac{x^2}{x^2 + z^2} + \frac{z^2}{x^2 + z^2}} \, dA = 4\iint_{S_{xz}} \sqrt{2} \, dA$$

$$= 4\sqrt{2}(\text{Area of } S_{xz}) = 4\sqrt{2}\{\tfrac{1}{4}\pi(1)^2\} = \sqrt{2}\pi.$$

FIGURE 14.37 **FIGURE 14.38**

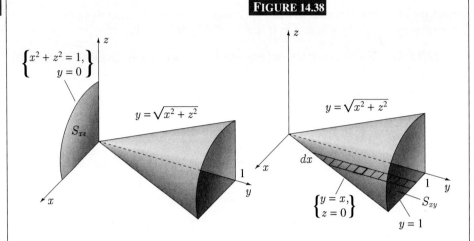

Method 2 Suppose instead that we project that part of the surface in the first octant onto the triangle S_{xy} in Figure 14.38. We write the equation of this part of the surface in the form $z = \sqrt{y^2 - x^2}$ and calculate

$$\text{Area} = 4\iint_{S_{xy}} \sqrt{1 + \left(\frac{\partial z}{\partial x}\right)^2 + \left(\frac{\partial z}{\partial y}\right)^2}\, dA$$

$$= 4\iint_{S_{xy}} \sqrt{1 + \left(\frac{-x}{\sqrt{y^2 - x^2}}\right)^2 + \left(\frac{y}{\sqrt{y^2 - x^2}}\right)^2}\, dA$$

$$= 4\iint_{S_{xy}} \sqrt{1 + \frac{x^2}{y^2 - x^2} + \frac{y^2}{y^2 - x^2}}\, dA = 4\sqrt{2} \iint_{S_{xy}} \frac{y}{\sqrt{y^2 - x^2}}\, dA.$$

To evaluate this double integral, it is advantageous to integrate first with respect to y:

$$\text{Area} = 4\sqrt{2} \int_0^1 \int_x^1 \frac{y}{\sqrt{y^2 - x^2}}\, dy\, dx = 4\sqrt{2} \int_0^1 \{\sqrt{y^2 - x^2}\}_x^1 dx$$

$$= 4\sqrt{2} \int_0^1 \sqrt{1 - x^2}\, dx.$$

If we now set $x = \sin\theta$, then $dx = \cos\theta\, d\theta$, and

$$\text{Area} = 4\sqrt{2} \int_0^{\pi/2} \cos\theta \cos\theta\, d\theta = 4\sqrt{2} \int_0^{\pi/2} \left\{\frac{1 + \cos 2\theta}{2}\right\} d\theta$$

$$= 2\sqrt{2} \left\{\theta + \frac{1}{2}\sin 2\theta\right\}_0^{\pi/2} = \sqrt{2}\,\pi.$$

∎

EXERCISES 14.6

In Exercises 1–6 find the required area.

1. The area of $2x + 3y + 6z = 1$ in the first octant

2. The area of $x + 2y - 3z + 4 = 0$ for which $x \leq 0$, $y \leq 0$ and $z \geq 0$

3. The area of $z = 1 - 4\sqrt{x^2 + y^2}$ above the xy-plane

4. The area of $z = \sqrt{2xy}$ cut out by the planes $x = 1$, $x = 2$, $y = 1$, $y = 3$

5. The area in the first octant cut out from the surface $z = x + y$ by the plane $x + 2y = 4$

6. The area of $z = x^{3/2} + y^{3/2}$ in the first octant cut off by the plane $x + y = 1$

In Exercises 7–12 set up, but do not evaluate, double iterated integrals to find the required area.

7. The area of $x^2 + y^2 + z^2 = 2$ inside the cone $z = \sqrt{x^2 + y^2}$

8. The area of $4x = y^2 + z^2$ cut off by $x = 4$

9. The area in the first octant cut from $y = xz$ by the cylinder $x^2 + z^2 = 1$

10. The area of $z = (x^2 + y^2)^2$ below $z = 4$

11. The area of $y = 1 - x^2 - 3z^2$ to the right of the xz-plane

12. The area of $z = \ln(1 + x + y)$ in the first octant cut off by $y = 1 - x^2$

13. Find the area of the surface $z = \ln x$ that projects onto the rectangle in the xy-plane bounded by the lines $x = 1$, $x = 2$, $y = 0$, $y = 2$

14. Verify that the area of the curved portion of a right-circular cone of radius r and height h is $\pi r \sqrt{r^2 + h^2}$.

In Exercises 15–19 set up, but do not evaluate, double iterated integrals to find the required area.

15. The area of $y = x^2 + z^2$ cut off by $y + z = 1$

16. The area of $y = z^2 + x$ inside $x^2 + y^2 = 1$

17. The area of $y^2 = z + x^2$ inside $x^2 + y^2 = 4$

18. The area of $z = x^3 + y^3$ that is in the first octant and between the planes $x + y = 1$ and $x + y = 2$

19. The area of $x^2 + y^2 = z^2 + 1$ between the planes $z = 1$ and $z = 4$

20. Find the area of that part of the surface $z = 2x^2 + 3y$ bounded by the planes $x = 2$, $y = 0$, and $y = x$.

SECTION 14.7

Double Iterated Integrals in Polar Coordinates

So far we have used only double iterated integrals in x and y to evaluate double integrals. But for some problems this is not convenient. For instance, the double integral of a continuous function $f(x, y)$ over the region R in Figure 14.39 requires three double iterated integrals in x and y. In other words, a subdivision of R into rectangles by coordinate lines $x = $ constant and $y = $ constant is simply not convenient for this region. For such an area, polar coordinates are more useful.

FIGURE 14.39

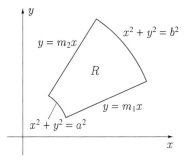

Polar coordinates with the origin as pole and the positive x-axis as polar axis are defined by

$$x = r\cos\theta, \qquad y = r\sin\theta$$

(see Section 10.2). We wish to obtain double iterated integrals in polar coordinates that represent the double integral of $f(x,y)$ over the region R in Figure 14.39. To do this we return to Definition 14.3 for a double integral, and choose a subdivision of R into sub-areas convenient to polar coordinates. When using Cartesian coordinates we drew coordinate lines $x =$ constant and $y =$ constant. When using polar coordinates we draw coordinate curves $r =$ constant and $\theta =$ constant. In particular, we subdivide R by a network of $n+1$ circles $r = r_i$, where

$$a = r_0 < r_1 < r_2 < \cdots < r_{n-1} < r_n = b,$$

and $m+1$ radial lines $\theta = \theta_j$, where

$$c = \theta_0 < \theta_1 < \theta_2 < \cdots < \theta_{m-1} < \theta_m = d.$$

(Figure 14.40). If ΔA_{ij} represents the area bounded by the circles $r = r_{i-1}$ and $r = r_i$ and the radial lines $\theta = \theta_{j-1}$ and $\theta = \theta_j$ (Figure 14.41), then it is straightforward to show that

$$\Delta A_{ij} = \frac{1}{2}(r_i^2 - r_{i-1}^2)(\theta_j - \theta_{j-1}).$$

If we set $\Delta r_i = r_i - r_{i-1}$ and $\Delta \theta_j = \theta_j - \theta_{j-1}$, then

$$\Delta A_{ij} = \frac{1}{2}(r_i + r_{i-1})(r_i - r_{i-1})(\theta_j - \theta_{j-1})$$
$$= \left[\frac{r_i + r_{i-1}}{2}\right] \Delta r_i \, \Delta \theta_j. \qquad (14.43)$$

FIGURE 14.40 **FIGURE 14.41**

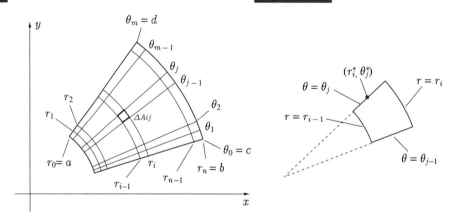

Our next task in using Definition 14.3 for the double integral of $f(x,y)$ over R is to choose a star-point in each ΔA_{ij}. If we select

$$(r_i^*, \theta_j^*) = \left(\frac{r_i + r_{i-1}}{2}, \theta_j\right),$$

then

$$\Delta A_{ij} = r_i^* \, \Delta r_i \, \Delta \theta_j,$$

and by Definition 14.3,

$$\iint_R f(x,y)\,dA = \lim_{\|\Delta A_{ij}\| \to 0} \sum_{j=1}^{m} \sum_{i=1}^{n} f(r_i^* \cos \theta_j^*, r_i^* \sin \theta_j^*) \Delta A_{ij}$$

$$= \lim_{\substack{\|\Delta r_i\| \to 0 \\ \|\Delta \theta_j\| \to 0}} \sum_{j=1}^{m} \sum_{i=1}^{n} f(r_i^* \cos \theta_j^*, r_i^* \sin \theta_j^*) r_i^* \, \Delta r_i \, \Delta \theta_j.$$

If we take the limit first as $||\Delta r_i|| \to 0$ and then as $||\Delta \theta_j|| \to 0$, we obtain the double iterated integral

$$\iint_R f(x,y)\,dA = \int_c^d \int_a^b f(r\cos\theta, r\sin\theta)\,r\,dr\,d\theta. \qquad (14.44)$$

Reversing the order of taking limits reverses the order of the iterated integrals:

$$\iint_R f(x,y)\,dA = \int_a^b \int_c^d f(r\cos\theta, r\sin\theta)\,r\,d\theta\,dr. \qquad (14.45)$$

For the region R of Figure 14.40, then, there are two double iterated integrals in polar coordinates representing the double integral of $f(x,y)$ over R.

We have interpreted double iterated integrals in Cartesian coordinates as integrations over horizontal or vertical strips. Double iterated integrals in polar coordinates can also be interpreted geometrically. Take, for instance, equation 14.44. A double iterated integral in polar coordinates implies a subdivision of the region R into areas as shown in Figure 14.41. Let us denote small variations in r and θ for a representative piece of area at position (r, θ) by dr and $d\theta$ (Figure 14.42). If dr and $d\theta$ are very small (as is implied in the definition of the double integral), then this piece of area is almost rectangular with an approximate area of $(r\,d\theta)\,dr$. In polar coordinates, then, we think of dA in 14.44 as being replaced by

$$dA = r\,dr\,d\theta. \qquad (14.46)$$

Each such area at (r, θ) is multiplied by the value of $f(x,y)$ at (r,θ) to give the product

$$f(r\cos\theta, r\sin\theta)\,r\,dr\,d\theta.$$

The inner integral

$$\int_a^b f(r\cos\theta, r\sin\theta)\,r\,dr\,d\theta$$

with respect to r holds θ constant and is therefore interpreted as a summation over the small areas in a wedge $d\theta$ from $r = a$ to $r = b$. The θ-integration then adds over all wedges starting at $\theta = c$ and ending at $\theta = d$. Limits on θ therefore identify positions of first and last wedges.

FIGURE 14.42

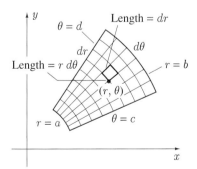

If the order of integration is reversed (equation 14.45), then the inner integral

$$\int_c^d f(r\cos\theta, r\sin\theta)\,r\,d\theta\,dr$$

holds r constant. We interpret this as an addition over the small areas in a ring dr, where the limits indicate that each and every ring starts on the curve $\theta = c$ and ends on the curve $\theta = d$. The outer r-integration is an addition over all rings with the first ring at $r = a$ and the last at $r = b$.

Double iterated integrals in polar coordinates for more general regions are now quite simple. For the region R of Figure 14.43,

$$\iint_R f(x,y)\,dA = \int_\alpha^\beta \int_{g(\theta)}^{h(\theta)} f(r\cos\theta, r\sin\theta)\, r\, dr\, d\theta.$$

FIGURE 14.43

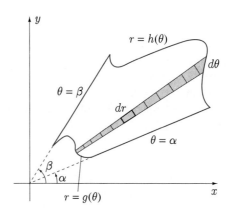

EXAMPLE 14.15

Find the area inside the circle $x^2 + y^2 = 4x$ and outside the circle $x^2 + y^2 = 1$.

SOLUTION If R is the region bounded by these circles and above the x-axis (Figure 14.44), then the required area is

$$2 \iint_R dA.$$

Since the curves intersect in the first quadrant at a point where $\theta = \overline{\theta} = \mathrm{Cos}^{-1}(\tfrac{1}{4})$, then

$$\text{Area} = 2\int_0^{\overline{\theta}} \int_1^{4\cos\theta} r\, dr\, d\theta = 2\int_0^{\overline{\theta}} \left\{\frac{r^2}{2}\right\}_1^{4\cos\theta} d\theta = \int_0^{\overline{\theta}} (16\cos^2\theta - 1)\, d\theta$$

$$= \int_0^{\overline{\theta}} \left\{16\left(\frac{1+\cos 2\theta}{2}\right) - 1\right\} d\theta = \int_0^{\overline{\theta}} (7 + 8\cos 2\theta)\, d\theta = \{7\theta + 4\sin 2\theta\}_0^{\overline{\theta}}$$

$$= 7\overline{\theta} + 4\sin 2\overline{\theta} = 7\,\mathrm{Cos}^{-1}(\tfrac{1}{4}) + 8\cos\overline{\theta}\sin\overline{\theta}$$

$$= 7\,\mathrm{Cos}^{-1}(\tfrac{1}{4}) + 8(\tfrac{1}{4})\sqrt{1 - \tfrac{1}{16}} = 7\,\mathrm{Cos}^{-1}(\tfrac{1}{4}) + \sqrt{15}/2.$$

FIGURE 14.44

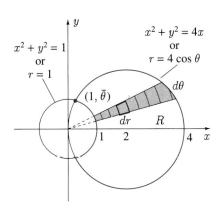

EXAMPLE 14.16

Evaluate the double iterated integral

$$\int_0^1 \int_0^{\sqrt{-x^2+x}} y^2 \, dy \, dx.$$

SOLUTION The limits identify the region of integration as the interior of the semicircle in Figure 14.45. The integrand suggests an interpretation of the integral as the second moment of area of this semicircle about the x-axis. Since the semicircle R in Figure 14.46 has exactly the same second moment about the x-axis, we can state that

$$\int_0^1 \int_0^{\sqrt{-x^2+x}} y^2 \, dy \, dx = \iint_R y^2 \, dA = \int_0^\pi \int_0^{1/2} (r^2 \sin^2 \theta) r \, dr \, d\theta$$

$$= \int_0^\pi \left\{ \frac{r^4}{4} \sin^2 \theta \right\}_0^{1/2} d\theta = \frac{1}{64} \int_0^\pi \sin^2 \theta \, d\theta$$

$$= \frac{1}{128} \int_0^\pi (1 - \cos 2\theta) \, d\theta = \frac{1}{128} \left\{ \theta - \frac{\sin 2\theta}{2} \right\}_0^\pi = \frac{\pi}{128}.$$

FIGURE 14.45

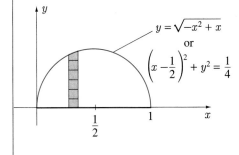

FIGURE 14.46

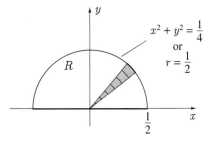

EXAMPLE 14.17

Find the centroid of the region R in Figure 14.47.

SOLUTION Evidently, $\bar{x} = 0$, and the area of the region is $A = (\pi b^2 - \pi a^2)/2$. Since

$$A\bar{y} = \iint_R y\,dA = \int_0^\pi \int_a^b (r\sin\theta)r\,dr\,d\theta = \int_0^\pi \left\{\frac{r^3}{3}\sin\theta\right\}_a^b d\theta$$

$$= \frac{1}{3}(b^3 - a^3)\int_0^\pi \sin\theta\,d\theta$$

$$= \frac{1}{3}(b^3 - a^3)\{-\cos\theta\}_0^\pi = \frac{2}{3}(b^3 - a^3),$$

if follows that

$$\bar{y} = \frac{2}{3}(b^3 - a^3)\frac{2}{\pi(b^2 - a^2)} = \frac{4}{3\pi}\frac{b^2 + ab + a^2}{a + b}.$$

FIGURE 14.47

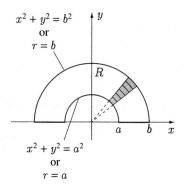

EXAMPLE 14.18

Find the area of that portion of the sphere $x^2 + y^2 + z^2 = 2$ inside the cone $z = \sqrt{x^2 + y^2}$.

SOLUTION If S is that portion of the sphere that is inside the cone and also in the first octant (Figure 14.48), then the required area is four times that of S; i.e.,

$$\text{Area} = 4\iint_{S_{xy}} \sqrt{1 + \left(\frac{\partial z}{\partial x}\right)^2 + \left(\frac{\partial z}{\partial y}\right)^2}\,dA,$$

where S_{xy} is the projection of S on the xy-plane. The curve of intersection of the cone and the sphere has equations

$$x^2 + y^2 + z^2 = 2, \qquad\qquad z = 1,$$

or equivalently

$$z = \sqrt{x^2 + y^2}, \qquad\qquad x^2 + y^2 = 1.$$

Consequently, S_{xy} is the interior of the quarter-circle $x^2 + y^2 \leq 1$, $x \geq 0$, $y \geq 0$. On S,
$$\frac{\partial z}{\partial x} = -\frac{x}{z} \quad \text{and} \quad \frac{\partial z}{\partial y} = -\frac{y}{z},$$
so that
$$\text{Area} = 4 \iint_{S_{xy}} \sqrt{1 + \frac{x^2}{z^2} + \frac{y^2}{z^2}} \, dA = 4 \iint_{S_{xy}} \sqrt{\frac{x^2 + y^2 + z^2}{z^2}} \, dA$$
$$= 4 \iint_{S_{xy}} \sqrt{\frac{2}{z^2}} \, dA = 4\sqrt{2} \iint_{S_{xy}} \frac{1}{\sqrt{2 - x^2 - y^2}} \, dA.$$

If we now use polar coordinates to evaluate this double integral, we have
$$\text{Area} = 4\sqrt{2} \int_0^{\pi/2} \int_0^1 \frac{1}{\sqrt{2 - r^2}} r \, dr \, d\theta = 4\sqrt{2} \int_0^{\pi/2} \{-\sqrt{2 - r^2}\}_0^1 \, d\theta$$
$$= 4\sqrt{2}(\sqrt{2} - 1) \int_0^{\pi/2} d\theta = 4\sqrt{2}(\sqrt{2} - 1) \frac{\pi}{2} = 2\sqrt{2}\pi(\sqrt{2} - 1).$$

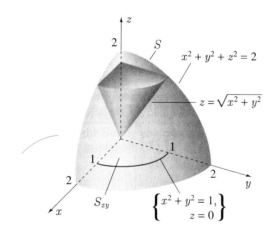

FIGURE 14.48

EXERCISES 14.7

In Exercises 1–5 evaluate the double integral of the function over the region R.

1. $f(x, y) = e^{x^2 + y^2}$ where R is bounded by $x^2 + y^2 = a^2$

2. $f(x, y) = x$ where R is bounded by $x = \sqrt{2y - y^2}$, $x = 0$

3. $f(x, y) = \sqrt{x^2 + y^2}$ where R is bounded by $y = \sqrt{9 - x^2}$, $y = x$, $x = 0$

4. $f(x, y) = 1/\sqrt{x^2 + y^2}$ where R is the region outside $x^2 + y^2 = 4$ and inside $x^2 + y^2 = 4x$

5. $f(x, y) = \sqrt{1 + 2x^2 + 2y^2}$ where R is bounded by $x^2 + y^2 = 1$, $x^2 + y^2 = 4$

Evaluate the double iterated integral in Exercises 6 and 7.

6. $\displaystyle\int_0^1 \int_0^{\sqrt{1-x^2}} \sqrt{x^2 + y^2} \, dy \, dx$

7. $\displaystyle\int_{-\sqrt{2}}^0 \int_{-y}^{\sqrt{4-y^2}} x^2 \, dx \, dy$

In Exercises 8–12 find the area bounded by the curves.

8. Outside $x^2 + y^2 = 9$ and inside $x^2 + y^2 = 2\sqrt{3}y$

9. $r = 9(1 + \cos\theta)$

10. $r = \cos 3\theta$

11. Common to $r = 2$ and $r^2 = 9\cos 2\theta$

12. Common to $r = 1 + \sin\theta$ and $r = 2 - 2\sin\theta$

13. Find the centroid of the area bounded by the curves $y = x$, $y = -x$, $x = \sqrt{2 - y^2}$

14. Find the second moment of area for a circular plate of radius R about any diameter.

15. A water tank in the form of a right-circular cylinder with radius R and length h has its axis horizontal. If it is full, what is the force due to water pressure on each end?

In Exercises 16–18 find the area of the surface.

16. The area of $z = x^2 + y^2$ below $z = 4$

17. The area of $x^2 + y^2 + z^2 = 4$ inside $x^2 + y^2 = 1$

18. The area of $z = xy$ inside $x^2 + y^2 = 9$

19. Prove that the area of a sphere of radius R is $4\pi R^2$.

20. Find the area of the hyperbolic paraboloid $z = x^2 - y^2$ between the cylinders $x^2 + y^2 = 1$ and $x^2 + y^2 = 4$

In Exercises 21 and 22 find the volume of the solid of revolution obtained by rotating the area bounded by the curve about the line.

21. $r = \cos^2\theta$ about the x-axis

22. $r = 1 + \sin\theta$ about the y-axis

23. Find the area bounded by the curve $(x^2 + y^2)^3 = 4a^2 x^2 y^2$.

24. Find the volume of the solid of revolution when a circle of radius R is rotated about a tangent line.

25. A circular plate of radius R (Figure 14.49) has a uniform charge distribution of ρ coulombs per square metre. If P is a point directly above the centre of the plate and dA is a small area on the plate, then the potential at P due to dA is given by

$$\frac{1}{4\pi\epsilon_0} \frac{\rho dA}{s},$$

where s is the distance from P to dA.

(a) Show that, in terms of polar coordinates, the potential V at P due to the entire plate is

$$V = \frac{\rho}{4\pi\epsilon_0} \int_{-\pi}^{\pi} \int_0^R \frac{r}{\sqrt{r^2 + d^2}} dr\, d\theta,$$

where d is the distance from P to the centre of the plate.

(b) Evaluate the double iterated integral to find V.

FIGURE 14.49

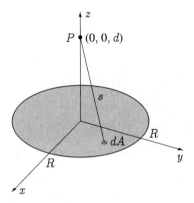

26. Use Coulomb's law (see Example 12.9 in Section 12.3) to find the force on a charge q at point P due to the charge on the plate in Exercise 25. What happens to this force as the radius of the plate gets very large?

In Exercises 27–29 find the area bounded by the curves.

27. $(x^2 + y^2)^2 = 2xy$

28. Inside both $r = 6\cos\theta$ and $r = 4 - 2\cos\theta$

29. $r = \cos^2\theta \sin\theta$

30. Find the centroid of the cardioid $r = 1 + \cos\theta$.

31. Find the second moment of area about the x-axis for the area bounded by $r^2 = 9\cos 2\theta$.

32. Evaluate the double integral of

$$f(x,y) = \sqrt{\frac{1 - x^2 - y^2}{1 + x^2 + y^2}}$$

over the area inside the circle $x^2 + y^2 = 1$.

33. Figure 14.50 illustrates a piece of an artery or vein with circular cross-section (radius R). The speed of blood flowing through the blood vessel is not uniform because of the viscosity of the blood and friction at the walls. Poiseuille's law states that for laminar blood flow, the speed v of blood at a distance r from the centre of the vessel is given by

$$v = \frac{P}{4nL}(R^2 - r^2),$$

where P is the pressure difference between the ends of the vessel, L is the length of the vessel, and n is the viscosity of the blood. Find the amount of blood flowing over a cross-section of the blood vessel per unit time.

FIGURE 14.50

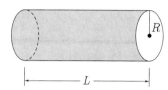

34. Find the area of that part of the sphere $x^2 + y^2 + z^2 = a^2$ inside $(x^2 + y^2)^2 = a^2(x^2 - y^2)$.

35. Find the area of that portion of the surface $x^2 + z^2 = a^2$ cut out by $x^2 + y^2 = a^2$.

36. A very important integral in statistics is

$$I = \int_0^\infty e^{-x^2}\,dx.$$

To evaluate the integral we set

$$I = \int_0^\infty e^{-y^2}\,dy,$$

and then multiply these two equations. Do this to prove that $I = \sqrt{\pi}/2$.

37. Use the result of Exercise 36 to evaluate the gamma function

$$\Gamma(n) = \int_0^\infty x^{n-1} e^{-x}\,dx$$

at $n = 1/2$.

SECTION 14.8

Triple Integrals and Triple Iterated Integrals

Triple integrals are defined in much the same way as double integrals. Suppose $f(x, y, z)$ is a function defined in some region V of space that has finite volume (Figure 14.51). We divide V into n subregions of volumes $\Delta V_1, \Delta V_2, \ldots, \Delta V_n$ in any manner whatsoever, and in each subregion ΔV_i ($i = 1, \ldots, n$) we choose an arbitrary point (x_i^*, y_i^*, z_i^*). We then form the sum

$$f(x_1^*, y_1^*, z_1^*)\Delta V_1 + f(x_2^*, y_2^*, z_2^*)\Delta V_2 + \cdots + f(x_n^*, y_n^*, z_n^*)\Delta V_n$$
$$= \sum_{i=1}^n f(x_i^*, y_i^*, z_i^*)\Delta V_i. \qquad (14.47)$$

If this sum approaches a limit as the number of subregions becomes increasingly large and every subregion shrinks to a point, we call the limit the triple integral of $f(x, y, z)$ over the region V and denote it by

$$\iiint_V f(x, y, z)\,dV = \lim_{\|\Delta V_i\| \to 0} \sum_{i=1}^n f(x_i^*, y_i^*, z_i^*)\Delta V_i. \qquad (14.48)$$

FIGURE 14.51

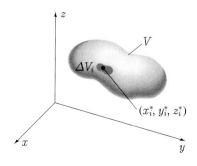

As in the case of double integrals, we require that this limit be independent of the manner of subdivision of V and the choice of star-points in the subregions. This is guaranteed for continuous functions by the following theorem.

Theorem 14.2

Let S be a piecewise-smooth surface that encloses a region V with finite volume. If $f(x,y,z)$ is continuous inside and on S, then the triple integral of $f(x,y,z)$ over V exists.

Properties analogous to those in equations 14.4–14.7 hold for triple integrals, although we will not list the first three here. Corresponding to equation 14.7, the volume of a region V is given by the triple integral

$$\text{Volume of } V = \iiint_V dV. \qquad (14.49)$$

We evaluate triple integrals with triple iterated integrals. If we use Cartesian coordinates there are six possible triple iterated integrals of a function $f(x,y,z)$, corresponding to the six permutations of the product of the differentials dx, dy, and dz:

$$dz\,dy\,dx, \quad dz\,dx\,dy, \quad dx\,dz\,dy, \quad dx\,dy\,dz, \quad dy\,dx\,dz, \quad dy\,dz\,dx.$$

The general triple iterated integral of $f(x,y,z)$ with respect to z, y, and x is of the form

$$\int_a^b \int_{g_1(x)}^{g_2(x)} \int_{h_1(x,y)}^{h_2(x,y)} f(x,y,z)\,dz\,dy\,dx. \qquad (14.50)$$

Because the first integration with respect to z holds x and y constant, the limits on z may therefore depend on x and y. Similarly, the second integration with respect to y holds x constant, and the limits may be functions of x.

EXAMPLE 14.19

Evaluate the triple iterated integral

$$\int_0^1 \int_0^{x^2} \int_{xy}^{x+y} xyz\,dz\,dy\,dx.$$

SOLUTION

$$\int_0^1 \int_0^{x^2} \int_{xy}^{x+y} xyz\,dz\,dy\,dx = \int_0^1 \int_0^{x^2} \left\{\frac{xyz^2}{2}\right\}_{xy}^{x+y} dy\,dx$$

$$= \frac{1}{2}\int_0^1 \int_0^{x^2} \{xy(x+y)^2 - xy(xy)^2\}dy\,dx$$

$$= \frac{1}{2}\int_0^1 \int_0^{x^2} (x^3y + 2x^2y^2 + xy^3 - x^3y^3)\,dy\,dx$$

$$= \frac{1}{2} \int_0^1 \left\{ \frac{x^3 y^2}{2} + \frac{2x^2 y^3}{3} + \frac{xy^4}{4} - \frac{x^3 y^4}{4} \right\}_0^{x^2} dx$$

$$= \frac{1}{24} \int_0^1 (6x^7 + 8x^8 + 3x^9 - 3x^{11}) \, dx$$

$$= \frac{1}{24} \left\{ \frac{3x^8}{4} + \frac{8x^9}{9} + \frac{3x^{10}}{10} - \frac{x^{12}}{4} \right\}_0^1 = \frac{19}{270}.$$

■

Because of the analogy between double and triple integrals, we accept without proof that triple integrals can be evaluated with triple iterated integrals. We must, however, examine how triple iterated integrals bring about the summations represented by triple integrals, for it is only by thoroughly understanding this process that we can obtain limits for triple iterated integrals.

In Section 14.3 we discussed in considerable detail the evaluation of double integrals by means of double iterated integrals. In particular, we showed that double iterated integrals in Cartesian coordinates represent the subdivision of an area into small rectangles by coordinate lines $x =$ constant and $y =$ constant. The first integration creates a summation over rectangles in a strip, and the second integration adds over all strips. It is fairly straightforward to generalize these ideas to triple integrals.

Consider evaluating the triple integral

$$\iiint_V f(x, y, z) \, dV$$

over the region V in Figure 14.52 bounded above by the surface $z = h(x, y)$, below by the area R in the xy-plane, and on the sides by a cylindrical wall standing on the curve bounding R.

FIGURE 14.52

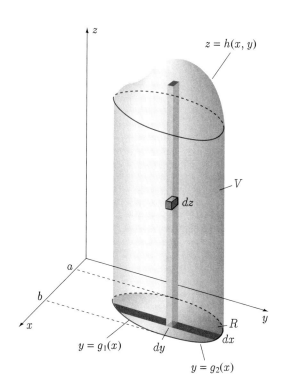

The choice of a triple iterated integral in Cartesian coordinates to evaluate this triple integral implies a subdivision of V into small rectangular parallelepipeds (boxes for short) by means of coordinate planes x = constant, y = constant, and z = constant. The dimensions of a representative box at position (x, y, z) in V are denoted by dx, dy, and dz, with resulting volume $dx\, dy\, dz$. If we decide on a triple iterated integral with respect to z, y, and x, then the first integration on z holds x and y constant. This integration therefore adds the quantities

$$f(x,y,z)\,dz\,dy\,dx$$

over boxes in a vertical column of cross-sectional dimensions dx and dy. Lower and upper limits on z identify where each and every column starts and stops, and must consequently be 0 and $h(x,y)$:

$$\int_0^{h(x,y)} f(x,y,z)\,dz\,dy\,dx.$$

Since this integration produces a function of x and y alone, the remaining integration with respect to y and x is essentially a double iterated integral in the xy-plane. These integrations must account for all columns in V and therefore the area in the xy-plane over which this double iterated integral is performed is the area R upon which all columns in V stand. Since the y-integration adds inside a strip in the y-direction and limits identify where all strips start and stop, they must therefore be $g_1(x)$ and $g_2(x)$. We have now

$$\int_{g_1(x)}^{g_2(x)} \int_0^{h(x,y)} f(x,y,z)\,dz\,dy\,dx.$$

Finally, the x-integration adds over all strips and the limits are $x = a$ and $x = b$:

$$\iiint_V f(x,y,z)\,dV = \int_a^b \int_{g_1(x)}^{g_2(x)} \int_0^{h(x,y)} f(x,y,z)\,dz\,dy\,dx.$$

Suppose now that V is the region bounded above by the surface $z = h_2(x,y)$ and below by $z = h_1(x,y)$ (Figure 14.53). In this case, the limits on the first integration with respect to z are $h_1(x,y)$ and $h_2(x,y)$ since every column starts on the surface $z = h_1(x,y)$ and ends on the surface $z = h_2(x,y)$:

$$\int_{h_1(x,y)}^{h_2(x,y)} f(x,y,z)\,dz\,dy\,dx.$$

For the volume of Figure 14.52 we interpreted the final two integrations as a double iterated integral in the xy-plane over the area R from which all columns emanated. For the present volume we interpret R as the area in the xy-plane onto which all columns project. We obtain then

$$\iiint_V f(x,y,z)\,dV = \int_a^b \int_{g_1(x)}^{g_2(x)} \int_{h_1(x,y)}^{h_2(x,y)} f(x,y,z)\,dz\,dy\,dx.$$

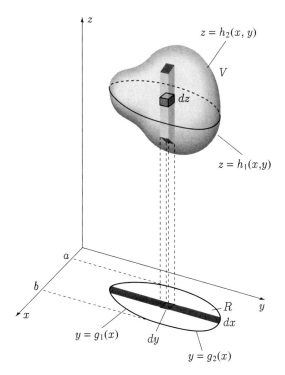

FIGURE 14.53

Schematically we have obtained the following interpretation for the limits of triple iterated integrals in Cartesian coordinates:

$$\int_{\substack{\text{position of} \\ \text{first strip}}}^{\substack{\text{position of} \\ \text{last strip}}} \int_{\substack{\text{where every} \\ \text{strip starts}}}^{\substack{\text{where every} \\ \text{strip stops}}} \int_{\substack{\text{where every} \\ \text{column starts}}}^{\substack{\text{where every} \\ \text{column stops}}} f(x,y,z) \begin{Bmatrix} dz\, dy\, dx \\ dz\, dx\, dy \\ dx\, dz\, dy \\ dx\, dy\, dz \\ dy\, dx\, dz \\ dy\, dz\, dx \end{Bmatrix}.$$

EXAMPLE 14.20

Set up the six triple iterated integrals in Cartesian coordinates for the triple integral of a function $f(x, y, z)$ over the region V in the first octant bounded by the surfaces

$$y^2 + z^2 = 1, \quad y = x, \quad z = 0, \quad x = 0.$$

SOLUTION The triple integral of $f(x, y, z)$ over V is given by each of the following triple iterated integrals (see Figure 14.54):

$$\int_0^1 \int_x^1 \int_0^{\sqrt{1-y^2}} f(x,y,z)\,dz\,dy\,dx,$$

$$\int_0^1 \int_0^y \int_0^{\sqrt{1-y^2}} f(x,y,z)\,dz\,dx\,dy,$$

$$\int_0^1 \int_0^{\sqrt{1-z^2}} \int_0^y f(x,y,z)\,dx\,dy\,dz,$$

$$\int_0^1 \int_0^{\sqrt{1-y^2}} \int_0^y f(x,y,z)\,dx\,dz\,dy,$$

$$\int_0^1 \int_0^{\sqrt{1-x^2}} \int_x^{\sqrt{1-z^2}} f(x,y,z)\,dy\,dz\,dx,$$

$$\int_0^1 \int_0^{\sqrt{1-z^2}} \int_x^{\sqrt{1-z^2}} f(x,y,z)\,dy\,dx\,dz.$$

FIGURE 14.54

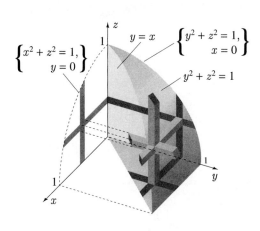

EXAMPLE 14.21

Set up triple iterated integrals with respect to z, y, and x for the triple integral of the function $f(x,y,z) = x^2 y \sin z$ over the region V bounded by the surfaces

$$z = \sqrt{y}, \quad y + z = 2, \quad x = 0, \quad z = 0, \quad x = 2.$$

SOLUTION The problem requires triple iterated integrals first with respect to z, then with respect to y and x. Since some columns end on the parabolic cylinder $z = \sqrt{y}$ (Figure 14.55) and others on the plane $y + z = 2$, we require two iterated integrals:

$$\iiint_V x^2 y \sin z \, dV = \int_0^2 \int_0^1 \int_0^{\sqrt{y}} x^2 y \sin z \, dz \, dy \, dx$$
$$+ \int_0^2 \int_1^2 \int_0^{2-y} x^2 y \sin z \, dz \, dy \, dx.$$

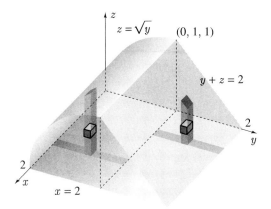

FIGURE 14.55

EXERCISES 14.8

In Exercises 1–12 evaluate the triple integral over the region.

1. $\iiint_V (x^2 z + ye^x) \, dV$ where V is bounded by $x = 0$, $x = 1$, $y = 1$, $y = 2$, $z = 0$, $z = 1$

2. $\iiint_V x \, dV$ where V is bounded by $x = 0$, $y = 0$, $z = 0$, $x + y + z = 4$

3. $\iiint_V \sin(y + z) \, dV$ where V is bounded by $z = 0$, $y = 2x$, $y = 0$, $x = 1$, $z = x + 2y$

4. $\iiint_V xy \, dV$ where V is enclosed by $z = \sqrt{1 - x^2 - y^2}$, $z = 0$

5. $\iiint_V dV$ where V is bounded by $x^2 + y^2 = 1$, $x^2 + z^2 = 1$

6. $\iiint_V (x^2 + 2z) \, dV$ where V is bounded by $z = 0$, $y + z = 4$, $y = x^2$

7. $\iiint_V x^2 y^2 z^2 \, dV$ where V is bounded by $z = 1 + y$, $y + z = 1$, $x = 1$, $x = 0$, $z = 0$

8. $\iiint_V xyz \, dV$ where V is the first octant volume cut out by $z = x^2 + y^2$, $z = \sqrt{x^2 + y^2}$

9. $\iiint_V dV$ where V is bounded by $z = x^2$, $y + z = 4$, $y = 0$

10. $\iiint_V (x + y + z) \, dV$ where V is bounded by $x = 0$, $x = 1$, $z = 0$, $y + z = 2$, $y = z$

11. $\iiint_V xyz \, dV$ where V is bounded by $z = 1$, $z = x^2/4 + y^2/9$

12. $\iiint_V x^2 y \, dV$ where V is the volume in the first octant bounded by $z = 1$, $z = x^2/4 + y^2/9$

13. Set up the six triple iterated integrals in Cartesian coordinates for the triple integral of a function $f(x, y, z)$ over the volume enclosed by the surfaces $y = 1 - x^2$, $z = 0$, and $y = z$.

In Exercises 14–17 set up, but do not evaluate, a triple iterated integral for the triple integral.

14. $\iiint_V (x^2 + y^2 + z^2) \, dV$ where V is bounded by $z = \sqrt{1 - x^2 - y^2}$, $z = x^2$

15. $\iiint_V xz \sin(x + y) \, dV$ where V is bounded by $y^2 = 1 + 4x^2 + 4z^2$, $y = \sqrt{4 + x^2}$

16. $\iiint_V xyz \, dV$ where V is bounded by $z = x^2 + 4y^2$, $2x + 8y + z = 4$

17. $\iiint_V x^2 y^2 z^2 \, dV$ where V is bounded by $x = y^2 + z^2$, $x + 1 = (y^2 + z^2)^2$

In Exercises 18–23 evaluate the triple integral over the region.

18. $\iiint_V (y + x^2) \, dV$ where V is bounded by $x + z^2 = 1$, $z = x + 1$, $y = 1$, $y = -1$

19. $\iiint_V (xy + z) \, dV$ where V is bounded by $y + z = 1$, $z = 2y$, $z = y$, $x = 0$, $x = 3$

20. $\iiint_V dV$ where V is bounded by $z = 0$, $x^2 + y^2 = 1$, $x + y + z = 2$

21. $\iiint_V dV$ where V is bounded by $z = x^2 + y^2$, $z = 4 - x^2 - y^2$

22. $\iiint_V (x + y + z) \, dV$ where V is bounded by $2z = y^2 - x^2$, $z = 1 - x^2$

23. $\iiint_V |yz| \, dV$ where V is bounded by $z^2 = 1 + x^2 + y^2$, $z = \sqrt{4 - x^2 - y^2}$

24. Set up, but do not evaluate, triple iterated integrals to evaluate the triple integral of the function $f(x, y, z) = x^2 + y^2 + z^2$ over the volume bounded by the surfaces $x^2 + y^2 = z^2 + 1$, $2z = \sqrt{x^2 + y^2}$, $z = 0$.

SECTION 14.9

Volumes

Because the volume of a region V is represented by triple integral 14.49 and triple integrals are evaluated by means of triple iterated integrals, it follows that volumes can be evaluated with triple iterated integrals. For example, to evaluate the volume of the region in Figure 14.53 using a triple iterated integral in x, y, and z, we subdivide V into boxes of dimensions dx, dy, and dz and therefore of volume

$$dz \, dy \, dx.$$

Integration with respect to z adds these volumes in the z-direction to give the volume of a vertical column (Figure 14.56):

$$\int_{h_1(x,y)}^{h_2(x,y)} dz \, dy \, dx.$$

The limits indicate that all columns start on the surface $z = h_1(x, y)$ and end on the surface $z = h_2(x, y)$. Integration with respect to y now adds the volumes of columns that project onto a strip in the y-direction:

$$\int_{g_1(x)}^{g_2(x)} \int_{h_1(x,y)}^{h_2(x,y)} dz \, dy \, dx,$$

where the limits indicate that all strips start on the curve $y = g_1(x)$ and end on the curve $y = g_2(x)$. Evidently this integration yields the volume of a slab as shown in Figure 14.56. Finally, integration with respect to x adds the volumes of all such slabs in V:

$$\int_a^b \int_{g_1(x)}^{g_2(x)} \int_{h_1(x,y)}^{h_2(x,y)} dz\, dy\, dx,$$

where a and b designate the positions of the first and last strips.

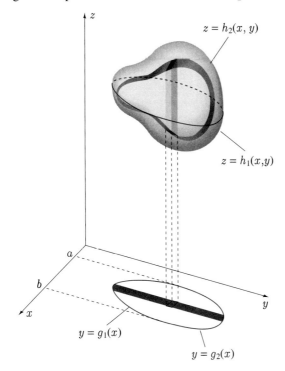

FIGURE 14.56

EXAMPLE 14.22

Find the volume bounded by the planes $z = x + y$, $y = 2x$, $z = 0$, $x = 0$, $y = 2$.

SOLUTION If we use vertical columns (Figure 14.57), we have

$$\text{Volume} = \int_0^1 \int_{2x}^2 \int_0^{x+y} dz\, dy\, dx = \int_0^1 \int_{2x}^2 (x+y)\, dy\, dx = \int_0^1 \left\{ xy + \frac{y^2}{2} \right\}_{2x}^2 dx$$

$$= 2\int_0^1 (1 + x - 2x^2)\, dx = 2\left\{ x + \frac{x^2}{2} - \frac{2x^3}{3} \right\}_0^1 = \frac{5}{3}.$$

FIGURE 14.57

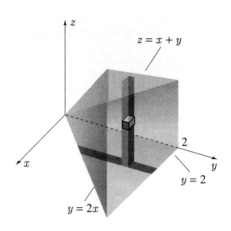

EXAMPLE 14.23

Find the volume in the first octant cut from the cylinder $x^2 + z^2 = 4$ by the plane $y + z = 6$.

SOLUTION With columns in the y-direction (Figure 14.58), we have

$$\text{Volume} = \int_0^2 \int_0^{\sqrt{4-x^2}} \int_0^{6-z} dy \, dz \, dx = \int_0^2 \int_0^{\sqrt{4-x^2}} (6-z) \, dz \, dx$$

$$= \int_0^2 \left\{ 6z - \frac{z^2}{2} \right\}_0^{\sqrt{4-x^2}} dx$$

$$= \int_0^2 \left\{ 6\sqrt{4-x^2} - \frac{1}{2}(4-x^2) \right\} dx.$$

In the first term we set $x = 2 \sin \theta$, from which we get $dx = 2 \cos \theta \, d\theta$, and

$$\text{Volume} = 6 \int_0^{\pi/2} (2 \cos \theta) 2 \cos \theta \, d\theta + \left\{ -2x + \frac{x^3}{6} \right\}_0^2$$

$$= 12 \int_0^{\pi/2} (1 + \cos 2\theta) \, d\theta - \frac{8}{3}$$

$$= 12 \left\{ \theta + \frac{1}{2} \sin 2\theta \right\}_0^{\pi/2} - \frac{8}{3} = 6\pi - \frac{8}{3}.$$

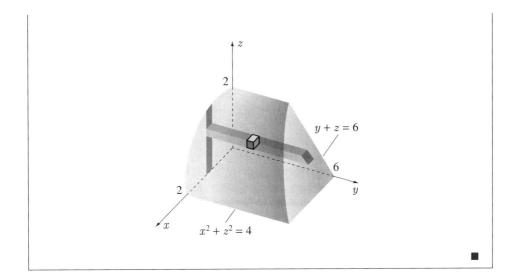

FIGURE 14.58

Had we used iterated integrals with respect to $z, y,$ and x in this example, we would have had the two integrals

$$\text{Volume} = \int_0^2 \int_0^{6-\sqrt{4-x^2}} \int_0^{\sqrt{4-x^2}} dz\, dy\, dx + \int_0^2 \int_{6-\sqrt{4-x^2}}^6 \int_0^{6-y} dz\, dy\, dx.$$

EXERCISES 14.9

In Exercises 1–17 find the volume of the region bounded by the surfaces.

1. $y = x^2$, $y = 1$, $z = 0$, $z = 4$
2. $x = z^2$, $z = x^2$, $y = 0$, $y = 2$
3. $x = 3z$, $z = 3x$, $y = 1$, $y = 0$, $x = 2$
4. $x + y + z = 6$, $y = 4 - x^2$, $z = 0$, $y = 0$
5. $z = x^2 + y^2$, $y = x^2$, $y = 4$, $z = 0$
6. $x + y + z = 4$, $y = 3z$, $x = 0$, $y = 0$
7. $x^2 + y^2 = 4$, $y^2 + z^2 = 4$
8. $y = x^2 - 1$, $y = 1 - x^2$, $x + z = 1$, $z = 0$
9. $z = 16 - x^2 - 4y^2$, $x + y = 1$, $z = 16$, $x = 0$, $y = 0$ (in the first octant)
10. $z = x^2 + y^2$, $x = 1$, $z = 0$, $x = y$, $x = 2y$
11. $z = 1 - x^2 - y^2$, $z = 0$
12. $x - z = 0$, $x + z = 3$, $y + z = 1$, $z = y + 1$, $z = 0$
13. $x + y + z = 2$, $x^2 + y^2 = 1$, $z = 0$
14. $y + z = 1$, $z = 2y$, $z = y$, $x = 0$, $x + y + z = 4$
15. $x^2 + 4y^2 = z$, $x^2 + 4y^2 = 12 - 2z$
16. $y = 1 - z^2$, $y = z^2 - 1$, $x = 1 - z^2$, $x = z^2 - 1$
17. $x + 3y + 2z = 6$, $z = 0$, $y = x$, $y = 2x$
18. Find the volume in the first octant bounded by the plane $2x + y + z = 2$ and inside the cylinder $y^2 + z^2 = 1$.
19. A pyramid has a square base with side length b and has height h at its centre.
    (a) Find its volume by taking cross-sections parallel to the base (see Section 7.8).
    (b) Find its volume using triple integrals.

The average value of a function $f(x, y, z)$ over a region with volume V is defined as

$$\overline{f} = \frac{1}{V} \iiint_V f(x, y, z)\, dV.$$

In Exercises 20–22 find the average value of the function over the region.

20. $f(x, y, z) = xy$ over the region bounded by the surfaces $x = 0$, $y = 0$, $z = 0$, $x + y + z = 1$
21. $f(x, y, z) = x + y + z$ over the region in the first octant bounded by the surfaces $z = 9 - x^2 - y^2$, $z = 0$ and for which $0 \leq x \leq 1$, $0 \leq y \leq 1$

22. $f(x, y, z) = x^2 + y^2 + z^2$ over the region bounded by the surfaces $x = 0$, $x = 1$, $y + z = 2$, $y = 2$, $z = 2$

23. Find the volume bounded by the surfaces $z = x^2 - y^2$, and $z = 4 - 2(x^2 + y^2)$.

24. Verify that the surfaces $z = x^2 - y^2$ and $z = 4 - x^2 - y^2$ do not bound a finite volume.

25. Find the volume bounded by the surfaces $x + z = 2$, $z = 0$, $4y^2 = x(2 - z)$.

26. Find the volume bounded by the surfaces $z = (x - 1)^2 + y^2$, $2x + z = 2$.

27. Find the volume inside the ellipsoid $x^2/a^2 + y^2/b^2 + z^2/c^2 = 1$.

28. The bottom and sides of a boat are defined by the surface equation $x = 10(1 - y^2 - z^2)$, $0 \le x \le 10$, where all dimensions are in metres.
 (a) Find the volume of water displaced by the boat when the water level on the side of the boat is d metres below the top of the boat.
 (b) Archimedes' principle states that the buoyant force on an object when immersed or partially immersed in a fluid is equal to the weight of the fluid displaced by the object. Find the maximum weight of the boat and contents just before sinking.

29. Find the volume inside all three surfaces $x^2 + y^2 = a^2$, $x^2 + z^2 = a^2$, $y^2 + z^2 = a^2$.

SECTION 14.10

Centres of Mass and Moments of Inertia

In this section we discuss centres of mass and moments of inertia for three-dimensional objects of density $\rho(x, y, z)$ (mass per unit volume). If we divide the object occupying region V in Figure 14.59 into small volumes dV, then the amount of mass in dV is $\rho \, dV$. The triple integral

$$M = \iiint_V \rho \, dV \qquad (14.51)$$

adds the masses of all such volumes (of ever-decreasing size) to produce the total mass M of the object.

FIGURE 14.59

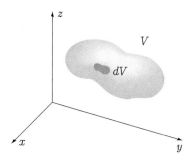

Corresponding to equations 14.29 and 14.30 for first moments of planar masses about the coordinate axes are the following formulas for first moments of the object about the coordinate planes:

$$\text{First moment of object about } yz\text{-plane} = \iiint_V x\rho \, dV, \qquad (14.52)$$

$$\text{First moment of object about } xz\text{-plane} = \iiint_V y\rho \, dV, \qquad (14.53)$$

$$\text{First moment of object about } xy\text{-plane} = \iiint_V z\rho \, dV. \qquad (14.54)$$

The centre of mass of the object is defined as that point $(\bar{x}, \bar{y}, \bar{z})$ at which a particle of mass M would have the same first moments about the coordinate planes as the object itself. Since the first moments of M at $(\bar{x}, \bar{y}, \bar{z})$ about the coordinate planes are $M\bar{x}, M\bar{y}$, and $M\bar{z}$, it follows that we can use the equations

$$M\bar{x} = \iiint_V x\rho \, dV, \qquad (14.55)$$

$$M\bar{y} = \iiint_V y\rho \, dV, \qquad (14.56)$$

$$M\bar{z} = \iiint_V z\rho \, dV, \qquad (14.57)$$

to solve for $(\bar{x}, \bar{y}, \bar{z})$ once M and the integrals on the right have been evaluated.

If we use a triple iterated integral with respect to z, y, and x to evaluate 14.57, say, for the object in Figure 14.56, then

$$\iiint_V z\rho \, dV = \int_a^b \int_{g_1(x)}^{g_2(x)} \int_{h_1(x,y)}^{h_2(x,y)} z\rho \, dz \, dy \, dx.$$

The quantity $z\rho \, dz \, dy \, dx$ is the first moment about the xy-plane of the mass in an elemental box of dimensions dx, dy, and dz. The z-integration then adds these moments over boxes in the z-direction to give the first moment about the xy-plane of the mass in a vertical column:

$$\int_{h_1(x,y)}^{h_2(x,y)} z\rho \, dz \, dy \, dx.$$

The y-integration then adds the first moments of columns that project onto a strip in the y-direction:

$$\int_{g_1(x)}^{g_2(x)} \int_{h_1(x,y)}^{h_2(x,y)} z\rho \, dz \, dy \, dx.$$

This quantity therefore represents the first moment about the xy-plane of the slab in Figure 14.56. Finally, the x-integration adds first moments of all such slabs to give the total first moment of V about the xy-plane:

$$\int_a^b \int_{g_1(x)}^{g_2(x)} \int_{h_1(x,y)}^{h_2(x,y)} z\rho \, dz \, dy \, dx.$$

EXAMPLE 14.24

Find the centre of mass of an object of constant density ρ if it is bounded by the surfaces $z = 1 - y^2$, $x = 0$, $z = 0$, $x = 2$.

SOLUTION From the symmetry of the object (Figure 14.60), we see that $\bar{x} = 1$ and $\bar{y} = 0$. Now

$$M = 2 \int_0^2 \int_0^1 \int_0^{1-y^2} \rho \, dz \, dy \, dx = 2\rho \int_0^2 \int_0^1 (1 - y^2) \, dy \, dx$$

$$= 2\rho \int_0^2 \left\{ y - \frac{y^3}{3} \right\}_0^1 dx = \frac{4\rho}{3} \int_0^2 dx = \frac{8\rho}{3};$$

and

$$M\bar{z} = 2\int_0^2\int_0^1\int_0^{1-y^2} z\rho\,dz\,dy\,dx = 2\rho\int_0^2\int_0^1\left\{\frac{z^2}{2}\right\}_0^{1-y^2} dy\,dx$$

$$= \rho\int_0^2\int_0^1 (1-2y^2+y^4)\,dy\,dx$$

$$= \rho\int_0^2\left\{y-\frac{2y^3}{3}+\frac{y^5}{5}\right\}_0^1 dx = \frac{8\rho}{15}\int_0^2 dx = \frac{16\rho}{15}.$$

Thus,

$$\bar{z} = \frac{16\rho}{15}\cdot\frac{3}{8\rho} = \frac{2}{5}.$$

FIGURE 14.60

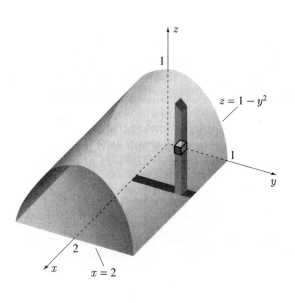

In view of our discussion on first moments in this section and on moments of inertia of thin plates in Section 14.5, it should not be necessary to give a full treatment of moments of inertia of three-dimensional objects. Instead, we simply note that the distances from a point (x,y,z) to the x-, y-, and z-axes are, respectively, $\sqrt{y^2+z^2}$, $\sqrt{x^2+z^2}$, and $\sqrt{x^2+y^2}$. If follows, then, that if an object of density $\rho(x,y,z)$ occupies a region V of space:

$$\text{Moment of inertia of object about } x\text{-axis} = \iiint_V (y^2+z^2)\rho\,dV, \quad (14.58)$$

$$\text{Moment of inertia of object about } y\text{-axis} = \iiint_V (x^2+z^2)\rho\,dV, \quad (14.59)$$

$$\text{Moment of inertia of object about } z\text{-axis} = \iiint_V (x^2+y^2)\rho\,dV. \quad (14.60)$$

EXAMPLE 14.25

Find the moment of inertia of a right-circular cylinder of constant density ρ about its axis.

SOLUTION Let the length and radius of the cylinder be h and r and choose the coordinate system in Figure 14.61. The required moment of inertia about the z-axis is four times the moment of inertia of that part of the cylinder in the first octant. Hence,

$$I = 4\int_0^r \int_0^{\sqrt{r^2-x^2}} \int_0^h (x^2+y^2)\rho\, dz\, dy\, dx = 4\rho h \int_0^r \int_0^{\sqrt{r^2-x^2}} (x^2+y^2)\, dy\, dx$$

$$= 4\rho h \int_0^r \left\{ x^2 y + \frac{y^3}{3} \right\}_0^{\sqrt{r^2-x^2}} dx = 4\rho h \int_0^r \left\{ x^2\sqrt{r^2-x^2} + \frac{1}{3}(r^2-x^2)^{3/2} \right\} dx$$

$$= \frac{4\rho h}{3} \int_0^r \{ r^2\sqrt{r^2-x^2} + 2x^2\sqrt{r^2-x^2} \} dx.$$

To evaluate this definite integral we set $x = r\sin\theta$, which implies that $dx = r\cos\theta\, d\theta$, and

$$I = \frac{4\rho h}{3} \int_0^{\pi/2} \{ r^2 r\cos\theta + 2r^2\sin^2\theta\, r\cos\theta \} r\cos\theta\, d\theta$$

$$= \frac{4\rho h r^4}{3} \int_0^{\pi/2} \{ \cos^2\theta + 2\sin^2\theta\cos^2\theta \} d\theta$$

$$= \frac{4\rho h r^4}{3} \int_0^{\pi/2} \left\{ \frac{1+\cos 2\theta}{2} + \frac{1-\cos 4\theta}{4} \right\} d\theta$$

$$= \frac{4\rho h r^4}{3} \left\{ \frac{3\theta}{4} + \frac{\sin 2\theta}{4} - \frac{\sin 4\theta}{16} \right\}_0^{\pi/2} = \frac{\rho h r^4 \pi}{2}.$$

FIGURE 14.61

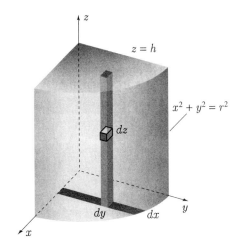

EXERCISES 14.10

In Exercises 1–5 the surfaces bound a solid object of constant density. Find its centre of mass.

1. $z = x^2 + y^2$, $z = 0$, $x = 0$, $y = 0$, $x = 1$, $y = 1$
2. $x + y + z = 1$, $x = 0$, $y = 0$, $z = 0$
3. $z = x^2$, $y + z = 4$, $y = 0$
4. $y = 4 - x^2$, $z = 0$, $y = z$
5. $x + y + z = 4$, $y = 3z$, $x = 0$, $y = 0$

In Exercises 6–10 the surfaces bound a solid object of constant density ρ. Find its moment of inertia about the line.

6. $x = 0$, $y = 0$, $z = 0$, $x = 1$, $y = 1$, $z = 1$ about the x-axis
7. $z = 2x$, $z = 0$, $y = 0$, $y = 2$, $x = 3$ about the y-axis
8. $x + y + z = 2$, $y = 0$, $x = 0$, $0 \leq z \leq \sqrt{1-y}$ about the x-axis
9. $z = xy$, $x^2 + y^2 = 1$, $z = 0$ (first octant) about the z-axis
10. $y + z = 2$, $x + z = 2$, $x = 0$, $y = 0$, $z = 0$ about the z-axis

11. Find the first moment about the xy-plane of a solid of constant density ρ if it is bounded by the surfaces $x = z$, $x + z = 0$, $z = 2$, $y = 0$, $y = 2$.

In Exercises 12–14 the surfaces bound a solid object of constant density. Find its centre of mass.

12. $y = x^3$, $x = y^2$, $z = 1 + x^2 + y^2$, $z = -x^2 - y^2$
13. $z = x^2$, $x + z = 2$, $z = y$, $y = 0$
14. $y + z = 0$, $y - z = 0$, $x + z = 0$, $x - z = 0$, $z = 2$

In Exercises 15–17 the surfaces bound a solid object of constant density ρ. Find its moment of inertia about the line.

15. $z = 4 - x^2$, $x + z + 2 = 0$, $y = 0$, $y = 2$ about the y-axis
16. $x^2 + z^2 = a^2$, $x^2 + y^2 = a^2$ about the x-axis
17. $x + y - z = 0$, $x = 3y$, $3y = 2x$, $x = 3$, $z = 0$ about the z-axis

18. Find the first moment about the plane $x + y + z = 1$ of a solid object of constant density ρ if it is bounded by the surfaces $x + 2y + 4z = 12$, $x = 0$, $y = 0$, $z = 0$.

19. Prove the parallel axis theorem for solid objects: The moment of inertia of a uniform solid about a line is equal to the moment of inertia about a parallel line through the centre of mass of the solid plus the mass multiplied by the square of the distance between the lines.

20. Find the centre of mass of a uniform solid in the first octant bounded by the ellipsoid $x^2/a^2 + y^2/b^2 + z^2/c^2 = 1$.

21. Let $\hat{\mathbf{v}} = (v_x, v_y, v_z)$ be a unit vector with its tail at the origin. Show that the moment of inertia I of any solid object occupying volume V about the line containing $\hat{\mathbf{v}}$ can be expressed in the form

$$I = v_x^2 I_x + v_y^2 I_y + v_z^2 I_z - 2 v_x v_y I_{xy} - 2 v_y v_z I_{yz} - 2 v_z v_x I_{xz}$$

where I_x, I_y, and I_z are moments of inertia about the x-, y-, and z-axes, and

$$I_{xy} = \iiint_V xy\rho\, dV, \quad I_{yz} = \iiint_V yz\rho\, dV,$$

$$I_{xz} = \iiint_V xz\rho\, dV.$$

22. Find the moment of inertia of a uniform solid sphere of radius R about any tangent line.

SECTION 14.11

Triple Iterated Integrals in Cylindrical Coordinates

In Section 14.7 we saw that polar coordinates are sometimes more convenient than Cartesian coordinates in evaluating double integrals. It should come as no surprise, then, that other coordinate systems can simplify the evaluation of triple integrals. Two of the most common are cylindrical and spherical coordinates.

Cylindrical coordinates are useful in problems involving an axis of symmetry. They are based on a Cartesian coordinate along the axis of symmetry and polar coordinates in a plane perpendicular to the axis of symmetry. If the z-axis is the axis of symmetry and polar coordinates are defined in the xy-plane with the origin as pole and the positive x-axis as polar axis, then cylindrical coordinates and Cartesian coordinates are related by the equations

$$x = r\cos\theta, \quad y = r\sin\theta, \quad z = z \qquad (14.61)$$

(see Figure 14.62). Recall that r can be expressed in terms of x and y by

$$r = \sqrt{x^2 + y^2}, \qquad (14.62\text{a})$$

and θ is defined implicitly by the equations

$$\cos\theta = \frac{x}{\sqrt{x^2 + y^2}}, \quad \sin\theta = \frac{y}{\sqrt{x^2 + y^2}}. \qquad (14.62\text{b})$$

FIGURE 14.62

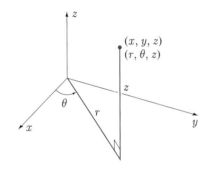

To use cylindrical coordinates in the evaluation of triple integrals, we must express equations of surfaces in terms of these coordinates. But this is very simple, for if $F(x, y, z) = 0$ is the equation of a surface in Cartesian coordinates, then to express this equation in cylindrical coordinates we substitute from equations 14.61: $F(r\cos\theta, r\sin\theta, z) = 0$. For example, the right-circular cylinder $x^2 + y^2 = 9$, which has the z-axis as its axis of symmetry, has the very simple equation $r = 3$ in cylindrical coordinates. The right-circular cone $z = \sqrt{x^2 + y^2}$ also has the z-axis as its axis of symmetry, and its equation in cylindrical coordinates takes the simple form $z = r$.

Suppose that we are to evaluate the triple integral of a continuous function $f(x, y, z)$,

$$\iiint_V f(x, y, z)\, dV,$$

over some region V of space. The choice of a triple iterated integral in cylindrical coordinates implies a subdivision of V into small volumes dV by means of coordinate surfaces $r = $ constant, $\theta = $ constant, and $z = $ constant (Figure 14.63). Surfaces $r = $ constant are right-circular cylinders coaxial with the z-axis; surfaces $\theta = $ constant are planes containing the z-axis and therefore perpendicular to the xy-plane; and surfaces $z = $ constant are planes parallel to the xy-plane.

FIGURE 14.63

FIGURE 14.64

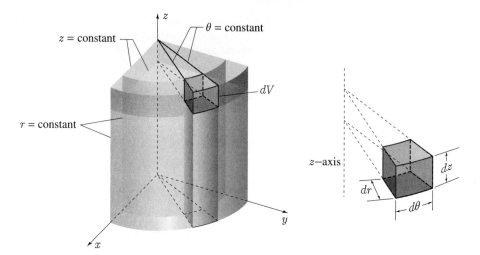

If we denote small variations in r, θ, and z for the element dV by $dr, d\theta$, and dz (Figure 14.64), then the volume of the element is approximately $(r\, dr\, d\theta)\, dz$, where $r\, dr\, d\theta$ is the polar cross-sectional area parallel to the xy-plane. Hence in cylindrical coordinates we set

$$dV = r\, dz\, dr\, d\theta. \qquad (14.63)$$

The integrand $f(x, y, z)$ is expressed in cylindrical coordinates as $f(r\cos\theta, r\sin\theta, z)$. It remains only to affix appropriate limits to the triple iterated integral, and these will, of course, depend on which of the six possible iterated integrals in cylindrical coordinates we choose. The most commonly used triple iterated integral is with respect to z, r, and θ, and in this case the z-integration adds the quantities

$$f(r\cos\theta, r\sin\theta, z)r\, dz\, dr\, d\theta$$

in a vertical column, where r and θ are constant. The limits therefore identify the surfaces on which each and every column starts and stops, and generally depend on r and θ:

$$\int_{h_1(r,\theta)}^{h_2(r,\theta)} f(r\cos\theta, r\sin\theta, z)r\, dz\, dr\, d\theta.$$

The remaining integrations with respect to r and θ perform additions over the area in the xy-plane onto which all vertical columns project. Since r and θ are simply polar coordinates, the r-integration adds over small areas in a wedge and the θ-integration adds over all wedges. The triple iterated integral with respect to z, r, and θ therefore has the form

$$\int_a^b \int_{g_1(\theta)}^{g_2(\theta)} \int_{h_1(r,\theta)}^{h_2(r,\theta)} f(r\cos\theta, r\sin\theta, z)r\, dz\, dr\, d\theta.$$

We comment on the geometric aspects of these additions more fully in the following examples.

EXAMPLE 14.26

Find the volume inside both the sphere $x^2 + y^2 + z^2 = 2$ and the cylinder $x^2 + y^2 = 1$.

SOLUTION The required volume is eight times the first octant volume shown in Figure 14.65. If we use cylindrical coordinates, the volume of an elemental piece is $r\,dz\,dr\,d\theta$. A z-integration adds these pieces to give the volume in a vertical column:

$$\int_0^{\sqrt{2-r^2}} r\,dz\,dr\,d\theta,$$

where the limits indicate that for the volume in the first octant all columns start on the xy-plane (where $z = 0$) and end on the sphere (where $r^2 + z^2 = 2$). An r-integration now adds the volumes of all columns that stand on a wedge:

$$\int_0^1 \int_0^{\sqrt{2-r^2}} r\,dz\,dr\,d\theta,$$

where the limits indicate that all wedges start at the origin (where $r = 0$) and end on the curve $x^2 + y^2 = 1$ (or $r = 1$) in the xy-plane. This integration therefore yields the volume of a slice (Figure 14.65). Finally, the θ-integration adds the volumes of all such slices

$$\int_0^{\pi/2} \int_0^1 \int_0^{\sqrt{2-r^2}} r\,dz\,dr\,d\theta,$$

where the limits 0 and $\pi/2$ identify the positions of the first and last wedges, respectively, in the first quadrant. We obtain the required volume, then, as

$$8 \int_0^{\pi/2} \int_0^1 \int_0^{\sqrt{2-r^2}} r\,dz\,dr\,d\theta = 8 \int_0^{\pi/2} \int_0^1 r\sqrt{2-r^2}\,dr\,d\theta$$

$$= 8 \int_0^{\pi/2} \left\{-\frac{1}{3}(2-r^2)^{3/2}\right\}_0^1 d\theta$$

$$= \frac{8}{3}(2\sqrt{2}-1) \int_0^{\pi/2} d\theta = \frac{4\pi}{3}(2\sqrt{2}-1).$$

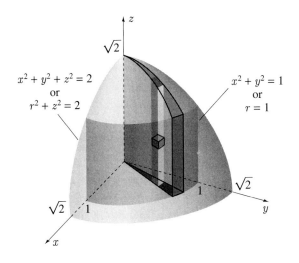

FIGURE 14.65

Had we used a triple iterated integral with respect to $z, y,$ and x in Example 14.26, we would have obtained

$$\text{Volume} = 8 \int_0^1 \int_0^{\sqrt{1-x^2}} \int_0^{\sqrt{2-x^2-y^2}} dz\, dy\, dx.$$

To appreciate the value of cylindrical coordinates, try to evaluate this triple iterated integral.

As further evidence of the value of cylindrical coordinates, repeat Examples 14.23 and 14.25 using cylindrical coordinates.

EXAMPLE 14.27

Evaluate the triple iterated integral

$$I = \int_0^1 \int_0^{\sqrt{1-y^2}} \int_0^{x^2+y^2} y^2 \, dz\, dx\, dy.$$

SOLUTION The first two integrations with respect to z and x are quite straightforward, giving

$$I = \frac{1}{3} \int_0^1 y^2 (1 + 2y^2) \sqrt{1 - y^2}\, dy.$$

To evaluate this definite integral we could make a trigonometric substitution $y = \sin\theta$. (Try it.) Alternatively, we could consider using cylindrical coordinates on the original triple iterated integral. We first obtain the region over which integration is being performed. The limits on z indicate that all columns begin on the xy-plane and end on the paraboloid $z = x^2 + y^2$. The limits on x indicate that in the xy-plane all strips begin on the y-axis ($x = 0$) and end on the curve $x = \sqrt{1 - y^2}$ (i.e., the circle $x^2 + y^2 = 1$). The first and last strips are at $y = 0$ and $y = 1$, respectively. These facts determine the region of integration as that region in the first octant bounded by the paraboloid $z = x^2 + y^2$ and the right-circular cylinder $x^2 + y^2 = 1$ (Figure 14.66a). If we use a triple iterated integral with respect to z, r and θ (Figure 14.66b), then

$$I = \int_0^{\pi/2} \int_0^1 \int_0^{r^2} r^2 \sin^2\theta\, r\, dz\, dr\, d\theta = \int_0^{\pi/2} \int_0^1 r^5 \sin^2\theta\, dr\, d\theta = \int_0^{\pi/2} \frac{1}{6} \sin^2\theta\, d\theta$$

$$= \frac{1}{6} \int_0^{\pi/2} \left\{ \frac{1 - \cos 2\theta}{2} \right\} d\theta = \frac{1}{12} \left\{ \theta - \frac{1}{2} \sin 2\theta \right\}_0^{\pi/2} = \frac{\pi}{24}.$$

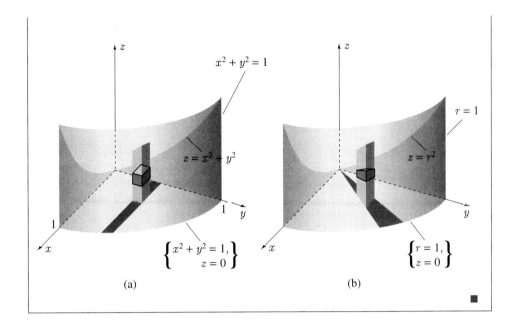

FIGURE 14.66

EXERCISES 14.11

In Exercises 1–10 find the equation for the surface in cylindrical coordinates. Sketch each surface and indicate whether it is symmetric about the z-axis.

1. $x^2 + y^2 + z^2 = 4$
2. $x^2 + y^2 = 1$
3. $y^2 + z^2 = 6$
4. $x + y = 5$
5. $z = 2\sqrt{x^2 + y^2}$
6. $z = x^2$
7. $x^2 + 4y^2 = 4$
8. $4z = x^2 + y^2$
9. $y = x$
10. $x^2 + y^2 = 1 + z^2$

In Exercises 11–15 find the volume bounded by the surfaces.

11. $z = \sqrt{x^2 + y^2}$, $x^2 + y^2 = 4$, $z = 0$
12. $z = \sqrt{2 - x^2 - y^2}$, $z = x^2 + y^2$
13. $z = xy$, $x^2 + y^2 = 1$, $z = 0$
14. $z = x^2 + y^2$, $z = 4 - x^2 - y^2$
15. $x + y + z = 2$, $x^2 + y^2 = 1$, $z = 0$

16. Find the volume inside the sphere $x^2 + y^2 + z^2 = 4$ but outside the cylinder $x^2 + y^2 = 1$.

17. Find the centre of mass of a uniform hemispherical solid.

18. Set up the six triple iterated integrals in cylindrical coordinates for the triple integral of a function $f(x, y, z)$ over the volume bounded by the surfaces $z = 1 + x^2 + y^2$, $x^2 + y^2 = 9$, $z = 0$.

19. Find the moment of inertia of a uniform right-circular cylinder of radius R and height h **(a)** about its axis; **(b)** about a line through the centre of its base and perpendicular to its axis.

20. Find the moment of inertia of a uniform sphere of radius R about any line through its centre.

In Exercises 21–24 evaluate the triple iterated integral.

21. $\displaystyle\int_0^3 \int_0^{\sqrt{9-x^2}} \int_0^{\sqrt{x^2+y^2}} dz\,dy\,dx$

22. $\displaystyle\int_0^9 \int_0^{\sqrt{81-y^2}} \int_0^{\sqrt{81-x^2-y^2}} \frac{1}{\sqrt{x^2+y^2}} dz\,dx\,dy$

23. $\displaystyle\int_0^4 \int_0^{\sqrt{4y-y^2}} \int_0^{y+x^2} dz\,dx\,dy$

24. $\displaystyle\int_0^{\sqrt{3}/2} \int_{5-\sqrt{21-y^2}}^{\sqrt{1-y^2}} \int_0^{x^2+y^2} y\,dz\,dx\,dy$

25. Find the centre of mass for the uniform solid bounded by the surfaces $x^2 + y^2 = 2x$, $z = \sqrt{x^2 + y^2}$, $z = 0$.

26. Find the moment of inertia of a uniform right-circular cone of radius R and height h about its axis.

27. A casting is in the form of a sphere of radius b with two cylindrical holes of radius $a < b$ such that the axes of the holes pass through the centre of the sphere and intersect at right angles. What volume of metal is required for the casting?

In Exercises 28–38 find the volume described.

28. Bounded by $x^2 + y^2 - z^2 = 1$, $4z^2 = x^2 + y^2$

29. Bounded by $z = x^2 + y^2$, $z = 0$, $(x^2 + y^2)^2 = x^2 - y^2$

30. Bounded by $x^2 + y^2 + z^2 = 4$, $x^2 + y^2 + z^2 = 16$, $z = \sqrt{x^2 + y^2}$ (smaller piece)

31. Bounded by $x^2 + y^2 + z^2 = 1$, $y = x$, $x = 2y$, $z = 0$ (in the first octant)

32. Inside both $2x^2 + 2y^2 + z^2 = 8$ and $x^2 + y^2 = 1$

33. Bounded by $z^2 = (x^2 + y^2)^2$, $x^2 + y^2 = 2y$

34. Inside $x^2 + y^2 + z^2 = a^2$ but outside, $x^2 + y^2 = ay$

35. Bounded by $z = 0$, $x^2 + y^2 = 1$, $z = e^{-x^2-y^2}$

36. Inside $x^2 + y^2 + z^2 = 9$ but outside $x^2 + y^2 = 1 + z^2$

37. Cut off from $z = x^2 + y^2$ by $z = x + y$

38. Inside $x^2 + y^2 + z^2 = 4$ and below $3z = x^2 + y^2$

39. Evaluate the triple integral of $\sqrt{x^2 + y^2 + z^2}$ over the volume bounded by $z = 3$ and $z = \sqrt{x^2 + y^2}$.

40. Evaluate the triple integral of $f(y, z) = |yz|$ over the volume bounded by $z^2 = 1 + x^2 + y^2$ and $z = \sqrt{4 - x^2 - y^2}$.

41. A tumbler in the form of a right-circular cylinder of radius R and height h is full of water. As the axis of the tumbler is tilted from the vertical, water pours over the side. Find the volume of water remaining in the tumbler as a function of the angle between the vertical and the axis of the tumbler.

42. Use cylindrical coordinates to find the volume of the torus $(\sqrt{x^2 + y^2} - a)^2 + z^2 = b^2$, $b < a$.

SECTION 14.12

Triple Iterated Integrals in Spherical Coordinates

Spherical coordinates are useful in solving problems concerning figures that are symmetric about a point. If the origin is that point, then spherical coordinates $(\mathcal{R}, \theta, \phi)$ are related to Cartesian coordinates (x, y, z) by the equations

$$\begin{aligned} x &= \mathcal{R} \sin\phi \cos\theta, \\ y &= \mathcal{R} \sin\phi \sin\theta, \\ z &= \mathcal{R} \cos\phi, \end{aligned} \tag{14.64}$$

which are illustrated in Figure 14.67. As is the case for polar and cylindrical coordinates, without restrictions on $\mathcal{R}, \theta$, and ϕ, each point in space has many sets of spherical coordinates. The positive value of its spherical coordinate $\mathcal{R}$ is given by

$$\mathcal{R} = \sqrt{x^2 + y^2 + z^2}. \tag{14.65}$$

The θ-coordinates in cylindrical and spherical coordinates are identical, so that no simple formula for θ in terms of x, y, and z exists. The ϕ-coordinate is the angle between the positive z-axis and the line joining the origin to the point (x, y, z), and that value of ϕ in the range $0 \leq \phi \leq \pi$ is determined by the formula

$$\phi = \mathrm{Cos}^{-1}\left(\frac{z}{\sqrt{x^2 + y^2 + z^2}}\right). \tag{14.66}$$

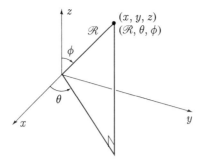

FIGURE 14.67

To transform equations $F(x,y,z) = 0$ of surfaces from Cartesian to spherical coordinates, we substitute from equations 14.64:

$$F(\mathcal{R}\sin\phi\cos\theta, \mathcal{R}\sin\phi\sin\theta, \mathcal{R}\cos\phi) = 0.$$

For example, the sphere $x^2 + y^2 + z^2 = 4$ is symmetric about its centre, and its equation in spherical coordinates is simply $\mathcal{R} = 2$. For the right-circular cone $z = \sqrt{x^2 + y^2}$, we write

$$\mathcal{R}\cos\phi = \sqrt{\mathcal{R}^2\sin^2\phi\cos^2\theta + \mathcal{R}^2\sin^2\phi\sin^2\theta} = \mathcal{R}\sin\phi.$$

Consequently, $\tan\phi = 1$ or $\phi = \pi/4$; i.e., $\phi = \pi/4$ is the equation of the cone in spherical coordinates.

Suppose that we are to evaluate the triple integral of a function $f(x,y,z)$,

$$\iiint_V f(x,y,z)\,dV,$$

over some region V of space. The choice of a triple iterated integral in spherical coordinates implies a subdivision of V into small volumes by means of coordinate surfaces $\mathcal{R} =$ constant, $\theta =$ constant, and $\phi =$ constant (Figure 14.68). Surfaces $\mathcal{R} =$ constant are spheres centred at the origin; surfaces $\theta =$ constant are planes containing the z-axis; and surfaces $\phi =$ constant are right-circular cones symmetric about the z-axis with the origin as apex.

FIGURE 14.68

FIGURE 14.69

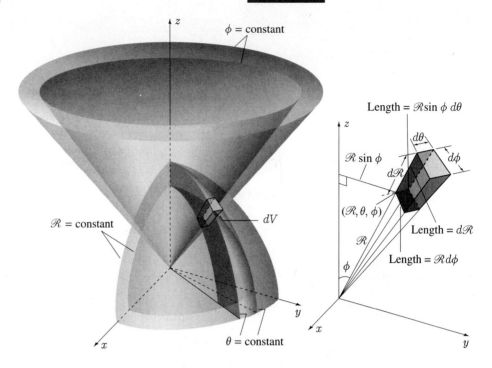

If we denote small variations in $\mathcal{R}, \theta$, and ϕ for the element dV by $d\mathcal{R}, d\theta$, and $d\phi$ (Figure 14.69) and approximate dV by a rectangular parallelepiped with dimensions $d\mathcal{R}, \mathcal{R}\, d\phi$, and $\mathcal{R}\sin\phi\, d\theta$, then

$$dV = (\mathcal{R}\sin\phi\, d\theta)(\mathcal{R}\, d\phi)\, d\mathcal{R} = \mathcal{R}^2 \sin\phi\, d\mathcal{R}\, d\theta\, d\phi. \qquad (14.67)$$

The integrand $f(x,y,z)$ is expressed in spherical coordinates as

$$f(\mathcal{R}\sin\phi\cos\theta, \mathcal{R}\sin\phi\sin\theta, \mathcal{R}\cos\phi).$$

It remains only to affix appropriate limits to the triple iterated integral, and these limits depend on which of the six possible triple iterated integrals in spherical coordinates that we choose. If we use a triple iterated integral with respect to $\mathcal{R}, \phi$, and θ, then it is of the form

$$\iiint_V f(x,y,z)\, dV = \int_a^b \int_{g_1(\theta)}^{g_2(\theta)} \int_{h_1(\theta,\phi)}^{h_2(\theta,\phi)} f(\mathcal{R}\sin\phi\cos\theta, \mathcal{R}\sin\phi\sin\theta, \mathcal{R}\cos\phi)\mathcal{R}^2 \sin\phi\, d\mathcal{R}\, d\phi\, d\theta.$$

The geometric interpretations of the additions represented by these integrations with respect to $\mathcal{R}, \phi$, and θ are left to the examples.

EXAMPLE 14.28

Find the volume of a sphere.

SOLUTION The equation of a sphere of radius R centred at the origin is $x^2 + y^2 + z^2 = R^2$ or, in spherical coordinates, $\mathcal{R} = R$. The volume of this sphere is eight times the first octant volume shown in Figure 14.70. If we use spherical coordinates, the volume of an elemental piece is

$$\mathcal{R}^2 \sin\phi \, d\mathcal{R} \, d\phi \, d\theta.$$

An $\mathcal{R}$-integration adds these volumes for constant ϕ and θ to give the volume in a "spike,"

$$\int_0^R \mathcal{R}^2 \sin\phi \, d\mathcal{R} \, d\phi \, d\theta,$$

where the limits indicate that all spikes start at the origin (where $\mathcal{R} = 0$) and end on the sphere (where $\mathcal{R} = R$). A ϕ-integration now adds the volumes of spikes for constant θ. This yields the volume of a slice

$$\int_0^{\pi/2} \int_0^R \mathcal{R}^2 \sin\phi \, d\mathcal{R} \, d\phi \, d\theta,$$

where the limits indicate that all slices in the first octant start on the z-axis (where $\phi = 0$) and end on the xy-plane (where $\phi = \pi/2$). Finally, the θ-integration adds the volumes of all such slices

$$\int_0^{\pi/2} \int_0^{\pi/2} \int_0^R \mathcal{R}^2 \sin\phi \, d\mathcal{R} \, d\phi \, d\theta,$$

where the limits 0 and $\pi/2$ identify the positions of the first and last slices, respectively, in the first octant. We obtain the required volume as

$$8 \int_0^{\pi/2} \int_0^{\pi/2} \int_0^R \mathcal{R}^2 \sin\phi \, d\mathcal{R} \, d\phi \, d\theta = 8 \int_0^{\pi/2} \int_0^{\pi/2} \left\{ \frac{\mathcal{R}^3}{3} \sin\phi \right\}_0^R d\phi \, d\theta$$

$$= \frac{8R^3}{3} \int_0^{\pi/2} \{-\cos\phi\}_0^{\pi/2} d\theta$$

$$= \frac{8R^3}{3} \{\theta\}_0^{\pi/2} = \frac{4}{3}\pi R^3.$$

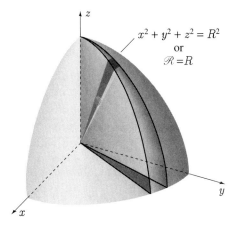

FIGURE 14.70

EXAMPLE 14.29

Find the centre of mass of a solid object of constant density if it is in the shape of a right-circular cone.

SOLUTION Let the cone have altitude h and base radius R. Then its mass is

$$M = \frac{1}{3}\pi R^2 h\rho,$$

where ρ is the density of the object. If we place axes as shown in Figure 14.71, then $\bar{x} = \bar{y} = 0$; i.e., the centre of mass is on the axis of symmetry of the cone. To find $\bar{z}$ we offer three solutions.

Method 1 If we use Cartesian coordinates, the equation of the surface of the cone is of the form $z = k\sqrt{x^2 + y^2}$. Since $(0, R, h)$ is a point on the cone, $h = kR$, and therefore $k = h/R$. Now

$$M\bar{z} = 4\int_0^R \int_0^{\sqrt{R^2-x^2}} \int_{k\sqrt{x^2+y^2}}^h z\rho \, dz \, dy \, dx = 4\rho \int_0^R \int_0^{\sqrt{R^2-x^2}} \left\{\frac{z^2}{2}\right\}_{k\sqrt{x^2+y^2}}^h dy \, dx$$

$$= 2\rho \int_0^R \int_0^{\sqrt{R^2-x^2}} \{h^2 - k^2(x^2 + y^2)\} dy \, dx$$

$$= 2\rho \int_0^R \left\{h^2 y - k^2 x^2 y - \frac{k^2 y^3}{3}\right\}_0^{\sqrt{R^2-x^2}} dx$$

$$= 2\rho \int_0^R \left\{h^2 \sqrt{R^2 - x^2} - k^2 x^2 \sqrt{R^2 - x^2} - \frac{k^2}{3}(R^2 - x^2)^{3/2}\right\} dx.$$

If we set $x = R\sin\theta$, then $dx = R\cos\theta \, d\theta$, and

$$M\bar{z} = 2\rho \int_0^{\pi/2} \left\{h^2 R\cos\theta - k^2(R^2 \sin^2\theta) R\cos\theta - \frac{k^2}{3}R^3 \cos^3\theta\right\} R\cos\theta \, d\theta$$

$$= 2\rho R^2 \int_0^{\pi/2} \left\{\frac{h^2}{2}(1 + \cos 2\theta) - \frac{k^2 R^2}{8}(1 - \cos 4\theta) \right.$$
$$\left. - \frac{k^2 R^2}{12}\left(1 + 2\cos 2\theta + \frac{1 + \cos 4\theta}{2}\right)\right\} d\theta$$

$$= \rho R^2 \left\{h^2\left(\theta + \frac{\sin 2\theta}{2}\right) - \frac{k^2 R^2}{4}\left(\theta - \frac{\sin 4\theta}{4}\right) \right.$$
$$\left. - \frac{k^2 R^2}{6}\left(\frac{3\theta}{2} + \sin 2\theta + \frac{\sin 4\theta}{8}\right)\right\}_0^{\pi/2}$$

$$= \rho R^2 \left\{\frac{\pi h^2}{2} - \frac{k^2 R^2 \pi}{8} - \frac{k^2 R^2 \pi}{8}\right\} = \frac{\pi R^2 h^2 \rho}{4}.$$

Thus,

$$\bar{z} = \frac{\pi R^2 h^2 \rho}{4} \cdot \frac{3}{\pi R^2 h\rho} = \frac{3h}{4}.$$

FIGURE 14.71

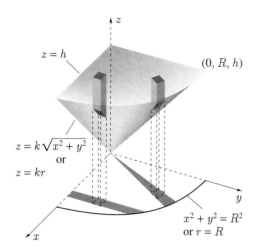

Method 2 If we use cylindrical coordinates, the equation of the surface of the cone is $z = kr$. From Figure 14.71, we see that

$$M\bar{z} = 4 \int_0^{\pi/2} \int_0^R \int_{kr}^h z\rho\, r\, dz\, dr\, d\theta = 4\rho \int_0^{\pi/2} \int_0^R \left\{\frac{rz^2}{2}\right\}_{kr}^h dr\, d\theta$$

$$= 2\rho \int_0^{\pi/2} \int_0^R r(h^2 - k^2 r^2)\, dr\, d\theta = 2\rho \int_0^{\pi/2} \left\{\frac{r^2 h^2}{2} - \frac{k^2 r^4}{4}\right\}_0^R d\theta$$

$$= \rho \left(R^2 h^2 - \frac{k^2 R^4}{2}\right) \{\theta\}_0^{\pi/2} = \rho \left(R^2 h^2 - \frac{k^2 R^4}{2}\right) \frac{\pi}{2} = \frac{\pi R^2 h^2 \rho}{4}.$$

Again, then, $\bar{z} = 3h/4$.

Method 3 If we use spherical coordinates, then the equation of the surface of the cone is

$$\phi = \text{Cos}^{-1} \frac{h}{\sqrt{h^2 + R^2}} = \phi_1.$$

From Figure 14.72, we see that

$$M\bar{z} = 4 \int_0^{\pi/2} \int_0^{\phi_1} \int_0^{h\sec\phi} (\mathcal{R}\cos\phi)\rho \mathcal{R}^2 \sin\phi\, d\mathcal{R}\, d\phi\, d\theta$$

$$= 4\rho \int_0^{\pi/2} \int_0^{\phi_1} \left\{\frac{\mathcal{R}^4}{4} \sin\phi\cos\phi\right\}_0^{h\sec\phi} d\phi\, d\theta$$

$$= \rho h^4 \int_0^{\pi/2} \int_0^{\phi_1} \frac{\sin\phi}{\cos^3\phi} d\phi\, d\theta = \rho h^4 \int_0^{\pi/2} \left\{\frac{1}{2\cos^2\phi}\right\}_0^{\phi_1} d\theta$$

$$= \frac{\rho h^4}{2} \left\{\frac{1}{\cos^2\phi_1} - 1\right\} \{\theta\}_0^{\pi/2} = \frac{\pi \rho h^4}{4} \left\{\frac{h^2 + R^2}{h^2} - 1\right\} = \frac{\pi h^2 R^2 \rho}{4}.$$

Again, $\bar{z} = 3h/4$.

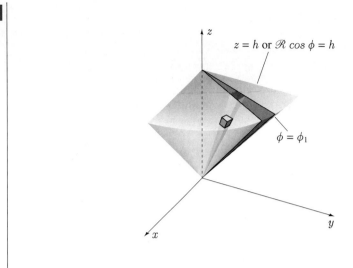

FIGURE 14.72

EXERCISES 14.12

In Exercises 1–7 find the equation of the surface in spherical coordinates. Sketch the surface.

1. $x^2 + y^2 + z^2 = 4$
2. $x^2 + y^2 = 1$
3. $3z = \sqrt{x^2 + y^2}$
4. $4z = x^2 + y^2$
5. $y = x$
6. $x^2 + y^2 = 1 + z^2$
7. $z = -2\sqrt{x^2 + y^2}$

In Exercises 8–12 find the volume described.

8. Bounded by $z = \sqrt{x^2 + y^2}$, $z = \sqrt{1 - x^2 - y^2}$
9. Bounded by $z = 1$, $z = \sqrt{4 - x^2 - y^2}$
10. Bounded by $x^2 + y^2 + z^2 = 1$, $y = x$, $y = 2x$, $z = 0$ (in the first octant)
11. Inside $x^2 + y^2 + z^2 = 2$ but outside $x^2 + y^2 = 1$
12. Bounded by $z = 2\sqrt{x^2 + y^2}$, $x^2 + y^2 = 4$, $z = 0$

13. Find the centre of mass of a uniform hemispherical solid.

14. Find the moment of inertia of a uniform solid sphere of radius R about any line through its centre.

15. Find the first moment about the yz-plane of the uniform solid in the first octant bounded by the surfaces $x^2 + y^2 + z^2 = 4$, $x^2 + y^2 + z^2 = 9$, $y = 0$ and $y = \sqrt{3}x$.

16. A solid sphere of radius R and centre at the origin has a continuous charge distribution throughout. If the density of the charge is $\rho(x, y, z) = k\sqrt{x^2 + y^2 + z^2}$ coulombs per cubic metre, find the total charge in the sphere.

17. Set up the six triple iterated integrals in spherical coordinates for the triple integral of a function $f(x, y, z)$ over the volume in the first octant under the sphere $x^2 + y^2 + z^2 = 2$ and inside the cylinder $x^2 + y^2 = 1$.

Evaluate the triple iterated integral in Exercises 18 and 19.

18. $\int_0^9 \int_0^{\sqrt{81-y^2}} \int_0^{\sqrt{81-x^2-y^2}} \frac{1}{x^2 + y^2 + z^2} \, dz \, dx \, dy$

19. $\int_0^1 \int_0^{\sqrt{1-x^2}} \int_{\sqrt{x^2+y^2}}^{\sqrt{2-x^2-y^2}} dz \, dy \, dx$

20. Find a formula for the smaller volume bounded by $x^2 + y^2 + z^2 = R^2$ and $z = k\sqrt{x^2 + y^2}$ $(k > 0)$.

21. Find the volume bounded by $(x^2 + y^2 + z^2)^2 = x$.

22. (a) Use Archimedes' principle to determine the density of a spherical ball if it floats half submerged in water.
    (b) What force is required to keep the ball with its centre at a depth of one-half the radius of the ball?

23. Find the volume bounded by the surface $(x^2 + y^2 + z^2)^2 = 2z(x^2 + y^2)$.

24. A sphere of radius R carries a uniform charge distribution of ρ coulombs per cubic metre (Figure 14.73). If P is a point on the z-axis (distance $d > R$ from the centre of the sphere) and dV is a small element of volume of the sphere, then the potential at P due to dV is given by

$$\frac{1}{4\pi\epsilon_0}\frac{\rho dV}{s},$$

where s is the distance from P to dV.

(a) Show that in terms of spherical coordinates the potential V at P due to the entire sphere is

$$V = \frac{\rho}{4\pi\epsilon_0}\int_{-\pi}^{\pi}\int_{0}^{\pi}\int_{0}^{R}\frac{\mathcal{R}^2\sin\phi}{\sqrt{\mathcal{R}^2+d^2-2\mathcal{R}d\cos\phi}}d\mathcal{R}\,d\phi\,d\theta.$$

(b) Because this iterated integral is very difficult to evaluate, we replace ϕ with the variable $s = \sqrt{\mathcal{R}^2+d^2-2\mathcal{R}d\cos\phi}$. Show that with this change

$$V = \frac{\rho}{4\pi\epsilon_0 d}\int_{-\pi}^{\pi}\int_{0}^{R}\int_{d-\mathcal{R}}^{d+\mathcal{R}}\mathcal{R}\,ds\,d\mathcal{R}\,d\theta.$$

(c) Evaluate the integral in (b) to verify that $V = Q/(4\pi\epsilon_0 d)$, where Q is the total charge on the sphere.

FIGURE 14.73

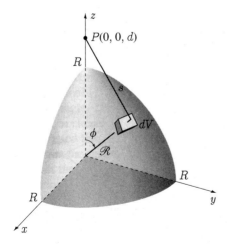

25. A sphere of constant density ρ and radius R is located at the origin (Figure 14.74). If a mass m is situated at a point P on the z-axis (distance $d > R$ from the centre of the sphere) and dV is a small element of volume of the sphere, then according to Newton's universal law of gravitation, the z-component of the force on m due to the mass in dV is given by

$$-\frac{Gm\rho dV\cos\psi}{s^2},$$

where G is a constant and s is the distance between P and dV.

(a) Show that in spherical coordinates the total force on m due to the entire sphere has z-component

$$F_z = -\frac{Gm\rho}{2d}\int_{-\pi}^{\pi}\int_{0}^{\pi}\int_{0}^{R}\left(\frac{s^2+d^2-\mathcal{R}^2}{s^3}\right)\mathcal{R}^2\sin\phi\,d\mathcal{R}\,d\phi\,d\theta.$$

(b) Use the transformation in Exercise 24(b) to write F_z in the form

$$F_z = -\frac{Gm\rho}{2d^2}\int_{-\pi}^{\pi}\int_{0}^{R}\int_{d-\mathcal{R}}^{d+\mathcal{R}}\mathcal{R}\left(\frac{s^2+d^2-\mathcal{R}^2}{s^2}\right)ds\,d\mathcal{R}\,d\theta,$$

and show that $F_z = -GmM/d^2$ where M is the total mass of the sphere.

FIGURE 14.74

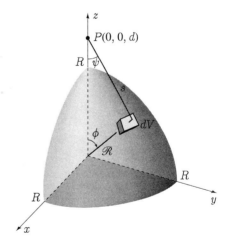

26. A homogeneous solid is bounded by two concentric spheres of radii a and b ($a < b$). Verify that the force that this layer exerts on a point mass at any point interior to the shell vanishes.

SECTION 14.13

Derivatives of Definite Integrals

If a function $f(x,y)$ of two independent variables is integrated with respect to y from $y = a$ to $y = b$, the result depends on x. Suppose we denote this function by $F(x)$:

$$F(x) = \int_a^b f(x,y)\,dy. \tag{14.68}$$

To calculate the derivative $F'(x)$ of this function, we should first integrate with respect to y and then differentiate with respect to x. The following theorem indicates that differentiation can be done first and integration later.

Theorem 14.3

If the partial derivative $\partial f/\partial x$ of $f(x,y)$ is continuous on a rectangle $a \leq y \leq b$, $c \leq x \leq d$, then for $c < x < d$

$$\frac{d}{dx}\int_a^b f(x,y)\,dy = \int_a^b \frac{\partial f(x,y)}{\partial x}\,dy. \tag{14.69}$$

Proof If we define

$$g(x) = \int_a^b \frac{\partial f(x,y)}{\partial x}\,dy$$

as the right-hand side of equation 14.69, then this function is defined for $c \leq x \leq d$. In fact, because $\partial f/\partial x$ is continuous, $g(x)$ is also continuous. We can therefore integrate $g(x)$ with respect to x from $x = c$ to any value of x in the interval $c \leq x \leq d$:

$$\int_c^x g(x)\,dx = \int_c^x \int_a^b \frac{\partial f(x,y)}{\partial x}\,dy\,dx.$$

This double iterated integral represents the double integral of $\partial f/\partial x$ over the rectangle in Figure 14.75, and if we reverse the order of integration, we have

$$\int_c^x g(x)\,dx = \int_a^b \int_c^x \frac{\partial f(x,y)}{\partial x}\,dx\,dy = \int_a^b \{f(x,y)\}_c^x\,dy$$
$$= \int_a^b \{f(x,y) - f(c,y)\}\,dy = \int_a^b f(x,y)\,dy - \int_a^b f(c,y)\,dy.$$

Because the second integral on the right is independent of x, if we differentiate this equation with respect to x, we get

$$\frac{d}{dx}\int_c^x g(x)\,dx = \frac{d}{dx}\int_a^b f(x,y)\,dy.$$

But Theorem 6.7 gives

$$g(x) = \frac{d}{dx}\int_a^b f(x,y)\,dy.$$

This completes the proof.

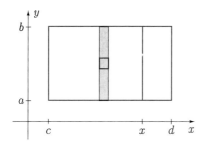

FIGURE 14.75

The limits on the integral in 14.68 need not be numerical constants; so far as the integration with respect to y is concerned, x is constant, and therefore a and b could be functions of x:

$$F(x) = \int_{a(x)}^{b(x)} f(x,y)\,dy.$$

Indeed, this is precisely what does occur in the evaluation of double integrals by means of double iterated integrals. To differentiate $F(x)$ now is more complicated than in Theorem 14.3 because we must also account for the x's in $a(x)$ and $b(x)$. The chain rule can be used to develop a formula for $F'(x)$.

> **Theorem 14.4**
>
> (Leibnitz's rule). If the partial derivative $\partial f/\partial x$ is continuous on the area bounded by the curves $y = a(x)$, $y = b(x)$, $x = c$, and $x = d$, then
>
> $$\frac{d}{dx}\int_{a(x)}^{b(x)} f(x,y)\,dy = \int_{a(x)}^{b(x)} \frac{\partial f(x,y)}{\partial x}\,dy + f[x, b(x)]\frac{db}{dx} - f[x, a(x)]\frac{da}{dx}. \quad (14.70)$$

Proof If $F(x) = \int_{a(x)}^{b(x)} f(x,y)\,dy$, then the schematic diagram

gives the chain rule

$$F'(x) = \left(\frac{\partial F}{\partial x}\right)_{a,b} + \left(\frac{\partial F}{\partial b}\right)_{a,x}\frac{db}{dx} + \left(\frac{\partial F}{\partial a}\right)_{b,x}\frac{da}{dx}.$$

The first term is precisely the situation covered in Theorem 14.3, and therefore

$$\left(\frac{\partial F}{\partial x}\right)_{a,b} = \int_a^b \frac{\partial f(x,y)}{\partial x}\,dy.$$

Since Theorem 6.7 indicates that

$$\frac{d}{db}\int_a^b f(y)\,dy = f(b),$$

it follows that

$$\frac{\partial}{\partial b}\int_a^b f(x,y)\,dy = f(x,b).$$

In other words,

$$\left(\frac{\partial F}{\partial b}\right)_{a,x} = f(x,b).$$

Furthermore,

$$\left(\frac{\partial F}{\partial a}\right)_{b,x} = \frac{\partial}{\partial a}\int_a^b f(x,y)\,dy = -\frac{\partial}{\partial a}\int_b^a f(x,y)\,dy = -f(x,a).$$

Substitution of these facts into the chain rule now gives Leibnitz's rule.

Our derivation of Leibnitz's rule for the differentiation of a definite integral that depends on a parameter (x in this case) shows that the first term accounts for those x's in the integrand, and the second and third terms for the x's in the upper and lower limits. The following geometric interpretation of Leibnitz's rule emphasizes this same point.

Suppose the function $f(x,y)$ has only positive values so that the surface $z = f(x,y)$ lies completely above the xy-plane (Figure 14.76). The equations $y = a(x)$ and $y = b(x)$ describe cylindrical walls standing on the curves $y = a(x)$, $z = 0$ and $y = b(x)$, $z = 0$ in the xy-plane. Were we to slice through the surface $z = f(x,y)$ with a plane $x =$ constant, then an area would be defined in this plane bounded on the top by $z = f(x,y)$, on the sides by $y = a(x)$ and $y = b(x)$, and on the bottom by the xy-plane. This area is clearly defined by the definite integral

$$\int_{a(x)}^{b(x)} f(x,y)\,dy$$

in Leibnitz's rule, and as x varies so too does the area. Note, in particular, that as the plane varies, the area changes, not only because the height of the surface $z = f(x,y)$ changes but also because the width of the area varies (i.e., the two cylindrical walls are not a constant distance apart). The first term in Leibnitz's rule accounts for the vertical variation, whereas the remaining two terms represent variations due to the fluctuating width.

FIGURE 14.76

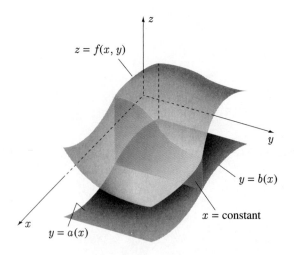

EXAMPLE 14.30

If $F(x) = \int_x^{2x}(y^3 \ln y + x^3 e^y)\,dy$, find $F'(x)$.

SOLUTION With Leibnitz's rule, we have

$$F'(x) = \int_x^{2x}(3x^2 e^y)\,dy + \{(2x)^3 \ln(2x) + x^3 e^{2x}\}(2) - \{x^3 \ln x + x^3 e^x\}(1)$$

$$= \{3x^2 e^y\}_x^{2x} + 16x^3(\ln 2 + \ln x) + 2x^3 e^{2x} - x^3 \ln x - x^3 e^x$$

$$= 3x^2 e^{2x} - 3x^2 e^x + (16 \ln 2)x^3 + 15x^3 \ln x + 2x^3 e^{2x} - x^3 e^x$$

$$= 3x^2 e^x(e^x - 1) + x^3(16 \ln 2 + 15 \ln x + 2e^{2x} - e^x).$$

∎

In Exercise 6 you are asked to evaluate the integral defining $F(x)$ in this example and then to differentiate the resulting function with respect to x. It will be clear, then, that for this example Leibnitz's rule simplifies the calculations considerably.

EXAMPLE 14.31

Evaluate

$$\int_0^1 \frac{y^x - 1}{\ln y}\,dy \quad \text{for } x > -1.$$

SOLUTION We use Leibnitz's rule in this example to avoid finding an antiderivative for $(y^x - 1)/\ln y$. If we set

$$F(x) = \int_0^1 \frac{y^x - 1}{\ln y}\,dy$$

and use Leibnitz's rule, we have

$$F'(x) = \int_0^1 \frac{\partial}{\partial x}\left\{\frac{y^x - 1}{\ln y}\right\}dy = \int_0^1 \frac{y^x \ln y}{\ln y}\,dy = \int_0^1 y^x\,dy = \left\{\frac{y^{x+1}}{x+1}\right\}_0^1 = \frac{1}{x+1}.$$

It follows, therefore, that $F(x)$ must be of the form

$$F(x) = \ln(x+1) + C.$$

But from the definition of $F(x)$ as an integral, it is clear that $F(0) = 0$, and hence $C = 0$. Thus,

$$\int_0^1 \frac{y^x - 1}{\ln y}\,dy = \ln(x+1).$$

∎

Note that this problem originally had nothing whatsoever to do with Leibnitz's rule. But by introducing the rule, we were able to find a simple formula for $F'(x)$, and this immediately led to $F(x)$. This can be a very useful technique for evaluating definite integrals that depend on a parameter.

EXAMPLE 14.32

When a shell is fired from the artillery gun in Figure 14.77, the barrel recoils along a well-lubricated guide and its motion is braked by a battery of heavy springs. We set up a coordinate system where $x = 0$ represents the firing position of the gun when no stretch or compression exists in the springs. Suppose that when the gun is fired, the horizontal component of the force causing recoil is $g(t)$ (t = time). If the mass of the gun is m and the effective spring constant for the battery of springs is k, then Newton's second law states that

$$m\frac{d^2x}{dt^2} = g(t) - kx,$$

or

$$m\frac{d^2x}{dt^2} + kx = g(t).$$

This differential equation can be solved by a method called **variation of parameters**, and the solution is

$$x(t) = A\cos\sqrt{\frac{k}{m}}t + B\sin\sqrt{\frac{k}{m}}t + \frac{1}{\sqrt{mk}}\int_0^t g(u)\sin\left(\sqrt{\frac{k}{m}}(t-u)\right)du,$$

where A and B are arbitrary constants. We do not discuss differential equations until Chapter 16, but with Leibnitz's rule it is possible to verify that this function is indeed a solution. Do so.

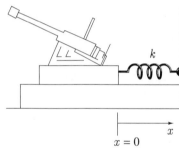

FIGURE 14.77

SOLUTION To differentiate the integral we rewrite Leibnitz's rule in terms of the variables of this problem:

$$\frac{d}{dt}\int_{a(t)}^{b(t)} f(t,u)\,du = \int_{a(t)}^{b(t)} \frac{\partial f(t,u)}{\partial t}\,du + f[t,b(t)]\frac{db}{dt} - f[t,a(t)]\frac{da}{dt}.$$

If we now apply this formula to the definite integral in $x(t)$ where

$$f(t,u) = g(u)\sin\left[\sqrt{\frac{k}{m}}(t-u)\right],$$

then

$$\frac{dx}{dt} = -\sqrt{\frac{k}{m}}A\sin\sqrt{\frac{k}{m}}t + \sqrt{\frac{k}{m}}B\cos\sqrt{\frac{k}{m}}t$$
$$+ \frac{1}{\sqrt{mk}}\left\{\int_0^t g(u)\sqrt{\frac{k}{m}}\cos\left(\sqrt{\frac{k}{m}}(t-u)\right)du + g(t)\sin\left(\sqrt{\frac{k}{m}}(t-t)\right)\right\}$$

$$= \sqrt{\frac{k}{m}} \left\{ -A \sin \sqrt{\frac{k}{m}} t + B \cos \sqrt{\frac{k}{m}} t \right\} + \frac{1}{m} \int_0^t g(u) \cos\left(\sqrt{\frac{k}{m}}(t-u)\right) du.$$

We now use Leibnitz's rule once more to find d^2x/dt^2:

$$\frac{d^2x}{dt^2} = \sqrt{\frac{k}{m}} \left\{ -\sqrt{\frac{k}{m}} A \cos \sqrt{\frac{k}{m}} t - \sqrt{\frac{k}{m}} B \sin \sqrt{\frac{k}{m}} t \right\}$$

$$+ \frac{1}{m} \left\{ \int_0^t -g(u) \sqrt{\frac{k}{m}} \sin\left(\sqrt{\frac{k}{m}}(t-u)\right) du + g(t) \cos\left(\sqrt{\frac{k}{m}}(t-t)\right) \right\}$$

$$= -\frac{k}{m} \left\{ A \cos \sqrt{\frac{k}{m}} t + B \sin \sqrt{\frac{k}{m}} t \right\}$$

$$- \sqrt{\frac{k}{m^3}} \int_0^t g(u) \sin\left(\sqrt{\frac{k}{m}}(t-u)\right) du + \frac{1}{m} g(t)$$

$$= -\frac{k}{m} \left\{ A \cos \sqrt{\frac{k}{m}} t + B \sin \sqrt{\frac{k}{m}} t + \frac{1}{\sqrt{mk}} \int_0^t g(u) \sin\left(\sqrt{\frac{k}{m}}(t-u)\right) du \right\}$$

$$+ \frac{1}{m} g(t)$$

$$= -\frac{k}{m} x(t) + \frac{1}{m} g(t).$$

In other words, the function $x(t)$ satisfies the differential equation

$$m \frac{d^2x}{dt^2} = -kx(t) + g(t),$$

and the proof is complete. ∎

This example illustrates that Leibnitz's rule is essential to the manipulation of solutions of differential equations that are represented as definite integrals.

EXERCISES 14.13

In Exercises 1–5 use Leibnitz's rule to find the derivative of $F(x)$. Check your result by evaluating the integral and then differentiating.

1. $F(x) = \displaystyle\int_0^3 (x^2 y^2 + 3xy)\, dy$

2. $F(x) = \displaystyle\int_1^x (x^2/y^2 + e^y)\, dy$

3. $F(x) = \displaystyle\int_{x-1}^{x^2} (x^3 y + y^2 + 1)\, dy$

4. $F(x) = \displaystyle\int_{x^2}^{x^3-1} (x + y \ln y)\, dy$

5. $F(x) = \displaystyle\int_0^x \frac{y-x}{y+x}\, dy$

6. Find $F'(x)$ in Example 14.30 by first evaluating the definite integral.

7. Use the result of Example 14.31 to prove that

$$\int_0^1 \frac{x^p - x^q}{\ln x}\, dx = \ln\left(\frac{p+1}{q+1}\right),$$

provided $p > -1$ and $q > -1$.

8. Find $F'(x)$ if $F(x) = \displaystyle\int_{\sin x}^{e^x} \sqrt{1+y^3}\, dy$.

In Exercises 9–11 use Leibnitz's rule to verify that the function $y(x)$ satisfies the differential equation.

9. $y(x) = \dfrac{1}{x^2}\displaystyle\int_0^x t^2 f(t)\, dt$; $x\dfrac{dy}{dx} + 2y = xf(x)$

10. $y(x) = \dfrac{1}{2}\displaystyle\int_0^x f(t)(e^{x-t} - e^{t-x})\, dt$; $\dfrac{d^2y}{dx^2} - y = f(x)$

11. $y(x) = \dfrac{1}{\sqrt{2}}\displaystyle\int_0^x e^{2(t-x)}\sin[\sqrt{2}(x-t)]f(t)\, dt$;
$\dfrac{d^2y}{dx^2} + 4\dfrac{dy}{dx} + 6y = f(x)$

12. Given that

$$\int_0^b \frac{1}{1+ax}\, dx = \frac{1}{a}\ln(1+ab),$$

find a formula for $\displaystyle\int_0^b \frac{x}{(1+ax)^2}\, dx$.

13. Given that

$$\int \frac{1}{\sqrt{a^2-x^2}}\, dx = \text{Sin}^{-1}\left(\frac{x}{a}\right) + C,$$

find a formula for $\displaystyle\int \frac{1}{(a^2-x^2)^{3/2}}\, dx$.

14. Given that

$$\int \frac{1}{a^2+x^2}\, dx = \frac{1}{a}\text{Tan}^{-1}\left(\frac{x}{a}\right) + C,$$

find a formula for $\displaystyle\int \frac{1}{(a^2+x^2)^3}\, dx$.

15. Use the result that

$$\int_0^{\pi/2} \frac{1}{a^2\cos^2 x + b^2\sin^2 x}\, dx = \frac{\pi}{2|ab|}$$

to find a formula for $\displaystyle\int_0^{\pi/2} \frac{1}{(a^2\cos^2 x + b^2\sin^2 x)^2}\, dx$.

In Exercises 16 and 17 use Leibnitz's rule to evaluate the integral.

16. $\displaystyle\int_0^\pi \frac{\ln(1 + a\cos x)}{\cos x}\, dx$ where $|a| < 1$

17. $\displaystyle\int_0^\infty \frac{\text{Tan}^{-1}(ax)}{x(1+x^2)}\, dx$ where $a > 0$

18. (a) What is the domain of the function
$$F(x) = \int_0^9 \ln(1 - x^2 y^2)\, dy?$$ What is $F(0)$?
(b) Find $F'(x)$ by Leibnitz's rule. What is $F'(0)$?
(c) Show that the graph of the function $F(x)$ is concave downward for all x in its domain of definition.

19. Laplace's equation for a function $u(r,\theta)$ in polar coordinates is

$$\frac{\partial^2 u}{\partial r^2} + \frac{1}{r}\frac{\partial u}{\partial r} + \frac{1}{r^2}\frac{\partial^2 u}{\partial \theta^2} = 0.$$

If the values of $u(r,\theta)$ are specified on the circle $r = R$ as $u(R,\phi)$, $-\pi < \phi \le \pi$, then Poisson's integral formula states that the value of $u(r,\theta)$ interior to this circle is defined by

$$u(r,\theta) = \frac{R^2 - r^2}{2\pi}\int_{-\pi}^{\pi} \frac{u(R,\phi)}{R^2 + r^2 - 2rR\cos(\theta-\phi)}\, d\phi.$$

Show that this function does indeed satisfy Laplace's equation.

SUMMARY

The definite integral of a function $f(x)$ from $x = a$ to $x = b$ is a limit of a sum

$$\int_a^b f(x)\,dx = \lim_{\|\Delta x_i\| \to 0} \sum_{i=1}^n f(x_i^*)\,\Delta x_i.$$

In this chapter we extended this idea to define double integrals of functions $f(x, y)$ over regions in the xy-plane and triple integrals of functions $f(x, y, z)$ over regions of space. Each is once again the limit of a sum:

$$\iint_R f(x, y)\,dA = \lim_{\|\Delta A_i\| \to 0} \sum_{i=1}^n f(x_i^*, y_i^*)\,\Delta A_i,$$

$$\iiint_V f(x, y, z)\,dV = \lim_{\|\Delta V_i\| \to 0} \sum_{i=1}^n f(x_i^*, y_i^*, z_i^*)\,\Delta V_i.$$

To evaluate double integrals we use double iterated integrals in Cartesian or polar coordinates. Which is the more useful in a given problem depends on the shape of the region R and the form of the function $f(x, y)$. For instance, circles centred at the origin and straight lines through the origin are represented very simply in polar coordinates, and therefore a region R with these curves as boundaries immediately suggests the use of polar coordinates. On the other hand, curves that can be described in the form $y = f(x)$, where $f(x)$ is a polynomial, a rational function, or a transcendental function, often suggest using double iterated integrals in Cartesian coordinates. Each integration in a double iterated integral can be interpreted geometrically, and through these interpretations it is a simple matter to find appropriate limits for the integrals. In particular, for a double iterated integral in Cartesian coordinates, the inner integration is over the rectangles in a strip (horizontal or vertical), and the outer integration adds over all strips. In polar coordinates, the inner integral is inside either a wedge or a ring, and the outer integral adds over all wedges or rings.

To evaluate triple integrals we use triple iterated integrals in Cartesian, cylindrical, or spherical coordinates. Once again the limits on these integrals can be determined by interpreting the summations geometrically. For example, the first integration in a triple iterated integral in Cartesian coordinates is always over the boxes in a column, the second over the rectangles inside a strip, and the third over all strips.

We used double integrals to find plane areas, volumes of solids of revolution, centroids, moments of inertia, fluid forces, and areas of surfaces. We dealt with these same applications (with the exception of surface area) in Chapter 7 using the definite integral, but with some difficulty: Volumes required two methods, shells and washers; centroids required an averaging formula for the first moment of a rectangle that has its length perpendicular to the axis about which a moment is required; moments of inertia needed a "one-third cubed formula" for rectangles with lengths perpendicular to the axis about which the moment of inertia is required; and fluid forces required horizontal rectangles. On the other hand, double integrals eliminate these difficulties but, more importantly, provide a unified approach to all applications.

In Section 14.13 we used double integrals to verify Leibnitz's rule for differentiating definite integrals that depend on a parameter.

Key Terms and Formulas

In reviewing this chapter, you should be able to define or discuss the following key terms:

Double integral
Double iterated integral
Area
Volume of a solid of revolution
Fluid pressure
Centre of mass
Moment of inertia
Surface area
Double integrals in polar coordinates

Triple integral
Triple iterated integral
Volume
Cylindrical coordinates
Triple integrals in cylindrical coordinates
Spherical coordinates
Triple integrals in spherical coordinates
Leibnitz's rule

REVIEW EXERCISES

In Exercises 1–21 evaluate the integral over the region.

1. $\iint_R (2x + y)\, dA$ where R is bounded by $y = x$, $y = 0$, $x = 2$

2. $\iiint_V xyz\, dV$ where V is bounded by $y = z$, $x = 0$, $y = 3$, $x = 1$, $z = 0$

3. $\iint_R x^3 y^2\, dA$ where R is bounded by $x = 1$, $x = -1$, $y = 2$, $y = -2$

4. $\iiint_V (x^2 - y^3)\, dV$ where V is bounded by $z = xy$, $z = 0$, $x = 1$, $y = 1$

5. $\iiint_V (x^2 - y^2)\, dV$ where V is the region in Exercise 4

6. $\iint_R y\, dA$ where R is bounded by $y = (x - 1)^2$, $y = x + 1$

7. $\iint_R xy^2\, dA$ where R is bounded by $x = 2 - 2y^2$, $x = -y^2$

8. $\iint_R x^2 y\, dA$ where R is the region of Exercise 7

9. $\iiint_V (x^2 + y^2 + z^2)\, dV$ where V is bounded by $z = x$, $z = -x$, $y = 0$, $y = 1$, $z = 2$

10. $\iint_R (xy - x^2 y^2)\, dA$ where R is bounded by $y = 2x^2$, $y = 4 - 2x^2$

11. $\iint_R x \sin y\, dA$ where R is bounded by $x = \sqrt{1 - y}$, $x = 0$, $y = 0$

12. $\iiint_V (x + y + z)\, dV$ where V is bounded by $z = 1 - x^2 - y^2$, $z = 0$

13. $\iint_R xe^y\, dA$ where R is bounded by $x = 0$, $y = 5$, $y = 2x + 1$

14. $\iiint_V dV$ where V is bounded by $z^2 = x^2 + y^2$, $x^2 + y^2 = 4$

15. $\iint_R (x + y)\, dA$ where R is bounded by $y = 1 - x^2$, $y = 1 - \sqrt{x}$, $y = x - 1$

16. $\iiint_V (x^2 + y^2 + z^2)\, dV$ where V is bounded by $z = \sqrt{1 - x^2 - y^2}$, $z = 0$

17. $\iint_R \dfrac{x}{x + y}\, dA$ where R is bounded by $y = x - 1$, $x = 2$, $y = 0$

18. $\iint_R (x^2 + y^2)\, dA$ where R is bounded by $x^2 + y^2 = 2x$

19. $\iiint_V \dfrac{x^2}{z^2}\, dV$ where V is bounded by $z = 1$, $z = \sqrt{4 - x^2 - y^2}$

20. $\iint_R \dfrac{1}{x^2 + y^2}\, dA$ where R is bounded by $y = x$, $x = 1$, $x = 2$, $y = 0$

860 CHAPTER 14 MULTIPLE INTEGRALS

21. $\iint_R (x^2 - y^2)\, dA$ where R is bounded by $y = x$, $y = x - 1$, $y = 5 - 2x$, $y = 14 - 2x$

22. If R represents the region of the xy-plane in Figure 14.78, what double integrals represent the following:
    (a) the area of R
    (b) the volumes of the solids of revolution when R is rotated about the lines $x = 2$ and $y = -4$
    (c) the first moments of area of R about the lines $x = 1$ and $y = -1$
    (d) the second moments of area of R about the lines $x = -1$ and $y = 4$
    (e) the total charge on R if it carries a charge per unit area $\sigma(x, y)$
    (f) the total mass if R is a plate with mass per unit area $\rho(x, y)$
    (g) the probability of an electron being in R if the probability of the electron being in unit area at point (x, y) is $P(x, y)$

FIGURE 14.78

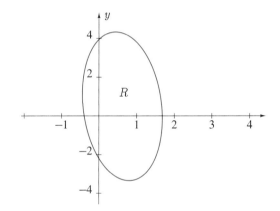

23. Find the volume bounded by the surfaces $z = 0$, $z = 1 - 2e^{-x^2 - y^2}$.

24. Find the area bounded by the curve $x^2(4 - x^2) = y^2$.

25. Find the centroid of the area bounded by the curves $4x = 4 - y^2$, $x = \sqrt{8 - 2y^2}$.

26. Find the volume bounded by the surfaces $y + z = 2$, $y - z = 2$, $y = x^2 + 1$.

27. If the viewing glass on a deep-sea diver's helmet is circular with diameter 10 centimetres, what is the force on the glass when the centre of the glass is 50 m below the surface?

28. Find the area of that part of $z = 1 - 4x^2 - 4y^2$ above the xy-plane.

29. Find the area bounded by the curve $x^4(4 - x^2) = y^2$.

30. Find the volumes of the solids of revolution when the area in Exercise 24 is rotated about the x- and y-axes.

31. Find the moment of inertia about the z-axis of the solid bounded by the surfaces $z = \sqrt{x^2 + y^2}$, $z = 2 - \sqrt{x^2 + y^2}$ if its density ρ is constant.

32. Find the second moment of area about the y-axis of the area in the first quadrant bounded by the curve $x^3 + y^3 = 1$.

33. Find the centre of mass of a uniform solid bounded by the surfaces $z = 1 + x^2 + y^2$, $x^2 + y^2 = 1$, $z = 0$.

34. Find the average value of the function $f(x, y) = x^2 + y^2$ over the region bounded by the curve $x^2 + y^2 = 4$.

35. Find the force due to water pressure on each side of the vertical parallelogram in Figure 14.79. All measurements are in metres.

FIGURE 14.79

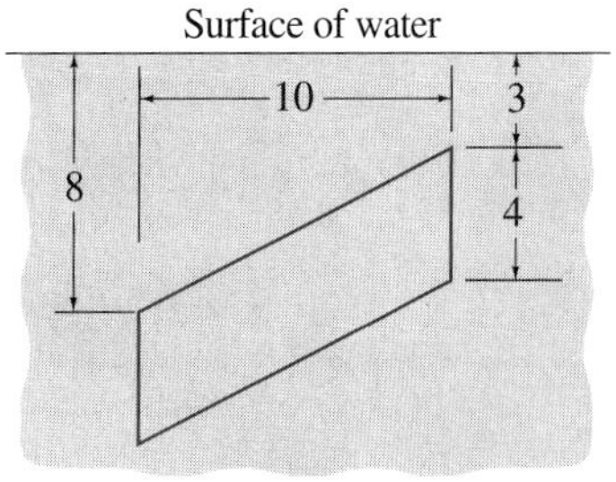

36. Find the moment of inertia about the z-axis of a uniform solid bounded by the surfaces $z = 0$, $z = 1$, $x^2 + y^2 = 1 + z^2$.

37. Find the average value of the function $f(x, y, z) = x + y + z$ over the region bounded by the surfaces $-x + 2y + 2z = 4$, $x + y + z = 2$, $y = 0$, $z = 0$.

38. Find the centroid of the bifolium $r = \sin\theta \cos^2\theta$.

39. Show that
$$y(x) = e^{-3x}(C_1 \cos x + C_2 \sin x) + \int_0^x f(t) e^{3(t-x)} \sin(x - t)\, dt$$
is a solution of the differential equation
$$\frac{d^2 y}{dx^2} + 6\frac{dy}{dx} + 10y = f(x) \text{ for any constants } C_1 \text{ and } C_2.$$

40. Find the centre of mass of the uniform solid in the first octant common to the cylinders $x^2 + z^2 = 1$, $y^2 + z^2 = 1$.

41. Find the area of that part of $z = \ln(x^2 + y^2)$ between the cylinders $x^2 + y^2 = 1$ and $x^2 + y^2 = 4$.

42. Find the volume bounded by the surface $\sqrt{x^2 + z^2} = y(2 - y)$.

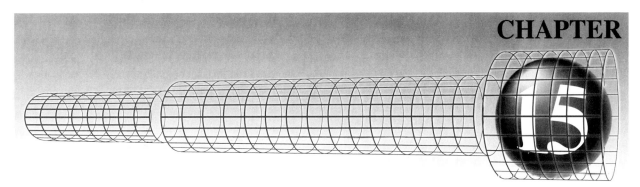

CHAPTER 15

Vector Calculus

In Sections 12.6 and 12.7 we considered vectors whose components are functions of a single variable. In particular, if an object moves along a curve C and C is defined parametrically by $x = x(t)$, $y = y(t)$, $z = z(t)$, $t \geq 0$, where t is time, then the components of the position, velocity, and acceleration vectors are functions of time:

$$\mathbf{r} = \mathbf{r}(t) = x(t)\hat{\mathbf{i}} + y(t)\hat{\mathbf{j}} + z(t)\hat{\mathbf{k}},$$

$$\mathbf{v} = \frac{d\mathbf{r}}{dt} = \frac{dx}{dt}\hat{\mathbf{i}} + \frac{dy}{dt}\hat{\mathbf{j}} + \frac{dz}{dt}\hat{\mathbf{k}},$$

$$\mathbf{a} = \frac{d\mathbf{v}}{dt} = \frac{d^2x}{dt^2}\hat{\mathbf{i}} + \frac{d^2y}{dt^2}\hat{\mathbf{j}} + \frac{d^2z}{dt^2}\hat{\mathbf{k}}.$$

If the object has constant mass m and is subjected to a force $\mathbf{F}$, which is given as a function of time t, $\mathbf{F} = \mathbf{F}(t)$, then Newton's second law expresses the acceleration of the object as $\mathbf{a} = \mathbf{F}(t)/m$. This equation can then be integrated to yield the velocity and position of the object as functions of time. Unfortunately, what often happens is that we do not know $\mathbf{F}$ as a function of time. Instead, we know that if the object were at such and such a position, then the force on it would be such and such; i.e., we know $\mathbf{F}$ as a function of position. For example, suppose a positive charge q is placed at the origin in space (Figure 15.1), and a second positive charge Q is placed at position (x, y, z). According to Coulomb's law, the force on Q due to q is

FIGURE 15.1

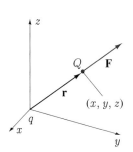

$$\mathbf{F} = \frac{qQ}{4\pi\epsilon_0|\mathbf{r}|^3}\mathbf{r} = \frac{qQ}{4\pi\epsilon_0(x^2 + y^2 + z^2)^{3/2}}(x\hat{\mathbf{i}} + y\hat{\mathbf{j}} + z\hat{\mathbf{k}}).$$

If we allow Q to move under the influence of this force, we will not know $\mathbf{F}$ as a function of time, but rather as a function of position. This makes Newton's second law much more difficult to deal with, but it is in fact the normal situation. Most forces are represented as a function of position rather than time. Besides electrostatic forces, consider, for instance, spring forces, gravitational forces, and fluid forces — all of these are functions of position. Forces that are functions of position are examples of vectors that are functions of position. In this chapter we study vectors that are functions of position. In particular, we integrate them along curves and over surfaces.

SECTION 15.1

Vector Fields

Domains for functions $f(x)$ of one variable are open, closed, half-open, or half-closed intervals on the x-axis. Domains for functions $f(x, y)$ of two variables are sets of points in the xy-plane, and domains for functions $f(x, y, z)$ are sets of points in xyz-space.

In order to state definitions and theorems for multivariable functions in this chapter, we require corresponding definitions of open and closed sets of points. We define them for the xy-plane; analogous definitions for space can be found in Exercise 10.

Consider a set S of points in the xy-plane. A point P in the plane is called an **interior point** of S if there exists a circle centred at P that contains only points of S. A point Q is called an **exterior point** of S if there exists a circle centred at Q that contains no point of S. A point R is called a **boundary point** of S if every circle with centre R contains at least one point in S and at least one point not in S. For example, consider the set of points $S_1 : 4x^2 + 9y^2 < 36$ (Figure 15.2). The fact that we have dotted the ellipse indicates that these points are not in S. Every point inside the ellipse is an interior point of S_1; every point outside the ellipse is an exterior point, and every point on the ellipse is a boundary point. For the set $S_2 : 4x^2 + 9y^2 \geq 36$ (Figure 15.3), every point outside the ellipse is interior to S_2, every point inside the ellipse is exterior to S_2, and every point on the ellipse is a boundary point.

FIGURE 15.2

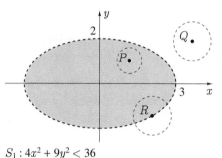

$S_1 : 4x^2 + 9y^2 < 36$

FIGURE 15.3

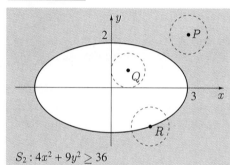

$S_2 : 4x^2 + 9y^2 \geq 36$

A set of points S is said to be **open** if all points in S are interior points. Alternatively, a set is open if it contains none of its boundary points. A set is said to be **closed** if it contains all of its boundary points. Set S_1 above is open; set S_2 is closed. If to S_1 we add the points on the upper half of the ellipse (Figure 15.4), this set, call it S_3, is neither open nor closed. It contains some of its boundary points but not all of them.

FIGURE 15.4   **FIGURE 15.5**

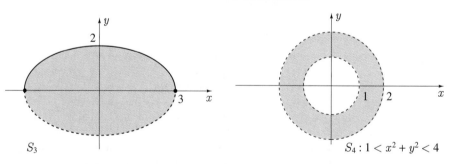

S_3 $S_4 : 1 < x^2 + y^2 < 4$

A set S is said to be **connected** if every pair of points in S can be joined by a piecewise smooth curve lying entirely within S. Sets S_1, S_2, and S_3 are all connected. Set S_4 in Figure 15.5 is also connected. Set S_5 in Figure 15.6 is not connected; it consists of two disjoint pieces.

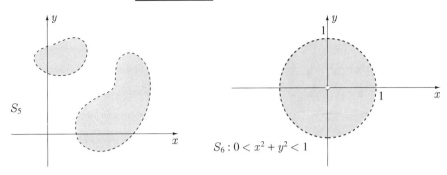

FIGURE 15.6   **FIGURE 15.7**

$S_6 : 0 < x^2 + y^2 < 1$

A **domain** is an open, connected set. Domain is perhaps a poor choice of words, as it might be confused with "domain of a function," but it has become accepted terminology. Context always makes it clear which interpretation is intended. Sets S_1 and S_4 are domains; sets S_2, S_3, and S_5 are not.

A domain is said to be **simply-connected** if every closed curve in the domain contains in its interior only points of the domain. In essence, a simply-connected domain has no holes. Domain S_1 is simply-connected, S_4 is not. The set $S_6 : 0 < x^2 + y^2 < 1$ (Figure 15.7) is a domain, but it is not simply-connected.

Analogous definitions for sets of points in space can be found in Exercise 10. One must be somewhat more careful here as there are different "kinds of holes." See Exercises 13 and 15.

When it is not necessary to indicate the particular characteristics of a set of points, we often use the word "region."

Many vectors are functions of position. We call such vectors **vector fields**. To be precise, we say that **F** is a vector field in a region D if **F** assigns a vector to each point in D. If D is a region of space, then **F** assigns a vector $\mathbf{F}(x, y, z)$ to each point in D. If P, Q, and R are the components of $\mathbf{F}(x, y, z)$, then each of these is a function of x, y and z:

$$\mathbf{F} = \mathbf{F}(x, y, z) = P(x, y, z)\hat{\mathbf{i}} + Q(x, y, z)\hat{\mathbf{j}} + R(x, y, z)\hat{\mathbf{k}}. \quad (15.1)$$

If D is a region of the xy-plane, then

$$\mathbf{F} = \mathbf{F}(x, y) = P(x, y)\hat{\mathbf{i}} + Q(x, y)\hat{\mathbf{j}}. \quad (15.2)$$

In Chapter 12 we stressed the fact that the tail of a vector could be placed at any point whatsoever. What was important was relative positions of tip and tail. For vector fields, we almost always place the tail of the vector associated with a point at that point.

Vector fields are essential to the study of most areas in the physical sciences. Moments ($\mathbf{M} = \mathbf{r} \times \mathbf{F}$) in mechanics (see Exercise 43 in Section 12.5) depend on the position of **F**, and are therefore vector fields. The electric field intensity **E**, the electric displacement **D**, the magnetic induction **B**, and the current density **J** are all important vector fields in electromagnetic theory. The heat flux vector **q** is the basis for the study of heat conduction.

We have also encountered vector fields that are of geometric importance. For example, the gradient of a scalar function $f(x, y, z)$ is a vector field,

$$\operatorname{grad} f = \nabla f = \frac{\partial f}{\partial x}\hat{\mathbf{i}} + \frac{\partial f}{\partial y}\hat{\mathbf{j}} + \frac{\partial f}{\partial z}\hat{\mathbf{k}}. \quad (15.3)$$

It assigns the vector ∇f to each point (x, y, z) in some region of space. We have seen that ∇f points in the direction in which $f(x, y, z)$ increases most rapidly, and its magnitude $|\nabla f|$ is the rate of increase in that direction. In addition, we know that if $f(x, y, z) = c$ is the equation of a surface that passes through a point (x, y, z), then ∇f at that point is normal to the surface.

Often we write

$$\nabla f = \left(\frac{\partial}{\partial x}\hat{\mathbf{i}} + \frac{\partial}{\partial y}\hat{\mathbf{j}} + \frac{\partial}{\partial z}\hat{\mathbf{k}} \right) f$$

and regard

$$\nabla = \frac{\partial}{\partial x}\hat{\mathbf{i}} + \frac{\partial}{\partial y}\hat{\mathbf{j}} + \frac{\partial}{\partial z}\hat{\mathbf{k}}$$

as a vector differential operator, called the **del operator**. It operates on a scalar function to produce a vector field, its gradient. As an operator ∇ should never stand alone, but should always be followed by something on which to operate. Because of this ∇ should not itself be considered a vector, in spite of the fact that it has the form of a vector. It is a differential operator, and must therefore operate on something. In the remainder of this section we use the del operator to define two extremely useful operations on vector fields: the **divergence** and the **curl**.

Definition 15.1

If $\mathbf{F}(x, y, z) = P(x, y, z)\hat{\mathbf{i}} + Q(x, y, z)\hat{\mathbf{j}} + R(x, y, z)\hat{\mathbf{k}}$ is a vector field in a region D, then the **divergence** of $\mathbf{F}$ is a scalar field in D defined by

$$\text{div}\,\mathbf{F} = \nabla \cdot \mathbf{F} = \frac{\partial P}{\partial x} + \frac{\partial Q}{\partial y} + \frac{\partial R}{\partial z}, \qquad (15.4)$$

provided that the partial derivatives exist at each point in D.

Is it clear why we use the notation $\nabla \cdot \mathbf{F}$, in spite of the fact that ∇ is not a vector in the true sense of the word?

EXAMPLE 15.1

Calculate $\nabla \cdot \mathbf{F}$ if

(a) $\mathbf{F} = 2xy\hat{\mathbf{i}} + z\hat{\mathbf{j}} + x^2 \cos(yz)\hat{\mathbf{k}}$.

(b) $\mathbf{F} = \dfrac{qQ}{4\pi\epsilon_0 |\mathbf{r}|^3}\mathbf{r}$, where $\mathbf{r} = x\hat{\mathbf{i}} + y\hat{\mathbf{j}} + z\hat{\mathbf{k}}$.

SOLUTION (a) $\nabla \cdot \mathbf{F} = \dfrac{\partial}{\partial x}(2xy) + \dfrac{\partial}{\partial y}(z) + \dfrac{\partial}{\partial z}(x^2 \cos(yz)) = 2y - x^2 y \sin(yz)$.

(b) For the derivative of the x-component of $\mathbf{F}$ with respect to x, we calculate

$$\frac{\partial}{\partial x}\left\{ \frac{qQx}{4\pi\epsilon_0 (x^2 + y^2 + z^2)^{3/2}} \right\} = \frac{qQ}{4\pi\epsilon_0}\left\{ \frac{1}{(x^2 + y^2 + z^2)^{3/2}} - \frac{3x^2}{(x^2 + y^2 + z^2)^{5/2}} \right\}$$

$$= \frac{qQ}{4\pi\epsilon_0}\left\{ \frac{-2x^2 + y^2 + z^2}{(x^2 + y^2 + z^2)^{5/2}} \right\}.$$

With similar results for the remaining two derivatives, we obtain

$$\nabla \cdot \mathbf{F} = \frac{qQ}{4\pi\epsilon_0}\left\{ \frac{-2x^2 + y^2 + z^2}{(x^2 + y^2 + z^2)^{5/2}} + \frac{x^2 - 2y^2 + z^2}{(x^2 + y^2 + z^2)^{5/2}} + \frac{x^2 + y^2 - 2z^2}{(x^2 + y^2 + z^2)^{5/2}} \right\}$$

$$= 0.$$

Any physical interpretation of $\nabla \cdot \mathbf{F}$ will depend on the interpretation of $\mathbf{F}$. The following discussion describes the interpretation of the divergence of a certain vector field in the theory of fluid flow. If a gas flows through a region D of space, then it flows with some velocity $\mathbf{v}$ past point $P(x, y, z)$ in D at time t (Figure 15.8). Consider a unit area A around P perpendicular to $\mathbf{v}$. If at time t the density of gas at P is ρ, then the vector $\rho\mathbf{v}$ represents the mass of gas flowing through A per unit time. At each point P in D, the direction of $\rho\mathbf{v}$ tells us the direction of gas flow, and its length indicates the mass of gas flowing in that direction. For changing conditions, each of ρ and $\mathbf{v}$ depend not only on position (x, y, z) but also on time t:

$$\rho\mathbf{v} = \rho(x, y, z, t)\mathbf{v}(x, y, z, t).$$

The vector $\rho\mathbf{v}$, then, is a vector field that also depends on time.

In fluid dynamics it is shown that at each point in D, $\rho\mathbf{v}$ must satisfy

FIGURE 15.8

$$\nabla \cdot (\rho\mathbf{v}) = -\frac{\partial \rho}{\partial t},$$

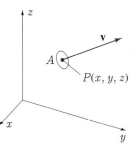

called the **equation of continuity** (see also Section 15.11). It is this equation that gives us an interpretation of the divergence of $\rho\mathbf{v}$. Density is the mass per unit volume of gas. If $\partial\rho/\partial t$ is positive, then density is increasing. This means that more mass must be entering unit volume than leaving it. Similarly, if $\partial\rho/\partial t$ is negative, more mass is leaving than entering. Since $\nabla \cdot (\rho\mathbf{v})$ is the negative of $\partial\rho/\partial t$, it follows that $\nabla \cdot (\rho\mathbf{v})$ must be a measure of how much more gas is leaving unit volume than entering. We can see, then, that the word "divergence" is appropriately chosen for this application.

The curl of a vector field is defined as follows.

Definition 15.2

If $\mathbf{F}(x, y, z) = P(x, y, z)\hat{\mathbf{i}} + Q(x, y, z)\hat{\mathbf{j}} + R(x, y, z)\hat{\mathbf{k}}$ is a vector field in a region D, then the **curl** of $\mathbf{F}$ is a vector function in D defined by

$$\text{curl } \mathbf{F} = \nabla \times \mathbf{F} = \begin{vmatrix} \hat{\mathbf{i}} & \hat{\mathbf{j}} & \hat{\mathbf{k}} \\ \frac{\partial}{\partial x} & \frac{\partial}{\partial y} & \frac{\partial}{\partial z} \\ P & Q & R \end{vmatrix}$$
$$= \left(\frac{\partial R}{\partial y} - \frac{\partial Q}{\partial z}\right)\hat{\mathbf{i}} + \left(\frac{\partial P}{\partial z} - \frac{\partial R}{\partial x}\right)\hat{\mathbf{j}} + \left(\frac{\partial Q}{\partial x} - \frac{\partial P}{\partial y}\right)\hat{\mathbf{k}}, \quad (15.5)$$

provided that the partial derivatives exist at each point in D.

In representing curl $\mathbf{F}$ in the form of a determinant we agree to expand the determinant along the first row.

EXAMPLE 15.2

Calculate the curls of the vector fields in Example 15.1.

SOLUTION (a)

$$\nabla \times \mathbf{F} = \begin{vmatrix} \hat{\mathbf{i}} & \hat{\mathbf{j}} & \hat{\mathbf{k}} \\ \dfrac{\partial}{\partial x} & \dfrac{\partial}{\partial y} & \dfrac{\partial}{\partial z} \\ 2xy & z & x^2 \cos(yz) \end{vmatrix}$$

$$= (-x^2 z \sin(yz) - 1)\hat{\mathbf{i}} + (-2x\cos(yz))\hat{\mathbf{j}} + (-2x)\hat{\mathbf{k}}$$

(b)

$$\nabla \times \mathbf{F} = \dfrac{qQ}{4\pi\epsilon_0} \begin{vmatrix} \hat{\mathbf{i}} & \hat{\mathbf{j}} & \hat{\mathbf{k}} \\ \dfrac{\partial}{\partial x} & \dfrac{\partial}{\partial y} & \dfrac{\partial}{\partial z} \\ \dfrac{x}{|\mathbf{r}|^3} & \dfrac{y}{|\mathbf{r}|^3} & \dfrac{z}{|\mathbf{r}|^3} \end{vmatrix}$$

The x-component of $\nabla \times \mathbf{F}$ is $qQ(4\pi\epsilon_0)$ multiplied by

$$\dfrac{\partial}{\partial y}\left(\dfrac{z}{|\mathbf{r}|^3}\right) - \dfrac{\partial}{\partial z}\left(\dfrac{y}{|\mathbf{r}|^3}\right) = \dfrac{\partial}{\partial y}\left(\dfrac{z}{(x^2+y^2+z^2)^{3/2}}\right)$$
$$- \dfrac{\partial}{\partial z}\left(\dfrac{y}{(x^2+y^2+z^2)^{3/2}}\right)$$
$$= \dfrac{-3yz}{(x^2+y^2+z^2)^{5/2}} + \dfrac{3yz}{(x^2+y^2+z^2)^{5/2}} = 0.$$

Similar results for the y- and z-components give $\nabla \times \mathbf{F} = \mathbf{0}$. ∎

If a vector $\rho\mathbf{v}$ is defined as above for gas flow through a region D, then the curl of $\rho\mathbf{v}$ describes the tendency for the motion of the gas to be circular rather than flowing in a straight line (see Section 15.11). This suggests why the term "curl" is used. With this interpretation, the following definition seems reasonable.

Definition 15.3

A vector field $\mathbf{F}$ is said to be **irrotational** in a region D if in D

$$\nabla \times \mathbf{F} = \mathbf{0}. \tag{15.6}$$

Applications of divergence and curl extend far beyond the topic of fluid dynamics. Both concepts are indispensable in many areas of applied mathematics such as electromagnetism, continuum mechanics, and heat conduction, to name a few.

The del operator ∇ operates on a scalar field to produce the gradient of the scalar field and on a vector field to give the divergence and the curl of the vector field. The following list of properties for the del operator is straightforward to verify (see Exercise 43).

If f and g are scalar fields that have first partial derivatives in a region D, and if $\mathbf{F}$ and $\mathbf{G}$ are vector fields in D with components that have first partial derivatives, then:

$$\nabla(f+g) = \nabla f + \nabla g, \tag{15.7}$$
$$\nabla \cdot (\mathbf{F}+\mathbf{G}) = \nabla \cdot \mathbf{F} + \nabla \cdot \mathbf{G}, \tag{15.8}$$

$$\nabla \times (\mathbf{F} + \mathbf{G}) = \nabla \times \mathbf{F} + \nabla \times \mathbf{G}, \tag{15.9}$$
$$\nabla(fg) = f\nabla g + g\nabla f, \tag{15.10}$$
$$\nabla \cdot (f\mathbf{F}) = \nabla f \cdot \mathbf{F} + f\nabla \cdot \mathbf{F}, \tag{15.11}$$
$$\nabla \times (f\mathbf{F}) = \nabla f \times \mathbf{F} + f\nabla \times \mathbf{F}, \tag{15.12}$$
$$\nabla \cdot (\mathbf{F} \times \mathbf{G}) = \mathbf{G} \cdot (\nabla \times \mathbf{F}) - \mathbf{F} \cdot (\nabla \times \mathbf{G}), \tag{15.13}$$
$$\nabla \times (\nabla f) = \mathbf{0}, \tag{15.14}$$
$$\nabla \cdot (\nabla \times \mathbf{F}) = 0. \tag{15.15}$$

For properties (15.14) and (15.15), we assume that f and $\mathbf{F}$ have continuous second-order partial derivatives in D. A typical way to verify these identities is to reduce each side of the identity to the same quantity. For example, to verify 15.13, we set $\mathbf{F} = P\hat{\mathbf{i}} + Q\hat{\mathbf{j}} + R\hat{\mathbf{k}}$ and $\mathbf{G} = L\hat{\mathbf{i}} + M\hat{\mathbf{j}} + N\hat{\mathbf{k}}$. Then

$$\mathbf{F} \times \mathbf{G} = (QN - RM)\hat{\mathbf{i}} + (RL - PN)\hat{\mathbf{j}} + (PM - QL)\hat{\mathbf{k}};$$

thus,

$$\nabla \cdot (\mathbf{F} \times \mathbf{G}) = \frac{\partial}{\partial x}(QN - RM) + \frac{\partial}{\partial y}(RL - PN) + \frac{\partial}{\partial z}(PM - QL)$$
$$= Q\frac{\partial N}{\partial x} + N\frac{\partial Q}{\partial x} - R\frac{\partial M}{\partial x} - M\frac{\partial R}{\partial x} + R\frac{\partial L}{\partial y} + L\frac{\partial R}{\partial y} - P\frac{\partial N}{\partial y}$$
$$- N\frac{\partial P}{\partial y} + P\frac{\partial M}{\partial z} + M\frac{\partial P}{\partial z} - Q\frac{\partial L}{\partial z} - L\frac{\partial Q}{\partial z}$$
$$= L\left(\frac{\partial R}{\partial y} - \frac{\partial Q}{\partial z}\right) + M\left(\frac{\partial P}{\partial z} - \frac{\partial R}{\partial x}\right) + N\left(\frac{\partial Q}{\partial x} - \frac{\partial P}{\partial y}\right)$$
$$+ P\left(\frac{\partial M}{\partial z} - \frac{\partial N}{\partial y}\right) + Q\left(\frac{\partial N}{\partial x} - \frac{\partial L}{\partial z}\right) + R\left(\frac{\partial L}{\partial y} - \frac{\partial M}{\partial x}\right).$$

On the other hand,

$$\mathbf{G} \cdot (\nabla \times \mathbf{F}) - \mathbf{F} \cdot (\nabla \times \mathbf{G}) = \mathbf{G} \cdot \begin{vmatrix} \hat{\mathbf{i}} & \hat{\mathbf{j}} & \hat{\mathbf{k}} \\ \partial/\partial x & \partial/\partial y & \partial/\partial z \\ P & Q & R \end{vmatrix} - \mathbf{F} \cdot \begin{vmatrix} \hat{\mathbf{i}} & \hat{\mathbf{j}} & \hat{\mathbf{k}} \\ \partial/\partial x & \partial/\partial y & \partial/\partial z \\ L & M & N \end{vmatrix}$$
$$= (L, M, N) \cdot \left(\frac{\partial R}{\partial y} - \frac{\partial Q}{\partial z}, \frac{\partial P}{\partial z} - \frac{\partial R}{\partial x}, \frac{\partial Q}{\partial x} - \frac{\partial P}{\partial y}\right)$$
$$- (P, Q, R) \cdot \left(\frac{\partial N}{\partial y} - \frac{\partial M}{\partial z}, \frac{\partial L}{\partial z} - \frac{\partial N}{\partial x}, \frac{\partial M}{\partial x} - \frac{\partial L}{\partial y}\right)$$
$$= L\left(\frac{\partial R}{\partial y} - \frac{\partial Q}{\partial z}\right) + M\left(\frac{\partial P}{\partial z} - \frac{\partial R}{\partial x}\right) + N\left(\frac{\partial Q}{\partial x} - \frac{\partial P}{\partial y}\right)$$
$$- P\left(\frac{\partial N}{\partial y} - \frac{\partial M}{\partial z}\right) - Q\left(\frac{\partial L}{\partial z} - \frac{\partial N}{\partial x}\right)$$
$$- R\left(\frac{\partial M}{\partial x} - \frac{\partial L}{\partial y}\right),$$

and this is the same expression as for $\nabla \cdot (\mathbf{F} \times \mathbf{G})$.

Given a scalar function $f(x, y, z)$ it is straightforward to calculate its gradient ∇f. Conversely, given the gradient of a function ∇f, it is possible to find the function $f(x, y, z)$. For example, if the vector field

$$\mathbf{F} = (3x^2yz + z^2)\hat{\mathbf{i}} + (x^3z + 2y)\hat{\mathbf{j}} + (x^3y + 2xz + 1)\hat{\mathbf{k}}$$

is known to be the gradient of some function $f(x, y, z)$, then

$$\frac{\partial f}{\partial x}\hat{\mathbf{i}} + \frac{\partial f}{\partial y}\hat{\mathbf{j}} + \frac{\partial f}{\partial z}\hat{\mathbf{k}} = (3x^2yz + z^2)\hat{\mathbf{i}} + (x^3z + 2y)\hat{\mathbf{j}} + (x^3y + 2xz + 1)\hat{\mathbf{k}}.$$

Since two vectors are equal if and only if they have identical components, we can say that

$$\frac{\partial f}{\partial x} = 3x^2yz + z^2, \quad \frac{\partial f}{\partial y} = x^3z + 2y, \quad \frac{\partial f}{\partial z} = x^3y + 2xz + 1.$$

Integration of the first of these with respect to x, holding y and z constant, implies that $f(x, y, z)$ must be of the form

$$f(x, y, z) = x^3yz + xz^2 + v(y, z)$$

for some function $v(y, z)$. To determine $v(y, z)$, we substitute this $f(x, y, z)$ into the second equation,

$$x^3z + \frac{\partial v}{\partial y} = x^3z + 2y,$$

and this implies that

$$\frac{\partial v}{\partial y} = 2y.$$

Consequently,

$$v(y, z) = y^2 + w(z)$$

for some $w(z)$, and therefore

$$f(x, y, z) = x^3yz + xz^2 + y^2 + w(z).$$

We now know both the x- and y-dependence of $f(x, y, z)$. To find $w(z)$ we substitute into the equation for $\partial f/\partial z$:

$$x^3y + 2xz + \frac{dw}{dz} = x^3y + 2xz + 1.$$

This equation requires

$$\frac{dw}{dz} = 1,$$

from which we have

$$w(z) = z + C, \quad C = \text{a constant}.$$

Thus,

$$f(x, y, z) = x^3yz + xz^2 + y^2 + z + C,$$

and this represents all functions that have a gradient equal to the given vector $\mathbf{F}$.

A much more difficult question is to determine whether a given vector field $\mathbf{F}(x, y, z)$ is the gradient of some scalar function $f(x, y, z)$. In the vast majority of cases, we can say that if the above procedure fails, then the answer is no. For instance, if $\mathbf{F} = x^2y\hat{\mathbf{i}} + xy\hat{\mathbf{j}} + z\hat{\mathbf{k}}$, and we attempt to find a function $f(x, y, z)$ so that $\nabla f = \mathbf{F}$, then

$$\frac{\partial f}{\partial x} = x^2y, \quad \frac{\partial f}{\partial y} = xy, \quad \frac{\partial f}{\partial z} = z.$$

The first implies that

$$f(x, y, z) = \frac{x^3y}{3} + v(y, z),$$

and when this is substituted into the second, we get

$$\frac{x^3}{3} + \frac{\partial v}{\partial y} = xy, \quad \text{or} \quad \frac{\partial v}{\partial y} = xy - \frac{x^3}{3}.$$

But this is an impossible situation since v is to be a function of y and z only. How then could its derivative depend on x? Although this type of argument will suffice in most examples, it is really not a satisfactory mathematical answer. The following theorem gives a test by which to determine whether a given vector function is the gradient of some scalar function.

> **Theorem 15.1**
>
> Suppose that the components $P(x,y,z)$, $Q(x,y,z)$, and $R(x,y,z)$ of $\mathbf{F} = P\hat{\mathbf{i}} + Q\hat{\mathbf{j}} + R\hat{\mathbf{k}}$ have continuous first partial derivatives in a domain D. If there exists a function $f(x,y,z)$ defined in D such that $\nabla f = \mathbf{F}$, then $\nabla \times \mathbf{F} = \mathbf{0}$. Conversely, if D is simply-connected, and $\nabla \times \mathbf{F} = \mathbf{0}$ in D, then there exists a function $f(x,y,z)$ such that $\nabla f = \mathbf{F}$ in D.

It is obvious that if $\mathbf{F} = \nabla f$, then $\nabla \times \mathbf{F} = \mathbf{0}$, for this is the result of equation 15.14. To prove the converse result requires Stokes's theorem from Section 15.10, and a proof is therefore delayed until that time. Notice that in the converse result, the domain (open, connected set) must be simply-connected. This is our first encounter with a situation where the nature of a region is important to the result.

In the special case in which $\mathbf{F}$ is a vector field in the xy-plane, the equation $\nabla \times \mathbf{F} = \mathbf{0}$ is still the condition for existence of a function $f(x,y)$ such that $\nabla f = \mathbf{F} = P\hat{\mathbf{i}} + Q\hat{\mathbf{j}}$, but the condition reduces to

$$\mathbf{0} = \begin{vmatrix} \hat{\mathbf{i}} & \hat{\mathbf{j}} & \hat{\mathbf{k}} \\ \partial/\partial x & \partial/\partial y & \partial/\partial z \\ P & Q & 0 \end{vmatrix} = \left(\frac{\partial Q}{\partial x} - \frac{\partial P}{\partial y} \right) \hat{\mathbf{k}},$$

or

$$\frac{\partial Q}{\partial x} = \frac{\partial P}{\partial y}. \tag{15.16}$$

EXAMPLE 15.3

Find, if possible, a function $f(x,y)$ such that

$$\nabla f = \left(\frac{x^3 - 2y^2}{x^3 y} \right) \hat{\mathbf{i}} + \left(\frac{y^2 - x^3}{x^2 y^2} \right) \hat{\mathbf{j}}.$$

SOLUTION We first note that

$$\frac{\partial}{\partial x} \left(\frac{y^2 - x^3}{x^2 y^2} \right) = \frac{\partial}{\partial x} \left(\frac{1}{x^2} - \frac{x}{y^2} \right) = \frac{-2}{x^3} - \frac{1}{y^2}$$

and

$$\frac{\partial}{\partial y} \left(\frac{x^3 - 2y^2}{x^3 y} \right) = \frac{\partial}{\partial y} \left(\frac{1}{y} - \frac{2y}{x^3} \right) = \frac{-1}{y^2} + \frac{-2}{x^3}.$$

Since the components of $\mathbf{F}$ are undefined whenever $x = 0$ or $y = 0$, we can state that in any simply-connected domain that does not contain points on either of the axes, there is a function $f(x,y)$ such that $\nabla f = \mathbf{F}$. To find $f(x,y)$, we set

$$\frac{\partial f}{\partial x} = \frac{1}{y} - \frac{2y}{x^3}, \quad \frac{\partial f}{\partial y} = \frac{1}{x^2} - \frac{x}{y^2}.$$

From the first equation, we have

$$f(x,y) = \frac{x}{y} + \frac{y}{x^2} + v(y),$$

which substituted into the second equation gives us

$$-\frac{x}{y^2} + \frac{1}{x^2} + \frac{dv}{dy} = \frac{1}{x^2} - \frac{x}{y^2}.$$

Consequently,

$$\frac{dv}{dy} = 0 \quad \text{or} \quad v(y) = C,$$

and therefore

$$f(x,y) = \frac{x}{y} + \frac{y}{x^2} + C.$$

EXERCISES 15.1

In Exercises 1–9 determine whether the set of points in the xy-plane is open, closed, connected, a domain, and/or a simply-connected domain.

1. $x^2 + (y+1)^2 < 4$
2. $x^2 + (y-3)^2 \leq 4$
3. $0 < x^2 + (y-1)^2 < 16$
4. $1 < (x-4)^2 + (y+1)^2 \leq 9$
5. $x > 3$
6. $y \leq -2$
7. $2(x-1)^2 - (y+2)^2 < 16$
8. All points satisfying $x^2 + y^2 < 1$ or $(x-2)^2 + y^2 < 1$
9. $4(x+1)^2 + 9(y-2)^2 > 20$
10. Give definitions of the following for sets of points in xyz-space: interior, exterior, and boundary points; open, closed, and connected sets; domain and simply-connected domain.

In Exercises 11–19 determine whether the set of points in space is open, closed, connected, a domain and/or a simply-connected domain.

11. $x^2 + y^2/4 + z^2/9 < 1$
12. $z \geq x^2 + y^2$
13. $x^2 + y^2 + z^2 > 0$
14. $1 < x^2 + y^2 + z^2 < 4$, $x \geq 0$, $y \geq 0$, $z \geq 0$
15. $x^2 + y^2 > 0$
16. $|z| > 0$
17. $z^2 > x^2 + y^2$
18. $z^2 > x^2 + y^2 - 1$
19. $z^2 < x^2 + y^2 - 1$
20. Prove that the only nonempty set in the xy-plane that is both open and closed is the whole plane.

In Exercises 21–40 calculate the required quantity.

21. ∇f if $f(x,y,z) = 3x^2y - y^3z^2$
22. ∇f if $f(x,y,z) = (x^2 + y^2 + z^2)^{-1/2}$
23. ∇f if $f(x,y) = \text{Tan}^{-1}(y/x)$
24. ∇f at $(1,2)$ if $f(x,y) = x^3y - 2x\cos y$
25. ∇f at $(1,-1,4)$ if $f(x,y,z) = e^{xyz}$
26. $\nabla \cdot \mathbf{F}$ if $\mathbf{F}(x,y,z) = 2xe^y\hat{\mathbf{i}} + 3x^2z\hat{\mathbf{j}} - 2x^2yz\hat{\mathbf{k}}$

27. $\nabla \cdot \mathbf{F}$ if $\mathbf{F}(x,y) = x\ln y\hat{\mathbf{i}} - y^3 e^x \hat{\mathbf{j}}$

28. $\nabla \cdot \mathbf{F}$ if $\mathbf{F}(x,y,z) = \sin(x^2 + y^2 + z^2)\hat{\mathbf{i}} + \cos(y+z)\hat{\mathbf{j}}$

29. $\nabla \cdot \mathbf{F}$ if $\mathbf{F}(x,y) = e^x \hat{\mathbf{i}} + e^y \hat{\mathbf{j}}$

30. $\nabla \cdot \mathbf{F}$ at $(1,1,1)$ if $\mathbf{F}(x,y,z) = x^2 y^3 \hat{\mathbf{i}} - 3xy\hat{\mathbf{j}} + z^2 \hat{\mathbf{k}}$

31. $\nabla \cdot \mathbf{F}$ at $(-1,3)$ if $\mathbf{F}(x,y) = (x+y)^2(\hat{\mathbf{i}} + \hat{\mathbf{j}})$

32. $\nabla \cdot \mathbf{F}$ if $\mathbf{F}(x,y,z) = (x\hat{\mathbf{i}} + y\hat{\mathbf{j}} + z\hat{\mathbf{k}})/\sqrt{x^2+y^2+z^2}$

33. $\nabla \cdot \mathbf{F}$ if $\mathbf{F}(x,y) = \text{Cot}^{-1}(xy)\hat{\mathbf{i}} + \text{Tan}^{-1}(xy)\hat{\mathbf{j}}$

34. $\nabla \times \mathbf{F}$ if $\mathbf{F}(x,y,z) = x^2 z \hat{\mathbf{i}} + 12xyz\hat{\mathbf{j}} + 32y^2 z^4 \hat{\mathbf{k}}$

35. $\nabla \times \mathbf{F}$ if $\mathbf{F}(x,y) = xe^y \hat{\mathbf{i}} - 2xy^2 \hat{\mathbf{j}}$

36. $\nabla \times \mathbf{F}$ if $\mathbf{F}(x,y,z) = x^2 \hat{\mathbf{i}} + y^2 \hat{\mathbf{j}} + z^2 \hat{\mathbf{k}}$

37. $\nabla \times \mathbf{F}$ at $(1,-1,1)$ if $\mathbf{F}(x,y,z) = xz^3 \hat{\mathbf{i}} - 2x^2 yz\hat{\mathbf{j}} + 2yz^4 \hat{\mathbf{k}}$

38. $\nabla \times \mathbf{F}$ at $(2,0)$ if $\mathbf{F}(x,y) = y\hat{\mathbf{i}} - x\hat{\mathbf{j}}$

39. $\nabla \times \mathbf{F}$ if $\mathbf{F}(x,y,z) = \ln(x+y+z)(\hat{\mathbf{i}} + \hat{\mathbf{j}} + \hat{\mathbf{k}})$

40. $\nabla \times \mathbf{F}$ if $\mathbf{F}(x,y) = \text{Sec}^{-1}(x+y)\hat{\mathbf{i}} + \text{Csc}^{-1}(y+x)\hat{\mathbf{j}}$

41. If $\mathbf{F} = x^2 y \hat{\mathbf{i}} - 2xz\hat{\mathbf{j}} + 2yz\hat{\mathbf{k}}$, find $\nabla \times (\nabla \times \mathbf{F})$.

42. (a) Verify that Laplace's equation 13.12 can be expressed in the form $\nabla \cdot \nabla f = 0$.
    (b) Show that $f(x,y,z) = (x^2 + y^2 + z^2)^{-1/2}$ satisfies Laplace's equation.

43. Prove properties 15.7–15.12, 15.14, and 15.15.

44. A gas is moving through some region D of space. If we follow a particular particle of the gas, it traces out some curved path C : $x = x(t), y = y(t), z = z(t)$, where t is time. If $\rho(x,y,z,t)$ is the density of gas at any point in D at time t, then along C we can express density in terms of t only, $\rho = \rho[x(t), y(t), z(t), t]$. Show that along C,
$$\frac{d\rho}{dt} = \frac{\partial \rho}{\partial t} + \nabla \rho \cdot \frac{d\mathbf{r}}{dt},$$
where $\mathbf{r} = x\hat{\mathbf{i}} + y\hat{\mathbf{j}} + z\hat{\mathbf{k}}$.

In Exercises 45–49 find all functions $f(x,y)$ such that ∇f is equal to the vector field.

45. $\mathbf{F}(x,y) = 2xy\hat{\mathbf{i}} + x^2 \hat{\mathbf{j}}$

46. $\mathbf{F}(x,y) = (3x^2 y^2 + 3)\hat{\mathbf{i}} + (2x^3 y + 2)\hat{\mathbf{j}}$

47. $\mathbf{F}(x,y) = e^y \hat{\mathbf{i}} + (xe^y + 4y^2)\hat{\mathbf{j}}$

48. $\mathbf{F}(x,y) = (x+y)^{-1}(\hat{\mathbf{i}} + \hat{\mathbf{j}})$

49. $\mathbf{F}(x,y) = -xy(1 - x^2 y^2)^{-1/2}(y\hat{\mathbf{i}} + x\hat{\mathbf{j}})$

In Exercises 50–55 find all functions $f(x,y,z)$ such ∇f is equal to the vector field.

50. $\mathbf{F}(x,y,z) = x\hat{\mathbf{i}} + y\hat{\mathbf{j}} + z\hat{\mathbf{k}}$

51. $\mathbf{F}(x,y,z) = yz\hat{\mathbf{i}} + xz\hat{\mathbf{j}} + (yx - 3)\hat{\mathbf{k}}$

52. $\mathbf{F}(x,y,z) = (1 + x + y + z)^{-1}(\hat{\mathbf{i}} + \hat{\mathbf{j}} + \hat{\mathbf{k}})$

53. $\mathbf{F}(x,y,z) = (2x/y^2 + 1)\hat{\mathbf{i}} - (2x^2/y^3)\hat{\mathbf{j}} - 2z\hat{\mathbf{k}}$

54. $\mathbf{F}(x,y,z) = (1 + x^2 y^2)^{-1}(y\hat{\mathbf{i}} + x\hat{\mathbf{j}}) + z\hat{\mathbf{k}}$

55. $\mathbf{F}(x,y,z) = (3x^2 y + yz + 2xz^2)\hat{\mathbf{i}} + (xz + x^3 + 3z^2 - 6y^2 z)\hat{\mathbf{j}} + (2x^2 z + 6yz - 2y^3 + xy)\hat{\mathbf{k}}$

56. (a) Find constants a, b, and c in order that the vector field
$$\mathbf{F} = (x^2 + 2y + az)\hat{\mathbf{i}} + (bx - 3y - z)\hat{\mathbf{j}} + (4x + cy + 2z)\hat{\mathbf{k}}$$
be irrotational.
(b) If $\mathbf{F}$ is irrotational, find a scalar function $f(x,y,z)$ such that $\nabla f = \mathbf{F}$.

57. A vector field $\mathbf{F}(x,y,z)$ is said to be *solenoidal* if $\nabla \cdot \mathbf{F} = 0$.
(a) Are either of $\mathbf{F} = (2x^2 + 8xy^2 z)\hat{\mathbf{i}} + (3x^3 y - 3xy)\hat{\mathbf{j}} - (4y^2 z^2 + 2x^3 z)\hat{\mathbf{k}}$ or $xyz^2 \mathbf{F}$ solenoidal?
(b) Show that $\nabla f \times \nabla g$ is solenoidal for arbitrary functions $f(x,y,z)$ and $g(x,y,z)$ which have continuous second partial derivatives.

58. Associated with every electric field is a scalar function $V(x,y,z)$ called potential. It is defined by $\mathbf{E} = -\nabla V$, where $\mathbf{E}$ is a vector field called the electric field intensity. In addition, if a point charge Q is placed at a point (x,y,z) in the electric field, then the force $\mathbf{F}$ on Q is $\mathbf{F} = Q\mathbf{E}$.
(a) If the force on Q due to a charge q at the origin is
$$\mathbf{F} = \frac{qQ}{4\pi\epsilon_0 |\mathbf{r}|^3}\mathbf{r},$$
where $\mathbf{r} = x\hat{\mathbf{i}} + y\hat{\mathbf{j}} + z\hat{\mathbf{k}}$ and ϵ_0 is a constant, find $V(x,y,z)$ for the field due to q.
(b) If the entire xy-plane is given a uniform charge density σ units of charge per unit area, it is found that the force on a charge Q placed z units above the plane is $\mathbf{F} = [Q\sigma/(2\epsilon_0)]\hat{\mathbf{k}}$. Find the potential V for the electric field due to this charge distribution.

59. Show that if $\mathbf{v} = \boldsymbol{\omega} \times \mathbf{r}$, where $\boldsymbol{\omega}$ is a constant vector, and $\mathbf{r} = x\hat{\mathbf{i}} + y\hat{\mathbf{j}} + z\hat{\mathbf{k}}$, then $\boldsymbol{\omega} = (1/2)(\nabla \times \mathbf{v})$.

60. Show that if a function $f(x,y,z)$ satisfies Laplace's equation 13.12, then its gradient is both irrotational and solenoidal.

61. Theorem 15.1 indicates that a vector field $\mathbf{F}$ is the gradient of some scalar field if $\nabla \times \mathbf{F} = \mathbf{0}$. Sometimes a given vector field $\mathbf{F}$ is the curl of another field $\mathbf{v}$; that is $\mathbf{F} = \nabla \times \mathbf{v}$. The following theorem indicates when this is the case: Let D be the interior of a sphere in which the components of a vector field $\mathbf{F}$ have continuous first partial derivatives. Then there exists a vector field $\mathbf{v}$ defined in D such that $\mathbf{F} = \nabla \times \mathbf{v}$ if and only if $\nabla \cdot \mathbf{F} = 0$ in D. In other words, $\mathbf{F}$ is the curl of a vector field if and only if $\mathbf{F}$ is solenoidal.

(a) Show that if $\mathbf{F} = P\hat{\mathbf{i}} + Q\hat{\mathbf{j}} + R\hat{\mathbf{k}}$ is solenoidal, then the components of $\mathbf{v} = L\hat{\mathbf{i}} + M\hat{\mathbf{j}} + N\hat{\mathbf{k}}$ would have to satisfy the equations

$$P = \frac{\partial N}{\partial y} - \frac{\partial M}{\partial z}, \quad Q = \frac{\partial L}{\partial z} - \frac{\partial N}{\partial x}, \quad R = \frac{\partial M}{\partial x} - \frac{\partial L}{\partial y}.$$

(b) Show that the vector field $\mathbf{v}(x, y, z)$ defined by

$$\mathbf{v}(x, y, z) = \int_0^1 t\mathbf{F}(tx, ty, tz) \times (x, y, z)\, dt$$

satisfies these equations. In other words, this formula defines a possible vector $\mathbf{v}$. Is it unique?

(c) Show that if $\mathbf{F}$ satisfies the property that $\mathbf{F}(tx, ty, tz) = t^n \mathbf{F}(x, y, z)$, then

$$\mathbf{v}(x, y, z) = \frac{1}{n+2} \mathbf{F} \times \mathbf{r}, \qquad \mathbf{r} = x\hat{\mathbf{i}} + y\hat{\mathbf{j}} + z\hat{\mathbf{k}}.$$

In Exercises 62–64 verify that the vector field is solenoidal, and then use the formulas in Exercise 61 to find a vector field $\mathbf{v}$ such that $\mathbf{F} = \nabla \times \mathbf{v}$.

62. $\mathbf{F} = x\hat{\mathbf{i}} + y\hat{\mathbf{j}} - 2z\hat{\mathbf{k}}$

63. $\mathbf{F} = (1+x)\hat{\mathbf{i}} - (x+z)\hat{\mathbf{k}}$

64. $\mathbf{F} = 2x^2\hat{\mathbf{i}} - y^2\hat{\mathbf{j}} + (2yz - 4xz)\hat{\mathbf{k}}$

SECTION 15.2

Line Integrals

Just as definite integrals, double integrals, and triple integrals are defined as limits of sums, so too are line and surface integrals. The only difference is that line integrals are applied to functions defined along curves, and surface integrals involve functions defined on surfaces.

A curve C in space is defined parametrically by three functions

$$C: \begin{array}{l} x = x(t), \\ y = y(t), \\ z = z(t), \end{array} \quad \alpha \le t \le \beta \qquad (15.17)$$

where α and β specify initial and final points A and B of the curve, respectively (Figure 15.9). The direction of a curve is from initial to final point, and in Section 12.7 we agreed to parametrize a curve using parameters that increase in the direction of the curve.

FIGURE 15.9

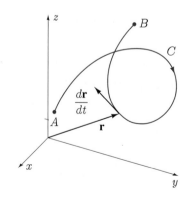

The curve is said to be continuous if each of the functions $x(t)$, $y(t)$, and $z(t)$ is continuous (implying that C is at no point separated). It is said to be smooth if each of these functions has a continuous first derivative; geometrically, this means that the tangent vector $d\mathbf{r}/dt = x'(t)\hat{\mathbf{i}} + y'(t)\hat{\mathbf{j}} + z'(t)\hat{\mathbf{k}}$ turns gradually or smoothly along C. A continuous curve that is not smooth but can be divided into a finite number of smooth subcurves is said to be piecewise smooth.

Suppose a function $f(x,y,z)$ is defined along a curve C joining A to B (Figure 15.10). We divide C into n subcurves of lengths $\Delta s_1, \Delta s_2, \ldots, \Delta s_n$ by any $n-1$ consecutive points $A = P_0, P_1, P_2, \ldots, P_{n-1}, P_n = B$, whatsoever. On each subcurve of length $\Delta s_i (i = 1, \ldots, n)$ we choose an arbitrary point $P_i^*(x_i^*, y_i^*, z_i^*)$. We then form the sum

$$f(x_1^*, y_1^*, z_1^*)\Delta s_1 + f(x_2^*, y_2^*, z_2^*)\Delta s_2 + \cdots + f(x_n^*, y_n^*, z_n^*)\Delta s_n$$
$$= \sum_{i=1}^{n} f(x_i^*, y_i^*, z_i^*)\Delta s_i. \quad (15.18)$$

FIGURE 15.10

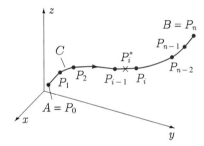

If this sum approaches a limit as the number of subcurves becomes increasingly large and the length of every subcurve approaches zero, we call the limit the line integral of $f(x,y,z)$ along the curve C, and denote it by

$$\int_C f(x,y,z)\,ds = \lim_{\|\Delta s_i\| \to 0} \sum_{i=1}^{n} f(x_i^*, y_i^*, z_i^*)\Delta s_i. \quad (15.19)$$

A more appropriate name might be curvilinear integral, rather than line integral, but line integral has become the accepted terminology. We must simply regard the word "line" as meaning "curved line" rather than "straight line."

For definition 15.19 to be useful we must demand that the limit be independent of the manner of subdivision of C and choice of star-points on the subcurves. Theorem 15.2 indicates that for continuous functions defined on smooth curves, this requirement is indeed satisfied.

Theorem 15.2

Let $f(x,y,z)$ be continuous on a smooth curve C of finite length, $C: x = x(t)$, $y = y(t)$, $z = z(t)$, $\alpha \leq t \leq \beta$. Then the line integral of $f(x,y,z)$ along C exists and can be evaluated by means of the following definite integral:

$$\int_C f(x,y,z)\,ds = \int_\alpha^\beta f[x(t), y(t), z(t)]\sqrt{\left(\frac{dx}{dt}\right)^2 + \left(\frac{dy}{dt}\right)^2 + \left(\frac{dz}{dt}\right)^2}\,dt. \quad (15.20)$$

It is not necessary to memorize 15.20 as a formula, since it is the result obtained by expressing x, y, z, and ds in terms of t and interpreting that result as a definite integral with respect to t. To be more explicit, recall from equation 12.74 that when length along a curve is measured from its initial point, then an infinitesimal length ds along C corresponding to an increment dt in t is given by

$$ds = \sqrt{(dx)^2 + (dy)^2 + (dz)^2} = \sqrt{\left\{\left(\frac{dx}{dt}\right)^2 + \left(\frac{dy}{dt}\right)^2 + \left(\frac{dz}{dt}\right)^2\right\}(dt)^2}$$

$$= \sqrt{\left(\frac{dx}{dt}\right)^2 + \left(\frac{dy}{dt}\right)^2 + \left(\frac{dz}{dt}\right)^2}\, dt.$$

If we substitute this into the left-hand side of 15.20 and at the same time use the equations for C to express $f(x, y, z)$ in terms of t, then

$$\int_C f(x, y, z)\, ds = \int_C f[x(t), y(t), z(t)] \sqrt{\left(\frac{dx}{dt}\right)^2 + \left(\frac{dy}{dt}\right)^2 + \left(\frac{dz}{dt}\right)^2}\, dt.$$

But if we now interpret the right-hand side of this equation as the definite integral of $f[x(t), y(t), z(t)]\sqrt{(dx/dt)^2 + (dy/dt)^2 + (dz/dt)^2}$ with respect to t and affix limits $t = \alpha$ and $t = \beta$ that identify end points of C, we obtain 15.20.

To evaluate a line integral, then, we simply express $f(x, y, z)$ and ds in terms of some parameter along C and evaluate the resulting definite integral. If equations for C are given in the form $C : y = y(x),\ z = z(x),\ x_A \leq x \leq x_B$, then x is a convenient parameter, and equation 15.20 takes the form

$$\int_C f(x, y, z)\, ds = \int_{x_A}^{x_B} f[x, y(x), z(x)] \sqrt{1 + \left(\frac{dy}{dx}\right)^2 + \left(\frac{dz}{dx}\right)^2}\, dx. \quad (15.21)$$

Similar expressions exist if either y or z is a convenient parameter.

When C is piecewise smooth rather than smooth, the line integral along C is found by evaluating the line integral along each smooth subcurve and adding the results.

EXAMPLE 15.4

Evaluate the line integral of $f(x, y, z) = 8x + 6xy + 30z$ from $A(0, 0, 0)$ to $B(1, 1, 1)$

(a) along the straight line joining A to B with parametrization

$$C_1 : \quad x = t, \quad y = t, \quad z = t, \quad 0 \leq t \leq 1.$$

(b) along the straight line in **(a)** with parametrization

$$C_1 : \quad x = -1 + \frac{t}{2}, \quad y = -1 + \frac{t}{2}, \quad z = -1 + \frac{t}{2}, \quad 2 \leq t \leq 4.$$

(c) along the curve (Figure 15.11)

$$C_2 : \quad x = t, \quad y = t^2, \quad z = t^3, \quad 0 \leq t \leq 1.$$

SOLUTION (a)

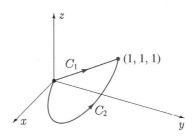

FIGURE 15.11

$$\int_{C_1}(8x+6xy+30z)\,ds = \int_0^1 (8t+6t^2+30t)\sqrt{(1)^2+(1)^2+(1)^2}\,dt$$
$$= \sqrt{3}\int_0^1 (38t+6t^2)\,dt = \sqrt{3}\{19t^2+2t^3\}_0^1$$
$$= 21\sqrt{3}.$$

(b)

$$\int_{C_1}(8x+6xy+30z)\,ds = \int_2^4 \{8(-1+t/2)+6(-1+t/2)^2+30(-1+t/2)\}$$
$$\times \sqrt{(1/2)^2+(1/2)^2+(1/2)^2}\,dt$$
$$= \frac{\sqrt{3}}{2}\int_2^4 \{38(-1+t/2)+6(-1+t/2)^2\}\,dt$$
$$= \frac{\sqrt{3}}{2}\{38(-1+t/2)^2+4(-1+t/2)^3\}_2^4$$
$$= \frac{\sqrt{3}}{2}\{38+4\} = 21\sqrt{3}.$$

(c)

$$\int_{C_2}(8x+6xy+30z)\,ds = \int_0^1 (8t+6t^3+30t^3)\sqrt{(1)^2+(2t)^2+(3t^2)^2}\,dt$$
$$= \int_0^1 (8t+36t^3)\sqrt{1+4t^2+9t^4}\,dt$$
$$= \{\tfrac{2}{3}(1+4t^2+9t^4)^{3/2}\}_0^1 = \tfrac{2}{3}(14\sqrt{14}-1).$$

■

Parts (a) and (b) of this example suggest that the value of a line integral does not depend on the particular parametrization of the curve used in its evaluation. This is indeed true, and should perhaps be expected since definition 15.20 makes no reference whatsoever to parametrization of the curve. For a proof of this fact see Exercise 37. Different parameters normally lead to different definite integrals, but they all give exactly the same value for the line integral. Parts (a) and (c) illustrate that a line integral does depend on the curve joining the points A and B; i.e., the value of the line integral may change if the curve C joining A and B changes.

EXAMPLE 15.5

Evaluate the line integral of $f(x,y) = x^2 + y^2$ once clockwise around the circle $x^2 + y^2 = 4$, $z = 0$.

SOLUTION If we use the parametrization

$$C: \quad x = 2\cos t, \quad y = -2\sin t, \quad z = 0, \quad 0 \leq t \leq 2\pi$$

(Figure 15.12), then

$$\int_C (x^2 + y^2)\,ds = \int_0^{2\pi} (4)\sqrt{\left(\frac{dx}{dt}\right)^2 + \left(\frac{dy}{dt}\right)^2}\,dt$$

$$= 4\int_0^{2\pi} \sqrt{(-2\sin t)^2 + (-2\cos t)^2}\,dt = 8\int_0^{2\pi} dt = 16\pi.$$

FIGURE 15.12

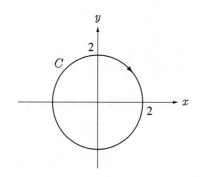

The value of this line integral, as well as many other line integrals in the xy-plane, can be given a geometric interpretation. Suppose a function $f(x,y)$ is positive along a curve C in the xy-plane. If at each point of C we draw a vertical line of height $z = f(x,y)$, then a vertical wall is constructed as shown in Figure 15.13. Since ds is an elemental piece of length along C, the quantity $f(x,y)\,ds$ can be interpreted as approximately the area of the vertical wall projecting onto ds. Because the line integral

$$\int_C f(x,y)\,ds,$$

like all integrals, is a limit-summation process, we interpret the value of this line integral as the total area of the vertical wall. Correct as this interpretation is, it really is of little use in the evaluation of line integrals and, in addition, the interpretation is valid only if the curve C along which the line integral is performed is contained in a plane.

FIGURE 15.13

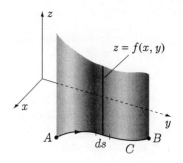

Because a line integral is a limit-summation, it should be obvious that the line integral

$$\int_C ds \qquad (15.22)$$

represents the length of the curve C. If C is a curve in the xy-plane, we substitute $ds = \sqrt{(dx)^2 + (dy)^2}$, and if C is a curve in space, then $ds = \sqrt{(dx)^2 + (dy)^2 + (dz)^2}$. This agrees with the results of equations 12.67 and 12.74.

We make one last point about notation. To indicate that a line integral is being evaluated around a closed curve, we usually draw a circle on the integral sign, as follows:

$$\oint_C f(x, y, z)\, ds.$$

Such would be the case for the curve of intersection of the cylinder $x^2 + y^2 = 1$ and the plane $x + z = 1$ (Figure 15.14).

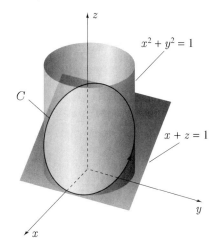

FIGURE 15.14

When C is a closed curve in the xy-plane that does not cross itself, we indicate the direction along C by an arrowhead on the circle. For the curves shown in Figure 15.15 we write

$$\oint_{C_1} f(x, y)\, ds \quad \text{and} \quad \oint_{C_2} f(x, y)\, ds.$$

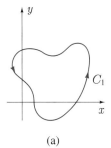
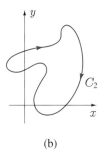

FIGURE 15.15

(a)   (b)

EXERCISES 15.2

In Exercises 1–6 evaluate the line integral.

1. $\int_C x\,ds$ where C is the curve $y = x^2$, $z = 0$ from $(0,0,0)$ to $(1,1,0)$

2. $\oint_C (x^2 + y^2)\,ds$ once around the square C in the xy-plane with vertices $(\pm 1, 1)$ and $(\pm 1, -1)$

3. $\oint_C (2 + x - 2xy)\,ds$ once around the circle $x^2 + y^2 = 4$, $z = 0$

4. $\int_C (x^2 + yz)\,ds$ along the straight line from $(1,2,-1)$ to $(3,2,5)$

5. $\int_C xy\,ds$ where C is the first octant part of $x^2 + y^2 = 1$, $x^2 + z^2 = 1$ from $(1,0,0)$ to $(0,1,1)$

6. $\int_C x^2 yz\,ds$ where C is the curve $z = x+y$, $x+y+z = 1$ from $(1, -1/2, 1/2)$ to $(-3, 7/2, 1/2)$

7. Prove that the length of the circumference of a circle is 2π multiplied by the radius.

8. A spring has six coils in the form of the helix
$$x = 3\cos t, \quad y = 3\sin t, \quad z = 3t/(4\pi), \quad 0 \leq t \leq 12\pi,$$
where all dimensions are in centimetres. Find the length of the spring.

9. Use the parametric equations $x = \cos^3\theta$, $y = \sin^3\theta$, $0 \leq \theta < 2\pi$, to sketch the astroid $x^{2/3} + y^{2/3} = 1$ in the xy-plane. At each point (x,y) on the astroid, a vertical line is drawn with height $z = x^2 + y^2$, thus forming a cylindrical wall. Find the area of the wall.

In Exercises 10–16 evaluate the line integral.

10. $\int_C xz\,ds$ along the first octant part of $y = x^2$, $z + y = 1$ from $(0,0,1)$ to $(1,1,0)$

11. $\int_C (x+y)^5\,ds$ along $C: x = t + 1/t$, $y = t - 1/t$ from $(2, 0)$ to $(17/4, 15/4)$

12. $\int_C x\sqrt{y+z}\,ds$ where C is that part of the curve $3x + 2y + 3z = 6$, $x - 2y + 4z = 5$ from $(1,0,1)$ to $(0, 9/14, 11/7)$

13. $\int_C xy\,ds$ where C is the curve $x = 1 - y^2$, $z = 0$ from $(1,0,0)$ to $(0,1,0)$

14. $\int_C (x+y)z\,ds$ where C is the curve $y = x$, $z = 1 + y^4$ from $(-1,-1,2)$ to $(1,1,2)$

15. $\int_C \frac{1}{y+z}\,ds$ where C is the curve $y = x^2$, $z = x^2$ from $(1,1,1)$ to $(2,4,4)$

16. $\int_C (2y + 9z)\,ds$ where C is the curve $z = xy$, $x = y^2$ from $(0,0,0)$ to $(4, -2, -8)$

17. (a) If the curve C in Figure 15.16 is rotated around the y-axis, show that the area of the surface that it traces out is represented by the line integral
$$\int_C 2\pi x\,ds.$$

(b) If C is rotated around the x-axis, what line integral represents the area of the surface traced out?

FIGURE 15.16

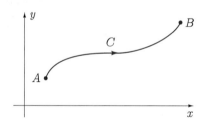

In Exercises 18–20, use the method in Exercise 17 to find the area of the surface traced out when the curve is rotated about the line.

18. $y = x^3$, $1 \leq x \leq 2$, about $y = 0$

19. $24xy = x^4 + 48$, $1 \leq x \leq 2$, about $x = 0$

20. $8y^2 = x^2(1 - x^2)$ about $y = 0$

21. Find the length of the parabola $y = x^2$ from $(0,0)$ to $(1,1)$.

In Exercises 22 and 23 find a definite integral which can be used to evaluate the line integral. Use power series to approximate the definite integral accurate to three decimals.

22. $\int_C xy\,ds$ where C is the curve $y = x^3$, $z = 0$ from $(0,0,0)$ to $(1/2, 1/8, 0)$

23. $\int_C e^{-(x+y-2)^2} \, ds$ where C is the curve $x + y + z = 2$, $y + 2z = 3$ from $(-1, 3, 0)$ to $(0, 1, 1)$

The average value of a function $f(x, y, z)$ defined along a curve C is defined as the value of the line integral of the function along the curve divided by the length of the curve. In Exercises 24–27 find the average value of the function along the curve.

24. $f(x, y) = x^2 y^2$ along $C: x^2 + y^2 = 4$, $z = 0$

25. $f(x, y, z) = x^2 + y^2 + z^2$ along $C: x = \cos t$, $y = \sin t$, $z = t$, $0 \leq t \leq \pi$

26. $f(x, y, z) = xyz$ along $C: z = x^2$, $y = x^2$ from $(0, 0, 0)$ to $(1, 1, 1)$

27. $f(x, y) = y$ along $C: y = x^3/4 + 1/(3x)$ from $(1, 7/12)$ to $(2, 13/16)$

28. At each point on the curve $(x^2 + y^2)^2 = x^2 - y^2$ a vertical line is drawn with height equal to the distance from the point to the origin. Find the area of the vertical wall so formed.

29. During a sleet storm, a power line between two poles at positions $x = \pm 20$ hangs in the shape $y = 40 \cosh(x/40) - 10$, where all distances are measured in metres. Ice accumulates more heavily on the middle part of the line than at the ends. In fact, the combined mass of ice and line per unit length in the x-direction at position x is given in kg/m by the formula $\rho(x) = 1 - |x|/40$. Find the total mass of the line.

In Exercises 30–33 use the fact that in polar coordinates small lengths along a curve can be expressed in the form $ds = \sqrt{r^2 + (dr/d\theta)^2} \, d\theta$ (see formula 10.14) to evaluate the line integral.

30. $\int_C \dfrac{x}{\sqrt{x^2 + y^2}} \, ds$ where C is the first quadrant part of the limaçon $r = 2 - \sin \theta$ starting from the point on the x-axis

31. $\oint_C (x^2 + y^2) \, ds$ where C is the cardioid $r = 1 + \cos \theta$

32. $\int_C xy \, ds$ where C is the spiral $r = e^\theta$ from $\theta = 0$ to $\theta = 2\pi$

33. $\oint_C \cos^3 2\theta \, ds$ around the first quadrant loop of the lemniscate $r^2 = \sin 2\theta$

In Exercises 34 and 35 find a definite integral which can be used to evaluate the line integral. Use Simpson's rule with 10 equal subdivisions to approximate the definite integral.

34. $\int_C (x^2 y + z) \, ds$ where C is the curve $z = x^2 + y^2$, $y + x = 1$ from $(-1, 2, 5)$ to $(1, 0, 1)$

35. $\oint_C x^2 y^2 \, ds$ where C is the ellipse $4x^2 + 9y^2 = 36$, $z = 0$

36. Find the surface area of the torus obtained by rotating the circle $(x - a)^2 + y^2 = b^2$ $(a > b)$ about the y-axis.

37. Show that the value of a line integral is independent of the parameter used to specify the curve.

SECTION 15.3

Line Integrals Involving Vector Functions

There are many ways in which $f(x, y, z)$ in the line integral

$$\int_C f(x, y, z) \, ds \qquad (15.23)$$

can arise. According to equation 15.22 we must choose $f(x, y, z) = 1$ in order to find the length of the curve C; Exercise 17 in Section 15.2 indicates that for the areas of the surfaces traced out when a curve in the xy-plane is rotated about the y- and x-axes, we must choose $f(x, y)$ equal to $2\pi x$ and $2\pi y$, respectively.

The most important and common type of line integral occurs when $f(x, y, z)$ is specified as the tangential component of some given vector field $\mathbf{F}(x, y, z)$ defined along C; i.e., $f(x, y, z)$ itself is not given, but $\mathbf{F}$ is, and to find $f(x, y, z)$ we must calculate the tangential component of $\mathbf{F}$ along C. By the tangential component of $\mathbf{F}(x, y, z)$ along C we mean the component of $\mathbf{F}$ along that tangent vector to C which points in the same direction as C (Figure 15.17).

FIGURE 15.17

FIGURE 15.18

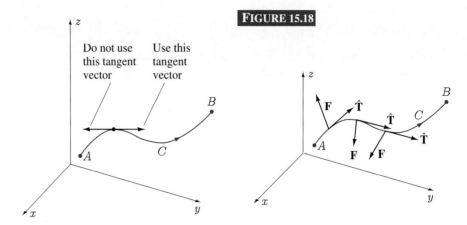

In Section 12.8 we saw that if s is a measure of length along a curve C from A to B (Figure 15.18), and if s is chosen equal to zero at A, then a unit tangent vector pointing in the direction of motion along C is

$$\hat{\mathbf{T}} = \frac{d\mathbf{r}}{ds}. \qquad (15.24)$$

Consequently, if $f(x, y, z)$ is the tangential component of $\mathbf{F}(x, y, z)$ along C, then

$$f(x, y, z) = \mathbf{F} \cdot \hat{\mathbf{T}} = \mathbf{F} \cdot \frac{d\mathbf{r}}{ds}, \qquad (15.25)$$

and we can write that

$$\int_C f(x, y, z)\,ds = \int_C \mathbf{F} \cdot \hat{\mathbf{T}}\,ds = \int_C \mathbf{F} \cdot \frac{d\mathbf{r}}{ds}\,ds = \int_C \mathbf{F} \cdot d\mathbf{r}. \qquad (15.26)$$

If the components of the vector field $\mathbf{F}(x, y, z)$ are

$$\mathbf{F}(x, y, z) = P(x, y, z)\hat{\mathbf{i}} + Q(x, y, z)\hat{\mathbf{j}} + R(x, y, z)\hat{\mathbf{k}}, \qquad (15.27)$$

then

$$\int_C \mathbf{F} \cdot d\mathbf{r} = \int_C (P\,dx + Q\,dy + R\,dz),$$

and if the parentheses are omitted, we have

$$\int_C \mathbf{F} \cdot d\mathbf{r} = \int_C P\,dx + Q\,dy + R\,dz. \qquad (15.28)$$

This discussion has shown that when the integrand $f(x, y, z)$ of a line integral is specified as the tangential component of $\mathbf{F} = P\hat{\mathbf{i}} + Q\hat{\mathbf{j}} + R\hat{\mathbf{k}}$ along C, the product $f(x, y, z)\,ds$ can be replaced by the sum of products $P\,dx + Q\,dy + R\,dz$:

$$\int_C f(x, y, z)\,ds = \int_C \mathbf{F} \cdot d\mathbf{r} = \int_C P\,dx + Q\,dy + R\,dz. \qquad (15.29)$$

According to the results of Section 15.2, evaluation of this line integral can be accomplished by expressing $P\,dx + Q\,dy + R\,dz$ in terms of any parametric representation of C and evaluating the resulting definite integral.

EXAMPLE 15.6

Evaluate
$$\int_C \frac{z}{y}dx + (x^2 + y^2 + z^2)dz,$$

where C is the first octant intersection of $x^2 + y^2 = 1$ and $z = 2x + 4$ joining $(1, 0, 6)$ to $(0, 1, 4)$.

SOLUTION If we choose the parametrization

$$x = \cos t, \quad y = \sin t, \quad z = 2\cos t + 4, \quad 0 \leq t \leq \pi/2,$$

for C (Figure 15.19), then

$$\int_C \frac{z}{y}dx + (x^2 + y^2 + z^2)dz$$

$$= \int_0^{\pi/2} \left\{ \left(\frac{2\cos t + 4}{\sin t}\right)(-\sin t\, dt) \right.$$

$$\left. + (\cos^2 t + \sin^2 t + 4\cos^2 t + 16\cos t + 16)(-2\sin t\, dt) \right\}$$

$$= -2\int_0^{\pi/2} \{\cos t + 2 + 17\sin t + 4\cos^2 t \sin t + 16\cos t \sin t\}dt$$

$$= -2\left\{\sin t + 2t - 17\cos t - \frac{4\cos^3 t}{3} + 8\sin^2 t\right\}_0^{\pi/2} = -2\pi - \frac{164}{3}.$$

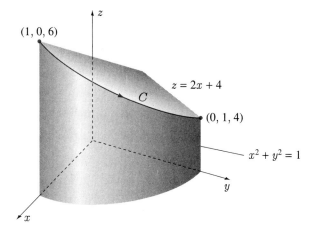

FIGURE 15.19

EXAMPLE 15.7

Evaluate
$$\oint_C y^2\, dx + x^2\, dy,$$

where C is the closed curve shown in Figure 15.20.

FIGURE 15.20

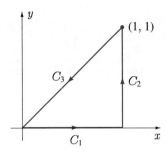

SOLUTION If we start at the origin and denote the three straight line paths by

$$C_1: \quad y = 0, \quad 0 \leq x \leq 1; \quad C_2: \quad x = 1, \quad 0 \leq y \leq 1;$$
$$C_3: \quad y = x = 1 - t, \quad 0 \leq t \leq 1,$$

then

$$\oint_C y^2\,dx + x^2\,dy = \int_{C_1} y^2\,dx + x^2\,dy + \int_{C_2} y^2\,dx + x^2\,dy + \int_{C_3} y^2\,dx + x^2\,dy$$

$$= \int_0^1 0\,dx + x^2 0 + \int_0^1 y^2 0 + 1\,dy$$

$$+ \int_0^1 (1-t)^2(-dt) + (1-t)^2(-dt)$$

$$= \{y\}_0^1 + \left\{\frac{2}{3}(1-t)^3\right\}_0^1 = 1 - \frac{2}{3} = \frac{1}{3}.$$

■

EXAMPLE 15.8

If a curve C has initial and final points A and B, then the curve that traces the same points but has initial and final points B and A is denoted by $-C$. Show that

$$\int_{-C} \mathbf{F} \cdot d\mathbf{r} = -\int_C \mathbf{F} \cdot d\mathbf{r}. \tag{15.30}$$

SOLUTION When 15.17 are parametric equations for C, then parametric equations for $-C$ are

$$-C: \quad x = x(t), \quad y = y(t), \quad z = z(t), \quad \beta \geq t \geq \alpha.$$

To obtain an increasing parameter along $-C$, we set $u = -t$, in which case

$$-C: \quad x = x(-u), \quad y = y(-u), \quad z = z(-u), \quad -\beta \leq u \leq -\alpha.$$

If $\mathbf{F} = P\hat{\mathbf{i}} + Q\hat{\mathbf{j}} + R\hat{\mathbf{k}}$, then the value of the line integral along $-C$ can be expressed as a definite integral with respect to u:

$$\int_{-C} \mathbf{F} \cdot d\mathbf{r} = \int_{-C} P\,dx + Q\,dy + R\,dz$$

$$= \int_{-\beta}^{-\alpha} \left\{ P[x(-u), y(-u), z(-u)] \frac{dx}{du} \right.$$

$$+ Q[x(-u), y(-u), z(-u)]\frac{dy}{du} + R[x(-u), y(-u), z(-u)]\frac{dz}{du}\bigg\} du.$$

If we now change variables of integration by setting $t = -u$, then

$$\frac{dx}{du} = \frac{dx}{dt}\frac{dt}{du} = -\frac{dx}{dt},$$

and similarly for dy/du and dz/du. Consequently,

$$\int_{-C} \mathbf{F} \cdot d\mathbf{r} = \int_{\beta}^{\alpha} \bigg\{ P[x(t), y(t), z(t)]\left(-\frac{dx}{dt}\right) + Q[x(t), y(t), z(t)]\left(-\frac{dy}{dt}\right)$$
$$+ R[x(t), y(t), z(t)]\left(-\frac{dz}{dt}\right)\bigg\}(-dt)$$
$$= -\int_{\alpha}^{\beta} \bigg\{ P[x(t), y(t), z(t)]\frac{dx}{dt} + Q[x(t), y(t), z(t)]\frac{dy}{dt}$$
$$+ R[x(t), y(t), z(t)]\frac{dz}{dt}\bigg\} dt$$
$$= -\int_{C} \mathbf{F} \cdot d\mathbf{r}.$$

According to this example when the direction along a curve is reversed, the value of a line integral of the form 15.28 along the new curve is the negative of its value along the original curve. This is because the signs of dx, dy, and dz are reversed when the direction along C is reversed. This is not the case for line integral 15.20; ds does not change sign when the direction along C is reversed.

Line integrals of the form 15.28 are singled out for special consideration because they repeatedly arise in physical problems. For example, suppose $\mathbf{F}$ represents a force, and we consider finding the work done by this force as a particle moves along a curve C from A to B. We begin by dividing C into n subcurves of lengths Δs_i as shown in Figure 15.21.

FIGURE 15.21

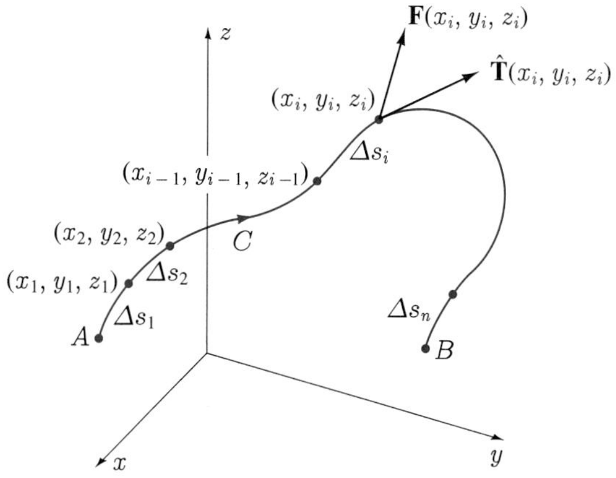

If $\mathbf{F}(x, y, z)$ is continuous along C, then along any given Δs_i, $\mathbf{F}(x, y, z)$ does not vary greatly (provided, of course, that Δs_i is small). If we approximate $\mathbf{F}(x, y, z)$ along

Δs_i by its value $\mathbf{F}(x_i, y_i, z_i)$ at the final point (x_i, y_i, z_i) of Δs_i, then an approximation to the work done by $\mathbf{F}$ along Δs_i is $\mathbf{F}(x_i, y_i, z_i) \cdot \hat{\mathbf{T}}(x_i, y_i, z_i)\Delta s_i$. An approximation to the total work done by $\mathbf{F}$ along C is therefore

$$\sum_{i=1}^{n} \mathbf{F}(x_i, y_i, z_i) \cdot \hat{\mathbf{T}}(x_i, y_i, z_i)\Delta s_i.$$

To obtain the exact value of the work done by $\mathbf{F}$ along C we take the limit of this sum as the number of subdivisions becomes larger and larger and each Δs_i approaches zero. But this is precisely the definition of the line integral of $\mathbf{F} \cdot \hat{\mathbf{T}}$, and we therefore write

$$W = \int_C \mathbf{F} \cdot \hat{\mathbf{T}} \, ds = \int_C \mathbf{F} \cdot d\mathbf{r}. \qquad (15.31)$$

This interpretation of a line integral as the work done by a force $\mathbf{F}$ is extremely important, and we will return to it in Section 15.5.

EXAMPLE 15.9

The force of repulsion between two positive point charges, one of size q and the other of size unity, has magnitude $q/(4\pi\epsilon_0 r^2)$, where ϵ_0 is a constant and r is the distance between the charges. The potential V at any point P due to charge q is defined as the work required to bring the unit charge to P from an infinite distance along the straight line joining q and P. Find V.

SOLUTION If that part of the line joining q and P from infinity to P is denoted by C (Figure 15.22), then

$$V = \int_C \mathbf{F} \cdot d\mathbf{r} = -\int_{-C} \mathbf{F} \cdot d\mathbf{r}$$

(see equation 15.30), where $\mathbf{F}$, the force necessary to overcome the electrostatic repulsion, is given by

$$\mathbf{F} = \frac{-q}{4\pi\epsilon_0 x^2}\hat{\mathbf{i}}.$$

Along $-C$, $d\mathbf{r} = dx\hat{\mathbf{i}}$, and therefore V can be evaluated by the (improper) definite integral

$$V = -\int_r^\infty \frac{-q}{4\pi\epsilon_0 x^2}\,dx = -\left\{\frac{q}{4\pi\epsilon_0 x}\right\}_r^\infty = \frac{q}{4\pi\epsilon_0 r}.$$

FIGURE 15.22

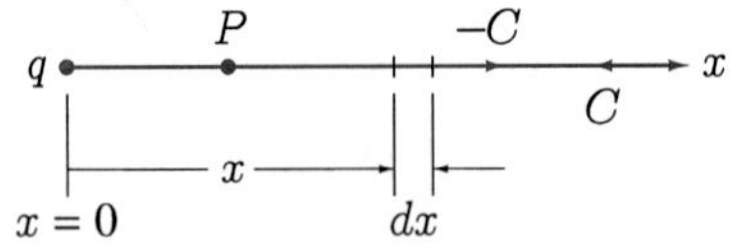

If the vector field $\mathbf{F}$ in equation 15.27 has an x-component that is only a function of x, $P = P(x)$, and if the curve C is a portion of the x-axis from $x = a$ to $x = b$, then

$$\int_C \mathbf{F} \cdot d\mathbf{r} = \int_a^b P(x)\,dx.$$

This equation indicates that definite integrals with respect to x can be regarded as line integrals along the x-axis.

EXERCISES 15.3

In Exercises 1–10 evaluate the line integral.

1. $\int_C x\,dx + x^2 y\,dy$ where C is the curve $y = x^3$, $z = 0$ from $(-1,-1,0)$ to $(2,8,0)$

2. $\int_C x\,dx + yz\,dy + x^2\,dz$ where C is the curve $y = x$, $z = x^2$ from $(-1,-1,1)$ to $(2,2,4)$

3. $\int_C x\,dx + (x+y)\,dy$ where C is the curve $x = 1 + y^2$ from $(2,1)$ to $(2,-1)$

4. $\int_C x^2\,dx + y^2\,dy + z^2\,dz$ where C is the curve $x + y = 1$, $x + z = 1$ from $(-2,3,3)$ to $(1,0,0)$

5. $\int_C (y + 2x^2z)\,dx$ where C is the curve $x = y^2$, $z = x^2$ from $(4,-2,16)$ to $(1,1,1)$

6. $\oint_C x^2 y\,dx + (x-y)\,dy$ once counterclockwise around the curve bounding the area described by the curves $x = 1 - y^2$, $y = x + 1$

7. $\int_C y^2\,dx + x^2\,dy$ where C is the semicircle $x = \sqrt{1-y^2}$ from $(0,1)$ to $(0,-1)$

8. $\int_C y\,dx + x\,dy + z\,dz$ where C is the curve $z = x^2 + y^2$, $x + y = 1$ from $(1,0,1)$ to $(-1,2,5)$

9. $\oint_C x^2 y\,dy + z\,dx$ where C is the curve $x^2 + y^2 = 1$, $x + y + z = 1$ directed so that x decreases when y is positive

10. $\oint_C y^2\,dx + x^2\,dy$ once clockwise around the curve $|x| + |y| = 1$

11. Find the work done by the force $\mathbf{F} = x^2 y\hat{\mathbf{i}} + x\hat{\mathbf{j}}$ as a particle moves from $(1,0)$ to $(6,5)$ along the straight line joining these points.

12. Consider the line integral $\int_C xy\,dx + x^2\,dy$, where C is the quarter-circle $x^2 + y^2 = 9$ from $(3,0)$ to $(0,3)$. Show that for each of the following parametrizations of C the value of the line integral is the same:
    (a) $x = 3\cos t$, $y = 3\sin t$, $0 \leq t \leq \pi/2$
    (b) $x = \sqrt{9 - y^2}$, $0 \leq y \leq 3$

13. Evaluate the line integral $\int_C xy\,dx + x\,dy$ from $(-5,3,0)$ to $(4,0,0)$ along each of the following curves:
    (a) the straight line joining the points
    (b) $x = 4 - y^2$, $z = 0$
    (c) $3y = x^2 - 16$, $z = 0$

14. Find the work done by a force $\mathbf{F} = x\hat{\mathbf{i}} + y\hat{\mathbf{j}}$ on a particle as it moves once counterclockwise around the ellipse $b^2 x^2 + a^2 y^2 = a^2 b^2$, $z = 0$.

In Exercises 15–19 evaluate the line integral.

15. $\int_C \dfrac{1}{yz}\,dx$ where C is the curve $z = \sqrt{1 - x^2}$, $y = \sqrt{1 - x^2}$ from $(1/\sqrt{2}, 1/\sqrt{2}, 1/\sqrt{2})$ to $(-1\sqrt{2}, 1/\sqrt{2}, 1/\sqrt{2})$

16. $\oint_C (x^2 + 2y^2)\,dy$ twice clockwise around the circle $(x - 2)^2 + y^2 = 1$, $z = 0$

17. $\int_C y\,dx - y(x-1)\,dy + y^2 z\,dz$ where C is the first octant intersection of $x^2 + y^2 + z^2 = 4$ and $(x - 1)^2 + y^2 = 1$ from $(2,0,0)$ to $(0,0,2)$

18. $\int_C x^2 y\,dx + y\,dy + \sqrt{1 - x^2}\,dz$ where C is the curve $y - 2z^2 = 1$, $z = x + 1$ from $(0,3,1)$ to $(1,9,2)$

19. $\int_C x\,dx + xy\,dy + 2\,dz$ where C is the curve $x + 2y + z = 4$, $4x + 3y + 2z = 13$ from $(2,-1,4)$ to $(3,1,-1)$

20. Evaluate the line integral
$$\int_C \dfrac{x^3}{(1+x^4)^3}\,dx + y^2 e^y\,dy + \dfrac{z}{\sqrt{1+z^2}}\,dz$$
where C is the series of line segments joining successively the points $(0,-1,1)$, $(1,-1,1)$, $(1,0,1)$, and $(1,0,2)$.

21. One end of a spring (with constant k) is fixed at point D in Figure 15.23. The other end is moved along the x-axis from A to B. If the spring is stretched an amount l at A, find the work done against the spring.

FIGURE 15.23

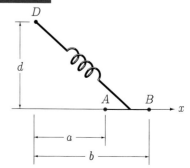

22. Two positive charges q_1 and q_2 are placed at positions $(5,5)$ and $(-2,3)$ respectively, in the xy-plane. A third positive charge q_3 is moved along the x-axis from $x = 1$ to $x = -1$. Find the total work done by the electrostatic forces of q_1 and q_2 on q_3.

In Exercises 23–26 set up a definite integral to evaluate the line integral. Use Simpson's rule with 10 equal subdivisions to approximate the definite integral.

23. $\displaystyle\int_C xy\,dx + xy^2\,dy$ where C is the curve $z = 0$,
    $y = 1/\sqrt{1 + x^3}$, from $(0,1,0)$ to $(2,1/3,0)$

24. $\displaystyle\int_C xz\,dx + \tan x\,dy + e^{xy}\,dz$ where C is the curve
    $x = y^2$, $z = y^3$ from $(1,-1,-1)$ to $(1,1,1)$

25. $\displaystyle\int_C \sqrt{1 + y^2}\,dz + zy\,dy$ where C is the curve
    $y = \cos^3 t$, $z = \sin^3 t$, $x = 0$, $0 \le t \le \pi/2$

26. $\displaystyle\int_C xyz\,dy$ where C is the curve $x = (1 - t^2)/(1 + t^2)$,
    $y = t(1 - t^2)/(1 + t^2)$, $z = t$, $-1 \le t \le 1$

In Exercises 27 and 28 evaluate the line integral along the polar coordinate curve.

27. $\displaystyle\oint_C y\,dx$ where C is the cardioid $r = 1 - \cos\theta$

28. $\displaystyle\int_C y\,dx + x\,dy$ where C is the curve $r = \theta$, $0 \le \theta \le \pi$

29. Suppose a gas is flowing through a region D of space. At each point $P(x,y,z)$ in D and time t, the gas has a certain velocity $\mathbf{v}(x,y,z,t)$. If C is a closed curve in D, the line integral
$$\Gamma = \oint_C \mathbf{v} \cdot d\mathbf{r}$$
is called the circulation of the flow for the curve C. If C is the circle $x^2 + y^2 = r^2$, $z = 1$ (directed clockwise as viewed from the origin), calculate Γ for the following flow vectors:
(a) $\mathbf{v}(x,y,z) = (x\hat{\mathbf{i}} + y\hat{\mathbf{j}} + z\hat{\mathbf{k}})/(x^2 + y^2 + z^2)^{3/2}$
(b) $\mathbf{v}(x,y,z) = -y\hat{\mathbf{i}} + x\hat{\mathbf{j}}$

30. We have shown that given a line integral 15.28, it is always possible to write it uniquely in form 15.23, where $f = \mathbf{F} \cdot \hat{\mathbf{T}}$. Show that the converse is not true; that is, given $f(x,y,z)$, there does not exist a unique $\mathbf{F}(x,y,z)$ such that $\mathbf{F} \cdot d\mathbf{r} = f(x,y,z)\,ds$.

31. Explain why the line integral
$$\oint_C f(x)\,dx + g(y)\,dy + h(z)\,dz$$
must have value zero when $f(x)$, $g(y)$, and $h(z)$ are continuous functions in some domain containing C.

32. The cycloid $x = a(\theta - \sin\theta)$, $y = a(1 - \cos\theta)$ (Figure 15.24) is the curve traced out by a fixed point on the circumference of a circle of radius a rolling along the x-axis (θ being the angle through which the point has rotated). Suppose the point is acted on by a force of unit magnitude directed toward the centre of the rolling circle.
   (a) Find the work done by the force as the point moves from $\theta = 0$ to $\theta = \pi$.
   (b) How much of the work in (a) is done by the vertical component of the force?

FIGURE 15.24

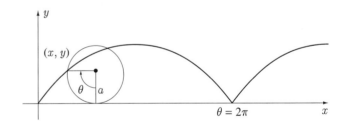

SECTION 15.4

Independence of Path

In Sections 15.2 and 15.3 we illustrated that the value of a line integral joining two points usually depends on the curve joining the points. In this section we show that certain line integrals have the same value for all curves joining the same two points. We formalize this idea in the following definition.

Definition 15.4

A line integral $\int \mathbf{F} \cdot d\mathbf{r}$ is said to be **independent of path** in a domain D if for each pair of points A and B in D, the value of the line integral

$$\int_C \mathbf{F} \cdot d\mathbf{r}$$

is the same for all piecewise-smooth paths C in D from A to B.

The value of such a line integral for given $\mathbf{F}$ will then depend only on the end points A and B. Note that we speak of independence of path only for the special class of line integrals of the form $\int \mathbf{F} \cdot d\mathbf{r}$. The question we must now ask is "How do we determine whether a given line integral is independent of path?" One answer is contained in the following theorem.

Theorem 15.3

Suppose $P(x,y,z)$, $Q(x,y,z)$, and $R(x,y,z)$ are continuous functions in some domain D. The line integral

$$\int \mathbf{F} \cdot d\mathbf{r} = \int P\,dx + Q\,dy + R\,dz$$

is independent of path in D if and only if there exists a function $\phi(x,y,z)$ defined in D such that

$$\nabla \phi = \mathbf{F} = P\hat{\mathbf{i}} + Q\hat{\mathbf{j}} + R\hat{\mathbf{k}}. \qquad (15.32)$$

Essentially, then, a line integral is independent of path if $\mathbf{F}$ is the gradient of some scalar function.

Proof Suppose first of all that in D there exists a function $\phi(x,y,z)$ such that $\nabla \phi = P\hat{\mathbf{i}} + Q\hat{\mathbf{j}} + R\hat{\mathbf{k}}$. If

$$C: \quad x = x(t), \quad y = y(t), \quad z = z(t), \quad \alpha \leq t \leq \beta$$

is any smooth curve in D from A to B, then

$$\int_C \mathbf{F} \cdot d\mathbf{r} = \int_C P\,dx + Q\,dy + R\,dz = \int_\alpha^\beta \left\{ \frac{\partial \phi}{\partial x}\frac{dx}{dt} + \frac{\partial \phi}{\partial y}\frac{dy}{dt} + \frac{\partial \phi}{\partial z}\frac{dz}{dt} \right\} dt.$$

But the term in braces is the chain rule for the derivative of the composite function $\phi[x(t), y(t), z(t)]$, and we can therefore write that

$$\int_C \mathbf{F} \cdot d\mathbf{r} = \int_\alpha^\beta \frac{d\phi}{dt}\,dt = \{\phi[x(t), y(t), z(t)]\}_\alpha^\beta$$
$$= \phi[x(\beta), y(\beta), z(\beta)] - \phi[x(\alpha), y(\alpha), z(\alpha)]$$
$$= \phi(x_B, y_B, z_B) - \phi(x_A, y_A, z_A).$$

(The same result is obtained even when C is piecewise smooth rather than smooth.) Because this last expression does not depend on the curve C taken from A to B, it follows that the line integral is independent of path in D.

Conversely, suppose now that the line integral

$$\int \mathbf{F} \cdot d\mathbf{r} = \int P\,dx + Q\,dy + R\,dz$$

is independent of path in D, and A is chosen as some fixed point in D. If $P(x, y, z)$ is any other point in D (Figure 15.25), and C is a piecewise-smooth curve in D from A to P, then the line integral

$$\phi(x, y, z) = \int_C \mathbf{F} \cdot d\mathbf{r}$$

defines a single-valued function $\phi(x, y, z)$ in D, and the value of $\phi(x, y, z)$ is the same for all piecewise-smooth curves from A to P.

FIGURE 15.25

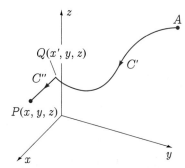

Consider a curve C composed of two parts: a straight-line portion C'' parallel to the x-axis from a fixed point $Q(x', y, z)$ to $P(x, y, z)$, and any other piecewise-smooth curve C' in D from A to Q. Then

$$\phi(x, y, z) = \int_{C'} \mathbf{F} \cdot d\mathbf{r} + \int_{C''} \mathbf{F} \cdot d\mathbf{r}.$$

Now along C'', y and z are both constant, and therefore

$$\phi(x, y, z) = \int_{C'} \mathbf{F} \cdot d\mathbf{r} + \int_{x'}^{x} P(t, y, z)\,dt.$$

The partial derivative of this function with respect to x is

$$\frac{\partial \phi}{\partial x} = \frac{\partial}{\partial x} \int_{C'} \mathbf{F} \cdot d\mathbf{r} + \frac{\partial}{\partial x} \int_{x'}^{x} P(t, y, z)\,dt,$$

but because $Q(x', y, z)$ is fixed,

$$\frac{\partial}{\partial x} \int_{C'} \mathbf{F} \cdot d\mathbf{r} = 0.$$

Consequently,

$$\frac{\partial \phi}{\partial x} = \frac{\partial}{\partial x} \int_{x'}^{x} P(t, y, z)\,dt = P(x, y, z).$$

By choosing other curves with straight-line portions parallel to the y- and z-axes, we can also show that $\partial \phi / \partial y = Q$ and $\partial \phi / \partial z = R$. Thus, $\mathbf{F} = \nabla \phi$, and this completes the proof.

This theorem points out that it is very simple to evaluate a line integral that is independent of path. We state this in the following corollary.

Corollary 1 When a line integral is independent of path in a domain D, and A and B are points in D, then

$$\int_C \mathbf{F} \cdot d\mathbf{r} = \phi(x_B, y_B, z_B) - \phi(x_A, y_A, z_A),$$

where $\nabla \phi = \mathbf{F}$, for every piecewise smooth curve C in D from A to B.

EXAMPLE 15.10

Evaluate $\int_C 2xy\,dx + x^2\,dy + 2z\,dz$, where C is the first octant intersection of $x^2 + y^2 = 1$ and $z = 2x + 4$ from $(0, 1, 4)$ to $(1, 0, 6)$.

SOLUTION Since $\nabla(x^2 y + z^2) = 2xy\hat{\mathbf{i}} + x^2\hat{\mathbf{j}} + 2z\hat{\mathbf{k}}$, the line integral is independent of path everywhere, and

$$\int_C 2xy\,dx + x^2\,dy + 2z\,dz = \left\{ x^2 y + z^2 \right\}_{(0,1,4)}^{(1,0,6)} = 36 - 16 = 20.$$

∎

The following corollary is also an immediate consequence of Theorem 15.3.

Corollary 2 The line integral $\int \mathbf{F} \cdot d\mathbf{r}$ is independent of path in a domain D if and only if

$$\oint_C \mathbf{F} \cdot d\mathbf{r} = 0 \qquad (15.33)$$

for every closed path in D.

Theorem 15.3 states that a necessary and sufficient condition for line integral 15.28 to be independent of path is the existence of a function $\phi(x, y, z)$ such that $\nabla\phi = \mathbf{F}$. For most problems it is quite obvious whether such a function $\phi(x, y, z)$ exists; for a few, however, it is not at all clear whether $\phi(x, y, z)$ exists or not. Since much time could be wasted searching for $\phi(x, y, z)$ (when in fact it does not exist), it would be helpful to have a test that states *a priori* whether $\phi(x, y, z)$ does indeed exist. Such a test is contained in Theorem 15.1. It states that $\mathbf{F}$ is the gradient of some scalar function $\phi(x, y, z)$ if $\nabla \times \mathbf{F} = \mathbf{0}$. When this result is combined with Theorem 15.3, we obtain this important theorem.

Theorem 15.4

Let D be a domain in which $P(x, y, z)$, $Q(x, y, z)$, and $R(x, y, z)$ have continuous first derivatives. If the line integral $\int \mathbf{F} \cdot d\mathbf{r} = \int P\,dx + Q\,dy + R\,dz$ is independent of path in D, then $\nabla \times \mathbf{F} = \mathbf{0}$ in D. Conversely, if D is simply-connected, and $\nabla \times \mathbf{F} = \mathbf{0}$ in D, then the line integral is independent of path in D.

We have in Theorem 15.4 a simple test to determine whether a given line integral is independent of path: We see whether the curl of $\mathbf{F}$ is zero. Evaluation of a line integral that is independent of path still requires the function $\phi(x, y, z)$, but it is at least nice to know that ϕ exists before searching for it.

Theorems 15.1, 15.3, and 15.4 have identified an important equivalence, at least in simply-connected domains:

1. $\int \mathbf{F} \cdot d\mathbf{r}$ is independent of path in D;

2. $\mathbf{F} = \nabla \phi$ for some function $\phi(x,y,z)$ defined in D;

3. $\nabla \times \mathbf{F} = \mathbf{0}$ in D.

Theorem 15.1 states that (2) and (3) are equivalent; Theorem 15.3 verifies the equivalence of (1) and (2); and these two imply the equivalence of (1) and (3) (Theorem 15.4).

For line integrals in the xy-plane, this equivalence is still valid except that $\nabla \times \mathbf{F} = \mathbf{0}$ can be stated more simply as

$$\frac{\partial Q}{\partial x} = \frac{\partial P}{\partial y} \qquad (15.34)$$

(see equation 15.16).

EXAMPLE 15.11

Evaluate $\int_C 2xye^z\,dx + (x^2 e^z + y)\,dy + (x^2 ye^z - z)\,dz$ along the straight line C from $(0,1,2)$ to $(2,1,-8)$.

SOLUTION

Method 1 Parametric equations for the straight line are

$$C: \quad x = 2t, \quad y = 1, \quad z = 2 - 10t, \quad 0 \le t \le 1.$$

If I is the value of the line integral, then

$$I = \int_0^1 \{2(2t)(1)e^{2-10t}(2\,dt) + [(2t)^2 e^{2-10t} + 1](0)$$
$$\quad + [(2t)^2(1)e^{2-10t} - 2 + 10t](-10\,dt)\}$$

$$= \int_0^1 \{8e^2(t - 5t^2)e^{-10t} + 20 - 100t\}dt$$

$$= 8e^2 \left\{(t - 5t^2)\frac{e^{-10t}}{-10}\right\}_0^1 + \frac{4e^2}{5}\int_0^1 (1 - 10t)e^{-10t}dt + 10\{2t - 5t^2\}_0^1$$

$$= 8e^2 \left\{\frac{2}{5}e^{-10}\right\} - 30 + \frac{4}{5}e^2 \left\{(1 - 10t)\frac{e^{-10t}}{-10}\right\}_0^1 + \frac{2e^2}{25}\int_0^1 -10e^{-10t}dt$$

$$= \frac{16}{5}e^{-8} - 30 + \frac{4}{5}e^2 \left\{\frac{9}{10}e^{-10} + \frac{1}{10}\right\} + \frac{2e^2}{25}\left\{e^{-10t}\right\}_0^1 = 4e^{-8} - 30.$$

Method 2 It is evident that

$$\nabla\left(x^2 ye^z + \frac{y^2}{2} - \frac{z^2}{2}\right) = 2xye^z\hat{\mathbf{i}} + (x^2 e^z + y)\hat{\mathbf{j}} + (x^2 ye^z - z)\hat{\mathbf{k}},$$

and hence the line integral is independent of path. Its value is therefore

$$I = \left\{x^2 ye^z + \frac{y^2}{2} - \frac{z^2}{2}\right\}_{(0,1,2)}^{(2,1,-8)} = \left\{4e^{-8} + \frac{1}{2} - 32\right\} - \left\{\frac{1}{2} - 2\right\} = 4e^{-8} - 30.$$

Method 3 Since

$$\nabla \times (2xye^z\hat{\mathbf{i}} + (x^2e^z + y)\hat{\mathbf{j}} + (x^2ye^z - z)\hat{\mathbf{k}}) = \begin{vmatrix} \hat{\mathbf{i}} & \hat{\mathbf{j}} & \hat{\mathbf{k}} \\ \partial/\partial x & \partial/\partial y & \partial/\partial z \\ 2xye^z & x^2e^z + y & x^2ye^z - z \end{vmatrix}$$

$$= (x^2e^z - x^2e^z)\hat{\mathbf{i}}$$
$$+ (2xye^z - 2xye^z)\hat{\mathbf{j}}$$
$$+ (2xe^z - 2xe^z)\hat{\mathbf{k}} = \mathbf{0},$$

the line integral is independent of path. Thus there exists a function $\phi(x, y, z)$ such that

$$\nabla \phi = 2xye^z\hat{\mathbf{i}} + (x^2e^z + y)\hat{\mathbf{j}} + (x^2ye^z - z)\hat{\mathbf{k}}$$

or

$$\frac{\partial \phi}{\partial x} = 2xye^z, \quad \frac{\partial \phi}{\partial y} = x^2e^z + y, \quad \frac{\partial \phi}{\partial z} = x^2ye^z - z.$$

Integration of the first of these equations yields

$$\phi(x, y, z) = x^2ye^z + K(y, z).$$

Substitution of this function into the left-hand side of the second equation gives

$$x^2e^z + \frac{\partial K}{\partial y} = x^2e^z + y,$$

which implies that

$$\frac{\partial K}{\partial y} = y.$$

Consequently,

$$K(y, z) = \frac{y^2}{2} + L(z),$$

and we know both the x- and y-dependence of ϕ:

$$\phi(x, y, z) = x^2ye^z + \frac{y^2}{2} + L(z).$$

To obtain the z-dependence contained in $L(z)$, we substitute into the left-hand side of the third equation to get

$$x^2ye^z + \frac{dL}{dz} = x^2ye^z - z,$$

from which we have

$$\frac{dL}{dz} = -z.$$

Hence, $L(z) = -z^2/2 + C$ (C a constant), and

$$\phi(x, y, z) = x^2ye^z + \frac{y^2}{2} - \frac{z^2}{2} + C.$$

Finally, then, we have

$$I = \left\{ x^2ye^z + \frac{y^2}{2} - \frac{z^2}{2} \right\}_{(0,1,2)}^{(2,1,-8)} = 4e^{-8} - 30.$$

∎

Method 1 is one of "brute force." The function $x^2 y e^z + y^2/2 - z^2/2$ in Method 2 was obtained by observation. Method 3 is the systematic procedure suggested in Section 15.1 for finding the function $\phi(x,y,z)$.

EXAMPLE 15.12

Evaluate
$$I = \int_C \left(\frac{x^3 - 2y^2}{x^3 y}\right) dx + \left(\frac{y^2 - x^3}{x^2 y^2}\right) dy + 2z^2\, dz,$$
where C is the curve $y = x^2$, $z = x - 1$ from $(1,1,0)$ to $(2,4,1)$.

SOLUTION If we set
$$\mathbf{F} = \left(\frac{x^3 - 2y^2}{x^3 y}\right)\hat{\mathbf{i}} + \left(\frac{y^2 - x^3}{x^2 y^2}\right)\hat{\mathbf{j}} + 2z^2\hat{\mathbf{k}},$$

it is straightforward to show that in any simply-connected domain not containing points on the xz- or yz-planes, $\nabla \times \mathbf{F} = \mathbf{0}$, and therefore the line integral is independent of path. Thus there exists a function $\phi(x,y,z)$ such that $\nabla \phi = \mathbf{F}$. If we write
$$\mathbf{F} = \left(\frac{1}{y} - \frac{2y}{x^3}\right)\hat{\mathbf{i}} + \left(\frac{1}{x^2} - \frac{x}{y^2}\right)\hat{\mathbf{j}} + 2z^2\hat{\mathbf{k}},$$

then it is evident that
$$\phi(x,y,z) = \frac{x}{y} + \frac{y}{x^2} + \frac{2z^3}{3}.$$

Thus,
$$I = \left\{\frac{x}{y} + \frac{y}{x^2} + \frac{2z^3}{3}\right\}_{(1,1,0)}^{(2,4,1)} = \left\{\frac{1}{2} + 1 + \frac{2}{3}\right\} - \{1 + 1\} = \frac{1}{6}.$$
∎

EXAMPLE 15.13

In thermodynamics the state of a gas is described by four variables — pressure P, temperature T, internal energy U, and volume V. These variables are related by two equations of state,
$$F(P,T,U,V) = 0 \quad \text{and} \quad G(P,T,U,V) = 0,$$
so that two of the variables are independent and two are dependent. If U and V are chosen as independent variables, then $T = T(U,V)$ and $P = P(U,V)$. An experimental law called the second law of thermodynamics states that the line integral
$$\int_C \frac{1}{T} dU + \frac{P}{T} dV$$
is independent of path in the UV-plane. Show that the second law can be expressed in the differential form
$$T\frac{\partial P}{\partial U} - P\frac{\partial T}{\partial U} + \frac{\partial T}{\partial V} = 0.$$

SOLUTION According to equation 15.34, the line integral is independent of path if and only if

$$\frac{\partial}{\partial U}\left(\frac{P}{T}\right) = \frac{\partial}{\partial V}\left(\frac{1}{T}\right),$$

or

$$0 = \frac{T\frac{\partial P}{\partial U} - P\frac{\partial T}{\partial U}}{T^2} + \frac{1}{T^2}\frac{\partial T}{\partial V};$$

that is,

$$0 = T\frac{\partial P}{\partial U} - P\frac{\partial T}{\partial U} + \frac{\partial T}{\partial V},$$

and the proof is complete.

Since the above line integral is independent of path, there exists a function $S(U, V)$ such that

$$\frac{\partial S}{\partial U} = \frac{1}{T}, \quad \frac{\partial S}{\partial V} = \frac{P}{T},$$

and the value of the line integral is given by

$$\int_C \frac{1}{T}dU + \frac{P}{T}dV = S(B) - S(A),$$

where C joins A and B. This function, called **entropy**, plays a key role in the field of thermodynamics. ∎

EXERCISES 15.4

In Exercises 1–10 show that the line integral is independent of path, and evaluate it.

1. $\int_C xy^2\, dx + x^2 y\, dy$ where C is the curve $y = x^2$, $z = 0$ from $(0,0,0)$ to $(1,1,0)$

2. $\int_C (3x^2 + y)\, dx + x\, dy$ where C is the straight line from $(2,1,5)$ to $(-3,2,4)$

3. $\int_C 2xe^y\, dx + (x^2 e^y + 3)\, dy$ where C is the curve $y = \sqrt{1-x^2}$, $z = 0$ from $(1,0,0)$ to $(-1,0,0)$

4. $\int_C 3x^2 yz\, dx + x^3 z\, dy + (x^3 y - 4z)\, dz$ where C is the curve $x^2 + y^2 + z^2 = 3$, $y = x$ from $(-1,-1,1)$ to $(1,1,-1)$

5. $\int_C -\frac{y}{z}\sin x\, dx + \frac{1}{z}\cos x\, dy - \frac{y}{z^2}\cos x\, dz$ where C is the helix $x = 2\cos t$, $y = 2\sin t$, $z = t$ from $(2,0,2\pi)$ to $(2,0,4\pi)$

6. $\oint_C y\cos x\, dx + \sin x\, dy$ once clockwise around the circle $x^2 + y^2 - 2x + 4y = 7$, $z = 0$

7. $\int_C x^2\, dx + y^2\, dy + z^2\, dz$ where C is the curve $x + y = 1$, $x + z = 1$ from $(-2,3,3)$ to $(1,0,0)$

8. $\int_C y\, dx + x\, dy + z\, dz$ where C is the curve $z = x^2 + y^2$, $x + y = 1$ from $(1,0,1)$ to $(-1,2,5)$

9. $\int_C \frac{1}{y}dx - \frac{x}{y^2}dy + dz$ where C is the curve $y = x^2 + 1$, $x + y + z = 2$ from $(0,1,1)$ to $(3,10,-11)$

10. $\int_C 3x^2 y^3\, dx + 3x^3 y^2\, dy$ where C is the curve $y = e^x$ from $(0,1)$ to $(1,e)$

11. Show that if $f(x)$, $g(y)$, and $h(z)$ have continuous first derivatives, then the line integral

$$\int_C f(x)\, dx + g(y)\, dy + h(z)\, dz$$

is independent of path.

12. If $\nabla \times \mathbf{F} = \mathbf{0}$ in a domain D which is not simply-connected, can you conclude that the line integral $\int_C \mathbf{F} \cdot d\mathbf{r}$ is not independent of path in D? Explain.

In Exercises 13–18 evaluate the line integral.

13. $\int_C zye^{xy}\, dx + zxe^{xy}\, dy + (e^{xy} - 1)\, dz$ where C is the curve $y = x^2$, $z = x^3$ from $(1,1,1)$ to $(2,4,8)$.

14. $\oint_C y(\tan x + x\sec^2 x)\, dx + x\tan x\, dy + dz$ once around the circle $x^2 + y^2 = 1$, $z = 0$.

15. $\int_C \left(\dfrac{1+y^2}{x^3}\right) dx - \left(\dfrac{y+x^2 y}{x^2}\right) dy + z\, dz$ where C is the broken line joining successively $(1,0,0)$, $(25,2,3)$ and $(5,2,1)$.

16. $\oint_C \dfrac{zy\, dx - xz\, dy + xy\, dz}{y^2}$ where C is the curve $x^2 + z^2 = 1$, $y + z = 2$.

17. $\int_C -\dfrac{1}{x^2}\operatorname{Tan}^{-1} y\, dx + \dfrac{1}{x+xy^2}\, dy$ where C is the curve $x = y^2 + 1$ from $(2,-1)$ to $(10,3)$.

18. $\int_C \dfrac{1}{(x-3)^2(y+5)}\, dx + \dfrac{1}{(x-3)(y+5)^2}\, dy + \dfrac{1}{z+4}\, dz$ where C is the curve $x = y = z$ from $(0,0,0)$ to $(2,2,2)$.

19. Evaluate $\oint_C \dfrac{-y\, dx + x\, dy}{x^2 + y^2}$

    (a) once counterclockwise around the circle $x^2 + y^2 = 1$, $z = 0$.

    (b) once counterclockwise around the circle $(x-2)^2 + y^2 = 1$, $z = 0$.

20. Evaluate $\int_C \dfrac{y}{x^2+y^2}\, dx - \dfrac{x}{x^2+y^2}\, dy$ where C is the series of line segments joining successively the points $(1,0)$, $(1,1)$, $(-1,1)$, and $(-1,0)$.

21. Is the line integral
$$\int_C \dfrac{x}{\sqrt{x^2+y^2}}\, dx + \dfrac{y}{\sqrt{x^2+y^2}}\, dy$$
independent of path in the domain consisting of the xy-plane with the origin removed? Is the line integral
$$\int_C \dfrac{x}{\sqrt{x^2+y^2+z^2}}\, dx + \dfrac{y}{\sqrt{x^2+y^2+z^2}}\, dy + \dfrac{z}{\sqrt{x^2+y^2+z^2}}\, dz$$
independent of path in the domain consisting of xyz-space with the origin removed?

22. In which of the following domains is the line integral
$$\int_C \dfrac{y\, dx - x\, dy}{x^2+y^2}$$
independent of path: (a) $x > 0$ (b) $x < 0$ (c) $y > 0$ (d) $y < 0$ (e) $x^2 + y^2 > 0$

23. The second law of thermodynamics states that the line integral $I = \int_C T^{-1}(dU + P\, dV)$ is independent of path in the UV-plane (see Example 15.13).

    (a) The equations of state for an ideal gas are $PV = nRT$, $U = f(T)$, where n and R are constants and $f(T)$ is some given function. Because of these, it is more convenient to choose T and V as independent variables and to express P and U in terms of T and V. If this is done, show that
    $$I = \int kT^{-1}\, dT + nRV^{-1}\, dV \quad \text{where } k = dU/dT.$$

    (b) Since the line integral is independent of path, there exists a function $S(T,V)$, called entropy, such that
    $$\int_C \dfrac{k}{T}\, dT + \dfrac{nR}{V}\, dV = S(B) - S(A),$$
    where C is any curve joining points A and B. Show in the case that k is constant that $S = k\ln T + nR\ln V + S_0$, where S_0 is a constant.

24. Evaluate $\oint_C (2xye^{x^2 y} + x^2 y)\, dx + x^2 e^{x^2 y}\, dy$ once clockwise around the ellipse $x^2 + 4y^2 = 4$, $z = 0$.

25. A spring has one end fixed at point P. The other end is moved along the curve $y = f(x)$ from $(x_0, f(x_0))$ to $(x_1, f(x_1))$. If the initial and final stretches in the spring are a and b ($b > a$), what work is done against the spring?

26. Electrostatic forces due to point charges and gravitational forces due to point masses are examples of inverse square force fields — force fields of the form $\mathbf{F} = k\hat{\mathbf{r}}/|\mathbf{r}|^2$, where k is a constant and $\mathbf{r} = x\hat{\mathbf{i}} + y\hat{\mathbf{j}} + z\hat{\mathbf{k}}$.

    (a) Is the line integral representing work done by such a force field independent of path?

    (b) What is the work done by $\mathbf{F}$ in moving a particle from (x_1, y_1, z_1) to (x_2, y_2, z_2)?

SECTION 15.5
Energy and Conservative Force Fields

Suppose a force field $\mathbf{F}(x,y,z)$ acts throughout some domain D of space. In physics and engineering, many force fields satisfy the following definition.

> **Definition 15.5**
>
> A force field $\mathbf{F}(x,y,z)$ is said to be **conservative** in a domain D if the line integral $\int \mathbf{F} \cdot d\mathbf{r}$ is independent of path in D.

Since the value of $\int_C \mathbf{F} \cdot d\mathbf{r}$ can be interpreted as the work done by $\mathbf{F}$ along C, a force field is conservative if the work done by it is independent of path taken from one point to another. According to the results of Section 15.4, we can also state that a force field $\mathbf{F}$ is conservative if and only if there exists a function $\phi(x,y,z)$ such that $\mathbf{F} = \nabla \phi$.

It is customary to associate a potential energy function $U(x,y,z)$ with a conservative force field $\mathbf{F}$. This function assigns to each point (x,y,z) a potential energy in such a way that if a particle moves from point A to point B, then the difference in potential energy $U(A) - U(B)$ is precisely the work done by $\mathbf{F}$; i.e., if C is any curve joining A and B, then

$$U(A) - U(B) = \int_C \mathbf{F} \cdot d\mathbf{r}. \qquad (15.35)$$

If $\int \mathbf{F} \cdot d\mathbf{r} > 0$, then the potential energy at A is greater than the potential energy at B; if $\int \mathbf{F} \cdot d\mathbf{r} < 0$, then the potential energy at B is greater than that at A. To find $U(x,y,z)$, we use the fact that because $\mathbf{F}$ is conservative, there exists a function $\phi(x,y,z)$ such that $\mathbf{F} = \nabla \phi$, and

$$\int_C \mathbf{F} \cdot d\mathbf{r} = \phi(B) - \phi(A).$$

If follows, then, that $U(x,y,z)$ must satisfy the equation

$$U(A) - U(B) = \phi(B) - \phi(A),$$

or

$$U(A) + \phi(A) = U(B) + \phi(B).$$

Since A and B are arbitrary points in D, this last equation states that the value of the function $U(x,y,z) + \phi(x,y,z)$ is the same at every point in the force field,

$$U(x,y,z) + \phi(x,y,z) = C,$$

where C is a constant. Thus,

$$U(x,y,z) = -\phi(x,y,z) + C. \qquad (15.36)$$

Equation 15.36 shows that the force field $\mathbf{F}$ defines a potential energy function $U(x,y,z)$ up to an additive constant. (This seems reasonable in that ϕ itself is defined only to an additive constant.) Because $U = -\phi + C$ and $\mathbf{F} = \nabla \phi$, we can also regard U as being defined by the equation

$$\mathbf{F} = -\nabla U. \qquad (15.37)$$

The advantage of this equation is that it defines U directly, not through the function ϕ.

For a conservative force field $\mathbf{F}$, then, we define a potential energy function $U(x,y,z)$ by equation 15.37. If a particle moves from A to B, then the work done by $\mathbf{F}$ is

$$W = U(A) - U(B); \tag{15.38}$$

in other words, work done by a conservative force field is equal to loss in potential energy.

On the other hand, if a particle moves under the action of a force $\mathbf{F}$ (and only $\mathbf{F}$), be it conservative or nonconservative, then it does so according to Newton's second law,

$$\mathbf{F} = \frac{d}{dt}(m\mathbf{v}) = m\frac{d\mathbf{v}}{dt},$$

where m is the mass of the particle (assumed constant), $\mathbf{v}$ is its velocity, and t is time. The action of $\mathbf{F}$ produces motion along some curve C, and the work done by $\mathbf{F}$ along this curve from A to B is

$$W = \int_C \mathbf{F} \cdot d\mathbf{r} = \int_\alpha^\beta m\frac{d\mathbf{v}}{dt} \cdot \frac{d\mathbf{r}}{dt} dt = \int_\alpha^\beta m\frac{d\mathbf{v}}{dt} \cdot \mathbf{v}\, dt = \int_\alpha^\beta \frac{d}{dt}\left\{\frac{1}{2}m\mathbf{v} \cdot \mathbf{v}\right\} dt$$

$$= \left\{\frac{1}{2}m\mathbf{v} \cdot \mathbf{v}\right\}_\alpha^\beta = \left\{\frac{1}{2}m|\mathbf{v}|^2\right\}_\alpha^\beta.$$

Thus if $K(x,y,z) = \frac{1}{2}m|\mathbf{v}|^2$ represents kinetic energy of the particle, the work done by $\mathbf{F}$ is equal to the gain in kinetic energy of the particle,

$$W = K(B) - K(A) \tag{15.39}$$

(and this is true for any force $\mathbf{F}$ as long as $\mathbf{F}$ is the total resultant force producing motion).

If the total force producing motion is a conservative force field $\mathbf{F}$ we have two expressions 15.38 and 15.39 for the work done as a particle moves from one point to another under the action of $\mathbf{F}$. If we equate them, we have

$$U(A) - U(B) = K(B) - K(A),$$

or

$$U(A) + K(A) = U(B) + K(B). \tag{15.40}$$

We have shown then that if a particle moves under the action of a conservative force field *only*, the sum of the kinetic and potential energies at B must be the same as the sum of the kinetic and potential energies at A. In other words, if E is the total energy of the particle, kinetic plus potential, then

$$E(A) = E(B). \tag{15.41}$$

Since B can be any point along the path of the particle, it follows that when a particle moves under the action of a conservative force field, and only a conservative force field, then at every point along its trajectory

$$E = \text{a constant.} \tag{15.42}$$

This is the **law of conservation of energy** for a conservative force field.

EXAMPLE 15.14

Show that the electrostatic force due to a point charge is conservative, and determine a potential energy function for the field.

SOLUTION The electrostatic force on a charge Q due to a charge q is

$$\mathbf{F} = \frac{qQ}{4\pi\epsilon_0 |\mathbf{r}|^3}\mathbf{r},$$

where $\mathbf{r}$ is the vector from q to Q. If we choose a Cartesian coordinate system with q at the origin (Figure 15.26), then

$$\mathbf{F} = \frac{qQ}{4\pi\epsilon_0 (x^2 + y^2 + z^2)^{3/2}}(x\hat{\mathbf{i}} + y\hat{\mathbf{j}} + z\hat{\mathbf{k}}).$$

Since

$$\nabla\left\{\frac{-qQ}{4\pi\epsilon_0 (x^2 + y^2 + z^2)^{1/2}}\right\} = \mathbf{F},$$

the force field is conservative, and possible potential energy functions are

$$U(x,y,z) = \frac{qQ}{4\pi\epsilon_0 (x^2 + y^2 + z^2)^{1/2}} + C = \frac{qQ}{4\pi\epsilon_0 r} + C,$$

where $r = |\mathbf{r}|$. In electrostatics it is customary to choose $U(x,y,z)$ so that $\lim_{r\to\infty} U = 0$, in which case $C = 0$, and

$$U(x,y,z) = \frac{qQ}{4\pi\epsilon_0 r}.$$

In addition, if V is defined as the potential energy per unit test charge Q, then

$$V(x,y,z) = \frac{U}{Q} = \frac{q}{4\pi\epsilon_0 r}.$$

This result agrees with that in Example 15.9.

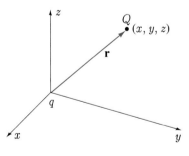

FIGURE 15.26

EXERCISES 15.5

In Exercises 1–5 determine whether the force field is conservative. Identify each conservative force field, and find a potential energy function.

1. $\mathbf{F}(x,y,z) = \dfrac{q_1 q_2}{4\pi\epsilon_0} \dfrac{x\hat{\mathbf{i}} + y\hat{\mathbf{j}} + z\hat{\mathbf{k}}}{(x^2 + y^2 + z^2)^{3/2}}$

2. $\mathbf{F}(x,y) = mx\hat{\mathbf{i}} + xy\hat{\mathbf{j}}$, m is a constant

3. $\mathbf{F}(x) = -kx\hat{\mathbf{i}}$, k is a constant

4. $\mathbf{F}(x,y,z) = -mg\hat{\mathbf{k}}$, m and g are constants

5. $\mathbf{F}(x,y,z) = GMm\dfrac{x\hat{\mathbf{i}} + y\hat{\mathbf{j}} + z\hat{\mathbf{k}}}{(x^2 + y^2 + z^2)^{3/2}}$, G, M, and m are constants

6. Suppose that $\mathbf{F}(x,y,z)$ is a conservative force field in some domain D, and $U(x,y,z)$ is a potential energy function associated with $\mathbf{F}$. The surfaces $U(x,y,z) = C$, where C is a constant, are called *equipotential surfaces*. Through each point P in D there is one and only one such equipotential surface for $\mathbf{F}$. Show that at P the force $\mathbf{F}$ is normal to the equipotential surface through P.

7. Draw the equipotential surfaces for the forces in Exercises 1, 4, and 5.

8. One end of a spring with unstretched length L is fixed at the origin in space. If the other end is at point (x,y,z) (all coordinates in metres), what is the force exerted by the spring? Is this force conservative?

9. Explain why friction is not conservative.

10. (a) When students in Universityland leave their houses, a supernatural power attracts them to university in such a way that the magnitude of the force at any point is inversely proportional to the square of the distance from university. This force acts until they are 100 m from university and then it disappears. Is this force conservative?

    (b) If someone diverts the power so that the force attracts students to the local donut shop, is this force conservative?

11. A force field $\mathbf{F}(x,y,z)$ is said to be radially symmetric about the origin if it can be written in the form

$$\mathbf{F}(x,y,z) = f\left(\sqrt{x^2 + y^2 + z^2}\right)\mathbf{r}, \qquad \mathbf{r} = x\hat{\mathbf{i}} + y\hat{\mathbf{j}} + z\hat{\mathbf{k}},$$

for some function f. We often write in such a case that

$$\mathbf{F}(x,y,z) = f(r)\mathbf{r}, \quad \text{where } r = |\mathbf{r}| = \sqrt{x^2 + y^2 + z^2}.$$

(a) Use Theorem 15.4 to show that such a force is conservative in suitably defined domains (provided $f(r)$ has a continuous first derivative).

(b) If A and B are the points in Figure 15.27 joined by the curve C, show that

$$\int_C \mathbf{F} \cdot d\mathbf{r} = \int_a^b rf(r)\,dr,$$

where the limits a and b are the distances from the origin to A and B.

(c) Have we discussed any radially symmetric force fields in this chapter?

FIGURE 15.27

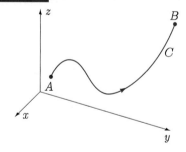

SECTION 15.6

Green's Theorem

Line integrals in the xy-plane are of the form $\int_C f(x,y)\,ds$, and in the special case that $f(x,y)$ is the tangential component of some vector field $\mathbf{F}(x,y) = P(x,y)\hat{\mathbf{i}} + Q(x,y)\hat{\mathbf{j}}$ along C, they take the form

$$\int_C \mathbf{F} \cdot d\mathbf{r} = \int_C P(x,y)\,dx + Q(x,y)\,dy. \qquad (15.43)$$

We now show that when C is a closed curve, line integral 15.43 can usually be replaced by a double integral. The precise result is contained in the following theorem.

Theorem 15.5

(Green's theorem). Let C be a piecewise-smooth, closed curve in the xy-plane that does not intersect itself and that encloses a region R (Figure 15.28). If $P(x,y)$ and $Q(x,y)$ have continuous first partial derivatives in a domain D containing C and R, then

$$\oint_C P\,dx + Q\,dy = \iint_R \left(\frac{\partial Q}{\partial x} - \frac{\partial P}{\partial y} \right) dA. \qquad (15.44)$$

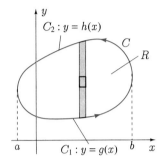

FIGURE 15.28

Proof We first consider a fairly simple region R. In particular, suppose every line parallel to the x- and y-axes that intersects C does so in at most two points (Figure 15.28). Then C can be subdivided into an upper and a lower part,

$$C_2: \quad y = h(x) \quad \text{and} \quad C_1: \quad y = g(x).$$

If we first consider the second term on the right of equation 15.44, we have

$$\iint_R -\frac{\partial P}{\partial y}\,dA = \int_a^b \int_{g(x)}^{h(x)} -\frac{\partial P}{\partial y}\,dy\,dx = \int_a^b \left\{ -P \right\}_{g(x)}^{h(x)} dx$$

$$= \int_a^b \{P[x,g(x)] - P[x,h(x)]\}\,dx.$$

On the other hand, the first term on the left of 15.44 is

$$\oint_C P\,dx = \int_{C_1} P\,dx + \int_{C_2} P\,dx = \int_{C_1} P\,dx - \int_{-C_2} P\,dx,$$

and if we use x as a parameter along C_1 and $-C_2$, then

$$\oint_C P\,dx = \int_a^b P[x,g(x)]\,dx - \int_a^b P[x,h(x)]\,dx$$

$$= \int_a^b \{P[x,g(x)] - P[x,h(x)]\}\,dx.$$

We have shown therefore that

$$\oint_C P\,dx = \iint_R -\frac{\partial P}{\partial y}\,dA.$$

By subdividing C into two parts of the type $x = g(y)$ and $x = h(y)$ (where $g(y) \leq h(y)$), we can also show that

$$\oint_C Q\,dy = \iint_R \frac{\partial Q}{\partial x}\,dA.$$

Addition of these results gives Green's theorem for this C and R.

Now consider a more general region R such as that shown in Figure 15.29, which can be decomposed into n subregions R_i, each of which satisfies the condition that lines parallel to the coordinate axes intersect its boundary in at most two points. For each subregion R_i, Green's theorem gives

$$\oint_{C_i} P\,dx + Q\,dy = \iint_{R_i} \left(\frac{\partial Q}{\partial x} - \frac{\partial P}{\partial y} \right) dA.$$

If these results are added, we get

$$\sum_{i=1}^n \oint_{C_i} P\,dx + Q\,dy = \sum_{i=1}^n \iint_{R_i} \left(\frac{\partial Q}{\partial x} - \frac{\partial P}{\partial y} \right) dA.$$

Now, R is composed of the subregions R_i; thus the right-hand side of this equation is precisely the double integral over R. Figure 15.29 illustrates that when the line integrals over the C_i are added, contributions from ancillary (interior) curves cancel in pairs, leaving the line integral around C. This completes the proof.

FIGURE 15.29

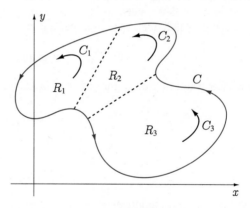

We omit a proof for even more general regions that cannot be divided into a finite number of these subregions. The interested reader should consult more advanced books.

EXAMPLE 15.15

Evaluate the line integral of Example 15.7.

SOLUTION By Green's theorem (see Figure 15.30), we have

$$\oint_C y^2\,dx + x^2\,dy = \iint_R (2x - 2y)\,dA = 2 \int_0^1 \int_0^x (x - y)\,dy\,dx$$

$$= 2 \int_0^1 \left\{ xy - \frac{y^2}{2} \right\}_0^x dx = \int_0^1 x^2\,dx = \left\{ \frac{x^3}{3} \right\}_0^1 = \frac{1}{3}.$$

FIGURE 15.30

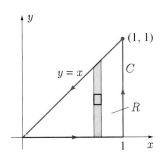

EXAMPLE 15.16

Evaluate

$$\oint_C (5x^2 + ye^x + y)\,dx + (e^x + e^y)\,dy,$$

where C is the circle $(x-1)^2 + y^2 = 1$.

SOLUTION By Green's theorem (see Figure 15.31), we have

$$\oint_C (5x^2 + ye^x + y)\,dx + (e^x + e^y)\,dy = \iint_R (e^x - e^x - 1)\,dA = -\iint_R dA$$
$$= -(\text{Area of } R) = -\pi(1)^2 = -\pi.$$

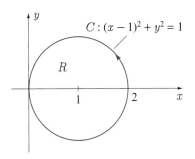

FIGURE 15.31

EXAMPLE 15.17

Show that the area of a region R is defined by each of the line integrals

$$\oint_C x\,dy = \oint_C -y\,dx = \frac{1}{2}\oint_C x\,dy - y\,dx,$$

where C is the boundary of R.

SOLUTION By Green's theorem, we have

$$\oint_C x\,dy = \iint_R 1\,dA = \text{Area of } R$$

and

$$\oint_C -y\,dx = \iint_R 1\,dA = \text{Area of } R.$$

The third expression for area is the average of these two equations. These formulas are of particular value when the curve C is defined parametrically. ∎

Green's theorem cannot be used to evaluate a line integral around a closed curve that contains a point at which either $P(x,y)$ or $Q(x,y)$ fails to have continuous first partial derivatives. For example, it cannot be used to evaluate the line integral in Exercise 19(a) of Section 15.4 because neither P nor Q is defined at the origin. Try it and compare the result to the correct answer (2π). A generalization of Green's theorem that can be useful in such situations is found in Exercise 30.

EXERCISES 15.6

In Exercises 1–11 use Green's theorem (if possible) to evaluate the line integral.

1. $\oint_C y^2\,dx + x^2\,dy$ where C is the circle $x^2 + y^2 = 1$

2. $\oint_C (x^2 + 2y^2)\,dy$ where C is the curve $(x-2)^2 + y^2 = 1$

3. $\oint_C x^2 e^y\,dx + (x+y)\,dy$ where C is the square with vertices $(\pm 1, 1)$ and $(\pm 1, -1)$

4. $\oint_C xy^3\,dx + x^2\,dy$ where C is the curve enclosing the area bounded by $x = \sqrt{1+y^2}$, $x = 2$

5. $\oint_C (x^3 + y^3)\,dx + (x^3 - y^3)\,dy$ where C is the curve enclosing the area bounded by $x = y^2 - 1$, $x = 1 - y^2$

6. $\oint_C 2\operatorname{Tan}^{-1}(y/x)\,dx + \ln(x^2 + y^2)\,dy$ where C is the circle $(x-4)^2 + (y-1)^2 = 2$

7. $\oint_C (3x^2y^3 + y)\,dx + (3x^3y^2 + 2x)\,dy$ where C is the boundary of the area enclosed by $x + y = 1$, $x = -1, y = -1$

8. $\oint_C (x^3 + y^3)\,dx + (x^3 - y^3)\,dy$ where C is the curve $2|x| + |y| = 1$

9. $\oint_C (x^2y^2 + 3x)\,dx + (2xy - y)\,dy$ where C is the boundary of the area enclosed by $x = 1 - y^2$ ($x \geq 0$), $y = x + 1$, $y + x + 1 = 0$

10. $\oint_C (xy^2 + 2x)\,dx + (x^2y + y + x^2)\,dy$ where C is the boundary of the area enclosed by $y^2 - x^2 = 4$, $x = 0, x = 3$

11. $\oint_C \dfrac{-y\,dx + x\,dy}{x^2 + y^2}$ where C is the circle $x^2 + y^2 = 1$

12. Show that Green's theorem can be expressed vectorially in the form

$$\oint_C \mathbf{F}\cdot d\mathbf{r} = \iint_R (\nabla \times \mathbf{F})\cdot \hat{\mathbf{k}}\,dA.$$

13. If a curve C is traced out in the direction defined by Green's theorem, it can be shown that a normal vector to C that always points to the outside of C is $\mathbf{n} = (dy, -dx)$. Show that Green's theorem can be written vectorially in the form

$$\oint_C \mathbf{F} \cdot \hat{\mathbf{n}}\, ds = \iint_R \nabla \cdot \mathbf{F}\, dA.$$

In Exercises 14–19 use the results of Example 15.17 to find the area enclosed by the curve.

14. $x^2/a^2 + y^2/b^2 = 1$

15. The strophoid $x = (1-t^2)/(1+t^2)$, $y = (t-t^3)/(1+t^2)$ (see Example 10.4)

16. The astroid $x = \cos^3\theta$, $y = \sin^3\theta$ (see Exercise 50 in Section 10.1)

17. The right loop of the curve of Lissajous (see Exercise 48 in Section 10.1)

18. The deltoid $x = 2\cos t + \cos 2t$, $y = 2\sin t - \sin 2t$

19. The droplet $x = 2\cos t - \sin 2t$, $y = \sin t$

20. **(a)** If C is the straight-line segment joining points $P_1(x_1, y_1)$ and $P_2(x_2, y_2)$ in the xy-plane, show that

$$\int_C x\, dy - y\, dx = x_1 y_2 - x_2 y_1.$$

(b) Let $P_1(x_1, y_1), P_2(x_2, y_2), \cdots, P_n(x_n, y_n)$ be the coordinates of a polygon labeled in the counterclockwise direction (Figure 15.32(a)). Use the result in (a) and Example 15.17 to derive the following formula for the area A of the polygon:

$$A = \frac{1}{2}\Big[(x_1 y_2 - x_2 y_1) + (x_2 y_3 - x_3 y_2) + \cdots$$
$$+ (x_{n-1} y_n - x_n y_{n-1}) + (x_n y_1 - x_1 y_n)\Big].$$

(c) Show that if the coordinates of the vertices are arranged in a vertical column with (x_1, y_1) repeated at the bottom (Figure 15.32(b)), then

$$A = \frac{1}{2}\Big\{[\text{sum of downward products to the right } (\searrow)]$$
$$- [\text{sum of downward products to the left } (\swarrow)]\Big\}.$$

FIGURE 15.32

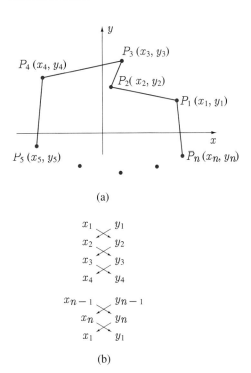

In Exercises 21–24 use the result of Exercise 20 to find the area of the polygon with the points as successive vertices.

21. $(1, 2), (-3, 2), (4, 1)$

22. $(2, -2), (1, -3), (-2, 1), (5, 6)$

23. $(3, 0), (1, 1), (2, 5), (-4, -4)$

24. $(0, 4), (-1, 0), (-2, 0), (-3, -4), (0, -5), (6, -2), (3, 0), (2, 2)$

In Exercises 25–29 evaluate the line integral.

25. $\oint_C (2xye^{x^2 y} + 3x^2 y)\, dx + x^2 e^{x^2 y}\, dy$ where C is the ellipse $x^2 + 4y^2 = 4$

26. $\oint_C (3x^2 y^3 - x^2 y)\, dx + (xy^2 + 3x^3 y^2)\, dy$ where C is the circle $x^2 + y^2 = 9$

27. $\oint_C -x^3 y^2\, dx + x^2 y^3\, dy$ where C is the right loop of $(x^2 + y^2)^{3/2} = x^2 - y^2$

28. $\int_C (x - y)(dx + dy)$ where C is the semicircular part of $x^2 + y^2 = 4$ above $y = x$ from $(-\sqrt{2}, -\sqrt{2})$ to $(\sqrt{2}, \sqrt{2})$

29. $\int_C (e^y - y\sin x)\,dx + (\cos x + xe^y)\,dy$ where C is the curve $x = 1 - y^2$ from $(0, -1)$ to $(0, 1)$

30. The result of this exercise is useful when the curve C in Green's theorem contains a point (or points) at which either P or Q fails to have continuous first partial derivatives (see Exercises 31–35).

    (a) Suppose a piecewise smooth curve C (Figure 15.33a) contains in its interior another piecewise smooth curve C', and that $P(x, y)$ and $Q(x, y)$ have continuous first partial derivatives in a domain containing C and C' and the area R between them. Prove that

    $$\oint_C P\,dx + Q\,dy + \oint_{C'} P\,dx + Q\,dy$$
    $$= \iint_R \left(\frac{\partial Q}{\partial x} - \frac{\partial P}{\partial y}\right) dA.$$

    Hint: Join C and C' by two curves such as those in Figure 15.33(a).

    (b) Extend this result to show that when C' is replaced by n distinct curves C_i (Figure 15.33b), and P and Q have continuous first partial derivatives in a domain containing C and the C_i and the area R between them,

    $$\oint_C P\,dx + Q\,dy + \oint_{C_1} P\,dx + Q\,dy +$$
    $$\cdots + \oint_{C_n} P\,dx + Q\,dy$$
    $$= \iint_R \left(\frac{\partial Q}{\partial x} - \frac{\partial P}{\partial y}\right) dA.$$

    (c) What can we conclude in (a) and (b) if $\partial Q/\partial x = \partial P/\partial y$ in R?

FIGURE 15.33

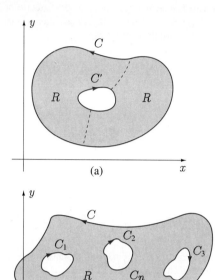

(a)

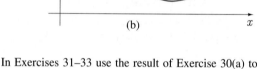

(b)

In Exercises 31–33 use the result of Exercise 30(a) to evaluate the line integral.

31. $\oint_C \dfrac{y\,dx - (x-1)\,dy}{(x-1)^2 + y^2}$ where C is the circle $x^2 + y^2 = 4$.

32. $\oint_C \dfrac{-x^2 y\,dx + x^3\,dy}{(x^2 + y^2)^2}$ where C is the ellipse $4x^2 + y^2 = 1$

33. $\oint_C \dfrac{-y\,dx + x\,dy}{x^2 + y^2}$ where C is the square with vertices $(\pm 2, 0)$ and $(0, \pm 2)$

34. Show that the line integral of Exercise 19 in Section 4 has value $\pm 2\pi$ for every piecewise smooth, closed curve enclosing the origin that does not intersect itself.

35. (a) In what domains is the line integral $\int_C \dfrac{x\,dx + y\,dy}{x^2 + y^2}$ independent of path?

    (b) Evaluate the integral clockwise around the curve $x^2 + y^2 - 2y = 1$.

In Exercises 36–38 assume that $P(x, y)$ and $Q(x, y)$ have continuous second partial derivatives in a domain containing R and C (Figure 15.34). Let the vector $\hat{\mathbf{n}} = (dy/ds, -dx/ds)$ be the outward pointing normal to C, and let

$$\frac{\partial P}{\partial n} = \nabla P \cdot \hat{\mathbf{n}}, \qquad \frac{\partial Q}{\partial n} = \nabla Q \cdot \hat{\mathbf{n}}$$

be the directional derivatives of P and Q in the direction $\hat{\mathbf{n}}$.

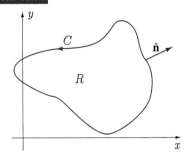

FIGURE 15.34

36. Show that
$$\oint_C \frac{\partial P}{\partial n} ds = \iint_R \nabla^2 P \, dA$$
where $\nabla^2 P = \frac{\partial^2 P}{\partial x^2} + \frac{\partial^2 P}{\partial y^2}$. Hint: See Exercise 13.
What can we conclude if $P(x,y)$ satisfies Laplace's equation in R?

37. Show that
$$\oint_C P \frac{\partial Q}{\partial n} ds = \iint_R P \nabla^2 Q \, dA + \iint_R \nabla P \cdot \nabla Q \, dA.$$
Hint: Use identity 15.11. This result is often called Green's first identity (in the plane).

38. Prove that
$$\oint_C \left(P \frac{\partial Q}{\partial n} - Q \frac{\partial P}{\partial n} \right) ds = \iint_R (P \nabla^2 Q - Q \nabla^2 P) \, dA.$$
This is often called Green's second identity (in the plane).

39. Find all possible values for the line integral
$$\oint_C \frac{-y \, dx + x \, dy}{x^2 + y^2}$$
for curves in the xy-plane not passing through the origin.

SECTION 15.7

Surface Integrals

Consider a function $f(x, y, z)$ defined on some surface S (Figure 15.35). We divide S into n subsurfaces of areas $\Delta S_1, \Delta S_2, \ldots, \Delta S_n$ in any manner whatsoever. On each subsurface $\Delta S_i (i = 1, \ldots, n)$ we choose an arbitrary point (x_i^*, y_i^*, z_i^*), and form the sum

$$f(x_1^*, y_1^*, z_1^*) \Delta S_1 + f(x_2^*, y_2^*, z_2^*) \Delta S_2 + \cdots + f(x_n^*, y_n^*, z_n^*) \Delta S_n$$

$$= \sum_{i=1}^n f(x_i^*, y_i^*, z_i^*) \Delta S_i. \quad (15.45)$$

If this sum approaches a limit as the number of subsurfaces becomes increasingly large and every subsurface shrinks to a point, we call the limit the surface integral of $f(x, y, z)$ over the surface S, and denote it by

$$\iint_S f(x, y, z) \, dS = \lim_{\|\Delta S_i\| \to 0} \sum_{i=1}^n f(x_i^*, y_i^*, z_i^*) \Delta S_i. \quad (15.46)$$

FIGURE 15.35

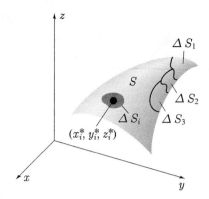

Surface integrals, like all integrals, are limit-summations. We think of dS as a small piece of area on S, and each dS is multiplied by the value of $f(x, y, z)$ for that area. All such products are then added together and the limit taken as the pieces of area shrink to points.

The following theorem guarantees the existence of the surface integral of a continuous function over a smooth surface.

> **Theorem 15.6**
>
> If a function $f(x, y, z)$ is continuous on a smooth surface S of finite area, then the surface integral of $f(x, y, z)$ over S exists. If S projects in a one-to-one fashion onto a region S_{xy} in the xy-plane, then the surface integral of $f(x, y, z)$ over S can be evaluated by means of the following double integral:
>
> $$\iint_S f(x, y, z)\, dS = \iint_{S_{xy}} f[x, y, g(x, y)] \sqrt{1 + \left(\frac{\partial z}{\partial x}\right)^2 + \left(\frac{\partial z}{\partial y}\right)^2}\, dA, \qquad (15.47)$$
>
> where $z = g(x, y)$ is the equation of S.

It is not necessary to memorize 15.47 as a formula, since it is the result obtained by expressing z and dS in terms of x and y and interpreting the result as a double integral over the projection of S in the xy-plane. Recall from Section 14.7 that when a surface S can be represented in the form $z = g(x, y)$, a small area dS on S is related to its projection dA in the xy-plane according to the formula

$$dS = \sqrt{1 + \left(\frac{\partial z}{\partial x}\right)^2 + \left(\frac{\partial z}{\partial y}\right)^2}\, dA. \qquad (15.48)$$

(Figure 15.36). If we substitute this into the left-hand side of 15.47 and at the same time use $z = g(x, y)$ to express $f(x, y, z)$ in terms of x and y, then

$$\iint_S f(x, y, z)\, dS = \iint_S f[x, y, g(x, y)] \sqrt{1 + \left(\frac{\partial z}{\partial x}\right)^2 + \left(\frac{\partial z}{\partial y}\right)^2}\, dA.$$

But if we now interpret the right-hand side of this equation as the double integral of $f[x, y, g(x, y)]\sqrt{1 + (\partial z/\partial x)^2 + (\partial z/\partial y)^2}$ over the projection S_{xy} of S onto the xy-plane, we obtain 15.47.

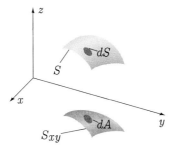

FIGURE 15.36

Note the analogy between equations 15.47 and 15.20. equation 15.20 states that the line integral on the left can be evaluated by means of the definite integral on the right. Equation 15.47 states that the surface integral on the left can be evaluated by means of the double integral on the right.

If a surface does not project one-to-one onto an area in the xy-plane, then one possibility is to subdivide it into parts, each of which projects one-to-one onto the xy-plane. The total surface integral over the surface is then the sum of the surface integrals over the parts. For example, if we require the surface integral of a function $f(x, y, z)$ over the sphere $S: x^2 + y^2 + z^2 = 1$ (Figure 15.37), then we could subdivide S into two hemispheres,

$$S_1: \quad z = \sqrt{1 - x^2 - y^2},$$
$$S_2: \quad z = -\sqrt{1 - x^2 - y^2},$$

each of which projects onto

$$S_{xy}: \quad x^2 + y^2 \leq 1.$$

Then,

$$\iint_S f(x, y, z) \, dS = \iint_{S_1} f(x, y, z) \, dS + \iint_{S_2} f(x, y, z) \, dS.$$

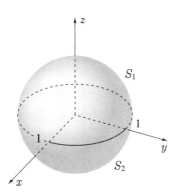

FIGURE 15.37

A second possibility is to project surfaces onto either the xz-plane or the yz-plane. If a surface projects one-to-one onto areas S_{xz} and S_{yz} in these planes, then

$$\iint_S f(x, y, z) \, dS = \iint_{S_{xz}} f[x, g(x, z), z] \sqrt{1 + \left(\frac{\partial y}{\partial x}\right)^2 + \left(\frac{\partial y}{\partial z}\right)^2} \, dA \quad (15.49)$$

and

$$\iint_S f(x, y, z) \, dS = \iint_{S_{yz}} f[g(y, z), y, z] \sqrt{1 + \left(\frac{\partial x}{\partial y}\right)^2 + \left(\frac{\partial x}{\partial z}\right)^2} \, dA. \quad (15.50)$$

We should also note that the area of a surface S is defined by the surface integral

$$\iint_S dS, \tag{15.51}$$

and if S projects one-to-one onto S_{xy}, then

$$\text{Area of } S = \iint_{S_{xy}} \sqrt{1 + \left(\frac{\partial z}{\partial x}\right)^2 + \left(\frac{\partial z}{\partial y}\right)^2} \, dA. \tag{15.52}$$

This is precisely formula 14.41.

The results in equations 15.47–15.52 were based on a smooth surface S. If S is piecewise smooth, rather than smooth, we simply subdivide S into smooth parts and apply each of these results to the smooth parts. The surface integral over S is then the summation of the surface integrals over its parts.

EXAMPLE 15.18

Evaluate $\iint_S (x + y + z) \, dS$, where S is that part of the plane $x + 2y + 4z = 8$ in the first octant.

SOLUTION The surface S projects one-to-one onto the triangle S_{xy} in the xy-plane shown in Figure 15.38. Since $z = (8 - x - 2y)/4$ on S,

$$\iint_S (x+y+z)\,dS = \iint_{S_{xy}} \left(x + y + 2 - \frac{x}{4} - \frac{y}{2}\right)\sqrt{1 + \left(\frac{\partial z}{\partial x}\right)^2 + \left(\frac{\partial z}{\partial y}\right)^2}\,dA$$

$$= \frac{1}{4}\iint_{S_{xy}} (3x + 2y + 8)\sqrt{1 + \left(-\frac{1}{4}\right)^2 + \left(-\frac{1}{2}\right)^2}\,dA$$

$$= \frac{\sqrt{21}}{16}\int_0^4 \int_0^{8-2y} (3x + 2y + 8)\,dx\,dy$$

$$= \frac{\sqrt{21}}{16}\int_0^4 \left\{\frac{3x^2}{2} + (2y + 8)x\right\}_0^{8-2y} dy$$

$$= \frac{\sqrt{21}}{8}\int_0^4 (80 - 24y + y^2)\,dy$$

$$= \frac{\sqrt{21}}{8}\left\{80y - 12y^2 + \frac{y^3}{3}\right\}_0^4 = \frac{56\sqrt{21}}{3}.$$

FIGURE 15.38

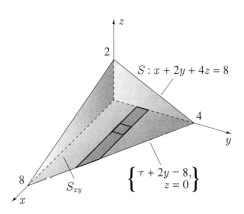

Evaluate $\iint_S z^2\, dS$ where S is the sphere $x^2 + y^2 + z^2 = 4$.

EXAMPLE 15.19

SOLUTION We divide S into two hemispheres (Figure 15.39),

$$S_1 : z = \sqrt{4 - x^2 - y^2}, \quad S_2 : z = -\sqrt{4 - x^2 - y^2}$$

each of which projects one-to-one onto the circle $S_{xy}: x^2 + y^2 \leq 4$, $z = 0$ in the xy-plane. For each hemisphere,

$$dS = \sqrt{1 + \left(\frac{\partial z}{\partial x}\right)^2 + \left(\frac{\partial z}{\partial y}\right)^2}\, dA$$

$$= \sqrt{1 + \frac{x^2}{4 - x^2 - y^2} + \frac{y^2}{4 - x^2 - y^2}}\, dA = \frac{2}{\sqrt{4 - x^2 - y^2}}\, dA,$$

and therefore

$$\iint_S z^2\, dS = \iint_{S_1} z^2\, dS + \iint_{S_2} z^2\, dS$$

$$= \iint_{S_{xy}} (4 - x^2 - y^2) \frac{2}{\sqrt{4 - x^2 - y^2}}\, dA$$

$$+ \iint_{S_{xy}} (4 - x^2 - y^2) \frac{2}{\sqrt{4 - x^2 - y^2}}\, dA$$

$$= 4 \iint_{S_{xy}} \sqrt{4 - x^2 - y^2}\, dA.$$

If we use polar coordinates to evaluate this integral over S_{xy}, then

$$\iint_S z^2\, dS = 4 \int_0^{2\pi} \int_0^2 \sqrt{4 - r^2}\, r\, dr\, d\theta$$

$$= 4 \int_0^{2\pi} \left\{-\frac{1}{3}(4 - r^2)^{3/2}\right\}_0^2 d\theta = \frac{32}{3} \{\theta\}_0^{2\pi} = \frac{64\pi}{3}.$$

FIGURE 15.39

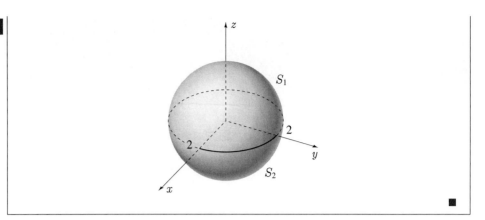

If parameters can be found to describe a surface, it may not be necessary to project the surface into one of the coordinate planes. Such is the case for a sphere centred at the origin. Formula 14.67 for the volume element in spherical coordinates indicates that an area element on the surface of a sphere of radius R can be expressed in terms of angles θ and ϕ (Figure 15.40) as

$$dS = R^2 \sin\phi \, d\phi \, d\theta. \tag{15.53}$$

With this choice of area element, the surface integral in Example 15.19 is evaluated as follows:

$$\iint_S z^2 \, dS = \int_0^{2\pi} \int_0^{\pi} (2\cos\phi)^2 4 \sin\phi \, d\phi \, d\theta = 16 \int_0^{2\pi} \int_0^{\pi} \cos^2\phi \sin\phi \, d\phi \, d\theta$$

$$= 16 \int_0^{2\pi} \left\{ -\frac{1}{3}\cos^3\phi \right\}_0^{\pi} d\theta = \frac{32}{3}\{\theta\}_0^{2\pi} = \frac{64\pi}{3}.$$

FIGURE 15.40

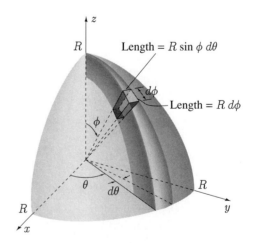

EXERCISES 15.7

In Exercises 1–8 evaluate the surface integral.

1. $\iint_S (x^2y + z)\, dS$ where S is the first octant part of
$$2x + 3y + z = 6$$

2. $\iint_S (x^2 + y^2)z\, dS$ where S is that part of $z = x + y$ cut out by $x = 0$, $y = 0$, $x + y = 1$

3. $\iint_S xyz\, dS$ where S is the surface of the cube
$$0 \le x \le 1, 0 \le y \le 1, 0 \le z \le 1$$

4. $\iint_S xy\, dS$ where S is the first octant part of
$$z = \sqrt{x^2 + y^2} \text{ cut out by } x^2 + y^2 = 1$$

5. $\iint_S \dfrac{1}{\sqrt{z - y + 1}}\, dS$ where S is the surface defined by
$$2z = x^2 + 2y,\ 0 \le x \le 1,\ 0 \le y \le 1$$

6. $\iint_S \sqrt{4y + 1}\, dS$ where S is the first octant part of
$$y = x^2 \text{ cut out by } 2x + y + z = 1$$

7. $\iint_S x^2 z\, dS$ where S is the surface $x^2 + y^2 = 1$ for
$$0 \le z \le 1$$

8. $\iint_S (x + y)\, dS$ where S is the surface bounding the volume enclosed by $x = 0$, $y = 0$, $z = 0$, $6x - 3y + 2z = 6$

9. Set up double iterated integrals for the surface integral of a function $f(x, y, z)$ over the surface defined by $z = 4 - x^2 - 4y^2$ ($x, y, z \ge 0$) if the surface is projected onto the xy-, the xz-, and the yz-planes.

10. Use a surface integral to find the area of the curved portion of a right-circular cone of radius R and height h.

In Exercises 11–17 evaluate the surface integral.

11. $\iint_S xyz^3\, dS$ where S is the surface defined by
$$x = y^2,\ 0 \le x \le 4,\ 0 \le z \le 1$$

12. $\iint_S xyz\, dS$ where S is the surface defined by
$$2y = \sqrt{9 - x},\ x \ge 0,\ 0 \le z \le 3$$

13. $\iint_S \dfrac{1}{\sqrt{2az - z^2}}\, dS$ where S is that part of $x^2 + y^2 + (z-a)^2 = a^2$ inside the cylinder $x^2 + y^2 = ay$, underneath the plane $z = a$, and in the first octant

14. $\iint_S z\, dS$ where S is that part of the surface $x^2 + y^2 - z^2 = 1$ between the planes $z = 0$ and $z = 1$

15. $\iint_S x^2 y^2\, dS$ where S is that part of $z = x^2 + y^2$ inside $x^2 + y^2 + z^2 = 2$

16. $\iint_S x^2\, dS$ where S is that part of $z = xy$ inside $x^2 + y^2 = 4$

17. $\iint_S z(y + x^2)\, dS$ where S is that part of $y = 1 - x^2$ bounded by $z = 0$, $z = 2$, and $y = 0$

In Exercises 18–22 evaluate the surface integral by projecting the surface into one of the coordinate planes and also by using area element 15.53.

18. $\iint_S dS$ where S is the sphere $x^2 + y^2 + z^2 = R^2$. Is this the formula for the area of a sphere?

19. $\iint_S x^2 z^2\, dS$ where S is the sphere $x^2 + y^2 + z^2 = 1$

20. $\iint_S (x^2 - y^2)\, dS$ where S is the hemisphere
$$z = \sqrt{9 - x^2 - y^2}$$

21. $\iint_S (x^2 + y^2)\, dS$ where S is the sphere
$$x^2 + y^2 + z^2 = R^2$$

22. $\iint_S \dfrac{1}{x^2 + y^2}\, dS$ where S is that part of the sphere $x^2 + y^2 + z^2 = 4R^2$ between the planes $z = 0$ and $z = R$

23. A viscous material is allowed to drip onto the sphere $x^2 + y^2 + z^2 = 1$ at the point $(0, 0, 1)$ (all dimensions in metres). The material spreads out evenly in all directions and runs down the sphere, becoming more and more viscous as it does so. The thickness of the material increases linearly with respect to angle ϕ (Figure 15.40) from 0.001 m at $(0, 0, 1)$ to 0.005 m at $(0, 0, -1)$. Find the volume of material on the sphere.

24. Show that if a surface S defined implicitly by the equation $F(x, y, z) = 0$ projects one-to-one onto the area S_{xy} in the xy-plane, then

$$\iint_S f(x, y, z)\, dS = \iint_{S_{xy}} f[x, y, g(x, y)] \dfrac{|\nabla F|}{|\partial F/\partial z|}\, dA.$$

25. **(a)** Find the area cut from the cones $z^2 = x^2 + y^2$ by the cylinder $x^2 + y^2 = 2x$.

(b) Find the area cut from the cylinder $x^2 + y^2 = 2x$ by the cones $z^2 = x^2 + y^2$.

SECTION 15.8

Surface Integrals Involving Vector Fields

The most important and common type of surface integral occurs when $f(x, y, z)$ in 15.46 is specified as the normal component of some given vector field $\mathbf{F}(x, y, z)$ defined on S. In other words, $f(x, y, z)$ itself is not given, but $\mathbf{F}$ is, and to find $f(x, y, z)$ we must calculate the component of $\mathbf{F}$ normal to S.

This presupposes that surfaces are two-sided and that a normal vector to a surface can be assigned in an unambiguous way. When this is possible, the surface is said to be **orientable**. All surfaces in this book are orientable, with the exception of the Möbius strip mentioned below.

Take a thin rectangular strip of paper and label its corners $A, B, C,$ and D (Figure 15.41a). Give the strip a half twist and join A and C, and B and D (Figure 15.41b).

FIGURE 15.41

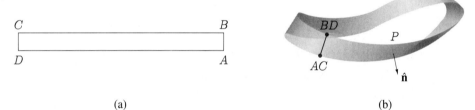

(a)   (b)

This surface, called a Möbius strip, cannot be assigned a unique normal vector which varies continuously over the surface. To illustrate, suppose at point P in Figure 15.41b, we assign a unit normal vector $\hat{\mathbf{n}}$ as shown. By moving once around the strip, we can vary the direction of $\hat{\mathbf{n}}$ continuously and arrive back at P with $\hat{\mathbf{n}}$ pointing in the opposite direction. This surface is said to have only one side, or to be nonorientable.

We consider only surfaces that are orientable, or have two sides, and can therefore be assigned a unit normal vector in an unambiguous way.

Suppose again that $\mathbf{F}(x, y, z)$ is a vector field defined on an (orientable) surface S and $f(x, y, z)$ is the component of $\mathbf{F}$ in one of the two normal directions to S. If $\hat{\mathbf{n}}$ is the unit normal vector to S in the specified direction, then $f(x, y, z) = \mathbf{F} \cdot \hat{\mathbf{n}}$, and

$$\iint_S f(x, y, z)\, dS = \iint_S \mathbf{F} \cdot \hat{\mathbf{n}}\, dS. \qquad (15.54)$$

EXAMPLE 15.20

Evaluate $\iint_S \mathbf{F} \cdot \hat{\mathbf{n}}\, dS$, where $\mathbf{F} = x^2 y \hat{\mathbf{i}} + xz \hat{\mathbf{j}}$ and $\hat{\mathbf{n}}$ is the upper normal to the surface $S : z = 4 - x^2 - y^2, z \geq 0$.

SOLUTION Since a normal vector to S is

$$\nabla(z - 4 + x^2 + y^2) = (2x, 2y, 1),$$

it follows that
$$\hat{\mathbf{n}} = \frac{(2x, 2y, 1)}{\sqrt{4x^2 + 4y^2 + 1}}.$$

Thus,
$$\mathbf{F} \cdot \hat{\mathbf{n}} = \frac{2x^3y + 2xyz}{\sqrt{4x^2 + 4y^2 + 1}}.$$

If we project S onto $S_{xy} : x^2 + y^2 \leq 4$ in the xy-plane (Figure 15.42), then

$$\iint_S \mathbf{F} \cdot \hat{\mathbf{n}} \, dS = \iint_{S_{xy}} \frac{2x^3y + 2xyz}{\sqrt{4x^2 + 4y^2 + 1}} \sqrt{1 + \left(\frac{\partial z}{\partial x}\right)^2 + \left(\frac{\partial z}{\partial y}\right)^2} \, dA$$

$$= \iint_{S_{xy}} \frac{2x^3y + 2xy(4 - x^2 - y^2)}{\sqrt{4x^2 + 4y^2 + 1}} \sqrt{1 + (-2x)^2 + (-2y)^2} \, dA$$

$$= 2 \iint_{S_{xy}} (4xy - xy^3) \, dA$$

$$= 2 \int_{-2}^{2} \int_{-\sqrt{4-x^2}}^{\sqrt{4-x^2}} (4xy - xy^3) \, dy \, dx$$

$$= 2 \int_{-2}^{2} \left\{ 2xy^2 - \frac{xy^4}{4} \right\}_{-\sqrt{4-x^2}}^{\sqrt{4-x^2}} dx$$

$$= 0.$$

FIGURE 15.42

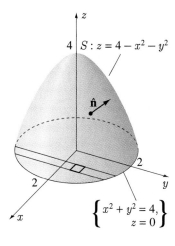

EXAMPLE 15.21

Evaluate
$$\oiint_S \mathbf{F} \cdot \hat{\mathbf{n}} \, dS,$$
where $\mathbf{F} = x\hat{\mathbf{i}} + y\hat{\mathbf{j}} + z\hat{\mathbf{k}}$ and $\hat{\mathbf{n}}$ is the unit outward-pointing normal to the surface S

enclosing the volume bounded by $x^2 + y^2 = 4, z = 0, z = 2$.

SOLUTION We can divide S into four parts (Figure 15.43):

$$S_1: \quad z = 0, \quad x^2 + y^2 \leq 4;$$
$$S_2: \quad z = 2, \quad x^2 + y^2 \leq 4;$$
$$S_3: \quad y = \sqrt{4-x^2}, \quad 0 \leq z \leq 2;$$
$$S_4: \quad y = -\sqrt{4-x^2}, \quad 0 \leq z \leq 2.$$

FIGURE 15.43

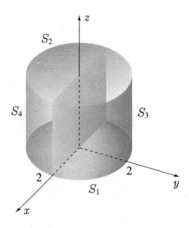

On $S_1, \hat{\mathbf{n}} = -\hat{\mathbf{k}}$; on $S_2, \hat{\mathbf{n}} = \hat{\mathbf{k}}$; and on S_3 and S_4,

$$\hat{\mathbf{n}} = \frac{\nabla(x^2 + y^2 - 4)}{|\nabla(x^2 + y^2 - 4)|} = \frac{(2x, 2y, 0)}{\sqrt{4x^2 + 4y^2}} = \frac{(x, y, 0)}{2}.$$

We know that S_1 and S_2 project onto $S_{xy} : x^2 + y^2 \leq 4$ in the xy-plane, and S_3 and S_4 project onto the rectangle $S_{xz} : -2 \leq x \leq 2, 0 \leq z \leq 2$ in the xz-plane. Consequently,

$$\oiint_S \mathbf{F} \cdot \hat{\mathbf{n}} \, dS = \iint_{S_1} -z \, dS + \iint_{S_2} z \, dS + \iint_{S_3} \left(\frac{x^2 + y^2}{2}\right) dS + \iint_{S_4} \left(\frac{x^2 + y^2}{2}\right) dS$$

$$= 0 + \iint_{S_{xy}} 2\sqrt{1} \, dA + \frac{1}{2} \iint_{S_{xz}} (x^2 + 4 - x^2) \sqrt{1 + \left(\frac{-x}{\sqrt{4-x^2}}\right)^2} \, dA$$

$$+ \frac{1}{2} \iint_{S_{xz}} (x^2 + 4 - x^2) \sqrt{1 + \left(\frac{x}{\sqrt{4-x^2}}\right)^2} \, dA$$

$$= 2 \iint_{S_{xy}} dA + 4 \iint_{S_{xz}} \frac{2}{\sqrt{4-x^2}} \, dA$$

$$= 2(\text{Area of } S_{xy}) + 8 \int_{-2}^{2} \int_{0}^{2} \frac{1}{\sqrt{4-x^2}} \, dz \, dx$$

$$= 2(4\pi) + 16 \int_{-2}^{2} \frac{1}{\sqrt{4-x^2}} \, dx.$$

If we set $x = 2 \sin \theta$, then $dx = 2 \cos \theta \, d\theta$ and

$$\oiint_S \mathbf{F} \cdot \hat{\mathbf{n}} \, dS = 8\pi + 16 \int_{-\pi/2}^{\pi/2} \frac{1}{2\cos\theta} 2\cos\theta \, d\theta = 8\pi + 16\left\{\theta\right\}_{-\pi/2}^{\pi/2} = 24\pi.$$

∎

Note that the notation $\oiint$ is similar to that for the line integrals for the surface integral over a closed surface (a closed surface being one that encloses a volume).

EXAMPLE 15.22

If a spherical object is submerged in a fluid with density ρ, it experiences a buoyant force due to fluid pressure. Show that the magnitude of this force is exactly the weight of the fluid displaced by the object.

SOLUTION Suppose the object is represented by the sphere $x^2 + y^2 + z^2 = R^2$ and the surface of the fluid by the plane $z = h$ ($h > R$) (Figure 15.44). If dS is a small area on the surface, then the force due to fluid pressure on dS has magnitude $P \, dS$, where P is pressure, and this force acts normal to the surface of the sphere. If $\hat{\mathbf{n}}$ is the unit inward-pointing normal to the sphere, then the force on dS is

$$(P \, dS)\hat{\mathbf{n}}.$$

Clearly, the resultant force on the sphere will be in the z-direction, the x- and y-components canceling because of the symmetry of the sphere. The z-component of the force on dS is $(P \, dS)\hat{\mathbf{n}} \cdot \hat{\mathbf{k}}$; thus the magnitude of the resultant force on the sphere is

$$\oiint_S P\hat{\mathbf{k}} \cdot \hat{\mathbf{n}} \, dS.$$

A normal to the surface is $\nabla(x^2 + y^2 + z^2 - R^2) = (2x, 2y, 2z)$, and therefore the unit inward-pointing normal is

$$\hat{\mathbf{n}} = \frac{-(x,y,z)}{\sqrt{x^2+y^2+z^2}} = -\frac{1}{R}(x,y,z).$$

With formula 15.53 for an area element on the sphere, and $P = 9.81\rho(h-z)$,

$$\oiint_S P\hat{\mathbf{k}} \cdot \hat{\mathbf{n}} \, dS = \oiint_S 9.81\rho(h-z)\left(-\frac{z}{R}\right) dS$$

$$= -\frac{9.81\rho}{R} \int_0^{2\pi} \int_0^{\pi} (h - R\cos\phi)(R\cos\phi) R^2 \sin\phi \, d\phi \, d\theta$$

$$= -9.81\rho R^2 \int_0^{2\pi} \left\{-\frac{h}{2}\cos^2\phi + \frac{R}{3}\cos^3\phi\right\}_0^{\pi} d\theta$$

$$= \frac{2(9.81)\rho R^3}{3} \{\theta\}_0^{2\pi}$$

$$= \frac{4}{3}\pi R^3 (9.81\rho).$$

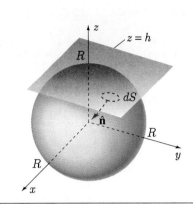

FIGURE 15.44

EXERCISES 15.8

In Exercises 1–16 evaluate the surface integral.

1. $\iint_S (x\hat{\mathbf{i}} + z\hat{\mathbf{k}}) \cdot \hat{\mathbf{n}}\, dS$ where S is the first octant part of $x + y + z = 3$, and $\hat{\mathbf{n}}$ is the unit normal to S with positive z-component

2. $\iint_S (yz^2\hat{\mathbf{i}} + ye^x\hat{\mathbf{j}} + x\hat{\mathbf{k}}) \cdot \hat{\mathbf{n}}\, dS$ where S is defined by $y = x^2$, $0 \le y \le 4$, $0 \le z \le 1$, and $\hat{\mathbf{n}}$ is the unit normal to S with positive y-component

3. $\iint_S (x\hat{\mathbf{i}} + y\hat{\mathbf{j}} + z\hat{\mathbf{k}}) \cdot \hat{\mathbf{n}}\, dS$ where S is the hemisphere $z = \sqrt{1 - x^2 - y^2}$, and $\hat{\mathbf{n}}$ is its upper normal

4. $\iint_S (yz\hat{\mathbf{i}} + zx\hat{\mathbf{j}} + xy\hat{\mathbf{k}}) \cdot \hat{\mathbf{n}}\, dS$ where S is that part of the surface $z = x^2 + y^2$ cut out by the planes $x = 1$, $x = -1$, $y = 1$, $y = -1$, and $\hat{\mathbf{n}}$ is the unit lower normal to S

5. $\oiint_S (z\hat{\mathbf{i}} - x\hat{\mathbf{j}} + y\hat{\mathbf{k}}) \cdot \hat{\mathbf{n}}\, dS$ where S is the surface enclosing the volume defined by $z = \sqrt{4 - x^2 - y^2}$, $z = 0$, and $\hat{\mathbf{n}}$ is the unit outer normal to S

6. $\iint_S (x\hat{\mathbf{i}} + y\hat{\mathbf{j}}) \cdot \hat{\mathbf{n}}\, dS$ where S is that part of the surface $z = \sqrt{x^2 + y^2}$ below $z = 1$, and $\hat{\mathbf{n}}$ is the unit normal to S with negative z-component

7. $\iint_S (xyz\hat{\mathbf{i}} - x\hat{\mathbf{j}} + z\hat{\mathbf{k}}) \cdot \hat{\mathbf{n}}\, dS$ where S is the smaller part of $x^2 + y^2 = 9$ cut out by $z = 0$, $z = 2$, $y = |x|$, and $\hat{\mathbf{n}}$ is the unit normal to S with positive y-component

8. $\iint_S (x^2 y\hat{\mathbf{i}} + xy\hat{\mathbf{j}} + z\hat{\mathbf{k}}) \cdot \hat{\mathbf{n}}\, dS$ where S is defined by $z = 2 - x^2 - y^2$, $z \ge 0$, and $\hat{\mathbf{n}}$ is the unit normal to S with negative z-component

9. $\oiint_S (yz\hat{\mathbf{i}} + xz\hat{\mathbf{j}} + xy\hat{\mathbf{k}}) \cdot \hat{\mathbf{n}}\, dS$ where S is the surface enclosing the volume defined by $x = 0$, $x = 2$, $z = 0$, $z = y$, $y + z = 2$, and $\hat{\mathbf{n}}$ is the unit outer normal to S

10. $\iint_S (x\hat{\mathbf{i}} + y\hat{\mathbf{j}}) \cdot \hat{\mathbf{n}}\, dS$ where S is the surface $x^2 + y^2 + z^2 = 4$, $z \ge 1$, and $\hat{\mathbf{n}}$ is the unit upper normal to S

11. $\oiint_S (x^2\hat{\mathbf{i}} + y^2\hat{\mathbf{j}} + z^2\hat{\mathbf{k}}) \cdot \hat{\mathbf{n}}\, dS$ where S is the sphere $x^2 + y^2 + z^2 = a^2$, and $\hat{\mathbf{n}}$ is the unit outer normal to S

12. $\iint_S (y\hat{\mathbf{i}} - x\hat{\mathbf{j}} + \hat{\mathbf{k}}) \cdot \hat{\mathbf{n}}\, dS$ where S is the smaller surface cut from the sphere $x^2 + y^2 + z^2 = 1$ by the plane $y + z = 1$, and $\hat{\mathbf{n}}$ is the unit upper normal to S

13. $\oiint_S \mathbf{F} \cdot \hat{\mathbf{n}}\, dS$ where $\mathbf{F} = (z^2 - x)\hat{\mathbf{i}} - xy\hat{\mathbf{j}} + 3z\hat{\mathbf{k}}$, S is the surface enclosing the volume defined by $z = 4 - y^2$, $x = 0$, $x = 3$, $z = 0$, and $\hat{\mathbf{n}}$ is the unit outer normal to S

14. $\iint_S (x^2\hat{\mathbf{i}} + xy\hat{\mathbf{j}} + xz\hat{\mathbf{k}}) \cdot \hat{\mathbf{n}}\, dS$ where S is that part of the surface $z = \sqrt{4 + y^2 - x^2}$ in the first octant cut out by the planes $y = 0$, $y = 1$, $x = 0$, $z = 0$, and $\hat{\mathbf{n}}$ is the unit normal to S with positive z-component.

15. $\iint_S (x^2\hat{\mathbf{i}} + yz\hat{\mathbf{j}} - x\hat{\mathbf{k}}) \cdot \hat{\mathbf{n}}\, dS$ where S is that part of the surface $x = yz$ in the first octant cut out by $y^2 + z^2 = 1$, and $\hat{\mathbf{n}}$ is the unit normal to S with positive x-component

16. $\oiint_S (yx\hat{\mathbf{i}} + y^2\hat{\mathbf{j}} + yz\hat{\mathbf{k}}) \cdot \hat{\mathbf{n}}\, dS$ where S is the ellipsoid $x^2 + y^2/4 + z^2 = 1$, and $\hat{\mathbf{n}}$ is the unit outer normal to S

17. Show that if a surface S projects one-to-one onto an area S_{xy} in the xy-plane, then

$$\iint_S (P\hat{\mathbf{i}} + Q\hat{\mathbf{j}} + R\hat{\mathbf{k}}) \cdot \hat{\mathbf{n}}\, dS = \pm \iint_{S_{xy}} \left(-P\frac{\partial z}{\partial x} - Q\frac{\partial z}{\partial y} + R\right) dA$$

the $\pm$ depending on whether $\hat{\mathbf{n}}$ is the upper or lower normal to S. What are corresponding formulas when S projects one-to-one onto areas S_{yz} and S_{xz} in the yz- and xz-coordinate planes?

18. Evaluate $\iint_S (y\hat{\mathbf{i}} - x\hat{\mathbf{j}} + z\hat{\mathbf{k}}) \cdot \hat{\mathbf{n}}\, dS$ where
   (a) S is that part of $z = 9 - x^2 - y^2$ cut out by $z = 2y$,
   (b) S is that part of $z = 2y$ cut out by $z = 9 - x^2 - y^2$, and $\hat{\mathbf{n}}$ is the unit upper normal to S in each case. Hint: Use polar coordinates with pole at $(0, -1)$.

19. Show that if a surface S, defined implicitly by the equation $G(x, y, z) = 0$, projects one-to-one onto the area S_{xy} in the xy-plane, then

$$\iint_S \mathbf{F} \cdot \hat{\mathbf{n}}\, dS = \pm \iint_{S_{xy}} \frac{\mathbf{F} \cdot \nabla G}{|\partial G/\partial z|}\, dA.$$

20. A circular tube $S: x^2 + z^2 = 1,\ 0 \leq y \leq 2$ is a model for a part of an artery. Blood flows through the artery and the force per unit area at any point on the arterial wall is given by

$$\mathbf{F} = e^{-y}\hat{\mathbf{n}} + \frac{1}{y^2 + 1}\hat{\mathbf{j}},$$

where $\hat{\mathbf{n}}$ is the unit outer normal to the arterial wall. Blood diffuses through the wall in such a way that if dS is a small area on S, the amount of diffusion through dS in one second is $\mathbf{F} \cdot \hat{\mathbf{n}}\, dS$. Find the total amount of blood leaving the entire wall per second.

21. A beam of light travelling in the positive y-direction has circular cross-section $x^2 + z^2 \leq a^2$. It strikes a surface $S: x^2 + y^2 + z^2 = a^2,\ y \geq a/2$. The intensity of the beam is given by

$$\mathbf{I} = \frac{e^{-t}}{y^2}\hat{\mathbf{j}}, \qquad \text{where } t \text{ is time.}$$

The absorption of light by a small area dS on S in time dt is $\mathbf{I} \cdot \hat{\mathbf{n}}\, dS\, dt$ where $\hat{\mathbf{n}}$ is the unit normal to S at dS.
   (a) Find the total absorption over S in time dt.
   (b) Find the total absorption over S from time $t = 0$ to time $t = 5$.

SECTION 15.9

The Divergence Theorem

In this section and in Section 15.10 we show that relationships may exist between line integrals and surface integrals and between surface integrals and triple integrals. In Section 15.11 we indicate how useful these relationships can be in the fields of electromagnetic theory and fluid dynamics.

The divergence theorem relates certain surface integrals over surfaces that enclose volumes to triple integrals over the enclosed volume. More precisely, we have the following.

Theorem 15.7

(Divergence theorem) Let S be a piecewise-smooth surface enclosing a region V (Figure 15.45). Let $\mathbf{F}(x,y,z) = L(x,y,z)\hat{\mathbf{i}} + M(x,y,z)\hat{\mathbf{j}} + N(x,y,z)\hat{\mathbf{k}}$ be a vector field whose components L, M, and N have continuous first partial derivative in a domain containing S and V. If $\hat{\mathbf{n}}$ is the unit outer normal to S, then

$$\oiint_S \mathbf{F} \cdot \hat{\mathbf{n}}\, dS = \iiint_V \nabla \cdot \mathbf{F}\, dV, \tag{15.55a}$$

or

$$\oiint_S (L\hat{\mathbf{i}} + M\hat{\mathbf{j}} + N\hat{\mathbf{k}}) \cdot \hat{\mathbf{n}}\, dS = \iiint_V \left(\frac{\partial L}{\partial x} + \frac{\partial M}{\partial y} + \frac{\partial N}{\partial z} \right) dV. \tag{15.55b}$$

FIGURE 15.45

FIGURE 15.46

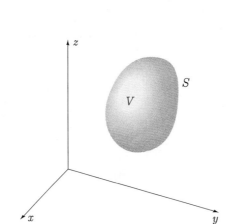

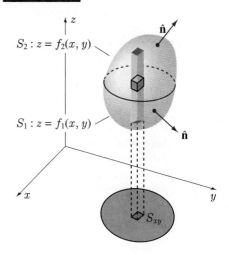

Proof We consider first of all a surface S for which any line parallel to any coordinate axis intersects S in at most two points (Figure 15.46). We can then divide S into an upper and a lower portion, $S_2 : z = f_2(x,y)$ and $S_1 : z = f_1(x,y)$, both of which have the same projection S_{xy} in the xy-plane. We consider the third term in the surface integral on the left-hand side of equation (15.55b):

$$\oiint_S N\hat{\mathbf{k}} \cdot \hat{\mathbf{n}}\, dS = \iint_{S_1} N\hat{\mathbf{k}} \cdot \hat{\mathbf{n}}\, dS + \iint_{S_2} N\hat{\mathbf{k}} \cdot \hat{\mathbf{n}}\, dS.$$

On S_1,

$$\hat{\mathbf{n}} = \frac{\left(\dfrac{\partial f_1}{\partial x}, \dfrac{\partial f_1}{\partial y}, -1 \right)}{\sqrt{1 + \left(\dfrac{\partial f_1}{\partial x} \right)^2 + \left(\dfrac{\partial f_1}{\partial y} \right)^2}},$$

and on S_2,
$$\hat{\mathbf{n}} = \frac{-\left(\frac{\partial f_2}{\partial x}, \frac{\partial f_2}{\partial y}, -1\right)}{\sqrt{1 + \left(\frac{\partial f_2}{\partial x}\right)^2 + \left(\frac{\partial f_2}{\partial y}\right)^2}}.$$

Consequently,
$$\oiint_S N\hat{\mathbf{k}} \cdot \hat{\mathbf{n}} \, dS = \iint_{S_1} \frac{-N}{\sqrt{1 + \left(\frac{\partial f_1}{\partial x}\right)^2 + \left(\frac{\partial f_1}{\partial y}\right)^2}} \, dS + \iint_{S_2} \frac{N}{\sqrt{1 + \left(\frac{\partial f_2}{\partial x}\right)^2 + \left(\frac{\partial f_2}{\partial y}\right)^2}} \, dS$$

$$= \iint_{S_{xy}} \frac{-N[x,y,f_1(x,y)]}{\sqrt{1 + \left(\frac{\partial f_1}{\partial x}\right)^2 + \left(\frac{\partial f_1}{\partial y}\right)^2}} \sqrt{1 + \left(\frac{\partial f_1}{\partial x}\right)^2 + \left(\frac{\partial f_1}{\partial y}\right)^2} \, dA$$

$$+ \iint_{S_{xy}} \frac{N[x,y,f_2(x,y)]}{\sqrt{1 + \left(\frac{\partial f_2}{\partial x}\right)^2 + \left(\frac{\partial f_2}{\partial y}\right)^2}} \sqrt{1 + \left(\frac{\partial f_2}{\partial x}\right)^2 + \left(\frac{\partial f_2}{\partial y}\right)^2} \, dA$$

$$= \iint_{S_{xy}} \{N[x,y,f_2(x,y)] - N[x,y,f_1(x,y)]\} \, dA.$$

On the other hand,
$$\iiint_V \frac{\partial N}{\partial z} dV = \iint_{S_{xy}} \left\{ \int_{f_1(x,y)}^{f_2(x,y)} \frac{\partial N}{\partial z} dz \right\} dA = \iint_{S_{xy}} \left\{ N \right\}_{f_1(x,y)}^{f_2(x,y)} dA$$

$$= \iint_{S_{xy}} \{N[x,y,f_2(x,y)] - N[x,y,f_1(x,y)]\} \, dA.$$

We have shown then that
$$\oiint_S N\hat{\mathbf{k}} \cdot \hat{\mathbf{n}} \, dS = \iiint_V \frac{\partial N}{\partial z} dV.$$

Projections of S onto the xz- and yz-planes lead in a similar way to
$$\oiint_S M\hat{\mathbf{j}} \cdot \hat{\mathbf{n}} \, dS = \iiint_V \frac{\partial M}{\partial y} dV \quad \text{and} \quad \oiint_S L\hat{\mathbf{i}} \cdot \hat{\mathbf{n}} \, dS = \iiint_V \frac{\partial L}{\partial x} dV.$$

By adding these three results, we obtain the divergence theorem for $\mathbf{F}$ and S.

The proof can be extended to general surfaces for which lines parallel to the coordinate axes intersect the surfaces in more than two points. Indeed, most volumes V bounded by surfaces S can be subdivided into n subvolumes V_i whose bounding surfaces S_i do satisfy this condition (Figure 15.47). For each such subvolume the divergence theorem is now known to apply:
$$\oiint_{S_i} \mathbf{F} \cdot \hat{\mathbf{n}} \, dS = \iiint_{V_i} \nabla \cdot \mathbf{F} \, dV, \quad i = 1, \ldots, n.$$

If these n equations are then added together, we have
$$\sum_{i=1}^{n} \oiint_{S_i} \mathbf{F} \cdot \hat{\mathbf{n}} \, dS = \sum_{i=1}^{n} \iiint_{V_i} \nabla \cdot \mathbf{F} \, dV.$$

The right side is precisely the triple integral of $\nabla \cdot \mathbf{F}$ over V since the V_i constitute V. Figure 15.47 illustrates that when the surface integrals over the S_i are added, contributions from the auxiliary (interior) surfaces cancel in pairs, and the remaining surface integrals add to give the surface integral of $\mathbf{F} \cdot \hat{\mathbf{n}}$ over S. This completes the proof.

FIGURE 15.47

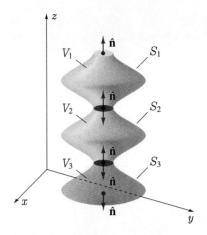

We omit a proof for even more general surfaces that cannot be subdivided into a finite number of subsurfaces of this type. The interested reader should consult more advanced books.

EXAMPLE 15.23

Verify the divergence theorem for $\mathbf{F} = x^2\hat{\mathbf{i}} + yz\hat{\mathbf{j}} + x\hat{\mathbf{k}}$ if V is the volume bounded by the surfaces $x + y + z = 1$, $x = 0$, $y = 0$, $z = 0$.

SOLUTION The surface S bounding V must be subdivided into four parts:

$$S_1: z = 0, \quad S_2: x = 0, \quad S_3: y = 0, \quad S_4: x + y + z = 1,$$

as shown in Figure 15.48. The surface integral of $\mathbf{F} \cdot \hat{\mathbf{n}}$ over S is then the sum of the surface integrals over these four parts:

$$\iint_{S_1} \mathbf{F} \cdot \hat{\mathbf{n}} \, dS = \iint_{S_1} \mathbf{F} \cdot (-\hat{\mathbf{k}}) \, dS = \iint_{S_1} -x \, dS$$

$$= -\iint_{S_{xy}} x\sqrt{1} \, dA$$

$$= -\int_0^1 \int_0^{1-x} x \, dy \, dx = -\int_0^1 \left\{ xy \right\}_0^{1-x} dx$$

$$= -\int_0^1 x(1-x) \, dx$$

$$= -\left\{ \frac{x^2}{2} - \frac{x^3}{3} \right\}_0^1$$

$$= -\frac{1}{6};$$

$$\iint_{S_2} \mathbf{F} \cdot \hat{\mathbf{n}}\,dS = \iint_{S_2} \mathbf{F} \cdot (-\hat{\mathbf{i}})\,dS = \iint_{S_2} -x^2\,dS = 0;$$

$$\iint_{S_3} \mathbf{F} \cdot \hat{\mathbf{n}}\,dS = \iint_{S_3} \mathbf{F} \cdot (-\hat{\mathbf{j}})\,dS = \iint_{S_3} -yz\,dS = 0;$$

$$\iint_{S_4} \mathbf{F} \cdot \hat{\mathbf{n}}\,dS = \iint_{S_4} \mathbf{F} \cdot \frac{(1,1,1)}{\sqrt{3}}\,dS = \frac{1}{\sqrt{3}} \iint_{S_4} (x^2 + yz + x)\,dS$$

$$= \frac{1}{\sqrt{3}} \iint_{S_{xy}} \{x^2 + y(1-x-y) + x\}\sqrt{1 + (-1)^2 + (-1)^2}\,dA$$

$$= \int_0^1 \int_0^{1-x} \{(x^2 + x) + y(1-x) - y^2\}\,dy\,dx$$

$$= \int_0^1 \left\{ (x^2 + x)y + (1-x)\frac{y^2}{2} - \frac{y^3}{3} \right\}_0^{1-x} dx$$

$$= \int_0^1 \left\{ x - x^3 + \frac{1}{6}(1-x)^3 \right\} dx$$

$$= \left\{ \frac{x^2}{2} - \frac{x^4}{4} - \frac{1}{24}(1-x)^4 \right\}_0^1$$

$$= \frac{7}{24}.$$

Adding these results gives us

$$\oiint_S \mathbf{F} \cdot \hat{\mathbf{n}}\,dS = -\frac{1}{6} + 0 + 0 + \frac{7}{24} = \frac{1}{8}.$$

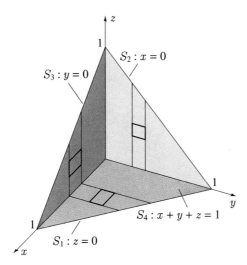

FIGURE 15.48

Alternatively, we have

$$\iiint_V \nabla \cdot \mathbf{F}\, dV = \iiint_V (2x+z)\, dV$$

$$= \int_0^1 \int_0^{1-x} \int_0^{1-x-y} (2x+z)\, dz\, dy\, dx$$

$$= \int_0^1 \int_0^{1-x} \left\{2xz + \frac{z^2}{2}\right\}_0^{1-x-y} dy\, dx$$

$$= \frac{1}{2}\int_0^1 \int_0^{1-x} \{4x(1-x-y) + (1-x-y)^2\} dy\, dx$$

$$= \frac{1}{2}\int_0^1 \left\{4x\left(y - xy - \frac{y^2}{2}\right) - \frac{1}{3}(1-x-y)^3\right\}_0^{1-x} dx$$

$$= \frac{1}{2}\int_0^1 \left\{2x(1-x)^2 + \frac{1}{3}(1-x)^3\right\} dx$$

$$= \frac{1}{2}\int_0^1 \left\{2x - 4x^2 + 2x^3 + \frac{1}{3}(1-x)^3\right\} dx$$

$$= \frac{1}{2}\left\{x^2 - \frac{4x^3}{3} + \frac{x^4}{2} - \frac{1}{12}(1-x)^4\right\}_0^1$$

$$= \frac{1}{8}.$$

We have therefore verified the divergence theorem for this **F** and S. ∎

EXAMPLE 15.24

Use the divergence theorem to evaluate the surface integral of Example 15.21.

SOLUTION By the divergence theorem (Figure 15.43), we have

$$\oiint_S \mathbf{F} \cdot \hat{\mathbf{n}}\, dS = \iiint_V \nabla \cdot \mathbf{F}\, dV$$

$$= \iiint_V (1+1+1)\, dV$$

$$= 3\iiint_V dV$$

$$= 3(\text{Volume of } V) = 3(4\pi)(2) = 24\pi.$$

∎

EXAMPLE 15.25

Evaluate

$$\oiint_S (x^3\hat{\mathbf{i}} + y^3\hat{\mathbf{j}} + z^3\hat{\mathbf{k}}) \cdot \hat{\mathbf{n}}\, dS,$$

where $\hat{\mathbf{n}}$ is the unit inner normal to the surface bounding the volume defined by the surfaces $y = x, z = 0, z = 2, y = 0, x = 4$.

SOLUTION By the divergence theorem (see Figure 15.49), we have

$$\oiint_S (x^3\hat{\mathbf{i}} + y^3\hat{\mathbf{j}} + z^3\hat{\mathbf{k}}) \cdot \hat{\mathbf{n}}\, dS = -\iiint_V \nabla \cdot (x^3\hat{\mathbf{i}} + y^3\hat{\mathbf{j}} + z^3\hat{\mathbf{k}})\, dV$$

$$= -\iiint_V (3x^2 + 3y^2 + 3z^2)\, dV$$

$$= -3\int_0^4 \int_0^x \int_0^2 (x^2 + y^2 + z^2)\, dz\, dy\, dx$$

$$= -3\int_0^4 \int_0^x \left\{ (x^2 + y^2)z + \frac{z^3}{3} \right\}_0^2 dy\, dx$$

$$= -2\int_0^4 \int_0^x \{3(x^2 + y^2) + 4\}\, dy\, dx$$

$$= -2\int_0^4 \{y^3 + (3x^2 + 4)y\}_0^x\, dx$$

$$= -2\int_0^4 (4x^3 + 4x)\, dx = -2\{x^4 + 2x^2\}_0^4$$

$$= -576.$$

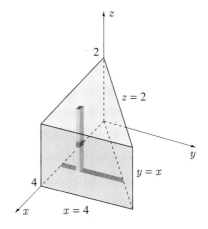

FIGURE 15.49

EXAMPLE 15.26

Prove Archimedes' principle, which states that when an object is submerged in a fluid, it experiences a buoyant force equal to the weight of the fluid displaced.

SOLUTION Suppose the surface of the fluid is taken as the xy-plane ($z = 0$), and the object occupies a region V with bounding surface S (Figure 15.50). The force due to fluid pressure P on a small area dS on S is $(P\, dS)\hat{\mathbf{n}}$, where $\hat{\mathbf{n}}$ is the unit inner normal to S at dS. If ρ is the density of the fluid, then $P = -9.81\rho z$, and the

force on dS is
$$(-9.81\rho z\, dS)\hat{\mathbf{n}}.$$

The resultant buoyant force is in the positive z-direction (the x- and y-components canceling), so that we require only the z-component of this force,
$$(-9.81\rho z\, dS)\hat{\mathbf{n}}\cdot\hat{\mathbf{k}}.$$

The total buoyant force must therefore have z-component
$$\oiint_S (-9.81\rho z)\hat{\mathbf{n}}\cdot\hat{\mathbf{k}}\, dS = \oiint_S (-9.81\rho z\hat{\mathbf{k}})\cdot\hat{\mathbf{n}}\, dS = \oiint_S (9.81\rho z\hat{\mathbf{k}})\cdot(-\hat{\mathbf{n}})\, dS,$$

where $-\hat{\mathbf{n}}$ is the unit outer normal to S. If we now use the divergence theorem, we have
$$\oiint_S (-9.81\rho z)\hat{\mathbf{n}}\cdot\hat{\mathbf{k}}\, dS = \iiint_V \nabla\cdot(9.81\rho z\hat{\mathbf{k}})\, dV$$
$$= \iiint_V 9.81\rho\, dV = 9.81\rho\,(\text{Volume of } V),$$

and this is the weight of the fluid displaced by the object. ∎

FIGURE 15.50

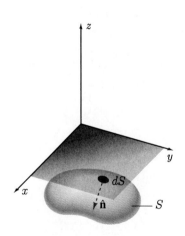

EXERCISES 15.9

In Exercises 1–12 use the divergence theorem to evaluate the surface integral.

1. $\oiint_S (x\hat{\mathbf{i}} + y\hat{\mathbf{j}} - 2z\hat{\mathbf{k}}) \cdot \hat{\mathbf{n}}\, dS$ where S is the surface bounding the volume defined by $z = 2x^2 + y^2$, $x^2 + y^2 = 3$, $z = 0$, and $\hat{\mathbf{n}}$ is the unit outer normal to S

2. $\oiint_S (x^2\hat{\mathbf{i}} + y^2\hat{\mathbf{j}} + z^2\hat{\mathbf{k}}) \cdot \hat{\mathbf{n}}\, dS$ where S is the sphere $x^2 + y^2 + z^2 = a^2$, and $\hat{\mathbf{n}}$ is the unit outer normal to S

3. $\oiint_S (yz\hat{\mathbf{i}} + xz\hat{\mathbf{j}} + xy\hat{\mathbf{k}}) \cdot \hat{\mathbf{n}}\, dS$ where S is the surface enclosing the volume defined by $x = 0$, $x = 2$, $z = 0$, $z = y$, $y + z = 2$, and $\hat{\mathbf{n}}$ is the unit outer normal to S

4. $\oiint_S [(z^2 - x)\hat{\mathbf{i}} - xy\hat{\mathbf{j}} + 3z\hat{\mathbf{k}}] \cdot \hat{\mathbf{n}}\, dS$ where S is the surface enclosing the volume defined by $z = 4 - y^2$, $x = 0$, $x = 3$, $z = 0$, and $\hat{\mathbf{n}}$ is the unit outer normal to S

5. $\oiint_S \mathbf{F} \cdot \hat{\mathbf{n}}\, dS$ where $\mathbf{F} = (x^2y\hat{\mathbf{i}} + y^2z\hat{\mathbf{j}} + z^2x\hat{\mathbf{k}})/2$, S is the surface bounding the volume in the first octant defined by $x = 0$, $y = 0$, $z = 1$, $z = 0$, $x^2 + y^2 = 1$, and $\hat{\mathbf{n}}$ is the unit inner normal to S

6. $\oiint_S (x\hat{\mathbf{i}} + y\hat{\mathbf{j}} + 2z\hat{\mathbf{k}}) \cdot \hat{\mathbf{n}}\, dS$ where S is the surface bounding the volume defined by $z = 2x^2 + y^2$, $x^2 + y^2 = 3$, $z = 0$, and $\hat{\mathbf{n}}$ is the unit outer normal to S

7. $\oiint_S (z\hat{\mathbf{i}} - x\hat{\mathbf{j}} + y\hat{\mathbf{k}}) \cdot \hat{\mathbf{n}}\, dS$ where S is the surface enclosing the volume defined by $z = \sqrt{4 - x^2 - y^2}$, $z = 0$, and $\hat{\mathbf{n}}$ is the unit outer normal to S

8. $\oiint_S (2x^2y\hat{\mathbf{i}} - y^2\hat{\mathbf{j}} + 4xz^2\hat{\mathbf{k}}) \cdot \hat{\mathbf{n}}\, dS$ where S is the surface enclosing the volume in the first octant defined by $y^2 + z^2 = 9$, $x = 2$, and $\hat{\mathbf{n}}$ is the unit outer normal to S

9. $\oiint_S (yx\hat{\mathbf{i}} + y^2\hat{\mathbf{j}} + yz\hat{\mathbf{k}}) \cdot \hat{\mathbf{n}}\, dS$ where S is the ellipsoid $x^2 + y^2/4 + z^2 = 1$, and $\hat{\mathbf{n}}$ is the unit outer normal to S

10. $\oiint_S (x^3\hat{\mathbf{i}} + y^3\hat{\mathbf{j}} - z^3\hat{\mathbf{k}}) \cdot \hat{\mathbf{n}}\, dS$ where S is the surface enclosing the volume defined by $z = 6 - x^2 - y^2$, $z = \sqrt{x^2 + y^2}$, and $\hat{\mathbf{n}}$ is the unit outer normal to S

11. $\oiint_S (y\hat{\mathbf{i}} - xy\hat{\mathbf{j}} + zy^2\hat{\mathbf{k}}) \cdot \hat{\mathbf{n}}\, dS$ where S is the surface enclosing the volume defined by $y^2 - x^2 - z^2 = 4$, $y = 4$, and $\hat{\mathbf{n}}$ is the unit inner normal to S

12. $\oiint_S (xy\hat{\mathbf{i}} + z^2\hat{\mathbf{k}}) \cdot \hat{\mathbf{n}}\, dS$ where S is the surface enclosing the volume in the first octant bounded by the planes $z = 0$, $y = x$, $y = 2x$, $x + y + z = 6$, and $\hat{\mathbf{n}}$ is the unit outer normal to S

In Exercises 13–15 use the divergence theorem to evaluate the surface integral. In each case an additional surface must be introduced in order to enclose a volume.

13. $\iint_S (x\hat{\mathbf{i}} + y\hat{\mathbf{j}} + z\hat{\mathbf{k}}) \cdot \hat{\mathbf{n}}\, dS$ where S is the top half of the ellipsoid $x^2 + 4y^2 + 9z^2 = 36$, and $\hat{\mathbf{n}}$ is the unit upper normal to S

14. $\iint_S (xy\hat{\mathbf{i}} - yz\hat{\mathbf{j}} + x^2z\hat{\mathbf{k}}) \cdot \hat{\mathbf{n}}\, dS$ where S is that part of the cone $z = \sqrt{x^2 + y^2}$ below $z = 2$, and $\hat{\mathbf{n}}$ is the unit normal to S with positive z-component

15. $\iint_S (y^2 e^z\hat{\mathbf{i}} - xy\hat{\mathbf{j}} + z\hat{\mathbf{k}}) \cdot \hat{\mathbf{n}}\, dS$ where S is that part of $z = 4 - x^2 - y^2$ cut out by $z = 2y$, and $\hat{\mathbf{n}}$ is the unit upper normal to S

16. Show that if $\hat{\mathbf{n}}$ is the unit outer normal to a surface S, then the region enclosed by S has volume

$$V = \frac{1}{3} \oiint_S \mathbf{r} \cdot \hat{\mathbf{n}}\, dS, \qquad \mathbf{r} = x\hat{\mathbf{i}} + y\hat{\mathbf{j}} + z\hat{\mathbf{k}}.$$

17. If $\hat{\mathbf{n}}$ is the unit outer normal to a surface S that encloses a region V, show that the area of S can be expressed in the form

$$\text{Area}(S) = \iiint_V \nabla \cdot \hat{\mathbf{n}}\, dV.$$

18. How would you prove Archimedes' principle in the case that an object is only partially submerged? (See Example 15.26.)

In Exercises 19–21 evaluate the surface integral.

19. $\oiint_S [(x+y)\hat{\mathbf{i}} + y^3\hat{\mathbf{j}} + x^2z\hat{\mathbf{k}}] \cdot \hat{\mathbf{n}}\, dS$ where S is the surface enclosing the volume defined by $x^2 + y^2 - z^2 = 1$, $2z^2 = x^2 + y^2$, and $\hat{\mathbf{n}}$ is the unit outer normal to S

20. $\oiint_S [(x+y)^2\hat{\mathbf{i}} + x^2y\hat{\mathbf{j}} - x^2z\hat{\mathbf{k}}] \cdot \hat{\mathbf{n}}\, dS$ where $\hat{\mathbf{n}}$ is the unit inner normal to the surface S enclosing the volume defined by $z^2 = (1 - x^2 - 2y^2)^2$

21. $\iint_S [(y^3 + x^2 y)\hat{\mathbf{i}} + (x^3 - xy^2)\hat{\mathbf{j}} + z\hat{\mathbf{k}}] \cdot \hat{\mathbf{n}} \, dS$ where $\hat{\mathbf{n}}$ is the unit upper normal to the surface
$S: z = \sqrt{1 - x^2 - y^2}$

22. If V is a region bounded by a closed surface S, and $\mathbf{B} = \nabla \times \mathbf{A}$, show that

$$\oiint_S \mathbf{B} \cdot \hat{\mathbf{n}} \, dS = 0.$$

23. Is Green's theorem related to the divergence theorem? (See Exercise 13 in Section 15.6.)

In Exercises 24–26 assume that $P(x, y, z)$ and $Q(x, y, z)$ have continuous first and second partial derivatives in a domain containing a closed surface S and its interior V. Let $\hat{\mathbf{n}}$ be the unit outer normal to S.

24. Show that

$$\oiint_S \nabla P \cdot \hat{\mathbf{n}} \, dS = \iiint_V \nabla^2 P \, dV,$$

where $\nabla^2 P = \dfrac{\partial^2 P}{\partial x^2} + \dfrac{\partial^2 P}{\partial y^2} + \dfrac{\partial^2 P}{\partial z^2}$. What can we conclude if $P(x, y, z)$ satisfies Laplace's equation in V?

25. Show that

$$\oiint_S P \nabla Q \cdot \hat{\mathbf{n}} \, dS = \iiint_V (P \nabla^2 Q + \nabla P \cdot \nabla Q) \, dV.$$

This result is called *Green's first identity*.

26. Prove that

$$\oiint_S (P \nabla Q - Q \nabla P) \cdot \hat{\mathbf{n}} \, dS = \iiint_V (P \nabla^2 Q - Q \nabla^2 P) \, dV.$$

This result is called *Green's second identity*.

27. Compare Exercises 24–26 with Exercises 36–38 in Section 15.6.

28. Let S be a closed surface, and let $\hat{\mathbf{n}}$ be the unit outer normal to S. If $\mathbf{r}_0 = x_0\hat{\mathbf{i}} + y_0\hat{\mathbf{j}} + z_0\hat{\mathbf{k}}$ is the position vector of some fixed point P_0, show that

$$\oiint_S \frac{\mathbf{r} - \mathbf{r}_0}{|\mathbf{r} - \mathbf{r}_0|^3} \cdot \hat{\mathbf{n}} \, dS = \begin{cases} 0 & \text{if } S \text{ does not enclose } P_0 \\ 4\pi & \text{if } S \text{ does enclose } P_0. \end{cases}$$

29. In this problem we use the result of Exercise 28 to prove Gauss's law for electrostatic fields. Let S be a closed surface containing n point charges q_i ($i = 1, \ldots, n$) at points $\mathbf{r}_i = (x_i, y_i, z_i)$ in its interior. According to Coulomb's law, the electric field $\mathbf{E}$ at a point $\mathbf{r} = (x, y, z)$ due to this charge distribution is defined by $\mathbf{E} = -\nabla V$, where V, the potential at $\mathbf{r}$, is

$$V(x, y, z) = \sum_{i=1}^{n} \frac{q_i}{4\pi\epsilon_0 |\mathbf{r} - \mathbf{r}_i|}.$$

(a) Verify that

$$\mathbf{E} = \sum_{i=1}^{n} \frac{q_i(\mathbf{r} - \mathbf{r}_i)}{4\pi\epsilon_0 |\mathbf{r} - \mathbf{r}_i|^3}.$$

(b) Now use the result of Exercise 28 to show that if $\hat{\mathbf{n}}$ is the unit outer normal to S, then

$$\oiint_S \mathbf{E} \cdot \hat{\mathbf{n}} \, dS = \frac{Q}{\epsilon_0},$$

where Q is the total charge inside S. This result is Gauss's law.

SECTION 15.10

Stokes's Theorem

Stokes's theorem relates certain line integrals around closed curves to surface integrals over surfaces that have the curves as boundaries.

> **Theorem 15.8**
>
> (Stokes's theorem). Let C be a closed, piecewise-smooth, curve that does not intersect itself and let S be a piecewise-smooth, orientable surface with C as boundary (Figure 15.51). Let $\mathbf{F}(x,y,z) = P(x,y,z)\hat{\mathbf{i}} + Q(x,y,z)\hat{\mathbf{j}} + R(x,y,z)\hat{\mathbf{k}}$ be a vector field whose components P, Q, and R have continuous first partial derivatives in a domain that contains S and C. Then
>
> $$\oint_C \mathbf{F} \cdot d\mathbf{r} = \iint_S (\nabla \times \mathbf{F}) \cdot \hat{\mathbf{n}} \, dS, \qquad (15.56\text{a})$$
>
> or
>
> $$\oint_C P\,dx + Q\,dy + R\,dz$$
> $$= \iint_S \left\{ \left(\frac{\partial R}{\partial y} - \frac{\partial Q}{\partial z}\right)\hat{\mathbf{i}} + \left(\frac{\partial P}{\partial z} - \frac{\partial R}{\partial x}\right)\hat{\mathbf{j}} + \left(\frac{\partial Q}{\partial x} - \frac{\partial P}{\partial y}\right)\hat{\mathbf{k}} \right\} \cdot \hat{\mathbf{n}} \, dS, \quad (15.56\text{b})$$
>
> where $\hat{\mathbf{n}}$ is the unit normal to S chosen in the following way: If when moving along C the surface S is on the left-hand side, then $\hat{\mathbf{n}}$ must be chosen as the unit normal on that side of S. On the other hand, if when moving along C, the surface is on the right, then $\hat{\mathbf{n}}$ must be chosen on the opposite side of S.

FIGURE 15.51

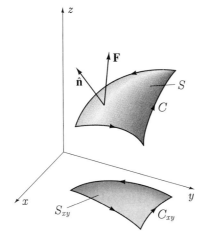

Proof We first consider a surface S that projects in a one-to-one fashion onto each of the three coordinate planes. Because S projects one-to-one onto some region S_{xy} in the xy-plane, we can take the equation for S in the form $z = f(x,y)$. If the direction along C is as indicated in Figure 15.51, and C_{xy} is the projection of C on the xy-plane, then

$$\oint_C P\,dx = \oint_{C_{xy}} P[x,y,f(x,y)]\,dx.$$

If we use Green's theorem on this line integral around C_{xy}, we have

$$\oint_C P\,dx = \iint_{S_{xy}} -\left\{\frac{\partial P}{\partial y} + \frac{\partial P}{\partial z}\frac{\partial z}{\partial y}\right\} dA.$$

On the other hand, since a unit normal to S is

$$\hat{\mathbf{n}} = \frac{\left(-\dfrac{\partial z}{\partial x}, -\dfrac{\partial z}{\partial y}, 1\right)}{\sqrt{1 + \left(\dfrac{\partial z}{\partial x}\right)^2 + \left(\dfrac{\partial z}{\partial y}\right)^2}},$$

it also follows that

$$\iint_S \left(\frac{\partial P}{\partial z}\hat{\mathbf{j}} - \frac{\partial P}{\partial y}\hat{\mathbf{k}}\right) \cdot \hat{\mathbf{n}} \, dS = \iint_S \frac{-\dfrac{\partial P}{\partial z}\dfrac{\partial z}{\partial y} - \dfrac{\partial P}{\partial y}}{\sqrt{1 + \left(\dfrac{\partial z}{\partial x}\right)^2 + \left(\dfrac{\partial z}{\partial y}\right)^2}} \, dS$$

$$= \iint_{S_{xy}} \frac{-\dfrac{\partial P}{\partial z}\dfrac{\partial z}{\partial y} - \dfrac{\partial P}{\partial y}}{\sqrt{1 + \left(\dfrac{\partial z}{\partial x}\right)^2 + \left(\dfrac{\partial z}{\partial y}\right)^2}} \sqrt{1 + \left(\frac{\partial z}{\partial x}\right)^2 + \left(\frac{\partial z}{\partial y}\right)^2} \, dA$$

$$= \iint_{S_{xy}} -\left\{\frac{\partial P}{\partial y} + \frac{\partial P}{\partial z}\frac{\partial z}{\partial y}\right\} dA.$$

We have shown, then, that

$$\oint_C P \, dx = \iint_S \left(\frac{\partial P}{\partial z}\hat{\mathbf{j}} - \frac{\partial P}{\partial y}\hat{\mathbf{k}}\right) \cdot \hat{\mathbf{n}} \, dS.$$

By projecting C and S onto the xz- and yz-planes, we can show similarly that

$$\oint_C R \, dz = \iint_S \left(\frac{\partial R}{\partial y}\hat{\mathbf{i}} - \frac{\partial R}{\partial x}\hat{\mathbf{j}}\right) \cdot \hat{\mathbf{n}} \, dS$$

and

$$\oint_C Q \, dy = \iint_S \left(\frac{\partial Q}{\partial x}\hat{\mathbf{k}} - \frac{\partial Q}{\partial z}\hat{\mathbf{i}}\right) \cdot \hat{\mathbf{n}} \, dS.$$

Addition of these three results gives Stokes's theorem for $\mathbf{F}$ and S.

The proof can be extended to general curves and surfaces that do not project in a one-to-one fashion onto all three coordinate planes. Most surfaces S with bounding curves C can be subdivided into n subsurfaces S_i with bounding curves C_i that do satisfy this condition (Figure 15.52). For each such subsurface, Stokes's theorem applies

$$\oint_{C_i} \mathbf{F} \cdot d\mathbf{r} = \iint_{S_i} (\nabla \times \mathbf{F}) \cdot \hat{\mathbf{n}} \, dS, \quad i = 1, \ldots, n.$$

If these n equations are now added together, we have

$$\sum_{i=1}^{n} \oint_{C_i} \mathbf{F} \cdot d\mathbf{r} = \sum_{i=1}^{n} \iint_{S_i} (\nabla \times \mathbf{F}) \cdot \hat{\mathbf{n}} \, dS.$$

Since the S_i constitute S, the right-hand side of this equation is the surface integral of $(\nabla \times \mathbf{F}) \cdot \hat{\mathbf{n}}$ over S. Figure 15.52 illustrates that when the line integrals over the C_i are added, contributions from the auxiliary (interior) curves cancel in pairs, and the remaining line integrals give the line integral of $\mathbf{F} \cdot d\mathbf{r}$ along C. This completes the proof.

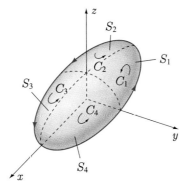

FIGURE 15.52

For general surfaces that cannot be divided into a finite number of subsurfaces of this type, the reader should consult a more advanced book.

Note that Green's theorem is a special case of Stokes's theorem. For if $\mathbf{F} = P(x,y)\hat{\mathbf{i}} + Q(x,y)\hat{\mathbf{j}}$, and C is a closed curve in the xy-plane, then by Stokes's theorem

$$\oint_C P\,dx + Q\,dy = \iint_S \left(\frac{\partial Q}{\partial x} - \frac{\partial P}{\partial y}\right) \hat{\mathbf{k}} \cdot \hat{\mathbf{n}}\,dS,$$

where S is any surface for which C is the boundary. If we choose S as that part of the xy-plane bounded by C, then $\hat{\mathbf{n}} = \hat{\mathbf{k}}$ and

$$\oint_C P\,dx + Q\,dy = \iint_S \left(\frac{\partial Q}{\partial x} - \frac{\partial P}{\partial y}\right) dA.$$

With Stokes's theorem, it is straightforward to verify the sufficiency half of Theorem 15.1. Suppose that the curl of a vector field $\mathbf{F}$ vanishes in a simply-connected domain D. If C is any piecewise smooth, closed curve in D, then there exists a piecewise smooth surface S in D with C as boundary. By Stokes's theorem,

$$\oint_C \mathbf{F} \cdot d\mathbf{r} = \iint_S (\nabla \times \mathbf{F}) \cdot \hat{\mathbf{n}}\,dS = 0.$$

According to Corollary 2 of Theorem 15.3, the line integral is independent of path in D, and the theorem itself implies the existence of a function $f(x, y, z)$ such $\nabla f = \mathbf{F}$.

EXAMPLE 15.27

Verify Stokes's theorem if $\mathbf{F} = x^2\hat{\mathbf{i}} + x\hat{\mathbf{j}} + xyz\hat{\mathbf{k}}$, and S is that part of the sphere $x^2 + y^2 + z^2 = 4$ above the plane $z = 1$.

SOLUTION If we choose $\hat{\mathbf{n}}$ as the upper normal to S, then C, the boundary of S, must be traversed in the direction shown in Figure 15.53. (If $\hat{\mathbf{n}}$ is chosen as the lower normal, then C must be traversed in the opposite direction.) Since parametric equations for C are

$$x = \sqrt{3}\cos t, \quad y = \sqrt{3}\sin t, \quad z = 1, \quad 0 \leq t \leq 2\pi,$$

then

$$\oint_C \mathbf{F} \cdot d\mathbf{r} = \oint_C x^2 \, dx + x \, dy + xyz \, dz$$
$$= \int_0^{2\pi} \{3\cos^2 t(-\sqrt{3}\sin t \, dt) + \sqrt{3}\cos t(\sqrt{3}\cos t \, dt)\}$$
$$= \int_0^{2\pi} \left\{-3\sqrt{3}\cos^2 t \sin t + \frac{3}{2}(1+\cos 2t)\right\} dt$$
$$= \left\{\sqrt{3}\cos^3 t + \frac{3t}{2} + \frac{3\sin 2t}{4}\right\}_0^{2\pi}$$
$$= 3\pi.$$

FIGURE 15.53

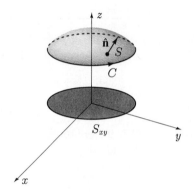

Alternatively,

$$\iint_S (\nabla \times \mathbf{F}) \cdot \hat{\mathbf{n}} \, dS = \iint_S (xz\hat{\mathbf{i}} - yz\hat{\mathbf{j}} + \hat{\mathbf{k}}) \cdot \frac{(2x, 2y, 2z)}{\sqrt{4x^2 + 4y^2 + 4z^2}} dS$$
$$= \iint_S \frac{x^2 z - y^2 z + z}{\sqrt{x^2 + y^2 + z^2}} dS$$
$$= \iint_{S_{xy}} \frac{z(x^2 - y^2 + 1)}{2} \sqrt{1 + \left(\frac{\partial z}{\partial x}\right)^2 + \left(\frac{\partial z}{\partial y}\right)^2} \, dA$$
$$= \frac{1}{2} \iint_{S_{xy}} z(x^2 - y^2 + 1) \sqrt{1 + (-x/z)^2 + (-y/z)^2} \, dA$$
$$= \frac{1}{2} \iint_{S_{xy}} z(x^2 - y^2 + 1) \frac{\sqrt{x^2 + y^2 + z^2}}{z} dA$$
$$= \iint_{S_{xy}} (x^2 - y^2 + 1) \, dA.$$

If we use polar coordinates to evaluate this double integral over $S_{xy} : x^2 + y^2 \leq 3$, we have

$$\iint_S (\nabla \times \mathbf{F}) \cdot \hat{\mathbf{n}}\, dS = \int_{-\pi}^{\pi} \int_0^{\sqrt{3}} (r^2 \cos^2 \theta - r^2 \sin^2 \theta + 1)\, r\, dr\, d\theta$$

$$= \int_{-\pi}^{\pi} \left\{ \frac{r^4}{4}(\cos^2 \theta - \sin^2 \theta) + \frac{r^2}{2} \right\}_0^{\sqrt{3}} d\theta$$

$$= \int_{-\pi}^{\pi} \left\{ \frac{9}{4} \cos 2\theta + \frac{3}{2} \right\} d\theta$$

$$= \left\{ \frac{9}{8} \sin 2\theta + \frac{3\theta}{2} \right\}_{-\pi}^{\pi} = 3\pi.$$

∎

EXAMPLE 15.28

Evaluate

$$\oint_C 2xy^3\, dx + 3x^2 y^2\, dy + (2z + x)\, dz,$$

where C is the series of line segments joining $A(2,0,0)$ to $B(0,1,0)$ to $D(0,0,1)$ to A.

SOLUTION By Stokes's theorem,

$$\oint_C 2xy^3\, dx + 3x^2 y^2\, dy + (2z + x)\, dz = \iint_S \nabla \times (2xy^3, 3x^2 y^2, 2z + x) \cdot \hat{\mathbf{n}}\, dS,$$

where S is any surface with C as boundary. If we choose S as the flat triangle bounded by C (Figure 15.54), then a normal vector to S is

$$\overrightarrow{BD} \times \overrightarrow{BA} = \begin{vmatrix} \hat{\mathbf{i}} & \hat{\mathbf{j}} & \hat{\mathbf{k}} \\ 0 & -1 & 1 \\ 2 & -1 & 0 \end{vmatrix} = (1, 2, 2),$$

and therefore

$$\hat{\mathbf{n}} = \frac{(1, 2, 2)}{3}.$$

Since

$$\nabla \times (2xy^3, 3x^2 y^2, 2z + x) = \begin{vmatrix} \hat{\mathbf{i}} & \hat{\mathbf{j}} & \hat{\mathbf{k}} \\ \partial/\partial x & \partial/\partial y & \partial/\partial z \\ 2xy^3 & 3x^2 y^2 & 2z + x \end{vmatrix} = -\hat{\mathbf{j}},$$

it follows that

$$\nabla \times (2xy^3, 3x^2 y^2, 2z + x) \cdot \hat{\mathbf{n}} = -\hat{\mathbf{j}} \cdot \frac{(1, 2, 2)}{3} = -\frac{2}{3}$$

and

$$\oint_C 2xy^3\, dx + 3x^2 y^2\, dy + (2z + x)\, dz = \iint_S -\frac{2}{3}\, dS = -\frac{2}{3}(\text{Area of } S).$$

But from equation (12.41), the area of triangle S is

$$\frac{1}{2} |\overrightarrow{BD} \times \overrightarrow{BA}| = \frac{1}{2} |(1, 2, 2)| = \frac{3}{2}.$$

Finally, then,
$$\oint_C 2xy^3\,dx + 3x^2y^2\,dy + (2z+x)\,dz = -\frac{2}{3}\left(\frac{3}{2}\right) = -1.$$

FIGURE 15.54

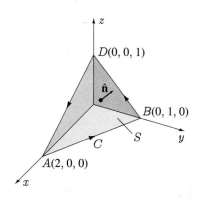

EXERCISES 15.10

In Exercises 1–14 use Stokes's theorem to evaluate the line integral.

1. $\oint_C x^2 y\,dx + y^2 z\,dy + z^2 x\,dz$ where C is the curve $z = x^2 + y^2$, $x^2 + y^2 = 4$, directed counterclockwise as viewed from the origin

2. $\oint_C y^2\,dx + xy\,dy + xz\,dz$ where C is the curve $x^2 + y^2 = 2y$, $y = z$, directed so that y increases when x is positive

3. $\oint_C (xyz + 2yz)\,dx + xz\,dy + 2xy\,dz$ where C is the curve $z = 1$, $x^2 + y^2 + z^2 = 4$, directed clockwise as viewed from the origin

4. $\oint_C (2xy + y)\,dx + (x^2 + xy - 3y)\,dy + 2xz\,dz$ where C is the curve $z = \sqrt{x^2 + y^2}$, $z = 4$

5. $\oint_C x^2\,dx + y^2\,dy + (x^2 + y^2)\,dz$ where C is the boundary of that part of the plane $x + y + z = 1$ in the first octant, directed counterclockwise as viewed from the origin

6. $\oint_C y\,dx + x\,dy + (x^2 + y^2 + z^2)\,dz$ where C is the curve $x^2 + y^2 = 1$, $z = xy$, directed clockwise as viewed from the point $(0, 0, 1)$

7. $\oint_C zy^2\,dx + xy\,dy + (y^2 + z^2)\,dz$ where C is the curve $x^2 + z^2 = 9$, $y = \sqrt{x^2 + z^2}$, directed counterclockwise as viewed from the origin

8. $\oint_C y\,dx + z\,dy + x\,dz$ where C is the curve $x + y = 2b$, $x^2 + y^2 + z^2 = 2b(x + y)$, directed clockwise as viewed from the origin

9. $\oint_C y^2\,dx + (x + y)\,dy + yz\,dz$ where C is the curve $x^2 + y^2 = 2$, $x + y + z = 2$, directed clockwise as viewed from the origin

10. $\oint_C (x + y)^2\,dx + (x + y)^2\,dy + yz^3\,dz$ where C is the curve $z = \sqrt{x^2 + y^2}$, $(x - 1)^2 + y^2 = 1$

11. $\oint_C xy\,dx - zx\,dy + yz\,dz$ where C is the boundary of that part of $z = x + y$ in the first octant cut off by $x + y = 1$, directed counterclockwise as viewed from the point $(0, 0, 1)$

12. $\oint_C y^3\,dx - x^3\,dy + xyz\,dz$ where C is the curve $x^2 + y^2 = z^2 + 3$, $z = 3 - \sqrt{x^2 + y^2}$, directed clockwise as viewed from the origin

13. $\oint_C z(x + y)^2\,dx + (y - x)^2\,dy + z^2\,dz$ where C is the smooth curve of intersection of the surfaces

$x^2 + z^2 = a^2$, $y^2 + z^2 = a^2$ which has a portion in the first octant, directed so that z decreases in the first octant

14. $\oint_C -2y^3 x^2\, dx + x^3 y^2\, dy + z\, dz$ where C is the curve
$x^2 + y^2 + z^2 = 4$, $x^2 + 4y^2 = 4$, directed so that x decreases along that part of the curve in the first octant

15. Evaluate the line integral $\oint_C 2x^2 y\, dx - yz\, dy + xz\, dz$
where C is the curve $x^2 + y^2 + z^2 = 4$, $z = \sqrt{3}(x^2 + y^2)$, directed clockwise as viewed from the origin, in four ways:

   (a) directly as a line integral

   (b) using Stokes's theorem with S as that part of $z = \sqrt{4 - x^2 - y^2}$ bounded by C

   (c) using Stokes's theorem with S as that part of $z = \sqrt{3(x^2 + y^2)}$ bounded by C

   (d) using Stokes's theorem with S as that part of $z = \sqrt{3}$ bounded by C

16. Let S_1 be that part of $x^2 + y^2 + z^2 = 1$ above the xy-plane and S_2 be that part of $z = 1 - x^2 - y^2$ above the xy-plane. Show that if $\hat{\mathbf{n}}_1$ and $\hat{\mathbf{n}}_2$ are the unit upper normals to these surfaces, and $\mathbf{F}$ is a vector field defined on both S_1 and S_2, then

$$\iint_{S_1} (\nabla \times \mathbf{F}) \cdot \hat{\mathbf{n}}_1\, dS = \iint_{S_2} (\nabla \times \mathbf{F}) \cdot \hat{\mathbf{n}}_2\, dS.$$

SECTION 15.11

Flux and Circulation

In many branches of engineering and physics we encounter the concepts of **flux** and **circulation**. In this section we discuss the relationships between flux and divergence and between circulation and curl. We find a physical setting previously mentioned in Section 15.1 most useful in developing these ideas.

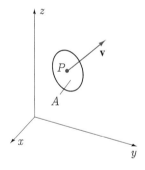

FIGURE 15.55

Fluid Flow

Consider a gas flowing through a region D of space. At time t and point $P(x, y, z)$ in D, gas flows through P with some velocity $\mathbf{v}(x, y, z, t)$. If A is a unit area around P perpendicular to $\mathbf{v}$ (Figure 15.55) and $\rho(x, y, z, t)$ is the density of the gas at P, then the amount of gas crossing A per unit time is $\rho \mathbf{v}$. At every point P in D, then, the vector $\rho \mathbf{v}$ is such that its direction $\mathbf{v}$ gives the velocity of gas flow, and its magnitude $\rho |\mathbf{v}|$ describes the mass of gas flowing in that direction per unit time.

Consider now some surface S in D (Figure 15.56). If $\hat{\mathbf{n}}$ is a unit normal to S, then $\rho \mathbf{v} \cdot \hat{\mathbf{n}}$ is the component of $\rho \mathbf{v}$ normal to the surface S. If dS is an element of area on S, then $\rho \mathbf{v} \cdot \hat{\mathbf{n}}\, dS$ describes the mass of gas flowing through dS per unit time. Consequently,

$$\iint_S \rho \mathbf{v} \cdot \hat{\mathbf{n}}\, dS$$

is the mass of gas flowing through S per unit time. This quantity is called the **flux** for the surface S.

If S is a closed surface (Figure 15.57) and $\hat{\mathbf{n}}$ is the unit outer normal to S, then

$$\oiint_S \rho \mathbf{v} \cdot \hat{\mathbf{n}}\, dS$$

is the mass of gas flowing out of the surface S per unit time. If this flux (for closed S) is positive, then there is a net outward flow of gas through S (i.e., more gas is leaving the volume bounded by S than is entering); if the flux is negative, the net flow is inward.

If we apply the divergence theorem to the flux integral over the closed surface S, we have

$$\oiint_S \rho \mathbf{v} \cdot \hat{\mathbf{n}}\, dS = \iiint_V \nabla \cdot (\rho \mathbf{v})\, dV. \tag{15.57}$$

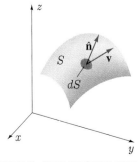

FIGURE 15.56

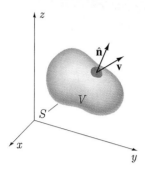

FIGURE 15.57

Now the flux (on the left-hand side of this equation) is the mass of gas per unit time leaving S. In order for the right-hand side to represent the same quantity, $\nabla \cdot (\rho \mathbf{v})$ must be interpreted as the mass of gas leaving unit volume per unit time, because then $\nabla \cdot (\rho \mathbf{v}) \, dV$ represents the mass per unit time leaving dV, and the triple integral is the mass per unit time leaving V.

We have obtained, therefore, an interpretation of the divergence of $\rho \mathbf{v}$. The divergence of $\rho \mathbf{v}$ is the flux per unit volume per unit time at a point: the mass of gas leaving unit volume in unit time. We can use this idea of flux to derive the "equation of continuity" for fluid flow. The triple integral

$$\iiint_V \frac{\partial \rho}{\partial t} \, dV = \frac{\partial}{\partial t} \iiint_V \rho \, dV$$

measures the time rate of change of mass in a volume V. If this triple integral is positive, then there is a net inward flow of mass; if it is negative, the net flow is outward. We conclude, therefore, that this triple integral must be the negative of the flux for the volume V; i.e.,

$$\iiint_V \frac{\partial \rho}{\partial t} \, dV = - \iiint_V \nabla \cdot (\rho \mathbf{v}) \, dV,$$

or

$$\iiint_V \left(\nabla \cdot (\rho \mathbf{v}) + \frac{\partial \rho}{\partial t} \right) dV = 0.$$

If $\nabla \cdot (\rho \mathbf{v})$ and $\partial \rho / \partial t$ are continuous functions, then this equation can hold for arbitrary volume V only if

$$\nabla \cdot (\rho \mathbf{v}) + \frac{\partial \rho}{\partial t} = 0. \tag{15.58}$$

This equation, called the **equation of continuity**, expresses conservation of mass.

If C is a closed curve in the flow region D, then the **circulation** of the flow for the curve C is defined by

$$\Gamma = \oint_C \mathbf{v} \cdot d\mathbf{r}. \tag{15.59}$$

To obtain an intuitive feeling for Γ, we consider two very simple two-dimensional flows. First suppose $\mathbf{v} = x\hat{\mathbf{i}} + y\hat{\mathbf{j}}$, so that all particles of gas flow along radial lines directed away from the origin (Figure 15.58). In this case, the line integral defining Γ is independent of path and $\Gamma = 0$ for any curve whatsoever.

Second, suppose $\mathbf{v} = -y\hat{\mathbf{i}} + x\hat{\mathbf{j}}$, so that all particles of the gas flow counterclockwise around circles centred at the origin (Figure 15.59). In this case Γ does not generally vanish. In particular, if C is the circle $x^2 + y^2 = r^2$, then $\mathbf{v}$ and $d\mathbf{r}$ are parallel, and

$$\Gamma = \oint_C \mathbf{v} \cdot d\mathbf{r} = \oint_C |\mathbf{v}| ds = \oint_C \sqrt{y^2 + x^2} \, ds = \oint_C r \, ds = 2\pi r^2.$$

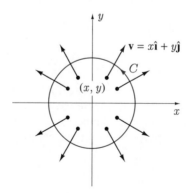

FIGURE 15.58

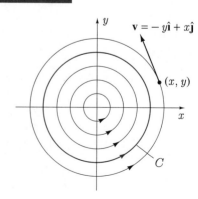

FIGURE 15.59

These two flow patterns indicate perhaps that circulation is a measure of the tendency for the flow to be circulatory. If we apply Stokes's theorem to the circulation integral for the closed curve C, we have

$$\oint_C \mathbf{v} \cdot d\mathbf{r} = \iint_S (\nabla \times \mathbf{v}) \cdot \hat{\mathbf{n}} \, dS, \tag{15.60}$$

where S is any surface in the flow with boundary C. If the right-hand side of this equation is to represent the circulation for C also, then $(\nabla \times \mathbf{v}) \cdot \hat{\mathbf{n}} \, dS$ must be interpreted as the circulation for the curve bounding dS (or simply for dS itself). Then the addition process of the surface integral (Figure 15.60) gives the circulation around C, the circulation around all internal boundaries canceling. But if $(\nabla \times \mathbf{v}) \cdot \hat{\mathbf{n}} \, dS$ is the circulation for dS, then it follows that $(\nabla \times \mathbf{v}) \cdot \hat{\mathbf{n}}$ must be the circulation for unit area perpendicular to $\hat{\mathbf{n}}$. Thus $\nabla \times \mathbf{v}$ describes the circulatory nature of the flow $\mathbf{v}$, its component $(\nabla \times \mathbf{v}) \cdot \hat{\mathbf{n}}$ in any direction describes the circulation for unit area perpendicular to $\hat{\mathbf{n}}$.

FIGURE 15.60

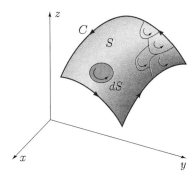

Electromagnetic Theory

The concepts of flux and circulation also play a prominent role in electromagnetic theory. For example, suppose a dielectric contains a charge distribution of density ρ (charge per unit volume). This charge produces an electric field represented by the electric displacement vector $\mathbf{D}$. If S is a surface in the dielectric, then the flux of $\mathbf{D}$ through S is defined as

$$\iint_S \mathbf{D} \cdot \hat{\mathbf{n}} \, dS,$$

and, in the particular case in which S is closed, as

$$\oiint_S \mathbf{D} \cdot \hat{\mathbf{n}} \, dS.$$

Gauss's law states that this flux integral must be exactly equal to the total charge enclosed by S. If V is the region enclosed by S, we can write

$$\oiint_S \mathbf{D} \cdot \hat{\mathbf{n}} \, dS = \iiint_V \rho \, dV. \tag{15.61}$$

On the other hand, if we apply the divergence theorem to the flux integral, we have

$$\oiint_S \mathbf{D} \cdot \hat{\mathbf{n}} \, dS = \iiint_V \nabla \cdot \mathbf{D} \, dV.$$

Consequently,
$$\iiint_V \nabla \cdot \mathbf{D}\, dV = \iiint_V \rho\, dV,$$
or
$$\iiint_V (\nabla \cdot \mathbf{D} - \rho)\, dV = 0.$$

If $\nabla \cdot \mathbf{D}$ and ρ are continuous functions, then the only way this equation can hold for arbitrary volume V in the dielectric is if

$$\nabla \cdot \mathbf{D} = \rho. \qquad (15.62)$$

This is the first of Maxwell's equations for electromagnetic fields.

Another of Maxwell's equations can be obtained using Stokes's theorem. The flux through a surface S of a magnetic field $\mathbf{B}$ is defined by

$$\iint_S \mathbf{B} \cdot \hat{\mathbf{n}}\, dS.$$

If $\mathbf{B}$ is a changing field, then an induced electric field intensity $\mathbf{E}$ is created. Faraday's induction law states that the time rate of change of the flux of $\mathbf{B}$ through S must be equal to the negative of the line integral of $\mathbf{E}$ around the boundary C of S:

$$\oint_C \mathbf{E} \cdot d\mathbf{r} = -\frac{\partial}{\partial t} \iint_S \mathbf{B} \cdot \hat{\mathbf{n}}\, dS. \qquad (15.63)$$

But Stokes's theorem applied to the line integral also gives

$$\oint_C \mathbf{E} \cdot d\mathbf{r} = \iint_S (\nabla \times \mathbf{E}) \cdot \hat{\mathbf{n}}\, dS.$$

It follows, therefore, that if S is stationary,

$$\iint_S (\nabla \times \mathbf{E}) \cdot \hat{\mathbf{n}}\, dS = \iint_S -\frac{\partial \mathbf{B}}{\partial t} \cdot \hat{\mathbf{n}}\, dS,$$

or

$$\iint_S \left(\nabla \times \mathbf{E} + \frac{\partial \mathbf{B}}{\partial t} \right) \cdot \hat{\mathbf{n}}\, dS = 0.$$

Once again, this equation can hold for arbitrary surfaces S, if $\nabla \times \mathbf{E}$ and $\partial \mathbf{B}/\partial t$ are continuous, only if

$$\nabla \times \mathbf{E} = -\frac{\partial \mathbf{B}}{\partial t}. \qquad (15.64)$$

SUMMARY

When a vector is a function of position, it becomes susceptible to the operations of differentiation and integration. In this chapter we developed various ways of differentiating and integrating vector fields beginning with the operations of divergence and curl. The divergence of a vector field is a scalar field, and the curl of a vector field is another vector field. Both are extremely useful in applied mathematics. Mathematically, the curl appeared in our discussion of independence of path for line integrals and in Stokes's theorem; we saw its physical importance in our study of fluid flow and electromagnetic theory. We introduced the divergence of a vector field in our discussion of the divergence theorem and in the same applications as those for the curl.

The line integral of a function $f(x,y,z)$ along a curve C is defined in the same way as a definite integral, a double integral, or a triple integral— that is, the limit of a sum,

$$\int_C f(x,y,z)\,ds = \lim_{\|\Delta s_i\| \to 0} \sum_{i=1}^{n} f(x_i^*, y_i^*, z_i^*)\,\Delta s_i.$$

The most important type of line integral occurs when $f(x,y,z)$ is the tangential component of a vector field $\mathbf{F} = P\hat{\mathbf{i}} + Q\hat{\mathbf{j}} + R\hat{\mathbf{k}}$ defined along C, and in this case we write

$$\int_C f(x,y,z)\,ds = \int_C \mathbf{F} \cdot d\mathbf{r} = \int_C P\,dx + Q\,dy + R\,dz.$$

We developed three methods for evaluating line integrals:

1. Express all parts of the line integral in terms of any parameter along C and evaluate the resulting definite integral. All line integrals can be evaluated in this way, but often methods (2) and (3) lead to much simpler calculations.

2. If a line integral is independent of path, then we can evaluate it by taking the difference in values of a function ϕ (where $\nabla \phi = \mathbf{F}$) at the ends of the curve.

3. If C is a closed curve, we can sometimes use Stokes's theorem to replace a line integral with a simpler surface integral. In this regard, Green's theorem is a special case of Stokes's theorem.

If the line integral of a force field $\mathbf{F}$ is independent of path, the force field is said to be conservative. Associated with every conservative force field is a potential function U such that the work done by $\mathbf{F}$ along a curve C from A to B is equal to the difference in U at A and B. In addition, motion of an object in a conservative force field is always characterized by an exchange of potential energy for kinetic energy in such a way that the sum of the two energies is always a constant value.

Surface integrals are also limits of sums,

$$\iint_S f(x,y,z)\,dS = \lim_{\|\Delta S_i\| \to 0} \sum_{i=1}^{n} f(x_i^*, y_i^*, z_i^*)\,\Delta S_i,$$

and the most important type of surface integral occurs when $f(x,y,z)$ is the normal component of a vector field $\mathbf{F}$ on S:

$$\iint_S f(x,y,z)\,dS = \iint_S \mathbf{F} \cdot \hat{\mathbf{n}}\,dS.$$

We suggested two methods for the evaluation of surface integrals:

1. Project S onto some region R in one of the coordinate planes, express all parts of the integral in terms of coordinates in that plane, and evaluate the resulting double integral over R.

2. If S is closed, it could be advantageous to replace a surface integral with the triple integral of $\nabla \cdot \mathbf{F}$ over the volume bounded by S (the divergence theorem).

Key Terms and Formulas

In reviewing this chapter, you should be able to define or discuss the following key terms:

Vector field
Scalar field
Interior point of region
Exterior point of region
Boundary point of region
Open set
Closed set
Connected set
Domain
Simply-connected domain
Del operator
Divergence of a vector field
Curl of a vector field
Irrotational vector field

Solenoidal vector field
Line integral
Path independence of a line integral
Conservative vector field
Green's theorem
Work
Potential energy
Law of conservation of Energy
Surface integral
Orientable surface
Divergence theorem
Stokes's theorem
Flux of a vector field
Circulation of a vector field

REVIEW EXERCISES

In Exercises 1–10 calculate the quantity.

1. ∇f if $f(x,y,z) = x^2 y^3 - xy + z$
2. $\nabla \cdot \mathbf{F}$ if $\mathbf{F}(x,y) = x^3 y \hat{\mathbf{i}} - (x^2/y)\hat{\mathbf{j}}$
3. $\nabla \times \mathbf{F}$ if $\mathbf{F}(x,y) = \sin(xy)\hat{\mathbf{i}} + \cos(xy)\hat{\mathbf{j}} + xy\hat{\mathbf{k}}$
4. $\nabla \times \mathbf{F}$ if $\mathbf{F}(x,y,z) = (x+y+z)(\hat{\mathbf{i}} + \hat{\mathbf{j}} + \hat{\mathbf{k}})$
5. ∇f if $f(x,y,z) = \ln(x^2 + y^2 + z^2)$
6. $\nabla \cdot \mathbf{F}$ if $\mathbf{F}(x,y,z) = ye^x \hat{\mathbf{i}} + ze^y \hat{\mathbf{j}} + xe^z \hat{\mathbf{k}}$
7. $\nabla \times \mathbf{F}$ if $\mathbf{F}(x,y,z) = xyz\hat{\mathbf{j}}$
8. ∇f if $f(x,y) = \mathrm{Sin}^{-1}(x+y)$
9. $\nabla \cdot \mathbf{F}$ if $\mathbf{F}(y,z) = yz\hat{\mathbf{i}} - (y^2 + z^2)\hat{\mathbf{j}} + y^2 z^2 \hat{\mathbf{k}}$
10. $\nabla \times \mathbf{F}$ if $\mathbf{F}(x,y,z) = \mathrm{Cot}^{-1}(xyz)\hat{\mathbf{i}}$

In Exercises 11–30 evaluate the integral.

11. $\int_C y\,ds$ where C is the curve $y = x^3$ from $(-1,-1)$ to $(2,8)$

12. $\iint_S (x^2 + yz)\,dS$ where S is that part of $x + y + z = 2$ in the first octant

13. $\iint_S (x\hat{\mathbf{i}} + y\hat{\mathbf{j}}) \cdot \hat{\mathbf{n}}\,dS$ where S is that part of $z = x^2 + y^2$ bounded by the surfaces $x = \pm 1$, $y = \pm 1$, and $\hat{\mathbf{n}}$ is the lower normal

14. $\iint_S (x\hat{\mathbf{i}} + y\hat{\mathbf{j}}) \cdot \hat{\mathbf{n}}\,dS$ where S is that part of $z = x^2 + y^2$ below $z = 1$, and $\hat{\mathbf{n}}$ is the lower normal

15. $\oint_C x\,dx + y\,dy - z^2\,dz$ where C is the curve $x^2 + y^2 = 1$, $y = z$

16. $\int_C xy\,dx + xz\,dz$ where C is the curve $y = \sqrt{1+x^2}$, $z = \sqrt{2 - x^2 - y^2}$, from $(1/\sqrt{2}, \sqrt{3/2}, 0)$ to $(-1/\sqrt{2}, \sqrt{3/2}, 0)$

17. $\oint_C 2xy^3\,dx + (3x^2 y^2 + 2xy)\,dy$ where C is the curve $(x-1)^2 + y^2 = 1$

18. $\oint_C 2xy^3\,dx + (3x^2 y^2 + x^2)\,dy$ where C is the curve $(x-1)^2 + y^2 = 1$

19. $\iint_S (x^2 \hat{\mathbf{i}} + y^2 \hat{\mathbf{j}} + z^2 \hat{\mathbf{k}}) \cdot \hat{\mathbf{n}}\,dS$ where S is the surface bounding the volume enclosed by $y = z$, $y + z = 2$, $x = 0$, $x = 1$, $z = 0$, and $\hat{\mathbf{n}}$ is the outer normal to S

20. $\iint_S (x^2 + y^2)\,dS$ where S is that part of $x^2 + y^2 + z^2 = 6$ inside $z = x^2 + y^2$

21. $\iint_S (x^2 + y^2)\hat{\mathbf{i}} \cdot \hat{\mathbf{n}}\,dS$ where S is that part of $x^2 + y^2 + z^2 = 6$ inside $z = x^2 + y^2$, and $\hat{\mathbf{n}}$ is the upper normal to S

22. $\oint_C (x^2\hat{\mathbf{i}} + y\hat{\mathbf{j}} - xz\hat{\mathbf{k}}) \cdot d\mathbf{r}$ where C is the curve
    $x^2 + y^2 = 1$, $z = x + 1$, directed clockwise as viewed from the origin

23. $\oint_C (xy\hat{\mathbf{i}} + z\hat{\mathbf{j}} - x^2\hat{\mathbf{k}}) \cdot d\mathbf{r}$ where C is the curve in Exercise 22

24. $\oint_C y\,dx + 2x\,dy - 3z^2\,dz$ where C is the curve
    $y = \sqrt{1 + z^2 - x^2}$, $x^2 + z^2 = 1$, directed counterclockwise as viewed from the origin

25. $\oint_C (xy + 4x^3y^2)\,dx + (z + 2x^4y)\,dy + (z^5 + x^2z^2)\,dz$
    where C is the curve $x^2 + z^2 = 4$, $x^2 + y^2 = 4$, $y = z$, directed counterclockwise as viewed from a point far up the positive z-axis

26. $\iint_S (x^2yz\hat{\mathbf{i}} - x^2yz\hat{\mathbf{j}} - xyz^2\hat{\mathbf{k}}) \cdot \hat{\mathbf{n}}\,dS$ where S is that part of $z = 1 - \sqrt{x^2 - y^2}$ above the xy-plane, and $\hat{\mathbf{n}}$ is the upper normal

27. $\iint_S dS$ where S is that part of $z = x^2 - y^2$ inside $x^2 + y^2 = 4$

28. $\iint_S y\,dS$ where S is that part of $x = y^2 + 1$ in the first octant which is under $x + z = 2$

29. $\oint_C (ye^{xy} + xy^2e^{xy})\,dx + (xe^{xy} + x^2ye^{xy} + x^3y)\,dy$
    where C is the curve $x^2 + y^2 = 2y$, $z = 0$

30. $\iint_S (x\hat{\mathbf{i}} + y\hat{\mathbf{j}}) \cdot \hat{\mathbf{n}}\,dS$ where S is that part of $z^2 - x^2 - y^2 = 1$ between the planes $z = 0$ and $z = 2$, and $\hat{\mathbf{n}}$ is the lower normal

31. Let S be that part of the sphere $x^2 + y^2 + z^2 = 1$ that lies above the parabolic cylinder $2z = x^2$. Set up, but do not evaluate, double iterated integrals to calculate the surface integral
    $$\iint_S x^2y^2z^2\,dS$$
    by projecting S onto (a) the xy-coordinate plane (b) the yz-coordinate plane (c) the xz-coordinate plane.

32. If $\mathbf{r} = x\hat{\mathbf{i}} + y\hat{\mathbf{j}} + z\hat{\mathbf{k}}$, show that $\nabla(|\mathbf{r}|^n) = n|\mathbf{r}|^{n-2}\mathbf{r}$.

33. Verify that $\nabla \times (\nabla \times \mathbf{F}) = \nabla(\nabla \cdot \mathbf{F}) - \nabla^2\mathbf{F}$, where $\nabla^2 = \partial^2/\partial x^2 + \partial^2/\partial y^2 + \partial^2/\partial z^2$.

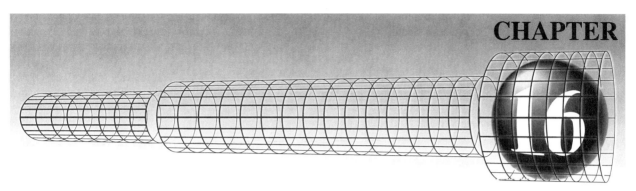

CHAPTER 16

Differential Equations

Differential equations serve as models in many areas of applied mathematics — physics, chemistry, economics, medicine, and engineering to name a few. In this chapter we discuss some of the methods for solving first-order and simple second-order equations. We also give a fairly thorough treatment of linear differential equations. We include a wide variety of applications to illustrate the relevance of differential equations in applied mathematics.

SECTION 16.1

Introduction

A **differential equation** is an equation that must be solved for an unknown function. What distinguishes a differential equation from other equations is the fact that it contains derivatives of the unknown function. For example, each of the following equations is a differential equation in y as a function of x:

$$\frac{dy}{dx} + \frac{2}{3}y = 9.81, \qquad (16.1)$$

$$\frac{d^2 y}{dx^2} = k\sqrt{1 + \left(\frac{dy}{dx}\right)^2}, \qquad (16.2)$$

$$x\frac{d^2 y}{dx^2} + \frac{dy}{dx} + xy = 0, \qquad (16.3)$$

$$\frac{d^4 y}{dx^4} - k^4 y = 0. \qquad (16.4)$$

Equation 16.1 is used to determine the position of a paratrooper who falls under the influences of gravity and an air resistance that is proportional to velocity (see Example 16.8); equation 16.2 describes the shape of a hanging cable (Exercise 11 in Section 16.4); equation 16.3, called Bessel's differential equation of order zero, is found in heat flow and vibration problems; and equation 16.4 is used to determine the deflection of beams.

Definition 16.1

The **order** of a differential equation is the order of the highest derivative in the equation.

Of the four differential equations 16.1–16.4, the first is first order, the second and third are second order, and the last is fourth order.

We have considered quite a number of differential equations in Chapters 3, 5, 8, and 9. In Section 5.4 we concentrated on describing physical situations by means of differential equations; in Chapters 3 and 8 we verified that particular combinations of transcendental functions satisfied certain differential equations; and in Chapter 9 we used our integration techniques to solve various equations. Almost all of these differential equations were based on applications, many from physics and engineering, but also some from geometry and other fields such as ecology, chemistry, and psychology. Because of this association with applications, differential equations are almost always accompanied by subsidiary conditions called **initial** or **boundary conditions**. For example, if a mass m falls from rest under gravity and is acted on by a force due to air resistance that is proportional to its instantaneous velocity, then the differential equation that describes its velocity $v(t)$ as a function of time t is

$$m\frac{dv}{dt} = -kv + mg, \quad \text{where } k \text{ is a constant.} \tag{16.5}$$

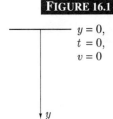

FIGURE 16.1

$y = 0$,
$t = 0$,
$v = 0$

This is simply a statement of Newton's second law, where dv/dt is the vertical component of the acceleration of m, and $-kv + mg$ is the vertical component of the total force on m due to gravity (mg) and air resistance ($-kv$). If distance y is chosen as positive downward (Figure 16.1), both g and k are positive. Furthermore, if we choose time $t = 0$ at the instant m is dropped, then because m falls from rest, the condition $v(0) = 0$ must be added to the differential equation. In other words, the real problem is to find the solution of the differential equation that also satisfies the initial condition:

$$m\frac{dv}{dt} = -kv + mg, \quad v(0) = 0. \tag{16.6}$$

This is the form in which applied mathematicians find differential equations—the differential equation is accompanied by subsidiary conditions that express physical requirements of the solution. It is not difficult to show that the solution of equation 16.6 is

$$v(t) = \frac{mg}{k} - \frac{mg}{k}e^{-kt/m}. \tag{16.7}$$

(All we need do to verify this is to substitute the function into the differential equation to see that it does indeed satisfy the equation. It is clear that it does satisfy the initial condition.)

If we change the initial condition to $v(0) = v_0$, so that the initial velocity of m has vertical component v_0, then the solution becomes

$$v(t) = \frac{mg}{k} - \left(\frac{mg}{k} - v_0\right)e^{-kt/m}. \tag{16.8}$$

In other words, every solution of differential equation 16.5 can be written in the form

$$v(t) = \frac{mg}{k} + Ce^{-kt/m}, \tag{16.9}$$

and when we impose the initial condition $v(0) = v_0$, then $C = v_0 - mg/k$.

In a similar way, equation 16.4 is normally accompanied by four boundary conditions such as perhaps

$$y(0) = y(L) = 0;$$
$$y''(0) = y''(L) = 0.^\dagger$$

It can be shown that every solution of equation 16.4 can be expressed in the form

$$y(x) = C_1 e^{kx} + C_2 e^{-kx} + C_3 \sin(kx) + C_4 \cos(kx), \qquad (16.10)$$

where $C_1, C_2, C_3,$ and C_4 are arbitrary constants, and when the boundary conditions are applied, these constants must satisfy the four equations

$$0 = C_1 + C_2 + C_4,$$
$$0 = C_1 e^{kL} + C_2 e^{-kL} + C_3 \sin(kL) + C_4 \cos(kL),$$
$$0 = C_1 + C_2 - C_4,$$
$$0 = C_1 e^{kL} + C_2 e^{-kL} - C_3 \sin(kL) - C_4 \cos(kL).$$

We have stated that every solution of 16.5 can be written in the form 16.9, and every solution of 16.4 can be expressed as 16.10. Note that 16.9 contains one arbitrary constant whereas 16.10 has four, but in both cases the number of arbitrary constants is exactly the same as the order of the differential equation. We might suspect that every solution of an nth-order differential equation can be expressed as a function involving n arbitrary constants. For many differential equations this is indeed true, but unfortunately it is not true for all equations. As an illustration, consider the equation

$$\frac{d^2 y}{dx^2} = \left(\frac{dy}{dx}\right)^2. \qquad (16.11)$$

In Example 16.6 we apply standard techniques for solving differential equations to obtain the solution $y(x) = C_1 - \ln(C_2 + x)$, which contains two arbitrary constants C_1 and C_2. This two-parameter family of solutions does not, however, contain all solutions of the differential equation, for no choice of C_1 and C_2 will give the perfectly acceptable solution $y(x) \equiv 1$. This solution is not particularly interesting, but it is nonetheless a solution that is not contained within the two-parameter family. Such a solution is called a **singular solution**. We have illustrated that a solution that contains the same number of arbitrary constants as the order of the differential equation may or may not contain all solutions of the differential equation. In spite of this unfortunate circumstance, there do exist large classes of differential equations for which a solution with the same number of arbitrary constants as the order of the equation does indeed represent all possible solutions. Because of this we make the following definition.

> **Definition 16.2**
>
> A **general solution** of a differential equation is a solution that contains the same number of arbitrary constants as the order of the differential equation.

Consequently, in order for a function to be a general solution of a differential equation, it must first be a solution, and second, contain the requisite number of arbitrary constants.

Once again we emphasize that a general solution of a differential equation may or may not contain all possible solutions. We will point out important cases in which a general solution does indeed contain all solutions.

---

† In this chapter it is frequently convenient to use the notation $y', y'', y''', \ldots$ to represent $dy/dx, d^2y/dx^2, d^3y/dx^3$, etc. In this notation $y''(a)$ is the second derivative of y evaluated at $x = a$. In addition, we denote the solution of a differential equation in y as a function of x by $y(x)$.

EXAMPLE 16.1

Show that $y(x) = C_1 \cos 2x + C_2 \sin 2x$ is a general solution of the differential equation

$$\frac{d^2 y}{dx^2} + 4y = 0.$$

SOLUTION If we substitute this function into the left-hand side of the equation, we have

$$\frac{d^2 y}{dx^2} + 4y = \{-4C_1 \cos 2x - 4C_2 \sin 2x\} + 4\{C_1 \cos 2x + C_2 \sin 2x\} = 0,$$

and the function is indeed a solution. Because it contains two arbitrary constants, the order of the differential equation, it is a general solution. ∎

Definition 16.3

A **particular solution** of a differential equation is a solution that contains no arbitrary constants.

It follows, therefore, that particular solutions can be obtained by assigning specific values to the arbitrary constants in a general solution. For example, $y(x) = 5 - \ln(3 + x)$ is a particular solution of differential equation 16.11, as is $y(x) = -\ln x$, both being obtained from $y(x) = C_1 - \ln(C_2 + x)$. On the other hand, the singular solution $y(x) = 10$ is also a particular solution, but it cannot be obtained from this general solution.

EXAMPLE 16.2

Find a particular solution of the differential equation

$$5\frac{d^3 y}{dx^3} + 3\frac{d^2 y}{dx^2} + 2y = 4.$$

SOLUTION In Section 16.10 we develop systematic techniques for finding general and particular solutions for differential equations such as this. But clearly those techniques are not needed here, since a simple glance tells us that $y(x) = 2$ is a solution. ∎

Many differential equations are immediately solvable (or, as we often say, immediately integrable). For example, to solve a differential equation of the form

$$\frac{dy}{dx} = M(x), \qquad (16.12)$$

where $M(x)$ is given, we integrate both sides of the equation with respect to x to obtain the general solution

$$y(x) = \int M(x)\,dx + C. \qquad (16.13)$$

We have called this *the* general solution rather than *a* general solution because in this case the general solution does indeed contain all solutions of the differential equation.

This result is easily extended to the nth-order equation

$$\frac{d^n y}{dx^n} = M(x), \quad n \text{ a positive integer.} \qquad (16.14)$$

We integrate successively n times to obtain the general solution

$$y(x) = \int \cdots \int M(x) \, dx \cdots dx + C_1 + C_2 x + \cdots + C_n x^{n-1}. \qquad (16.15)$$

EXERCISES 16.1

In Exercises 1–10 show that the function is a general solution of the differential equation.

1. $y(x) = 2 + Ce^{-x^2}$; $\dfrac{dy}{dx} + 2xy = 4x$

2. $y(x) = \dfrac{x}{1 + Cx}$; $\dfrac{dy}{dx} = \dfrac{y^2}{x^2}$

3. $y(x) = \dfrac{x^3}{2} + Cx^3 e^{1/x^2}$; $x^3 \dfrac{dy}{dx} + (2 - 3x^2)y = x^3$

4. $y(x) = C_1 \sin 3x + C_2 \cos 3x$; $\dfrac{d^2 y}{dx^2} + 9y = 0$

5. $y(x) = \dfrac{C_1^2 e^{2x} + 1}{2C_1 e^x} + C_2$; $\left(\dfrac{d^2 y}{dx^2}\right)^2 = 1 + \left(\dfrac{dy}{dx}\right)^2$

6. $y(x) = C_1 e^{2x} \cos\left(\dfrac{x}{\sqrt{2}}\right) + C_2 e^{2x} \sin\left(\dfrac{x}{\sqrt{2}}\right)$;
$2\dfrac{d^2 y}{dx^2} - 8\dfrac{dy}{dx} + 9y = 0$

7. $y(x) = C_1 \cos 2x + C_2 \sin 2x + C_3 \cos x + C_4 \sin x$;
$\dfrac{d^4 y}{dx^4} + 5\dfrac{d^2 y}{dx^2} + 4y = 0$

8. $y(x) = \left(C_1 + C_2 x - \dfrac{x^2}{4}\right)e^{4x}$;
$2\dfrac{d^2 y}{dx^2} - 16\dfrac{dy}{dx} + 32y = -e^{4x}$

9. $y(x) = C_1 \cos(2 \ln x) + C_2 \sin(2 \ln x) + \dfrac{1}{4}$;
$x^2 \dfrac{d^2 y}{dx^2} + x\dfrac{dy}{dx} + 4y = 1$

10. $y(x) = C_1 \dfrac{\sin x}{\sqrt{x}} + C_2 \dfrac{\cos x}{\sqrt{x}}$;
$x^2 \dfrac{d^2 y}{dx^2} + x\dfrac{dy}{dx} + \left(x^2 - \dfrac{1}{4}\right)y = 0$

In Exercises 11–14 find a particular solution of the differential equation in Exercise 4 that satisfies the conditions.

11. $y(0) = 1$, $y'(0) = 6$

12. $y(0) = 2$, $y(\pi/2) = 3$

13. $y(\pi/12) = 0$, $y'(\pi/12) = 1$

14. $y(1) = 1$, $y(2) = 2$

In Exercises 15–20 find a general solution of the differential equation.

15. $\dfrac{dy}{dx} = 6x^2 + 2x$

16. $\dfrac{dy}{dx} = \dfrac{1}{9 + x^2}$

17. $\dfrac{d^2 y}{dx^2} = 2x + e^x$

18. $\dfrac{d^2 y}{dx^2} = x \ln x$

19. $\dfrac{d^3 y}{dx^3} = \dfrac{1}{3x^5}$

20. $\dfrac{dy}{dx} = y$

21. **(a)** A boy initially at O (Figure 16.2) walks along the edge of a swimming pool (the y-axis) towing his sailboat by a string of length L. If the boat starts at Q and the string always remains straight, show that the equation of the curved path $y = y(x)$ followed by the boat must satisfy the differential equation

$$\dfrac{dy}{dx} = -\dfrac{\sqrt{L^2 - x^2}}{x}.$$

(b) Solve this differential equation for $y(x)$.

FIGURE 16.2

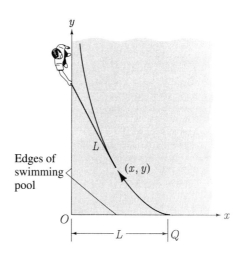

SECTION 16.1 INTRODUCTION

22. Show that $y(x) = -(x^2 + C)^{-1}$ is a general solution of the differential equation

$$\frac{dy}{dx} = 2xy^2.$$

Find a singular solution.

23. Show that $y(x) = 1 - (x^3 + C)^{-1}$ is a general solution of the differential equation

$$\frac{dy}{dx} = 3x^2(y-1)^2.$$

Find a singular solution.

24. (a) Verify that $y(x) = Ce^{2x}$ is a general solution of the differential equation

$$\frac{dy}{dx} = 2y.$$

(b) Draw the one-parameter family of curves defined by this general solution.

(c) Show that there is a particular solution that passes through any given point (x_0, y_0), and that this solution can be obtained by choosing C appropriately.

25. (a) Verify that a general solution of the differential equation

$$2x \frac{dy}{dx} = y$$

is defined implicitly by the equation $y^2 = Cx$.

(b) Draw the one-parameter family of curves defined by this equation.

(c) Show that, with the exception of points on the y-axis, there is a particular solution that passes through any given point (x_0, y_0), and that this solution can be obtained by specifying C appropriately.

26. (a) Draw the one-parameter family of curves defined by the general solution in Exercise 22.

(b) Show that, with the exception of points on the x-axis, there is a solution passing through any given point in the xy-plane.

27. Consider the differential equation

$$\frac{dy}{dx} = \frac{1}{x^2}.$$

(a) Find a solution that satisfies the condition $y(1) = 1$.

(b) Find a solution that satisfies the condition $y(-1) = 2$.

(c) Find a solution that satisfies the conditions in both (a) and (b).

(d) Draw the curves defined by the general solution to explain why this is possible.

SECTION 16.2

Separable Differential Equations

In this section and the next, we consider first-order differential equations that can be written in the form

$$\frac{dy}{dx} = F(x, y). \qquad (16.16)$$

Since any function $F(x, y)$ can always be considered as the quotient of two other functions,

$$F(x, y) = \frac{M(x, y)}{N(x, y)},$$

equation 16.16 can also be written in the equivalent form

$$N(x, y)\, dy = M(x, y)\, dx. \qquad (16.17)$$

Depending on the form of $F(x, y)$ (or $M(x, y)$ and $N(x, y)$) various methods can be used to obtain the unknown function $y(x)$. Two of the more important techniques are considered here and in Section 16.3; others are discussed in the exercises.

Differential equation 16.16 is said to be separable if it can be expressed in the form

$$\frac{dy}{dx} = \frac{M(x)}{N(y)}; \qquad (16.18)$$

that is, if dy/dx is equal to a function of x divided by a function of y. Equivalently, a differential equation is said to be separable if it can be written in the form

$$N(y)\,dy = M(x)\,dx. \qquad (16.19)$$

When a differential equation is written in this way, it is said to be separated—separated in the sense that the x- and y-variables appear on opposite sides of the equation. For a separated equation we can write therefore that

$$N(y)\frac{dy}{dx} = M(x), \qquad (16.20)$$

and if we integrate both sides with respect to x, we have

$$\int N(y)\frac{dy}{dx}\,dx = \int M(x)\,dx + C. \qquad (16.21)$$

Cancellation of differentials on the left leads to the solution

$$\int N(y)\,dy = \int M(x)\,dx + C. \qquad (16.22)$$

What we mean by saying that 16.21 and 16.22 represent general solutions of 16.19 is that any function defined *implicitly* by 16.21 or 16.22 is a solution of 16.19. For example, if we divide the differential equation

$$3x^3y^2\,dx + (xy^3 - xy^2)\,dy = 0$$

by xy^2, it becomes separated:

$$(1-y)\,dy = 3x^2\,dx.$$

According to 16.21 we should divide by dx and integrate both sides with respect to x:

$$\int (1-y)\frac{dy}{dx}\,dx = \int 3x^2\,dx = x^3 + C.$$

The antidifferentiation on the left must be interpreted as "implicit antidifferentiation," asking for that function which when differentiated with respect to x gives $(1-y)\,dy/dx$. Since an antiderivative is $y - y^2/2$, a general solution of the differential equation is

$$y - \frac{y^2}{2} = x^3 + C.$$

Were we to use 16.22 after separation (instead of 16.21), we would write

$$\int (1-y)\,dy = \int 3x^2\,dx + C,$$

and integrate for a general solution

$$y - \frac{y^2}{2} = x^3 + C.$$

As indicated above, by saying that $y - y^2/2 = x^3 + C$ represents a general solution of the original differential equation, we mean that any function defined implicitly by $y - y^2/2 = x^3 + C$ is a solution.

EXAMPLE 16.3

Find general solutions for the following differential equations:

(a) $2x^3 y^2 \, dx = xy^3 \, dy$

(b) $\dfrac{dy}{dx} = \dfrac{y \sin x + y^3 \sin x}{(1+y^2)^2}$

SOLUTION

(a) If we divide the differential equation by xy^2, we obtain

$$y \, dy = 2x^2 \, dx,$$

which is separated. A general solution is therefore defined implicitly by

$$\int y \, dy = \int 2x^2 \, dx + C,$$

or

$$\dfrac{y^2}{2} = \dfrac{2x^3}{3} + C.$$

Note that $y(x) = 0$ is also a solution, but it cannot be obtained by specifying C. In other words, $y(x) = 0$ is a singular solution. We removed this solution when we divided the original equation by xy^2.

(b) Since

$$\dfrac{dy}{dx} = \dfrac{y \sin x (1+y^2)}{(1+y^2)^2} = \dfrac{y \sin x}{1+y^2},$$

the differential equation can be separated:

$$\sin x \, dx = \dfrac{1+y^2}{y} dy = \left(\dfrac{1}{y} + y\right) dy.$$

A general solution is therefore defined implicitly by

$$\int \sin x \, dx + C = \int \left(\dfrac{1}{y} + y\right) dy,$$

or,

$$-\cos x + C = \ln|y| + \dfrac{y^2}{2}.$$

∎

EXAMPLE 16.4

Find the function (in explicit form) that satisfies the differential equation

$$e^{2x+y} dx - 2e^{x-y} dy = 0, \quad y(0) = 1.$$

SOLUTION If we divide the equation by e^{x+y}, we have

$$e^x \, dx = 2e^{-2y} dy.$$

A general solution is defined implicitly by

$$e^x = -e^{-2y} + C.$$

The condition $y(0) = 1$ requires $e^0 = -e^{-2} + C$, which implies that $C = 1 + e^{-2}$. The required solution is therefore defined implicitly by

$$e^x = -e^{-2y} + e^{-2} + 1,$$

and from this equation, we have

$$e^{-2y} = e^{-2} + 1 - e^x, \quad \text{or} \quad y(x) = -\tfrac{1}{2} \ln(e^{-2} + 1 - e^x).$$

∎

EXERCISES 16.2

In Exercises 1–10 find a general solution of the differential equation.

1. $y^2 \, dx - x^2 \, dy = 0$

2. $\dfrac{dy}{dx} + 2xy = 4x$

3. $2xy \, dx + (x^2 + 1) \, dy = 0$

4. $\dfrac{dy}{dx} = 3y + 2$

5. $3(y^2 + 2) \, dx = 4y(x - 1) \, dy$

6. $(x^2 y + x^2) \, dx + (xy^2 - y^2) \, dy = 0$

7. $\dfrac{dy}{dx} = -\dfrac{\cos y}{\sin x}$

8. $(x^2 y e^x - y) \, dx + xy^3 \, dy = 0$

9. $(x^2 y^2 \sec x \tan x + xy^2 \sec x) \, dx + xy^3 \, dy = 0$

10. $\dfrac{dy}{dx} = \dfrac{1 + y^2}{1 + x^2}$

In Exercises 11–15 find a solution of the differential equation that also satisfies the given condition.

11. $2y \, dx + (x + 1) \, dy = 0$, $y(1) = 2$

12. $(xy + y) \, dx - (xy - x) \, dy = 0$, $y(1) = 2$

13. $\dfrac{dy}{dx} = e^{x+y}$, $y(0) = 0$

14. $\dfrac{dy}{dx} = 2x(1 + y^2)$, $y(2) = 4$

15. $\dfrac{dy}{dx} = \dfrac{\sin^2 y}{\cos^2 x}$, $y(0) = \pi/2$

16. A girl lives 6 km from school. She decides to travel to school so that her speed is always proportional to the square of her distance from the school.
    (a) Find her distance from school at any time.
    (b) When does she reach school?

17. Find a general solution for the differential equation

$$\dfrac{dy}{dx} = -\dfrac{1 + y^3}{xy^2 + x^3 y^2}.$$

18. When a container of water at temperature 80 °C is placed in a room at temperature 20 °C, Newton's law of cooling states that the time rate of change of the temperature of the water is proportional to the difference between the temperature of the water and room temperature. If the water cools to 60 °C in 2 min, find a formula for its temperature as a function of time.

19. A thermometer reading 23 °C is taken outside where the temperature is −20 °C. If the reading drops to 0 °C in 4 min, when will it read −19 °C?

20. The amount of a drug such as penicillin injected into the body is used up at a rate proportional to the amount still present. If a dose decreases by 5% in the first hour, when does it decrease to one-half its original amount?

21. When a deep-sea diver inhales air, his body tissues absorb extra amounts of nitrogen. Suppose the diver enters the water at time $t = 0$, drops very quickly to depth d, and remains at this depth for a very long time. The amount N of nitrogen in his body tissues increases as he remains at this depth until a maximum amount $\overline{N}$ is reached. The time rate of change of N is proportional to the difference $\overline{N} - N$. Show that if N_0 is the amount of nitrogen in his body tissues when he enters the water, then

$$N = N_0 e^{-kt} + \overline{N}(1 - e^{-kt}),$$

where $k > 0$ is a constant.

22. When a substance such as glucose is administered intravenously into the bloodstream, it is used up by the body at a rate proportional to the amount present at that time. If it is added at a constant rate of R units per unit time, and A_0 is the amount present in the bloodstream when the intravenous feeding begins, find a formula for the amount in the bloodstream at any time.

23. Find the equations for all curves that satisfy the condition that the normal at any point on the curve, and the line joining the point to the origin, form an isosceles triangle with the x-axis as base.

24. When two substances A and B are brought together in one solution, they react to form a third substance C in such a way that one gram of A reacts with one gram of B to produce two grams of C. The rate at which C is formed is proportional to the amounts of A and B still present in the solution. If 10 g of A and 15 g of B are originally brought together, find a formula for the amount of C present in the mixture at any time.

25. What is the solution to Exercise 24 when the initial amounts of A and B are both 10 g?

26. A first-order differential equation in $y(x)$ is said to be *homogeneous* if it can be written in the form

$$\frac{dy}{dx} = f\left(\frac{y}{x}\right).$$

Show that the change of dependent variable $v = y/x$ yields a differential equation in $v(x)$ that is always separable.

27. State a condition on the functions M and N in equation 16.17 in order that the differential equation be homogeneous in the sense of Exercise 26.

In Exercises 28–33 show that the differential equation is homogeneous, and use the change of variable in Exercise 26 to find a general solution.

28. $(y^2 - x^2)\, dx + xy\, dy = 0$

29. $2x\, dy - 2y\, dx = \sqrt{x^2 + 4y^2}\, dx$

30. $\dfrac{dy}{dx} = \dfrac{y + x}{y - x}$

31. $x\, dy - y\, dx = x \cos(y/x)\, dx$

32. $\dfrac{dy}{dx} = \dfrac{x^2 e^{-y/x} + y^2}{xy}$

33. $(x^2 y + y^3)\, dx + x^3\, dy = 0$

34. If a curve passes through the point $(1, 2)$ and is such that the length of that part of the tangent line at (x, y) from (x, y) to the y-axis is equal to the y-intercept of the tangent line, find the equation of the curve.

35. Find in explicit form a function that satisfies the differential equation $dy/dx = \csc y$ and the following conditions: **(a)** $y(0) = \pi/4$ **(b)** $y(0) = 7\pi/4$

36. In a chemical reaction, one molecule of trypsinogen yields one molecule of trypsin. In order for the reaction to take place, an initial amount of trypsin must be present. Suppose that the initial amount is y_0. Thereafter, the rate at which trypsinogen is changed into trypsin is proportional to the product of the amounts of each chemical in the reaction. Find a formula for the amount of trypsin if the initial amount of trypsinogen is A.

37. If a first-order differential equation can be written in the form

$$\frac{dy}{dx} = f(ax + by),$$

where a and b are constants, show that the change of dependent variable $v = ax + by$ always gives a differential equation in $v(x)$ that is separable.

Use the method of Exercise 37 to find a general solution of the differential equation in Exercises 38–41.

38. $\dfrac{dy}{dx} = x + y$

39. $\dfrac{dy}{dx} = (x + y)^2$

40. $\dfrac{dy}{dx} = \dfrac{1}{2x + 3y}$

41. $\dfrac{dy}{dx} = \sin^2(x - y)$

42. A certain chemical dissolves in water at a rate proportional to the product of the amount of undissolved chemical and the difference between the concentrations in a saturated solution and the existing concentration in the solution. A saturated solution contains 25 g of chemical in 100 mL of solution.
   (a) If 50 g of chemical are added to 200 mL of water, find a formula for the amount of chemical dissolved as a function of time. Sketch its graph.
   (b) Repeat (a) if the 50 g of chemical are added to 100 mL of water.
   (c) Repeat (a) if 10 g of chemical are added to 100 mL of water.

43. Snow has been falling for some time when a snowplow starts plowing the highway. The plow begins at 12:00 and travels 2 km during the first hour and 1 km during the second hour. Make reasonable assumptions to find out when the snow started falling.

44. In order to perform a one-hour operation on a dog a veterinarian anesthetizes the dog with sodium pentobarbital. During the operation, the dog's body breaks down the drug at a rate proportional to the amount still present, and only half an original dose remains after 5 h. If the dog has mass 20 kg, and 20 mg of sodium pentobarbital per kilogram of body mass are required to maintain surgical anesthesia, what original dose is required?

45. Explain what the cancellation of differentials in proceeding from 16.21 to 16.22 really means.

46. Solve the differential equation

$$(x^3 y^4 + 2xy^4)\,dx + (x - xy^6)\,dy = 0, \quad y(1) = 1.$$

47. Two substances A and B react to form a third substance C in such a way that 2 g of A react with 1 g of B to produce 3 g of C. The rate at which C is formed is proportional to the amounts of A and B still present in the mixture. Find the amount of C present in the mixture as a function of time when the original amounts of A and B brought together at time $t = 0$ are as follows. Sketch graphs of all three functions on the same axes.

(a) 20 g and 10 g
(b) 20 g and 5 g
(c) 20 g and 20 g

48. A cylindrical tank of radius r and height H has a vertical axis and no top. It is originally (time $t = 0$) full of water, and for $t > 0$, water leaks out through a hole in its bottom.

(a) To find the speed at which water exits through the hole consider the situation when the depth of water is h. In a small time dt, the depth drops by an amount dh. Use the fact that during dt, the potential energy of the small disc of water with thickness $|dh|$ at height h above the bottom of the cylinder is converted into kinetic energy of the same volume exiting through the hole with speed v to show that $v = \sqrt{2gh}$ where $g = 9.81$. This formula describes an ideal situation where all potential energy is converted into kinetic energy. In less than ideal situations, it is customary to assume that $v = c\sqrt{2gh}$, where c is a constant and $0 < c < 1$.

(b) Show that the rate of change of the volume of water in the cylinder is given by the two expressions

$$\frac{dV}{dt} = \pi r^2 \frac{dh}{dt} \quad \text{and} \quad \frac{dV}{dt} = -Av$$

where A is the cross-sectional area of the hole, and hence deduce that

$$\frac{dh}{dt} = -\frac{\sqrt{2g}\,cA}{\pi r^2}\sqrt{h}.$$

This is sometimes called Torricelli's law.

(c) Find h as a function of t and thereby determine a formula for how long it takes for the cylinder to empty.

49. Repeat Exercise 48 for an open-topped right-circular cone of radius r and height H (with hole at the vertex).

50. A bird is due east of its nest a distance L away and at the same height above the ground as the nest. Wind is blowing due north at speed v. If the bird flies horizontally with constant speed V always pointing straight at its nest, what is the equation of the curve that it follows. Take the nest at the origin and the x- and y-directions as east and north.

51. Find the equation of the curve that passes through $(1, 1)$ and is such that the tangent and normal lines at any point (x, y) make with the x-axis a triangle whose area is equal to the slope of the tangent line at (x, y).

SECTION 16.3

Linear First-Order Differential Equations

A first-order differential equation that can be written in the form

$$\frac{dy}{dx} + P(x)y = Q(x) \qquad (16.23)$$

is said to be linear. We will explain the significance of the adjective "linear" in Section 16.7. To illustrate how to solve such differential equations, consider the equation

$$\frac{dy}{dx} + \frac{1}{x}y = 1.$$

If we multiply both sides by x, we have

$$x\frac{dy}{dx} + y = x.$$

But note now that the left side is the derivative of the product xy; i.e.,

$$\frac{d}{dx}(xy) = x\frac{dy}{dx} + y.$$

In other words, we can write the differential equation in the form

$$\frac{d}{dx}(xy) = x,$$

and integration immediately gives a general solution

$$xy = \frac{x^2}{2} + C.$$

This is the principle behind all linear first-order equations: Multiply the equation by a function of x in order that the left side is expressible as the derivative of a product. To show that this is always possible, we turn now to the general equation 16.23. If the equation is multiplied by a function $\mu(x)$,

$$\mu\frac{dy}{dx} + \mu P(x)y = \mu Q(x). \tag{16.24}$$

This equation is equivalent to 16.23 in the sense that $y(x)$ is a solution of 16.23 if and only if it is a solution of 16.24. We ask whether it is possible to find μ so that the left side of 16.24 can be written as the derivative of the product μy; that is, can we find $\mu(x)$ so that

$$\mu\frac{dy}{dx} + \mu P(x)y = \frac{d}{dx}(\mu y)?$$

If we expand the right side, μ must satisfy

$$\mu\frac{dy}{dx} + \mu P(x)y = \mu\frac{dy}{dx} + y\frac{d\mu}{dx},$$

from which we get

$$\mu P = \frac{d\mu}{dx} \quad \text{or} \quad \frac{d\mu}{\mu} = P\,dx.$$

Thus μ must satisfy a separated differential equation, one solution of which is

$$\ln|\mu| = \int P(x)\,dx, \quad \text{or} \quad \mu = \pm e^{\int P(x)\,dx}.$$

We have shown then that if 16.23 is multiplied by the factor

$$e^{\int P(x)\,dx} \quad (\text{or by } -e^{\int P(x)\,dx}),$$

then the differential equation becomes

$$e^{\int P(x)\,dx}\frac{dy}{dx} + P(x)ye^{\int P(x)\,dx} = Q(x)e^{\int P(x)\,dx},$$

and the left side can be expressed as the derivative of a product:

$$\frac{d}{dx}\{ye^{\int P(x)\,dx}\} = Q(x)e^{\int P(x)\,dx}.$$

Integration now gives a general solution

$$ye^{\int P(x)\,dx} = \int Q(x)e^{\int P(x)\,dx}\,dx + C. \qquad (16.25)$$

The quantity $e^{\int P(x)\,dx}$ is called an **integrating factor** for equation 16.23 because when equation 16.23 is multiplied by this factor, it becomes immediately integrable.

In summary, if linear differential equation 16.23 is multiplied by the function $e^{\int P(x)\,dx}$, then the left side of the equation becomes the derivative of the product of y and this function. In this form the equation can immediately be integrated to give a general solution. Given a specific linear equation to solve, we have three choices on how to proceed:

(1) Memorize 16.25 as a formula.

(2) Memorize that an integrating factor is $e^{\int P(x)\,dx}$.

(3) Develop the integrating factor in every example.

We feel that (2) is a reasonable compromise in that we must memorize something, but not too much, and we must do some calculations, but not too many.

EXAMPLE 16.5

Find general solutions for the following differential equations:

(a) $\dfrac{dy}{dx} + xy = x$

(b) $(y - x\sin x)\,dx + x\,dy = 0$

(c) $\cos x \dfrac{dy}{dx} + y\sin x = 1$

SOLUTION

(a) An integrating factor for this linear equation is

$$e^{\int x\,dx} = e^{x^2/2}.$$

If we multiply the equation by this integrating factor, we have

$$e^{x^2/2}\dfrac{dy}{dx} + yxe^{x^2/2} = xe^{x^2/2},$$

or,

$$\dfrac{d}{dx}\{ye^{x^2/2}\} = xe^{x^2/2}.$$

Integration yields

$$ye^{x^2/2} = \int xe^{x^2/2}\,dx = e^{x^2/2} + C,$$

or,

$$y(x) = 1 + Ce^{-x^2/2}.$$

(b) If we write the differential equation in the form

$$\dfrac{dy}{dx} + \dfrac{y}{x} = \sin x,$$

then we see that it is linear first order. An integrating factor is therefore

$$e^{\int 1/x\,dx} = e^{\ln|x|} = |x|.$$

If we multiply the differential equation by this factor, we get

$$|x|\frac{dy}{dx} + \frac{|x|}{x}y = |x|\sin x.$$

If $x > 0$, then we write

$$x\frac{dy}{dx} + \frac{x}{x}y = x\sin x,$$

whereas if $x < 0$,

$$-x\frac{dy}{dx} - \frac{x}{x}y = -x\sin x.$$

In either case, however, the equation simplifies to

$$x\frac{dy}{dx} + y = x\sin x,$$

or

$$\frac{d}{dx}(xy) = x\sin x.$$

Integration now gives

$$xy = \int x\sin x\,dx = -x\cos x + \sin x + C.$$

Finally, then,

$$y(x) = -\cos x + \frac{\sin x}{x} + \frac{C}{x}.$$

(c) An integrating factor for the linear equation

$$\frac{dy}{dx} + y\tan x = \frac{1}{\cos x}$$

is

$$e^{\int \tan x\,dx} = e^{\ln|\sec x|} = |\sec x|.$$

For either $\sec x < 0$ or $\sec x > 0$, we obtain

$$\frac{1}{\cos x}\frac{dy}{dx} + y\frac{\sin x}{\cos^2 x} = \frac{1}{\cos^2 x},$$

or,

$$\frac{d}{dx}\left\{\frac{y}{\cos x}\right\} = \sec^2 x.$$

Integration now yields
$$\frac{y}{\cos x} = \tan x + C,$$
and hence
$$y(x) = \sin x + C \cos x.$$

∎

EXERCISES 16.3

In Exercises 1–12 find a general solution for the differential equation.

1. $\dfrac{dy}{dx} + 2xy = 4x$

2. $\dfrac{dy}{dx} + \dfrac{2}{x}y = 6x^3$

3. $(2y - x)\,dx + dy = 0$

4. $\dfrac{dy}{dx} + y \cot x = 5e^{\cos x}$

5. $(x^2 + 2xy)\,dx + (x^2 + 1)\,dy = 0$

6. $(x+1)\dfrac{dy}{dx} - 2y = 2(x+1)$

7. $\dfrac{1}{x}\dfrac{dy}{dx} - \dfrac{y}{x^2} = \dfrac{1}{x^3}$

8. $(y + e^{2x})\,dx = dy$

9. $\dfrac{dy}{dx} + y = 2\cos x$

10. $x^3\dfrac{dy}{dx} + (2 - 3x^2)y = x^3$

11. $\dfrac{dy}{dx} + \dfrac{y}{x \ln x} = x^2$

12. $(-2y \cot 2x - 1 + 2x \cot 2x + 2 \csc 2x)\,dx + dy = 0$

In Exercises 13–15 solve the differential equation.

13. $\dfrac{dy}{dx} + 3x^2 y = x^2$, $y(1) = 2$

14. $(-e^x \sin x + y)\,dx + dy = 0$, $y(0) = -1$

15. $\dfrac{dy}{dx} + \dfrac{x^3 y}{x^4 + 1} = x^7$, $y(0) = 1$

16. Find a general solution for the differential equation $(y^3 - x)\,dy = y\,dx$.

17. A differential equation of the form
$$\frac{dy}{dx} + P(x)y = y^n Q(x)$$
is called a *Bernoulli equation*. Show that the change of dependent variable $z = y^{1-n}$ gives
$$\frac{dz}{dx} + (1-n)Pz = (1-n)Q,$$
a linear first-order equation in $z(x)$.

In Exercises 18–22 use the change of variable in Exercise 17 to find a general solution for the differential equation.

18. $\dfrac{dy}{dx} + y = y^2 e^x$

19. $\dfrac{dy}{dx} + \dfrac{y}{x} = \dfrac{y^2}{x^2}$

20. $\dfrac{dy}{dx} - y + (x^2 + 2x)y^2 = 0$

21. $x\,dy + y\,dx = x^3 y^5\,dx$

22. $\dfrac{dy}{dx} + y \tan x = y^4 \sin x$

23. Repeat Exercise 22 in Section 16.2 if the glucose is added at a rate $R(t)$ that is a function of time t.

24. A tank has 100 L of solution containing 4 kg of sugar. A mixture with 10 g of sugar per litre of solution is added at a rate of 200 mL per minute. At the same time, 100 mL of well-stirred mixture are removed each minute. Find the amount of sugar in the tank as a function of time.

25. Repeat Exercise 24 if 300 mL of mixture are removed each minute.

26. A tank originally contains 1000 L of water in which 5 kg of salt has been dissolved. A mixture containing 2 kg of salt for each 100 L of solution is added to the tank at 10 mL/s. At the same time, the well-stirred mixture in the tank is removed at the rate of 5 mL/s. Find the amount of salt in the tank as a function of time. Sketch a graph of the function.

27. Repeat Exercise 26 if the mixture is removed at 10 mL/s. What is the limit of the amount of salt in the tank for large time?

28. Repeat Exercise 26 if the mixture is removed at 20 mL/s.

29. (a) The current I in the LR-circuit in Figure 16.3 must satisfy the differential equation

$$L\frac{dI}{dt} + RI = E.$$

If $E(t) = E_0 \sin(\omega t)$, $t \geq 0$, where E_0 and ω are constants, solve this differential equation for $I(t)$, and show that the solution can be written in the form

$$I(t) = Ae^{-Rt/L} + \frac{E_0}{Z}\sin(\omega t - \phi),$$

where A is an arbitrary constant and

$$Z = \sqrt{R^2 + \omega^2 L^2}, \qquad \phi = \text{Tan}^{-1}\left(\frac{\omega L}{R}\right).$$

(b) What is the value of A if the current in the circuit at time $t = 0$ when $E(t)$ is connected is I_0?

FIGURE 16.3

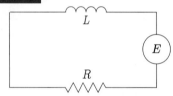

30. (a) The current I in the RC-circuit in Figure 16.4 must satisfy the differential equation

$$R\frac{dI}{dt} + \frac{I}{C} = \frac{dE}{dt}.$$

If $E(t)$ is as in Exercise 29, show that the solution can be written in the form

$$I(t) = Ae^{-t/(RC)} + \frac{E_0}{Z}\sin(\omega t - \phi),$$

where A is an arbitrary constant and

$$Z = \sqrt{R^2 + \frac{1}{\omega^2 C^2}}, \qquad \phi = \text{Tan}^{-1}\left(-\frac{1}{\omega CR}\right).$$

(b) What is the value of A if the current in the circuit at time $t = 0$ when $E(t)$ is connected is I_0?

FIGURE 16.4

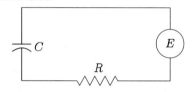

31. Repeat Exercise 21 in Section 16.2 given that the diver descends slowly to the bottom. Assume that his descent is at a constant rate over a time interval of length T. Assume also that maximum pressure $\overline{P}$ is proportional to depth below the surface.

SECTION 16.4

Second-Order Equations Reducible to Two First-Order Equations

We now consider two types of second-order differential equations that can be reduced to a pair of first-order equations.

Type I: Dependent Variable Missing

If a second-order differential equation in $y(x)$ is explicitly independent of y, then it is of the form

$$F(x, y', y'') = 0. \qquad (16.26)$$

In such a case we set

$$v = \frac{dy}{dx} \quad \text{and} \quad \frac{dv}{dx} = \frac{d^2y}{dx^2}. \qquad (16.27)$$

If we substitute these into 16.26, we obtain

$$F(x, v, v') = 0,$$

a first-order differential equation in $v(x)$. If we can solve this equation for $v(x)$, we can then integrate $dy/dx = v(x)$ for $y(x)$.

EXAMPLE 16.6

Find general solutions for the following differential equations:
(a) $xy'' - y' = 0$
(b) $y'' = (y')^2$

SOLUTION

(a) Since y is explicitly missing, we substitute $v = y'$ and $v' = y''$:

$$x\frac{dv}{dx} - v = 0.$$

Variables are now separable,

$$\frac{dv}{v} = \frac{dx}{x},$$

and a solution for $v(x)$ is

$$\ln|v| = \ln|x| + C \quad \text{or} \quad v = Dx \quad (D = \pm e^C).$$

Because $v = dy/dx$,

$$\frac{dy}{dx} = Dx,$$

and we can integrate for

$$y(x) = \frac{D}{2}x^2 + E$$
$$= Fx^2 + E \quad (F = D/2).$$

(b) Since y is explicitly missing, we substitute $v = y'$ and $v' = y''$ to get

$$\frac{dv}{dx} = v^2.$$

Variables are again separable,

$$\frac{dv}{v^2} = dx,$$

and a solution for $v(x)$ is

$$-\frac{1}{v} = x + C.$$

Consequently,
$$v = \frac{dy}{dx} = -\frac{1}{x+C}.$$

Integration now yields
$$y(x) = -\ln|x+C| + D.$$

Type II: Independent Variable Missing

If a second-order differential equation in $y(x)$ is explicitly independent of x, then it is of the form
$$F(y, y', y'') = 0. \qquad (16.28)$$

In such a case we set
$$\frac{dy}{dx} = v \quad \text{and} \quad \frac{d^2y}{dx^2} = \frac{dv}{dx} = \frac{dv}{dy}\frac{dy}{dx} = v\frac{dv}{dy}. \qquad (16.29)$$

When we substitute these into 16.28, we obtain
$$F\left(y, v, v\frac{dv}{dy}\right) = 0,$$

a first-order differential equation in $v(y)$. If we can solve this equation for $v(y)$, we can separate $dy/dx = v(y)$ and integrate for $y(x)$.

EXAMPLE 16.7

Find general solutions for the following differential equations:
(a) $yy'' + (y')^2 = 1$
(b) $y'' = (y')^2$

SOLUTION

(a) Since x is explicitly missing, we substitute
$$y' = v \quad \text{and} \quad y'' = \frac{dv}{dx} = \frac{dv}{dy}\frac{dy}{dx} = v\frac{dv}{dy}$$

to get
$$yv\frac{dv}{dy} + v^2 = 1.$$

If variables are separated, we have
$$\frac{v\,dv}{v^2 - 1} = -\frac{dy}{y},$$

and a general solution for $v(y)$ is
$$\tfrac{1}{2}\ln|v^2 - 1| = -\ln|y| + C.$$

Thus,
$$|v^2 - 1| = \frac{e^{2C}}{y^2},$$

from which we have
$$\frac{dy}{dx} = v = \frac{\pm\sqrt{D + y^2}}{y} \quad (D = \pm e^{2C}).$$

We separate variables again to get
$$\frac{y\,dy}{\sqrt{y^2 + D}} = \pm dx,$$

and obtain an implicit definition of the solution $y(x)$,
$$\sqrt{y^2 + D} = \pm x + E.$$

(b) Since x is explicitly missing, we again substitute from 16.29 to obtain
$$v\frac{dv}{dy} = v^2.$$

If variables are separated,
$$\frac{dv}{v} = dy,$$

and a solution for $v(y)$ is
$$\ln|v| = y + C.$$

Thus,
$$|v| = e^{y+C}$$

and
$$\frac{dy}{dx} = v = De^y \quad (D = \pm e^C).$$

We separate variables again,
$$e^{-y}\,dy = D\,dx,$$

and find an implicit definition of the solution $y(x)$:
$$-e^{-y} = Dx + E.$$

∎

In each of Examples 16.6 and 16.7 we solved the differential equation $y'' = (y')^2$ since both the independent variable x and the dependent variable y are missing. Although the solutions appear different, each is easily derivable from the other.

Let us summarize the results of this section. Substitutions 16.27 for differential equation 16.26 with the dependent variable missing, replace the second-order differential equation with two first-order equations: a first-order equation in $v(x)$, followed by a first-order equation in $y(x)$. Contrast this with the method for equation 16.28 with the independent variable missing. Substitutions 16.29 again replace the second-order equation with two first-order equations. However, the first first-order equation is in $v(y)$, so that for this equation, y is the independent variable rather than the dependent variable. The second first-order equation is again one for $y(x)$.

EXERCISES 16.4

In Exercises 1–10 find a general solution for the differential equation.

1. $xy'' + y' = 4x$
2. $2yy'' = 1 + (y')^2$
3. $y'' = y' + 2x$
4. $x^2 y'' = (y')^2$
5. $y'' \sin x + y' \cos x = \sin x$
6. $y'' = [1 + (y')^2]^{3/2}$
7. $y'' + 4y = 0$
8. $y'' = yy'$
9. $y'' + (y')^2 = 1$
10. $(y'')^2 = 1 + (y')^2$

11. When a flexible cable of uniform mass per unit length hangs between two fixed points A and B (Figure 16.5), the differential equation that describes the shape of the cable is

$$\frac{d^2 y}{dx^2} = k \sqrt{1 + \left(\frac{dy}{dx}\right)^2},$$

where k is a constant. Solve this equation for the curve $y = y(x)$.

FIGURE 16.5

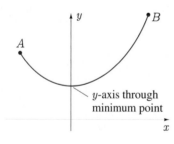

12. A hawk at position $(L, 0)$ (Figure 16.6) spots a pigeon at the origin flying with speed v in the positive y-direction. The hawk immediately takes off after the pigeon with speed $V > v$, always heading directly toward the pigeon. After time t, the pigeon is at position $P(0, vt)$. If the equation of the pursuit curve of the hawk is $y = y(x)$, then during time t, the hawk travels distance Vt along this curve. But distance along this curve can be calculated by means of the definite integral

$$\int_x^L \sqrt{1 + \left(\frac{dy}{dx}\right)^2} \, dx.$$

(a) Show that $y(x)$ must satisfy the integro–differential equation

$$x \frac{dy}{dx} - y = \frac{v}{V} \int_L^x \sqrt{1 + \left(\frac{dy}{dx}\right)^2} \, dx.$$

(b) Differentiate this equation to obtain the second-order differential equation

$$x \frac{d^2 y}{dx^2} = \frac{v}{V} \sqrt{1 + \left(\frac{dy}{dx}\right)^2}.$$

(c) Solve this differential equation for the pursuit curve of the hawk.

(d) Show that the hawk catches the pigeon a distance $vVL/(V^2 - v^2)$ up the y-axis.

FIGURE 16.6

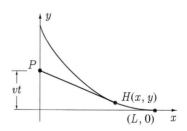

SECTION 16.5

Newtonian Mechanics

One of the most important applications of differential equations is in the study of moving particles and objects. In classical mechanics, motion is governed by Newton's second law 12.104, which states that when an object of constant mass m is subjected to a force $\mathbf{F}$, the resultant acceleration is described by

$$\mathbf{F} = m\mathbf{a}. \qquad (16.30)$$

When $\mathbf{F}$ is given, this is an algebraic equation giving acceleration $\mathbf{a}$. If we substitute $\mathbf{a} = d\mathbf{v}/dt$, we obtain a first-order differential equation

$$\mathbf{F} = m\frac{d\mathbf{v}}{dt} \qquad (16.31)$$

for velocity $\mathbf{v}$ as a function of time t. If we substitute $\mathbf{a} = d^2\mathbf{r}/dt^2$, we obtain a second-order differential equation for position $\mathbf{r}$ as a function of time:

$$\mathbf{F} = m\frac{d^2\mathbf{r}}{dt^2}. \qquad (16.32)$$

In practice it is seldom this simple. Often $\mathbf{F}$ is not given as a function of time, but as a function of position, or velocity, or some other variable. In such cases we may have to change the dependent or independent variable, or both, in order to solve the differential equation.

In this section we are concerned with motion in one direction only. If $r(t), v(t), a(t)$, and F are the components of $\mathbf{r}(t), \mathbf{v}(t), \mathbf{a}(t)$, and $\mathbf{F}$ in this direction, then equations 16.30–16.32 take the forms

$$F = ma, \qquad (16.33)$$
$$F = m\frac{dv}{dt}, \qquad (16.34)$$
$$F = m\frac{d^2r}{dt^2}, \qquad (16.35)$$

respectively.

EXAMPLE 16.8

A paratrooper and his parachute have mass 100 kg. At the instant his parachute opens, the paratrooper is traveling vertically downward at 10 m/s. Air resistance on the chute varies directly as the instantaneous velocity and is 800 N when the paratrooper's velocity is 12 m/s. Find the position and velocity of the paratrooper as functions of time.

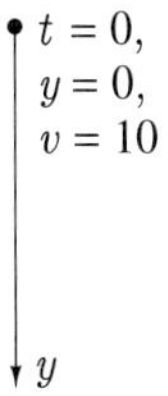

SOLUTION Let us measure y as positive downward, taking $y = 0$ and time $t = 0$ at the instant the paratrooper opens his parachute (Figure 16.7). If $F_{\mathbf{a}}$ is the vertical component of the force of air resistance, then

$$F_{\mathbf{a}} = -kv.$$

Since $F_{\mathbf{a}} = -800$ when $v = 12$, it follows that $-800 = -12k$ or $k = 200/3$. Since the total force on the paratrooper during the fall has component $100g - 200v/3$

FIGURE 16.7

- $t = 0$,
  $y = 0$,
  $v = 10$

↓ y

($g = 9.81$), Newton's second law gives

$$100\frac{dv}{dt} = 100g - \frac{200}{3}v, \quad t \geq 0.$$

Note that the force on the right is not a function of t so that the equation is not immediately integrable. It is however separable,

$$\frac{dv}{2v - 3g} = -\frac{dt}{3},$$

and a general solution for $v(t)$ is defined implicitly by

$$\frac{1}{2}\ln|2v - 3g| = -\frac{t}{3} + C.$$

An explicit solution is found by solving this equation for v:

$$v(t) = \frac{3g}{2} + De^{-2t/3} \quad (D = \pm e^{2C}/2).$$

Since $v = 10$ when $t = 0$,

$$10 = \frac{3g}{2} + D \quad \text{or} \quad D = 10 - \frac{3g}{2}.$$

The velocity of the paratrooper as a function of time is therefore

$$v(t) = \frac{3g}{2} + \left(10 - \frac{3g}{2}\right)e^{-2t/3}.$$

Because $v(t) = dy/dt$, we set

$$\frac{dy}{dt} = \frac{3g}{2} + \left(10 - \frac{3g}{2}\right)e^{-2t/3},$$

and this equation is immediately integrable to

$$y(t) = \frac{3g}{2}t - \frac{3}{2}\left(10 - \frac{3g}{2}\right)e^{-2t/3} + E.$$

Since $y(0) = 0$,

$$0 = -\frac{3}{2}\left(10 - \frac{3g}{2}\right) + E \quad \text{or} \quad E = 15 - \frac{9g}{4}.$$

Consequently, the distance fallen by the paratrooper as a function of time is given by

$$y(t) = \frac{3g}{2}t - \left(15 - \frac{9g}{4}\right)e^{-2t/3} + \left(15 - \frac{9g}{4}\right) = 14.7t + 7.07(e^{-2t/3} - 1).$$

Note that the velocity of the paratrooper does not increase indefinitely. In fact, as time passes a limiting velocity is approached:

$$\lim_{t\to\infty} v(t) = \lim_{t\to\infty}\left\{\frac{3g}{2} + \left(10 - \frac{3g}{2}\right)e^{-2t/3}\right\} = \frac{3g}{2} = 14.7 \text{ m/s}.$$

This is called the terminal velocity of the paratrooper; it is a direct result of the assumption that air resistance is proportional to instantaneous velocity. ∎

Whenever we move an object such as the block in Figure 16.8 over a surface, there is resistance to the motion. This resistance, called **friction**, is due to the fact that the interface between the block and the surface is not smooth; each surface is inherently rough, and this roughness retards the motion of one surface over the other. In effect, a force opposing the motion of the block is created, and this force is called the **force of friction**. Many experiments have been performed to obtain a functional representation for this force. It turns out that when the block in Figure 16.8 slides along a horizontal surface, the magnitude of the force of friction opposing the motion is given by

$$|\mathbf{F}| = \mu m g, \qquad (16.36)$$

where m is the mass of the block, g is the acceleration due to gravity, and μ is a constant called the **coefficient of kinetic friction**. In other words, the force of friction is directly proportional to the weight mg of the block. We caution the reader that this result is valid for the situation shown in Figure 16.8, but it may not be valid for other configurations (say perhaps for an inclined plane).

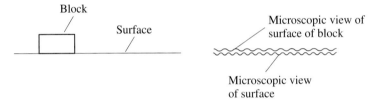

FIGURE 16.8

EXAMPLE 16.9

A block of mass 2 kg is given initial speed 5 m/s along a horizontal surface. If the coefficient of kinetic friction between the block and surface is $\mu = 0.25$, how far does the block slide before stopping?

FIGURE 16.9

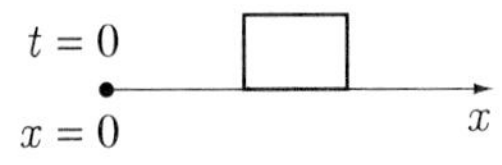

SOLUTION Let us measure x as positive in the direction of motion (Figure 16.9), taking $x = 0$ and $t = 0$ at the instant the block is released. The x-component of the force of friction on the block is

$$F = -0.25(2)g = -\frac{g}{2} \quad (g = 9.81).$$

According to Newton's second law,

$$2\frac{dv}{dt} = -\frac{g}{2},$$

from which we get

$$v(t) = -\frac{g}{4}t + C.$$

Since $v(0) = 5$, it follows that $C = 5$, and

$$v(t) = -\frac{g}{4}t + 5.$$

But $v = dx/dt$, and hence

$$\frac{dx}{dt} = -\frac{g}{4}t + 5.$$

Integration gives

$$x(t) = -\frac{g}{8}t^2 + 5t + D.$$

Because we chose $x = 0$ at time $t = 0$, D must also be zero, and

$$x(t) = -\frac{g}{8}t^2 + 5t.$$

The block comes to rest when $v = 0$; i.e., when

$$-\frac{g}{4}t + 5 = 0, \text{ or, } t = \frac{20}{g}.$$

The position of the block at this time is

$$x = -\frac{g}{8}\left(\frac{20}{g}\right)^2 + 5\left(\frac{20}{g}\right) = 5.10.$$

The block therefore slides 5.10 m before stopping.

EXERCISES 16.5

1. A car of mass 1500 kg starts from rest at an intersection. The engine exerts a constant force of 3000 N, and air friction causes a resistive force whose magnitude in newtons is equal to the speed of the car in metres per second. Find the speed of the car and its distance from the intersection after 10 s.

2. You are called on as an expert to testify in a traffic accident hearing. The question concerns the speed of a car that made an emergency stop with brakes locked and wheels sliding. The skid mark on the road measured 9 m. If you assume that the coefficient of kinetic friction between the tires and road was less than one, what can you say about the speed of the car before the brakes were applied? Are you testifying for the prosecution or the defence?

3. A boat and its contents have mass 250 kg. Water exerts a resistive force on the motion of the boat that is proportional to the instantaneous speed of the boat and is 200 N when the speed is 30 km/h.
   (a) If the boat starts from rest and the engine exerts a constant force of 250 N in the direction of motion, find the speed of the boat as a function of time.

   (b) What is the limiting speed of the boat?

4. (a) In Example 16.8 it was necessary to solve the equation
   $$\frac{1}{2}\ln|2v - 3g| = -\frac{t}{3} + C$$
   for $v = v(t)$ and use the condition $v(0) = 10$ to evaluate C. We chose first to solve for $v(t)$ and then to evaluate the constant. Show the details of this analysis.

   (b) Instead, first use the condition $v(0) = 10$ to evaluate C, and then solve the equation for $v(t)$.

5. A body of weight W falls from rest under gravity. It is acted on by a resistive force that is proportional to the square root of the speed at any instant. If the magnitude of this resistive force is F^* when the speed is v^*, find an expression for its terminal speed.

6. A spring (with constant k) is attached on one end to a wall and on the other end to a mass M (Figure 16.10). The mass is set into motion along the x-axis by pulling it a distance x_0 to the right of the position it would occupy were the spring unstretched and given speed

v_0 to the left. During the subsequent motion, there is a frictional force between M and the horizontal surface with coefficient of kinetic friction equal to μ.

(a) Show that the differential equation describing the motion of M is

$$M\frac{d^2x}{dt^2} = -kx + \mu Mg, \quad x(0) = x_0, \quad v(0) = -v_0,$$

if we take $t = 0$ at the instant that motion is initiated. When is this equation valid?

(b) Since t is explicitly missing from the equation in (a), show that it can be rewritten in the form

$$Mv\frac{dv}{dx} = -kx + \mu Mg, \quad v(x_0) = -v_0,$$

and that therefore

$$\frac{k}{2}(x_0^2 - x^2) = \frac{M}{2}(v^2 - v_0^2) + \mu Mg(x_0 - x).$$

Interpret each of the terms in this equation physically.

(c) If x^* represents the position at which M comes to rest for the first time, use the equation in (b) to determine x^* as a function of μ, M, g, k, x_0, and v_0. Discuss the possibilities of x^* being positive, negative, and zero.

FIGURE 16.10

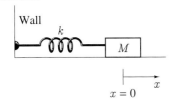

7. When a body falls in air, it is acted on by gravity and also by a force due to air resistance that is proportional to its instantaneous speed.

   (a) If the body is initially projected upward, find its velocity as a function of time, and sketch a graph of the function.

   (b) Repeat (a) if the body is projected downward with velocity less than its terminal velocity.

   (c) Repeat (b) if the initial velocity is greater than its terminal velocity.

8. A 1-kg rock is thrown vertically upward with speed 20 m/s. Air resistance to its motion when measured in newtons has magnitude equal to one-tenth its speed in metres per second. Find the maximum height attained by the rock.

9. A stone of mass 100 g is thrown vertically upward with speed 20 m/s. Air exerts a resistive force on the stone proportional to its instantaneous speed, and has magnitude 1/10 N when the speed of the stone is 10 m/s. Find an equation defining the time when the stone returns to its projection point. Solve this equation numerically, and compare with the time taken if air resistance is neglected.

10. (a) A 1-kg mass falls under the influence of gravity. It is also acted on by air resistance which is proportional to the square of its velocity and is 5 N when its velocity is 50 m/s. If the velocity of the mass has magnitude 20 m/s at time $t = 0$, find a formula for its velocity as a function of time.

    (b) Sketch a graph of the velocity function. Is there a terminal velocity?

11. Repeat Exercise 10 if the initial velocity has magnitude 100 m/s.

12. For how long does the mass in Exercise 10 rise if it is thrown upward with velocity 20 m/s?

13. A mass m falls under the influence of gravity and air resistance which is proportional to the square of velocity. If the speed of the mass is v_0 at time $t = 0$, find a formula for its speed as a function of time.

14. (a) If the mass in Exercise 13 is thrown upward with speed v_0, find a formula for speed on its ascent. Is this formula also valid for its speed when it begins to fall?

    (b) Find a formula for its height. How high does it rise?

15. A mass m slides from rest down a frictionless plane inclined at angle α to the horizontal. Find a formula for the time taken to travel a distance D down the plane. What is its speed at this time?

16. Find formulas for speed and distance travelled for the mass in Exercise 15 if air resistance proportional to velocity also acts on the mass.

17. In Example 7.33 of Section 7.9 we derived the escape velocity of a projectile from the earth's surface based on energy principles. In this exercise we obtain the same result using differential equations. When a projectile of mass m, fired from the earth's surface, is a distance r from the centre of the earth, the magnitude of the force of attraction on it is given by Newton's universal law of gravitation, $F = GMm/r^2$, where M is the mass of the earth and G is a constant. Use Newton's second law and a substitution corresponding to 16.29 to find the velocity of the projectile as a function of r. What minimum initial velocity guarantees that the projectile escapes the gravitational field of the earth?

18. A huge cannon fires a projectile with initial velocity v_0 directly toward the moon (Figure 16.11). When the projectile is a distance r above the earth's surface, the force of attraction of the earth on the projectile has

magnitude
$$\frac{GMm}{(r+R)^2},$$

where G is a constant, R is the radius of the earth M is the mass of the earth, and m is the mass of the projectile. At this point the moon's gravitational attraction has magnitude

$$\frac{GM^*m}{(a+R^*-r)^2},$$

where M^* is the mass of the moon.

(a) Show that if only the above two forces are considered to act on the projectile, then the differential equation describing its motion is

$$\frac{d^2r}{dt^2} = -\frac{gR^2}{(r+R)^2} + \frac{g^*R^{*2}}{(a+R^*-r)^2},$$

where g and g^* are gravitational accelerations on the surfaces of the earth and the moon.

(b) Prove that the velocity of the projectile at a distance r above the surface of the earth is defined by

$$v^2 = \frac{2gR^2}{r+R} + \frac{2g^*R^{*2}}{a+R^*-r} + v_0^2 - 2gR - \frac{2g^*R^{*2}}{a+R^*}.$$

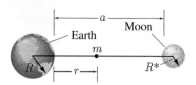

FIGURE 16.11

19. Newton's second law states that if an object of variable mass $m(t)$ is subjected to a force $\mathbf{F}(t)$, then

$$\frac{d}{dt}(m\mathbf{v}) = \mathbf{F},$$

where t is time and $\mathbf{v}$ is the velocity of the object. A uniform chain of length 3 m and mass 6 kg is held by one end so that the other end just touches the floor. If the chain is released, find the velocity of the falling chain as a function of the length of chain still falling. How fast does the end hit the floor?

SECTION 16.6

Population Dynamics

The rate of growth of a country's population depends on many factors: the birth rate, death rate, immigration rate, size of the country, availability of food, state of technology, natural disasters, etc. Developing a mathematical formula that would account for all possible factors would be impossible. Some factors, on the other hand, are much more important to the growth of a population than others. Perhaps we could develop a mathematical model that would initially account for the important factors, and that could be refined at a later stage to incorporate minor factors. Our problem, then, is to develop a model that predicts populations of countries with some reasonable amount of accuracy.

On the basis of the census figures of a country such as the United States, it is possible to calculate the number of births per year in the populaton. If we divide this by the existing size of the population, we get what is called the relative birth rate b of the population (or the individual birth rate). It is an average figure, but it is found that for many populations the relative birth rate is reasonably constant over long periods of time. In a similar way, it is possible to define a relative death rate d (death due to natural causes), and the constant $k = b - d$ is called the relative growth rate of the population. It is the increase in population per individual per year that can be attributed solely to births and deaths. On the other hand, if $N = N(t)$ represents the size of the population at any time t (in years), then

$$\frac{1}{N}\frac{dN}{dt}$$

is the total change in population per individual per year, and should therefore account for all factors that affect the size of the population. If we were to decide that the most important factor in determining the size of a population were its birth and death rates,

we would write
$$\frac{1}{N}\frac{dN}{dt} = k,$$
or,
$$\frac{dN}{dt} = kN. \tag{16.37}$$

This is a mathematical model (called the **Malthusian model**) that attempts to describe the size of a population in terms only of its birth and death rates.

Because 16.37 is a separable differential equation
$$\frac{1}{N}dN = k\,dt,$$
we obtain a general solution
$$\ln|N| = kt + D.$$
Because N is never negative, we can state that
$$N = Ce^{kt} \quad (C = e^D). \tag{16.38}$$

In Table 16.1 and Figure 16.12 we show approximate census figures for the United States from 1790 to 1970. Note that the population figure in 1940 is uncharacteristic of the previous growth from 1900 to 1930 (and history makes the reason clear). Suppose we attempt to fit exponential curve 16.38 to the population between 1790 and 1860. One way to do this is to demand that $N(1790) = 3.93$ and $N(1860) = 31.44$:
$$3.93 = Ce^{1790k}, \quad 31.44 = Ce^{1860k}.$$

The solution of these equations for C and k is $C = 3.17 \times 10^{-23}$, $k = 0.0297$. Consequently,
$$N = 3.17 \times 10^{-23} e^{0.0297t},$$
and this curve is shown in Figure 16.12. It is indeed an excellent description of the United States population in the years between 1790 and 1860.

TABLE 16.1

Year	1790	1800	1810	1820	1830	1840	1850	1860	1870	1880
Population (millions)	3.93	5.31	7.24	9.64	12.87	17.07	23.19	31.44	38.56	50.16

Year	1890	1900	1910	1920	1930	1940	1950	1960	1970
Population (millions)	62.95	76.00	91.97	105.71	122.78	131.67	150.70	179.32	203.19

FIGURE 16.12

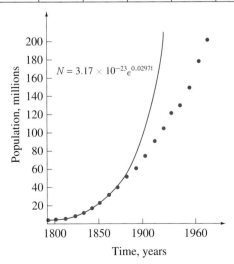

It is equally obvious, however, that this particular exponential function is not an adequate representation of the population beyond 1860. In fact no exponential function can describe the census points in Table 16.1. This implies that after 1860 other factors besides simply pure birth and death processes become important in determining the size of the population. One possible explanation is that because 16.37 predicts exponential growth, it must assume that there are sufficient resources (area, food, water, etc.) to sustain any level of population. This is unrealistic in the long term because there is a limit to the size of the population that any environment can support. This limit, called the **carrying capacity** C of the environment, is such that as the size of the population draws closer to C its growth rate must slow down. One way to incorporate this into model 16.37 is to set

$$\frac{dN}{dt} = kN\left(1 - \frac{N}{C}\right) \qquad (16.39)$$

(called the **logistic model**). For small N, $1 - N/C$ is approximately equal to one, and N will experience exponential growth, but as N approaches C, this factor diminishes and the growth rate decreases.

To solve this differential equation, we again separate variables,

$$k\,dt = \frac{C\,dN}{N(C-N)} = \left\{\frac{1}{N} + \frac{1}{C-N}\right\}dN,$$

and obtain a general solution

$$kt + E = \ln N - \ln(C - N).$$

We now solve this equation for $N(t)$ by first exponentiating to obtain

$$\frac{N}{C - N} = De^{kt} \quad (D = e^E),$$

and cross-multiplying to get

$$N = (C - N)De^{kt}.$$

Thus,

$$N(t) = \frac{CDe^{kt}}{1 + De^{kt}} = \frac{C}{1 + Fe^{-kt}} \quad (F = 1/D), \qquad (16.40)$$

and this is called the **logistic growth function** (or **logistic curve**). If we use the census figures of 1790, 1880, and 1970 to determine the three constants F, k, and C, we obtain

$$N(t) = \frac{257}{1 + 3.54 \times 10^{25} e^{-0.0305t}}.$$

This function and the census figures of Table 16.1 are shown in Figure 16.13. Note that this logistic function predicts an upper limit of 257 million people for the United States, not very realistic in light of today's population. It has resulted from assuming that the population of the United States can reasonably be described by a logistic function that passes through the census points in 1790, 1880, and 1970. This implies, then, that such a representation is still not an adequate description of the United States population; other factors besides birth and death rates and carrying capacity must be incorporated. Some of these are introduced in the exercises.

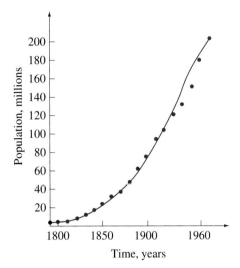

FIGURE 16.13

EXERCISES 16.6

1. Bacteria in a culture increase at a rate proportional to the number present. If the original number of bacteria increases by 50% in one-half hour, how long does it take to quintuple the original number?

2. Two rodents begin a colony. During the next ten months they are permitted to reproduce and their numbers are recorded in the following table.

Time t (months)	0	2	5	7	10
Number N of rodents	2	5	15	35	110

   (a) Find a formula for $N(t)$ based on the Malthusian model and use the conditions $N(0) = 2$ and $N(2) = 5$ to determine the unknown constants in the function.
   (b) Repeat (a) but use the conditions $N(0) = 2$ and $N(10) = 110$ to evaluate the constants.
   (c) Sketch both functions in (a) and (b) and the data points on the same graph.

3. If we modify the Malthusian model for predicting the population of a country to incorporate a constant immigration rate of λ individuals per unit time, then the model becomes
   $$\frac{dN}{dt} = kN + \lambda.$$
   Solve the differential equation for $N(t)$.

4. Draw the logistic curve 16.40 indicating any relative extrema and points of inflection. Assume that $N(0) = N_0 < C/2$.

5. An initial population of 100 inhabitants enters an area with carrying capacity 100 000. In the first year, the population increases to 120. Assume that the population follows logistic growth.

   (a) Determine the population as an explicit function of time.
   (b) How long does it take the population to reach 95 000?

6. (a) A country institutes an immigration policy that states that the immigration rate shall be proportional to the difference between the carrying capacity C and the present size N of the population. Show that a mathematical model to describe $N(t)$ is
   $$\frac{dN}{dt} = -\frac{k}{C}(N - C)(N + C\lambda/k),$$
   where k is the relative growth rate of the population attributed only to birth and death rates, and λ is a positive constant.
   (b) Solve this differential equation, and show that
   $$N(t) = \frac{C - (FC\lambda/k)e^{-(k+\lambda)t}}{1 + Fe^{-(k+\lambda)t}}.$$

7. (a) For some species, if the population drops below a certain level, the species will become extinct. Given that m is the minimum viable population for such a species, one way to incorporate this condition into the logistic model is to write
   $$\frac{dN}{dt} = kN\left(1 - \frac{N}{C}\right)\left(1 - \frac{m}{N}\right).$$
   Solve this differential equation subject to the condition $N(0) = N_0$ to obtain
   $$N(t) = \frac{[(C - N_0)/(N_0 - m)]me^{-[(C-m)/C]kt} + C}{1 + [(C - N_0)/(N_0 - m)]e^{-[(C-m)/C]kt}}.$$

(b) Use the result in (a) to show that if $N_0 < m$, then extinction must occur.

Logistic model 16.39 incorporates the term $1 - N/C$ in order to decrease dN/dt for N approaching C. In Exercises 8–10, we show that there are other ways to do this.

8. The Gompertz model multiplies the right-hand side of the Malthusian model by the factor e^{-lt} ($l > 0$ a constant):

$$\frac{dN}{dt} = kNe^{-lt}.$$

Solve this differential equation for $N(t)$, and sketch the curve $N = N(t)$. Does it predict an upper limit for $N(t)$?

9. The monomolecular model multiplies the right-hand side of the Malthusian model by the factor $be^{-kt}/(1-be^{-kt})$ ($0 < b < 1$ a constant):

$$\frac{dN}{dt} = kN\frac{be^{-kt}}{1 - be^{-kt}}.$$

Solve this differential equation for $N(t)$, and sketch the curve $N = N(t)$. Does it predict an upper limit for $N(t)$?

10. The Von Bertalanffy model is much like the monomolecular model,

$$\frac{dN}{dt} = kN\frac{3be^{-kt}}{1 - be^{-kt}}.$$

Solve this differential equation for $N(t)$, and sketch the curve $N = N(t)$. Does it predict an upper limit for $N(t)$?

11. Verify that if the constants C, k, and F in the logistic function 16.40 are evaluated by using the census figures of 1790, 1880, and 1970 for the United States, then $C = 257$, $F = 3.54 \times 10^{25}$, and $k = 0.0305$.

SECTION 16.7

Linear Differential Equations

Throughout Chapters 5, 8, and 9 and Section 16.1–16.6, we have stressed the use of differential equations in solving applied problems. We have considered examples from such diverse areas as engineering, geometry, ecology, and psychology, hoping thereby to illustrate how valuable differential equations can be in modeling physical situations mathematically. Perhaps the most important type of differential equation is the **linear differential equation**.

Definition 16.4

A differential equation of the form

$$a_0(x)\frac{d^n y}{dx^n} + a_1(x)\frac{d^{n-1} y}{dx^{n-1}} + a_2(x)\frac{d^{n-2} y}{dx^{n-2}} + \cdots + a_{n-1}(x)\frac{dy}{dx} + a_n(x)y = F(x), \quad (16.41)$$

where $a_0(x) \not\equiv 0$, is called a **linear nth-order differential equation**.

Note in particular that none of the derivatives of $y(x)$ are multiplied together, nor are they squared or cubed or taken to any other power, nor do they appear as the argument of any transcendental function. All we see is a function of x multiplying y, plus a function of x multiplying the first derivative of y, plus a function of x multiplying the second derivative of y, and so on.

If $n = 1$, equation 16.41 reduces to

$$a_0(x)\frac{dy}{dx} + a_1(x)y = F(x),$$

and at any point at which $a_0(x) \neq 0$, we can divide to obtain

$$\frac{dy}{dx} + \frac{a_1(x)}{a_0(x)}y = \frac{F(x)}{a_0(x)}.$$

If we set $P(x) = a_1(x)/a_0(x)$ and $Q(x) = F(x)/a_0(x)$, we have

$$\frac{dy}{dx} + P(x)y = Q(x);$$

i.e., every linear first-order differential equation can be expressed in this form. We discussed equations of this type in Section 16.3, where it was shown that such equations have a general solution

$$y(x) = e^{-\int P(x)dx} \left\{ \int Q(x) e^{\int P(x)dx} dx + C \right\}.$$

In other words, we already know how to solve linear first-order differential equations, and therefore our discussion in the next five sections is directed primarily at second- and higher-order equations. Keep in mind, however, that all results are also valid for first-order linear equations.

Because equation 16.41 is so cumbersome, we introduce notation to simplify its representation. In particular, if we use the notation $D = d/dx$, $D^2 = d^2/dx^2$, etc., we can write

$$a_0(x)D^n y + a_1(x)D^{n-1}y + a_2(x)D^{n-2}y + \cdots + a_{n-1}(x)Dy + a_n(x)y = F(x), \quad (16.42)$$

or,

$$\{a_0(x)D^n + a_1(x)D^{n-1} + a_2(x)D^{n-2} + \cdots + a_{n-1}(x)D + a_n(x)\}y = F(x). \quad (16.43)$$

The quantity in braces is called a **differential operator**; it operates on whatever follows it—in this case, y. It is a "differential" operator because it operates by taking derivatives. Because the operator involves only x's and D's, we denote it by

$$\phi(x, D) = a_0(x)D^n + a_1(x)D^{n-1} + a_2(x)D^{n-2} + \cdots + a_{n-1}(x)D + a_n(x). \quad (16.44)$$

The general linear nth-order differential equation can then be represented very simply by

$$\phi(x, D)y = F(x). \quad (16.45)$$

For example, the linear second-order differential equation

$$xy'' + y' + xy = 0$$

is called Bessel's differential equation of order zero. In operator notation we write

$$(xD^2 + D + x)y = 0,$$

or,

$$\phi(x, D)y = 0,$$

where $\phi(x, D) = xD^2 + D + x$.

For the differential equation

$$y'' + 2y' - 3y = e^{-x},$$

we write

$$\phi(D)y = e^{-x},$$

where $\phi(D) = D^2 + 2D - 3$.

We now indicate the meaning of the term "linear." Suppose that L is an operator that operates on each function $y(x)$ in some set S. For example, L might be the operation that multiplies each function by 5, or perhaps squares each function, or perhaps differentiates each function. It is said to be a linear operator if it satisfies the following definition.

Definition 16.5

An operator L is said to be **linear** on a set of functions S if for any two functions $y_1(x)$ and $y_2(x)$ in S, and any constant c,

$$L(y_1 + y_2) = Ly_1 + Ly_2, \qquad (16.46\text{a})$$
$$L(cy_1) = c(Ly_1). \qquad (16.46\text{b})$$

Many of the operations in calculus are therefore linear. For instance, taking limits is a linear operation, as is differentiation, antidifferentiation, and taking definite integrals. On the other hand, taking the square root of a positive function is not a linear operation, since

$$L(y_1 + y_2) = \sqrt{y_1 + y_2} \neq L(y_1) + L(y_2).$$

It is not difficult to show that the differential operator $\phi(x, D)$ in 16.44 is linear; that is,

$$\phi(x, D)(y_1 + y_2) = \phi(x, D)y_1 + \phi(x, D)y_2,$$
$$\phi(x, D)(cy_1) = c[\phi(x, D)y_1],$$

and because of this, differential equation 16.45 is also said to be linear. The following two differential equations are examples of equations that are not linear:

$$\frac{d^2y}{dx^2} + y^2 = x, \qquad \frac{d^2y}{dx^2}\frac{dy}{dx} = e^x.$$

In particular, if we substitute $y_1(x) + y_2(x)$ into the left-hand side of the first equation, we obtain

$$\frac{d^2}{dx^2}(y_1 + y_2) + (y_1 + y_2)^2 = \frac{d^2y_1}{dx^2} + \frac{d^2y_2}{dx^2} + y_1^2 + 2y_1y_2 + y_2^2.$$

If we substitute $y_1(x)$, then $y_2(x)$, and then add the results, we find a different expression:

$$\frac{d^2y_1}{dx^2} + y_1^2 + \frac{d^2y_2}{dx^2} + y_2^2.$$

EXERCISES 16.7

In Exercises 1–10 prove either that the operator L is linear or that it is not linear. In each case assume that the set S of functions on which L operates is the set of all functions on which L can operate. For instance, in Exercise 6, assume that S is the set of all functions $y(x)$ that have a first derivative dy/dx.

1. L multiplies functions $y(x)$ by 5

2. L multiplies functions $y(x)$ by $15x$

3. L adds the fixed function $z(x)$ to functions $y(x)$

4. L takes the limit of functions $y(x)$ as x approaches 3

5. L takes the limit of functions $y(x)$ as x approaches infinity

6. L takes the first derivative of functions $y(x)$ with respect to x

7. L takes the third derivative of functions $y(x)$ with respect to x

8. L takes the antiderivative of functions $y(x)$ with respect to x

9. L takes the definite integral of functions $y(x)$ with respect to x from $x = -1$ to $x = 4$

10. L takes the cube root of functions $y(x)$

In Exercises 11–20 determine whether the differential equation is linear or nonlinear. Write those equations that are linear in operator notation 16.45.

11. $2x\dfrac{d^2y}{dx^2} + x^3y = x^2 + 5$

12. $2x\dfrac{d^2y}{dx^2} + x^3y = x^2 + 5y$

13. $2x\dfrac{d^2y}{dx^2} + x^3y = x^2 + 5y^2$

14. $x\dfrac{d^3y}{dx^3} + 3x\dfrac{d^2y}{dx^2} - 2\dfrac{dy}{dx} + y = 10 \sin x$

15. $x\dfrac{d^3y}{dx^3} + 3x\dfrac{d^2y}{dx^2} - 2\dfrac{dy}{dx} + y^2 = 10 \sin x$

16. $y\dfrac{d^3y}{dx^3} + 3x\dfrac{d^2y}{dx^2} - 2\dfrac{dy}{dx} + y = 10 \sin x$

17. $y'' - 3y' - 2y = 9 \sec^2 x$

18. $yy'' + 3y' - 2y = e^x$

19. $\sqrt{1 + y'} + x^2 = 4$

20. $y'''' + y'' - y = \ln x$

21. The Laplace transform of a function $y(t)$ is defined as the function $L(y) = \int_0^\infty e^{-st} y(t)\, dt$, provided the improper integral converges. Show that if S is the set of all functions that have a Laplace transform, then the operation of taking the Laplace transform is linear on S.

22. The finite Fourier cosine transform of a function $y(x)$ is defined as $L(y) = \int_0^{2\pi} y(x) \cos nx\, dx$, where n is a nonnegative integer, provided the definite integral exists. Show that if S is the set of all functions that have a finite Fourier cosine transform, then the operation of taking the transform is linear on S.

SECTION 16.8

Homogeneous Linear Differential Equations

Two classes of linear differential equations present themselves: those for which $F(x) \equiv 0$, and those for which $F(x) \not\equiv 0$. In this section and the next we consider equations for which $F(x) \equiv 0$; we will discuss the more difficult class, in which $F(x) \not\equiv 0$, in Section 16.10. First let us name each of these classes of linear differential equations.

Definition 16.6

A linear differential equation $\phi(x, D)y = F(x)$ is said to be **homogeneous** if $F(x) \equiv 0$, and **nonhomogeneous** otherwise.

The meaning of "homogeneous" to describe a property of linear differential equation in this definition is totally different from the meaning in Exercise 16.2-26.

The fundamental idea behind the solution of all linear differential equations is the following theorem.

Theorem 16.1 (Superposition principle). *If $y_1(x), y_2(x), \ldots, y_m(x)$ are solutions of a homogeneous linear differential equation*

$$\phi(x, D)\, y = 0,$$

then so too is any linear combination thereof,

$$C_1 y_1(x) + C_2 y_2(x) + \cdots + C_m y_m(x)$$

(for arbitrary constants $C_1, C_2, \ldots, C_m$).

Proof The proof requires only linearity of the operator $\phi(x, D)$, for

$$\phi(x, D)[C_1 y_1(x) + \cdots + C_m y_m(x)] = C_1[\phi(x, D) y_1] + \cdots + C_m[\phi(x, D) y_m]$$
$$= 0 + 0 + \cdots + 0 = 0.$$

Solutions of a linear differential equation that are linearly combined to produce other solutions are said to be superposed—consequently, the name "superposition principle" for Theorem 16.1. For example, it is straightforward to verify that $y_1(x) = e^{4x}$ and $y_2(x) = e^{-3x}$ are solutions of the homogeneous equation $y'' - y' + 12y = 0$. The superposition principle then states that for any constants C_1 and C_2, the function $y(x) = C_1 y_1(x) + C_2 y_2(x) = C_1 e^{4x} + C_2 e^{-3x}$ must also be a solution. But because $y(x)$ contains two arbitrary constants, it is not only a solution, it is a general solution of the differential equation.

Similarly, superposition of the three solutions e^x, xe^x, and $x^2 e^x$ of the linear differential equation $y''' - 3y'' + 3y' - y = 0$ gives a general solution $y(x) = C_1 e^x + C_2 xe^x + C_3 x^2 e^x$.

Now we begin to see the importance of the superposition principle. If we can find n solutions $y_1(x), y_2(x), \ldots, y_n(x)$ of an nth-order homogeneous linear differential equation, then a general solution is

$$y(x) = C_1 y_1(x) + C_2 y_2(x) + \cdots + C_n y_n(x).$$

In other words, all that we need do is find n solutions; the superposition principle will do the rest. There is a problem, however, if we take things a little too literally. For instance, $y_1(x) = e^{4x}$ and $y_2(x) = 10 e^{4x}$ are both solutions of $y'' - y' + 12y = 0$. By superposition, so too then is $y(x) = C_1 y_1 + C_2 y_2 = C_1 e^{4x} + 10 C_2 e^{4x}$. But is it a general solution? The answer is no, because we could write $y(x) = (C_1 + 10 C_2) e^{4x}$, and by setting $C_3 = C_1 + 10 C_2$, we have $y(x) = C_3 e^{4x}$. Superposition of the solutions e^{4x} and $10 e^{4x}$ has not therefore led to a general solution, and the reason is that they are essentially the same solution: $y_2(x)$ is $y_1(x)$ multiplied by a constant. Superposition does not therefore lead to a solution with two arbitrary constants.

In a similar way, $y_1(x) = e^x$, $y_2(x) = xe^x$, and $y_3(x) = 2e^x - 3xe^x$ are all solutions of $y''' - 3y'' + 3y' - y = 0$, and therefore so is $y(x) = C_1 y_1 + C_2 y_2 + C_3 y_3$. But because we can write

$$y(x) = C_1 e^x + C_2 xe^x + C_3(2e^x - 3xe^x)$$
$$= (C_1 + 2C_3) e^x + (C_2 - 3C_3) xe^x$$
$$= C_4 e^x + C_5 xe^x,$$

$y(x)$ is not a general solution. This is a direct result of the fact that $y_3(x)$ is a linear combination of the solutions $y_1(x)$ and $y_2(x)$; it is twice $y_1(x)$ minus three times $y_2(x)$.

Our problem seems to come down to this: If we have n solutions of an nth-order homogeneous linear differential equation, how can we determine whether superposition leads to a (general) solution that contains n arbitrary constants? Our examples have suggested that if any one of the solutions is a linear combination of the others, then a general solution is not obtained, and this is indeed true. If one of the solutions is a linear combination of the others, we say that the n solutions are **linearly dependent**; if no solution is a linear combination of the others, we say that the n solutions are **linearly independent**. We summarize our results in the following theorem.

Theorem 16.2

If $y_1(x), y_2(x), \ldots, y_n(x)$ are n linearly independent solutions of an nth-order homogeneous linear differential equation, then $y(x) = C_1 y_1(x) + C_2 y_2(x) + \cdots + C_n y_n(x)$ is a general solution of the differential equation.

What we should now do is devise a test to determine whether a set of n solutions is linearly independent or linearly dependent. For most examples, no test is really necessary; it is obvious whether one of the solutions can be written as a linear combination of the others. For those rare occasions when it is not obvious, Exercise 10 describes a test that can be used to determine whether functions are linearly independent.

In summary, the superposition principle states that solutions of a homogeneous linear differential equation can be superposed to produce other solutions. If n linearly independent solutions of an nth-order equation are superposed, a general solution is obtained. This is the importance of the superposition principle. We need not devise a method that will take us directly to a general solution; we need a method for finding n linearly independent solutions—the superposition principle will do the rest. Unfortunately, for completely general coefficients $a_i(x)$ in $\phi(x, D)$ (see equation 16.44), it is impossible to give a method that will always yield n linearly independent solutions of a homogeneous equation. There is, however, one special case of great practical importance in which it is always possible to produce n linearly independent solutions in a very simple way. This special case occurs when the coefficients $a_i(x)$ are all constants a_i, and this is the subject of Section 16.9.

EXAMPLE 16.10

If $y_1(x) = \cos 3x$ and $y_2(x) = \sin 3x$ are solutions of the differential equation $y'' + 9y = 0$, find a general solution.

SOLUTION Since $y_1(x)$ and $y_2(x)$ are linearly independent solutions (one is not a constant times the other), a general solution can be obtained by superposition:

$$y(x) = C_1 \cos 3x + C_2 \sin 3x.$$

∎

EXAMPLE 16.11

Given that $y_1(x) = e^{2x} \cos x$ and $y_2(x) = e^{2x} \sin x$ are solutions of the homogeneous linear differential equation $y'' - 4y' + 5y = 0$, find that solution which satisfies the conditions $y(\pi/4) = 1, y(\pi/3) = 2$.

SOLUTION By superposition, a general solution of the differential equation is

$$y(x) = C_1 e^{2x} \cos x + C_2 e^{2x} \sin x = e^{2x}(C_1 \cos x + C_2 \sin x).$$

To satisfy the conditions $y(\pi/4) = 1$ and $y(\pi/3) = 2$, we have

$$1 = e^{\pi/2}(C_1/\sqrt{2} + C_2/\sqrt{2}), \quad 2 = e^{2\pi/3}(C_1/2 + \sqrt{3}C_2/2).$$

Thus C_1 and C_2 are defined by the pair of equations

$$C_1 + C_2 = \sqrt{2}\,e^{-\pi/2}, \quad C_1 + \sqrt{3}\,C_2 = 4e^{-2\pi/3},$$

the solution of which is

$$C_2 = \frac{4e^{-2\pi/3} - \sqrt{2}\,e^{-\pi/2}}{\sqrt{3} - 1} = 0.2713, \quad C_1 = \sqrt{2}\,e^{-\pi/2} - C_2 = 0.0227.$$

The required solution is therefore

$$y(x) = e^{2x}(0.0227 \cos x + 0.2713 \sin x). \quad \blacksquare$$

We pointed out in Section 16.1 that a general solution of a differential equation might not contain all solutions of the equation. This is not true for linear differential equations. It can be shown that if $y(x)$ is a general solution of a linear differential equation on an interval I, then every solution of the differential equation on I can be obtained by specifying particular values for the arbitrary constants in $y(x)$. Here then is a very important class of differential equations for which a general solution contains all possible solutions.

EXERCISES 16.8

In Exercises 1–8 show that the functions are solutions of the differential equation. Check that the differential equation is linear and homogeneous, and then find a general solution.

1. $y'' + y' - 6y = 0$; $y_1(x) = e^{2x}$, $y_2(x) = e^{-3x}$

2. $y' + y \tan x = 0$; $y_1(x) = \cos x$

3. $y'''' + 5y'' + 4y = 0$; $y_1(x) = \cos 2x$, $y_2(x) = \sin 2x$, $y_3(x) = \cos x$, $y_4(x) = \sin x$

4. $2y'' - 16y' + 32y = 0$; $y_1(x) = 3e^{4x}$, $y_2(x) = -2xe^{4x}$

5. $y''' - 3y'' + 2y' = 0$; $y_1(x) = 10$, $y_2(x) = 3e^x$, $y_3(x) = 4e^{2x}$

6. $2y'' - 8y' + 9y = 0$; $y_1(x) = e^{2x}\cos(x/\sqrt{2})$, $y_2(x) = e^{2x}\sin(x/\sqrt{2})$

7. $x^2 y'' + xy' + (x^2 - 1/4)y = 0$; $y_1(x) = (\sin x)/\sqrt{x}$, $y_2(x) = (\cos x)/\sqrt{x}$

8. $x^2 y'' + xy' + 4y = 0$; $y_1(x) = \cos(2\ln x)$, $y_2(x) = \sin(2\ln x)$

9. Show that $y_1(x) = -2/(x+1)$ and $y_2(x) = -2/(x+2)$ are both solutions of the differential equation $y'' = yy'$. Is $y(x) = y_1(x) + y_2(x)$ a solution? Explain.

10. We stated in this section that n functions $y_1(x), \ldots, y_n(x)$ are linearly dependent if at least one of the functions can be expressed as a linear combination of the others; they are linearly independent if none of the functions is a linear combination of the others. Another way of saying this is as follows: Functions $y_1(x), \ldots, y_n(x)$ are linearly dependent on an interval I if there exist constants $C_1, \ldots, C_n$, not all zero, such that on I

$$C_1 y_1(x) + \cdots + C_n y_n(x) \equiv 0.$$

If this equation can be satisfied only with $C_1 = C_2 = \cdots = C_n = 0$, the functions are linearly independent. In this exercise we give a test to determine whether functions are linearly independent or dependent. If

$y_1(x), \ldots, y_n(x)$ have derivatives up to and including order $n-1$ on the interval I, we define the Wronskian of the functions as the $n \times n$ determinant:

$$W(y_1, \ldots, y_n) = \begin{vmatrix} y_1 & y_2 & \cdots & y_n \\ y_1' & y_2' & \cdots & y_n' \\ y_1'' & y_2'' & \cdots & y_n'' \\ \vdots & \vdots & \ddots & \vdots \\ y_1^{(n-1)} & y_2^{(n-1)} & \cdots & y_n^{(n-1)} \end{vmatrix}.$$

Show that if $y_1, \ldots, y_n$ are linearly dependent on I, then $W(y_1, \ldots, y_n) \equiv 0$ on I. It follows then that if there exists at least one point in I at which $W(y_1, \ldots, y_n) \neq 0$, the functions $y_1, \ldots, y_n$ are linearly independent on I.

In Exercises 11–15 use the method of Exercise 10 to determine whether the functions are linearly dependent or independent on the interval.

11. $\{1, x, x^2\}$ on $-\infty < x < \infty$

12. $\{x, 2x - 3x^2, x^2\}$ on $-\infty < x < \infty$

13. $\{\sin x, \cos x\}$ on $0 \leq x \leq 2\pi$

14. $\{x, xe^x, x^2 e^x\}$ on $0 \leq x \leq 1$

15. $\{x \sin x, e^{2x}\}$ on $-\infty < x < \infty$

SECTION 16.9

Homogeneous Linear Differential Equations with Constant Coefficients

We now consider homogeneous linear differential equations

$$a_0 D^n y + a_1 D^{n-1} y + a_2 D^{n-2} y + \cdots + a_{n-1} D y + a_n y = 0, \quad (16.47)$$

where the coefficients $a_0, a_1, \ldots, a_n$ are all constants. In operator notation we write

$$\phi(D) y = 0, \quad (16.48\text{a})$$

where

$$\phi(D) = a_0 D^n + a_1 D^{n-1} + \cdots + a_{n-1} D + a_n. \quad (16.48\text{b})$$

The superposition principle states that a general solution of equation 16.47 is $y(x) = C_1 y_1(x) + \cdots + C_n y_n(x)$, provided $y_1(x), \ldots, y_n(x)$ are any n linearly independent solutions of the equation. Our problem then is to devise a technique for finding n linearly independent solutions; to illustrate a possible procedure, we first consider three examples of second-order equations.

Our first example is

$$y'' + 2y' - 3y = 0.$$

It is not unreasonable to expect that perhaps for some value of m, $y(x) = e^{mx}$ might be a solution of this equation. After all, the equation says that the second derivative of the function must be equal to three times the function minus twice its first derivative. Since the exponential function reproduces itself when differentiated, perhaps m can be chosen to produce this combination. To see whether this is possible, we substitute $y = e^{mx}$ into the differential equation, and find that if $y = e^{mx}$ is to be a solution, then

$$m^2 e^{mx} + 2m e^{mx} - 3 e^{mx} = 0;$$

this implies that

$$0 = m^2 + 2m - 3 = (m+3)(m-1).$$

Thus $y = e^{mx}$ is a solution if m is chosen as either 1 or -3; i.e., $y_1 = e^x$ and $y_2 = e^{-3x}$ are solutions of the differential equation. Since they are linearly independent, a general solution is $y(x) = C_1 e^x + C_2 e^{-3x}$.

For our second example, we take
$$y'' + 2y' + y = 0.$$

Since exponentials worked in the first example, we once again try a solution of the form $y(x) = e^{mx}$. If we substitute into the differential equation, we obtain
$$m^2 e^{mx} + 2me^{mx} + e^{mx} = 0,$$
which implies that
$$0 = m^2 + 2m + 1 = (m+1)^2.$$

Thus $y_1 = e^{-x}$ is a solution, but unfortunately it is the only solution that we obtain as a result of our guess. We need a second linearly independent solution $y_2(x)$ in order to obtain a general solution. Clearly, no other exponential will work. Perhaps if we multiplied e^{-x} by another function, we might find a second solution; in other words, perhaps there is a solution of the form $y(x) = v(x)e^{-x}$ for some $v(x)$. To see, we again substitute into the differential equation,
$$0 = \{v''e^{-x} - 2v'e^{-x} + ve^{-x}\} + 2\{v'e^{-x} - ve^{-x}\} + ve^{-x} = v''e^{-x}.$$

Consequently, $v'' = 0$, and this implies that $v(x) = Ax + B$, for any constants A and B. In particular, if $A = 1$ and $B = 0$, $v(x) = x$, and $y_2(x) = v(x)e^{-x} = xe^{-x}$ is also a solution of the differential equation. By superposition, then, we find that a general solution is $y(x) = C_1 e^{-x} + C_2 xe^{-x} = (C_1 + C_2 x)e^{-x}$. Note that if we had set $A = 0$ and $B = 1$, then $v(x) = 1$, and the solution $y(x) = v(x)e^{-x}$ would have been $y_1(x)$. Further, if we had simply set $y(x) = v(x)e^{-x} = (Ax + B)e^{-x}$, we would have found the general solution.

Our third example is
$$y'' + 2y' + 10y = 0.$$

As in the previous two examples, if we assume a solution $y = e^{mx}$, then
$$m^2 e^{mx} + 2me^{mx} + 10e^{mx} = 0,$$
or
$$0 = m^2 + 2m + 10.$$

The solutions of this quadratic equation are the complex numbers
$$m = \frac{-2 \pm \sqrt{4 - 40}}{2} = -1 \pm 3i.$$

This means that there is no real exponential $y = e^{mx}$ that will satisfy the differential equation. If, however, we form complex exponentials $y_1(x) = e^{(-1+3i)x}$ and $y_2(x) = e^{(-1-3i)x}$ and superpose these solutions, then
$$y(x) = Ae^{(-1+3i)x} + Be^{(-1-3i)x}$$
must also be a solution. When we use Euler's identity for complex exponentials, $e^{i\theta} = \cos\theta + i\sin\theta$, we can write $y(x)$ in the form
$$\begin{aligned}y(x) &= Ae^{-x}e^{3xi} + Be^{-x}e^{-3xi}\\ &= Ae^{-x}(\cos 3x + i\sin 3x) + Be^{-x}(\cos 3x - i\sin 3x)\\ &= e^{-x}[(A+B)\cos 3x + i(A-B)\sin 3x]\\ &= e^{-x}(C_1 \cos 3x + C_2 \sin 3x),\end{aligned}$$

where $C_1 = A + B$ and $C_2 = i(A - B)$. In other words, the function $y(x) = e^{-x}(C_1 \cos 3x + C_2 \sin 3x)$ is a general solution of the differential equation, and it has been derived from the complex roots $m = -1 \pm 3i$ of the equation $m^2 + 2m + 10 = 0$. Note that what multiplies x in the exponential is the real part of these complex numbers, and what multiplies x in the trigonometric functions is the imaginary part. In Exercise 17 we show that this solution can also be derived without complex numbers, but we feel that in general the use of complex numbers is the best method.

In each of these examples we guessed $y = e^{mx}$ as a possible solution. We then substituted into the differential equation to obtain an algebraic equation for m. In each case the equation for m was

$$\phi(m) = 0;$$

that is, take the operator $\phi(D)$, replace D by m, and set the polynomial equal to zero. This is not a peculiarity of these examples for it is straightforward to show that for any homogeneous linear equation 16.48, if we assume a solution of the form $y = e^{mx}$, then m must satisfy the equation $\phi(m) = 0$. We name this equation in the following definition.

Definition 16.7

With every linear differential equation that has constant coefficients $\phi(D)y = F(x)$, we associate an equation

$$\phi(m) = 0 \qquad (16.49)$$

called the **auxiliary equation**.

To summarize, in each of the examples we assumed a solution $y = e^{mx}$ and found that m had to satisfy the auxiliary equation $\phi(m) = 0$. From the roots of the auxiliary equation we obtained solutions of the differential equation, and superposition then led to a general solution. We have found what could be a general procedure for determining a general solution of every homogeneous linear differential equation with constant coefficients. This is indeed true so that every such differential equation can be solved in exactly the same way. But if the procedure is the same in every case, surely we can set down rules that eliminate the necessity of tediously repeating these steps in every example. This we do in the following theorem.

Theorem 16.3

If $\phi(m) = 0$ is the auxiliary equation associated with the homogeneous linear differential equation $\phi(D)y = 0$, then there are two possibilities:

(i) $\phi(m) = 0$ has a real root m of multiplicity k. Then a solution of the differential equation is

$$(C_1 + C_2 x + \cdots + C_k x^{k-1})e^{mx}. \qquad (16.50\text{a})$$

(ii) $\phi(m) = 0$ has a pair of complex conjugate roots $a \pm bi$ each of multiplicity k. Then a solution of the differential equation is

$$e^{ax}[(C_1 + C_2 x + \cdots + C_k x^{k-1}) \cos bx \\ + (D_1 + D_2 x + \cdots + D_k x^{k-1}) \sin bx]. \qquad (16.50\text{b})$$

A general solution of the differential equation is obtained by superposing all solutions in **(i)** and **(ii)**.

For a proof of this theorem see Exercise 20. Let us now apply the theorem to our previous examples. The auxiliary equation for $y'' + 2y' - 3y = 0$ is

$$0 = m^2 + 2m - 3 = (m+3)(m-1)$$

with solutions $m = 1$ and $m = -3$. If we now use part **(i)** of Theorem 16.3 with two real roots, each of multiplicity 1, a general solution of the differential equation is

$$y(x) = C_1 e^x + C_2 e^{-3x}.$$

The auxiliary equation for $y'' + 2y' + y = 0$ is

$$0 = m^2 + 2m + 1 = (m+1)^2$$

with solutions $m = -1$ and $m = 1$. Part **(i)** of Theorem 16.3 with a single real root of multiplicity 2 gives the general solution

$$y(x) = (C_1 + C_2 x) e^{-x}.$$

The auxiliary equation for $y'' + 2y' + 10y = 0$ is

$$0 = m^2 + 2m + 10$$

with solutions $m = -1 \pm 3i$. Part **(ii)** of Theorem 16.3 with a pair of complex conjugate roots, each of multiplicity 1, gives the general solution

$$y(x) = e^{-x}(C_1 \cos 3x + C_2 \sin 3x).$$

EXAMPLE 16.12 Find a general solution for the differential equation

$$y''' - 3y'' - 4y' + 6y = 0.$$

SOLUTION The auxiliary equation is

$$0 = m^3 - 3m^2 - 4m + 6 = (m-1)(m^2 - 2m - 6)$$

with solutions $m = 1$ and $m = 1 \pm \sqrt{7}$. A general solution of the differential equation is therefore

$$y(x) = C_1 e^x + C_2 e^{(1+\sqrt{7})x} + C_3 e^{(1-\sqrt{7})x}.$$

EXAMPLE 16.13 Find a general solution for $y''' - y = 0$.

SOLUTION The auxiliary equation is

$$0 = m^3 - 1 = (m-1)(m^2 + m + 1)$$

with solutions $m = 1$ and $m = -(1/2) \pm (\sqrt{3}/2)i$. A general solution of the differential equation is therefore

$$y(x) = C_1 e^x + e^{-x/2}[C_2 \cos(\sqrt{3}x/2) + C_3 \sin(\sqrt{3}x/2)].$$

EXAMPLE 16.14

If the roots of the auxiliary equation $\phi(m) = 0$ are

$$3, 3, 3, \pm 2i, -2, 1 \pm \sqrt{3}, -4 \pm i, -4 \pm i,$$

find a general solution of the differential equation $\psi(D)y = 0$.

SOLUTION A general solution is

$$y(x) = (C_1 + C_2 x + C_3 x^2)e^{3x} + C_4 \cos 2x + C_5 \sin 2x + C_6 e^{-2x}$$
$$+ C_7 e^{(1+\sqrt{3})x} + C_8 e^{(1-\sqrt{3})x} + e^{-4x}[(C_9 + C_{10}x)\cos x$$
$$+ (C_{11} + C_{12}x)\sin x].$$

EXERCISES 16.9

In Exercises 1–12 find a general solution for the homogeneous differential equation.

1. $y'' + y' - 6y = 0$
2. $2y'' - 16y' + 32y = 0$
3. $2y'' + 16y' + 82y = 0$
4. $y'' + 2y' - 2y = 0$
5. $y'' - 4y' + 5y = 0$
6. $y''' - 3y'' + y' - 3y = 0$
7. $y'''' + 2y'' + y = 0$
8. $y''' - 6y'' + 12y' - 8y = 0$
9. $3y''' - 12y'' + 18y' - 12y = 0$
10. $y'''' + 5y'' + 4y = 0$
11. $y''' - 3y'' + 2y' = 0$
12. $y'''' + 16y = 0$

In Exercises 13–16 find a homogeneous linear differential equation that has the function as general solution.

13. $y(x) = C_1 e^x + (C_2 + C_3 x)e^{-4x}$
14. $y(x) = e^{-2x}(C_1 \cos 4x + C_2 \sin 4x)$
15. $y(x) = C_1 + C_2 e^{\sqrt{3}x} + C_3 e^{-\sqrt{3}x}$
16. $y(x) = e^x(C_1 + C_2 x)\cos\sqrt{2}x + e^x(C_3 + C_4 x)\sin\sqrt{2}x$

17. Show that if we assume that $y(x) = e^{ax} \sin bx$ is a solution of the differential equation $y'' + 2y' + 10y = 0$, then a and b must be equal to -1 and ± 3 respectively. Verify that for this a and b, $y(x) = e^{ax} \cos bx$ is also a solution, and therefore a general solution is $y(x) = e^{-x}(C_1 \cos 3x + C_2 \sin 3x)$.

18. The equation $y''' + ay'' + by' + cy = 0$, where a, b, and c are constants, has solution $y(x) = C_1 e^{-x} + e^{-2x}(C_2 \sin 4x + C_3 \cos 4x)$. Find a, b, and c.

19. Show that if p is constant and $f(x)$ is differentiable, then

$$D\{e^{px} f(x)\} = e^{px}\{(D+p)f(x)\}.$$

Now use mathematical induction to prove that if $f(x)$ is k times differentiable, then

$$D^k\{e^{px} f(x)\} = e^{px}\{(D+p)^k f(x)\}.$$

Finally verify that

$$\phi(D)\{e^{px} f(x)\} = e^{px}\{\phi(D+p)f(x)\},$$

a result called the *operator shift theorem*.

20. (a) If m_0 is a real root of multiplicity k for the auxiliary equation $\phi(m) = 0$, show that the operator $\phi(D)$ can be expressed in the form

$$\phi(D) = (D - m_0)^k \psi(D),$$

where $\psi(D)$ is a polynomial in D. Now use the operator shift theorem of Exercise 19 to verify that $(C_1 + C_2 x + \cdots + C_k x^{k-1})e^{m_0 x}$ is a solution of $\phi(D)y = 0$.

(b) If $a \pm bi$ are complex conjugate roots each of multiplicity k for the auxiliary equation $\phi(m) = 0$, show that $\phi(D)$ can be expressed in the form

$$\phi(D) = (D - a - bi)^k (D - a + bi)^k \psi(D),$$

where $\psi(D)$ is a polynomial in D. Now use the operator shift theorem of Exercise 19 to verify that 16.50b is a solution of $\phi(D)y = 0$.

21. If M, β, and k are all positive constants, find a general solution for the linear differential equation

$$M\frac{d^2 x}{dt^2} + \beta \frac{dx}{dt} + kx = 0.$$

SECTION 16.10

Nonhomogeneous Linear Differential Equations with Constant Coefficients

The general nonhomogeneous linear differential equation with constant coefficients is

$$\phi(D)y = F(x), \qquad (16.51\text{a})$$

where

$$\phi(D) = a_0 D^n + a_1 D^{n-1} + \cdots + a_{n-1} D + a_n. \qquad (16.51\text{b})$$

It is natural to ask whether we can use the results of Section 16.9 concerning homogeneous equations with constant coefficients to solve nonhomogeneous problems. Fortunately the answer is yes, as shown by the following definition.

Definition 16.8 With every nonhomogeneous linear differential equation with constant coefficients

$$\phi(D)y = F(x),$$

we associate a homogeneous equation

$$\phi(D)y = 0, \qquad (16.52)$$

called the **homogeneous (reduced, or complimentary)** equation associated with $\phi(D)y = F(x)$.

We now prove the following theorem.

Theorem 16.4 A general solution of the linear differential equation $\phi(D)y = F(x)$ is $y(x) = y_h(x) + y_p(x)$, where $y_h(x)$ is a general solution of the associated homogeneous equation, and $y_p(x)$ is any particular solution of the given equation.

Proof Since $\phi(D)$ is a linear operator,

$$\phi(D)(y_h + y_p) = \phi(D)y_h + \phi(D)y_p = 0 + F(x) = F(x),$$

so that $y_h + y_p$ is indeed a solution of the given differential equation. Because $y_h(x)$ is a general solution of the associated homogeneous equation, it contains the requisite number of arbitrary constants for $y(x)$ to be a general solution of $\phi(D)y = F(x)$, and this completes the proof.

We note in passing that Theorem 16.4 is also valid for linear differential equations with variable coefficients.

Theorem 16.4 indicates that the discussion of nonhomogeneous differential equations can be divided into two separate parts. First we find a general solution $y_h(x)$ of the associated homogeneous equation 16.52, and this can be done using the results of Section 16.9. To this we add any particular solution $y_p(x)$ of 16.51. We present two methods for finding a particular solution: (1) the method of undetermined coefficients, and (2) the method of operators. Both methods apply in general only to differential equations in which $F(x)$ is a power (x^n, n a nonnegative integer), an exponential (e^{px}), a sine ($\sin px$), a cosine ($\cos px$), and/or any sums or products thereof.

Method of Undetermined Coefficients for a Particular Solution

As stated above, the method of undetermined coefficients is to be used only when $F(x)$ in equation 16.51 is of the form $x^n, e^{px}, \sin px, \cos px$, and/or sums or products thereof. For example, if

$$y'' + y' - 6y = e^{4x},$$

then the method essentially says that Ae^{4x} is the simplest function that could conceivably yield e^{4x} when substituted into the left-hand side of the differential equation. Consequently, it is natural to assume $y_p = Ae^{4x}$ and attempt to determine the unknown coefficient A. Substitution of this function into the differential equation gives

$$16 Ae^{4x} + 4 Ae^{4x} - 6 Ae^{4x} = e^{4x}.$$

If we divide by e^{4x}, then

$$14 A = 1 \quad \text{and} \quad A = \tfrac{1}{14}.$$

A particular solution is therefore $y_p = e^{4x}/14$.

Before stating a general rule, we illustrate a few more possibilities in the following example.

EXAMPLE 16.15

Find a particular solution of $y'' + y' - 6y = F(x)$ in each case.

(a) $F(x) = 6x^2 + 2x + 3$
(b) $F(x) = 2 \sin 2x$
(c) $F(x) = xe^{-x} - e^{-x}$

SOLUTION

(a) Since terms in x^2, x, and constants yield terms in x^2, x, and constants when substituted into the left-hand side of the differential equation, we attempt to find a particular solution of the form $y_p = Ax^2 + Bx + C$. Substitution into the differential equation gives

$$(2A) + (2Ax + B) - 6(Ax^2 + Bx + C) = 6x^2 + 2x + 3,$$

or,
$$(-6A)x^2 + (2A - 6B)x + (2A + B - 6C) = 6x^2 + 2x + 3.$$

But this equation can hold for all values of x only if coefficients of corresponding powers of x are identical (see Exercise 35 in Section 3.7). Equating coefficients then gives
$$-6A = 6,$$
$$2A - 6B = 2,$$
$$2A + B - 6C = 3.$$

These imply that $A = -1, B = -2/3, C = -17/18$, and
$$y_p = -x^2 - \frac{2x}{3} - \frac{17}{18}.$$

(b) Since terms in $\sin 2x$ and $\cos 2x$ yield terms in $\sin 2x$ when substituted into the left-hand side of the differential equation, we assume that $y_p = A\sin 2x + B\cos 2x$. Substitution into the differential equation gives

$$(-4A\sin 2x - 4B\cos 2x) + (2A\cos 2x - 2B\sin 2x)$$
$$- 6(A\sin 2x + B\cos 2x) = 2\sin 2x,$$

or,
$$(-10A - 2B)\sin 2x + (2A - 10B)\cos 2x = 2\sin 2x.$$

Equating coefficients gives
$$-10A - 2B = 2,$$
$$2A - 10B = 0.$$

These imply that $A = -5/26, B = -1/26$, and hence
$$y_p = -\frac{1}{26}(5\sin 2x + \cos 2x).$$

(c) Since terms in xe^{-x} and e^{-x} yield terms in xe^{-x} and e^{-x} when substituted into the left-hand side of the differential equation, we assume that $y_p = Axe^{-x} + Be^{-x}$. Substitution into the differential equation gives

$$(Axe^{-x} - 2Ae^{-x} + Be^{-x}) + (-Axe^{-x} + Ae^{-x} - Be^{-x})$$
$$- 6(Axe^{-x} + Be^{-x}) = xe^{-x} - e^{-x},$$

or,
$$(-6A)xe^{-x} + (-A - 6B)e^{-x} = xe^{-x} - e^{-x}.$$

Equating coefficients yields
$$-6A = 1,$$
$$-A - 6B = -1.$$

These imply that $A = -1/6, B = 7/36$, and hence $y_p = -(1/6)xe^{-x} + (7/36)e^{-x}$. ∎

The following rule encompasses each part of this example.

> **Rule 1**
>
> If a term of $F(x)$ consists of a power (x^n), an exponential (e^{px}), a sine ($\sin px$), a cosine ($\cos px$), or any product thereof, assume as a part of y_p a constant multiplied by that term plus a constant multiplied by any linearly independent function arising from it by differentiation.

For Example 16.15(a), since $F(x)$ contains the term $6x^2$, we assume y_p contains Ax^2. Differentiation of Ax^2 yields a term in x and a constant so that we form $y_p = Ax^2 + Bx + C$. No new terms for y_p are obtained from the terms $2x$ and 3 in $F(x)$.

For Example 16.15(b), we assume that y_p contains $A \sin 2x$ to account for the term $2 \sin 2x$ in $F(x)$. Differentiation of $A \sin 2x$ gives a linearly independent term in $\cos 2x$ so that we form $y_p = A \sin 2x + B \cos 2x$.

For Example 16.15(c), since $F(x)$ contains the term xe^{-x}, we assume that y_p contains Axe^{-x}. Differentiation of Axe^{-x} yields a term in e^{-x} so that we form $y_p = Axe^{-x} + Be^{-x}$. No new terms for y_p are obtained from the term $-e^{-x}$ in $F(x)$.

EXAMPLE 16.16

What is the form of the particular solution predicted by Rule 1 for the differential equation
$$y'' + 15y' - 6y = x^2 e^{4x} + x + x \cos x?$$

SOLUTION Rule 1 suggests that
$$y_p = Ax^2 e^{4x} + Bxe^{4x} + Ce^{4x} + Dx + E + Fx \cos x + Gx \sin x + H \cos x + I \sin x.$$
∎

Unfortunately, exceptions to Rule 1 do occur. For the differential equation $y'' + y = \cos x$, Rule 1 would predict $y_p = A \cos x + B \sin x$. If we substitute this into the differential equation we obtain the absurd identity $0 = \cos x$, and certainly no equations to solve for A and B. This result could have been predicted had we first calculated $y_h(x)$. The auxiliary equation $m^2 + 1 = 0$ has solutions $m = \pm i$ so that $y_h(x) = C_1 \cos x + C_2 \sin x$. Since y_p as suggested by Rule 1 is precisely y_h with different names for the constants, then certainly $y_p'' + y_p = 0$. Suppose as an alternative we multiply this y_p by x, and assume that $y_p = Ax \cos x + Bx \sin x$. Substitution into the differential equation now gives
$$-2A \sin x + 2B \cos x = \cos x.$$
Identification of coefficients requires $A = 0, B = 1/2$, and $y_p = (1/2)x \sin x$.

This example suggests that if y_p predicted by Rule 1 is already contained in y_h, then a modification of y_p is necessary. A precise statement of the situation is given in the following rule.

> **Rule 2**
>
> Suppose that a term in $F(x)$ is of the form $x^n f(x)$ (n a nonnegative integer). Suppose further that $f(x)$ can be obtained from $y_h(x)$ by specifying values for the arbitrary constants. If this term in y_h results from a root of the auxiliary equation of multiplicity k, then corresponding to $x^n f(x)$, assume as a part of y_p the term $Ax^k(x^n f(x)) = Ax^{n+k} f(x)$, plus a constant multiplied by any linearly independent function arising from it by differentiation.

To use this rule we first require $y_h(x)$. Then, and only then, can we decide on the form of $y_p(x)$. As an illustration, consider the following example.

EXAMPLE 16.17

Find a general solution for $y''' - y = x^3 e^x$.

SOLUTION In Example 16.13 we solved the auxiliary equation to obtain $m = 1$ and $m = -1/2 \pm (\sqrt{3}/2)i$, from which we formed

$$y_h(x) = C_1 e^x + e^{-x/2}[C_2 \cos(\sqrt{3}x/2) + C_3 \sin(\sqrt{3}x/2)].$$

Now $x^3 e^x$ is x^3 times e^x, and e^x can be obtained from y_h by specifying $C_1 = 1$ and $C_2 = C_3 = 0$. Since this term results from the root $m = 1$ of multiplicity 1, we assume that y_p contains $Ax(x^3 e^x) = Ax^4 e^x$. Differentiation of this function gives terms in $x^3 e^x, x^2 e^x, xe^x$, and e^x. We therefore take

$$y_p = Ax^4 e^x + Bx^3 e^x + Cx^2 e^x + Dxe^x.$$

(We do not include a term in e^x since it is already in y_h.) Substitution into the differential equation and simplification gives

$$(12A)x^3 e^x + (36A + 9B)x^2 e^x + (24A + 18B + 6C)xe^x$$
$$+ (6B + 6C + 3D)e^x = x^3 e^x.$$

Equating coefficients gives

$$12A = 1,$$
$$36A + 9B = 0,$$
$$24A + 18B + 6C = 0,$$
$$6B + 6C + 3D = 0.$$

These imply that $A = 1/12, B = -1/3, C = 2/3, D = -2/3$, and

$$y_p = \tfrac{1}{12}x^4 e^x - \tfrac{1}{3}x^3 e^x + \tfrac{2}{3}x^2 e^x - \tfrac{2}{3}xe^x.$$

A general solution of the differential equation is therefore

$$y(x) = C_1 e^x + e^{-x/2}[C_2 \cos(\sqrt{3}x/2) + C_3 \sin(\sqrt{3}x/2)]$$
$$+ \frac{x^4 e^x}{12} - \frac{x^3 e^x}{3} + \frac{2x^2 e^x}{3} - \frac{2xe^x}{3}.$$

EXAMPLE 16.18

If the roots of the auxiliary equation $\phi(m) = 0$ for the differential equation $\phi(D)y = x^2 - 2\sin x + xe^{-2x}$ are $\pm i, -2, -2, -2, 4$, and 4, what is the form of y_p predicted by the method of undetermined coefficients?

SOLUTION From the roots of the auxiliary equation, we can form

$$y_h(x) = C_1 \cos x + C_2 \sin x + (C_3 + C_4 x + C_5 x^2)e^{-2x} + (C_6 + C_7 x)e^{4x}.$$

Corresponding to the term x^2 in $F(x)$, Rule 1 requires that y_p contain $Ax^2 + Bx + C$. Because $-2\sin x$ can be obtained from $y_h(x)$ by specifying $C_2 = -2$, $C_1 = C_3 = C_4 = C_5 = C_6 = C_7 = 0$, and this term results from the roots $m = \pm i$, each of multiplicity 1, Rule 2 suggests that y_p contain $Dx \sin x + Ex \cos x$. (We do not include terms in $\sin x$ and $\cos x$ since they are already in y_h.) Finally, xe^{-2x} is x times e^{-2x}, and this function can be obtained from $y_h(x)$ by setting $C_3 = 1$, and $C_1 = C_2 = C_4 = C_5 = C_6 = C_7 = 0$. Because this term results from the root $m = -2$ of multiplicity 3, y_p must contain $Fx^4 e^{-2x} + Gx^3 e^{-2x}$ (but not terms in $x^2 e^{-2x}$, xe^{-2x}, and e^{-2x} since they are in y_h). The total particular solution is therefore

$$y_p = Ax^2 + Bx + C + Dx \sin x + Ex \cos x + Fx^4 e^{-2x} + Gx^3 e^{-2x}.$$ ∎

Operator Method for a Particular Solution

This method, like that of undetermined coefficients, is designed only for functions $F(x)$ in equation 16.51 of the form $x^n, e^{px}, \sin px, \cos px$, and/or sums or products thereof. Essentially the operator method says that if

$$\phi(D)y = F(x), \qquad (16.53)$$

then

$$y = \frac{1}{\phi(D)} F(x). \qquad (16.54)$$

But there is a problem. What does it mean to say that

$$\frac{1}{\phi(D)} = \frac{1}{a_0 D^n + a_1 D^{n-1} + \cdots + a_{n-1} D + a_n} \qquad (16.55)$$

operates on $F(x)$? The operator method then depends on our explaining how $1/\phi(D)$ operates on $F(x)$. The simplest $\phi(D)$ is $\phi(D) = D$. In this case the differential equation is

$$Dy = F(x),$$

and the solution is

$$y(x) = \int F(x)\,dx.$$

If the solution is also to be represented in operator notation by

$$y(x) = \frac{1}{D}F(x),$$

then we must define the operator $1/D$ by

$$\frac{1}{D}F(x) = \int F(x)\,dx. \qquad (16.56)$$

If $1/D$ means to integrate, then $1/D^2$ must mean to integrate twice, $1/D^3$ to integrate three times, and so on.

We have stated that the method of operators is applicable in general only when $F(x)$ consists of powers, exponentials, and/or sines and cosines. Consider first the case in which $F(x)$ is a power x^n, in which case equations 16.53 and 16.54 become

$$\phi(D)y = x^n \tag{16.57}$$

and

$$y = \frac{1}{\phi(D)} x^n. \tag{16.58}$$

If we forget for the moment that D is a differential operator, and simply regard $1/\phi(D)$ as a rational function of a variable D, then we can express $1/\phi(D)$ as an infinite series of the form

$$\frac{1}{\phi(D)} = \frac{1}{D^k}\{b_0 + b_1 D + b_2 D^2 + \cdots\} \tag{16.59}$$

for some nonnegative integer k. (We will show how in a moment.) But this suggests that we write 16.58 in the form

$$y = \frac{1}{D^k}\{b_0 + b_1 D + b_2 D^2 + \cdots\} x^n. \tag{16.60}$$

If we now reinterpret D as d/dx, then the operator on the right will produce a polynomial in x. (Note that $D^m x^n = 0$ if $m > n$.) It turns out that if we ignore all arbitrary constants that result from the integrations (when $k \geq 1$), this polynomial is a solution of 16.57. In other words, when $F(x) = x^n$, a particular solution of 16.57 is

$$y_p = \frac{1}{D^k}\{b_0 + b_1 D + b_2 D^2 + \cdots\} x^n. \tag{16.61}$$

We now show how to expand $1/\phi(D)$ as a series of the form 16.59. Only two situations arise and each of these can be illustrated with a simple example. First suppose that $\phi(D) = D^2 + 4D + 5$. For $1/\phi(D)$ we write

$$\frac{1}{\phi(D)} = \frac{1}{D^2 + 4D + 5} = \frac{1}{5\left(1 + \frac{4D + D^2}{5}\right)},$$

and interpret the right-hand side (less the 5 outside the parentheses) as the sum of an infinite geometric series with common ratio $-(4D + D^2)/5$. When we write this series out, we have

$$\frac{1}{D^2 + 4D + 5} = \frac{1}{5}\left\{1 - \left(\frac{4D + D^2}{5}\right) + \left(\frac{4D + D^2}{5}\right)^2 - \cdots\right\}$$

$$= \frac{1}{5}\left\{1 - \frac{4}{5}D + \frac{11}{25}D^2 + \cdots\right\},$$

which is of the required form 16.59 with $k = 0$.

Power k in 16.59 is positive only when $\phi(D)$ has no constant term. For example, if $\phi(D) = D^4 - 2D^3 + 3D^2$, then we factor out D^2, and proceed as above:

$$\frac{1}{\phi(D)} = \frac{1}{D^2(3 - 2D + D^2)} = \frac{1}{3D^2\left(1 - \frac{2D - D^2}{3}\right)}$$

$$= \frac{1}{3D^2}\left\{1 + \left(\frac{2D - D^2}{3}\right) + \left(\frac{2D - D^2}{3}\right)^2 + \cdots\right\}$$

$$= \frac{1}{3D^2}\left\{1 + \frac{2}{3}D + \frac{1}{9}D^2 + \cdots\right\}.$$

EXAMPLE 16.19

Find a particular solution of the differential equation

$$y'' + 6y' + 4y = x^2 + 4.$$

SOLUTION We write

$$y_p = \frac{1}{D^2 + 6D + 4}(x^2 + 4) = \frac{1}{4\left(1 + \frac{6D + D^2}{4}\right)}(x^2 + 4)$$

$$= \frac{1}{4}\left\{1 - \left(\frac{6D + D^2}{4}\right) + \left(\frac{6D + D^2}{4}\right)^2 - \cdots\right\}(x^2 + 4).$$

Since $D^n(x^2 + 4) = 0$ if $n > 2$, we require only the constant term and terms in D and D^2:

$$y_p = \frac{1}{4}\left\{1 - \frac{3}{2}D + 2D^2 + \cdots\right\}(x^2 + 4)$$

$$= \frac{1}{4}(x^2 + 4) - \frac{3}{8}(2x) + \frac{1}{2}(2) = \frac{x^2}{4} - \frac{3x}{4} + 2.$$

EXAMPLE 16.20

Find a particular solution of the differential equation

$$y''' + 2y' = x^2 - x.$$

SOLUTION A particular solution is

$$y_p = \frac{1}{D^3 + 2D}(x^2 - x) = \frac{1}{D(2 + D^2)}(x^2 - x)$$

$$= \frac{1}{2D\left(1 + \frac{D^2}{2}\right)}(x^2 - x)$$

$$= \frac{1}{2D}\left\{1 - \frac{D^2}{2} + \cdots\right\}(x^2 - x)$$

$$= \frac{1}{2D}\left\{x^2 - x - \frac{1}{2}(2)\right\},$$

and since $1/D$ means to integrate with respect to x, we have

$$y_p = \frac{1}{2}\left\{\frac{x^3}{3} - \frac{x^2}{2} - x\right\} = \frac{x^3}{6} - \frac{x^2}{4} - \frac{x}{2}.$$

In summary, to evaluate 16.54 when $F(x)$ is a power x^n (or a polynomial), we expand $1/\phi(D)$ in a series of the form 16.59 and perform the indicated differentiations (and integrations if $k > 0$).

For all other cases of $F(x)$, we make use of a theorem called the **inverse operator shift theorem**. This theorem states that

$$\frac{1}{\phi(D)}\{e^{px} f(x)\} = e^{px} \frac{1}{\phi(D+p)} f(x) \qquad (16.62)$$

(see Exercise 16). The theorem enables us to shift the exponential e^{px} past the operator $1/\phi(D)$, but in so doing, the operator must be modified to $1/\phi(D+p)$. Equation 16.62 immediately yields $y_p(x)$ whenever $f(x)$ is a power x^n, n a nonnegative integer (or a polynomial). This is illustrated in the following example.

EXAMPLE 16.21 Find particular solutions for the following differential equations:

(a) $y'' + 3y' + 10y = x^2 e^{-x}$
(b) $y'' - 4y' + 4y = e^{2x}$
(c) $y''' + 2y'' + 3y' - y = e^{2x}$

SOLUTION

(a) Equation 16.62 gives

$$y_p = \frac{1}{D^2 + 3D + 10} x^2 e^{-x} = e^{-x} \frac{1}{(D-1)^2 + 3(D-1) + 10} x^2$$

$$= e^{-x} \frac{1}{D^2 + D + 8} x^2,$$

and we now proceed as in Example 16.19:

$$y_p = e^{-x} \frac{1}{8\left(1 + \frac{D + D^2}{8}\right)} x^2$$

$$= \frac{e^{-x}}{8}\left\{1 - \left(\frac{D + D^2}{8}\right) + \left(\frac{D + D^2}{8}\right)^2 - \cdots\right\} x^2$$

$$= \frac{e^{-x}}{8}\left\{1 - \frac{D}{8} - \frac{7D^2}{64} + \cdots\right\} x^2 = \frac{e^{-x}}{8}\left\{x^2 - \frac{x}{4} - \frac{7}{32}\right\}.$$

(b) Once again we use the inverse operator shift theorem to get

$$y_p = \frac{1}{D^2 - 4D + 4} e^{2x} = e^{2x} \frac{1}{(D+2)^2 - 4(D+2) + 4}(1)$$

$$= e^{2x} \frac{1}{D^2}(1) = \frac{x^2}{2} e^{2x},$$

since $1/D^2$ means to integrate twice.

(c) For this differential equation,

$$y_p = \frac{1}{D^3 + 2D^2 + 3D - 1}e^{2x} = e^{2x}\frac{1}{(D+2)^3 + 2(D+2)^2 + 3(D+2) - 1}(1)$$

$$= e^{2x}\frac{1}{D^3 + 8D^2 + 23D + 21}(1) = e^{2x}\frac{1}{21\left(1 + \frac{23D + 8D^2 + D^3}{21}\right)}(1)$$

$$= \frac{e^{2x}}{21}\{1 + \cdots\}(1) = \frac{e^{2x}}{21}. \quad \blacksquare$$

EXAMPLE 16.22

Find a general solution for $y''' - 3y'' + 3y' - y = 2x^2 e^x$.

SOLUTION The auxiliary equation is

$$0 = m^3 - 3m^2 + 3m - 1 = (m-1)^3$$

with solutions 1, 1, and 1. Thus,

$$y_h(x) = (C_1 + C_2 x + C_3 x^2)e^x.$$

A particular solution is

$$y_p(x) = \frac{2}{(D-1)^3}x^2 e^x = 2e^x\frac{1}{(D+1-1)^3}x^2 = 2e^x\frac{1}{D^3}x^2 = \frac{x^5}{30}e^x.$$

A general solution of the differential equation is therefore

$$y(x) = (C_1 + C_2 x + C_3 x^2)e^x + \frac{x^5}{30}e^x. \quad \blacksquare$$

EXAMPLE 16.23

Find a particular solution for the differential equation

$$y''' + 3y'' - 4y = xe^{-2x} + x^2.$$

SOLUTION

$$y_p = \frac{1}{D^3 + 3D^2 - 4}(xe^{-2x} + x^2)$$

$$= e^{-2x}\frac{1}{(D-2)^3 + 3(D-2)^2 - 4}x + \frac{1}{D^3 + 3D^2 - 4}x^2$$

$$= e^{-2x}\frac{1}{D^3 - 3D^2}x - \frac{1}{4\left(1 - \frac{3D^2 + D^3}{4}\right)}x^2$$

$$= e^{-2x}\frac{1}{D^2}\frac{1}{D-3}x - \frac{1}{4}\left\{1 + \left(\frac{3D^2 + D^3}{4}\right) + \cdots\right\}x^2$$

$$= e^{-2x}\frac{1}{D^2}\frac{1}{-3\left(1-\dfrac{D}{3}\right)}x - \frac{1}{4}\left(x^2 + \frac{3}{2}\right)$$

$$= \frac{e^{-2x}}{-3}\frac{1}{D^2}\left\{1 + \frac{D}{3} + \cdots\right\}x - \frac{x^2}{4} - \frac{3}{8}$$

$$= \frac{e^{-2x}}{-3}\frac{1}{D^2}\left\{x + \frac{1}{3}\right\} - \frac{x^2}{4} - \frac{3}{8}$$

$$= \frac{e^{-2x}}{-3}\left\{\frac{x^3}{6} + \frac{x^2}{6}\right\} - \frac{x^2}{4} - \frac{3}{8}. \quad\blacksquare$$

When $F(x)$ in 16.54 is of the form $x^n \sin px$ or $x^n \cos px$, n a nonnegative integer and p a constant, we introduce complex exponentials. Specifically, because

$$x^n e^{ipx} = x^n(\cos px + i\sin px),$$

we can write

$$x^n \cos px = \text{Real part of } x^n e^{ipx} = \text{Re}\{x^n e^{ipx}\}, \quad (16.63\text{a})$$
$$x^n \sin px = \text{Imaginary part of } x^n e^{ipx} = \text{Im}\{x^n e^{ipx}\}. \quad (16.63\text{b})$$

To operate on either of these functions by $1/\phi(D)$, we interchange the operations of $1/\phi(D)$ and taking real and imaginary parts:

$$\frac{1}{\phi(D)}\{x^n \cos px\} = \frac{1}{\phi(D)}\text{Re}\{x^n e^{ipx}\} = \text{Re}\left\{\frac{1}{\phi(D)}(x^n e^{ipx})\right\}, \quad (16.64\text{a})$$
$$\frac{1}{\phi(D)}\{x^n \sin px\} = \frac{1}{\phi(D)}\text{Im}\{x^n e^{ipx}\} = \text{Im}\left\{\frac{1}{\phi(D)}(x^n e^{ipx})\right\}. \quad (16.64\text{b})$$

We can now proceed by using inverse operator shift theorem 16.62.

EXAMPLE 16.24 Find particular solutions for the following differential equations:

(a) $y'' + y = \sin 2x$;
(b) $y'' + 4y = x^2 \cos x$;
(c) $y'' + 9y = \sin 3x$;
(d) $y'' + 4y = x \sin 2x$;
(e) $y'' + 2y' + 4y = e^{-x}\sin\sqrt{3}\,x$.

SOLUTION

(a) Equation 16.64b gives

$$y_p = \frac{1}{D^2 + 1}\sin 2x = \frac{1}{D^2 + 1}\{\text{Im}(e^{2ix})\} = \text{Im}\left\{\frac{1}{D^2 + 1}e^{2ix}\right\}.$$

We now use inverse operator shift theorem 16.62,

$$y_p = \text{Im}\left\{e^{2ix}\frac{1}{(D+2i)^2+1}(1)\right\}$$

$$= \text{Im}\left\{e^{2ix}\frac{1}{D^2+4iD-3}(1)\right\} = \text{Im}\left\{e^{2ix}\frac{1}{-3\left(1-\frac{4iD+D^2}{3}\right)}(1)\right\}$$

$$= \text{Im}\left\{\frac{e^{2ix}}{-3}\right\} = -\frac{1}{3}\sin 2x.$$

(b)

$$y_p = \frac{1}{D^2+4}x^2\cos x = \frac{1}{D^2+4}\text{Re}(x^2 e^{ix}) = \text{Re}\left\{\frac{1}{D^2+4}x^2 e^{ix}\right\}$$

$$= \text{Re}\left\{e^{ix}\frac{1}{(D+i)^2+4}x^2\right\} = \text{Re}\left\{e^{ix}\frac{1}{D^2+2iD+3}x^2\right\}$$

$$= \text{Re}\left\{e^{ix}\frac{1}{3\left(1+\frac{2iD+D^2}{3}\right)}x^2\right\}$$

$$= \text{Re}\left\{\frac{e^{ix}}{3}\left[1-\left(\frac{2iD+D^2}{3}\right)+\left(\frac{2iD+D^2}{3}\right)^2-\cdots\right]x^2\right\}$$

$$= \frac{1}{3}\text{Re}\left\{e^{ix}\left[1-\frac{2iD}{3}-\frac{7D^2}{9}+\cdots\right]x^2\right\} = \frac{1}{3}\text{Re}\left\{e^{ix}\left[x^2-\frac{4ix}{3}-\frac{14}{9}\right]\right\}$$

$$= \frac{1}{3}\left\{x^2\cos x + \frac{4x}{3}\sin x - \frac{14}{9}\cos x\right\}.$$

(c) For the differential equation $y'' + 9y = \sin 3x$, we have

$$y_p = \frac{1}{D^2+9}\sin 3x = \frac{1}{D^2+9}\text{Im}(e^{3ix}) = \text{Im}\left\{\frac{1}{D^2+9}e^{3ix}\right\}$$

$$= \text{Im}\left\{e^{3ix}\frac{1}{(D+3i)^2+9}(1)\right\} = \text{Im}\left\{e^{3ix}\frac{1}{D^2+6iD}(1)\right\}$$

$$= \text{Im}\left\{e^{3ix}\frac{1}{D}\frac{1}{D+6i}(1)\right\} = \text{Im}\left\{\frac{e^{3ix}}{6i}\frac{1}{D}\frac{1}{1+D/6i}(1)\right\}$$

$$= \text{Im}\left\{-\frac{i}{6}e^{3ix}\frac{1}{D}\left(1-\frac{D}{6i}+\cdots\right)(1)\right\} = \text{Im}\left\{-\frac{i}{6}e^{3ix}\frac{1}{D}(1)\right\}$$

$$= \text{Im}\left\{-\frac{i}{6}e^{3ix}x\right\} = -\frac{x}{6}\cos 3x.$$

(d) For the differential equation $y'' + 4y = x\sin 2x$, we get

$$y_p = \frac{1}{D^2 + 4} x \sin 2x = \frac{1}{D^2 + 4} \text{Im}(xe^{2ix}) = \text{Im}\left\{\frac{1}{D^2 + 4} xe^{2ix}\right\}$$

$$= \text{Im}\left\{e^{2ix}\frac{1}{(D+2i)^2 + 4}x\right\} = \text{Im}\left\{e^{2ix}\frac{1}{D^2 + 4iD}x\right\}$$

$$= \text{Im}\left\{e^{2ix}\frac{1}{D}\frac{1}{D+4i}x\right\} = \text{Im}\left\{\frac{1}{4i}e^{2ix}\frac{1}{D}\left(1 - \frac{D}{4i} + \cdots\right)x\right\}$$

$$= \text{Im}\left\{-\frac{i}{4}e^{2ix}\frac{1}{D}\left(x - \frac{1}{4i}\right)\right\} = \text{Im}\left\{-\frac{i}{4}e^{2ix}\left(\frac{x^2}{2} + \frac{ix}{4}\right)\right\}$$

$$= -\frac{x^2}{8}\cos 2x + \frac{x}{16}\sin 2x.$$

(e) For the differential equation $y'' + 2y' + 4y = e^{-x}\sin\sqrt{3}x$, we have

$$y_p = \frac{1}{D^2 + 2D + 4} e^{-x}\sin\sqrt{3}x = \frac{1}{D^2 + 2D + 4}\text{Im}(e^{-x}e^{\sqrt{3}ix})$$

$$= \text{Im}\left\{\frac{1}{D^2 + 2D + 4} e^{(-1+\sqrt{3}i)x}\right\}$$

$$= \text{Im}\left\{e^{(-1+\sqrt{3}i)x}\frac{1}{(D-1+\sqrt{3}i)^2 + 2(D-1+\sqrt{3}i) + 4}(1)\right\}$$

$$= \text{Im}\left\{e^{(-1+\sqrt{3}i)x}\frac{1}{D^2 + 2\sqrt{3}iD}(1)\right\} = \text{Im}\left\{e^{(-1+\sqrt{3}i)x}\frac{1}{D}\frac{1}{D+2\sqrt{3}i}(1)\right\}$$

$$= \text{Im}\left\{e^{(-1+\sqrt{3}i)x}\frac{1}{D}\frac{1}{2\sqrt{3}i}\right\} = \text{Im}\left\{-\frac{i}{2\sqrt{3}}xe^{(-1+\sqrt{3}i)x}\right\}$$

$$= -\frac{xe^{-x}}{2\sqrt{3}}\cos\sqrt{3}x.$$

■

EXAMPLE 16.25

Find a general solution of $y'' + 6y' + y = \sin 3x$.

SOLUTION The auxiliary equation is $m^2 + 6m + 1 = 0$ with solutions $m = -3 \pm 2\sqrt{2}$. Consequently,

$$y_h(x) = C_1 e^{(-3+2\sqrt{2})x} + C_2 e^{(-3-2\sqrt{2})x}.$$

A particular solution is

$$y_p = \frac{1}{D^2 + 6D + 1}\text{Im}(e^{3ix}) = \text{Im}\left\{\frac{1}{D^2 + 6D + 1}e^{3ix}\right\}$$

$$= \text{Im}\left\{e^{3ix}\frac{1}{(D+3i)^2 + 6(D+3i) + 1}(1)\right\}$$

$$= \text{Im}\left\{e^{3ix}\frac{1}{D^2 + (6+6i)D + (-8+18i)}(1)\right\} = \text{Im}\left\{e^{3ix}\frac{1}{-8+18i}\right\}$$

$$= \text{Im}\left\{e^{3ix}\frac{1}{-8+18i}\frac{-8-18i}{-8-18i}\right\} = -\frac{1}{194}(4\sin 3x + 9\cos 3x).$$

Finally, then, a general solution of the differential equation is

$$y(x) = C_1 e^{(-3+2\sqrt{2})x} + C_2 e^{(-3-2\sqrt{2})x} - \frac{1}{194}(4\sin 3x + 9\cos 3x).$$

∎

Examples 16.19–16.25 have illustrated that with identity 16.62 and the concept of series, we can obtain a particular solution for 16.53 whenever $F(x)$ is $x^n, e^{px}, \sin px, \cos px$, and/or any sums or products thereof. In summary:

When $F(x) = x^n$, use series 16.59 for $1/\phi(D)$.

In any other situation, use 16.62, and then series 16.59 for $1/\phi(D+p)$.

We make one final comment. We introduced complex exponentials e^{ipx} to handle terms involving $\sin px$ and $\cos px$. This is not the only way to treat trigonometric functions, since there do exist other methods that completely avoid complex numbers. Unfortunately, these methods require memorization of somewhat involved identities. Because of this, we prefer the use of complex numbers.

EXERCISES 16.10

In Exercises 1–11 find a particular solution for the differential equation, by both the method of operators and the method of undetermined coefficients. Find a general solution of the equation.

1. $2y'' - 16y' + 32y = -e^{4x}$
2. $y'' + 2y' - 2y = x^2 e^{-x}$
3. $y''' - 3y'' + y' - 3y = 3xe^x + 2$
4. $y'''' + 2y'' + y = \cos 2x$
5. $y''' - 6y'' + 12y' - 8y = 2e^{2x}$
6. $y'''' + 5y'' + 4y = e^{-2x}$
7. $y''' - 3y'' + 2y' = x^2 + e^{-x}$
8. $2y'' + 16y' + 82y = -2e^{2x}\sin x$
9. $y'' + y' - 6y = x + \cos x$
10. $y'' - 4y' + 5y = x\cos x$
11. $3y''' - 12y'' + 18y' - 12y = x^2 + 3x - 4$

In Exercises 12–15 state the form for the particular solution predicted by the method of undetermined coefficients. Do not evaluate the coefficients.

12. $y''' + 9y'' + 27y' + 27y = xe^{3x} + 2x\cos x$
13. $y''' + 4y'' + y' + 4y = xe^x \sin x$
14. $2y''' - 6y'' - 12y' + 16y = xe^x + 2x^3 - 4\cos x$
15. $2y'' - 4y' + 10y = 5e^x \sin 2x$
16. Use the operator shift theorem of Exercise 19 in Section 16.9 to verify the inverse operator shift theorem 16.62.

In Exercises 17 and 18 find a general solution for the differential equation.

17. $y'' + 2y' - 4y = \cos^2 x$

18. $2y'' - 4y' + 3y = \cos x \sin 2x$

In Exercises 19 and 20 find a solution for the differential equation.

19. $y'' - 3y' + 2y = 8x^2 + 12e^{-x}$, $y(0) = 0$, $y'(0) = 2$

20. $y'' + 9y = x(\sin 3x + \cos 3x)$, $y(0) = y'(0) = 0$

21. If J, k, and w are positive constants, find a general solution for $J\dfrac{d^4 y}{dx^4} + ky = w$.

22. The second-order linear differential equation

$$x^2 \frac{d^2 y}{dx^2} + ax\frac{dy}{dx} + by = F(x), \qquad a, b \text{ constants,}$$

is called the *Cauchy-Euler linear equation*. Because of the x^2- and x-factors, it does not have constant coefficients, and is therefore not immediately amenable to the techniques of this chapter. Show that if we make a change of independent variable $x = e^z$, then

$$x\frac{dy}{dx} = \frac{dy}{dz}, \qquad x^2\frac{d^2 y}{dx^2} = \frac{d^2 y}{dz^2} - \frac{dy}{dz},$$

and that as a result the Cauchy-Euler equation is transformed into a linear equation in $y(z)$ with constant coefficients.

In Exercises 23 and 24 use the technique of Exercise 22 to find a general solution for the differential equation.

23. $\dfrac{d^2 u}{dr^2} + \dfrac{1}{r}\dfrac{du}{dr} - \dfrac{u}{r^2} = 0$, $r > 0$

24. $x^2 y'' + xy' + 4y = 1$, $x > 0$

25. If M, β, k, A, and ω are all positive constants, find a particular solution of the linear differential equation

$$M\frac{d^2 x}{dt^2} + \beta\frac{dx}{dt} + kx = A\sin\omega t.$$

SECTION 16.11

Applications of Linear Differential Equations

Vibrating Mass-Spring Systems

In Figure 16.14 we have shown a mass M suspended vertically from a spring. If M is given an initial motion in the vertical direction (and the vertical direction only), then we expect M to oscillate up and down for some time. In this section we show how to describe these oscillations mathematically.

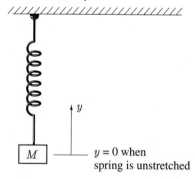

FIGURE 16.14

In order to describe the position of M as a function of time t, we must choose a vertical coordinate system. There are two natural places to choose the origin $y = 0$, one being the position of M when the spring is unstretched. Suppose we do this and choose y as positive upward. When M is a distance y away from the origin, the restoring force of the spring has y-component $-ky$ ($k > 0$). In addition, if g is the acceleration due to gravity ($g < 0$), then the force of gravity on M has y-component Mg. Finally, suppose the oscillations take place in a medium that exerts a damping force proportional to the instantaneous velocity of M. This damping force must therefore have a y-component of the form $-\beta(dy/dt)$, where β is a positive constant. The total force on M therefore has y-component

$$-ky + Mg - \beta\frac{dy}{dt},$$

and Newton's second law states that the acceleration d^2y/dt^2 of M must satisfy the equation

$$-ky + Mg - \beta\frac{dy}{dt} = M\frac{d^2y}{dt^2}.$$

Consequently, the differential equation that determines the position $y(t)$ of M relative to the unstretched position of the spring is

$$M\frac{d^2y}{dt^2} + \beta\frac{dy}{dt} + ky = Mg. \qquad (16.65)$$

The alternative possibility for describing oscillations is to attach M to the spring and slowly lower M until it reaches an equilibrium position. At this position, the restoring force of the spring is exactly equal to the force of gravity on the mass, and the mass, left by itself, will remain motionless. If s is the amount of stretch in the spring at equilibrium, and g is the acceleration due to gravity, then at equilibrium

$$ks + Mg = 0, \quad \text{where } s > 0 \text{ and } g < 0. \qquad (16.66)$$

Suppose we take the equilibrium position as $x = 0$ and x as positive upward (Figure 16.15). When M is a distance x away from its equilibrium position, the restoring force on M has x-component $k(s-x)$. The x-component of the force of gravity remains as Mg, and that of the damping force is $-\beta(dx/dt)$. Newton's second law therefore implies that

$$M\frac{d^2x}{dt^2} = k(s-x) + Mg - \beta\frac{dx}{dt},$$

or,

$$M\frac{d^2x}{dt^2} + \beta\frac{dx}{dt} + kx = Mg + ks.$$

But according to 16.66, $Mg + ks = 0$, and hence

$$M\frac{d^2x}{dt^2} + \beta\frac{dx}{dt} + kx = 0. \qquad (16.67)$$

This is the differential equation describing the displacement $x(t)$ of M relative to the equilibrium position of M.

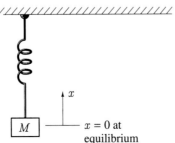

FIGURE 16.15

Note that both equations 16.65 and 16.67 are linear second-order differential equations with constant coefficients. The advantage of 16.67 is that it is homogeneous as well, and this is simply due to a convenient choice of dependent variable (x as opposed to y). Physically, we are saying that there are two parts to the spring force $k(s-x)$: a part ks and a part $-kx$. Gravity is always acting on M, and that part ks of the spring force is counteracting it in an attempt to restore the spring to its unstretched position. Because these forces always cancel, we might just as well eliminate both of them from our discussion. This would leave us $-kx$, and we therefore interpret $-kx$ as the *spring force attempting to restore the mass to its equilibrium position*.

If we choose equation 16.67 to describe the motion of M (and this equation is usually chosen over equation 16.65), we must remember three things: x is measured from equilibrium, $-kx$ is the spring force attempting to restore M to its equilibrium position, and gravity has been taken into account.

There are three basic ways to initiate the motion. First, we can move the mass away from its equilibrium position and then release it, giving it an initial displacement but no initial velocity. Second, we can strike the mass at the equilibrium position, imparting an initial velocity but no initial displacement. And finally, we can give the mass both an initial displacement and an initial velocity. Each of these methods adds two initial conditions to the differential equation.

To be complete we note that when (in addition to the forces already mentioned) there is an externally applied force acting on the mass that is represented as a function of time by $F(t)$, then equation 16.67 is modified to

$$M\frac{d^2x}{dt^2} + \beta\frac{dx}{dt} + kx = F(t). \qquad (16.68)$$

Perhaps, for example, M contains some iron, and $F(t)$ is due to a magnet directly below M that constantly exerts an attractive force on M.

EXAMPLE 16.26

A 2-kg mass is suspended vertically from a spring with constant 16 N/m. The mass is raised 10 cm above its equilibrium position and then released. If damping is ignored, find the amplitude, period, and frequency of the motion.

SOLUTION If we choose $x = 0$ at the equilibrium position of the mass and x positive upward (Figure 16.15), then the differential equation for the motion $x(t)$ of the mass is

$$2\frac{d^2x}{dt^2} = -16x,$$

or,

$$\frac{d^2x}{dt^2} + 8x = 0,$$

along with the initial conditions

$$x(0) = 1/10, \quad x'(0) = 0.$$

The auxiliary equation is $m^2 + 8 = 0$ with solutions $m = \pm 2\sqrt{2}i$. Consequently,

$$x(t) = C_1 \cos(2\sqrt{2}t) + C_2 \sin(2\sqrt{2}t).$$

The initial conditions require

$$1/10 = C_1, \quad 0 = 2\sqrt{2}C_2.$$

Thus,

$$x(t) = \frac{1}{10}\cos(2\sqrt{2}t).$$

The amplitude of the oscillations is 1/10 m, the period is $2\pi/(2\sqrt{2}) = \pi/\sqrt{2}$ s, and the frequency is $\sqrt{2}/\pi$ s^{-1}. A graph of this function (Figure 16.16) illustrates the oscillations of the mass about its equilibrium position. This is an example of **simple harmonic motion**.

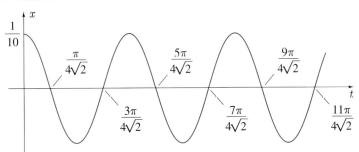

FIGURE 16.16

EXAMPLE 16.27

A 100-g mass is suspended vertically from a spring with constant 5 N/m. The mass is pulled 5 cm below its equilibrium position and given a velocity of 2 m/s upward. If, during the motion, the mass is acted on by a damping force in newtons numerically equal to one-twentieth the instantaneous velocity in metres per second, find the position of the mass at any time.

SOLUTION If we choose $x = 0$ at the equilibrium position of the mass and x positive upward (Figure 16.15), then the differential equation for the motion $x(t)$ of the mass is

$$\frac{1}{10}\frac{d^2x}{dt^2} = -5x - \frac{1}{20}\frac{dx}{dt},$$

or,

$$2\frac{d^2x}{dt^2} + \frac{dx}{dt} + 100x = 0,$$

along with the initial conditions

$$x(0) = -1/20, \quad x'(0) = 2.$$

The auxiliary equation is $2m^2 + m + 100 = 0$ with solutions

$$m = \frac{-1 \pm \sqrt{1-800}}{4} = \frac{-1 \pm \sqrt{799}\,i}{4}.$$

Consequently,

$$x(t) = e^{-t/4}[C_1 \cos(\sqrt{799}\,t/4) + C_2 \sin(\sqrt{799}\,t/4)].$$

The initial conditions require

$$-1/20 = C_1, \quad 2 = -C_1/4 + \sqrt{799}\,C_2/4,$$

from which we get

$$C_2 = \frac{159\sqrt{799}}{15\,980}.$$

Finally, then,

$$x(t) = e^{-t/4}\left[-\frac{1}{20}\cos\left(\frac{\sqrt{799}\,t}{4}\right) + \frac{159\sqrt{799}}{15\,980}\sin\left(\frac{\sqrt{799}\,t}{4}\right)\right]$$
$$= e^{-0.25t}[-0.05\cos(7.07t) + 0.28\sin(7.07t)].$$

The graph of this function in Figure 16.17 clearly indicates how the amplitude of the oscillations decreases in time.

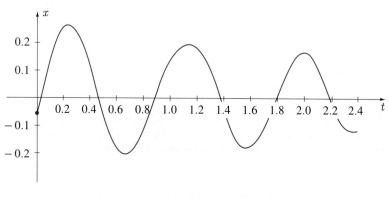

FIGURE 16.17

LCR Circuits

If a resistance R, an inductance L, and a capacitance C are connected in series with an electromotive force $E(t)$ (Figure 16.18) and the switch is closed, current flows in the circuit and charge builds up in the capacitor. If at any time t, Q is the charge on the capacitor and I is the current in the loop, then Kirchhoff's law states that

$$L\frac{dI}{dt} + RI + \frac{Q}{C} = E(t), \qquad (16.69)$$

where $L\,dI/dt$, RI, and Q/C represent the voltage drops across the inductor, the resistor, and the capacitor, respectively. If we substitute $I = dQ/dt$, then

$$L\frac{d^2Q}{dt^2} + R\frac{dQ}{dt} + \frac{1}{C}Q = E(t), \qquad (16.70)$$

a second-order linear differential equation for $Q(t)$.

Alternatively, if we differentiate this equation, we obtain

$$L\frac{d^2I}{dt^2} + R\frac{dI}{dt} + \frac{1}{C}I = E'(t), \qquad (16.71)$$

a second-order linear differential equation for $I(t)$.

FIGURE 16.18

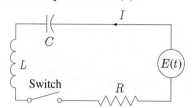

EXAMPLE 16.28

At time $t = 0$, a 25-Ω resistor, a 2-H inductor, and a 0.01-F capacitor are connected in series with a generator producing an alternating voltage of $10 \sin(5t)$, $t \geq 0$ (Figure 16.19). Find the charge on the capacitor and the current in the circuit if the capacitor is uncharged when the circuit is closed.

FIGURE 16.19

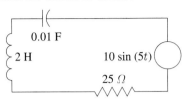

SOLUTION The differential equation for the charge Q on the capacitor is (see equation 16.70)

$$2 \frac{d^2 Q}{dt^2} + 25 \frac{dQ}{dt} + 100 Q = 10 \sin(5t),$$

to which we add the initial conditions

$$Q(0) = 0, \quad Q'(0) = I(0) = 0.$$

The auxiliary equation is $2m^2 + 25m + 100 = 0$ with solutions

$$m = \frac{-25 \pm \sqrt{625 - 800}}{4} = \frac{-25 \pm 5\sqrt{7}i}{4}.$$

Consequently, the general solution of the homogeneous equation is

$$Q_h(t) = e^{-25t/4}[C_1 \cos(5\sqrt{7}t/4) + C_2 \sin(5\sqrt{7}t/4)].$$

To find a particular solution of the nonhomogeneous equation by undetermined coefficients, we set

$$Q_p(t) = A \sin(5t) + B \cos(5t).$$

Substitution of this function into the differential equation gives

$$2\{-25 A \sin(5t) - 25 B \cos(5t)\} + 25\{5 A \cos(5t) - 5 B \sin(5t)\}$$

$$+ 100\{A \sin(5t) + B \cos(5t)\} = 10 \sin(5t).$$

This equation requires A and B to satisfy

$$50 A - 125 B = 10, \quad 125 A + 50 B = 0,$$

the solution of which is $A = 4/145$, $B = -10/145$. A particular solution is therefore

$$Q_p(t) = \frac{2}{145}[-5 \cos(5t) + 2 \sin(5t)]$$

and

$$Q(t) = Q_h(t) + Q_p(t)$$
$$= e^{-25t/4}\left[C_1 \cos\left(\frac{5\sqrt{7}t}{4}\right) + C_2 \sin\left(\frac{5\sqrt{7}t}{4}\right)\right]$$
$$+ \frac{2}{145}[2\sin(5t) - 5\cos(5t)].$$

The initial conditions require

$$0 = C_1 - \frac{10}{145}, \quad 0 = -\frac{25}{4}C_1 + \frac{5\sqrt{7}}{4}C_2 + \frac{20}{145},$$

and these imply that $C_1 = 10/145$, $C_2 = 34/(145\sqrt{7})$. Consequently,

$$Q(t) = \frac{e^{-25t/4}}{145\sqrt{7}}\left[10\sqrt{7}\cos\left(\frac{5\sqrt{7}t}{4}\right) + 34\sin\left(\frac{5\sqrt{7}t}{4}\right)\right]$$
$$+ \frac{2}{145}[2\sin(5t) - 5\cos(5t)].$$

The current in the circuit is

$$I(t) = \frac{dQ}{dt} = \left(-\frac{25}{4}\right)\frac{e^{-25t/4}}{145\sqrt{7}}\left[10\sqrt{7}\cos\left(\frac{5\sqrt{7}t}{4}\right) + 34\sin\left(\frac{5\sqrt{7}t}{4}\right)\right]$$
$$+ \frac{e^{-25t/4}}{145\sqrt{7}}\left[-\frac{175}{2}\sin\left(\frac{5\sqrt{7}t}{4}\right) + \frac{85\sqrt{7}}{2}\cos\left(\frac{5\sqrt{7}t}{4}\right)\right]$$
$$+ \frac{2}{145}[10\cos(5t) + 25\sin(5t)]$$
$$= -\frac{e^{-25t/4}}{29\sqrt{7}}\left[4\sqrt{7}\cos\left(\frac{5\sqrt{7}t}{4}\right) + 60\sin\left(\frac{5\sqrt{7}t}{4}\right)\right]$$
$$+ \frac{2}{29}[2\cos(5t) + 5\sin(5t)].$$

The solution $Q(t)$ contains two parts. The first two terms (containing the exponential $e^{-25t/4}$) are $Q_h(t)$ with the constants C_1 and C_2 determined by the initial conditions; the last two terms are $Q_p(t)$. We point this out because the two parts display completely different characteristics. For small t, both parts of $Q(t)$ are present and contribute significantly, but for very large t, the first two terms become negligible. In other words, after a long time, the charge $Q(t)$ on the capacitor is defined essentially by $Q_p(t)$. We call $Q_p(t)$ the steady-state part of the solution, and the two other terms in $Q(t)$ are called the transient part of the solution. Similarly, the first two terms in $I(t)$ are called the transient part of the current and the last two terms the steady-state part of the current.

Finally, note that the frequency of the steady-state part of either $Q(t)$ or $I(t)$ is exactly that of the forcing voltage $E(t)$. ∎

The similarity between differential equations 16.68 and 16.70 cannot go unmentioned:

$$M\frac{d^2x}{dt^2} + \beta\frac{dx}{dt} + kx = F(t),$$

$$L\frac{d^2Q}{dt^2} + R\frac{dQ}{dt} + \frac{1}{C}Q = E(t).$$

Each of the coefficients M, β, and k for the mechanical system has its analogue L, R, and $1/C$ in the electrical system. This suggests that LCR circuits might be used to model complicated physical systems subject to vibrations, and conversely, that mass-spring systems might represent complicated electrical systems.

EXERCISES 16.11

1. A 1-kg mass is suspended vertically from a spring with constant 16 N/m. The mass is pulled 10 cm below its equilibrium position, and then released. Find the position of the mass, relative to its equilibrium position, at any time if
   (a) damping is ignored.
   (b) a damping force in newtons equal to one-tenth the instantaneous velocity in metres per second acts on the mass.
   (c) a damping force in newtons equal to ten times the instantaneous velocity in metres per second acts on the mass.

2. A 200-g mass suspended vertically from a spring with constant 10 N/m is set into vibration by an external force in newtons given by $4\sin 10t$, $t \geq 0$. During the motion a damping force in newtons equal to $3/2$ the velocity of the mass in metres per second acts on the mass. Find the position of the mass as a function of time t.

3. A 0.001-F capacitor and a 2-H inductor are connected in series with a 20-V battery. If there is no charge on the capacitor before the battery is connected, find the current in the circuit as a function of time.

4. At time $t = 0$, a 0.02-F capacitor, a 100-Ω resistor, and 1-H inductor are connected in series. If the charge on the capacitor is initially 5 C, find its charge as a function of time.

5. A 5-H inductor and 20-Ω resistor are connected in series with a generator supplying an oscillating voltage of $10\sin 2t$, $t \geq 0$. What are the transient and steady-state currents in the circuit?

6. At time $t = 0$, a mass M is attached to the end of a hanging spring with constant k, and then released. Assuming that friction is negligible, find the subsequent displacement of the mass as a function of time.

7. A 0.5-kg mass sits on a table attached to a spring with constant 18 N/m (Figure 16.20). The mass is pulled so as to stretch the spring 5 cm and then released.

   (a) If friction between the mass and the table creates a force of 0.5 N that opposes motion, show that the differential equation determining motion is

   $$\frac{d^2x}{dt^2} + 36x = 1, \quad x(0) = 0.05, \quad x'(0) = 0.$$

   (b) Find where the mass comes to rest for the first time. Will it move from this position?

   FIGURE 16.20

   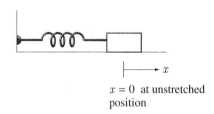

   $x = 0$ at unstretched position

8. Repeat Exercise 7 given that the mass is pulled 25 cm to the right.

9. At time $t = 0$ an uncharged 0.1-F capacitor is connected in series with 0.5-H inductor and a 3-Ω resistor. If the current in the circuit at this instant is 1 A, find the maximum charge that the capacitor stores.

10. Differential equation 16.67 describes the motion of a mass M at the end of a spring, taking damping proportional to velocity into account. Show each of the following.
    (a) If $\beta = 0$,

    $$x(t) = C_1\cos(\sqrt{k/M}t) + C_2\sin(\sqrt{k/M}t),$$

    called *simple harmonic motion*.

(b) If $\beta \neq 0$ and $\beta^2 - 4kM < 0$,
$$x(t) = e^{-\beta t/(2M)}(C_1 \cos \omega t + C_2 \sin \omega t),$$
where $\omega = \dfrac{\sqrt{4kM - \beta^2}}{2M}$,

called *damped oscillatory motion*.

(c) If $\beta \neq 0$ and $\beta^2 - 4kM > 0$,
$$x(t) = e^{-\beta t/(2M)}(C_2 e^{\omega t} + C_2 e^{-\omega t}),$$
where $\omega = \dfrac{\sqrt{\beta^2 - 4kM}}{2M}$,

called *overdamped motion*.

(d) If $\beta \neq 0$ and $\beta^2 - 4kM = 0$,
$$x(t) = (C_1 + C_2 t)e^{-\beta t/(2M)},$$

called *critically damped motion*.

11. A 100-g mass is suspended from a spring with constant 4000 N/m. At its equilibrium position, it is suddenly (time $t = 0$) given an upward velocity of 10 m/s. If an external force $3 \cos 200 t$, $t \geq 0$, acts on the mass, find its displacement as a function of time. Do you note anything strange?

12. A vertical spring having constant 64 N/m has a 1-kg mass attached to it. An external force $F(t) = 2 \sin 8t$, $t \geq 0$ is applied to the mass. If the mass is at rest at its equilibrium position at time $t = 0$, and damping is negligible, find the position of the mass as a function of time. What happens to the oscillations as time progresses? This phenomenon is called *resonance*; it occurs when the frequency of the forcing function is identical to the natural frequency of the undamped system.

13. A mass M is suspended from a vertical spring with constant k. If an external force $F(t) = A \cos \omega t$, $t \geq 0$, is applied to the mass, find the value of ω that causes resonance.

14. A $25/9$-H inductor, a 0.04-F capacitor, and a generator with voltage $15 \cos 3t$ are connected in series at time $t = 0$. Find the current in the circuit as a function of time. Does resonance occur?

15. A mass M is suspended from a vertical spring with constant k. When an external force $F(t) = A \cos \omega t$, $t \geq 0$ is applied to the mass, and a damping force proportional to velocity acts on the mass during its subsequent oscillations, the differential equation governing the motion of M is

$$M \frac{d^2 x}{dt^2} + \beta \frac{dx}{dt} + kx = A \cos \omega t.$$

(a) Show that steady-state oscillations of M are defined by
$$x(t) = \frac{A}{(k - M\omega^2)^2 + \beta^2 \omega^2} \times [(k - M\omega^2) \cos \omega t + \beta \omega \sin \omega t].$$

(b) Verify that this function can be expressed in the form
$$x(t) = \frac{A}{\sqrt{(k - M\omega^2)^2 + \beta^2 \omega^2}} \sin(\omega t + \phi),$$

where
$$\sin \phi = \frac{k - M\omega^2}{\sqrt{(k - M\omega^2)^2 + \beta^2 \omega^2}},$$
$$\cos \phi = \frac{\beta \omega}{\sqrt{(k - M\omega^2)^2 + \beta^2 \omega^2}}.$$

(c) If resonance is said to occur when the amplitude of the steady-state oscillations is a maximum, what value of ω yields resonance? What is the maximum amplitude?

16. (a) A cube L m on each side and with mass M kg floats half submerged in water. If it is pushed down slightly and then released, oscillations take place. Use Archimedes' principle to find the differential equation governing these oscillations. Assume no damping forces due to the viscosity of the water.

(b) What is the frequency of the oscillations?

17. A cylindrical buoy 20 cm in diameter floats partially submerged with its axis vertical. When it is depressed slightly and released, its oscillations have a period equal to 4 s. What is the mass of the buoy?

18. A sphere of radius R floats half submerged in water. It is set into vibration by pushing it down slightly and then releasing it. If y denotes the instantaneous distance of its centre below the surface, show that

$$\frac{d^2 y}{dt^2} = \frac{-3g}{2R^3}\left(R^2 y - \frac{y^3}{3}\right),$$

where g is the acceleration due to gravity.

19. A cable hangs over a peg, 10 m on one side and 15 m on the other. Find the time for it to slide off the peg
(a) if friction at the peg is negligible.
(b) if friction at the peg is equal to the weight of 1 m of cable.

SUMMARY

A differential equation is an equation that contains an unknown function and some of its derivatives, and the equation must be solved for this function. Depending on the form of the equation, various techniques may be used to find the solution. From this point of view, solving differential equations is much like evaluating antiderivatives: We must first recognize the technique appropriate to the particular problem at hand, and then proceed through the mechanics of the technique. It is important, then, to immediately recognize the type of differential equation under consideration.

Broadly speaking, we divided the differential equations we considered into two main groups: first-order equations together with simple second order equations, and linear differential equations. A first-order differential equation in $y(x)$ is said to be separable if it can be written in the form $N(y)\,dy = M(x)\,dx$, and a general solution for such an equation is defined implicitly by

$$\int N(y)\,dy = \int M(x)\,dx + C.$$

A differential equation of the form $dy/dx + P(x)y = Q(x)$ is said to be linear first order. If this equation is multiplied by the integrating factor $e^{\int P(x)\,dx}$, then the left side becomes the derivative of $ye^{\int P(x)\,dx}$, and the equation is immediately integrable.

If a second-order differential equation in $y(x)$ has either the dependent variable or the independent variable explicitly missing, it can be reduced to a pair of first-order equations. This is accomplished in the former case by setting $v = y'$ and $v' = y''$, and in the latter case by setting $v = y'$ and $y'' = v\,dv/dy$.

A differential equation in $y(x)$ is said to be linear if it is of the form

$$a_0 \frac{d^n y}{dx^n} + a_1 \frac{d^{n-1} y}{dx^{n-1}} + \cdots + a_{n-1} \frac{dy}{dx} + a_n y = F(x).$$

The general solution of such an equation is composed of two parts: $y(x) = y_h(x) + y_p(x)$. The function $y_h(x)$ is the general solution of the associated homogeneous equation obtained by replacing $F(x)$ by 0; $y_p(x)$ is any particular solution of the given equation whatsoever. In the special case that the a_i are constants, it is always possible to find $y_h(x)$. This is done by calculating all solutions of the auxiliary equation

$$a_0 m^n + a_1 m^{n-1} + \cdots + a_{n-1} m + a_n = 0,$$

a polynomial equation in m, and then using the rules of Theorem 16.3.

When the a_i are constants and $F(x)$ is a polynomial, an exponential, a sine, a cosine, or any sums or products thereof, we can find $y_p(x)$ either by undetermined coefficients or by operators. The method of undetermined coefficients is simply an intelligent way of guessing at $y_p(x)$ on the basis of $F(x)$ and $y_h(x)$. Note once again that $y_h(x)$ must be calculated before using the method of undetermined coefficients. The method of operators, on the other hand, is based on formal algebraic manipulations and the inverse operator shift theorem. It does not require prior calculation of $y_h(x)$ and, in its simplest form, uses complex numbers.

Key Terms and Formulas

In reviewing this chapter, you should be able to define or discuss the following key terms:

Differential equation
Order of a differential equation
Initial conditions
Boundary conditions
General solution
Particular solution
Singular solution
Separable differential equation
Homogeneous differential equation
Linear differential equation
Integrating factor
Bernoulli equation
Friction
Force of friction
Coefficient of kinetic friction
Malthusian population model

Carrying capacity
Logistic population model
Differential operator
Linear operator
Homogeneous linear differential equation
Superposition principle
Linearly independent solutions
Auxiliary equation
Method of undetermined coefficients
Operator method
Operator shift theorem
Inverse operator shift theorem
Cauchy-Euler equation
Simple harmonic motion
Resonance

REVIEW EXERCISES

In Exercises 1–20 find a general solution for the differential equation.

1. $x^2 \, dy - y \, dx = 0$
2. $(x+1) \, dx - xy \, dy = 0$
3. $\dfrac{dy}{dx} + 3xy = 2x$
4. $\dfrac{dy}{dx} + 4y = x^2$
5. $\dfrac{d^2 y}{dx^2} + 4 \dfrac{dy}{dx} + 3y = 2$
6. $\dfrac{d^2 y}{dx^2} + 3 \dfrac{dy}{dx} + 4y = 2$
7. $yy' = \sqrt{1+y^2}$
8. $\dfrac{d^2 y}{dx^2} + \dfrac{1}{x} \dfrac{dy}{dx} = x$
9. $y'' + 6y' + 3y = xe^x$
10. $\dfrac{dy}{dx} + 2xy = 2x^3$
11. $y^2 y'' = y'$
12. $y'' - 4y' + 4y = \sin x$
13. $y'' - 4y' + 4y = x^2 e^{2x}$
14. $y'' + 4y = \sin 2x$
15. $y'' + 4y' = x^2$
16. $2xy^2 \dfrac{dy}{dx} + (x+1)^2 y^3 = 0$
17. $\dfrac{d^3 y}{dx^3} + 3 \dfrac{d^2 y}{dx^2} + 3 \dfrac{dy}{dx} + y = 2e^{-x}$
18. $y'' + 2y' + 4y = e^{-x} \cos \sqrt{3} x$
19. $\dfrac{dy}{dx} = y \tan x + \cos x$
20. $(2y^2 + 3x) \, dy + dx = 0$

In Exercises 21–24 find the solution of the differential equation satisfying the given condition(s).

21. $y^2 \, dx + (x+1) \, dy = 0$, $y(0) = 3$
22. $y'' - 8y' - 9y = 2x + 4$, $y(0) = 3$, $y'(0) = 7$
23. $y'' + 9y = e^x$, $y(0) = 0$, $y(\pi/2) = 4$
24. $y' + \dfrac{2}{x} y = \sin x$, $y(1) = 1$

25. The quantity of radioactive material present in a sample decays at a rate proportional to the amount of the material in the sample (see Section 9.2). If one-quarter of a sample decays in 5 a, how long does it take for 90% of the sample to decay?

26. (a) A piece of wood rises from the bottom of a container of oil 1 m deep. If the wood has mass 0.5 g and volume 1 cm³, and the density of the oil is 0.9 g/cm³, show that Archimedes' principle

predicts a buoyant force due to fluid pressure of 8.829×10^{-3} N. What is the force on the piece of wood due to gravity and fluid pressure?

(b) The viscosity of the oil opposes motion by exerting a force equal (in newtons) to twice its velocity (in metres per second). Find the distance travelled by the wood as a function of time, assuming that it starts from rest on the bottom.

27. A 100-g mass is suspended vertically from a spring with constant 1 N/m. The mass is pulled 4 cm above its equilibrium position and then released. Find the position of the mass at any time if
    (a) damping is ignored.
    (b) a damping force in newtons equal to $1/5$ the instantaneous velocity of the mass in metres per second acts on the mass.
    (c) a damping force in newtons equal to $\sqrt{2/5}$ the instantaneous velocity of the mass in metres per second acts on the mass.

28. A 10-g stone is dropped over the side of a bridge 50 m above a river 10 m deep. Air resistance is negligible, but as the stone sinks in the water, its motion is retarded by a force in newtons equal to one-fifth the speed of the stone in metres per second. If the stone loses 10% of its speed in penetrating the surface of the water, find
    (a) its velocity as a function of time.
    (b) its position relative to the drop point as a function of time.
    (c) the time it takes to reach the bottom of the river from the point at which it was dropped.

    Assume that the water is stationary and that buoyancy due to Archimedes' principle may be neglected.

29. Repeat Exercise 28 but assume that, for the purpose of Archimedes' principle, the volume of the stone is 3 cm^3.

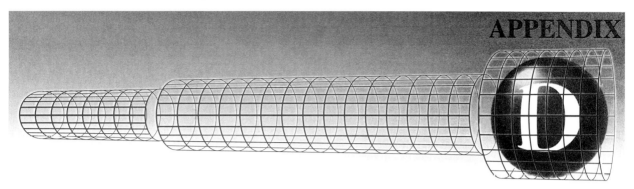

Answers to Even-numbered Exercises

CHAPTER 11

Exercises 11.1

2. Increasing, $V = 4$
4. Decreasing, $U = 9/16$, $V = 0$, $L = 0$
6. Not monotonic, $U = 1$, $V = -1$, $L = 0$
8. Decreasing, $U = 1/4$, $V = 0$, $L = 0$
10. Increasing, $U = \pi/2$, $V = \pi/4$, $L = \pi/2$
12. Not monotonic
14. Not monotonic, $U = 4$, $V = -4$, $L = 0$
16. False $\{n\}$ 18. False $\{-n\}$ 20. True
22. True   24. True   26. False $\{(-1)^n/n\}$
28. False $\{(-1)^n/n\}$ 30. False $\{(-1)^n\}$
32. False $\{-3^n\}$ 34. True
36. Not monotonic, $U = 9/4$, $V = -7$, $L = 0$
38. Decreasing, $U = 1/e$, $V = 0$, $L = 0$
40. Increasing, $U = \pi/2$, $V = 0$, $L = \pi/2$
42. Increasing, $U = 1$, $V = 0$, $L = 0.419241$
44. Increasing, $U = 5$, $V = 0$, $L = (\sqrt{21} + 1)/2$
46. Increasing, $U = 10$, $V = 3$, $L = (3 + \sqrt{29})/2$
48. Decreasing, $U = 4$, $V = 0$, $L = (7 - \sqrt{5})/2$
50. $(3n + 1)/n^2$ 52. $[1 + (-1)^{n+1}]/2$
54. 1   56. 0   58. 4
60. Decreasing, $U = 2$, $V = 0$, $L = (3 - \sqrt{5})/2$
62. Decreasing, $U = 1$, $V = 0$, $L = 1/2$
64. Decreasing, $U = 1$, $V = 0$, $L = 0$
66. Decreasing, $U = 2$, $V = 0$, $L = 0$
68. Increasing, $U = 1$, $V = 0$, $L = (3 - \sqrt{3})/2$
72. $V = 1/2$, $U = 1$

76. (a) 1, 1, 2, 3, 5, 8, 13, 21, 34, 55   (b) Increasing, $V = 1$, no upper bound, no limit   (e) $(\sqrt{5} + 1)/2$
80. $[5 + (-1)^n/2^{n-2}]/3$

Exercises 11.2

2. $-0.381\,966\,0$ 4. $-2.618\,034\,0$ 6. $3.044\,723\,1$
8. $1.214\,648\,0$ 10. $0.334\,734\,1$ 12. $-1.388\,792\,0$
14. (a) $40(0.99)^n$ m   (b) $4(0.981)^{-1/2}(0.99)^{n/2}$ s
16. $4P/3$, $16P/9$, $(4/3)^n P$, does not exist
18. 0.0625, 0.1125
20. $(\sqrt{3}P^2/36)[1 + (1/3) + (4/3^3) + (4^2/3^5) + \cdots + (4^{n-1}/3^{2n-1})]$
22. -0.4814 24. 2.9122 26. 3.3247
28. 0.7849 30. 0.7953

Exercises 11.3

2. $2/15$ 4. Diverges   6. $10\,804.5$
8. $-1/3$ 10. Diverges   12. $13/99$
14. $430\,162/9999$ 16. Diverges   18. 4
20. Diverges   22. 1   24. 804 s
26. $(\sqrt{3}P^2/180)[8 - 3(4/9)^n]$, $2\sqrt{3}P^2/45$
28. 2   30. $51/64$ 32. $2/3$
36. $600/11$ min after 10:00   38. (c) $A_0 e^{kT}/(e^{kT} - 1)$

Exercises 11.4

2. Diverges   4. Converges   6. Converges
8. Diverges   10. Converges   12. Converges
14. Converges   16. Converges   18. Diverges
20. Diverges   22. Diverges

24. Converges for $p > 1$, diverges for $p \leq 1$

Exercises 11.5

2. Converges 4. Converges 6. Diverges
8. Diverges 10. Converges 12. Converges
14. Converges 16. Diverges 18. Converges
20. Converges

Exercises 11.6

2. Converges conditionally 4. Converges absolutely
6. Diverges 8. Converges absolutely
10. Converges conditionally 12. Converges conditionally
14. Converges absolutely

Exercises 11.7

2. -0.9470 4. -0.0127 6. 1.07
8. -0.7 10. 4

Exercises 11.8

2. $-1 < x < 1$ 4. $-1/3 < x < 1/3$
6. $-4 < x < -2$ 8. $7/2 < x < 9/2$
10. $-1 < x < 1$ 12. $-1 \leq x < 1$
14. $-1/e \leq x \leq 1/e$ 16. $x = 0$
18. $-1/3 < x < 1/3$ 20. $-5^{1/3} < x < 5^{1/3}$
22. $-1 \leq x \leq 1$ 24. 1
26. $4/(4-x^3)$, $|x| < 4^{1/3}$
28. $(x-1)/(10-x)$, $-8 < x < 10$
30. (b) $-\infty < x < \infty$

Exercises 11.9

2. $\sum_{n=0}^{\infty} (5^n/n!)x^n$, $-\infty < x < \infty$
4. $(1/\sqrt{2})[1 + (x - \pi/4) - (x-\pi/4)^2/2! - (x-\pi/4)^3/3! + (x-\pi/4)^4/4! + \cdots]$, $-\infty < x < \infty$
6. $1/8 - (3/8^2)(x-2) + (3^2/8^3)(x-2)^3 - (3^3/8^4)(x-2)^4 + \cdots$, $-2/3 < x < 14/3$
12. (d) 0 (e) $x = 0$

Exercises 11.10

2. $\sum_{n=0}^{\infty} [(-1)^n/4^{n+1}]x^{2n}$, $|x| < 2$
4. $\sum_{n=0}^{\infty} [(-1)^n/(2n)!]x^{4n}$, $-\infty < x < \infty$
6. $\sum_{n=0}^{\infty} (5^n/n!)x^n$, $-\infty < x < \infty$
8. $\sum_{n=0}^{\infty} [1/(2n+1)!]x^{2n+1}$, $-\infty < x < \infty$

10. $1 + 9x/2 + 27x^2/8 + \sum_{n=3}^{\infty} \{(-1)^n (2n-5)!\,3^{n+1} / [2^{2n-3} n!(n-3)!]\}x^n$, $-1/2 \leq x \leq 1/3$
12. $x^4 + 3x^2 - 2x + 1$
14. $33 - 46(x+2) + 27(x+2)^2 - 8(x+2)^3 + (x+2)^4$
16. $\sum_{n=0}^{\infty} (e^n/n!)(x-3)^n$, $-\infty < x < \infty$
18. $\sqrt{3}(1 + x/6) + \sum_{n=2}^{\infty} \{4\sqrt{3}(-1)^{n-1}(2n-3)!/[12^n n!(n-2)!]\}x^n$, $|x| \leq 3$
20. $x - x^2/3 - \sum_{n=3}^{\infty} \{(2)(5)\cdots(3n-7)/[3^{n-1}(n-1)!]\}x^n$, $|x| \leq 1$
22. $1 + x^2/2 + 5x^4/24 + 61x^6/720$
24. $1 + \sum_{n=1}^{\infty} [(-1)^n 2^{2n-1}/(2n)!]x^{2n}$, $-\infty < x < \infty$
26. $x^2 + \sum_{n=1}^{\infty} \{(2n)!/[(2n+1)2^{2n}(n!)^2]\}x^{4n+2}$, $|x| < 1$
28. $\sqrt{2} \sum_{n=0}^{\infty} \{1/[2^n(2n+1)]\}x^{2n+1}$, $|x| < \sqrt{2}$
30. (a) $1/p$ (b) 6
32. $\sum_{n=0}^{\infty} \{(-1)^n \pi^{2n}/[(4n+1)2^{2n}(2n)!]\}x^{4n+1}$, $-\infty < x < \infty$;
$\sum_{n=0}^{\infty} \{(-1)^n \pi^{2n+1}/[(4n+3)2^{2n+1}(2n+1)!]\}x^{4n+3}$, $-\infty < x < \infty$
36. (a) $-1/2, 1/6, 0, -1/30, 0$

Exercises 11.11

2. $2/(1-x)^3$, $|x| < 1$ 4. $(x+1)/(1-x)^3$, $|x| < 1$
6. $-(1/x) \ln(1-x)$, $-1 \leq x < 1$
8. $-\ln(1+x^2)$, $-1 \leq x \leq 1$
10. $1/(x-x^2) - (1/x^2)\ln(1-x)$, $|x| < 1$
22. $-9/100$

Exercises 11.12

2. 4.2×10^{-10} 4. 4.2×10^{-10} 6. 1.4×10^{-9}
8. 0.115 10. 0.00625 12. 0.497
14. 0.133 16. 0.291 18. -0.122
20. $1/2$ 22. $1/2$ 24. 0
26. $1 - x^3/2 + 3x^6/8 - 5x^9/16$ 28. $1 - x^2$
30. $a_0 + a_1 \sum_{n=1}^{\infty} [(-1)^{n+1}/n!]x^n = C + De^{-x}$
32. $a_0 \sum_{n=0}^{\infty} [(-1)^n/(2n)!]x^n$
34. $a_1 \sum_{n=1}^{\infty} \{(-1)^{n+1}/[n!(n-1)!]\}x^n$
38. 2.44

Review Exercises

2. Decreasing, $U = 1$, $V = 0$, $L = 1/\sqrt{3}$
4. Increasing, $U = 100$, $V = 7$, $L = (31 + \sqrt{53})/2$
6. $|k| \leq 1$ 10. Converges 12. Converges
14. Converges conditionally

16. Diverges **18.** Converges **20.** Converges
22. Diverges **24.** Converges **26.** Converges
28. Converges conditionally **30.** $-2 \leq x \leq 2$
32. $-\infty < x < \infty$ **34.** $-4 < x < -2$
36. $-2^{-1/3} \leq x < 2^{-1/3}$
38. $e^5 \sum_{n=0}^{\infty} (1/n!) x^n$, $-\infty < x < \infty$
40. $\sum_{n=2}^{\infty} [(-1)^n 2^{n-1}/(n-1)] x^n$, $-1/2 < x \leq 1/2$
42. $(1/2) \sum_{n=0}^{\infty} (-1)^{n+1} (1 - 1/3^n) x^n$, $|x| < 1$
44. $x + \sum_{n=2}^{\infty} \{(-1)^n/[n(n-1)]\} x^n$, $-1 < x \leq 1$
46. $\sum_{n=0}^{18} [(-1)^n/n!] x^{2n}$
48. $\sum_{n=0}^{\infty} \{(-1)^n/[2^{2n}(2n)!]\} x^{2n} + \sum_{n=0}^{\infty} \{(-1)^n/[2^{2n+1}(2n+1)!]\} x^{2n+1}$

CHAPTER 12

Exercises 12.1

2. $\sqrt{33}$
4. $(0,0,0), (2,0,0), (0,2,0), (0,0,2), (2,2,0), (2,0,2), (0,2,2), (2,2,2)$
6. $\sqrt{29}, 5, 2\sqrt{5}, \sqrt{13}$ **8.** $5, 3, 4, 5$
12. $10x + 2y + 2z = 5$, plane
14. $(\sqrt{2}, \sqrt{2}, 5), (\sqrt{2} \pm 1/4, \sqrt{2} \pm 1/4, 5 - \sqrt{7}/4), (\sqrt{2} \pm 1/4, \sqrt{2} \pm 1/4, 9/2 - \sqrt{7}/4)$
16. (a) $(2, 1/2, -7/2)$ **(b)** $(5, 5, -5)$
20. $(11, 1, 7/5)$

Exercises 12.2

2.
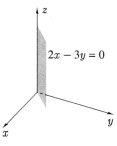

4.
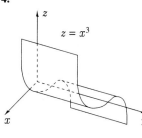

6.
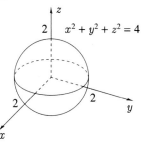

8.
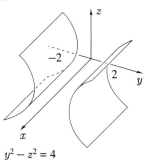

10.
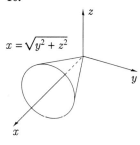

12.
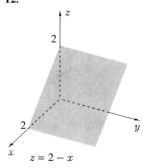

14.
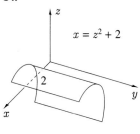

16.
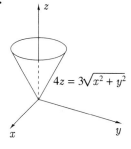

18.
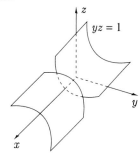

20.
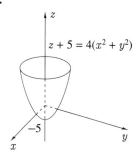

22.
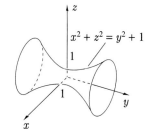

24.
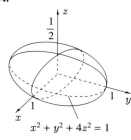

26.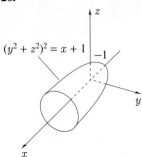
$(y^2+z^2)^2 = x+1$

28.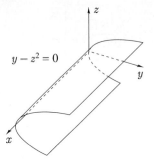
$y - z^2 = 0$

30.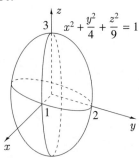
$x^2 + \dfrac{y^2}{4} + \dfrac{z^2}{9} = 1$

32.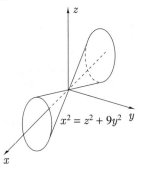
$x^2 = z^2 + 9y^2$

34.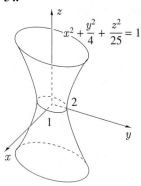
$x^2 + \dfrac{y^2}{4} + \dfrac{z^2}{25} = 1$

36.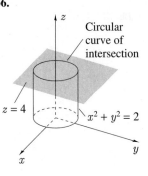
Circular curve of intersection
$z = 4$
$x^2 + y^2 = 2$

38.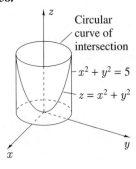
Circular curve of intersection
$x^2 + y^2 = 5$
$z = x^2 + y^2$

40.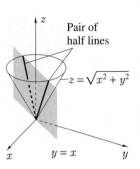
Pair of half lines
$z = \sqrt{x^2+y^2}$
$y = x$

42.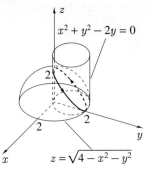
$x^2 + y^2 - 2y = 0$
$z = \sqrt{4 - x^2 - y^2}$

44.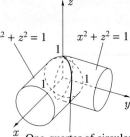
$y^2 + z^2 = 1$
$x^2 + z^2 = 1$
One-quarter of circular curve of intersection (Two such circles)

46.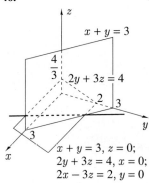
$x + y = 3$
$2y + 3z = 4$
$x + y = 3, z = 0;$
$2y + 3z = 4, x = 0;$
$2x - 3z = 2, y = 0$

48.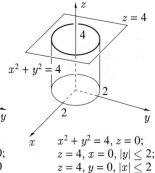
$z = 4$
$x^2 + y^2 = 4$
$x^2 + y^2 = 4, z = 0;$
$z = 4, x = 0, |y| \leq 2;$
$z = 4, y = 0, |x| \leq 2$

50.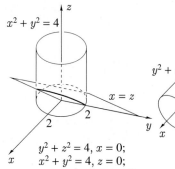
$x^2 + y^2 = 4$
$x = z$
$y^2 + z^2 = 4, x = 0;$
$x^2 + y^2 = 4, z = 0;$
$z = x, y = 0, |x| \leq 2$

52.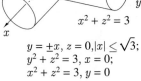
$y^2 + z^2 = 3$
$x^2 + z^2 = 3$
$y = \pm x, z = 0, |x| \leq \sqrt{3};$
$y^2 + z^2 = 3, x = 0;$
$x^2 + z^2 = 3, y = 0$

54.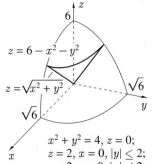
$z = 6 - x^2 - y^2$
$z = \sqrt{x^2 + y^2}$
$x^2 + y^2 = 4, z = 0;$
$z = 2, x = 0, |y| \leq 2;$
$z = 2, y = 0, |x| \leq 2$

56.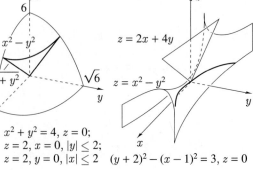
$z = 2x + 4y$
$z = x^2 - y^2$
$(y+2)^2 - (x-1)^2 = 3, z = 0$

58.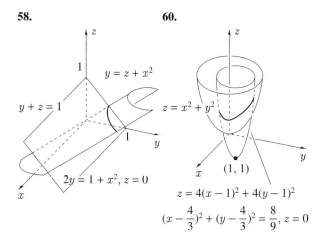
60.

62.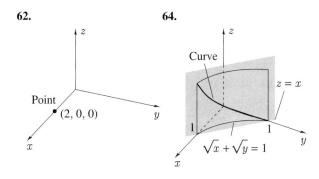
64.

66.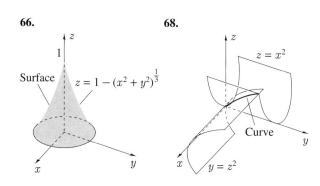
68.

70.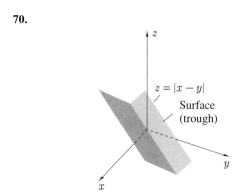

Exercises 12.3

2. $(2,6,8)$ **4.** $(-1/\sqrt{5}, 0, 2/\sqrt{5})$
6. $(-4\sqrt{5} - 8, -6, 8\sqrt{5} + 4)$
8. $(-6\sqrt{46} - 4\sqrt{5}, -12\sqrt{5}, 12\sqrt{46} - 24\sqrt{5})$
10. $(-6/\sqrt{17}, -3/\sqrt{17}, 6/\sqrt{17})$
12. $3\hat{\mathbf{i}} - 2\hat{\mathbf{j}}$
14. $(2/\sqrt{5} - 1/\sqrt{10})\hat{\mathbf{i}} + (1/\sqrt{5} + 3/\sqrt{10})\hat{\mathbf{j}}$
16. $(5, 0, 0)$ **18.** $(0, 3/\sqrt{5}, 6/\sqrt{5})$
20. $(1, 1, \ 3/2)$ **22.** $(5/4, 5\sqrt{2}/4, 5/4)$
24. $(2/\sqrt{3}, 2/\sqrt{3}, 2/\sqrt{3})$ **30.** $(3/\sqrt{17}, 12/\sqrt{17})$
32. $5G(-3/\sqrt{10} + 30\sqrt{11}/121, -10\sqrt{11}/121, -1/\sqrt{10} + 10\sqrt{11}/121)$ N
34. Linearly dependent **36.** Linearly dependent
42. $k_1(1 - l/\sqrt{x^2 + l^2})(-x, l) + k_2[1 - l/\sqrt{(L-x)^2 + l^2}](L-x, l)$

Exercises 12.4

2. $(-10, 15, -5)$ **4.** $4/\sqrt{14}$ **6.** -178
8. 1 **10.** 0 **12.** Yes
14. Yes **16.** 0.68 **18.** 2.20
20. π **22.** $2x + y - 2z + 5 = 0$
24. $6/\sqrt{13}$ **26.** $4/\sqrt{6}$ **28.** $23\sqrt{14}/28$
34. (a) $1.30, 1.01, 2.50$ **(b)** $\pi/2, 1.25, 2.82$
(c) $1.73, 1.89, 0.36$
36. $3/\sqrt{2}, 1/\sqrt{2}$ **38.** $\sqrt{5}, -\sqrt{14}/2, -\sqrt{70}/2$
40. $-3/10, 11/20$ **42.** $x - 2y + 3z = 6 \pm 2\sqrt{14}$
44. $k[1 + l(1 - \sqrt{5})]/2$ J **46.** $(\sqrt{2} - 1)GMm/(\sqrt{2}R)$

Exercises 12.5

2. $(48, 24, -42)$ **4.** $(2\sqrt{247})^{-1}(17, -23, -7)$
6. $(-41, 11, 28)$ **8.** $(17, -23, -7)$ **10.** $(-13, 19, 5)$
12. $\lambda(9, 0, 1)$ **14.** $\lambda(3, -1, 17)$
18. $(x, y, z) = (1, -1, 3) + t(2, 4, -3);\ x = 1 + 2t, y = -1 + 4t, z = 3 - 3t;\ (x-1)/2 = (y+1)/4 = (z-3)/(-3)$
20. $(x, y, z) = (2, -3, 4) + t(3, 5, -5);\ x = 2 + 3t, y = -3 + 5t, z = 4 - 5t;\ (x-2)/3 = (y+3)/5 = (z-4)/(-5)$
22. $(x, y, z) = (1, 3, 4) + t(0, 0, 1);\ x = 1, y = 3, z = 4 + t$
24. $(x, y, z) = (2, 0, 3) + t(1, 0, -2);\ x = 2 + t, y = 0, z = 3 - 2t;\ x - 2 = (z-3)/(-2), y = 0$
26. $(x, y, z) = (0, -5, 30) + t(1, 2, -11);\ x = t, y = -5 + 2t, z = 30 - 11t;\ x = (y+5)/2 = (z-30)/(-11)$
28. $x + 3y - 3z = 4$ **30.** $10x + 4y - 3z = 3$

32. $8x + 13y + 9z = 18$ **34.** $7x + 6y - 5z + 34 = 0$
36. No
40. $x = \sqrt{2} \pm 1/4$, $y = \sqrt{2} \pm 1/4$, $z = 9/2 - \sqrt{7}/4$, $\sqrt{7}x - z = \sqrt{14} - 5$, $\sqrt{7}x + z = \sqrt{14} + 5$, $\sqrt{7}y - z = \sqrt{14} - 5$, $\sqrt{7}y + z = \sqrt{14} + 5$
42. $\sqrt{154}$ **44.** $\sqrt{73/2}$ **46.** $2/\sqrt{6}$
48. $(-10, 5, -2)$ **50.** $(-5, 1, 4)$

Exercises 12.6

2. $-\infty < t < \infty$ **4.** $t > -4$ **6.** $\hat{\mathbf{i}} - 2t\hat{\mathbf{j}} + 2\hat{\mathbf{k}}$
8. $2t(4t^2 - 3)\hat{\mathbf{i}} + t^2(9 - 10t^2)\hat{\mathbf{j}} + 4t(4t^2 - 3)\hat{\mathbf{k}}$
10. $3t^2(4 - 5t^2)\hat{\mathbf{i}} + 4t(1 - 3t^2)\hat{\mathbf{j}} - 3t^2\hat{\mathbf{k}}$
12. $3\hat{\mathbf{i}} - 2(3t + 4)\hat{\mathbf{j}} + 6(1 + 4t)\hat{\mathbf{k}}$
14. $9t^2\hat{\mathbf{i}} + (6t - 20t^3)\hat{\mathbf{j}} + (30t^4 - 21t^2 + 6)\hat{\mathbf{k}}$
16. $(-15t^4 + 12t^2)\hat{\mathbf{i}} + (4t - 12t^3)\hat{\mathbf{j}} - 3t^2\hat{\mathbf{k}}$
18. $(3t^4/2 - 4t^2)\hat{\mathbf{i}} + (17t^3/3 - 12t^5/5)\hat{\mathbf{j}} + (3t^6 - 27t^4/4 + t^2)\hat{\mathbf{k}} + \mathbf{C}$
20. $(-14t^4 - 6t^3 - 42t^2 + 4t)\hat{\mathbf{i}} + (28t^5 + 84t^3 - 6t^2)\hat{\mathbf{j}} - (126t^4 + 42t^6 + 2t)\hat{\mathbf{k}}$

Exercises 12.7

2.

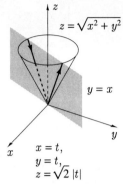

4.

6.

$x = t$, $y = t$, $z = \sqrt{2}|t|$
$\mathbf{r} = t\hat{\mathbf{i}} + t\hat{\mathbf{j}} + \sqrt{2}|t|\hat{\mathbf{k}}$

8.

$x = \cos t$, $y = 1 + \sin t$, $z = \sqrt{2 - 2\sin t}$ $\quad 0 \leq t < 2\pi$
$\mathbf{r} = \cos t\hat{\mathbf{i}} + (1 + \sin t)\hat{\mathbf{j}} + \sqrt{2 - 2\sin t}\hat{\mathbf{k}}$, $0 \leq t < 2\pi$

10.

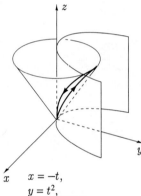

$x = -t$, $y = t^2$, $z = \sqrt{t^2 + t^4}$
$\mathbf{r} = -t\hat{\mathbf{i}} + t^2\hat{\mathbf{j}} + \sqrt{t^2 + t^4}\hat{\mathbf{k}}$

12.

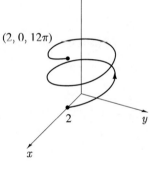

$(2, 0, 12\pi)$

14.

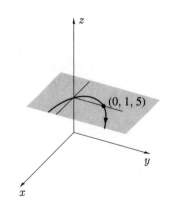

$(0, 1, 5)$

For problem 2:
$x = \sqrt{2}\cos t$, $y = \sqrt{2}\sin t$, $z = 4$; $0 \leq t < 2\pi$
$\mathbf{r} = \sqrt{2}\cos t\hat{\mathbf{i}} + \sqrt{2}\sin t\hat{\mathbf{j}} + 4\hat{\mathbf{k}}$, $0 \leq t < 2\pi$

For problem 4:
$x = \sqrt{5}\cos t$, $y = \sqrt{5}\sin t$, $z = 5$; $0 \leq t < 2\pi$
$\mathbf{r} = \sqrt{5}\cos t\hat{\mathbf{i}} + \sqrt{5}\sin t\hat{\mathbf{j}} + 5\hat{\mathbf{k}}$, $0 \leq t < 2\pi$

Exercises 12.8

2. $\mathbf{r} = t\hat{\mathbf{i}} + t^2\hat{\mathbf{j}} + t^3\hat{\mathbf{k}}$; $\hat{\mathbf{T}} = (\hat{\mathbf{i}} + 2t\hat{\mathbf{j}} + 3t^2\hat{\mathbf{k}})/\sqrt{1 + 4t^2 + 9t^4}$

4. $\mathbf{r} = -t\hat{\mathbf{i}} + (5 + t)\hat{\mathbf{j}} + (t^2 - t - 5)\hat{\mathbf{k}}$, $-5 \le t \le 0$; $\hat{\mathbf{T}} = [-\hat{\mathbf{i}} + \hat{\mathbf{j}} + (2t - 1)\hat{\mathbf{k}}]/\sqrt{4t^2 - 4t + 3}$

6. $(-2\hat{\mathbf{i}} + 3\hat{\mathbf{j}} + \hat{\mathbf{k}})/\sqrt{14}$ **8.** $(\hat{\mathbf{i}} - \hat{\mathbf{j}})/\sqrt{2}$

10. $-\hat{\mathbf{j}}$ **12.** $\sqrt{42}$

14. $(616\sqrt{616} - 157\sqrt{157})/459$

16. $\cos t\,\hat{\mathbf{i}} + \sin t\,\hat{\mathbf{j}}$

18. (a) $(0,0,0)$ (b) $2\hat{\mathbf{i}} + 2\hat{\mathbf{k}}$

Exercises 12.9

2. $[-(2t + 9t^3)\hat{\mathbf{i}} + (1 - 9t^4)\hat{\mathbf{j}} + (3t + 6t^3)\hat{\mathbf{k}}]/\sqrt{1 + 13t^2 + 54t^4 + 117t^6 + 81t^8}$; $(3t^2\hat{\mathbf{i}} - 3t\hat{\mathbf{j}} + \hat{\mathbf{k}})/\sqrt{1 + 9t^2 + 9t^4}$

4. $[(2t - 1)\hat{\mathbf{i}} + (1 - 2t)\hat{\mathbf{j}} + 2\hat{\mathbf{k}}]/\sqrt{8t^2 - 8t + 6}$; $(\hat{\mathbf{i}} + \hat{\mathbf{j}})/\sqrt{2}$

6. $-(5\hat{\mathbf{i}} + 3\hat{\mathbf{j}} + \hat{\mathbf{k}})/\sqrt{35}$; $(-\hat{\mathbf{j}} + 3\hat{\mathbf{k}})/\sqrt{10}$

8. $-(\hat{\mathbf{i}} + \hat{\mathbf{j}})/\sqrt{2}$; $-\hat{\mathbf{k}}$

10. $-(\hat{\mathbf{i}} + \hat{\mathbf{k}})/\sqrt{2}$; $(\hat{\mathbf{i}} - \hat{\mathbf{k}})/\sqrt{2}$

12. $\kappa = 0$, ρ undefined

14. $\kappa = 2e^t\sqrt{1 + e^{2t}}/(1 + 2e^{2t})^{3/2}$, $\rho = (1 + 2e^{2t})^{3/2}/(2e^t\sqrt{1 + e^{2t}})$

16. $\kappa = 1/[\sqrt{2}(1 + \cos^2 t)^{3/2}]$, $\rho = \sqrt{2}(1 + \cos^2 t)^{3/2}$

18. $\kappa = \sqrt{1 + 36t^4 + 16t^6}/[2(1 + t^2 + 4t^6)^{3/2}]$, $\rho = 2(1 + t^2 + 4t^6)^{3/2}/\sqrt{1 + 36t^4 + 16t^6}$

22. 0

24. (a) $-\sin t\,\hat{\mathbf{i}} + \cos t\,\hat{\mathbf{j}}$; $-(\cos t\,\hat{\mathbf{i}} + \sin t\,\hat{\mathbf{j}})$; $\hat{\mathbf{k}}$
(b) $2\sin 2t(\sin t - \cos t)$, $-4(\cos^3 t + \sin^3 t)$;
(c) $2\sin 2t(\sin t - \cos t)\hat{\mathbf{T}} - 4(\cos^3 t + \sin^3 t)\hat{\mathbf{N}}$

26. $(-\sin t\,\hat{\mathbf{i}} + \cos t\,\hat{\mathbf{j}} + \hat{\mathbf{k}})/\sqrt{2}$; $-(\cos t\,\hat{\mathbf{i}} + \sin t\,\hat{\mathbf{j}})$; $(\sin t\,\hat{\mathbf{i}} - \cos t\,\hat{\mathbf{j}} + \hat{\mathbf{k}})/\sqrt{2}$; $(1 - \cos t \sin t + \cos^2 t \sin^2 t)\hat{\mathbf{T}}/\sqrt{2} - \cos t(\cos t + \sin^3 t)\hat{\mathbf{N}} + (-\cos^2 t \sin^2 t + \cos \sin t + 1)\hat{\mathbf{B}}/\sqrt{2}$

Exercises 12.10

2. $[(t^2 - 1)\hat{\mathbf{i}} + (t^2 + 1)\hat{\mathbf{j}}]/t^2$; $\sqrt{2 + 2t^4}/t^2$; $2(\hat{\mathbf{i}} - \hat{\mathbf{j}})/t^3$

4. $2t\hat{\mathbf{i}} + 2e^t(t + 1)\hat{\mathbf{j}} - 2\hat{\mathbf{k}}/t^3$; $2\sqrt{t^2 + e^{2t}(t + 1)^2 + 1/t^6}$; $2\hat{\mathbf{i}} + 2e^t(t + 2)\hat{\mathbf{j}} + 6\hat{\mathbf{k}}/t^4$

6. $(t^4/4 + 1)\hat{\mathbf{i}} + (t^3 + 3t^2 + 12)\hat{\mathbf{j}}/6 - (t^5/5 + 1)\hat{\mathbf{k}}$

8. $4t/\sqrt{1 + 4t^2}$; $2/\sqrt{1 + 4t^2}$

12. (a) 0 (b) $56t^6$

14. $150x\hat{\mathbf{j}}$

18. (b) 7.79 km/s

20. $365\hat{\mathbf{i}} + 325\sqrt{3}\hat{\mathbf{j}}$ km/h; 670.9 km/h

22. 4.54 min

24. $(3/2, 3/2, 2)$ **26.** $(3, 0)$ **32.** Yes

36. (b) $(S/R)[(R - b\cos\theta)\hat{\mathbf{i}} + b\sin\theta\hat{\mathbf{j}}]$;
$(S/R)\sqrt{R^2 - 2bR\cos\theta + b^2}$; $(bS^2/R^2)(\sin\theta\hat{\mathbf{i}} + \cos\theta\hat{\mathbf{j}})$
(c) $(bS^2/R^2)|R\cos\theta - b|/\sqrt{R^2 - 2bR\cos\theta + b^2}$;
$(bS^2/R)\sin\theta/\sqrt{R^2 - 2bR\cos\theta + b^2}$

Review Exercises

2. -8 **4.** $(39, -33, -30)$

6. -2 **8.** $(-16, -24, -4)$

10. $(67/7, 21, -65/7)$

12.

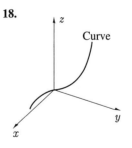

14.

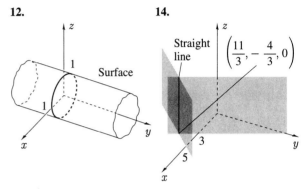

$\left(\dfrac{11}{3}, -\dfrac{4}{3}, 0\right)$

16. No points

18.

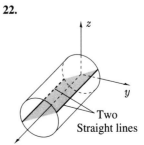

20.

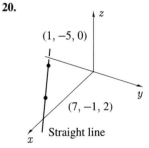

22.

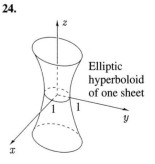

24.

Elliptic hyperboloid of one sheet

26.

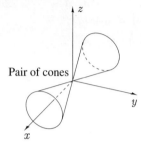
Pair of cones

28. $x = 6 + 5t$, $y = 6 - 2t$, $z = 2 + t$
30. $x = 1 + u$, $y = 3 + 2u$, $z = 2 + 3u$
32. $x = y + z$ **34.** $3x - 4y + z = 0$
36. $35/\sqrt{41}$ **38.** $2\sqrt{5}/15$ **40.** $\sqrt{59}$
42. $(2\cos t\,\hat{\imath} - 2\sin t\,\hat{\jmath} + \hat{k})/\sqrt{5}$; $-\sin t\,\hat{\imath} - \cos t\,\hat{\jmath}$;
$(\cos t\,\hat{\imath} - \sin t\,\hat{\jmath} - 2\hat{k})/\sqrt{5}$
44. $\hat{\imath} + 2t\hat{\jmath} + 2t\hat{k}$; $\sqrt{1+8t^2}$; $2(\hat{\jmath} + \hat{k})$; 0,
$\sqrt{1+8t^2}$; $2\sqrt{2}/\sqrt{1+8t^2}$, $8t/\sqrt{1+8t^2}$
46. (a) 4.46 m/s   (b) $0.226\hat{\imath} - \hat{\jmath}$ m   (c) 0.255 m from point on floor directly below point it left table
48. $k(1-x)[1 - 1/\sqrt{1+4(1-x)^2}]\hat{\imath} +$
$(k/2)[1 - 1/\sqrt{1+4(1-x)^2}]\hat{\jmath}$ N

CHAPTER 13

Exercises 13.1

4.
All points between the branches of the hyperbola $x^2 - y^2 = 1$

6.
All points in space except $(0, 0, 0)$

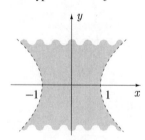

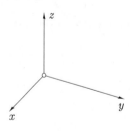

8.

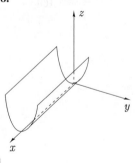

10.

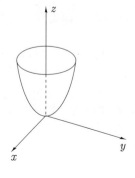

12.

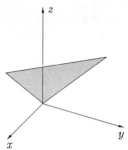

14.

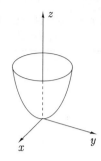

16.

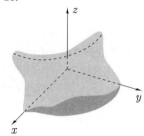

18.

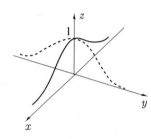

20.

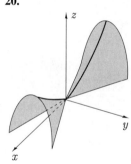

22.

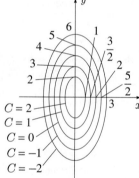

24.

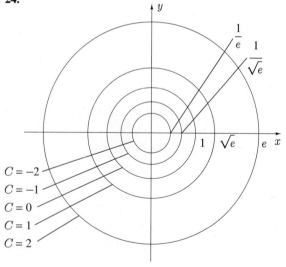

26. $lw(15 - lw)/(w + l)$

28. $8c|xy|\sqrt{1 - x^2/a^2 - y^2/b^2}$
30. $x \sin\theta(1 - x - x\sin\theta \csc\phi) + (x^2/2)\sin\theta(\cos\theta + \sin\theta \cot\phi)$
32. $30.25 + 11G + 17.5S$ cents. The domain consists of all points (G, S) satisfying the inequalities $200 \leq 37G + 31S \leq 300$, $145 \leq 22G + 85S \leq 170$.

Exercises 13.2

2. $3/5$ **4.** $-3/7$ **6.** 0
8. $\pi/2$ **10.** 1 **12.** 4
14. Does not exist   **16.** 0
18. Does not exist   **20.** Does not exist
22. $(0, 0)$ **24.** $x = 0; y = 0; z = 0$
26. $x = 0; y = 0; y = -x$ **28.** Does not exist
30. 1 **32.** (a) 1, no   (b) 1, yes
34. False

Exercises 13.3

2. $3y - 16x^3y^4$, $3x - 16x^4y^3$
4. $-x(x + 2y)/[y(x + y)^2]$, $x^2(x + 2y)/[y^2(x + y)^2]$
6. $y\cos(xy)$, $x\cos(xy)$
8. $x/\sqrt{x^2 + y^2}$, $y/\sqrt{x^2 + y^2}$
10. $4x\sec^2(2x^2 + y^2)$, $2y\sec^2(2x^2 + y^2)$
12. ye^{xy}, xe^{xy}
14. $2x/(x^2 + y^2)$, $2y/(x^2 + y^2)$
16. $ye^x\cos(ye^x)$, $e^x\cos(ye^x)$
18. $2xy\cos^2(x^2y)\sin(x^2y)/[1 - \cos^3(x^2y)]^{2/3}$, $x^2\cos^2(x^2y)\sin(x^2y)/[1 - \cos^3(x^2y)]^{2/3}$
20. $\tan\sqrt{x+y}/(2\sqrt{x+y})$, $\tan\sqrt{x+y}/(2\sqrt{x+y})$
22. $-2z/[1 + (x^2 + z^2)^2]$
24. 2
26. $-1/[1 + (1 + x + y + z)^2]$
28. $3x^2/y + \sin(yz/x) - (yz/x)\cos(yz/x)$
30. $z\mathrm{Sin}^{-1}(x/z)$ if $z > 0$; $z\mathrm{Sin}^{-1}(x/z) + xz/\sqrt{z^2 - x^2}$ if $z < 0$
34. (a) Yes   (b) No   (c) Yes   (d) No
36. No   **38.** (a) Yes   (b) No

Exercises 13.4

2. $2xyz\hat{\mathbf{i}} + x^2z\hat{\mathbf{j}} + x^2y\hat{\mathbf{k}}$
4. $(2xy + y^2)\hat{\mathbf{i}} + (x^2 + 2xy)\hat{\mathbf{j}}$
6. $(1 + x^2y^2z^2)^{-1}(yz\hat{\mathbf{i}} + xz\hat{\mathbf{j}} + xy\hat{\mathbf{k}})$
8. $e^{x+y+z}(\hat{\mathbf{i}} + \hat{\mathbf{j}} + \hat{\mathbf{k}})$
10. $-(x^2 + y^2 + z^2)^{-3/2}(x\hat{\mathbf{i}} + y\hat{\mathbf{j}} + z\hat{\mathbf{k}})$

12. $-\sin 1(\hat{\mathbf{i}} + \hat{\mathbf{j}} + \hat{\mathbf{k}})$
14. $-4e^{-8}(\hat{\mathbf{i}} + \hat{\mathbf{j}})$
22. ∇F not defined along the line $y = x$

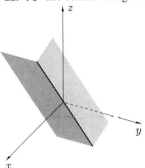

24. $x^2y - xy + C$ **26.** $xyz + y^2z + C$
28. $f(x, y) = g(x, y) + C$

Exercises 13.5

2. $-12x/y^4 + 72x^3y$ **4.** $x(1 + z + y + yz)e^{x+y+z}$
6. $e^{x+y} + 4/y^3$ **8.** 0
10. $(x^2 + y^2 - z^2)/(x^2 + y^2 + z^2)^2$
12. $2xy/(x^2 + y^2)^2$ **14.** $127/(756\sqrt{7})$
16. $8!y^9z^{10}$ **18.** $3y^2\cos(x + y^3)$
20. $-xy/(x^2y^2 - 1)^{3/2}$
24. Entire plane with $(0, 0)$ deleted
26. All space   **28.** Not harmonic
32. Curve of intersection of surface with plane $y = y_0$ is concave upward for $x_1 \leq x \leq x_2$
34. (a) Yes   (b) Yes
36. (b) $2xy + C$
38. $n = 0$, all space; $n = -1/2$, $x^2 + y^2 + z^2 > 0$

Exercises 13.6

2. $(2xe^y + y/x)(-s^2\sin t) + (x^2e^y + \ln x)\left[8t/\left((t^2 + 2s)\sqrt{(t^2 + 2s)^2 - 1}\right)\right]$
4. $xv^2[2yv(3u^2 + 2) + 2xuv/(u^2 + 1) + 3xye^u(u + 1)]$
6. $-2(\ln 3)y3^{x+2}\csc(r^2 + t)\cot(r^2 + t)$
8. $-2rst[5st/y^6 + 2yrt/(y^2 + z^2)^2 + 2rs/y^3]$
10. $6xve^t(v - 2)[1 - 2u/(x^2 - y^2)^2] + 8ye^{4t}[1 + 2u/(x^2 - y^2)^2]$
12. $2(t^2 + 2t)^2e^{2t} + (t^3 + 6t^2 + 6t + 2yt^2 + 4y + 8yt)e^t - 2$
14. $-(3y\sin v - 4x\cos v)^2\sin(xy) - (24\sin v\cos v + 3y\cos v + 4x\sin v)\cos(xy)$
18. $200\pi/3$ cm^3/min; no
20. -7.11×10^4 N/s

22. (a) Yes (b) No (c) Yes (d) No (e) Yes (f) No
(g) No (h) Yes
32. 18.77 kg/m^3/s
38. (c) $u = -[2c/(n+1)]k^{(n+1)/2} + D$

Exercises 13.7

2. $1/(x+y) - 1$
4. $[24x - \cos(x+y)]/[2y - 1 + \cos(x+y)]$
6. $-(2xz^2 + 3)/(2x^2z + y), -z/(2x^2z + y)$
8. $z(1 + y^2z^2)/(y - x - xy^2z^2), z/(x + xy^2z^2 - y)$
10. $[\cos(x+t) + \cos(x-t)]/[\cos(x-t) - \cos(x+t)]$
12. $(x+y)(3y^2 - 3x^2 - 5)/(3y^2z - 5z)$
14. 0
16. $e^x[(2t+1)\cos y/(3x^2 + e^x) + (y^2 + 2yt - 1)\sin y/(t^2 + 2yt + 1)]$
18. $[u^3 + u\cos(uv)]e^{-u}\cos v + [3u^2v + v\cos(uv)]e^{-u}\sin v$
20. $u(3v^2 - u^2)/[4(u^2 + v^2)^3]$
22. $-2t/(9y^2)$

Exercises 13.8

2. $1/\sqrt{5}$ **4.** $-2/\sqrt{17}$ **6.** $1/5$
8. $-13/\sqrt{29}$ **10.** $11/\sqrt{82}$ **12.** $-40/\sqrt{6}$
14. $(1, 4)$ **16.** $(1, -3, 2)$ **18.** $(1, 1)$
20. (a) $\pm(2, 1)$ (b) $(-1 \pm 2\sqrt{19}, 2 \pm \sqrt{19})$
(c) No direction
22. (a) Yes (b) No
24. $9t/[\sqrt{13}\sqrt{4 + 9t^2}], 0$
26. $(0, 0, 0), (1/4, \pm 1/2, 1/4)$
28. $\sqrt{65}/3$ **30.** -2
32. (a) $\pi/(\sqrt{2}\sqrt{8 - 4\pi + \pi^2}), \pi/\sqrt{4 + \pi^2}$ (b) $1/\sqrt{2}, 0$
(c) $1/\sqrt{2}, 1$

Exercises 13.9

2. $x = 1 + u, y = 1 + 2u, z = 1 + 3u$
4. $x = -2 + t, y = 4 - 4t, z = -2 + t$
6. $x = 1 - 2u, y = 5 + 2u, z = 1 + u$
8. $132x + 49y = 328, z = 0$
10. $x = 1 - u, y = u, z = u$
12. $x = 2 + t, y = -\sqrt{5} + (1/\sqrt{5})t, z = -1 - t$
14. $x = 4 + \sqrt{17}u, y = 1, z = \sqrt{17} + 4u$
16. $x = 12 + 16t, y = -14 - 31t, z = 2 + t$
18. $x = t, y = 1, z = 1 + t$
20. $x = 2\pi u, y = 2\pi + u, z = 4\pi + 2u$
22. $3x + 6y + z + 2 = 0$ **24.** $x + y + z = 4$

26. $x + y = 1$ **30.** $6\sqrt{2}$ **32.** 0
34. $x_0 x/a^2 + y_0 y/b^2 + z_0 z/c^2 = 1$
36. No points
38. $(0, 0, -1)$ and $x^2 + y^2 = 1/2, z = -1/2$

Exercises 13.10

2. $(0, 0)$ saddle point; $(1, 1)$ relative maximum
4. None
6. $(0, n\pi)$ saddle points
8. (x, x) relative minima
10. $(x, 0)$ saddle points; $(0, y) \; y > 0$ relative minima; $(0, y) \; y < 0$ relative maxima
12. $(0, 0)$ relative minimum; $(0, y) \; y \neq 0$ none of these
14. $(0, 0)$ relative maximum; $(0, \pm 1/\sqrt{2}), (\pm 1/\sqrt{2}, 0)$ saddle points; $(1/\sqrt{2}, \pm 1/\sqrt{2}), (-1/\sqrt{2}, \pm 1/\sqrt{2})$ relative minima
16. None
18. All points on the coordinate axes
22. (d) $2.45, 3.54$
24. $(0, 0)$ relative minimum; $(-1/3, \pm 2/9)$ saddle points
26. (a) $(2, \pm 2, \pm 2), (-2, \pm 2, \mp 2), (2, \mp 2, \mp 2), (-2, \mp 2, \pm 2)$

Exercises 13.11

2. $4, -1/3$ **4.** $3, -3$ **6.** $3, -2$
8. $(2\sqrt{2} + 5)/\sqrt{2}, (2\sqrt{2} - 5)/\sqrt{2}$
10. $(1, 1, -2)$ **12.** $(1/2, 1/2, 1/2)$ **14.** $2, -32$
16. $100(2/3)^{1/3} \times 100(2/3)^{1/3} \times 100(3/2)^{2/3}$ cm
20. $2a/\sqrt{3} \times 2b/\sqrt{3} \times 2c/\sqrt{3}$
24. $12/\sqrt{5}$ m, $(50\sqrt{5} - 27\pi)/(3\sqrt{5}\pi)$ m
26. $\|AB\| = 1/3$ m, $\theta = \pi/3$
28. $\pm 2\sqrt{3}/9$ **30.** $3\sqrt{3}r^2/4$

Exercises 13.12

2. $(8 \pm 9\sqrt{13})/2$ **4.** ± 27 **6.** $\pm\sqrt{32/27}$
8. $(\pm\sqrt{2} - 1)/2$ **18.** $33.12, -0.12$ **20.** $\pm 1/\sqrt{2}$
22. $(\pm 2\sqrt{14}, \mp\sqrt{14}), (\pm 2, \pm 4)$
24. $(3a/2, 3a/2)$ **26.** $1.171, 2.373$

Exercises 13.13

2. $y(1 + x^2y^2)^{-1}dx + x(1 + x^2y^2)^{-1}dy$
4. $[yz\cos(xyz) - 2xy^2z^2]dx + [xz\cos(xyz) - 2x^2yz^2]dy + [xy\cos(xyz) - 2x^2y^2z]dz$

6. $(1 - x^2 y^2)^{-1/2}(y\, dx + x\, dy)$
8. $(y + t)\, dx + (x + z)\, dy + (y + t)\, dz + (z + x)\, dt$
10. $2e^{x^2+y^2+z^2-t^2}(x\, dx + y\, dy + z\, dz - t\, dt)$
12. 3%

Exercises 13.14

2. $\frac{1}{3!}[f_{xxx}(c,d)(x-c)^3 + 3f_{xxy}(c,d)(x-c)^2(y-d) + 3f_{xyy}(c,d)(x-c)(y-d)^2 + f_{yyy}(c,d)(y-d)^3]$

4. $\sum_{n=0}^{\infty}\sum_{r=0}^{n}\frac{e^5(-1)^{n-r}2^r 3^{n-r}}{(n-r)!\, r!}(x-1)^r(y+1)^{n-r}$

6. $\sum_{n=1}^{\infty}\sum_{r=0}^{n}\frac{(-1)^{n+1}(n-1)!}{(n-r)!\, r!}x^{2r}y^{2n-2r}$

8. $[(x+1)-1]\sum_{n=0}^{\infty}(-1)^n y^{2n+2}$

10. $[72 + 12(x-2) + 24(y-1) - (x-2)^2 + 8(x-2)(y-1) - 4(y-1)^2]/(24\sqrt{3})$

12. $[2x + 2(y-1) + 3x^2 + 6x(y-1) + 3(y-1)^2]/2$

14. 0

16. $\sum_{n=0}^{\infty}\frac{1}{n!}\left[(x-c)\frac{\partial}{\partial x}+(y-d)\frac{\partial}{\partial y}\right]^n f(c,d)$

Review Exercises

2. $2(x^2 - y^2 + z^2)/(x^2+y^2+z^2)^2$
4. $(3-z^2)(1+z^2)/(1+2xz+2xz^3)$
6. $3(2x-ye^{xy})(t^2+1)+(2y-xe^{xy})(\ln t + 1)$
8. $(y + 3yv^2 - 2xuv)/(2u + 6uv^2 + 3v + vx^2)$
10. $e^t(t+1)(y-2x)+e^{-t}(1-t)(x-2y)$
12. $\cos\theta$
14. $(2x - 3x^2 y^2)\cos\theta - 2yx^3\sin\theta$
16. $6(6x^4 t^6 + 9x^5 + x^4 t^9 z^2 + 2xt^9 z^5 - 9t^{12}z^5 - 6xt^6 z^4 - 4x^5 t^3 z - 2x^2 t^3 z^4)/(x^4 t^8 z^4)$
18. $xz[(1-3xz^2)(1-6xz^2+6x^2 z^4)+(1-2xz^2)(1-3xz^2+6x^2 z^4)]/(x-3x^2 z^2)^3$
20. $2y(v^2 t - 1)(t^2 - 3/\sqrt{1-x^2 y^2}) + xv\sec^2 t\,(t^2 - 3/\sqrt{1-x^2 y^2}) + 2xyt$
26. $\sqrt{6}$ 28. $9/\sqrt{14}$ 30. 2
32. $x + 3y - 6z + 4 = 0$
34. $x = 2 + u,\ y = u,\ z = 6 + 4u$
36. $x = 1 + t,\ y = 1 - t,\ z = 1$
38. None
40. $(0,0)$ relative maximum; all points on $x^2 + y^2 = 1$ relative minima
42. $1/2,\ -1/2$
44. $(\pm\sqrt{(\sqrt{5}-1)/2},0),\ (0,\pm 1)$
46. $x = 100\,bcqr/(acpr+abpq+bcqr),\ y = 100\,acpr/(acpr+abpq+bcqr),\ z = 100\,abpq/(acpr+abpq+bcqr)$
48. $1/\sqrt{2} + \sqrt{2}(\pi+6)(x-1)/4 + (y-\pi/4)/\sqrt{2} + \sqrt{2}(24+14\pi-\pi^2)(x-1)^2/16 + \sqrt{2}(10-\pi)(x-1)(y-\pi/4)/4 - \sqrt{2}(y-\pi/4)^2/4 + \cdots$

CHAPTER 14

Exercises 14.1

2. 0   4. 3/4   6. $e^2(1-e)^2/2$
8. 0   10. 5/144   12. 0
14. -0.54 16. $(1-\ln 2)/2$ 18. $128\sqrt{2}/5$
20. $11\,664/35$ 22. $(1-2\sqrt{2})/12$ 24. 1/3
26. 8/189   28. $2(1-\sqrt{2})$
30. $[\sqrt{2} + \ln(\sqrt{2}+1)]/6$

Exercises 14.2

2. 128/15   4. $-621/140$ 6. 0
8. 0   10. 304/15   12. $\sqrt{13} - 7/2$
14. $(1-\cos 1)/2$ 16. $2(1-\sqrt{2})$ 18. $(5/4)\ln 5 - 1$
20. 4   22. Double   24. 0
26. Double   28. 0   30. 101/70
32. $(e^2-1)/(2e)$ 34. 24π 36. $e^2 - 2e - 1$
38. $(1-2\sqrt{2})/12$ 40. 1/3

Exercises 14.3

2. 343/6   4. 8   6. $1 + 3e^{-2}$
8. 343/6   10. 20/3   12. 16π
14. $1024\pi/15$ 16. $26\pi/15$ 18. 45π
20. $34\pi/3$ 22. 235/3   24. 7/6
26. $8\pi/3 - 2\sqrt{3}$ 28. $23\,328/35$ 30. $2 - \ln 3$
32. $16\pi(1+\sqrt{2})/15$ 34. $68\pi/9$
36. 5.38   38. $512\pi/(15\sqrt{5})$ 40. $7\sqrt{2}\pi/6$

Exercises 14.4

2. $256\rho g/5$ 4. $48\rho g$ 6. $370\rho g/3$
8. $250\rho g/3$ 10. 7.63×10^6 N   12. $5\sqrt{29}\rho g$
14. $10\rho g$ 16. $\pi\rho g r^3$ 18. 9.00×10^4 N

Exercises 14.5

2. $(0, 24/5)$ 4. $\bar{x} = \bar{y} = 9/(15 - 16\ln 2)$
6. $\bigl(8(2-\sqrt{2})/(3\pi),\, 8\sqrt{2}/(3\pi)\bigr)$
8. $(177/85, 9/17)$ 10. $(4/3, 5/3)$ 12. $32/3$

14. $603/10$ **16.** $48\rho/5$ **18.** $(0,192/205)$
20. $(-2/(8\sqrt{3}-1),0)$ **22.** $(-61/28,807/700)$
24. $\rho ab(a^2+b^2)/12$ **26.** $4761/140$
28. $81\sqrt{2}/40$

Exercises 14.6

2. $4\sqrt{14}/3$
4. $4(5\sqrt{3}-2\sqrt{6}-3+\sqrt{2})/3$
6. $(247\sqrt{13}+64)/1215$
8. $2\int_0^4\int_0^{\sqrt{16-y^2}}\sqrt{4+y^2+z^2}\,dz\,dy$
10. $4\int_0^{\sqrt{2}}\int_0^{\sqrt{2-x^2}}\sqrt{1+16(x^2+y^2)^3}\,dy\,dx$
12. $\int_0^1\int_0^{1-x^2}\sqrt{1+2/(1+x+y)^2}\,dy\,dx$
16. $\sqrt{2}\int_{-1/\sqrt{2}}^{1/\sqrt{2}}\int_x^{\sqrt{1-x^2}}\sqrt{2+1/(y-x)}\,dy\,dx$
18. $\int_0^1\int_{1-x}^{2-x}\sqrt{1+9x^4+9y^4}\,dy\,dx +$
$\int_1^2\int_0^{2-x}\sqrt{1+9x^4+9y^4}\,dy\,dx$
20. $(37\sqrt{74}-5\sqrt{10})/24$

Exercises 14.7

2. $2/3$ **4.** $4(3\sqrt{3}-\pi)/3$ **6.** $\pi/6$
8. $(3\sqrt{3}-\pi)/2$ **10.** $\pi/4$
12. $15\pi/4-19\sqrt{2}/3-(9/2)\mathrm{Sin}^{-1}(1/3)$
14. $\pi R^4/4$ **16.** $\pi(17\sqrt{17}-1)/6$
18. $2\pi(10\sqrt{10}-1)/3$ **20.** $(17\sqrt{17}-5\sqrt{5})\pi/6$
22. $8\pi/3$ **24.** $2\pi^2 R^3$
26. $[q\rho/(2\epsilon_0)](1-d/\sqrt{R^2+d^2}),q\rho/(2\epsilon_0)$
28. $9\pi-12\sqrt{3}$ **30.** $(5/6,0)$
32. $\pi(\pi-2)/2$ **34.** $2a^2(\pi+4-4\sqrt{2})$

Exercises 14.8

2. $32/3$ **4.** 0 **6.** $1024/21$
8. $1/96$ **10.** $11/6$ **12.** $48/35$
14. $4\int_0^{\sqrt{(\sqrt{5}-1)/2}}\int_{x^2}^{\sqrt{1-x^2-x^4}}\int_0^{\sqrt{1-x^2-y^2}}(x^2+y^2+z^2)\,dz\,dy\,dx$
16. $\int_{-5/2}^{1/2}\int_{-1-\sqrt{9-4(y+1)^2}}^{-1+\sqrt{9-4(y+1)^2}}\int_{x^2+4y^2}^{4-2x-8y}xyz\,dz\,dx\,dy$
18. $729/70$ **20.** 2π **22.** $\pi/3$
24. $4\int_0^1\int_0^{\sqrt{1-x^2}}\int_0^{\sqrt{x^2+y^2}/2}(x^2+y^2+z^2)\,dz\,dy\,dx$
$+4\int_0^1\int_{\sqrt{1-x^2}}^{\sqrt{4/3-x^2}}\int_{\sqrt{x^2+y^2-1}}^{\sqrt{x^2+y^2}/2}(x^2+y^2+z^2)\,dz\,dy\,dx$
$+4\int_1^{2/\sqrt{3}}\int_0^{\sqrt{4/3-x^2}}\int_{\sqrt{x^2+y^2-1}}^{\sqrt{x^2+y^2}/2}(x^2+y^2+z^2)\,dz\,dy\,dx$

Exercises 14.9

2. $2/3$ **4.** $704/15$ **6.** 8
8. $8/3$ **10.** $19/96$ **12.** $7/3$
14. $5/18$ **16.** $64/15$ **18.** $\pi/4-1/3$
20. $1/20$ **22.** $13/3$ **26.** $\pi/2$
28. (a) $(5/3)[3\pi/2-3\mathrm{Sin}^{-1}d-d\sqrt{1-d^2}(5-2d^2)]$
(b) $2500\pi g$

Exercises 14.10

2. $(1/4,1/4,1/4)$ **4.** $(0,16/7,8/7)$
6. $2\rho/3$ **8.** $773\rho/2520$ **10.** $64\rho/15$
12. $(6772/11\,847,7300/14\,001,1/2)$
14. $(0,0,3/2)$ **16.** $128\rho a^5/45$ **18.** $51\sqrt{3}\rho$
20. $(3a/8,3b/8,3c/8)$ **22.** $28\pi\rho R^5/15$

Exercises 14.11

2.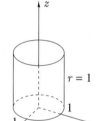

4.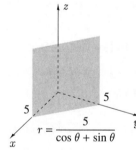

6.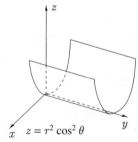

8.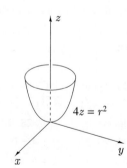

10.

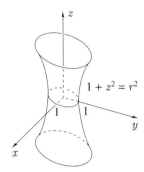

12. $(8\sqrt{2}-7)\pi/6$ **14.** 4π **16.** $4\sqrt{3}\pi$
18. $\int_0^{2\pi}\int_0^3\int_0^{1+r^2} f(r\cos\theta,r\sin\theta,z)\,r\,dz\,dr\,d\theta$,
$\int_0^3\int_0^{2\pi}\int_0^{1+r^2} f(r\cos\theta,r\sin\theta,z)\,r\,dz\,d\theta\,dr$,
$\int_0^3\int_0^{1+r^2}\int_0^{2\pi} f(r\cos\theta,r\sin\theta,z)\,r\,d\theta\,dz\,dr$,
$\int_0^1\int_0^3\int_0^{2\pi} f(r\cos\theta,r\sin\theta,z)\,r\,d\theta\,dr\,dz\ +$
$\int_1^{10}\int_{\sqrt{z-1}}^3\int_0^{2\pi} f(r\cos\theta,r\sin\theta,z)\,r\,d\theta\,dr\,dz$,
$\int_0^{2\pi}\int_0^1\int_0^3 f(r\cos\theta,r\sin\theta,z)\,r\,dr\,dz\,d\theta\ +$
$\int_0^{2\pi}\int_1^{10}\int_{\sqrt{z-1}}^3 f(r\cos\theta,r\sin\theta,z)\,r\,dr\,dz\,d\theta$
$\int_0^1\int_0^{2\pi}\int_0^3 f(r\cos\theta,r\sin\theta,z)\,r\,dr\,d\theta\,dz\ +$
$\int_1^{10}\int_0^{2\pi}\int_{\sqrt{z-1}}^3 f(r\cos\theta,r\sin\theta,z)\,r\,dr\,d\theta\,dz$
20. $8\pi\rho R^5/15$ **22.** $81\pi^2/8$ **24.** 0.084
26. $\pi h\rho R^4/10$ **28.** $4\sqrt{3}\pi/9$ **30.** $56\pi(2-\sqrt{2})/3$
32. $4\pi(8\sqrt{2}-3\sqrt{6})/3$ **34.** $2a^3(3\pi+4)/9$
36. $64\pi/3$ **38.** $15\pi/2$
40. $3\sqrt{6}/5$ **42.** $2\pi^2 ab^2$

Exercises 14.12

2. $\mathcal{R}\sin\phi=1$ (see figure for Exercise 14.11–2)
4. $\mathcal{R}=4\cot\phi\csc\phi$ (see figure for Exercise 14.11–8)
6. $\mathcal{R}^2=-\sec 2\phi$ (see figure for Exercise 14.11–10)
8. $(2-\sqrt{2})\pi/3$ **10.** $[4\text{Tan}^{-1}2-\pi]/12$
12. $32\pi/3$ **14.** $8\pi\rho R^5/15$ **16.** $k\pi R^4\,C$
18. $9\pi/2$ **20.** $(2\pi R^3/3)(1-k/\sqrt{1+k^2})$
22. (a) $\rho_b=\rho_w/2$ **(b)** $11\pi\rho_w g R^3/24$

Exercises 14.13

2. $2x-1+e^x$
4. $4x^3-3x^2-1+3x^2(x^3-1)\ln(x^3-1)-2x^3\ln(x^2)$
8. $e^x\sqrt{1+e^{3x}}-\cos x\sqrt{1+\sin^3 x}$
12. $(1/a^2)\ln(1+ab)-b/(a+a^2b)$
14. $[3/(8a^5)]\text{Tan}^{-1}(x/a)+x(3x^2+5a^2)/[8a^4(a^2+x^2)^2]+C$
16. $\pi\text{Sin}^{-1}a$
18. (a) $|x|<1/9,\ 0$
  (b) $18/x+(1/x^2)\ln[(1-9x)/(1+9x)],\ x\ne 0;\ 0$

Review Exercises

2. $81/16$ **4.** $1/40$ **6.** $36/5$
8. 0 **10.** $-4544/945$ **12.** $\pi/6$
14. $32\pi/3$ **16.** $2\pi/5$ **18.** $3\pi/2$
20. $(\pi\ln 2)/4$
22. (a) $\iint_R dA$
  (b) $\iint_R 2\pi(2-x)\,dA,\ \iint_R 2\pi(y+4)\,dA$
  (c) $\iint_R (x-1)\,dA,\ \iint_R (y+1)\,dA$
  (d) $\iint_R (x+1)^2\,dA,\ \iint_R (y-4)^2\,dA$
  (e) $\iint_R \sigma(x,y)\,dA$
  (f) $\iint_R \rho(x,y)\,dA$
  (g) $\iint_R P(x,y)\,dA$
24. $32/3$ **26.** $16/15$
28. $\pi(17\sqrt{17}-1)/96$ **30.** $128\pi/15,\ 4\pi^2$
32. $1/4$ **34.** 2
36. $14\rho\pi/15$ **38.** $(0,1/8)$
40. $(9\pi/64,9\pi/64,3/8)$ **42.** $16\pi/15$

CHAPTER 15

Exercises 15.1

2. Closed, connected **4.** Connected
6. Closed, connected **8.** Open
10. For interior, exterior, and boundary points replace circle with sphere in planar definitions. Open, closed, connected, and domain definitions are identical. A domain is simply-connected if every closed curve in the domain is the boundary of a surface that contains only points of the domain.
12. Closed, connected **14.** Connected
16. Open
18. Open, connected, simply-connected domain
22. $-(x^2+y^2+z^2)^{-3/2}(x\hat{\mathbf{i}}+y\hat{\mathbf{j}}+z\hat{\mathbf{k}})$
24. $(6-2\cos 2)\hat{\mathbf{i}}+(1+2\sin 2)\hat{\mathbf{j}}$
26. $2(e^y-x^2y)$
28. $2x\cos(x^2+y^2+z^2)-\sin(y+z)$
30. 1 **32.** $2/\sqrt{x^2+y^2+z^2}$
34. $4y(16z^4-3x)\hat{\mathbf{i}}+x^2\hat{\mathbf{j}}+12yz\hat{\mathbf{k}}$
36. 0 **38.** $-2\hat{\mathbf{k}}$
40. $-2\hat{\mathbf{k}}/[(x+y)\sqrt{(x+y)^2-1}]$
46. $x^3y^2+3x+2y+C$ **48.** $\ln|x+y|+C$
50. $(x^2+y^2+z^2)/2+C$ **52.** $\ln|1+x+y+z|+C$
54. $\text{Tan}^{-1}(xy)+z^2/2+C$
56. (a) $4,2,-1$ **(b)** $x^3/3+2xy+4xz-3y^2/2-yz+z^2+C$

58. (a) $q/(4\pi\epsilon_0|\mathbf{r}|) + C$ (b) $-\sigma z/(2\epsilon_0) + C$
62. $yz\hat{\mathbf{i}} - xz\hat{\mathbf{j}}$
64. $(1/4)[(4xyz - 3y^2z)\hat{\mathbf{i}} + (2xyz - 6x^2z)\hat{\mathbf{j}} + (2x^2y + xy^2)\hat{\mathbf{k}}]$

Exercises 15.2

2. $32/3$ **4.** $50\sqrt{10}/3$ **6.** $37\sqrt{2}/3$
8. $9\sqrt{1+16\pi^2}$ cm **10.** $37/80$ **12.** 0.78
14. 0 **16.** $(1 - 161\sqrt{161})/6$
18. $\pi(145\sqrt{145} - 10\sqrt{10})/27$
20. $\pi/2$ **22.** 0.007 **24.** 2
26. 0.242 **28.** π **30.** $(5\sqrt{5} - 1)/6$
32. $\sqrt{2}(1 - e^{6\pi})/13$ **34.** 17.08
36. $4\pi^2 ab$

Exercises 15.3

2. $51/4$ **4.** -15 **6.** $-99/140$
8. 10 **10.** 0 **12.** 9
14. 0 **16.** -8π **18.** $(768 + 5\pi)/20$
20. $67/32 + \sqrt{5} - \sqrt{2} - 5/e$
22. $(4\pi\epsilon_0)^{-1}[q_1 q_3(1/\sqrt{41} - 1/\sqrt{61}) + q_2 q_3(1/\sqrt{18} - 1/\sqrt{10})]$
24. 3.719 **26.** -4.26×10^{-4} **28.** 0
32. (a) $2a$ (b) 0

Exercises 15.4

2. -43 **4.** -2 **6.** 0
8. 10 **10.** e^3 **12.** No
14. 0 **16.** 0 **18.** $8/105 + \ln(3/2)$
20. $-\pi$
22. (a) Yes (b) Yes (c) Yes (d) Yes (e) No
24. 2π
26. (b) $k(1/\sqrt{x_1^2 + y_1^2 + z_1^2} - 1/\sqrt{x_2^2 + y_2^2 + z_2^2})$

Exercises 15.5

2. Not conservative **4.** mgz
8. $-k(\sqrt{x^2 + y^2 + z^2} - L)(x\hat{\mathbf{i}} + y\hat{\mathbf{j}} + z\hat{\mathbf{k}})/\sqrt{x^2 + y^2 + z^2}$, yes
10. (a) Yes (b) Yes

Exercises 15.6

2. 4π **4.** $2\sqrt{3}/5$ **6.** 0
8. $-3/8$ **10.** $4(13\sqrt{13} - 8)/3$
14. πab **16.** $3\pi/8$ **18.** 2π
22. 24 **24.** $77/2$ **26.** $81\pi/2$
28. -4π **32.** π

Exercises 15.7

2. $2\sqrt{3}/15$ **4.** $\sqrt{2}/8$ **6.** $(-61 + 44\sqrt{2})/5$
8. -3 **10.** $\pi R\sqrt{R^2 + h^2}$
12. $3(145^{5/2} - 361)/5120$ **14.** $\pi(3\sqrt{3} - 1)/3$
16. $(50\sqrt{5} + 2)\pi/15$ **18.** $4\pi R^2$
20. 0 **22.** $\pi \ln 3$

Exercises 15.8

2. $2e^2 - 10e^{-2}$ **4.** 0 **6.** $2\pi/3$
8. -2π **10.** $10\pi/3$ **12.** $\sqrt{2}\pi/4$
14. $2\sqrt{5} + 8\ln[(\sqrt{5} + 1)/2]$
16. 0 **18.** (a) 30π (b) -20π
20. $2\pi(1 - e^{-2})$

Exercises 15.9

2. 0 **4.** 16 **6.** 27π
8. 180 **10.** $-1328\pi/5$ **12.** $57/2$
14. $52\pi/5$ **20.** 0

Exercises 15.10

2. 0 **4.** $\pm 16\pi$ **6.** 0
8. $-2\sqrt{2}\pi b^2$ **10.** $\pm 2\pi$ **12.** -24π
14. 3π

Review Exercises

2. $3x^2 y + x^2/y^2$ **4.** 0
6. $ye^x + ze^y + xe^z$ **8.** $(\hat{\mathbf{i}} + \hat{\mathbf{j}})/\sqrt{1 - (x + y)^2}$
10. $x(-y\hat{\mathbf{j}} + z\hat{\mathbf{k}})/(1 + x^2 y^2 z^2)$
12. $2\sqrt{3}$ **14.** π
16. $\sqrt{2}/3$ **18.** 2π
20. $8\pi(18 - 7\sqrt{6})/3$ **22.** 0
24. 0 **26.** $-\pi/120$
28. $(25\sqrt{5} - 11)/120$ **30.** $8\pi/3$

CHAPTER 16

Exercises 16.1

12. $-3\sin 3x + 2\cos 3x$
14. $[(2\cos 3 - \cos 6)\sin 3x + (\sin 6 - 2\sin 3)\cos 3x]/\sin 3$

16. $(1/3)\operatorname{Tan}^{-1}(x/3) + C$

18. $(x^3/6)\ln x - 5x^3/36 + C_1 x + C_2$

20. Ce^x **22.** $y = 0$

24.(b)

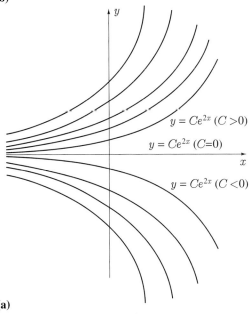

26.(a)

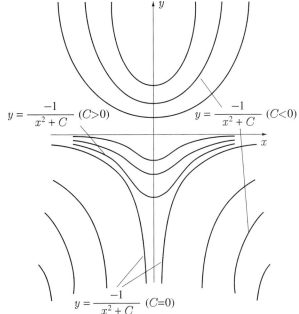

Exercises 16.2

2. $y(x) = 2 + Ce^{-x^2}$ **4.** $y(x) = Ce^{3x} - 2/3$

6. $y^2 + x^2 + 2(x - y) + 2\ln|y + 1| + 2\ln|x - 1| = C$

8. $y(x) = [C + 3\ln|x| + 3e^x(1 - x)]^{1/3}$

10. $y(x) = (x + C)/(1 - Cx)$

12. $xy = 2e^{y-x-1}$

14. $y(x) = [\tan(x^2 - 4) + 4]/[1 - 4\tan(x^2 - 4)]$

16. (a) $6(1 - 6kt)^{-1}$ km   **(b)** never

18. $20 + 60e^{-0.203t}$ **20.** 13.51 h

22. $(R/k)(1 - e^{-kt}) + A_0 e^{-kt}$

24. $60(1 - e^{kt})/(2 - 3e^{kt})$ g

28. $x^2(x^2 - 2y^2) = C$ **30.** $x^2 + 2xy - y^2 = C$

32. $(y - x)e^{y/x} = x\ln|x| + Cx$

34. $x^2 + y^2 = 5x$

36. $y_0(A + y_0)/[y_0 + Ae^{-k(A+y_0)t}]$

38. $y = Ce^x - x - 1$ **40.** $6y - 3\ln|4x + 6y + 3| = C$

42.(a)   **(b)**

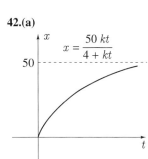 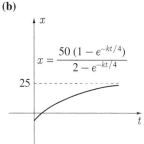

(c)

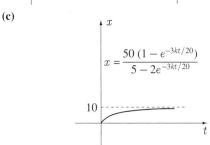

44. 459.5 mg   **46.** $y^3(x^3 + 6x - 5) = 1 + y^6$

48. $h = [\sqrt{H} - \sqrt{g}cAt/(\sqrt{2}\pi r^2)]^2$, $\pi r^2 \sqrt{2H}/(cA\sqrt{g})$

50. $y = (L/2)[(x/L)^{1-v/V} - (x/L)^{v/V+1}]$

Exercises 16.3

2. $y(x) = x^4 + C/x^2$ **4.** $y\sin x = C - 5e^{\cos x}$

6. $y(x) = C(x + 1)^2 - 2x - 2$

8. $y(x) = Ce^x + e^{2x}$

10. $y(x) = x^3/2 + Cx^3 e^{1/x^2}$

12. $y(x) = x + \cos 2x + C\sin 2x$

14. $y(x) = e^x(2\sin x - \cos x)/5 - 4e^{-x}/5$

16. $xy = C + y^4/4$ **18.** $y(x) = e^{-x}/(C - x)$

20. $x^2 y + Cye^{-x} = 1$

22. $y(x) = [(3/4)\cos x + C\sec^3 x]^{-1/3}$

24. $1000 + t + 3 \times 10^6/(1000 + t)$ g

26.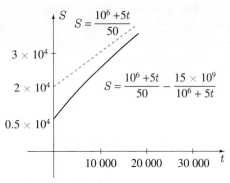

28.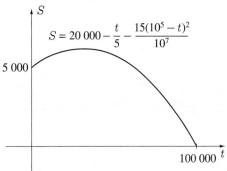

30. (b) $I_0 - E_0/(\omega CZ^2)$

Exercises 16.4

2. $2\sqrt{Cy-1} = \pm Cx + D$
4. $y(x) = Cx - C^2 \ln|x+C| + D$
6. $(x+C)^2 + (y+D)^2 = 1$
8. $D + x/2 = \{-1/y;$ or $C\text{Tan}^{-1}(Cy);$ or, $[1/(2C)] \ln|(y-C)/(y+C)|\}$
10. $y(x) = (1/2)(Ce^x + C^{-1}e^{-x}) + D$
12. (c) $LvV/(V^2 - v^2) + (LV/2)$
$\{[1/(V+v)](x/L)^{v/V+1} - [1/(V-v)](x/L)^{-v/V+1}\}$

Exercises 16.5

2. Less than 47.8 km/h; defence
6. (c) $x^* = (\mu Mg \pm \sqrt{\mu^2 M^2 g^2 + Mkv_0^2 + k^2 x_0^2 - 2k\mu Mx_0})/k$
8. 18.0 m
10. (a) $70.0(1 - 0.556\,e^{-0.280t})/(1 + 0.556\,e^{-0.280t})$ m/s

(b)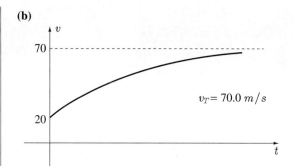

12. 1.99 s
14. (a) $v_T \tan[\text{Tan}^{-1}(v_0/v_T) - kv_T t/m]$ where $v_T = \sqrt{9.81m/k}$
(b) $(m/k) \ln|\cos[\text{Tan}^{-1}(v_0/v_T) - kv_T t/m]| + (m/k) \ln(\sqrt{v_0^2 + v_T^2}/v_T)$, $(m/k) \ln(\sqrt{v_0^2 + v_T^2}/v_T)$
16. $(9.81m \sin\alpha/k)(1 - e^{-kt/m})$, $(9.81m^2 \sin\alpha/k^2)(kt/m + e^{-kt/m} - 1)$

Exercises 16.6

2. (a) $2e^{[(1/2)\ln(5/2)]t}$ **(b)** $2e^{[(1/10)\ln 55]t}$

4. 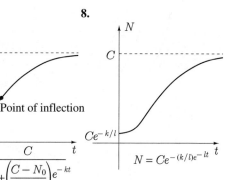   **8.**

10.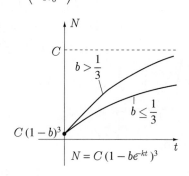

Exercises 16.7

2. Linear   **4.** Linear   **6.** Linear
8. Linear   **10.** Not linear   **12.** Linear

14. Linear   **16.** Not linear   **18.** Not linear
20. Linear

Exercises 16.8

2. $C \cos x$ **4.** $C_1 e^{4x} + C_2 x e^{4x}$
6. $e^{2x}[C_1 \cos(x/\sqrt{2}) + C_2 \sin(x/\sqrt{2})]$
8. $C_1 \cos(2 \ln x) + C_2 \sin(2 \ln x)$
12. Dependent   **14.** Independent

Exercises 16.9

2. $(C_1 + C_2 x)e^{4x}$
4. $C_1 e^{(-1+\sqrt{3})x} + C_2 e^{(-1-\sqrt{3})x}$
6. $C_1 e^{3x} + C_2 \cos x + C_3 \sin x$
8. $(C_1 + C_2 x + C_3 x^2)e^{2x}$
10. $C_1 \cos x + C_2 \sin x + C_3 \cos 2x + C_4 \sin 2x$
12. $e^{\sqrt{2}x}[C_1 \cos(\sqrt{2}x) + C_2 \sin(\sqrt{2}x)] + e^{-\sqrt{2}x}[C_3 \cos(\sqrt{2}x) + C_4 \sin(\sqrt{2}x)]$
14. $y'' + 4y' + 20y = 0$
16. $y'''' - 4y''' + 10y'' - 12y' + 9y = 0$
18. 5, 24, 20

Exercises 16.10

2. $-(x^2/3 + 2/9)e^{-x}$, $C_1 e^{(-1+\sqrt{3})x} + C_2 e^{-(1+\sqrt{3})x} - (x^2/3 + 2/9)e^{-x}$
4. $(1/9)\cos 2x$, $(C_1 + C_2 x)\cos x + (C_3 + C_4 x)\sin x + (1/9)\cos 2x$
6. $e^{-2x}/40$, $C_1 \cos x + C_2 \sin x + C_3 \cos 2x + C_4 \sin 2x + e^{-2x}/40$
8. $e^{2x}(\cos x - 5\sin x)/312$, $e^{-4x}(C_1 \cos 5x + C_2 \sin 5x) + e^{2x}(\cos x - 5\sin x)/312$
10. $[(2x+1)\cos x - 2(x+1)\sin x]/16$, $e^{2x}(C_1 \cos x + C_2 \sin x) + [(2x+1)\cos x - 2(x+1)\sin x]/16$
12. $Axe^{3x} + Be^{3x} + Cx\cos x + Dx\sin x + E\cos x + F\sin x$
14. $Ax^2 e^x + Bxe^x + Cx^3 + Dx^2 + Ex + F + G\cos x + H\sin x$
18. $e^x[C_1 \cos(x/\sqrt{2}) + C_2 \sin(x/\sqrt{2})] + (4\cos 3x - 5\sin 3x)/246 + (\sin x + 4\cos x)/34$
20. $-(1/108)\sin 3x + (x^2/12)(\sin 3x - \cos 3x) + (x/36)(\sin 3x + \cos 3x)$

Exercises 16.11

2. $e^{-15t/4}[(12/65)\cos(5\sqrt{23}t/4) + (20/(13\sqrt{23}))\sin(5\sqrt{23}t/4)] - (4/65)[3\cos(10t) + 2\sin(10t)]$ m
4. $-0.0253 e^{-99.50t} + 5.03 e^{-0.50t}$ C

6. $(M|g|/k)\cos(\sqrt{k/M}t)$ from equilibrium
8. (b) $x = -7/36$ m, yes
12. $(1/64)\sin 8t - (t/8)\cos 8t$ m
14. $(9/10)[3t\cos 3t - \sin 3t]A$, yes
16. (b) $0.705/\sqrt{L}$

Review Exercises

2. $y = \pm\sqrt{2(x + \ln|x|) + C}$
4. $y = Ce^{-4x} + (8x^2 - 4x + 1)/32$
6. $y = e^{-3x/2}[C_1 \cos(\sqrt{7}x/2) + C_2 \sin(\sqrt{7}x/2)] + 1/2$
8. $y = x^3/9 + C_1 \ln|x| + C_2$
10. $y = Ce^{-x^2} + x^2 - 1$
12. $y = (C_1 + C_2 x)e^{2x} + (3\sin x + 4\cos x)/25$
14. $y = C_1 \cos 2x + C_2 \sin 2x - (x/4)\cos 2x$
16. $xy^2 = Ce^{-2x-x^2/2}$
18. $e^{-x}[C_1 \cos(\sqrt{3}x) + C_2 \sin(\sqrt{3}x)] + (\sqrt{3}x/6)e^{-x}\sin(\sqrt{3}x)$
20. $x = Ce^{-3y} + (2/27)(-9y^2 + 6y - 2)$
22. $y = (424/405)e^{9x} + (11/5)e^{-x} - 2x/9 - 20/81$
24. $y = -\cos x + (2\sin x)/x + (2\cos x + 1 - \cos 1 - 2\sin 1)/x^2$
26. (a) 3.924×10^{-3} N   **(b)** $1.962 \times 10^{-3}t + 4.905 \times 10^{-7}(e^{-4000t} - 1)$ m
28. (a) In air, $v = 9.81t$; in water $v = 0.34335 + 1.502 \times 10^{29}e^{-20t}$
(b) In air, $y = 4.905t^2$; in water, $y = 50.30 + 0.34335t - 7.51 \times 10^{27}e^{-20t}$
(c) 28.25 s

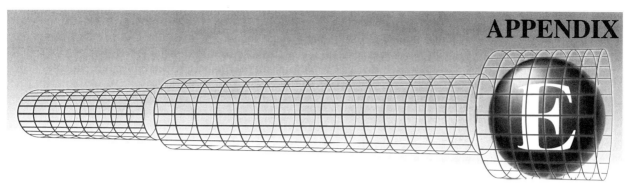

APPENDIX E

Additional Exercises

When you solve the exercises in a particular section of this text, you already know the nature of the problem and the techniques that should be used. For example, a problem in Exercises 13.6 will require chain rules, a problem in Exercises 14.11 will require cylindrical coordinates, etc. To ensure that you understand all parts of a chapter, we recommended the review exercises at the end of each chapter. They require you to first identify the particular section that is useful in solving a problem, and then to follow through with its application. Now that you have studied all the chapters in this book, you should test your understanding of multivariable calculus as a whole. The following exercises, which review all aspects of the course, will do this. They test your ability to recognize the essential nature of a problem — be it a maximum-minimum problem, a triple integral problem, an application of differential equations, a problem in infinite series, etc. — and they tax your organizational, analytical, and interpretive skills.

The first 200 or so exercises are a reasonably straightforward review of the many types of problems that you have already encountered. The next 40 are more difficult; they may require some originality on your part or some extra calculations. The last few are more challenging. Set your sights high; attempt as many problems as you can. Answers to selected exercises are found at the end of the appendix.

ADDITIONAL EXERCISES

In Exercises 1–20 determine whether the series converges or diverges. In the case of a series with both positive and negative terms, determine whether convergence is absolute or conditional.

1. $\sum_{n=1}^{\infty} \dfrac{n^2+1}{3n^3+5}$

2. $\sum_{n=1}^{\infty} \dfrac{(n^3+4n)^{2/3}}{n^4-3n^3+1}$

3. $\sum_{n=1}^{\infty} \dfrac{(-1)^{n+1} n^2}{2^n}$

4. $\sum_{n=1}^{\infty} \dfrac{(-1)^n}{\sqrt{n+1}}$

5. $\sum_{n=2}^{\infty} \dfrac{1}{n \ln n}$

6. $\sum_{n=1}^{\infty} n^{1/n}$

7. $\sum_{n=1}^{\infty} \dfrac{\sin(n+1)}{n^2+4}$

8. $\sum_{n=1}^{\infty} \dfrac{1+2^{-n}}{1+3^{-n}}$

9. $\sum_{n=1}^{\infty} \sqrt{\dfrac{n^2+n-1}{2n^4+n^3+5}}$

10. $\sum_{n=1}^{\infty} \left(\dfrac{n^2-2}{2n^2+5}\right)^n$

11. $\sum_{n=1}^{\infty} \dfrac{n^2}{n^{3n}}$

12. $\sum_{n=1}^{\infty} \dfrac{\operatorname{Tan}^{-1}(n^2+5)}{2n^2+4}$

13. $\sum_{n=1}^{\infty} \dfrac{1 \cdot 3 \cdot 5 \cdots (2n+1)}{2 \cdot 4 \cdot 6 \cdots (2n)} \left(\dfrac{1}{n}\right)$

14. $\sum_{n=1}^{\infty} \dfrac{(n!)^2}{(2n+1)!}$

15. $\sum_{n=1}^{\infty} \dfrac{\sqrt{n+\sqrt{1+n^2}}}{n^2}$

16. $\sum_{n=1}^{\infty} \left(\dfrac{1}{2} - \dfrac{1}{n+1}\right)^{10}$

17. $\sum_{n=1}^{\infty} \dfrac{3}{n(1+\ln n)^3}$

18. $\sum_{n=1}^{\infty} \dfrac{3^{-2n}+2^{-2n}}{4^{-n}+5^{-n}}$

19. $\sum_{n=1}^{\infty} \dfrac{(n-1)^n}{n^{n-1}}$

20. $\sum_{n=1}^{\infty} \operatorname{Sin}^{-1}\left(\dfrac{1}{n^2}\right)$

In Exercises 21–46 find a general solution for the differential equation.

21. $\dfrac{dy}{dx} = \dfrac{1-2x}{y^2}$

22. $\dfrac{dy}{dx} = -\sqrt{\dfrac{1-y^2}{1-x^2}}$

23. $\dfrac{dy}{dx} = \dfrac{1}{x+y^2}$

24. $(1+e^x)\dfrac{dy}{dx} + e^x y = \sin x$

25. $x\dfrac{d^2y}{dx^2} = \dfrac{dy}{dx} + \left(\dfrac{dy}{dx}\right)^3$

26. $\dfrac{d^2y}{dx^2} + \left(\dfrac{dy}{dx}\right)^2 + 1 = 0$

27. $y'' - 3y' - 4y = x^2 + e^{2x}$

28. $y'' + 4y = 3\cos 2x$

29. $y'' + 6y' + 9y = \sin x$

30. $y''' + 5y'' + 5y' + y = 2\sin 3x$

31. $\dfrac{dy}{dx} + \sin\left(\dfrac{x+y}{2}\right) = \sin\left(\dfrac{x-y}{2}\right)$

32. $x^2 \dfrac{dy}{dx} + (2x + x^2) y = e^x$

33. $\dfrac{d^2y}{dx^2} + \dfrac{dy}{dx} = \sin x$

34. $\dfrac{d^2y}{dx^2} + 12\dfrac{dy}{dx} + 2y = e^{-6x}$

35. $y''' - 3y'' + 4y = e^{-x} + 3x + 2$

36. $(x^2+1)\dfrac{d^2y}{dx^2} + 2x\dfrac{dy}{dx} + x^2 = 0$

37. $\cos^2 x \dfrac{dy}{dx} + y = 1$

38. $xe^{2y}\cos x\, dx - y^2\, dy = 0$

39. $y''' + 3y'' + 3y' + y = 2xe^{-x}$

40. $y''' + 3y'' + 3y' + y = e^{2x}$

41. $\sqrt{y}\dfrac{d^2y}{dx^2} = \dfrac{dy}{dx}$

42. $y'' + 6y' + 9y = x^2 e^{-3x}$

43. $\dfrac{d^2y}{dx^2} + 4y = 3\cos x + \sin x$

44. $y'' + 4y = 3x\cos 2x$

45. $\dfrac{d^2y}{dx^2} + 6\dfrac{dy}{dx} + 9y = x\sin x - x$

46. $x^2\dfrac{d^2y}{dx^2} - 2x\dfrac{dy}{dx} - 4y = 0$

In Exercises 47–66 sketch whatever is defined by the equation or equations in xyz-space.

47. $z^2 = 3x^2 + y^2$

48. $2x + 3y - 4z = 0$

49. $x^2 + y^2 = 4$, $y = 3x$

50. $x^2 + y^2 = 4$, $z = 5$

51. $x^2 + y^2 + z^2 = 4$, $y = 3x$

52. $x^2 + y^2 + z^2 = 4$, $z = 5$

53. $z = y + x^2$

54. $x + y^3 = 1$

55. $x^3 + y^3 + z^3 = 1$

56. $y^2 + 2z^2 - x^2 = 8$

57. $y^2 + 2z^2 - x^2 = 8$, $2z = x$

58. $y^2 + 2z^2 = 8$, $z = y$

59. $y = 4x^2$, $y = x$

60. $y = 4x^2$, $y = z$

61. $y = 4x^2$, $x = z$

62. $z = x^2 + 4y^2$

63. $z = x^2 + 4y^2$, $x + y + z = 1$

64. $z = x^2 + 4y^2$, $x + y + z = 1$, $x = 2y$

65. $z = x^2 + 4y^2$, $x + y + z = 1$, $x + y = 1$

66. $z = \sqrt{x^2 + y^2}$, $x^2 + y^2 = 2y$

67. Find the components of a vector with length 5 that makes angles of 1 and 0.8 radians with the positive x- and y-axes.

68. Is there a direction in which the vector $2\hat{\imath} - 3\hat{\jmath} + 4\hat{k}$ has a component of length 6?

69. If the components of a vector along the vectors $\mathbf{v} = 2\hat{\imath} - 3\hat{\jmath}$ and $\mathbf{w} = \hat{\imath} + \hat{\jmath}$ are 2 and -3, what are its Cartesian components?

In Exercises 70–73 find the equation of the plane.

70. containing the point $(1, 2, -3)$ and the line $3x + 4y - 2z = 5$, $x + y + 2z = 6$

71. containing the point $(2,-1,3)$ and perpendicular to the line $x = 2y = z + 2$

72. containing the points $(1,4,0)$, $(2,-2,1)$, and $(0,3,2)$

73. through the point of intersection of the plane $x - 2y + 3z = 4$ and the line $x = 1 - t$, $y = 2 + 3t$, $z = 2(1+t)$, and containing the line $(3x+1)/2 = (1-y)/3 = z$

In Exercises 74–75 find equations for the line.

74. through the point $(0,-1,-2)$ and perpendicular to the plane $2x - 3y + 5z = 6$

75. through the midpoint of the line segment joining $(-1,2,0)$ and $(3,2,2)$ and perpendicular to the plane containing the aforementioned line segment and the line $x = 3 + 3t$, $y = 2 - t$, $z = 2 + 4t$

76. Find the length of the curve $3x = y^2$, $9z = 2xy$ from $(0,0,0)$ to $(3,3,2)$.

77. Find a unit tangent vector to the curve $x = t^2 \cos 2t$, $y = t^2 \sin 2t$, $z = 1 - t$ at the point $(16\pi, 0, 1 - 4\pi)$.

78. Find a tangent vector at each point on the curve $x + y + 2z = 4$, $z = x^2 + y^2$ directed clockwise as viewed from the origin.

79. Find the unit vector that is normal to the curve $x = t^2$, $y = t^3 + t$, $z = 2t - 1$ at the point $(4, 10, 3)$ and has an x-component equal to the sum of its y- and z-components.

80. Show that the ratio of curvature to torsion for the curve $x = e^t \cos t$, $y = e^t \sin t$, $z = e^t$ is always $\sqrt{2}$.

81. Find the curvature of the curve $x = \ln(\cos t)$, $y = \ln(\sin t)$, $z = \sqrt{2}t$, $0 < t < \pi/2$.

82. Evaluate the following limits if they exist:

(a) $\lim_{(x,y) \to (0,0)} \dfrac{x^2 + y^2}{3x^2 - 4y^2}$

(b) $\lim_{(x,y) \to (1,3)} \dfrac{xy^2 + 2xy + x - y^2 - 2y - 1}{xy - 2x - y + 2}$

83. Along which straight line(s) approaching $(0,0)$ is the limit of the function $f(x,y) = (2x - 3y)/(4x + 2y)$ equal to 5?

In Exercises 84–88 find the indicated derivative.

84. $\partial z/\partial y$ if $z = \ln(y + \sqrt{x^2 + y^2})$

85. $\partial z/\partial x$ and $\partial z/\partial y$ if $z = \text{Tan}^{-1}\sqrt{x^y}$

86. $\partial z/\partial u$ if $z = \text{Sin}^{-1}\left(\dfrac{\sqrt{u^2 - v^2}}{\sqrt{u^2 + v^2}}\right)$

87. $\partial z/\partial x$ if $z = u^2 + u \sin v$, $u = \text{Tan}^{-1}(x + y)$, $v = \ln(e^x + e^y)$

88. $\partial u/\partial x)_y$ if $u^2 \sin x - vy = 3x + 2$, $u^3 \cos x + v^3 y - v = 3xy - 4yu$

89. Find all points at which the gradient of the function $f(x,y) = x^2 y + 3x$ is equal to (a) $2\hat{\imath} + 3\hat{\jmath}$ (b) $2\hat{\imath} - 3\hat{\jmath}$.

90. If u and v are functions of x, y, and z defined by $uv = 3x - 2y + z$, $v^2 = x^2 + y^2 + z^2$, show that $x\dfrac{\partial u}{\partial x} + y\dfrac{\partial u}{\partial y} + z\dfrac{\partial u}{\partial z} = 0$.

91. Find the following rates of change of the function $f(x,y,z) = x^2 yz - xy^3 z$:

(a) $\partial f / \partial x$

(b) $\partial f / \partial y$

(c) $\partial f / \partial z$

(d) $\partial^2 f / \partial x^2$ at $(1, -2, 3)$

(e) at $(1, -2, 3)$ toward the point $(3, 2, 4)$

(f) with respect to t along the curve $x = t^3 + 3t + 1$, $y = \sin t$, $z = 2 \cos t$

(g) at $(1, 0, 2)$ with respect to distance travelled along the curve in (f)

(h) at $(2, 2, 2)$ with respect to distance travelled along the curve $x^3 + y^3 + z^3 = 24$, $y = x$ directed so that z increases in the first octant

(i) at $(1, 1, 3)$ normal to the surface $x^2 + y + z = 5$

(j) at $(1, 2, -1/6)$ in any direction tangent to the surface $x^2 yz - xy^3 z = 1$

92. Find the tangent plane to the surface $x^2 y + xyz + z^2 = 5$ at the point $(2, 2, -1)$.

93. Find the point(s) where the curve $x + y + z = 1$, $y = x^2 - 4$ intersects the tangent plane to the surface $x^3 + y^3 + z^3 = 10$ at the point $(1, 1, 2)$.

94. Show that the tangent line to the curve $x = 18t - 3$, $y = 3t^2 + 1/t$, $z = t + 1/t^2$ at the point $(15, 4, 2)$ and the tangent plane to the surface $x^2 + y^2 - z^2 = 3x + 1$ at $(1, 2, -1)$ do not intersect.

In Exercises 95–97 find and classify the critical points of the function as yielding relative maxima, relative minima, or saddle points.

95. $f(x,y) = y\sqrt{1+x} + x\sqrt{1+y}$

96. $f(x,y) = xy(a - bx - cy)$ where a, b, and c are nonzero constants

97. $f(x,y) = (2x - x^2)(2y - y^2)$

98. Find the point in the xy-plane for which the sum of the squares of the distances from n fixed points $P_i(x_i, y_i)$, $(i = 1, \ldots, n)$ is a minimum.

99. Find the point in the xy-plane which minimizes the sum of the squares of the distances from the x-axis, the y-axis, and the line $Ax + By + C = 0$.

100. Show that a positively homogeneous function $f(x,y)$ of degree k can be expressed in the form $x^k F(y/x)$ when $x > 0$. What is the extension of this result for a positively homogeneous function $f(x,y,z)$?

In Exercises 101–147 evaluate the integral.

101. $\displaystyle\int_{-1}^{0} \int_{x^2}^{3x+1} (x+y)^2 \, dy \, dx$

102. $\displaystyle\int_{0}^{2} \int_{0}^{y} x \sin y \, dx \, dy$

103. $\displaystyle\int_{1}^{2} \int_{x}^{1} \frac{1}{(2x-y)^3} \, dy \, dx$

104. $\displaystyle\int_{0}^{1} \int_{0}^{y} x^2 e^{y^2} \, dx \, dy$

105. $\displaystyle\int_{0}^{2} \int_{0}^{1} |2x-y| \, dy \, dx$

106. $\displaystyle\int_{0}^{2} \int_{y}^{2} \frac{y}{\sqrt{x^2+y^2}} \, dx \, dy$

107. $\displaystyle\int_{0}^{2} \int_{0}^{\sqrt{4-x^2}} \sqrt{x^2+y^2} \, dy \, dx$

108. $\displaystyle\iint_{R} \frac{1}{(16-x^2-y^2)^{3/2}} \, dA$ where R is the region in the first quadrant bounded by the curves $y = x$, $y = \sqrt{9-x^2}$, $y = 0$

109. $\displaystyle\int_{-1}^{1} \int_{x}^{x^2} \int_{z+x}^{2} (x+y) \, dy \, dz \, dx$

110. $\displaystyle\int_{0}^{2} \int_{0}^{1} \int_{x}^{1} z \sin y^2 \, dy \, dx \, dz$

111. $\displaystyle\int_{0}^{2} \int_{0}^{\sqrt{4-x^2}} \int_{0}^{4-\sqrt{x^2+y^2}} (x+y)^2 \, dz \, dy \, dx$

112. $\displaystyle\int_{0}^{2} \int_{0}^{\sqrt{4-x^2}} \int_{-\sqrt{4-x^2-y^2}}^{\sqrt{4-x^2-y^2}} x^2 z^2 \, dz \, dy \, dx$

113. $\displaystyle\int_{C} (x^2 y - z) \, ds$ where C is the straight line from $(0,1,2)$ to $(-1,2,-3)$

114. $\displaystyle\int_{C} (x^2+y^2) \, dx + (y^2+z^2) \, dy + (z^2+x^2) \, dz$ where C is the curve $y = 2x^2$, $z+y = 0$ from $(1,2,-2)$ to $(-2,8,-8)$

115. $\displaystyle\int_{C} \frac{2xy}{z} \, dx + \frac{x^2}{z} \, dy - \frac{x^2 y}{z^2} \, dz$ where C is the shorter part of the curve $x^2+y^2 = 8$, $z = y$ from $(2,2,2)$ to $(-1,\sqrt{7},\sqrt{7})$

116. $\displaystyle\oint_{C} x^2 y \, dx - y^3 x \, dy$ where C is the boundary of the region defined by the lines $x = 1$, $y = 1$, $x + 2y = 5$

117. $\displaystyle\iint_{S} (x+y)^2 \, dS$ where S is that part of $z = x^2+y^2$ inside $x^2+y^2 = 4$

118. $\displaystyle\oiint_{S} (x^2+y^2+z^2) \, dS$ where S is the sphere $x^2+y^2+z^2 = a^2$

119. $\displaystyle\iint_{S} (x\hat{\mathbf{i}} + y\hat{\mathbf{j}} + z\hat{\mathbf{k}}) \cdot \hat{\mathbf{n}} \, dS$ where S is that part of $x+2y+3z = 6$ in the first octant and $\hat{\mathbf{n}}$ is the unit normal to S with positive y-component

120. $\displaystyle\oiint_{S} (x\hat{\mathbf{i}} + y^2\hat{\mathbf{j}} - z^3\hat{\mathbf{k}}) \cdot \hat{\mathbf{n}} \, dS$ where S is the surface bounding the volume $0 \le x \le 1$, $0 \le y \le 2$, $0 \le z \le 3$ and $\hat{\mathbf{n}}$ is the unit outer normal

121. $\displaystyle\oint_{C} xz \, dx - yz \, dy + z \, dz$ where C is the curve $x^2+y^2 = 4$, $y = z$ directed so that x decreases in the first octant

122. $\displaystyle\int_{0}^{\sqrt{3}} \int_{-\sqrt{4-y^2}}^{-y/\sqrt{3}} (x^2 - y^2) \, dx \, dy$

123. $\displaystyle\iiint_{V} xye^z \, dV$ where V is bounded by $z = x^2+y^2$, $z = 0$, $y = 0$, $x = 1$, $y = x$

124. $\displaystyle\iiint_{V} \sqrt{x^2+y^2} \, dV$ where V is the volume bounded by $z = \sqrt{x^2+y^2}$, $z = 2 - x^2 - y^2$

125. $\displaystyle\iiint_{V} (x^2+y^2+z^2)^2 \, dV$ where V is the volume between the spheres $x^2+y^2+z^2 = a^2$ and $x^2+y^2+z^2 = b^2$, and $a > b$

126. $\displaystyle\int_{C} (x-y) \, ds$ where C is the curve $z = x^2+y^2$, $y + x = 1$ from $(1,0,1)$ to $(3,-2,13)$

127. $\displaystyle\oint_{C} y \, dx + x^2 \, dy$ where C is the ellipse $x^2+4y^2 = 4$

128. $\displaystyle\int_{C} e^y \sin 2y \, dx + xe^y(\sin 2y + 2\cos 2y) \, dy$ where C is the curve $x = t^3 + 3t$, $y = 1+t^2$ from $t = -1$ to $t = 0$

129. $\displaystyle\iint_{S} (x^2 z^3 + y) \, dS$ where S is that part of

$3x + 6y + 2z = 6$ cut out by $x = 0$, $z = 0$, $2x + z = 2$

130. $\oint_C e^{xy}(1 + xy)\,dx + (x^2 e^{xy} + xy^2)\,dy$ around the ellipse $x^2 + 4y^2 = 4$

131. $\iint_S (x^2\hat{\mathbf{i}} + y^2\hat{\mathbf{j}} + z^2\hat{\mathbf{k}}) \cdot \hat{\mathbf{n}}\,dS$ where S is that part of $z = 4 - x^2 - y^2$ above the xy-plane, and $\hat{\mathbf{n}}$ is the unit upper normal to S

132. $\oiint_S (x^3\hat{\mathbf{i}} + y^3\hat{\mathbf{j}} + z^3\hat{\mathbf{k}}) \cdot \hat{\mathbf{n}}\,dS$ where S is the surface enclosing the volume $x^2 + y^2 \le 1$, $0 \le z \le 1$ and $\hat{\mathbf{n}}$ is the unit outer normal to S

133. $\oint_C yz^2\,dx + x^2z\,dy + xy^2\,dz$ where C is the smooth curve of intersection of the surfaces $x^2 + z^2 = a^2$, $y^2 + z^2 = a^2$ ($a > 0$) that has a portion in the first octant (directed so that z increases in the first octant)

134. $\iint_R \dfrac{y}{x}\,dA$ where R is the region inside $x^2 + y^2 = 2x$ and above the x-axis

135. $\iiint_V x^2yz\,dV$ where V is the volume in the first octant bounded by $z = 2x + 3y$, $z = 0$, $y = 0$, $x = 1$, $y = x^2 + 1$, $x = 0$

136. $\iiint_V x^2y^2z\,dV$ where V is the volume bounded by $x^2 + y^2 = z^2 + 1$, $z = 1$, $z = 0$

137. $\iiint_V (x^2 + y^2)\,dV$ where V is the volume inside the sphere $x^2 + y^2 + z^2 = 1$ and outside the cones $z^2 = x^2 + y^2$

138. $\int_C (x^2 + y^2)^{3/2}\,ds$ where C is the curve $x = y$, $z = x^2$ from $(-1, -1, 1)$ to $(1, 1, 1)$

139. $\oint_C x\,dy$ once around the polar coordinate curve $r = 4 + 3\sin\theta$

140. $\iint_S y^2(\hat{\mathbf{i}} + 2\hat{\mathbf{j}} + 3\hat{\mathbf{k}}) \cdot \hat{\mathbf{n}}\,dS$ where S is the surface $x^2 + y^2 = 1$, $0 \le z \le 1$ and $\hat{\mathbf{n}}$ is the unit normal pointing away from the z-axis

141. $\oiint_S (xz^2\hat{\mathbf{i}} + x^2y\hat{\mathbf{j}} + z\hat{\mathbf{k}}) \cdot \hat{\mathbf{n}}\,dS$ where S is the surface enclosing the volume bounded by $x^2 + z^2 = 4 + y^2$, $x^2 + z^2 = 3$ and $\hat{\mathbf{n}}$ is the unit outer normal to S

142. $\oint_C (x^2 + 3)\,dx + 2xyz\,dy - x^2y^2\,dz$ where C is the curve $x^2 + 4z^2 = 4$, $y = x^2$ directed clockwise as viewed from the negative y-axis

143. $\int_C \left(\dfrac{y}{x^2 + y^2} + z^2\right)dx - \dfrac{x}{x^2 + y^2}\,dy + 2xz\,dz$ where C is the upper half of the curve $y = 2x^2 - 1$, $x^2 + z^2 = 1$ from $(-1, 1, 0)$ to $(1, 1, 0)$

144. $\oint_C \sin y\,dx - \cos x\,dy$ where C is the curve $|x| + |y| = 1$

145. $\iint_S (y\hat{\mathbf{i}} + x\hat{\mathbf{j}} - \hat{\mathbf{k}}) \cdot \hat{\mathbf{n}}\,dS$ where S is the first octant part of $x^2 + y^2 + z^2 = 4$ above $z = y$ and $\hat{\mathbf{n}}$ is the unit lower normal to S

146. $\iint_S (x^3\hat{\mathbf{i}} + 3xz\hat{\mathbf{j}} - e^{x+y}\hat{\mathbf{k}}) \cdot \hat{\mathbf{n}}\,dS$ where S is the surface enclosing the volume bounded by $z^2 + 9y^2 = 16$, $x = 0$, $x = 4$ and $\hat{\mathbf{n}}$ is the unit inner normal to S

147. $\int_C \sin(x + y + z)\,dx$ where C is the boundary of that part of $2x + 3y + 6z = 6$ in the first octant directed counterclockwise as viewed from the origin

148. At what points is the function $f(x, y) = 1/(\sin^2\pi x + \sin^2\pi y)$ discontinuous?

149. Show that the function $f(x, y) = \text{Tan}^{-1}\left(\dfrac{x + y}{x - y}\right)$ satisfies $\dfrac{\partial z}{\partial x} + \dfrac{\partial z}{\partial y} = \dfrac{x - y}{x^2 + y^2}$.

150. Show that $f(x, y) = \ln(e^x + e^y)$ satisfies $\dfrac{\partial^2 f}{\partial x^2}\dfrac{\partial^2 f}{\partial y^2} - \left(\dfrac{\partial^2 f}{\partial x\partial y}\right)^2 = 0$.

151. Show that $f(x, y) = \sin x + F(\sin y - \sin x)$, where F is a differentiable function, satisfies $\dfrac{\partial f}{\partial y}\cos x + \dfrac{\partial f}{\partial x}\cos y = \cos x \cos y$.

152. In which direction(s) from the point $(2, 3)$ does the function $f(x, y) = x^2y - xy$ have a rate of change equal to 4?

153. Show that the tangent plane to the surface $z = xf(y/x)$, where f is a differentiable function, always passes through the origin.

154. Find the point in space that minimizes the sum of the squares of the distances from the coordinate planes and the plane $Ax + By + Cz + D = 0$.

155. Of all right-angled triangles with perimeter P find the one with largest area.

156. Show that for small $|x|$ and $|y|$,
$$\frac{\cos x}{\cos y} \approx 1 - \frac{1}{2}(x^2 - y^2).$$

157. Use double integrals to find the volumes of the solids of revolution formed when the region bounded by the curves $y = x^2 + 1$, $y = 2x + 1$ is rotated about the lines **(a)** $x = 0$ **(b)** $y = 0$ **(c)** $x = 4$ **(d)** $y = -2$ **(e)** $x + y = 1$

158. Repeat Exercise 157 for the region bounded by the curves $x = y(y+1)$, $x = -2y(y+1)$.

159. Find the centroid of the region bounded by the curves $x = 0$, $y = 0$, $y = 1$, $x = \sqrt{1 + y^2}$.

160. A flat vertical plate is immersed vertically in oil with density 900 kg/m³. It is a parallelogram with sides of lengths 2 m and 1 m with the longer sides parallel to the surface of the oil. If one of the longer sides is $1/2$ m below the surface and the parallel side is 1 m below the surface, use a double integral to find the force on each side of the plate due to the surrounding oil.

161. What is the force on the plate in Exercise 160 if the longer diagonal is vertical, and the uppermost vertex is in the surface of the oil?

162. Find the centre of mass of a uniform plate whose edges are defined by the curves $y = \sqrt{4 - x^2}$, $4y = 4 - x^2$.

In Exercises 163–166 set up, but do not evaluate, double iterated integrals for the area described.

163. the area of $x^3 + y^3 + z^3 = 1$ in the first octant

164. the area of $y = x + z^2$ inside the cylinder $x^2 + z^2 = 4$

165. the area of $x^2 + y^2 - z^2 = 9$ between $z = 1$ and $z = 2$

166. the area of $z = 1 - x^4 - y^4$ above the xy-plane

167. Find the volume of the solid of revolution when the area inside $r^2 = 4|\cos\theta|$ is rotated about the x-axis.

168. If A, B, and C are positive, and $D < 0$, find the volume of the tetrahedron formed by the plane $Ax + By + Cz + D = 0$ and the coordinate planes.

169. Show that the tangent plane to the surface $xyz = a$, where $a > 0$, cuts a constant volume from the first octant.

170. A solid object of uniform density ρ is bounded by the surfaces $z = 4x^2$, $z = 4$, $y = -1$, $y = 1$. Find its centre of mass and its moments of inertia about the coordinate axes.

171. Find the volume bounded by $z = x^2 + y^2 - \sqrt{x^2 + y^2}$ and the xy-plane.

172. Find the volume of the region inside the sphere $x^2 + y^2 + z^2 = a^2$ and outside the cones $z^2 = b^2(x^2 + y^2)$.

173. Given that $f(x,y,z) = xyz$ and $\mathbf{F}(x,y,z) = x^2\hat{\mathbf{i}} - yz\hat{\mathbf{j}} + xz\hat{\mathbf{k}}$, find **(a)** ∇f **(b)** $\nabla \cdot \mathbf{F}$ **(c)** $\nabla \times \mathbf{F}$ **(d)** $\mathbf{F} \times (\nabla \cdot \mathbf{F})\nabla f$ **(e)** $\nabla f \cdot \nabla \times (x^2 \mathbf{F})$ **(f)** $\nabla(f^2) \times (\nabla \times \nabla f)$ **(g)** $\nabla \cdot \mathbf{F}(\nabla f \times y\mathbf{F})$

174. Show that the gradient of the function $\text{Tan}^{-1}(y/x)$ is irrotational and solenoidal in any domain that does not contain points on the y-axis.

175. Show that vector fields of the form $\mathbf{F} = f(\sqrt{x^2 + y^2 + z^2})(x\hat{\mathbf{i}} + y\hat{\mathbf{j}} + z\hat{\mathbf{k}})$ are irrotational.

176. Show that the function $y = \frac{1}{3} \int_0^x f(t) e^{t-x} \sin 3(x-t) \, dt$ satisfies the differential equation
$$\frac{d^2 y}{dx^2} + 2\frac{dy}{dx} + 10y = f(x).$$

177. In Kintsch's model of choice behaviour, he states that
$$\sum_{n=2}^{\infty} (1-b)^{n-2} bs^n = \frac{bs^2}{1-(1-b)s}.$$

Is this true, and would you place any restrictions on b and s?

178. Evaluate $\lim_{n\to\infty} \frac{1}{n^2}(1 + 2 + 3 + \cdots + n)$.

179. Find the Maclaurin series for $f(x) = 1/(x^4 - x^2 - 2)$.

180. If $f(x)$ is a polynomial of degree 4, and it is known that $f(1) = 1$, $f'(1) = -2$, $f''(1) = 0$, $f'''(1) = 3$, and $f''''(1) = -5$, what is $f(-2)$?

181. Can the function $f(x,y) = xy/(x^2 + y^2)$, $(x,y) \neq (0,0)$ be made continuous at $(0,0)$?

182. Find all points at which the sum of the first partial derivatives of the function $f(x,y,z) = x^3 + y^3 - z^3$ is equal to **(a)** 3 **(b)** -3 **(c)** 0

183. Verify that $f(x,y) = e^x(x\cos y - y\sin y)$ is harmonic.

184. Suppose that $f(x,y) = xF(ax + by) + yG(ax + by)$ where a and b are constants and F and G are arbitrary twice differentiable functions. Show that
$$\frac{\partial^2 f}{\partial x^2} - 2\frac{\partial^2 f}{\partial x \partial y} + \frac{\partial^2 f}{\partial y^2} = 0 \text{ if and only if } a = b.$$

185. In which direction(s) from the point $(1, 2, 3)$ does the function $f(x,y,z) = xyz + x^2 + y^2 + z^2$ have a rate of change equal to -2?

186. Find all tangent planes to the ellipsoid $x^2/a^2 + y^2/b^2 + z^2/c^2 = 1$ which have equal intercepts on the coordinate axes.

187. Approximate $\int_0^{1/2} \frac{1}{1+x^4}\,dx$ using the first four terms of the Maclaurin series of the integrand.

188. Find the Maclaurin series for $f(x) = \ln[(1+x)^{1+x}] + \ln[(1-x)^{1-x}]$. What is the interval of convergence of the series?

189. What are the first three terms in the Maclaurin series for the function defined by the equation $y + \ln(1+xy) + x = 5$?

190. Solve the equation $\dfrac{dy}{dx} = y^2 - 4$ subject to
    (a) $y(0) = 1$ (b) $y(0) = -2$

191. Radioactive substances decay at a rate proportional to the amount of radioactive material present at any time. If 10% of a radioactive sample decays in 1 year, how long does it take for 50% to decay (called the half-life of the material)?

192. Find a continuous function satisfying $\dfrac{dy}{dx} + y = f(x)$,
    $y(0) = 1$, where $f(x) = \begin{cases} 3 & 0 \le x \le 2 \\ 0 & x > 2 \end{cases}$.

193. Verify that the length of the curve $4y = (z+x)^2$, $4y^2 + 3z^2 = 3x^2$ from $(0,0,0)$ to any point (x,y,z) where $x > 0$ is $\sqrt{2}\,x$. Hint: Regard x as the parameter and use implicit differentiation to find dy/dx and dz/dx.

194. Find all points at which the gradient of the function $f(x,y,z) = x^2 y^2 z^2 + xyz$ is equal to $2\hat{\imath} + 3\hat{\jmath} - \hat{k}$.

195. Evaluate $\lim\limits_{x \to 0} \dfrac{6x - x^3 - 6\sin x}{x^5}$.

196. Find constants b_n in order that the function $y = f(x) = \sum_{n=1}^{\infty} b_n \sin nx$ will satisfy the differential equation
    $$\frac{d^2 y}{dx^2} - 3y = \sum_{n=1}^{\infty} \frac{1}{n^2} \sin nx.$$
    Assume that the series for $f(x)$ can be differentiated term-by-term.

197. Show that a function $f(x)$ and the nth degree polynomial
    $$P_n(x) = f(c) + f'(c)(x-c) + \cdots + \frac{f^{(n)}(c)}{n!}(x-c)^n$$
    have the same first n derivatives at $x = c$.

198. If $f(x,y) = [F(ax-by) + G(ax+by)]/x$, where a and b are constants and F and G are twice differentiable functions, show that $b^2 \dfrac{\partial}{\partial x}\left(x^2 \dfrac{\partial f}{\partial x}\right) = a^2 x^2 \dfrac{\partial^2 f}{\partial y^2}$.

199. If y is defined as a function of x by $y = f(x,t)$, $F(x,y,t) = 0$, show that $\dfrac{dy}{dx} = \dfrac{f_x F_t - f_t F_x}{f_t F_y + F_t}$.

200. Find point(s) on the curve $x^2 + y^2 + z^2 = 1$, $y = z$ at which the rate of change of the function $f(x,y,z) = 3x + 2y - 2z$ with respect to length along the curve is equal to 1.

201. Find the second moment of area of the region bounded by the curves $y = 1 - x^2$, $x + y + 1 = 0$ about the lines (a) $y = 0$ (b) $x = 0$ (c) $y = 1$

202. A plate in the form of the region bounded by the curves $y = x^2$ and $y = 1$ (x and y in metres) is to be lowered into water until the force on each side is 10^5 N. Find the position of the horizontal side relative to the surface.

203. Repeat Exercise 202 if the force must be 1000 N.

204. Find the area of $z/3 = 1 - |x| - |y|$ above the xy-plane.

205. Find the area of that part of $y = 1 + 2\sqrt{x^2 + z^2}$ to the left of the plane $y = 2$.

206. Find the area inside each petal of the rose $r = a \sin n\theta$.

207. Find the volume bounded by the surfaces $3y + z = 6$, $3y + 2z = 12$, $x + y + z = 6$, $x = 0$, $y = 0$, $z = 0$.

208. Find the centre of mass of the uniform solid bounded by $4x^2 + y^2 = 4$, $z = 0$, $y + z = 2$.

209. Calculate $\dfrac{d^{20}}{dx^{20}}[(x^2+1)\sin x]$.

210. Prove that the function
    $$f(x) = \frac{x^2}{2} + \frac{1}{2} x\sqrt{x^2+1} + \ln\sqrt{x + \sqrt{x^2+1}}$$
    satisfies the differential equation $x\dfrac{dy}{dx} + \ln\left(\dfrac{dy}{dx}\right) = 2y$.

211. A man eats a diet of 2 500 calories per day; 1 200 of them go to basic metabolism (i.e. get used up automatically). He also spends 16 calories per day for each kilogram of body mass in exercise. Assume that the storage of calories as fat is 100% efficient and that 1 kilogram of fat contains 10 000 calories. Find his mass as a function of time, and sketch this function if initially his mass is (a) 100 kg (b) 70 kg. Is he on a reducing diet?

212. A model of learning proposed by Bush and Mosteller assumes that on the nth trial of a sequence of trials, the probability p_n of obtaining a certain response is related to the probability on the immediately preceding trial by $p_n = a + (1-a-b)p_{n-1}$ where a and b are constants that depend on the amounts of reward and inhibition. Prove that an explicit formula for p_n is
    $$p_n = p_0(1-a-b)^n + \frac{a}{a+b}[1-(1-a-b)^n]$$

where p_0 is the initial probability of response.

213. Evaluate $\displaystyle\lim_{n\to\infty} \dfrac{\frac{1}{4} + \frac{1}{8} + \frac{1}{16} + \cdots + \frac{1}{2^n}}{1 + \frac{2}{5} + \frac{4}{25} + \frac{8}{125} + \cdots + \left(\frac{2}{5}\right)^{n-1}}$.

214. Prove that if the amount of radioactive material in a sample is A_1 at time t_1 and the amount is A_2 at time t_2, where $t_2 > t_1$, then the half-life of the material is $(t_2 - t_1)\ln 2 / \ln(A_1/A_2)$. (See Exercise 191 for the rate of decay of radioactive materials.)

215. A 100 g mass falls from rest under gravity. It is also acted on by air resistance which is 2 N when the speed of the mass is 10 m/s. Find the velocity of the mass when the resistive force is proportional to **(a)** velocity **(b)** the square of velocity **(c)** the square root of velocity. Also find distance travelled in (a).

216. (a) Solve differential equation 16.39 if the term $1 - N/C$ is replaced by $1 - \ln N / \ln C$. Sketch a graph of $N(t)$.

(b) Does the solution in (a) or the logistic function 16.40 approach C more rapidly?

217. In Figure E.1 the four squares each have area one. A rectangle with one vertex at the origin, and two sides along the x- and y-axes has sides of length x and y. Find a formula for the amount of shaded area enclosed by this rectangle as a function of x and y.

FIGURE E.1

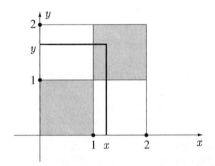

218. A function $f(x,y)$ satisfies Laplace's equation 13.11. It is also known to be radially symmetric; that is, when expressed in polar coordinates r and θ, it is independent of θ. Find the form of $f(x,y)$. (Hint: See Exercise 36 in Section 13.6.)

219. Show that if $\sum_{n=1}^{\infty} a_n^2$ and $\sum_{n=1}^{\infty} b_n^2$ are convergent series, then $\sum_{n=1}^{\infty} a_n b_n$ is absolutely convergent.

220. Find $f^{(6)}(0)$ if $f(x) = x^2/\sqrt{1 - x^2}$.

221. Find a formula for d^2y/dx^2 in terms of derivatives of $f(x,y)$ when $f(x,y) = 0$ defines y as a function of x.

222. Suppose the line $Ax + By + C = 0$ has nonzero x- and y-intercepts. Find the point in the xy-plane that minimizes the sum of the distances from the x-axis, the y-axis, and the line.

223. The area of a triangle with sides of length a, b, and c is $A = \sqrt{s(s-a)(s-b)(s-c)}$ where s is half the perimeter of the triangle. If the length of each side is increased by 1%, what is the approximate percentage increase in its area?

224. Of all right-angled triangles with area A, find the one with smallest perimeter.

225. Beer containing 6% alcohol is pumped into a vat containing 10 000 L of beer with 4% alcohol at 100 L per minute. If well-mixed beer is removed at the same rate, what is the percentage of alcohol in the vat as a function of time t. When does the beer contain 5% alcohol?

226. (a) Show that $y_1(x) = (\sin x)/\sqrt{x}$ is a solution of the differential equation

$$x^2\dfrac{d^2y}{dx^2} + x\dfrac{dy}{dx} + \left(x^2 - \dfrac{1}{4}\right)y = 0.$$

(b) Show that a second solution $y_2(x) = (\cos x)/\sqrt{x}$ can be obtained by setting $y_2(x) = v(x)y_1(x)$.

(c) What is the general solution of the differential equation?

227. Find the first three nonzero terms in the Maclaurin series for $\ln(\cos x)$.

228. Find the Maclaurin series for $\cos^4 x$.

229. Find the Maclaurin series for $x(2 + 3x)^{3/2}$.

230. A pyramid with rectangular horizontal base is inscribed in a sphere of radius r (Figure E.2). Find a formula for the volume in the pyramid in terms of x and y, where (x, y, z) are the coordinates of one vertex.

FIGURE E.2

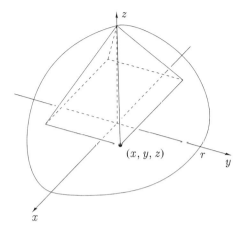

231. Of all planes passing through the point (a, b, c) in the first octant, find the one for which the tetrahedron formed by the plane and the three coordinate planes has the least possible volume.

232. Find the first moment of area of the region bounded by the curves $y = x^2$, $y = 2 - x^2$ about the line $x + y + 1 = 0$.

233. Find the area of $x^2 + y^2 + z^2 = 4$ between $z = \sqrt{x^2 + y^2}$ and $z = 2\sqrt{x^2 + y^2}$.

234. Find the first moment of area of the bifolium $r = \cos^2 \theta \sin \theta$ about the x-axis.

235. In Iraq an epidemic of methyl-mercury poisoning killed 459 people in 1972. Human beings became exposed to the poison when they ate homemade bread accidentally prepared from seed-wheat treated with methyl-mercurial fungicide. Symptoms appeared only after weeks of exposure. Assume for simplicity that a person takes a constant daily dose d of the poison and that a certain percentage p of the accumulated poison is excreted each day. Find a formula which relates the amount of poison stored in the body to the number of days.

236. A 100 g mass is attached to a vertical spring with constant 1000 N/m, and slowly lowered to its equilibrium position. If it is given an initial speed of 4 m/s downward, find its position as a function of time if a damping force proportional to velocity acts on the mass and has the following magnitude when speed is 2 m/s. **(a)** 20 N **(b)** 40 N **(c)** 80 N

237. Repeat Exercise 236 if a force $\sin 20t$, $t \geq 0$ also acts on the mass.

238. Find $\sqrt{4.000\,000\,000\,01}$ accurate to 20 decimal places.

239. **(a)** Show that if $y = (1-x)^{-a} e^{-ax}$, then
$$(1-x)\frac{dy}{dx} = axy.$$
(b) Now show that
$$(1-x)\frac{d^{n+1}y}{dx^{n+1}} - (n+ax)\frac{d^n y}{dx^n} - an\frac{d^{n-1}y}{dx^{n-1}} = 0.$$

240. Dropping leaflets by air is a means sometimes used in advertising to reach a large group of customers. Suppose it is assumed that a constant proportion λ of the leaflets actually survive a given time period T, while the remainder are lost, destroyed, or otherwise rendered unreadable. It is further assumed that each leaflet that survives until the ith time period will reach β people, on the average, during that period. Given that N leaflets are dropped, how many people are reached by the leaflets?

241. Find the Taylor series about $x = 1$ for the function $\sqrt[3]{7+x}$.

242. Use the Maclaurin series for the function $f(x) = (1+x)e^{-x} - (1-x)e^x$ to find the sum of the series $\sum_{n=1}^{\infty} n/(2n+1)!$.

243. Find the volumes bounded by the following surfaces:
(a) $4x^2 + y^2 = 4$, $4x + 3y + z = 12$, $z = 1$
(b) $4x^2 + y^2 = 4$, $4x + 3y + z = 12$, $z = 24$
(c) inside $4x^2 + y^2 = 4$, above $4x + 3y + z = 12$, under $z = 12$

244. Show that when a surface $z = f(r, \theta)$ in cylindrical coordinates projects one-to-one onto an area S_{xy} in the xy-plane, its area is given by
$$\iint_{S_{xy}} \sqrt{1 + \left(\frac{\partial z}{\partial r}\right)^2 + \frac{1}{r^2}\left(\frac{\partial z}{\partial \theta}\right)^2}\, r\, dr\, d\theta.$$

245. Use the result of Exercise 244 to find the area of the sphere $x^2 + y^2 + z^2 = 4$ inside the cylinder $x^2 + y^2 = 1$.

246. When chemicals A, B, and C are brought together, 1 g of A reacts with 2 g of B and 3 g of C to form 6 g of substance D. Initially there are 10 g of A, 20 g of B, and 30 g of C. Find the amount of D as a function of time under each of the following conditions:

(a) the rate at which D is formed is proportional to the product of the amounts of A, B, and C present in the reaction

(b) the rate at which D is formed is proportional to the product of the amounts of B and C present in the reaction.

247. **(a)** Find equations defining solutions to Exercise 246 if the initial amounts of A, B, and C are all 10 g.

(b) When is the amount of D equal to 15 g if there is 10 g after 5 min?

248. Inscribed in a circle of radius R is a square. A circle is then inscribed in the square, a square in the circle, and so on and so on. Find the limit of the sum of the areas of all the circles and the limit of the sum of the areas of all the squares.

249. How many terms in the Maclaurin series for the integrand guarantee a value for $\int_1^2 \frac{\sin 2x}{x} dx$ accurate to 10^{-3}?

250. Find a polynomial approximating the indefinite integral $\int \frac{1}{\sqrt[4]{1+x^4}} dx$ to within 10^{-6} on the interval $0 \leq x \leq 1$.

251. What is the minimum value of $f(x, y) = x^3 + y^3$ on that part of the line $Ax + By = C$ (where $A \geq 0$, $B \geq 0$, and $C \geq 0$) for which $x \geq 0$ and $y \geq 0$.

252. Find the moment of inertia about the line $y = 2x$ of a plate with uniform mass per unit area ρ if its edges are defined by $y = -x$, $y = x$, $x = 1$.

253. If $f(x, y)$ and its first partial derivatives are continuous in a simpy-connected domain containing a piecewise smooth, closed curve C that does not intersect itself, and $\partial f/\partial y = \partial f/\partial x$ inside C, what is the value of

$$\oint_C f(x, y)(dx + dy)?$$

254. Show that if $f(x, y)$ is harmonic in a simply-connected domain D and C is a closed, piecewise smooth curve in D, then

$$\oint_C \frac{\partial f}{\partial y} dx - \frac{\partial f}{\partial x} dy = 0.$$

255. If $r = \sqrt{x^2 + y^2 + z^2}$, and $f(r)$ is a continuous function, show that

$$\oiint_S f(r)\, dS = 4\pi a^2 f(a)$$

when S is the sphere $x^2 + y^2 + z^2 = a^2$.

256. Find a polynomial in x approximating $f(x) = \cos^3 x$ on $0 \leq x \leq 0.5$ with maximum error 10^{-6}.

257. Show that if $y_1(x)$ is a solution of the homogeneous, linear differential equation $y'' + P(x)y' + Q(x)y = 0$, then a second solution is

$$y_2(x) = y_1(x) \int \frac{e^{\int P(x)\, dx}}{[y_1(x)]^2} dx.$$

Hint: Let $y_2(x) = v(x) y_1(x)$.

258. (a) Use the transformation of Exercise 22 in Section 16.10 to solve $x^2 y'' - 4xy' + 4y = 0$.

 (b) Solve the differential equation by setting $y = x^m$ and finding appropriate values for m.

259. Use the techniques in Exercise 258 to solve $x^2 y'' + 3xy' + 4y = 0$.

260. Use the methods of Exercise 258(b) and Exercise 257 to solve $4x^2 y'' + 8xy' + y = 0$.

261. Assuming that the ellipsoid $x^2/a^2 + y^2/b^2 + z^2/c^2 = 1$ and the plane $Ax + By + Cz + D = 0$ do not intersect, find the points on the ellipsoid closest to and farthest from the plane.

262. Find the points on the ellipsoid $x^2 + 96 y^2 + 96 z^2 = 96$ closest to and farthest from the plane $3x + 4y + 12 z = 288$.

ANSWERS

2. Converges
4. Converges conditionally
6. Diverges
8. Diverges
10. Converges absolutely
12. Converges
14. Converges
16. Diverges
18. Diverges
20. Converges
22. $x\sqrt{1-y^2} + y\sqrt{1-x^2} = C$
24. $y = (C - \cos x)/(1 + e^x)$ **26.** $y = \ln|\cos(C-x)| + D$
28. $y = C_1 \cos 2x + (C_2 + 3x/4)\sin 2x$
30. $y = C_1 e^{-x} + C_2 e^{(\sqrt{3}-2)x} + C_3 e^{-(\sqrt{3}+2)x} + (3\cos 3x - 11\sin 3x)/260$
32. $y = (e^x + Ce^{-x})/(2x^2)$
34. $y = C_1 e^{(\sqrt{34}-6)x} + C_2 e^{-(\sqrt{34}+6)x} - (1/34)e^{-6x}$
36. $y = C\,\mathrm{Tan}^{-1} x - x^2/6 + (1/6)\ln(x^2+1) + D$
38. $4(x\sin x + \cos x) + e^{-2y}(2y^2 + 2y + 1) = C$
40. $y = (C_1 + C_2 x + C_3 x^2)e^{-x} + (1/27)e^{2x}$
42. $y = (C_1 + C_2 x + x^4/12)e^{-3x}$
44. $y = (C_1 + 3x/16)\cos 2x + (C_2 + 3x^2/8)\sin 2x$
46. $y = C_1 x^4 + C_2/x$

48.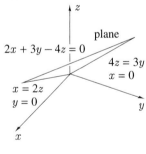

50.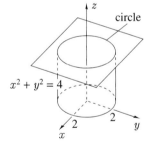

52.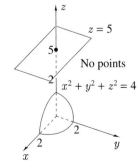

54.
Cubical cylinder

56. Elliptic hyperboloid

58. Two straight lines

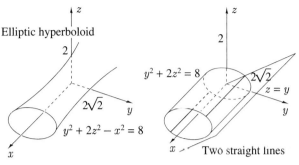

60. Curve

62. Elliptic paraboloid

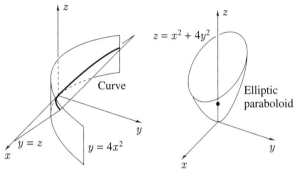

64. Two points

66. Curve

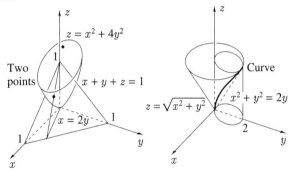

68. No
70. $13x + 16y + 2z = 39$
72. $11x + 3y + 7z = 23$
74. $x/2 = -(y+1)/3 = (z+2)/5$
76. 5
78. $-(1+4y)\hat{\mathbf{i}} + (4x+1)\hat{\mathbf{j}} + 2(y-x)\hat{\mathbf{k}}$
82. (a) Does not exist   (b) 16
84. $1/\sqrt{x^2 + y^2}$

86. $\sqrt{2}\,u|v|/[(u^2+v^2)\sqrt{u^2-v^2}]$

88. $[y(u^3\sin x+3y)+(1-3v^2y)(u^2\cos x-3)]/[2u\sin x(3v^2y-1)+y(3u^2\cos x+4y)]$

92. $3x+y+z=7$

96. $(0,0),(0,a/c),(a/b,0)$ give saddle points; $(a/(3b),a/(3c))$ gives a relative maximum if $abc>0$ and a relative minimum if $abc<0$

98. $((1/n)\sum_{i=1}^{n}x_i,(1/n)\sum_{i=1}^{n}y_i)$

102. $2\sin 2-\cos 2-1$ 104. $1/6$

106. $2(\sqrt{2}-1)$ 108. $(4-\sqrt{7})\pi/(16\sqrt{7})$

110. $1-\cos 1$ 112. $128\pi/105$

114. $618/5$ 116. $-28/5$

118. $4\pi a^4$ 120. -36

122. $\sqrt{3}$ 124. $13\pi/30$

126. $\sqrt{2}(51\sqrt{51}-3\sqrt{3})/12$

128. $4e^2\sin 4$ 130. $-\pi/2$

132. $5\pi/2$ 134. 1

136. $5\pi/64$ 138. $4(6\sqrt{3}+1)/15$

140. 0 142. 16π

144. $4(\cos 1-1)$ 146. $1024\pi/3$

148. (x,y) where x and y are integers

152. $(36\pm 2\sqrt{69},8\mp 9\sqrt{69})$

154. $x=-AD/[2(A^2+B^2+C^2)]$, $y=-BD/[2(A^2+B^2+C^2)]$, $z=-CD/[2(A^2+B^2+C^2)]$

158. (a) $2\pi/15$ (b) $\pi/2$ (c) $39\pi/10$ (d) $3\pi/2$
(e) $7\sqrt{2}\pi/10$

160. 6.62×10^3 N

162. $\bar{x}=0,\bar{y}=32/(15\pi-20)$

164. $4\sqrt{2}\int_0^2\int_0^{\sqrt{4-x^2}}\sqrt{1+2z^2}\,dz\,dx$

166. $4\int_0^1\int_0^{(1-x^4)^{1/4}}\sqrt{1+16x^6+16y^6}\,dy\,dx$

168. $-D^3/(6ABC)$

170. $\bar{x}=0,\bar{y}=0,\bar{z}=12/5$, $I_x=4832\rho/63$, $I_y=7904\rho/105$, $I_z=256\rho/45$

172. $(4/3)\pi a^3 b/\sqrt{1+b^2}$

178. $1/2$

180. $-187/8$

182. (a) $x^2+y^2=z^2+1$ (b) $x^2+y^2+1=z^2$
(c) $z^2=x^2+y^2$

186. $x+y+z=\pm\sqrt{a^2+b^2+c^2}$

188. $\sum_{n=1}^{\infty}x^{2n}/(2n^2-n)$, $-1\le x\le 1$

190. (a) $y=2(3-e^{4x})/(3+e^{4x})$ (b) $y=-2$

192. $y=3-2e^{-x}$ for $0\le x\le 2$, $(3e^2-2)e^{-x}$ for $x>2$

194. $(0.9848,0.6565,-1.9696)$

196. $-1/(n^4+3n^2)$

200. $(2\sqrt{2}/3,\pm\sqrt{2}/6,\pm\sqrt{2}/6)$, $(-2\sqrt{2}/3,\pm\sqrt{2}/6,\pm\sqrt{2}/6)$

202. 7.25 m below

204. $2\sqrt{19}$

206. $\pi a^2/(4n)$

208. $\bar{x}=0,\bar{y}=-1/2,\bar{z}=4/5$

216. (a)

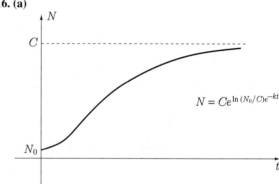

(b) logistic

218. $C\ln(x^2+y^2)+D$

220. 270

222. $(0,0)$

224. Sides of lengths $\sqrt{2A},\sqrt{2A},2\sqrt{A}$

226. (c) $C_1 y_1(x)+C_2 y_2(x)$

228. $1+\sum_{n=1}^{\infty}(-1)^n(2^{2n-1}+2^{4n-3})x^{2n}/(2n)!$

230. $(4/3)|xy|(r\pm\sqrt{r^2-x^2-y^2})$

232. $8\sqrt{2}/3$

234. $\pi/256$

236. (a) $(-2\sqrt{3}/75)e^{-50t}\sin 50\sqrt{3}\,t$ (b) $-4te^{-100t}$
(c) $(\sqrt{3}/150)[e^{-(200+100\sqrt{3})t}-e^{(-200+100\sqrt{3})t}]$

238. $2.000\,000\,000\,002\,500\,000\,00$

240. $\beta\lambda N/(1-\lambda)$

242. $1/(2e)$

246. (a) $60-60/\sqrt{7200kt+1}$ (b) $60-60/(60lt+1)$

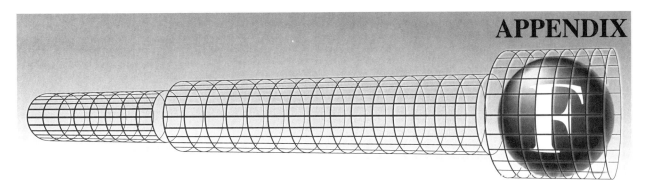

APPENDIX F

Determinants

A **determinant of order** n is n^2 numbers arranged in n rows and n columns and enclosed by two vertical lines. Thus,

$$\begin{vmatrix} 2 & 1 \\ 3 & 4 \end{vmatrix}, \quad \begin{vmatrix} -1 & 0 & 3 \\ 2 & 1 & 6 \\ 7 & -1 & 4 \end{vmatrix}, \quad \begin{vmatrix} 0 & 0 & 1 & 5 \\ 2 & -1 & 3 & 6 \\ -1 & -2 & -3 & 4 \\ 6 & 7 & 8 & 10 \end{vmatrix}$$

are determinants of orders 2, 3, and 4, respectively. The general determinant of order n is written in the form

$$D = \begin{vmatrix} a_{11} & a_{12} & a_{13} & \cdots & a_{1n} \\ a_{21} & a_{22} & a_{23} & \cdots & a_{2n} \\ \vdots & \vdots & \vdots & \ddots & \vdots \\ a_{n1} & a_{n2} & a_{n3} & \cdots & a_{nn} \end{vmatrix}. \quad \text{(F.1)}$$

The element in the ith row and j^{th} column is called the $(i,j)^{\text{th}}$ element and is denoted by a_{ij}. For brevity we write

$$D = |a_{ij}|_{n \times n} \quad \text{(F.2)}$$

to identify the general determinant of order n.

We wish to assign a value to every determinant, and to do this we first define what is meant by a minor and a cofactor of an element in a determinant.

> **Definition F.1**
> The **minor** M_{ij} of a_{ij} in $D = |a_{ij}|_{n \times n}$ is the determinant of order $n-1$ obtained by deleting the ith row and j^{th} column of D.

For example, if

$$D = \begin{vmatrix} -1 & 0 & 3 \\ 2 & 1 & 6 \\ 7 & -1 & 4 \end{vmatrix},$$

then

$$M_{11} = \begin{vmatrix} 1 & 6 \\ -1 & 4 \end{vmatrix}, \quad M_{23} = \begin{vmatrix} -1 & 0 \\ 7 & -1 \end{vmatrix}, \quad M_{31} = \begin{vmatrix} 0 & 3 \\ 1 & 6 \end{vmatrix}.$$

> **Definition F.2** The **cofactor** A_{ij} of a_{ij} in $D = |a_{ij}|_{n \times n}$ is $(-1)^{i+j} M_{ij}$.

If D is as in the previous paragraph, then

$$A_{11} = (-1)^{1+1} M_{11} = \begin{vmatrix} 1 & 6 \\ -1 & 4 \end{vmatrix}, \qquad A_{23} = (-1)^{2+3} M_{23} = -\begin{vmatrix} -1 & 0 \\ 7 & -1 \end{vmatrix}.$$

The following two rules now specify how to find the value of every determinant:
(1) The value of a determinant $D = |a_{11}|$ of order 1 is a_{11}.
(2) The value of a determinant $D = |a_{ij}|_{n \times n}$ of order n is obtained by choosing any line (row or column) and adding elements in that line, each multiplied by its cofactor.

If we select row i, then

$$D = a_{i1} A_{i1} + a_{i2} A_{i2} + \cdots + a_{in} A_{in} = \sum_{j=1}^{n} a_{ij} A_{ij}, \qquad (F.3\,a)$$

and if we select column j,

$$D = a_{1j} A_{1j} + a_{2j} A_{2j} + \cdots + a_{nj} A_{nj} = \sum_{i=1}^{n} a_{ij} A_{ij}. \qquad (F.3\,b)$$

Rule (2) defines a determinant of order n in terms of n determinants of order $n-1$ (the cofactors); each of these determinants is defined in terms of determinants of order $n-2$, and the process is continued until only determinants of order 1 are involved.

It is not clear that the *value of a determinant is independent of the line chosen in its evaluation*, but this is indeed the case. In other words, rules (1) and (2) define a unique value for every determinant.

The following result is essential to the speedy evaluation of determinants.

> **Theorem F.1** The value of a determinant of order 2 is
>
> $$D = \begin{vmatrix} a_{11} & a_{12} \\ a_{21} & a_{22} \end{vmatrix} = a_{11} a_{22} - a_{12} a_{21}. \qquad (F.4)$$

Proof If we expand D along its first row, we have

$$D = a_{11} A_{11} + a_{12} A_{12} = a_{11}(-1)^{1+1} M_{11} + a_{12}(-1)^{1+2} M_{12} = a_{11}(a_{22}) - a_{12}(a_{21}).$$

EXAMPLE F.1

Evaluate
$$\begin{vmatrix} 2 & 3 \\ -4 & 6 \end{vmatrix}.$$

SOLUTION By Theorem F.1,
$$\begin{vmatrix} 2 & 3 \\ -4 & 6 \end{vmatrix} = (2)(6) - (3)(-4) = 24.$$
∎

EXAMPLE F.2

Evaluate
$$D = \begin{vmatrix} 3 & -2 & 6 \\ 1 & 3 & 4 \\ 2 & -1 & 2 \end{vmatrix}.$$

SOLUTION If we expand along the first column, we have
$$D = 3(-1)^2 \begin{vmatrix} 3 & 4 \\ -1 & 2 \end{vmatrix} + 1(-1)^3 \begin{vmatrix} -2 & 6 \\ -1 & 2 \end{vmatrix} + 2(-1)^4 \begin{vmatrix} -2 & 6 \\ 3 & 4 \end{vmatrix}$$
$$= 3(6+4) - (-4+6) + 2(-8-18) = -24.$$
∎

Using equations F.3 to evaluate determinants is not always particularly easy. For instance, even for a determinant of order 5, it is necessary to evaluate 60 determinants of order 2. As a result, we now prove two theorems that are used to simplify determinants.

Theorem F.2

If any line of a determinant with value D has its elements multiplied by c, the new determinant has value cD.

Proof If the ith row of $D = |a_{ij}|_{n \times n}$ is multiplied by c, and the resulting determinant is expanded along this row, its value is
$$\sum_{j=1}^{n} c a_{ij} A_{ij} = c \sum_{j=1}^{n} a_{ij} A_{ij} = cD.$$

Theorem F.3

If a multiple of one line of a determinant with value D is added to a parallel line, the resulting determinant also has value D.

Proof Suppose c times the ith row is added to the kth row of $D = |a_{ij}|_{n \times n}$ to form

$$E = \begin{vmatrix} a_{11} & a_{12} & \cdots & a_{1n} \\ a_{21} & a_{22} & \cdots & a_{2n} \\ \vdots & \vdots & \ddots & \vdots \\ a_{i1} & a_{i2} & \cdots & a_{in} \\ \vdots & \vdots & \ddots & \vdots \\ a_{k1} + ca_{i1} & a_{k2} + ca_{i2} & \cdots & a_{kn} + ca_{in} \\ \vdots & \vdots & \ddots & \vdots \\ a_{n1} & a_{n2} & \cdots & a_{nn} \end{vmatrix}.$$

If we expand E along the kth row,

$$E = \sum_{j=1}^{n} (a_{kj} + ca_{ij}) A_{kj} = \sum_{j=1}^{n} a_{kj} A_{kj} + c \sum_{j=1}^{n} a_{ij} A_{kj}.$$

Now $\sum_{j=1}^{n} a_{kj} A_{kj} = D$, and $\sum_{j=1}^{n} a_{ij} A_{kj}$ is the value of the following determinant, which has identical i^{th} and k^{th} rows:

$$\begin{vmatrix} a_{11} & a_{12} & \cdots & a_{1n} \\ \vdots & \vdots & \ddots & \vdots \\ a_{i1} & a_{i2} & \cdots & a_{in} \\ \vdots & \vdots & \ddots & \vdots \\ a_{i1} & a_{i2} & \cdots & a_{in} \\ \vdots & \vdots & \ddots & \vdots \\ a_{n1} & a_{n2} & \cdots & a_{nn} \end{vmatrix} \begin{matrix} \\ \\ \leftarrow i^{\text{th}} \\ \\ \leftarrow k^{\text{th}} \\ \\ \end{matrix}.$$

In Exercise F.16 we show that a determinant with two identical parallel lines has value zero, and therefore $E = D$, which completes the proof.

This theorem is the key to evaluation of determinants. We use it to replace a determinant with an equivalent determinant that has a large number of zero elements. This makes evaluation by cofactors much simpler.

EXAMPLE F.3

Evaluate the determinant

$$D = \begin{vmatrix} 1 & 2 & -3 & 4 \\ 2 & 4 & 0 & -1 \\ 3 & 6 & 1 & 2 \\ 4 & 0 & 1 & 5 \end{vmatrix}.$$

SOLUTION Instead of immediately expanding the determinant according to equation F.3, we use Theorem F.3 to create zeros in column 3. We do this by first adding 3 times row 3 to row 1, and second, adding -1 times row 3 to row 4. The result is

$$D = \begin{vmatrix} 10 & 20 & 0 & 10 \\ 2 & 4 & 0 & -1 \\ 3 & 6 & 1 & 2 \\ 1 & -6 & 0 & 3 \end{vmatrix}.$$

We now expand D along the third column,

$$D = (1)(-1)^6 \begin{vmatrix} 10 & 20 & 10 \\ 2 & 4 & -1 \\ 1 & -6 & 3 \end{vmatrix},$$

and factor 10 from the first row and 2 from the second column,

$$D = 20 \begin{vmatrix} 1 & 1 & 1 \\ 2 & 2 & -1 \\ 1 & -3 & 3 \end{vmatrix}.$$

Finally we expand D along the first row to obtain

$$D = 20[1(3) - 1(7) + 1(-8)] = -240.$$

EXAMPLE F.4

Evaluate

$$D = \begin{vmatrix} 3 & 0 & 2 & -1 & 4 \\ 6 & 3 & 7 & -2 & 8 \\ 10 & 3 & 1 & 0 & 6 \\ 2 & 1 & 3 & -4 & 1 \\ 6 & -2 & -3 & -5 & 2 \end{vmatrix}.$$

SOLUTION By adding multiples of column 4 to columns 1, 3, and 5, we can create zeros in the first row:

$$D = \begin{vmatrix} 0 & 0 & 0 & -1 & 0 \\ 0 & 3 & 3 & -2 & 0 \\ 10 & 3 & 1 & 0 & 6 \\ -10 & 1 & -5 & -4 & -15 \\ -9 & -2 & -13 & -5 & -18 \end{vmatrix}.$$

Expansion along the first row gives

$$D = \begin{vmatrix} 0 & 3 & 3 & 0 \\ 10 & 3 & 1 & 6 \\ -10 & 1 & -5 & -15 \\ -9 & -2 & -13 & -18 \end{vmatrix}.$$

We now add -1 times column 2 to column 3,

$$D = \begin{vmatrix} 0 & 3 & 0 & 0 \\ 10 & 3 & -2 & 6 \\ -10 & 1 & -6 & -15 \\ -9 & -2 & -11 & -18 \end{vmatrix}.$$

Expansion along row 1 gives

$$D = 3(-1) \begin{vmatrix} 10 & -2 & 6 \\ -10 & -6 & -15 \\ -9 & -11 & -18 \end{vmatrix}.$$

We now factor -1 from rows 2 and 3, 2 from row 1, and 3 from column 3, and then expand along row 1,

$$D = -18 \begin{vmatrix} 5 & -1 & 1 \\ 10 & 6 & 5 \\ 9 & 11 & 6 \end{vmatrix} = -18[5(-19) + 1(15) + 1(56)] = 432.$$

■

Solution of Linear Equations by Cramer's Rule

To solve a pair of linear equations in two unknowns such as

$$2x + 3y = 6,$$
$$x - 4y = -1,$$

we can eliminate one of the unknowns, say y, solve the resulting equation for x, and then substitute this value of x into either of the original equations to obtain y. The result for the pair above is $x = 21/11$, $y = 8/11$. A similar procedure can be followed for three linear equations in three unknowns:

$$2x - 3y + 4z = 6,$$
$$x + 4y - 2z = 7,$$
$$3x - 2y + z = -2.$$

First one variable is eliminated, say x, to obtain two equations in the two unknowns y and z. These are then solved for y and z and substituted into one of the original equations to find x. The solution is $x = 3/7$, $y = 128/35$, $z = 141/35$.

A formula can be derived for the solution of linear equations, and the formula can be stated simply using determinants; this result is called **Cramer's rule**. We illustrate it for the second example above and then demonstrate its general validity. The coefficients of the unknowns are arranged in a determinant called the determinant of the system of equations. For the above system of three equations, it is

$$D = \begin{vmatrix} 2 & -3 & 4 \\ 1 & 4 & -2 \\ 3 & -2 & 1 \end{vmatrix}.$$

We now define three other determinants. They are obtained by replacing the first, second, and third columns in D by the coefficients on the right-hand sides of the equations:

$$D_x = \begin{vmatrix} 6 & -3 & 4 \\ 7 & 4 & -2 \\ -2 & -2 & 1 \end{vmatrix}, \quad D_y = \begin{vmatrix} 2 & 6 & 4 \\ 1 & 7 & -2 \\ 3 & -2 & 1 \end{vmatrix}, \quad D_z = \begin{vmatrix} 2 & -3 & 6 \\ 1 & 4 & 7 \\ 3 & -2 & -2 \end{vmatrix}.$$

Cramer's rule states that

$$x = \frac{D_x}{D}, \quad y = \frac{D_y}{D}, \quad z = \frac{D_z}{D}.$$

We check the first of these and leave it to the reader to verify the other two. If we expand D_x and D along their first rows, then

$$x = \frac{D_x}{D} = \frac{6(0) + 3(3) + 4(-6)}{2(0) + 3(7) + 4(-14)} = \frac{-15}{-35} = \frac{3}{7}.$$

For our first example of two linear equations in x and y, Cramer's rule gives

$$x = \frac{D_x}{D} = \frac{\begin{vmatrix} 6 & 3 \\ -1 & -4 \end{vmatrix}}{\begin{vmatrix} 2 & 3 \\ 1 & -4 \end{vmatrix}} = \frac{-21}{-11} = \frac{21}{11}, \qquad y = \frac{D_y}{D} = \frac{\begin{vmatrix} 2 & 6 \\ 1 & -1 \end{vmatrix}}{-11} = \frac{-8}{-11} = \frac{8}{11}.$$

To prove Cramer's rule we require the following theorem.

> **Theorem F.4**
>
> *If the elements in any line of a determinant are multiplied by the cofactors of corresponding elements of a distinct parallel line, the resulting sum is zero.*

Proof Suppose we take $D = |a_{ij}|_{n \times n}$, and construct a determinant E from D by replacing the kth row of D by its ith row:

$$E = \begin{vmatrix} a_{11} & a_{12} & \cdots & a_{1n} \\ \vdots & \vdots & \ddots & \vdots \\ a_{i1} & a_{i2} & \cdots & a_{in} \\ \vdots & \vdots & \ddots & \vdots \\ a_{i1} & a_{i2} & \cdots & a_{in} \\ \vdots & \vdots & \ddots & \vdots \\ a_{n1} & a_{n2} & \cdots & a_{nn} \end{vmatrix} \begin{matrix} \\ \\ \leftarrow i^{\text{th}} \\ \\ \leftarrow k^{\text{th}} \\ \\ \end{matrix}.$$

If we expand this determinant along its kth row, the net effect is to multiply the elements of row i by the cofactors of row k:

$$E = \sum_{j=1}^{n} a_{ij} A_{kj}.$$

However, because E has two identical parallel lines, its value must be zero (see Exercise 16). Thus,

$$\sum_{j=1}^{n} a_{ij} A_{kj} = 0 \qquad \text{whenever } i \neq k. \tag{F.5a}$$

A similar proof for columns gives

$$\sum_{i=1}^{n} a_{ij} A_{ik} = 0 \qquad \text{whenever } j \neq k. \tag{F.5b}$$

We now prove the following theorem.

Theorem F.5

(Cramer's rule) *A system of n linear equations in n unknowns $x_1, x_2, \ldots, x_n$,*

$$
\begin{aligned}
a_{11}x_1 + a_{12}x_2 + \cdots + a_{1n}x_n &= c_1, \\
a_{21}x_1 + a_{22}x_2 + \cdots + a_{2n}x_n &= c_2, \\
&\vdots \\
a_{n1}x_1 + a_{n2}x_2 + \cdots + a_{nn}x_n &= c_n,
\end{aligned}
\qquad (F.6)
$$

can be represented compactly in the form

$$
\sum_{j=1}^{n} a_{ij}x_j = c_i, \qquad i = 1, \ldots, n. \qquad (F.7)
$$

From the determinant $D = |a_{ij}|_{n \times n}$ of the system, we define n other determinants D_k by replacing the kth column of D by the column of constants $c_1, c_2, \ldots, c_n$. The solution of equations F.6 is then

$$
x_k = \frac{D_k}{D}, \qquad k = 1, \ldots, n, \qquad (F.8)
$$

provided $D \neq 0$.

Proof We multiply the first equation in F.6 by the cofactor A_{1k}, the second equation by A_{2k}, and so on, until the last equation is multiplied by A_{nk}. Symbolically, this is represented by multiplying the ith equation by A_{ik} to get

$$
\sum_{j=1}^{n} a_{ij}A_{ik}x_j = c_i A_{ik}, \qquad i = 1, \ldots, n.
$$

We now add all these equations together:

$$
\sum_{i=1}^{n}\sum_{j=1}^{n} a_{ij}A_{ik}x_j = \sum_{i=1}^{n} c_i A_{ik}.
$$

The right-hand side of this equation is the expansion of D_k along its kth column. Consequently,

$$
D_k = \sum_{j=1}^{n} \left(\sum_{i=1}^{n} a_{ij}A_{ik} \right) x_j.
$$

But according to Theorem F.4, the summation in parentheses is zero unless $j = k$; and when $j = k$, the result is

$$
D_k = \left(\sum_{i=1}^{n} a_{ik}A_{ik} \right) x_k = D x_k,
$$

or

$$
x_k = \frac{D_k}{D}.
$$

EXAMPLE F.5

Use Cramer's rule to solve

$$3x - 2y = -1,$$
$$x + 4y - 2z = 6,$$
$$3y + 4z = 7.$$

SOLUTION We have

$$D = \begin{vmatrix} 3 & -2 & 0 \\ 1 & 4 & -2 \\ 0 & 3 & 4 \end{vmatrix} = 3(22) + 2(4) = 74;$$

$$D_x = \begin{vmatrix} -1 & -2 & 0 \\ 6 & 4 & -2 \\ 7 & 3 & 4 \end{vmatrix} = -1(22) + 2(38) = 54;$$

$$D_y = \begin{vmatrix} 3 & -1 & 0 \\ 1 & 6 & -2 \\ 0 & 7 & 4 \end{vmatrix} = 3(38) + 1(4) = 118;$$

$$D_z = \begin{vmatrix} 3 & -2 & -1 \\ 1 & 4 & 6 \\ 0 & 3 & 7 \end{vmatrix} = 3(10) - 1(-11) = 41.$$

Hence, $x = 54/74 = 27/37$, $y = 118/74 = 59/37$, $z = 41/74$. ∎

EXERCISES

In Exercises 1–9 evaluate the determinant.

1. $\begin{vmatrix} 3 & 2 \\ -1 & 4 \end{vmatrix}$

2. $\begin{vmatrix} 1 & 0 \\ 0 & 1 \end{vmatrix}$

3. $\begin{vmatrix} -2 & -4 \\ -6 & -8 \end{vmatrix}$

4. $\begin{vmatrix} 1 & 2 & 3 \\ 2 & 4 & 6 \\ -1 & 3 & 0 \end{vmatrix}$

5. $\begin{vmatrix} -2 & 0 & 5 \\ 1 & 3 & 6 \\ -7 & 8 & 10 \end{vmatrix}$

6. $\begin{vmatrix} 10 & 20 & 30 \\ 16 & 32 & 64 \\ -1 & 2 & -3 \end{vmatrix}$

7. $\begin{vmatrix} 1 & 1 & 1 & -3 \\ 2 & 1 & 3 & 6 \\ 7 & -8 & 9 & 10 \\ 3 & 4 & -2 & 1 \end{vmatrix}$

8. $\begin{vmatrix} 3 & -2 & 1 & 6 \\ 4 & 5 & 2 & -1 \\ 0 & 0 & 3 & 2 \\ -1 & 3 & 4 & -1 \end{vmatrix}$

9. $\begin{vmatrix} 0 & 1 & 2 & 3 & 4 \\ -1 & 0 & 5 & 6 & 7 \\ -2 & -5 & 0 & 8 & 9 \\ -3 & -6 & -8 & 0 & 10 \\ -4 & -7 & -9 & -10 & 0 \end{vmatrix}$

In Exercises 10–15 use Cramer's rule to solve the system of equations.

10. $-3x + 4y = 2$, $x - 2y = 6$

11. $2x - 3y = -10$, $x + y = 0$

12. $4r - 2s = 6$, $3r - s = -1$

13. $2x - 3y + z = 2$, $6x - y + 2z = 4$, $x - y = 1$

14. $3z - 2y + x = 6$, $z + y + 4x = 2$, $-z - x + 2y = -1$

15. $2x - 3y + 4z + w = 1$, $x - 3y + 2w = 6$, $3y + 4z - w = 2$, $3x - y + z = 0$

16. **(a)** Use mathematical induction to verify that when two parallel lines of a determinant with value D are interchanged, the value of the new determinant is $-D$.

(b) Use part (a) to prove that a determinant with two identical parallel lines has value zero.

17. A determinant is said to be *skew-symmetric* if its elements satisfy the property $a_{ij} + a_{ji} = 0$. (The determinant in Exercise 9 is skew-symmetric.) Show that a skew-symmetric determinant of odd order has value zero.

18. Solve the system of equations
$$2x + 3y - 4z + w = 0,$$
$$x + y - 2z + 3w = 0,$$
$$2x - 3y + z - 2w = 0,$$
$$x + y - z + w = 0.$$

19. Can we use Cramer's rule to solve the system
$$2x + 3y - 4z + w = 0,$$
$$x + y - 2z + 3w = 0,$$
$$2x - 3y + z - 2w = 0,$$
$$5x + y - 5z + 2w = 0?$$

20. (a) Show that the equation of the straight line in the xy-plane through the two points (x_1, y_1) and (x_2, y_2) can be expressed in the form
$$\begin{vmatrix} x & y & 1 \\ x_1 & y_1 & 1 \\ x_2 & y_2 & 1 \end{vmatrix} = 0.$$

(b) Use the result in (a) to find the equation of the line through $(1, 1)$ and $(-2, 3)$.

21. (a) Show that the equation of the circle in the xy-plane through the three points (x_1, y_1), (x_2, y_2), and (x_3, y_3) can be expressed in the form
$$\begin{vmatrix} x^2 + y^2 & x & y & 1 \\ x_1^2 + y_1^2 & x_1 & y_1 & 1 \\ x_2^2 + y_2^2 & x_2 & y_2 & 1 \\ x_3^2 + y_3^2 & x_3 & y_3 & 1 \end{vmatrix} = 0.$$

(b) Use the result in (a) to find the circle through $(2, 1)$, $(-3, -3)$, and $(7, -5)$.

22. A parabola of the form $y = ax^2 + bx + c$ is to pass through the three points $(1, 0)$, $(2, 11)$, and $(-2, 1)$. Find a determinant that implicitly defines its equation.

23. Find a condition in the form of a determinant that serves as a test to determine whether a circle can be drawn through four given points, no three of which are collinear. (Hint: See Exercise 21.)

24. Evaluate
$$\begin{vmatrix} a+b & a & a & \cdots & a \\ a & a+b & a & \cdots & a \\ a & a & a+b & \cdots & a \\ \vdots & \vdots & \vdots & \ddots & \vdots \\ a & a & a & \cdots & a+b \end{vmatrix}_{n \times n}.$$

25. (a) Evaluate
$$\begin{vmatrix} a & b & b & b & \cdots & b & b \\ b & a & b & b & \cdots & b & b \\ b & b & a & b & \cdots & b & b \\ b & b & b & a & \cdots & b & b \\ \vdots & \vdots & \vdots & \vdots & \ddots & \vdots & \vdots \\ b & b & b & b & \cdots & a & b \\ b & b & b & b & \cdots & b & a \end{vmatrix}_{n \times n}.$$

(b) Use the result in (a) to solve the system of equations
$$ax_1 + bx_2 + bx_3 + \cdots + bx_n = 1,$$
$$bx_1 + ax_2 + bx_3 + \cdots + bx_n = 1,$$
$$bx_1 + bx_2 + ax_3 + \cdots + bx_n = 1,$$
$$\vdots$$
$$bx_1 + bx_2 + bx_3 + \cdots + ax_n = 1.$$

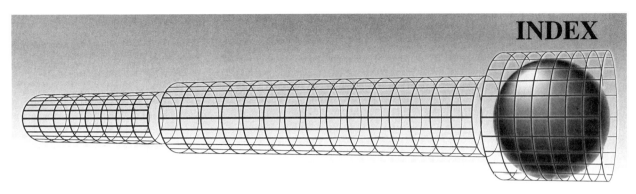

INDEX

Absolute convergence, 542
Absolute maxima and minima, 755
Acceleration, 676
    tangential and normal components, 681
    vector, 676
Addition of vectors, 615
Alternating series, 543
    approximating sum of, 547, 584
    convergence theorem for, 544
    harmonic, 544
Analytic geometry, three-dimensional, 595
Angle, direction, 635
    between lines, 626
    between planes, 635
    between vectors, 626
Angular momentum, 685
Angular speed, 683
Antiderivative, 234
    for function of more than one variable, 783
    for vector function, 652
Approximation. Refer to what is being approximated
Archimedes' principle, 850, 925, 927
Area, as a double integral, 793
    as a line integral, 904
    of a parallelogram, 645
    of a polygon, 905
    of a surface, 809
    of a surface of revolution, 880
    of a triangle in space, 644
Astroid, 880, 905
Auxiliary equation, 981
Average value of a function, 792, 833, 881

Bernoulli equation, 957
Bernoulli numbers, 580
Bessel functions, 558, 560, 579, 580, 589
Bifolium, 861
Binomial, coefficients, 520
    series, 575
    theorem, 520
Binormal vector, 667
Boundary conditions for differential equation, 944

Boundary point, 864
Bounded sequence, 511
Brachistochrone problem, 723

Cardioid, 685
Carrying capacity, 970
Cartesian, coordinates, 596
    components, 612
Cauchy-Euler linear equation, 998
Cauchy-Riemann equations, 704, 711, 712
Centre of curvature, 671
    mass, 803, 834
Centroid, 808
Chain rule, for multivariable function, 712
    for vector functions of one variable, 654
Circle of curvature, 671
Circulation, 888, 936
Closed curve, 657
    set, 864
    surface, 917
Coefficient of friction, 965
Coefficients, binomial, 520
    undetermined, 985
Cofactor in a determinant, 1042
Common ratio for geometric series, 524
Comparison test for series, 534
Completeness property for reals, 560
Complimentary equation, 984
Component, normal and tangential, 681
    scalar, 620
    vector, 620
    x, y, z, 610
Conditional convergence, 543
Cone, 606
Connected set, 864
Conservation of energy, law, of 898
Conservative force field, 897
Constraint, 764
Continuity, of a curve, 656
    equation of, 704, 867, 936
    of a function of two variables, 699
    of a vector function, 649
Continuous. *See* Continuity
Convergence, absolute, 542
    conditional, 543

    radius of, 554
    of a sequence, 511
    of a series, 526
Coordinate plane, 595
Coordinates, Cartesian for points in space, 596
    cylindrical, 839
    rectangular, 596
    spherical, 844
Coulomb's law, 621, 822, 863
Cramer's rule, 726, 1046
Critical point, 745
Critically damped motion, 1006
Cross product, 637
Curl of a vector field, 867
Curvature, centre of, 671
    circle of, 671
    of a curve, 671, 672
    radius of, 671, 672
Curve, closed, 657, 765
    continuous, 656
    curvature of, 671, 672
    direction of, 655
    length of, 663, 664
    parametric definition of, 655
    piecewise smooth, 661
    smooth, 661
Cycloid, 685, 723, 737, 888
Cylinder, 605, 606
Cylindrical coordinates, 839
Cylindrical shell method for volumes, 794

Damped oscillations, 1006
Decreasing sequence, 511
Del operator, 866
Deltoid, 905
Derivative(s), of composite functions, 117
    of several variables, 712
    of vector functions, 654
    directional, 731
    higher order, 708
    for implicitly defined functions, 724
    of an integral, 852
    partial, 701
    of a power series, 567, 571

symbols for, 701
of vector functions, 650
Determinant, 1041
Differential, 773
Differential equation, 943
   Bernoulli, 957
   boundary conditions for, 944
   complimentary, 984
   general solution of, 945
   homogeneous, 952
   initial conditions for, 944
   linear first order, 953
   linear n^{th} order, 972
      homogeneous, 975, 979
      nonhomogeneous, 975, 984
   order of, 943
   particular solution of, 946, 984
   power series solutions of, 587
   reduced, 984
   second order reducible to two
      first order, 958
   separable, 948
   singular solution of, 945
Direction, of a curve, 655, 874
   of a vector, 608
Direction angles, 635
Directional derivative, 731
   second, 735
Displacement vector, 648, 657, 675
Distance, between a point and a line in a
      plane, 630
   between a point and a line in space, 643
   between a point and a plane, 631, 635,
      763
   between two lines in a plane, 631
   between two lines in space, 647
   between two planes, 636
   between two points, 597
Divergence, of a sequence, 511
   of a series, 526
   of a vector field, 866, 936
Divergence theorem, 920
Domain, 865
   of a function, 691
Dot product, 624
Double integral, definition of, 782
   evaluation of, 786
   in polar coordinates, 815
   properties of, 782
Double iterated integral, 784
Droplet, 905

Electric circuits, 958, 1002
Electric displacement, 937
Electromagnetic theory, 937
Electrostatic potential, 707, 711, 822, 851,
      873, 899
Ellipsoid, 606
Elliptic, cone, 606

cylinder, 606
hyperboloid, 606
paraboloid, 606
Energy, law of conservation of, 898
   potential, 897
Entropy, 895, 896
Equation of continuity, 704, 867, 936
Equipotential surface, 900
Error, function, 579, 586
   truncation, 547, 584
Escape velocity, 967
Euler-Lagrange equations, 723
Euler's identity, 980
   theorem, 720
Exterior point, 864
Extrema, absolute, 755
   relative, 748, 749

Faraday's induction law, 938
Fibonacci sequence, 517
First moment, 803, 834
Fluid flow, 867, 935
Fluid pressure, 799
Flux, 935, 937
Force, moment of, 646, 685
   as a vector, 617
   work done by, 632
Force field, 897
Fourier transform, 975
Frenet-Serret formulas, 674
Fresnel integral, 579
Friction, 965
Function, average value of, 792, 833, 881
   continuous, 649, 699
   derivative of (see Derivative)
   differential of, 773
   domain of, 691
   homogeneous, 719
   implicit definition of, 724
   linearly independent, 977, 978

Gamma function, 823
Gauss's law, 928, 937
General solution of differential equation,
      945
Geometric series, 524, 526
Gompertz model, 972
Gradient vector, definition of, 705
   and directional derivative, 732
   and line integrals, 889
   and normals to surfaces, 740
Gravitational force, 851
Green's identities, in the plane, 907
   in space, 928
Green's theorem, 901

Hanging cable, 962
Harmonic, conjugates, 704, 711
   function, 710, 722

series, 525, 527
Heat equation, 712
Helix, 662, 880
Homogeneous, function, 719
   first order differential equation, 952
   linear differential equation, 975, 979
Hyperbolic, cylinder, 606
   paraboloid, 606
Hyperboloid, of one sheet, 606
   of two sheets, 606
Hypergeometric series, 560

Imaginary part of complex number, 981,
      994
Implicit differentiation, 724
Increasing sequence, 511
Independence of path, 888
Infinite series. See series
Initial condition for differential equation,
      944
Inner product, 624
Integral, derivative of, 852
   double (see Double integral)
   iterated, 784, 824
   line, 875
   surface, 907
   triple (see Triple integral)
Integral test, 531
Integrating factor, 955
Interior point, 864
Interval of convergence, 552
Intrinsic property of a curve, 666
Inverse operator shift theorem, 992
Inverse square force field, 896
Involute of a circle, 736
Irrotational vector field, 868
Iterated integral, 784, 824

Jacobian, 726

Kinetic energy, 591, 683, 898
Kirchhoff's law, 1002

Lagrange multiplier, 766
Lagrangian, 766
Laplace transform, 975
Laplace's equation, 710, 723, 873
   in polar coordinates, 722, 858
Least squares, 754
Legendre polynomials, 580
Leibnitz's rule, 853
Length, of a curve in Cartesian coordi-
      nates, 663, 664
   of a line segment in space, 597
   of a vector, 612
Level curves, 695
Limit comparison test, 535
Limit ratio test, 538

Limit root test, 539
Limits, definition of, 697
   of a sequence, 510, 514
   of a vector function, 649
Line(s), angle between, 626
   equations of in space, 640–641
   tangent, 737
Line integral, 875
Linear differential equation, 972
   applications, 998, 1002
   first order, 953
   n^{th} order, 972
      homogeneous, 975, 979
      nonhomogeneous, 975, 984
Linear operator, 974
Linearly dependent and independent
   functions, 977, 978
   vectors, 623
Logistic, function, 970
   model, 970
Lower bound for a sequence, 511

Maclaurin series, 563
Magnetic field, 938
Malthusian model, 969
Mass in relativity theory, 591
Mathematical model, 968
Maximum, absolute, 755
   relative, 748
Maxwell's equations, 938
Median of a triangle, 623
Midpoint, 598
Minimum, absolute, 755
   relative, 749
Minor in a determinant, 1041
Mobius strip, 914
Moment of a force, 646, 685,
Moment of inertia, 806, 836
Moments, first, 803, 834
   second, 806, 836
Momentum, 677
Monomolecular growth curve, 972
Monotonic sequence, 511
Multiplier, Lagrange, 766

Newtonian mechanics, 963
Newton's iterative procedure, 518, 523
Newton's law of cooling, 951
Newton's second law of motion, 677, 685,
   898, 963
Newton's universal law of gravitation, 622,
   683, 967
Nondecreasing sequence, 511
Nonhomogeneous linear differential equation, 975, 984
Nonincreasing sequence, 511
Nonnegative series, 531
Normal, component of acceleration, 681
   principal, 667
   vector, to a curve, 666, 667
      to a surface, 740, 742
Numerical integration, 585

Octant, 597
Open set, 864
Operator method, 989
Operator shift theorem, 983
Order of a differential equation, 943
Ordered triple, 596
Orientable, 914
Origin, 595
Outer product, 637
Overdamped motion, 1006

p-series, 533
Pappus, theorem of, 809
Parabolic cylinder, 606
Parallel axis theorem, 809, 838
Parallelepiped, 647
Parallelogram addition of vectors, 615
   law for lengths of diagonals, 635
Parametric equations, for a curve, 655
   for a line, 641
Partial derivative, 701
Partial sum of a series, 525
Particular solution, 946, 984
Perpendicular vectors, 626
Piecewise smooth, curve, 661
   surface, 743
Planck's law, 591
Plane, 627
   coordinate, 595
   equation of, 627
   normal vector to, 628
   tangent, 739
Poiseuille's law, 822
Poisson distribution, 579
Poisson's integral formula, 858
Polar coordinates, double integrals in, 815
   lengths of curves in, 881
Polygon, area of, 905
Population growth, 968
Position vector, 648, 657, 675
Potential, electrostatic, 707, 711, 822, 851,
   873, 899
   gravitational, 711
Potential energy, 897
Power series, 552
   addition of, 570
   binomial, 575
   differentiation of, 567, 571
   integration of, 567, 571
   interval of convergence of, 552
   multiplication of, 573
   radius of convergence of, 554
   sum of, 556, 580
Pressure, fluid, 799
Principal normal, 667

Product, dot, 624
   cross, 637
Projectile, 679
Projecting curves in planes, 607
Pursuit curve, 962

Quadric surface, 605

Radially symmetric force field, 900
Radius, of convergence, 554
   of curvature, 671, 672
Range of a projectile, 679, 684
Rate of change along a curve, 733
Rayleigh-Jeans law, 591
Real part of complex number, 981, 994
Rectangular coordinates in space, 596
Recursive sequence, 509
Reduced differential equation, 984
Relative maximum, 748
   minimum, 749
Relativity, mass in theory of, 591
Remainder in Taylor's formula, 562
Resolution of vectors, 628
Resonance, 1006
Resultant force, 618
Right-handed coordinate system, 596
Right-hand rule, 638

Saddle point, 749
Scalar, 608
Scalar, multiplication, 613
   product, 624
   triple product, 647
Second law of thermodynamics, 728, 730,
   894, 896
Second moment, 806, 836
Separable differential equation, 948
Sequence, 508
   bounded, 511
   explicitly defined, 509
   limit of, 510, 514
   monotonic, 511
   of partial sums, 525
   recursively defined, 509
Series, 524
   absolutely convergent, 542
   alternating (*see* Alternating series)
   approximating sum of, 547
   binomial, 575
   comparison test for, 534
   conditionally convergent, 543
   convergent, 526
   divergent, 526
   geometric, 524, 526
   harmonic, 525, 527
   integral test for, 531
   interval of convergence of, 552
   limit comparison test for, 535
   limit ratio test for, 538

limit root test for, 539
Maclaurin, 563
nonnegative, 531
nth term test for, 528
p-, 533
partial sum of, 525
power (see Power series)
radius of convergence of, 554
sum of, 526, 580
Taylor, 563, 776
Simple harmonic motion, 1001, 1005
Simply-connected domain, 865
Sine law, 647
Singular solution, 945
Skew-symetric determinant, 1050
Smooth, curve, 661
  surface, 742
Solenoidal vector field, 873, 874
Solid of revolution, 794
Speed, 676
Spherical coordinates, 844
Steady-state, 1004
Stokes's theorem, 929
Strophoid, 905
Successive substitution method, 520, 522, 523
Sum, partial, 525
  of power series, 556, 580
  of series of constants, 526, 580
  of vectors, 615
Superposition principle, 976
Surface, quadric, 605
  area, 809
  integral, 907
Symmetric equations for line, 641

Tangent, line to a curve, 737
  plane, 739
  vector to a curve, 659
Tangential component, of acceleration, 681

of velocity, 680
Taylor series, 563, 776
Taylor's remainder formula, 561
  integral form, 569
Terminal speed (velocity), 965
Tetrahedron, 598
Torricelli's law, 953
Torsion, 675
Torus, 844, 881
Traffic flow, 722
Transient, 1004
Triangle inequality, 623
Triangular addition of vectors, 615
Triple integral, 823
  in cylindrical coordinates, 839
  in spherical coordinates, 844
Triple iterated integral, 824
Trochoid, 685
Truncation error, 547, 584

Undetermined coefficients, 985
Unit tangent vector, 662, 663, 664
Unit vector, 612
Upper bound for sequence, 511

Vector(s), 608
  acceleration, 676
  addition of, 615
  angle between, 626
  binormal, 667
  components of, 609, 620, 629
  cross product of, 637
  derivative of, 650
  difference of, 616
  displacement, 648, 657, 675
  dot product of, 624
  equality of, 609, 611
  gradient, 705
  inner product of, 624
  length of, 612
  linearly dependent, 623

linearly independent, 623
  multiplication of scalar by, 613
  normal to a curve, 666, 667
  normal to a surface, 740, 742
  outer product of, 637
  perpendicular, 626
  position, 648, 657, 675
  principal normal, 667
  product, 637
  representation of curve, 657
  scalar product of, 624
  sum of, 615
  tangent, 659
  unit, 612
  unit tangent, 662, 663, 664
  vector product of, 637
  velocity, 675
  zero, 616
Vector field, 865
  conservative, 897
  curl of, 867
  divergence of, 866, 936
Velocity, instantaneous, 675
  escape, 967
  terminal, 965
Volume, of a parallelepiped, 647
  as a surface integral, 927
  by triple integrals, 830
Volume of solid of revolution, 794
  by cylindrical shell method, 794
  with double integral, 795
  by washer method, 794
Von Bertalanffy growth curve, 972

Washer method for volumes, 794
Wave equation, 711, 719
Work, 632, 886
Wronskian, 979

Zero vector, 616